Pressure:

$$\mathbf{1\ atm} = 101.3\ \text{kPa} = 1.013\ \text{bar} = 14.696\ \text{lbf/in}^2 = 33.89\ \text{ft of water}$$

$$= 29.92\ \text{in of mercury} = 1.033\ \text{kgf/cm}^2 = 10.33\ \text{m of water}$$

$$= 760\ \text{mm of mercury} = 760\ \text{torr}$$

$$\mathbf{1\ Pa} = \text{N/m}^2 = \text{kg/(m} \cdot \text{s}^2) = 10^{-5}\ \text{bar} = 1.450 \times 10^{-4}\ \text{lbf/in}^2$$

Viscosity:

$$\mathbf{1\ cP} = 0.01\ \text{P} = 0.01\ \text{g/(cm} \cdot \text{s}) = 0.001\ \text{kg/(m} \cdot \text{s}) = 0.001\ \text{Pa} \cdot \text{s}$$

$$= 6.72 \times 10^{-4}\ \text{lbm/(ft} \cdot \text{s}) = 2.42\ \text{lbm/(ft} \cdot \text{h}) = 2.09 \times 10^{-5}\ \text{lbf} \cdot \text{s/ft}^2$$

$$= 0.01\ \text{dyn} \cdot \text{s/cm}^2$$

Kinematic viscosity:

$$\mathbf{1\ cSt} = 0.01\ \text{St} = 0.01\ \text{cm}^2\text{/s} = 10^{-6}\ \text{m}^2\text{/s} = 1\ \text{cP/(g/cm}^3)$$

$$= 1.08 \times 10^{-5}\ \text{ft}^2\text{/s} = 1\text{cP/(62.4 lbm/ft}^3)$$

$T(K) = T°C + 273.15$

$T(°R) = T(°F) + 459.67$

$T(°R) = 1.8T(K)$

$T(°F) = 1.8T(°C) + 32$

FLUID MECHANICS FOR CHEMICAL ENGINEERS

McGraw-Hill Chemical Engineering Series

Building the Literature of a Profession

Fifteen prominent chemical engineers first met in New York more than 60 years ago to plan a continuing literature for their rapidly growing profession. From Industry came such pioneer practitioners as Leo H. Baekeland, Arthur D. Little, Charles L. Reese, John V. N. Dorr, M. C. Whitaker, and R. S. McBride. From the universities came such eminent educators as William H. Walker, Alfred H. White, D. D. Jackson, J. H. James, Warren K. Lewis, and Harry A. Curtis. H. C. Parmelee, then editor of *Chemical and Metallurgical Engineering*, served as chairman and was joined subsequently by S. D. Kirkpatrick as consulting editor.

After several meetings, this committee submitted its report to the McGraw-Hill Book Company in September 1925. In the report were detailed specifications for a correlated series of more than a dozen texts and reference books which have since become the McGraw-Hill Series in Chemical Engineering and which became the cornerstone of the chemical engineering curriculum.

From this beginning there has evolved a series of texts surpassing by far the scope and longevity envisioned by the founding Editorial Board. The McGraw-Hill Series in Chemical Engineering stands as a unique historical record of the development of chemical engineering education and practice. In the series one finds the milestones of the subject's evolution: industrial chemistry, stoichiometry, unit operations and processes, thermodynamics, kinetics, and transfer operations.

Chemical engineering is a dynamic profession, and its literature continues to evolve. McGraw-Hill, with its editor, B. J. Clark, and its consulting editors, remains committed to a publishing policy that will serve, and indeed lead, the needs of the chemical engineering profession during the years to come.

The Series

Bailey and Ollis: *Biochemical Engineering Fundamentals*
Bennett and Myers: *Momentum, Heat, and Mass Transfer*
Brodkey and Hershey: *Transport Phenomena: A Unified Approach*
Carberry: *Chemical and Catalytic Reaction Engineering*
Constantinides: *Applied Numerical Methods with Personal Computers*
Coughanowr: *Process Systems Analysis and Control*
de Nevers: *Fluid Mechanics for Chemical Engineers*
Douglas: *Conceptual Design of Chemical Processes*
Edgar and Himmelblau: *Optimization of Chemical Processes*
Gates, Katzer, and Schuit: *Chemistry of Catalytic Processes*
Holland: *Fundamentals of Multicomponent Distillation*
Holland and Liapis: *Computer Methods for Solving Dynamic Separation Problems*
Katz and Lee: *Natural Gas Engineering: Production and Storage*
King: *Separation Processes*
Lee: *Fundamentals of Microelectronics Processing*
Luyben: *Process Modeling, Simulation, and Control for Chemical Engineers*
McCabe, Smith, J. C., and Harriott: *Unit Operations of Chemical Engineering*
Mickley, Sherwood, and Reed: *Applied Mathematics in Chemical Engineering*
Nelson: *Petroleum Refinery Engineering*
Perry and Chilton (Editors): *Perry's Chemical Engineers' Handbook*
Peters: *Elementary Chemical Engineering*
Peters and Timmerhaus: *Plant Design and Economics for Chemical Engineers*
Reid, Prausnitz, and Rolling: *Properties of Gases and Liquids*
Smith, J. M.: *Chemical Engineering Kinetics*
Smith, J. M., and Van Ness: *Introduction to Chemical Engineering Thermodynamics*
Treybal: *Mass Transfer Operations*
Valle-Riestra: *Project Evaluation in the Chemical Process Industries*
Wei, Russell, and Swartzlander: *The Structure of the Chemical Processing Industries*
Wentz: *Hazardous Waste Management*

FLUID MECHANICS FOR CHEMICAL ENGINEERS

Second Edition

Noel de Nevers

Department of Chemical Engineering
University of Utah

McGraw Hill

Boston, Massachusetts Burr Ridge, Illinois
Dubuque, Iowa Madison, Wisconsin New York, New York
San Francisco, California St. Louis, Missouri

This book was set in Times Roman.
The editors were B. J. Clark and Margery Luhrs;
the production supervisor was Janelle S. Travers.
The cover was designed by John Hite.

McGraw-Hill

A Division of The ***McGraw-Hill*** *Companies*

FLUID MECHANICS FOR CHEMICAL ENGINEERS

11 12 13 14 15 16 17 18 19 BKM BKM 0 9 8 7 6 5 4 3

ISBN 0-07-016375-8

Library of Congress Cataloging-in-Publication Data

de Nevers, Noel, (date).
Fluid mechanics for chemical engineers / Noel de Nevers.—2nd ed.
p. cm.
Rev. ed. of: Fluid mechanics, 1970.
Includes index.
ISBN 0-07-016375-8
1. Fluid mechanics. 2. Chemical engineering. I. de Nevers, Noel, (date). Fluid mechanics. II. Title.
QC145.2.D42 1991
532—dc20 90-43155

CONTENTS

	Notation	xiii
	Preface	xix
Chapter 1	Introduction	1
1.1	What Is Fluid Mechanics?	1
1.2	What Good Is Fluid Mechanics?	3
1.3	Basic Ideas in Fluid Mechanics	4
1.4	Liquids and Gases	5
1.5	Properties of Fluids	6
1.6	Pressure	15
1.7	Force, Mass, and Weight	18
1.8	Units and Conversion Factors	18
1.9	Principles versus Techniques	25
1.10	Engineering Problems	26
1.11	Summary	28
Chapter 2	Fluid Statics	32
2.1	The Basic Equation of Fluid Statics	32
2.2	Pressure-Depth Relationships	35
2.3	Pressure Forces on Surfaces	39
2.4	Buoyancy	46
2.5	Pressure Measurement	49
2.6	Manometer-like Situations	55
2.7	Variable Gravity	58
2.8	Pressure in Accelerated Rigid-Body Motions	59
2.9	Thin-Walled Pressure Vessels	64
2.10	More Problems in Fluid Statics	65
2.11	Summary	65

Chapter 3 The Balance Equation and the Mass Balance 76

3.1 The Balance Equation 76
3.2 The Mass Balance 79
3.3 Steady-State Balances 80
3.4 The Steady-State Flow, One-Dimensional Mass Balance 82
3.5 Unsteady-State Mass Balances 84
3.6 Mass Balances for Mixtures 87
3.7 Mass Balances for Multidimensional Flows 89
3.8 Summary 90

Chapter 4 The First Law of Thermodynamics 94

4.1 Energy 94
4.2 Forms of Energy 95
4.3 Energy Transfer 98
4.4 The Energy Balance 99
4.5 Kinetic and Potential Energies 101
4.6 Internal Energy 103
4.7 The Work Term 106
4.8 Injection Work 107
4.9 Enthalpy 109
4.10 Restricted Forms 110
4.11 Some Common Machines and Processes 112
4.12 Unsteady-State Systems, Accumulation 117
4.13 Less Restricted Systems 121
4.14 Other Forms of Work and Energy 126
4.15 Limitations of the First Law 130
4.16 Summary 131

Chapter 5 Bernoulli's Equation 139

5.1 The Energy Balance for a Steady, Incompressible Flow 139
5.2 The Friction Heating Term 140
5.3 Zero Flow 143
5.4 The Head Form of Bernoulli's Equation 143
5.5 Diffusers and Sudden Expansions 144
5.6 Bernoulli's Equation for Gases 145
5.7 Torricelli's Equation and Its Variants 147
5.8 Bernoulli's Equation for Fluid Flow Measurement 151
5.9 Negative Absolute Pressures: Cavitation 161
5.10 Bernoulli's Equation for Unsteady Flows 163
5.11 Nonuniform Flows 166
5.12 Summary 168

Chapter 6 Fluid Friction in Steady, One-Dimensional Flow 178

6.1 The Pressure-Drop Experiment 179
6.2 Reynolds' Experiment 180

6.3	Laminar Flow	182
6.4	Turbulent Flow	188
6.5	The Three Friction Factor Problems	192
6.6	Some Comments about the Friction Factor Method and Turbulent Flow	198
6.7	More Convenient Methods	198
6.8	Computer Methods	205
6.9	Fitting Losses	206
6.10	Enlargements and Contractions	208
6.11	Fluid Friction in One-Directional Flow and Other Geometries	210
6.12	More Complex Problems Involving Bernoulli's Equation	214
6.13	Economic Pipe Diameter	218
6.14	Flow around Submerged Objects	222
6.15	Summary	229
Chapter 7	**The Momentum Balance**	241
7.1	Momentum	242
7.2	The Momentum Balance	243
7.3	Some Steady-Flow Applications of the Momentum Balance	247
7.4	Starting and Stopping Flows	259
7.5	Relative Velocities	262
7.6	A Very Brief Introduction to Aeronautical Engineering	266
7.7	The Angular Momentum Balance: Rotating Systems	270
7.8	The Momentum Balance for Three-Dimensional Flow	272
7.9	The Navier-Stokes Equations	275
7.10	Summary	279
Chapter 8	**One-Dimensional, High-Velocity Gas Flow**	289
8.1	The Speed of Sound	290
8.2	Steady, Frictionless, Adiabatic, One-Dimensional Flow of a Perfect Gas	294
8.3	Nozzle Choking	304
8.4	High-Velocity Gas Flow with Friction, Heating, or Both	305
8.5	Normal Shock Waves	311
8.6	Relative Velocities	314
8.7	Nozzles and Diffusers	316
8.8	Pitot Tubes for High-Velocity Gas Flow	320
8.9	Summary	322
Chapter 9	**Pumps, Compressors, and Turbines**	329
9.1	Positive-Displacement Pumps	329
9.2	Centrifugal Pumps	334
9.3	Positive-Displacement Compressors	339
9.4	Rotary Compressors	343
9.5	Compressor Efficiencies	345
9.6	Fluid Engines and Turbines	347
9.7	Fluid Engine and Turbine Efficiency	351
9.8	Summary	351

Chapter 10 Potential Flow 355

10.1 The History of Potential Flow and Boundary Layer 355
10.2 Streamlines 357
10.3 Potential Flow 358
10.4 Irrotational Flow 367
10.5 Stream Function 371
10.6 Bernoulli's Equation for Two-Dimensional, Perfect-Fluid, Irrotational Flows 375
10.7 Flow around a Cylinder 377
10.8 Separation 380
10.9 Summary 382

Chapter 11 The Boundary Layer 385

11.1 Prandtl's Boundary-Layer Equations 385
11.2 The Steady-Flow, Laminar Boundary Layer on a Flat Plate Parallel to the Flow 386
11.3 Turbulent Boundary Layers 395
11.4 Turbulent Flow in Pipes 396
11.5 The Steady, Turbulent Boundary Layer on a Flat Plate 400
11.6 The Successes of Boundary-Layer Theory 402
11.7 Summary 405

Chapter 12 Flow through Porous Media 410

12.1 Fluid Friction in Porous Media 412
12.2 Two-Fluid Cocurrent Flow in Porous Media 420
12.3 Countercurrent Flow in Porous Media 424
12.4 Simple Filter Theory 426
12.5 Fluidization 429
12.6 Summary 431

Chapter 13 Models, Dimensional Analysis, and Similitude 433

13.1 Models 433
13.2 Dimensionless Numbers 435
13.3 Finding the Dimensionless Numbers 436
13.4 Judgment, Guesswork, and Caution 446
13.5 Summary 447

Chapter 14 Gas-Liquid Flow 449

14.1 Vertical, Upward Gas-Liquid Flow 450
14.2 Horizontal Gas-Liquid Flow 454
14.3 Two-Phase Flow with Boiling 456
14.4 Summary 456

Chapter 15 Nonnewtonian Fluids 458

15.1 The Role of Structure in Nonnewtonian Behavior 458

15.2 Measurement and Description of Nonnewtonian Fluids 459
15.3 Laminar Flow of Nonnewtonian Fluids in Circular Tubes 462
15.4 Turbulent Flow of Nonnewtonian Fluids in Pipes 465
15.5 Summary 467

Chapter 16 Turbulence 469

16.1 Why Study Turbulence? 472
16.2 Turbulence Measurements 474
16.3 Free and Confined Turbulent Flows 476
16.4 Turbulent Kinetic Energy 476
16.5 Experimental and Mathematical Descriptions of Turbulent Flows 477
16.6 Reynolds Stresses 484
16.7 Turbulence Theories 487
16.8 Summary 487

Chapter 17 Surface Forces 489

17.1 Surface Tension and Surface Energy 490
17.2 Wetting and Contact Angle 491
17.3 Measurement of Surface Tension 492
17.4 Interfacial Tension 495
17.5 Forces due to Curved Surfaces 495
17.6 Some Example of Surface Force Effects 498
17.7 Summary 502

Appendixes

A Tables and Charts 507
A.1 Viscosities of Various Fluids at 1-atm Pressure 507
A.2 Pressure-Enthalpy Diagram for Freon-12 Refrigerant 508
A.3 Steel Pipe Dimensions: Capacities and Weights 509
A.4 Flow of Water through Schedule 40 Steel Pipe 513
A.5 Compressible-Flow Tables for $k = 1.4$ 515
A.6 Fluid Densities 521
A.7 Some Properties of Gases 522
A.8 Compressibility Factor 523
A.9 Values of the Universal Gas Constant 524
A.10 Some Properties of Liquids 524

B Proof that for a Fluid at Rest the Pressure Is the Same in All Directions 525

C The Hydraulic Jump Equations 527

D Properties of a Perfect Gas 529
D.1 Definitions 529
D.2 Isentropic Relations 530
D.3 Entropy Change 532

E The Area Ratio 533

F Normal Shock Waves 535

G Equations for Adiabatic, Zero-Clearance, Isentropic Compressors 539

H Proof that the Curves of Constant ϕ and of Constant ψ Are Perpendicular 541

References 543

Answers to Selected Problems 551

Index 555

NOTATION

a	acceleration	ft/s^2	m/s^2
a	some arbitrary direction or length (Sec. 2.1)	ft	m
a_x, a_y, a_z	x, y, and z components of acceleration	ft/s^2	m/s^2
a	resistance of filter medium (Sec. 12.4)	1/ft	1/m
a_c	centrifugal acceleration	ft/s^2	m/s^2
a, b, c, d	exponents in algebraic procedure (Sec. 13.3)	—	—
A	area or cross-sectional area perpendicular to flow	ft^2	m^2
A, B, C, D	arbitrary constants	various	various
A	independent variable (Sec. 13.3)	various	various
B	dependent variable (Sec. 13.3)	various	various
c	speed of light (Chap. 4 only)	ft/s	m/s
c	speed of sound	ft/s	m/s
c	concentration	lbm/ft^3	kg/m^3
C_d	drag coefficient (Sec. 6.13)	—	—
C_l	lift coefficient (Sec. 6.13)	—	—
C_f	integrated drag coefficient (Sec. 11.2)	—	—
C_f'	local drag coefficient (Sec. 11.2)	—	—
C_P	heat capacity at constant pressure	Btu/(lbm · °F) or Btu/(lbmol · °F)	J/(kg · K) or J/(mol · K)
C_v	orifice or venturi coefficient (Sec. 5.8)	—	—
C_V	heat capacity at constant volume	Btu/(lbm · °F) or Btu/(lbmol · °F)	J/(kg · K) or J/(mol · K)
CC	capital-cost factor (Sec. 6.12)	1/year	1/year
D	diameter	ft	m
D_p	particle diameter	ft	m
erf	gauss error function	—	—
E	energy	Btu or equivalent	J

E	voltage	V	V
E_s	surface energy	ft · lbf/ft^2	J/m^2
f	friction factor (Sec. 6.4)	—	—
f_{PM}	friction factor for porous medium (Sec. 12.1)	—	—
$f(n)$	spectrum function (Sec. 16.5)	1/Hz	1/Hz
°F	temperature or temperature interval in degrees Fahrenheit	°F	
F	force	lbf	N
$\mathcal{F}$	friction heating per unit mass	ft · lbf/lbm or equivalent	J/kg
F_x, F_y, F_z	x, y, and z components of force	lbf	N
F_θ	tangential component of force	lbf	N
$F_I, F_V, F_G, F_S, F_E, F_P$	inertia, viscous, gravity, surface, elastic, and pressure forces (Sec. 13.3)	lbf	N
$F(n)$	spectrum function (Sec. 16.5)	—	—
g	acceleration of gravity	ft/s^2	m/s^2
g_c	conversion factor $= 1 = 32.2$ lbm · ft/(lbf · s^2)	—	—
h	height or depth	ft	m
h	enthalpy per unit mass or mole	Btu/lbm or Btu/(lbmol)	J/kg or J/mol
h_c	centroid depth measured from free surface (Sec. 2.3)	ft	m
H	enthalpy ($U + PV$)	Btu	J
hp	horsepower	ft · lbf/s	
HR	hydraulic radius	ft	m
i, **j**, **k**	unit vectors in the x, y, and z directions	—	—
I	electric current ($d\mathcal{Q}/dt$)	A	A
I	moment of inertia	lbm · ft^2	kg · m^2
I_{sp}	specific impulse	lbf · s/lbm	N · s/kg
J_x, J_y, J_z	x, y, and z components of the electric current density (Sec. 10.3)	A/m^2	A/m^2
k	number of independent dimensions (Sec. 13.3)	—	—
k	ratio of specific heats C_P/C_V (Sec. 8.1)	—	—
k	thermal conductivity (Sec. 10.3)	Btu/(h · °F · ft)	W/(m · K)
k	permeability (Sec. 10.3 and Chap. 12)	ft^2	m^2
ke	kinetic energy per unit mass	Btu/lbm	J/kg
K	arbitrary constant in power law (Chap. 15)	lbf · s^n/ft^2	N · s^n/m^2
K	bulk modulus (Sec. 8.1)	lbf/in^2	Pa
K	resistance coefficient (Sec. 6.9)	—	—
KE	kinetic energy	Btu	J
l	length	ft	m
L	length	ft	m
L	angular momentum	lbm · ft^2/s	kg · m^2/s
L	scale of turbulence (Sec. 16.5)	ft	m
m	mass	lbm	kg

$\dot{m}$	mass flow rate	lbm/s	kg/s
M	molecular weight	lbm/(lbmol)	g/mol
$\mathcal{M}$	Mach number	—	—
n	number of independent variables	—	—
n	number of moles	lbmol	mol
n	arbitrary power in power law (Chap. 15)	—	—
n	frequency	cycles/s	Hz
N	$4f\,\Delta x/D$ (Sec. 8.4)	—	—
pe	potential energy per unit mass	Btu/lbm	J/kg
P	pressure	lbf/in^2	Pa
Po	power	ft · lbf/s	W
PE	potential energy	Btu	J
PC	pumping cost (Sec. 6.12)	\$/(yr · hp)	
PP	purchased-price factor for a pipe (Sec. 6.12)	\$/(in · ft)	
q	emission rate (Sec. 3.6)	lbm/s	kg/s
q_x, q_y, q_z	x, y, and z components of heat flux (Sec. 10.3)	Btu/(h · ft^2)	W/m^2
Q	volumetric flow rate	ft^3/s	m^3/s
Q	heat	Btu	J
$\mathcal{Q}$	charge	C	C
r	radius	ft	m
R	universal gas constant	See Table A.9	
R	correlation coefficient (Sec. 16.5)	—	—
R	radius of curvature (Chap. 17)	ft	m
$\mathcal{R}$	Reynolds number	—	—
$\mathcal{R}_p$	particle Reynolds number	—	—
$\mathcal{R}_{PM}$	Reynolds number for porous media	—	—
$\mathcal{R}_x$	Reynolds number based on distance from leading edge	—	—
s	entropy per unit mass or per mole	Btu/(lbm · °R) or Btu/(lbmol · °R)	J/(kg K) or J/(mol K)
s	cake compressibility coefficient (Sec. 12.4)	—	—
SG	specific gravity	—	—
t	time	s	s
t	wall thickness (Sec. 2.9)	ft	m
T	absolute temperature	°R or K	K
T	relative intensity of turbulence (Sec. 16.5)	—	—
u	internal energy per unit mass or per mole	Btu/lbm or Btu/(lbmol)	J/kg or J/mol
u^*	friction velocity (Sec. 11.4)	ft/s	m/s
u^+	V_x/u^* (Sec. 11.4)	—	—
U	internal energy	Btu	J
v	volume per unit mass	ft^3/lbm	m^3/kg
ν	fluctuating component of velocity (Chaps. 11 and 16)	ft/s	m/s
V	velocity	ft/s	m/s
V_x, V_y, V_z	x, y, and z components of velocity	ft/s	m/s
V_θ	tangential component of velocity	ft/s	m/s
V_r	radial velocity	ft/s	m/s

V_{av}	average velocity for a turbulent flow	ft/s	m/s
V_x	free-stream velocity	ft/s	m/s
V_s	superficial velocity (Sec. 12.1)	ft/s	m/s
V_I	interstitial velocity (Sec. 12.1)	ft/s	m/s
V_{mf}	minimum fluidizing velocity (Sec. 12.5)	ft/s	m/s
V	volume	ft^3	m^3
W	work	ft · lbf	J
W	weight	lbf	N
W	width	ft	m
W	volumetric solids content of slurry (Sec. 12.4)	—	—
$W_{a.o.}$	work, excluding injection work	ft · lbf	J
x, y, z	directions of coordinate axes, or lengths	ft	m
x	distance	ft	m
y	distance perpendicular to flow direction	ft	m
y^+	$(r_{wall} - r)u^*/\nu$ (Sec. 11.4)	—	—
z	elevation	ft	m
α	coefficient of thermal expansion	1/°F	1/K
α	specific resistance of filter cake (Sec. 12.4)	1/lbf	1/N
α	small angle	rad	rad
β	isothermal compressibility = 1/(bulk modulus)	$1/(lbf/in^2)$	1/Pa
γ	specific weight = ρg	lbf/ft^3	N/m^3
Γ	torque	ft · lbf	Nm
δ	boundary-layer thickness (Chap. 11)	ft	m
δ^*	displacement thickness (Sec. 11.2)	ft	m
ε	absolute roughness	ft	m
ε	porosity or void fraction or volume fraction of gas	—	—
ε	eddy (kinematic) viscosity	ft^2/s	m^2/s
ξ	vorticity = 2ω	1/s	1/s
η	efficiency	—	—
η	$y[V_x/(\nu x)]^{1/2}$ (Sec. 11.2)	—	—
η	viscosity (nonnewtonian fluids)	lbm/(ft · s) or cP	Pa · s
θ	angle	rad	rad
θ	momentum thickness (Sec. 11.2)	ft	m
θ	contact angle (Sec. 17.3)	rad	rad
μ	viscosity	lbm/(ft · s) or cP	Pa · s
ν	kinematic viscosity μ/ρ	ft^2/s or cSt	m^2/s
ρ	density	lbm/ft^3	kg/m^3
ρ	resistivity (Sec. 10.3)	Ω · m	Ω · m
σ	surface tension	lbf/ft	N/m
σ	stress	lbf/in^2	Pa
σ_{xx}	normal stress in x direction	lbf/in^2	Pa
τ	shear stress	lbf/in^2	Pa
τ_{xy}	shear stress in x direction on a face perpendicular to y axis	lbf/in^2	Pa

τ_{wall}	shear stress at a pipe wall	lbf/in^2	Pa
τ_0	shear stress at a solid surface	lbf/in^2	Pa
τ_{yield}	yield stress for Bingham fluid	lbf/in^2	Pa
ϕ	potential	ft^2/s for fluid flow	m^2/s for fluid flow
$\phi(t)$	arbitrary function of time (Sec. 16.2)	—	—
ψ	stream function	ft^2/s	m^2/s
ω	angular velocity	rad/s	rad/s

Superscripts

*	sonic condition (Chap. 8)

Subscripts

R	reservoir state in Chap. 8
S	isentropic condition (speed of sound)
1, 2	arbitrary states
x, y	conditions before and after normal shock in Chap. 8

PREFACE

This book represents an introduction to fluid mechanics for engineering students in their third academic year. It is based on the belief that engineering consists of applying scientific knowledge to find real solutions for problems of practical importance, and that the best way for a student to learn engineering is by solving such problems (in simplified form) under the supervision of competent teachers. Every effort has been made to keep the presentation clear and simple, with enough detailed examples that the instructor can assign problems without previous class discussion. A student is more likely to ask a pertinent question on some topic after working on a problem on that topic than after the material is presented only in a lecture or as a reading.

Throughout the text, emphasis is placed on the connection between physical reality and the mathematical models of reality, which we manipulate. The ultimate test of a mathematical solution is its ability to predict the results of future experiments. Because a mathematically correct consequence of inapplicable assumptions is often simply wrong, the text occasionally offers intentionally wrong solutions to caution the student.

Considerable attention is paid to the units of quantities in the equations because students usually have trouble with them, and because this reminds them that the symbols in our equations stand for real physical quantities.

A separate chapter is devoted to the balance equation. One might think that this is such a simple topic that it deserves only a few lines. However, it is a continual source of trouble to students. Furthermore, it is the most all-pervasive concept of chemical engineering, forming the basic mathematical framework for the application of the laws of thermodynamics, newtonian mechanics, stoichiometry, and for the study of chemically reacting systems.

The energy balance approach to fluid flow problems is developed before the momentum balance is introduced. This leads to a very simple and logical development of Bernoulli's equation and an intuitively satisfying treatment of fluid friction. In the undergraduate program at the University of Utah, the students study basic engineering thermodynamics before they are introduced to fluid mechanics; thus, Chap. 4 is merely a review for them.

Chapters 1 through 8 are the core of the book, covering all the basic ideas in fluid mechanics, and most of the problems are of interest to all chemical engineers. The last nine chapters introduce areas of fluid mechanics which are of great practical interest to some chemical engineers, which are not covered in an introductory course, for want of time. These chapters, which can be assigned as supplementary reading, or covered briefly in class, introduce students to the terminology and basic ideas of these fields and help them read related matters in the current literature. In our introductory course I spend one meeting each on Chaps. 9, 10, 11 and 13.

In revising the 1970 text, I have tried to take into account two major changes that have occurred in engineering since 1970. The first is the radical alteration of the teaching and practice of engineering caused by the spectacular increase in power and simultaneous decrease in cost of computers. As a result, many shortcut, approximate, and rule-of-thumb approaches are no longer needed because computers can provide rigorous or semirigorous solutions quickly and cheaply. The second change is the major commitment made by the engineering profession to switch to SI units, and the strong encouragement of engineering educators to aid in this conversion by teaching mostly or exclusively in SI units.

Computers do not make hand calculations unnecessary. No new or unfamiliar computer solution should be believed until manual plausibility checks have shown that the computer is indeed solving the problem we think it is solving and that its solution is physically reasonable. Furthermore, simply plugging values into available computer packages does not build physical insight, which is one of the most important tools of the successful engineer. Thus I believe that good pedagogy begins with hand solutions of simplified versions of the real problem, which build physical insight and some understanding of physical magnitudes, followed by computer solutions which can relax the simplifications and cover a wider variety of conditions, followed by manual plausibility checks on the computer solutions. I have tried to show this in the revised text.

After an initial rush of enthusiasm for SI, engineering educators seem to be deciding that the English system of units is not likely to vanish overnight. For this reason our students must become like educated Europeans, who speak more than one language fluently and can read and understand one or two additional languages. Our students must be fluent in SI and in the English system of units and must understand traditional metric and cgs, and be able to read and understand texts using the slug and the poundal. This second edition has a long discussion of these various systems of units; examples are presented in both SI and English units. This is unlikely to please purists of any persuasion, but it probably serves our students as well as any other approach and better than some.

My goal remains, as in the first edition, to present a text which average chemical engineering juniors can read and understand and from which they can attack a variety of meaningful problems. I have tried to help the student develop physical insight into the processes of fluid mechanics and develop the understanding that the equations on these pages truly describe what nature does. I have tried to choose examples from the students' own experiences, or which relate to things students can observe in their everyday lives. The home is a wonderful place to observe the principles of chemical engineering; good teachers help students interpret what they see in the home in terms of chemical engineering principles.

I thank the many secretaries who worked on the numerous drafts of the first edition—particularly Mrs. Diana Jennings Woodside. I also thank Anne Simmons and Heather Romney, who transcribed the first edition into computer-readable form for the second edition. I am also indebted to the many faculty and students who used the first edition and provided me with helpful comments and criticisms—especially Dr. Alan Fletcher, who used drafts of the first edition in class.

I would also like to thank the following reviewers for their many helpful comments and suggestions: Ron Darby, Texas A & M University; James Fair, University of Texas; William Schowalter, University of Illinois; and Matthew Tirrell, University of Minnesota.

Noel de Nevers

FLUID MECHANICS FOR CHEMICAL ENGINEERS

CHAPTER 1

INTRODUCTION

1.1 WHAT IS FLUID MECHANICS?

Mechanics is the study of forces and motions. Therefore, fluid mechanics is the study of forces and motions in fluids. But what is a fluid? We can all think of some things which obviously are fluids: air, water, gasoline, lubricating oil, and milk. We can also think of some things which obviously are not fluid: steel, diamonds, rubber bands, and paper. These we call solids. But there are some very interesting intermediate types of matter: Jell-o, peanut butter, cold cream, mayonnaise, toothpaste, roofing tar, library paste, bread dough, and automobile grease.

To decide what we mean by the word "fluid," first we have to consider the idea of shear stress. It is easiest to discuss shear stress in comparison with tensile stress and compressive stress; see Fig. 1.1.

In Fig. 1.1(a) a rope is holding up a weight. The weight exerts a force which tends to pull the rope apart. A stress is the ratio of the applied force to the area over which it is exerted. Thus the stress in the rope is the force exerted by the weight divided by the cross-sectional area of the rope. The force which tries to pull things apart is called a *tensile force*, and the stress it causes is called a *tensile stress*.

In Fig. 1.2(b) a steel column is holding up a weight. The weight exerts a force which tends to crush the column. This kind of force is called a *compressive force*, and the stress in the column, the force divided by the cross-sectional area of the column, is called a *compressive stress*.

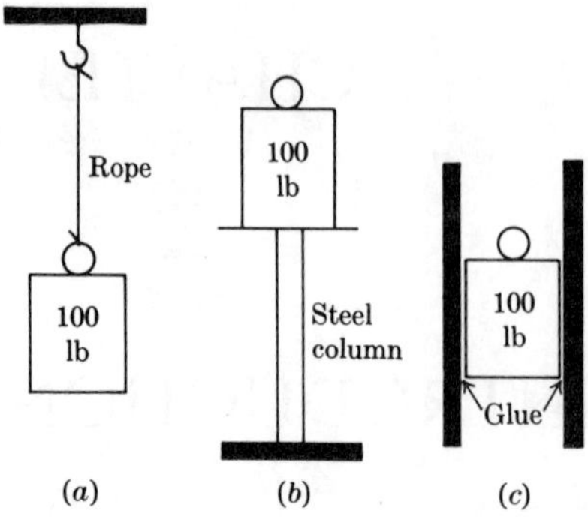

FIGURE 1.1
Comparison of tensile, compressive, and shear stresses. (*a*) The rope is in tensile stress; (*b*) the column is in compressive stress; (*c*) the glue is in shear stress.

In Fig. 1.1(*c*) some glue is holding up a weight. The weight exerts a force that tends to pull the weight down the walls and thus to *shear* the glue. This force, which tends to make one surface *slide* parallel to an adjacent surface, is called a *shear force*, and the stress in the glue, the force divided by the area of the glue joint, is called a *shear stress*.

A more detailed examination of these examples would show that all three kinds of stress were present in each case, but those we have identified are the main ones. (For more information on this topic, see any text on strength of materials.)

In our attempt to differentiate between fluids and solids, we can now say that solids are substances which can permanently resist very large shear forces. When subjected to a shear force solids move a short distance (elastic deformation), thereby setting up internal shear stresses which resist the external force, *and then they stop moving*. Materials that obviously are fluids cannot permanently resist a shear force, no matter how small. When subjected to a shear force, fluids start to move and *keep on moving as long as the force is applied*.

Substances intermediate between the two are materials which can permanently resist a small shear force but cannot permanently resist a large one. For example, if we put a blob of any obvious liquid on a vertical wall, gravity will make it run down the wall. If we attach a piece of steel or diamond securely to a wall, it will remain there, no matter how long we wait. If we attach some peanut butter to a wall, it will probably stay; but if we increase the shear stress on the peanut butter by spreading it with a knife, the peanut butter will flow just as a fluid would. We cannot, of course, spread steel with a knife.

If, as shown above, the relevant difference between peanut butter and steel is the magnitude of the shear stress that the material can resist, then the difference is one of degree, not of kind. At very high shear stresses, steel can be made to "flow like a fluid": this is called *plastic deformation* in books on the strength of materials. In the remainder of this book, we talk mostly about materials, such as air and water, which cannot permanently resist *any* shear force. However, it is well to keep our minds open to other possibilities of "fluid" behavior [1].†

1.2 WHAT GOOD IS FLUID MECHANICS?

The problems in fluid mechanics are basically no different from those in ordinary mechanics (the mechanics of solids) or in thermodynamics. Therefore, in principle, we can solve problems in fluid mechanics by the same methods used in mechanics or thermodynamics. However, for many problems involving the flow of fluids (or the movement of bodies through fluids), we use a combination of the problem-solving methods of mechanics and those of thermodynamics to obtain a solution. Furthermore, the methods that work for hydraulics problems (dams, canals, locks, river flow, etc.) are applicable, with slight modifications, to aerodynamics problems (airplanes, rockets, wind forces on bridges, etc.) and to problems of special interest to chemical engineers, such as the flow in chemical reactors, distillation columns, or polymer extrusion dies. Therefore, it makes sense to combine the study of this class of similar problems into one discipline, which we call *fluid mechanics*.

We may grasp the significance of fluid mechanics by considering the importance of fluids in our lives. Important fluids are the air we breathe, the water we drink, many of the foods we consume, most fuels that heat our houses or propel our vehicles, and various fluids in our bodies, such as blood, which make up our internal environment. Without some grasp of the behavior of fluids, we can have only a very limited understanding of the world around us.

Some of the subdivisions and applications of fluid mechanics are as follows:

1. Hydraulics: the flow of water in rivers, pipes, canals, pumps, turbines
2. Aerodynamics: the flow of air around airplanes, rockets, projectiles, structures
3. Meteorology: the flow of the atmosphere
4. Particle dynamics: the flow of fluids around particles, the interaction of particles and fluids (i.e., dust settling, slurries, pneumatic transport, fluidized beds, air pollutant particles)
5. Hydrology: the flow of water and waterborne pollutants in the ground
6. Reservoir mechanics: the flow of oil, gas, and water in petroleum reservoirs
7. Multiphase flow: coffee percolators, oil wells, carburetors, fuel injectors, combustion chambers, sprays

† Numbers in brackets refer to items listed in the References at the end of the book.

8. Combinations of fluid flow: with chemical reactions in combustion, with electromagnetic phenomena in magnetohydrodynamics, with mass transport in distillation or drying
9. Viscous dominated flows: lubrication, injection molding, wire coating, volcanoes, and continental drift

1.3 BASIC IDEAS IN FLUID MECHANICS

Fluid mechanics is based largely on working out the detailed consequences of four basic ideas:

1. The principle of the conservation of mass
2. The first law of thermodynamics (the principle of the conservation of energy)
3. The second law of thermodynamics
4. Newton's law of motion which may be summarized as $F = ma$

Each of these four ideas is a generalization of experimental data. No one of them can be deduced from the others or from any other prior principle. None can be "proved" mathematically. Rather, they stand on their ability to predict correctly the results of any experiment ever run to test them.

Sometimes in fluid mechanics we may start with these four ideas and the measured physical properties of the materials under consideration and proceed directly to solve mathematically for the desired forces, velocities, and so on. This is generally possible only in the case of very simple flows. The observed behavior of a great many fluid flows is too complex to be solved directly from these four principles, so we must resort to experimental tests. Through the use of techniques called *dimensional analysis* (Chap. 13) often we can use the results of one experiment to predict the results of a much different experiment. Thus, careful experimental work is very important in fluid mechanics. With the development of supercomputers, we are now able to solve many complex problems mathematically by using the methods outlined in Chaps. 10 and 11, which previously would have required experimental tests. As computers become faster and cheaper, we will probably see additional complex fluid mechanics problems solved on supercomputers. Ultimately, the computer solutions must be tested experimentally.

These four ideas are applied to fluid mechanics problems as follows: In Chap. 1 we discuss some of the measurable properties of fluids and some definitions. In Chap. 2 we apply Newton's law of motion to the particularly simple case of a fluid which is not moving. In Chap. 3 we explain and apply the principle of the conservation of mass. In Chap. 4 we consider the principle of the conservation of energy. In Chaps. 5 and 6 we apply the principle of the conservation of energy to a class of relatively simple fluid flows which includes many flows of great practical importance. In Chap. 7 we recast Newton's laws of motion into the form of a momentum balance, which is convenient for many fluid flow problems, and we show some applications. In Chap. 8 we apply the material of the preceding chapters to the flow of gases at high velocity, where several additional interesting phenomena are observed.

The first eight chapters form the core of the book. The remaining nine are intended as extra reading which will introduce the student to further topics in fluid mechanics. No new principles are introduced in these last chapters, but special topics and techniques in fluid mechanics are discussed. These chapters show the reader the connections between the basic material in Chaps. 1 to 8 and the sometimes special terminology and special ideas in other areas of fluid mechanics.

The students using this book should have completed a course in elementary thermodynamics. Chapters 3 and 4 serve as a review of matter previously covered; they are included because the principles involved are central to fluid mechanics. It is assumed that students are familiar with the second law of thermodynamics, which is used occasionally.

Remember that this entire book is devoted to the application of the four basic principles and the results of experimental tests to fluid-flow problems. Although the details can become quite involved, the basic ideas are few.

1.4 LIQUIDS AND GASES

Fluids come in two types: liquids and gases. On the molecular level, these are quite different. In liquids the molecules are close together and are held together by significant forces of attraction; in gases the molecules are relatively far apart and have very weak forces of attraction. As temperature and pressure increase, these differences become less and less, until the liquid and gas become identical at the critical temperature and pressure. The difference between the behavior of liquids and gases is most marked when these fluids are expanded. Suppose that some fluid completely fills the space below the piston in Fig. 1.2. When we raise the piston, the volume occupied by the fluid is increased. If the fluid is a gas, it expands readily, filling all the space vacated by the piston; gases can expand without limit to occupy space made available to them. But if the fluid is a liquid, then as the piston is raised, the liquid can

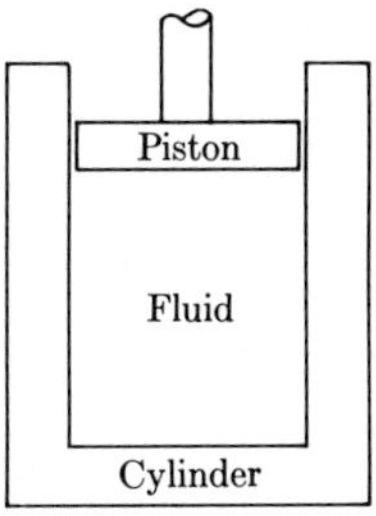

FIGURE 1.2
Piston and cylinder.

expand only a small amount and then it can expand no more. What fills the space between the piston and the liquid? Part of the liquid must turn into a gas by boiling, and this gas expands to fill the vacant space. This can be explained on the molecular level by saying that there is a maximum distance between molecules over which the attractive forces hold them together to form a liquid and that when the molecules separate more than this distance, they cease to behave as a liquid and behave instead as a gas.

Because of their closer molecular spacing, liquids normally have higher densities, viscosities, refractive indices, etc., than gases (see Prob. 1.2). In engineering this frequently leads to quite different behaviors of liquids and of gases, as we shall see.

1.5 PROPERTIES OF FLUIDS

Among those physical properties of fluids that are found in our calculations most often are density, viscosity, and surface tension.

A. Density

The *density* ρ is defined as the mass per unit volume:

$$\rho = \frac{m}{V} \tag{1.1}$$

We are all aware of the differences in density between various materials, such as lead and wood. How can we measure the density of a material? If we wish to know the density of a liquid, we can weigh a bottle of known volume (determine its mass), fill it with the liquid, weigh it again, and compute the density with the aid of Eq. 1.1. (This is a standard laboratory method of determining liquid density; the special weighing bottles designed for this purpose are called *pycnometers*.) If we want to know the density of a cubic block of a solid, we can measure the length of its sides, compute its volume, weigh it, and apply these results in Eq. 1.1.

Now suppose we want to determine the density of a piece of Swiss cheese. If we have a large block of the cheese, we can cut off a cube, measure its sides, compute its volume, weigh it, and then calculate its density. This is an average density, one that includes the density of the air in the holes in the cheese. As long as we are dealing with large pieces of cheese, it is a satisfactory density. Suppose, however, we want to find the density at some point inside a large block of the cheese. If we can cut open the cheese and if we find that the point in question is in the solid cheese and not in one of its holes, we can find the density easily enough; or if the point in question is in a hole, we can find the density of the air in the hole. But if the point is on the surface of a hole, the problem is more difficult. Then the density is discontinuous; see Fig. 1.3. In fact, there is no meaningful single value of the density at x.

Why this long discussion about the density of Swiss cheese? Because the world is full of holes! Atomic physics tells us that even in a solid bar of steel

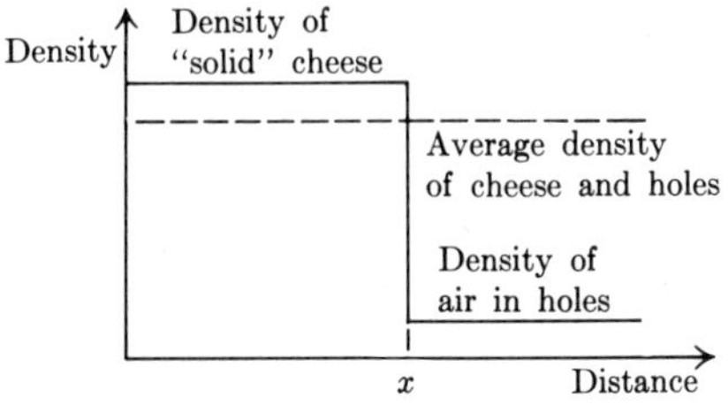

FIGURE 1.3
Density of Swiss cheese.

the space occupied by the electrons, protons, and neutrons is a very small fraction of the total space; the rest presumably is empty. Furthermore, even at the molecular level there are holes; in a typical gas, the space actually occupied by the individual gas molecules at any instant is a small fraction of the total space. Thus, in any attempt to speak of density at a given point, we are in the same trouble as we were in concerning the Swiss cheese. Therefore, we must restrict the definition of density to samples large enough to average out the holes. This causes no problem in fluid mechanics, because of the size of the samples normally used, but it indicates that the concept of density does not apply readily to samples of molecular and subatomic sizes.

In addition, we must be careful in defining the densities of composite materials. For example, a piece of reinforced concrete consists of several parts with different densities. In discussing such materials, we must distinguish between the *particle densities* of the individual pebbles or steel-reinforcing bars and the *bulk density* of the mixed mass. When we refer to bulk density, our sample must be large compared with the dimensions of one particle. Some examples of composite solid materials are cast iron, fiberglass-reinforced plastics, and wood. Some examples of composite liquids are slurries, such as muds, milkshakes, and toothpaste, and emulsions, such as homogenized milk, mayonnaise, and cold cream. Smokes and clouds behave as composite gases.

B. Specific Gravity

Specific gravity (SG) is defined as follows:

$$\text{SG} = \frac{\text{density}}{\text{density of water at specified temperature and pressure}} \tag{1.2}$$

This definition has the merit of being a ratio and hence a pure number that is independent of the system of units chosen. Ocassionally it leads to confusion, because some specific gravities are referred to water at 60°F, others to water at 70°F, and still others to water at 39°F = 4°C (all at a pressure of 1 atm). The differences are small but great enough to cause trouble.

If the temperature of the water is specified as 39°F (4°C), then the density of water is 1.000 g/cm^3. (The gram was defined to make this number come out 1.000.) Thus, if this basis of measurement is chosen, then specific gravities become numerically identical to densities expressed in grams per cubic centimeter or kilograms per liter or metric tons per cubic meter.

Many process industries use special scales of fluid density, which are usually referred to as *gravities*. Some are the American Petroleum Institute (API) gravity for oil and petroleum products (Prob. 1.5), Brix gravity for the sugar industry, and Baume gravity for sulfuric acid. Each scale is directly convertible to density, and conversion tables and formulas are widely available. Specific gravities of gases also are used; they are based on the density of air at 1 atm and a specific temperature (usually 4°C, sometimes 60 or 70°F).

Throughout this text we use specific gravities referred to water at 4°C. Thus a fluid with specific gravity of 0.8 is a fluid with density 0.8 g/cm^3.

C. Viscosity

The *viscosity* is a measure of resistance to flow. If we tip over a glass of water on the table, the water will spill out before we can stop it. If we tip over a jar of honey, we probably can set it upright again before much honey flows out; this is possible because the honey has much greater resistance to flow, more viscosity, than water. A more precise definition of viscosity is possible in terms of the following experiment.

Consider two long, solid plates separated by a thin film of fluid (see Fig. 1.4).† If we slide the upper plate steadily in the x direction with velocity V_0, a force is needed to overcome the friction in the fluid between the plates. This force is different for different velocities, plate sizes, fluids, and distances between the plates. We can eliminate the effect of different plate sizes, however, by measuring the force per unit area of the plate, which we define as the shear stress τ.

It has been demonstrated experimentally that at low values of V_0 the velocity profile in the fluid between the plates is linear, i.e.,

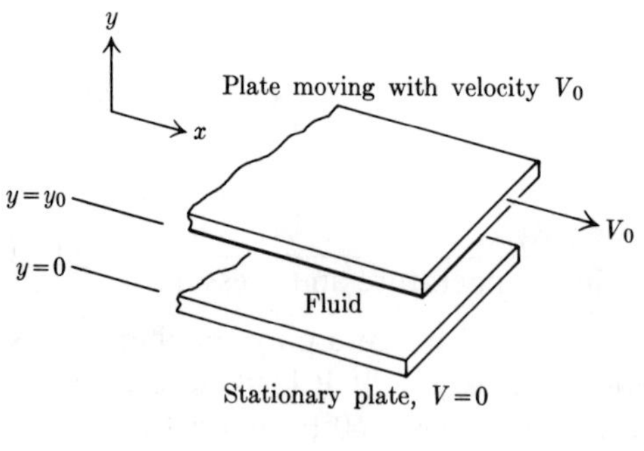

FIGURE 1.4
The sliding-plate experiment.

† This experiment is easy to grasp conceptually and mathematically but difficult to perform, because the fluid leaks out at the edges. Other experiments which are more complex mathematically but easier to perform actually are used to measure viscosities. Some are discussed in Sec. 6.3 and Chap. 15.

$$V = \frac{V_0 y}{y_0} \tag{1.3}$$

so that

$$\frac{dV}{dy} = \frac{V_0}{y_0} \tag{1.4}$$

It has also been demonstrated experimentally that for most fluids the results of this experiment can be shown most conveniently on a plot τ versus dV/dy (see Fig. 1.5). As shown here, dV/dy is simply a velocity divided by a distance. In more complex geometries, it is the limiting value of such a ratio at a point. It is commonly called the *shear rate*, *rate of strain*, and *rate of shear deformation*, which all mean exactly the same thing. Four different kinds of curve are shown as experimental results in the figure. All four are observed in nature. The behavior most common in nature is that represented by the straight line through the origin. This line is called *newtonian* because it is described by Newton's law of viscosity

$$\tau = \mu \frac{dV}{dy} \tag{1.5}$$

This equation says that the shear stress τ is linearly proportional to the velocity gradient dV/dy. It is also the definition of viscosity, because we can rearrange it as

$$\mu = \frac{\tau}{dV/dy} \tag{1.6}$$

Here μ is called the *viscosity* or the *coefficient of viscosity*.†

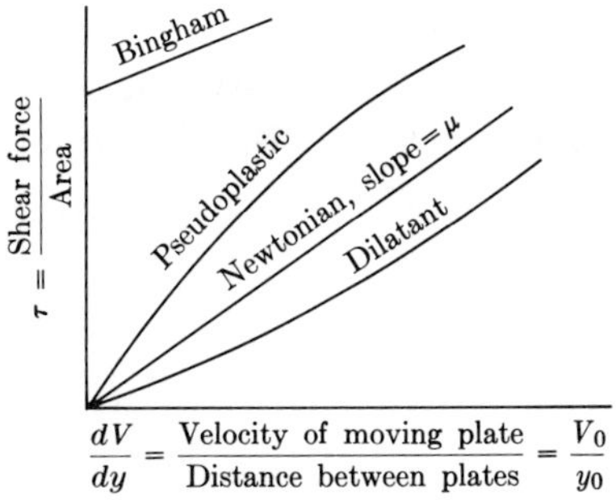

FIGURE 1.5
Possible outcomes of the sliding-plate experiment at constant temperature and pressure.

† We occasionally see this equation written with a minus sign in front of the τ. This is done so that the equation will have the same form as the heat conduction and mass diffusion equations [2]. Since the direction of the shear stress to which we assign the positive sign is arbitrary, we can introduce this minus sign and reverse our idea of the direction of τ, so that the result is always the same as in Eq. 1.6.

For fluids such as air, the value of μ is very low; therefore, their observed behavior is represented in Fig. 1.5 by a straight line through the origin, very close to the dV/dy axis. For fluids such as corn syrup, the value of μ is very large, and the straight line through the origin is close to the τ axis.

Fluids that exhibit this behavior in the sliding-plate experiment (i.e., fluids that obey Newton's law of viscosity) are called *newtonian fluids*. All others are called *nonnewtonian fluids*. Which fluids are newtonian? All gases are newtonian. All liquids for which we can write a simple chemical formula are newtonian, such as water, benzene, ethyl alcohol, carbon tetrachloride, and hexane. Most solutions of simple molecules are newtonian, such as aqueous solutions of inorganic salts and of sugar. Which fluids are nonnewtonian? Generally, nonnewtonian fluids are complex mixtures: slurries, pastes, gels, polymer solutions, etc. Most nonnewtonian fluids are composed of molecules or particles that are much larger than water molecules, such as the sand grains in a mud or the collagen molecules in gelatin, which are thousands or millions of times larger than water molecules.

In discussing nonnewtonian fluids, we must agree on what we mean by viscosity. If we retain the definition given by Eq. 1.6, then the viscosity can no longer be considered a constant independent of dV/dy for a given temperature, but must be considered a function of dV/dy. This is shown in Fig. 1.6. Here each of lines OA, OB, and OC have slope μ, so the viscosity is decreasing with increasing dV/dy. Using this definition, we can observe that there are three common types of nonnewtonian fluid (Fig. 1.5):

1. *Bingham fluids*, sometimes called *Bingham plastics*, resist a small shear stress indefinitely but flow easily under larger shear stresses. One may say that at low stresses the viscosity is infinite and that at higher stresses the viscosity decreases with increasing velocity gradient. Examples are bread dough, toothpaste, jellies, and some slurries.
2. *Pseudoplastic fluids* show a viscosity that decreases with increasing velocity gradient. Examples are most slurries, muds, polymer solutions, solutions of natural gums, and blood.

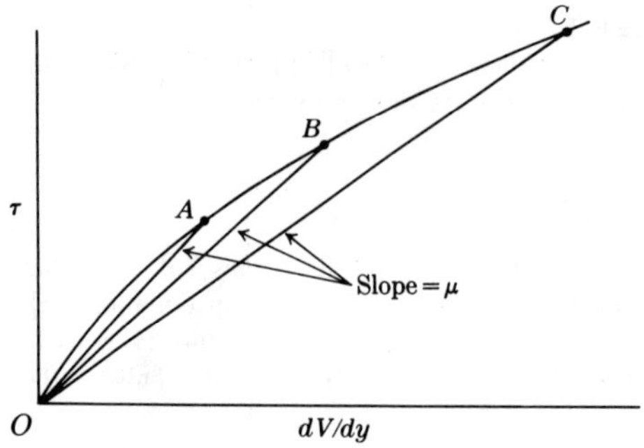

FIGURE 1.6
Viscosity of nonnewtonian fluids at constant temperature.

3. *Dilatant fluids* show a viscosity that increases with increasing velocity gradient. They are uncommon, but starch suspensions behave in this way.

So far, we have assumed that the curve of τ versus dV/dy is not a function of time or that, e.g., if we move the sliding plate at a constant speed, we will always require the same force. This is true of most fluids, but not all. A more complete picture is given in Fig. 1.7. There we see a constant dV/dy slice out of the solid constructed of τ versus dV/dy versus time. There are three possibilities:

1. The viscosity can remain constant with time, in which case the fluid is called *time-independent*.
2. The viscosity can decrease with time, in which case the fluid is called *thixotropic*.
3. The viscosity can increase with time, in which case the fluid is called *rheopectic*.

All newtonian fluids are time-independent, as are most nonnewtonian fluids. Many thixotropic fluids are known, of which almost all are slurries or solutions of polymers, and a few examples of rheopectic fluids are known.

In addition, some fluids can show not only the kinds of behavior represented in Figs. 1.6 and 1.7 but also elastic properties, which allow them to "spring back" when a shear force is released. These are called *viscoelastic fluids*. The most common example is the rubber cement sold at stationery stores. Its viscoelastic properties can be demonstrated most easily by starting to pour a little out of the bottle and then snapping it back into the bottle with a

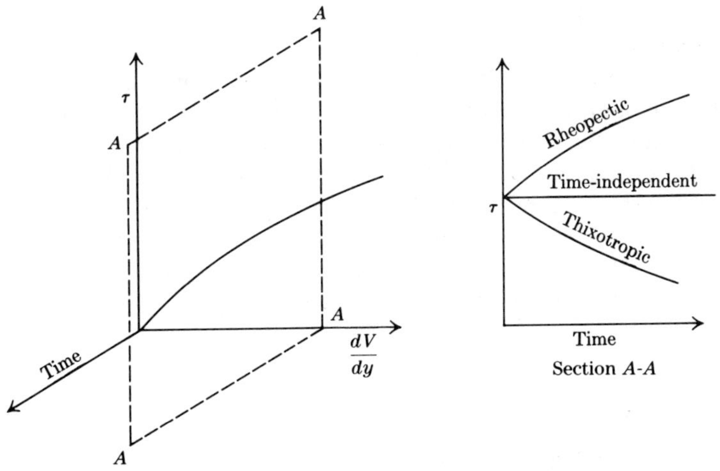

FIGURE 1.7
Time-dependent viscosity behavior at constant temperature.

quick jerk of the hand. The same can be done with an egg white, which is also viscoelastic. This is quite impossible with any ordinary fluid such as water.

These strange types of fluid behavior have considerable practical use. A good toothpaste should be a Bingham fluid, so that it can be easily squeezed out of the tube but will not drip off the toothbrush as water or honey would. A good paint should be thixotropic, so that in the can it will be very viscous and the pigment will not settle to the bottom, but when it is stirred, it will become less viscous and can be easily brushed onto a surface. In addition, the brushing should temporarily reduce the viscosity, so that the paint will flow sideways and fill in the brush marks (called *leveling* in the paint industry); then, as it stands, its viscosity should increase, so that it will not form drops and run down the wall. A good motor oil should be pseudoplastic, so that in the bearings, where the value of dV/dy is high, it will offer little frictional resistance and so that at all the gaskets and joints, where the value of dV/dy is low, it will be viscous and not leak through.

Most engineering applications of fluid flow involve water, air, gases, and simple fluids. Therefore, most fluid flow problems have to do with newtonian fluids, as do most of the problems in this book. Nonnewtonian fluids are important, however, precisely because of their nonnewtonian behavior; they are discussed in Chap. 15.

The viscosity of very simple gases, such as helium, can be calculated for all temperatures and pressures from the kinetic theory of gases by using only one experimental measurement [2]. For the viscosities of most gases and all liquids, several experimental data points are required, although ways of predicting the change in viscosity with changing temperature and pressure are available [2]. As a general rule, the viscosity of gases increases slowly with increasing temperature, and the viscosity of liquids decreases rapidly with increasing temperature. The viscosity of both gases and liquids is practically independent of pressure at low and moderate pressures.

The basic unit of viscosity is the *poise*, where $1\ \text{P} = 1\ \text{g}/(\text{cm}\cdot\text{s}) = 0.1\ \text{Pa}\cdot\text{s} = 6.72 \times 10^{-2}\ \text{lbm}/(\text{ft}\cdot\text{s})$. It is widely used for materials such as high-polymer solutions and molten polymers. However, it is too large a unit for most common fluids. By sheer coincidence the viscosity of pure water at about 68°F is 0.01 P; for that reason the common unit of viscosity in the United States is the *centipoise*, where $1\ \text{cP} = 0.01\ \text{P} = 0.01\ \text{g}/(\text{cm}\cdot\text{s}) = 6.72 \times 10^{-4}\ \text{lbm}/(\text{ft}\cdot\text{s}) = 0.001\ \text{Pa}\cdot\text{s}$. Hence, the viscosity of a fluid, expressed in centipoise, is the same as the ratio of its viscosity to that of water at room temperature. The viscosities of some common liquids and gases are shown in App. A.1.

D. Kinematic Viscosity

In many engineering problems, viscosity appears only in the relation of viscosity divided by density. Therefore, to save writing, we define

$$\text{Kinematic viscosity} = \nu = \frac{\mu}{\rho} \tag{1.7}$$

The most common unit of kinematic viscosity is the centistoke (cSt):

$$1\text{ cSt} = \frac{1\text{ cP}}{1\text{ g/cm}^3} = 1.08 \times 10^{-5}\,\frac{\text{ft}^2}{\text{s}} = 10^{-6}\,\frac{\text{m}^2}{\text{s}}$$

(The kinematic viscosity of water at room temperature is 1 cSt.) To avoid confusion over which viscosity is being used, some writers refer to the viscosity μ as the *absolute viscosity*. In Chap. 6 we will see some examples of the practical convenience of the kinematic viscosity.

E. Surface Tension

Liquids behave as if they were surrounded by a skin that tends to shrink, or contract, like a sheet of stretched rubber. This phenomenon is known as *surface tension*. It is seen in many everyday events, the most disheartening of which is the tendency of water, when poured slowly from a glass, to run down the edge of the glass (see Fig. 1.8).

Surface tension is caused by the attractive forces in liquids. All the molecules attract each other; those in the center are attracted equally in all directions, but those at the surface are drawn toward the center because there are no liquid molecules in the other direction to pull them outward (see Fig. 1.9). The effort of each molecule to get into the center causes the fluid to try to take a shape that would have the greatest number of molecules nearest the center, a sphere (Prob. 1.9). Any other shape has more surface per unit volume; therefore, regardless of the shape of a fluid, the attractive forces tend to pull the fluid into a sphere. The fluid thus tries to minimize its surface area.[†]

The tendency of a surface to contract can be measured with the device shown in Fig. 1.10. A wire frame with one movable side is dipped into a fluid

FIGURE 1.8
Disheartening effect of surface tension.

[†] An analogous situation in two dimensions is observable in the behavior of some army ants. They travel in large groups, and when viewed from above, the swarm often looks like a circle. The reason appears to be that the ants are attracted by the smell of other ants, and hence all try to go to where the smell is strongest, the center. The ants all stay in one plane, so the result is the plane figure with the smallest possible ratio of perimeter to area—a circle [3].

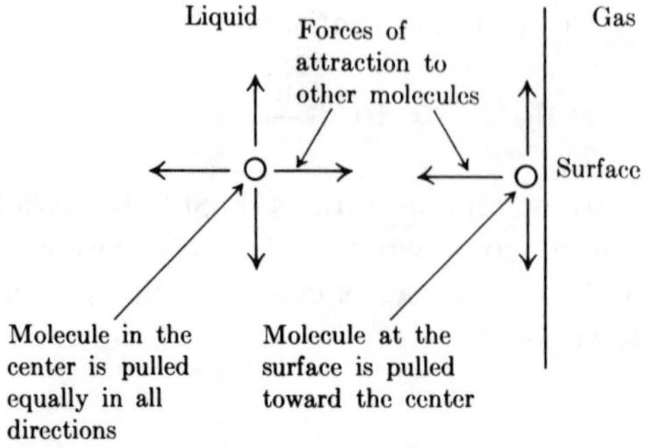

FIGURE 1.9
Surface tension is caused by the attractive forces between molecules.

and carefully removed with a film of fluid in the space formed by the frame. The film tries to take up a spherical shape, but since it adheres to the wire, it draws the movable part of the frame inward. The force necessary to resist this motion is measured by a weight. It is found experimentally that the ratio of the force to the length of the sliding part of the wire is always the same for a given fluid at a given temperature, regardless of the size of the apparatus. The film in the frame has two surfaces (front and back), so the force per unit length of *one* of the surfaces is exactly one-half the total measured force. The surface tension of the liquid is then defined as

$$\text{Surface tension} = \frac{\text{force of one film}}{\text{length}} \quad \text{or} \quad \sigma = \frac{F}{l} \tag{1.8}$$

This force is very slightly influenced by what the surrounding gas is—air or water vapor or some other gas. Typical values of the surface tension of liquid surfaces exposed to air are shown in Table 1.1.[†]

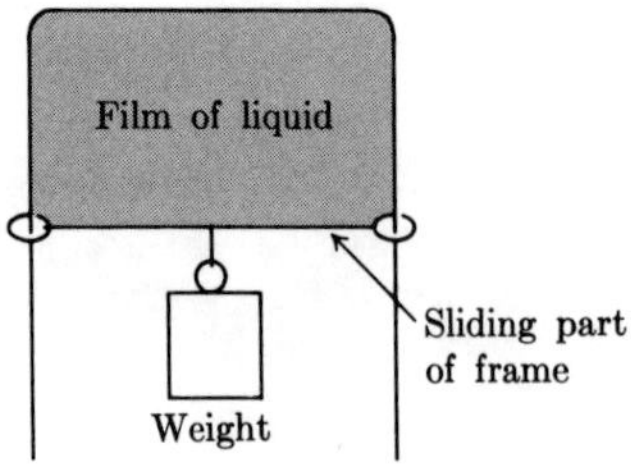

FIGURE 1.10
One way to measure surface tension.

[†] The device shown in Fig. 1.10 is easy to understand but not a very practical measuring device; more practical ones are discussed in Chap. 17.

TABLE 1.1
Surface tensions of pure fluids exposed to air at 68°F (20°C)

Fluid	Surface tension, lbf/in
Acetic acid	0.00016
Acetone	0.00014
Benzene	0.00016
Carbon tetrachloride	0.00015
Ethyl alcohol	0.00013
n-Octane	0.00012
Toluene	0.00016
Water	0.00043
Mercury	0.0030

Observe that most organic liquids have about the same surface tension (0.00015 lbf/in) while that of water is about 3 times as high and that of mercury 20 times as high.

We indicated that the fluid adheres to the solid in the apparatus in Fig. 1.10. Fluids adhere strongly to some solids but not to others. For example, water adheres strongly to glass but very weakly to polyethylene. This greatly complicates the whole subject of surface tensions; the phenomenon shown in Fig. 1.8 occurs much more often with glass, ceramic, or metal cups than with polyethylene or Teflon cups.

Two other effects due to surface tension are the capillary rise of liquids in small tubes and porous wicks (without which kerosene lanterns or copper sweat-solder fittings would not work at all) and the tendency of jets of liquid to break up into drops (as from a garden hose or diesel fuel injector). Surface tension effects are very important in systems involving large surface areas, such as emulsions (mayonnaise, cold cream, water-based paints) and multiphase flow through porous media (oil fields). We discuss the effects in Chap. 17; see also Refs. 4 and 5.

1.6 PRESSURE

Pressure is defined as a compressive stress, or compressive force per unit area. In a stationary fluid the compressive force per unit area is the same in all directions. In a solid or in a moving fluid the compressive force per unit area at some point is not necessarily the same in all directions. We can visualize why by considering what happens when we squeeze a rubber eraser between our fingers (see Fig. 1.11). As we squeeze the eraser, it becomes thinner and longer, as shown. If we analyze the stresses in the eraser, we find that the eraser is in compression in the y direction and in tension in the x direction. (This seems strange, but the eraser has been stretched in the x direction, and its elastic forces are trying to pull it back; hence the tension.) The contraction

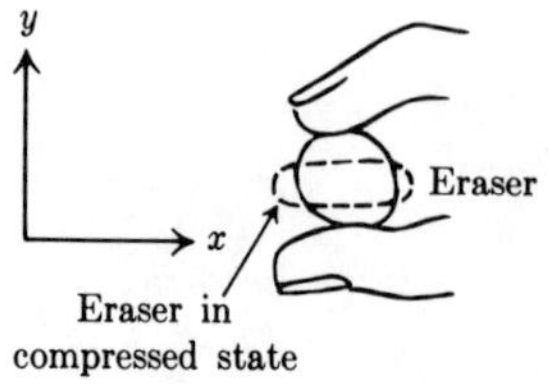

FIGURE 1.11
The response of an elastic solid to compression in one direction.

in one direction and expansion in another in an elastic solid are described in terms of Poisson's ratio, which is discussed in any text on strength of materials. Because the tensile and compressive forces are at right angles to each other, there is also a strong shear stress at 45° to the x axis.

Now, what would happen if we held our fingers in a cup of water and tried to squeeze the water between our fingers? Obviously, the water would run out from between our fingers, and our fingers would come together. Why? When we start to squeeze the water, it behaves as the eraser does: it sets up internal shear and tensile forces in the same directions as the eraser. However, ordinary fluids cannot permanently resist shear forces, so the water begins to flow and finally flows away. The eraser also flowed, until it had taken up a new shape, in which its internal tensile and shear resistance were enough to hold our fingers apart. Water cannot set up such resistance and so simply flows away.

If we really wanted to squeeze the water, we would put it in some container that would prevent its flowing out to the side. If we did this with the eraser, then as we compressed it from the top, it would press out on the sides of the container. So also does water.

The foregoing is a description of why the pressure at a point in a fluid at rest is the same in all directions. It is not a proof of that fact; for a proof see App. B.

What we mean by pressure is not as clear for a solid as for a liquid. The compressive stresses at a given point in a solid are not the same in all directions. The usual definition of pressure in a solid is as follows: Pressure at a point is the average of the compressive stresses measured in three perpendicular directions. Since, as we have seen, these three stresses are all the same in a fluid at rest, the two definitions are the same. For a fluid in motion the three perpendicular compressive stresses may not be the same. However, for this difference to be significant, the shear stresses must be very large, well outside the range of normal problems in fluid mechanics, Therefore, we normally extend the notion that pressure in a fluid at rest is the same in all directions to fluids in motion, with the reservation that at very high shear stresses (such as in the flow of metals or polymer melts through forming dies) this is not necessarily true. For polymer solutions and polymer melts the differences between the compressive stresses in directions at right angles to one another can be very significant and can lead to behavior quite different from that of simple fluids. For more on this subject, see Lodge [6].

In the solution of many problems, particularly those involving gases, it is most convenient to deal with pressures in an absolute sense, i.e., pressures relative to a compressive stress of zero; these are called *absolute pressures*. In the solution of many other problems, particularly those involving free surfaces, such as are encountered in hydraulics and situations with open tanks, it is more convenient to deal with pressures above an arbitrary datum, which is the local atmospheric pressure. Pressures relative to the local atmospheric pressure are called *gauge pressures*.

Because both systems of measurement are in common use, it is necessary to make clear which kind of pressure we mean when we write "a pressure of $15\,\text{lb/in}^2$." [This unit is also called psi (pounds per square inch). It is usual to say "15 psi absolute" or "15 psia" for absolute pressure and "15 psi gauge" or "15 psig" for gauge pressure.] The SI unit of pressure is the pascal, where $1\,\text{Pa} = 1\,\text{N/m}^2$. There does not seem to be a common set of abbreviations for pascal absolute and pascal gauge, so these must be written out.

Another two-datum situation is found in the measurement of elevation. Mountaintops, road routes, and rivers normally are surveyed relative to mean sea level, which serves as an "absolute" datum, but most buildings are designed and constructed relative to some local elevation (usually a marker in the street). See Fig. 1.12. In both cases, the most common measuring method gives answers in terms of the local datum. Most pressure gauges read the difference between the measured pressure and the local atmospheric pressure. For instance, the pressure gauge on the compressed air system in the figure would read 20 psig = 137.9 kPa gauge; the building height (by tape measure or transits) might be given as 100 ft = 30.5 m in elevation. Both such measurements usually involve negative values, based on the local datum; the basement has a negative elevation relative to the street, −30 ft = −9.15 m, and for the vacuum system, with a negative pressure relative to the atmosphere, −5 psig = −34.5 kPa gauge.

Negative elevations relative to sea level can exist; the Dead Sea, for instance, is about 1200 ft (366 m) below sea level. Can negative absolute

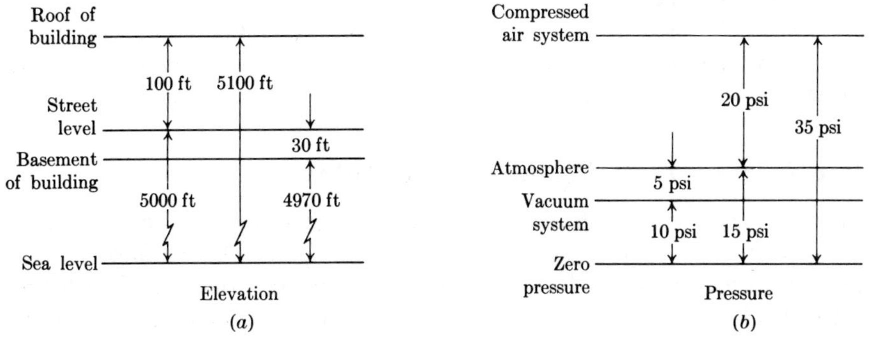

FIGURE 1.12
The relation between gauge and absolute pressure and a comparison with elevation measurements.

pressures exist? Certainly. A negative absolute pressure is a negative compressive stress, i.e., a tensile stress. These occur often in solids, very rarely in liquids, never in gases. They are rare because all liquids possess a finite vapor pressure. If the pressure of a liquid is reduced below its vapor pressure, the liquid boils and thus replaces the low pressure with the equilibrium vapor pressure of the liquid. However, this boiling never takes place spontaneously in an absolutely pure liquid [7], but rather occurs around small particles of impurities or at the wall of the container. (Most people have observed this phenomenon when they pour a cold carbonated drink into a glass; the bubbles form mostly at the edge of the glass, not in the bulk of the liquid. It can be shown dramatically by dropping some sugar into a cold, fresh glass of soft drink; do this over a sink!) Thus if a liquid is very pure and the surfaces of its container are very smooth, the liquid can exist in tension at a negative absolute pressure. This situation is unstable, and a slight disturbance can cause the liquid to boil [8].

1.7 FORCE, MASS, AND WEIGHT

In fluid mechanics often we are concerned with forces, masses, and weights. The problem of units of force and mass is discussed in the next section. An unbalanced force tends to make things change speed or direction. Most forces in the world are balanced by opposite forces (a building exerts a force on the ground, the ground exerts an equal and opposite force on the building, and neither moves). To make anything start moving or stop moving, we must exert an unbalanced force.

Mass is an indication of how much matter is present. The more matter, the more mass. (We may think of matter in any size, as bricks, molecules, atoms, nucleons, quarks, etc.) Mass is also an indicator of how hard it is to get some amount of matter moving or how hard it is to stop the matter once it is moving. We can all stop a baseball moving at 50 ft/s (15.2 m/s) and suffer little more damage than a possible sore hand. If we step in front of an automobile moving at that speed, we will certainly be killed. The automobile has much more mass; it is much harder to stop.

Weight is a force. It is the force that a body exerts due to the acceleration of gravity. When there is no gravity, there is no weight (e.g., in earth satellites there is no apparent gravity; this state is referred to as *weightlessness*).

1.8 UNITS AND CONVERSION FACTORS

Engineering is about real physical things, which can be measured and described in terms of the units of measure. Most engineering calculations involve these units of measure. It would be simple if there were only one set of such units that the whole world agreed on and used, but that is not the case today. In the United States, most measurements use the English system of units, based on the foot, pound, and degree Fahrenheit, but most of the world uses the metric (or SI) system of units based on the meter, kilogram, and degree Celsius. The metric system has been legally accepted in the United States since 1866, and it had been the declared policy of the U.S. government to convert to metric since 1975 [9]. Progress has been disappointingly slow.

The situation is similar with languages; it would be easier if we all spoke one language. But we do not; the world has many languages. Most educated Europeans speak at least two languages well and generally can read one or two more. Similarly, U.S. engineers must be fluent in English and in metric units, be able to understand older literature written in the centimeter-gram-second (cgs) system, and be familiar with variant English systems that use the poundal or the slug and with specialized industrial units, such as the 42-gal barrel for petroleum products or pressure differences expressed in inches of water. U.S. engineers must even deal with mixed systems, such as air pollutant emissions expressed in grams per mile or grains per cubic meter. Furthermore they must understand the differences between the common-use version of the metric system and SI; they will be better able to deal with those differences if they understand why the differences arise.

In fluid mechanics, most often we deal with dimensioned quantities, such as 12 ft/s(= 3.66 m/s), rather than with pure numbers, such as 12 or 3.66. To become competent at solving fluid mechanics problems, you must become virtually infallible in the handling of such units and their conversion factors. For most engineers the major sources of difficulties with units and conversion factors are carelessness and the simultaneous appearance of force and mass in the same equation.

A useful "system" for avoiding carelessness and consistently converting the dimensions of engineering quantities from one set of units to another has two rules:

1. *Always* (repeat, *always*) include the dimensions with any engineering quantity you write down.
2. Convert the dimensions you have written down to the dimensions you want in your answer by multiplying or dividing by 1.

Example 1.1. We are required to convert a speed of 327 mi/h to a speed in feet per second. The first step is to write the equation

$$\text{Speed} = 327 \text{ mi/h} \tag{1.9}$$

This is not the same as 327 km/h or 327. If we omit the dimensions, our equation is meaningless. We now write as an equation the definition of a mile:

$$1 \text{ mi} = 5280 \text{ ft} \tag{1.10}$$

If we divide both sides of this equation by 1 mi,

$$\frac{1 \text{ mi}}{1 \text{ mi}} = 1 = \frac{5280 \text{ ft}}{\text{mi}} \tag{1.11}$$

You may not be used to thinking of 5280 ft/mi being the same thing as 1, but this shows that they are the same. Similarly, we write the definition of an hour as

$$1\ \text{h} = 3600\ \text{s} \tag{1.12}$$

and divide both sides by 3600 s to find

$$\frac{3600\ \text{s}}{3600\ \text{s}} = 1 = \frac{1\ \text{h}}{3600\ \text{s}} \tag{1.13}$$

Again, you may not be used to thinking of 1 h/3600 s as the same thing as 1, but it is. Now we return to Eq. 1.9 and multiply both sides by 1 twice, choosing our equivalents of 1 from Eqs. 1.11 and 1.13:

$$\text{Speed} \cdot 1 \cdot 1 = \frac{327\ \text{mi}}{\text{h}} \cdot \frac{5280\ \text{ft}}{\text{mi}} \cdot \frac{\text{h}}{3600\ \text{s}} \tag{1.14}$$

We can now cancel the two 1s on the left side, because they do not change the value of "speed," and we can cancel the units that appear both above and below the line on the right side, to obtain

$$\begin{aligned}\text{Speed} &= \frac{327\ \cancel{\text{mi}}}{\cancel{\text{h}}} \cdot \frac{5280\ \text{ft}}{\cancel{\text{mi}}} \cdot \frac{\cancel{\text{h}}}{3600\ \text{s}} \\ &= \frac{327 \cdot 5280\ \text{ft}}{3600\ \text{s}} = 480\ \frac{\text{ft}}{\text{s}} = 146\ \frac{\text{m}}{\text{s}}\end{aligned} \tag{1.15}$$

■[†]

This was an easy example, one you could certainly solve without going into as much detail as shown here, but it illustrates the procedure to be used in more complicated problems.

Example 1.2. Suppose time equals 2.6 h. How many seconds in this? Again, we begin by writing it with dimension as an equation:

$$\text{Time} = 2.6\ \text{h} \tag{1.16}$$

We want to know its value in seconds, so we divide by 1:

$$\text{Time} = 2.6\ \text{h} \cdot \frac{3600\ \text{s}}{\text{h}} = 2.6 \cdot 3600\ \text{s} = 9380\ \text{s} \tag{1.17}$$

■

How did we know to multiply by 1 h/3600 s in Example 1.1 and to divide by 1 h/3600 s in Example 1.2? In each case we chose the value of 1 which allowed us to cancel the unwanted dimension. Three ideas are involved here:

[†] The ■ marks the end of an example.

1. Dimensions are treated as algebraic quantities and multiplied or divided accordingly.
2. Multiplying or dividing any quantity by 1 does not change its value.
3. Any dimensioned equation can be converted to 1 = 1 by dividing through by either side.

Using the last procedure, we can write

$$1 = \frac{60\text{ s}}{\text{min}} = \frac{12\text{ in}}{\text{ft}} = \frac{7000\text{ gr}}{\text{lbm}} = \frac{\text{mi}^2}{640\text{ acres}} = \frac{1\text{ Btu}}{252\text{ cal}} = \frac{1\text{ W}}{\text{VA}} = \text{etc}. \tag{1.18}$$

and as many other values of 1 as we like.

The previous examples did not involve the unit conversions that cause difficulties, the ones involving force and mass or thermal and mechanical energies. If everyone always used SI, we would never have those difficulties. In SI there is no difficulty with the units of force and mass; force is measured in newtons (N) and mass in kilograms (kg), and the only unit of energy is the mechanical-energy unit, the joule (J), where $1\text{ J} = \text{N} \cdot \text{m}$.

Unfortunately, in the English system (and in the traditional metric system as it is actually used by the public in Europe) there is difficulty with force-mass unit conversion. If one asks a typical European male what he weighs, he might well respond, "80 kilos," meaning 80 kg. If he were speaking in SI, he would not use the kilogram as a unit of weight, because weight is a force and the SI unit of force is the newton. He should respond, "784.6 newtons" because that is the weight of an 80-kg mass in a standard gravitational field of $9.807\text{ m/s}^2 = 32.17\text{ ft/s}^2$. It is hard enough to teach novice engineers the difference between weight and mass; it is probably impossible to get the general public to take the view that a mass of 80 kg does not exert a force of 80 kg. To make this come out right, we decide that there are really *two* kilogram units, the kilogram-mass (kgm) and the kilogram-force (kgf). We can define these so that 1 kgm exerts a force of 1 kgf at standard gravity. That is what most of the people in the world actually do. Similarly in the English system, we need two kinds of pounds: pound-mass (lbm) and pound-force (lbf). Again we define these so that 1 lbm has a weight of (exerts a force of) 1 lbf at standard gravity.

Why does this cause problems? Because the kgm and kgf look the same, so one is tempted to believe they are the same thing, and the lbm and the lbf look the same, so one is tempted to believe they are the same thing. That is wrong. It is a trap for the unwary. They are not the same. This leads to serious errors in engineering calculations.

Newton's second law of motion is

$$F = ma \tag{1.19}$$

where F is force, m is mass, and a is acceleration. The pound-force (lbf) is defined as that force which, acting on a mass of 1 lbm, produces an acceleration of 32.2 ft/s^2. Substituting this definition into the last equation, we find

$$1\ \text{lbf} = 1\ \text{lbm} \cdot 32.2\ \text{ft/s}^2 \tag{1.20}$$

Dividing both sides of this by 1 lbf, we find

$$1 = \frac{\text{lbf}}{\text{lbf}} = 32.2\ \frac{\text{lbm} \cdot \text{ft}}{\text{lbf} \cdot \text{s}^2} \tag{1.21}$$

Then if we make the mistake of canceling the lbm on top and the lbf on the bottom right-hand side, we will conclude that $1 = 32.2\ \text{ft/s}^2$. This is clearly wrong, and if we do it in a problem, we will find that the dimensions do not check and the numerical value of the answer will be wrong by a factor of 32.2 (if we use English units) or 9.8 (if we use metric units). Similarly, in the traditional metric system, we have

$$1\ \text{kgf} = 1\ \text{kgm} \cdot 9.8\ \text{m/s}^2 \tag{1.22}$$

and if we divide both sides by kgf,

$$1 = \frac{\text{kgf}}{\text{kgf}} = 9.8\ \frac{\text{kgm} \cdot \text{m}}{\text{kgf} \cdot \text{s}^2} \tag{1.23}$$

If we then cancel kgm on top and kgf on the bottom right, we will conclude that $1 = 9.8\ \text{m/s}^2$, which is equally absurd.

How can we get out of this difficulty? One way is to always work exclusively in SI. In that case kg will always mean kgm, and kgf will never appear. Instead the unit of force will always be the N = (1/9.8) kgf. However, then we will be unable to deal with the public, who speak (unintentionally) in kgf and lbf, or to understand those parts of the engineering literature that use kgf and lbf. The other way is to decide we must live with the kgf and lbf, and so we will have to regularly use the force-mass conversion factor whenever units of force and of mass occur in the same equation. This conversion factor has the values shown:

$$1 = 9.8\ \frac{\text{kgm} \cdot \text{m}}{\text{kgf} \cdot \text{s}^2} = 1.0\ \frac{\text{kgm} \cdot \text{m}}{\text{N} \cdot \text{s}^2} = 32.2\ \frac{\text{lbm} \cdot \text{ft}}{\text{lbf} \cdot \text{s}^2} \tag{1.24}$$

Furthermore, we must know some history to understand the older literature. First, we must know that many older textbooks and articles used the symbol g_c to stand for this force-mass conversion factor. So whenever we see a g_c in an equation, we must recognize it as a reminder that we must use the force-mass conversion factor. We must not confuse g_c, the force-mass conversion factor, with the acceleration of gravity g.

Second, we should recognize that engineers using English units have tried to evade this difficulty by inventing two new units: the slug (1 slug = 32.2 lbm) and the poundal (pdl) [1 pdl = (1/32.2) lbf]. Using these, we have the following force-mass conversion factors:

$$1 = 9.8\,\frac{\text{kgm}\cdot\text{m}}{\text{kgf}\cdot\text{s}^2} = 1.0\,\frac{\text{kgm}\cdot\text{m}}{\text{N}\cdot\text{s}^2}$$

$$= 32.2\,\frac{\text{lbm}\cdot\text{ft}}{\text{lbf}\cdot\text{s}^2} = \frac{\text{slug}\cdot\text{ft}}{\text{lbf}\cdot\text{s}^2} = \frac{\text{lbm}\cdot\text{ft}}{\text{pdl}\cdot\text{s}^2} \tag{1.25}$$

The kgf and the lbf have been around a long time, in spite of the efforts of scientists and engineers to replace them with the newton or the poundal. They survive because they seem natural to nonscientific users. Probably they will continue to be widely used, in spite of the efforts of the scientific community to replace them. Prudent engineers will learn to live with this fact, to use them when it seems appropriate, and to understand why they came about.

The second difficulty with units concerns mechanical and thermal units of energy. In SI the only unit of energy is the joule, where $1\text{ J} = 1\text{ N}\cdot\text{m}$. This is clearly a mechanical unit, the product of a force and a distance. If we are transferring thermal energy (e.g., heating our houses or our soup), it seems natural to base the measurements on the quantity of thermal energy required to raise the temperature of some reference substance by some finite temperature interval. In the English system, this quantity is the British thermal unit (Btu), which is the quantity of thermal energy required to raise the temperature of 1 lbm of water by 1°F. In the metric system the unit is the calorie (cal), which is the quantity of thermal energy required to raise the temperature of 1 g of water 1°C, or the kilocalorie (1 kcal = 1000 cal; this is the "calorie" used in describing the energy content of foods). If we wish to use the calorie or the Btu, then we need to convert from joules to calories or ft · lbf to Btu:

$$1 = \frac{\text{Btu}}{778\text{ ft}\cdot\text{lbf}} = \frac{\text{Btu}}{1055\text{ J}} = \frac{\text{cal}}{4.18\text{ J}} = \frac{\text{kcal}}{4180\text{ J}} = \frac{\text{kcal}}{4.18\text{ kJ}} \tag{1.26}$$

The Btu and the cal (or kcal) seem likely to continue in common usage; the Btu appears on almost all U.S. heating appliance and fuel bills (sometimes natural-gas bills use the therm (thm), where $1\text{ thm} = 10^5\text{ Btu}$), and the kilocalorie appears on numerous food products.

In summary, if you can do all your work in SI, you need never be concerned about force-mass conversions ($\text{N} = \text{kg}\cdot\text{m/s}^2$) or energy conversions ($\text{J} = \text{N}\cdot\text{m} = \text{W}\cdot\text{s}$). If you are confronted with problems (or literature, or current U.S. legal definitions) involving the kgf, lbf, cal, kcal, or Btu, you must follow the rules outlined above: Always write down the dimensions, treat the dimensions as algebraic quantities, and multiply by 1 as often as needed to get the quantities into the desired set of units, using the appropriate values of the force-mass conversion factor and the thermal-mechanical energy conversion factor. Even in SI, if you stray from the basic units (m, kg, s, A, K, mol, and cd), you will need conversion factors such as

$$1 = \frac{1000\text{ g}}{\text{kg}} = \frac{100\text{ cm}}{\text{m}} = \frac{1000\text{ mV}}{\text{V}} \tag{1.27}$$

Example 1.3. A mass of 10 lbm (4.54 kgm) is acted on by a force of 3.5 lbf (15.56 N or 1.59 kgf). What is the acceleration in feet per minute squared?

Rearranging Eq. 1.23, we obtain

$$a = \frac{F}{m} \tag{1.28}$$

Substituting, we obtain

$$a = \frac{3.5\ \text{lbf}}{10\ \text{lbm}} \tag{1.29}$$

Here we want the acceleration in feet per square minute, so we must multiply or divide by those equivalents of 1 that will convert the units:

$$a = \frac{3.5\ \text{lbf}}{10\ \text{lbm}} \cdot \frac{32.2\ \text{lbm}\cdot\text{ft}}{\text{lbf}\cdot\text{s}^2} \cdot \left(\frac{60\ \text{s}}{\text{min}}\right)^2$$

$$= \frac{3.5 \cdot 32.2 \cdot 60^2}{10} \frac{\text{ft}}{\text{min}^2} = 40{,}570\ \frac{\text{ft}}{\text{min}^2} \tag{1.30}$$

or

$$a = \frac{15.56\ \text{N}}{4.54\ \text{kg}} \cdot \frac{\text{kg}\cdot\text{m}}{\text{N}\cdot\text{s}^2} \cdot \frac{\text{ft}}{0.3048\ \text{m}} \cdot \left(\frac{60\ \text{s}}{\text{min}}\right)^2 = 40{,}480\ \frac{\text{ft}}{\text{min}^2} \tag{1.31}$$

or

$$a = \frac{1.59\ \text{kgf}}{4.54\ \text{kg}} \cdot \frac{9.8\ \text{kgm}\cdot\text{m}}{\text{kgf}\cdot\text{s}^2} \cdot \frac{\text{ft}}{0.3048} \cdot \left(\frac{60\ \text{s}}{\text{min}}\right)^2 = 40{,}540\ \frac{\text{ft}}{\text{min}^2} \tag{1.32}$$

You may verify that the difference between these three answers is due to round off error in the conversion factors used. If more figures had been carried (say, 1 kgf = 9.80650 N), the answers would have agreed exactly; but since we know the input data to only two significant figures, our best answer, in both cases, should be 40,500 ft/min^2. ■

Example 1.3 is the last example in this book to use the kgf. Clearly the method of dealing with kgm and kgf is just the same as the method of dealing with lbm and lbf. For the rest of this book, we use either lbm and lbf, or SI.

Example 1.4. An aluminum cell (Hall process) has a current of 50,000 A. Assuming it is 100 percent efficient, how much metallic aluminum does it produce per hour?

First we convert the current to gram equivalents per hour, using the necessary values of 1, one of which we take from Prob. 1.14:

$$I = 50{,}000\ \text{A} \cdot \frac{\text{C}}{\text{A}\cdot\text{s}} \cdot \frac{3600\ \text{s}}{\text{h}} \cdot \frac{\text{g equiv}}{96{,}500\ \text{C}} = 1870\ \frac{\text{g equiv}}{\text{h}} \tag{1.33}$$

For aluminum, $$27\ \text{g} = 1\ \text{mol} \tag{1.34}$$

and $$1\ \text{mol} = 3\ \text{g equiv} \tag{1.35}$$

Therefore,

$$I = 1870\,\frac{\text{g equiv}}{\text{h}} \cdot \frac{\text{mol}}{3\text{ g equiv}} \cdot \frac{27\text{ g}}{\text{mol}} \cdot \frac{\text{lbm}}{454\text{ g}}$$

$$= 37\text{ lbm/h} = 16.9\text{ kg/h} \tag{1.36}$$

■

In solving Example 1.4, we multiplied by 1 six times. Nonetheless, the procedure is simple and straightforward. Each multiplication by 1 gets rid of an undesired dimension and brings us closer to an answer in the desired units.

In Example 1.4, we saw that an apparently complex problem was really a simple conversion-of-units problem. In the course of our studies and our professional careers, we will have to convert units as quickly and as easily as we now add and subtract. It will be easiest if we develop the habit of following the two rules given at the start of Sec. 1.8:

1. Always include the dimensions with any engineering quantity you write down.
2. Convert the dimensions you have written down to the dimensions you want in your answer by multiplying or dividing by 1.

A short table of these conversion factors can be found inside the front cover. The American Society for Testing and Materials (ASTM) [10] has prepared a much longer and more useful table, which reveals some additional complexity. For example, there are five different calorie definitions in common use. The largest is 1.002 times the smallest. Only in the most careful work is this small a difference relevant. But if you are doing that kind of work, it is worthwhile to obtain, study, and use the tables in Ref. 10.

1.9 PRINCIPLES VERSUS TECHNIQUES

As discussed in Sec. 1.3, there are very few underlying ideas in fluid mechanics. With these few ideas we can solve a great variety of problems. In so doing, we can focus on either the application of principles or on the techniques of solving problems. The author recommends attention to the principles. In the 10 years following the author's graduation from college the engineering business was revolutionized by the digital computer, transistor, and space industry, among other things. None of these amounted to much in 1954 and they were not part of undergraduate courses.

All these technologies rigidly obey Newton's laws and the laws of thermodynamics. Students who learned "cookbook" techniques for solving problems in 1954 were not well prepared for the technologies that appeared during the next 10 years, but those who learned the basic ideas and how to apply them could adapt to any one of them. There is little reason to believe that the pace of technological change will be slower in the future. If we concentrate on learning techniques, we may be faced in a few years with "technical obsolescence," but we should have no such problem if we learn principles and their applications. The author believes that there will never be a surplus of people who really understand Newton's laws and the laws of thermodynamics.

1.10 ENGINEERING PROBLEMS

Although this book may fall into the hands of a practicing engineer, most readers will be college juniors; the following is addressed to them.

Engineering students start out in their first and second years by doing "plug-in" problems. That is, given a problem statement, they select the appropriate formula, either from the textbook or from memory, and "plug in" the data to obtain the final answer. In the third year, they begin to find problems that can be readily reduced to plug-ins or to problems involving two or more equations that require some manipulation to be put in plug-in form. Furthermore, engineering students may be exposed to problems that cannot be reduced to plug-ins and must be solved by trial and error. It is assumed that they can do simple plug-in problems (such as gas-law calculations) without hesitation.

Instructors of third-year students would like to assign more complicated or difficult problems but generally cannot because

1. The time required is too great—they cannot be done in the time that most students will devote to one homework problem.
2. The students would probably get intellectual indigestion. Therefore, at the third-year level most problems and examples in texts like this one are the plug-ins or can be readily reduced to plug-ins.

When the students start a senior laboratory or design course, they find their first real engineering problems. One may require 10 or 20 h of work and consist of 15 or 20 parts, each comparable to the problems and examples in this book. To deal with these problems, students break them up into pieces small enough to handle as plug-ins. The interesting and exciting part of engineering is often the task of deciding how to divide a problem into reasonable pieces and then how to reassemble the pieces into a recognizable whole so that they fit together properly.

In the examples and problems in this book, there are numerous simple plug-in problems. They are included because their solutions give the reader some feel for the numerical values involved in fluid mechanics. There are also more complex problems, in which two or more basic principles are involved (such as the mass balance and the energy balance). In these some manipulation is required to get the equations into plug-in form. This is the recommended procedure for solving such problems:

1. Be sure that you understand precisely what the problem is; in particular, know precisely what is being asked for.
2. Decide which physical laws relate what you know to what you want to find.
3. Write the working form of these laws (as discussed later), and rearrange them to get the symbol for the quantity you seek, standing alone to the left of the equal sign. In so doing, probably you will have to discard several terms in the physical-law equations. The discarding of each term corresponds to an assumption about the physical nature of the system (e.g., that a certain velocity is negligible). Thus a list of such terms dropped is a list of assumptions made in solving the problem.
4. When you have finished step 3, the problem is reduced to a plug-in. Insert the given data, check the units, and find the numerical value of the answer.
5. Check the answer for plausibility: Does it indicate negative masses, velocities greater than the speed of light, or efficiencies greater than 100 percent? Also reexamine the assumptions listed in step 3 to see whether they are consistent with the answer. If these checks are met, the answer probably is satisfactory.
6. If the problem is one that you may have to repeat with different data (such as the calculation of a fluid-flow rate from a measured pressure difference), then it may be worthwhile to see whether the answer can be put in a more convenient form, e.g., some general plot or diagram. Perhaps the problem occurs often enough to justify programming its solution on a personal computer or entering it in a spreadsheet program.

In all engineering we must consider the degree of precision needed. Voltaire's famous dictum "The perfect is the enemy of the good!" describes the situation of the engineer. We could always spend more engineering effort, do more testing, and thereby refine our design or calculation a bit more. But in any real problem the engineer's time is one of the limiting resources. We would all like the conditions that the famous architect Kobori Enshu demanded and received from the Japanese dictator Hideyoshi for the Katsura Villa: no limit on expense, no limit on time, and no client visits until the job is done. Many believe the result to be the greatest achievement of Japanese architecture and garden planning [11]. (If you are ever in Kyoto, visit it and decide for yourself.) But most engineers (and other professionals) are always working with limited time and limited budgets as well as clients who want intermediate progress reports. For us the goal is always to do the best possible, within the time, budget, and other constraints imposed by the client (or codes and regulations). So engineers must allocate their time well, handling routine matters swiftly and concentrating on those that are not routine and may be a source of trouble. Much of what you learn in this book is routine to practicing engineers. The goal of this book is that you learn not only to do those routine things but also learn the scientific basis of the solution of those routine problems. In so doing, you will learn how engineers and scientists have turned yesterday's difficult problems into today's routine ones. That will help you to develop the habits of mind that will turn today's difficult problems into tomorrow's routine problems.

You should also consider the degree of confidence in the solution to a problem. If in the calculation you used physical property data that is accurate to no more than ±5 percent, then it makes no sense to report the answer to 3 or more significant figures. If the solution presented required really speculative calculating approaches, the reader should be alerted to that fact.

In the problems at the end of each chapter, one or two need to be broken down into simpler ones before they can be solved. The practice gained in doing this is well worth the effort.

1.11 SUMMARY

1. Fluid mechanics is the study of the forces and motions in fluids.
2. Fluids are substances that move continually when subjected to a shear force as long as the force is applied. Solids deform slightly when subjected to a shear force and then stop moving and permanently resist the force. There are, however, intermediate types of substance; the distinction between solid and liquid is one of degree rather than of kind.
3. Fluid mechanics is based on Newton's laws of motion, the first two laws of thermodynamics, the principle of the conservation of matter, and careful experiments.
4. Gases have weak intermolecular attractions and expand without limit. Liquids have much stronger intermolecular attractions and can expand very little. With increasing temperature and pressure, the differences between liquids and gases gradually disappear.
5. Density is mass per unit volume. Specific gravity is density/(density of water at 4°C).
6. Viscosity is a measure of a fluid's resistance to flow. Most simple fluids are represented well by Newton's law of viscosity. The exceptions (nonnewtonian fluids) are generally complex mixtures, some of which have great practical significance. Kinematic viscosity is viscosity divided by density.
7. Surface tension is a measure of a liquid's tendency to take a spherical shape, caused by the mutual attraction of the liquid's molecules.
8. Pressure is the compressive force divided by the area. It is the same in all directions for a fluid at rest and practically the same in all directions for most moving fluids.
9. In handling the units (dimensions) in this text, always write down the units of any dimensioned quantity and then multiply or divide by 1 to obtain the desired units in the answer.

PROBLEMS

See the Common Units and Values for Problems and Examples inside the back cover.

1.1. In Sec. 1.3 the basic laws on which fluid mechanics rests are listed. How many basic laws of nature are not included in the list? To answer, make a list of what you consider to be the basic laws of nature. By *basic laws* we means laws that cannot be derived from other more basic ones; e.g., Galileo's "laws of falling bodies" can be derived from Newton's laws and so are not considered basic.

1.2. At low pressures there is a significant difference between the densities of liquids and of gases. For example, at 1 atm the densest gas known to the author is uranium hexafluoride, which has a molecular weight of 352; its normal boiling point is 56.2°C. Calculate its density in the gas phase at 1 atm and 56.2°C, assuming that it obeys the perfect-gas law. The least dense liquid known to the author is liquid hydrogen, which at its normal boiling point of 20 K has a density of $0.071\ g/cm^3$. Liquid helium also has a very low density, about $0.125\ g/cm^3$ (at 4 K). Excluding these remarkable materials, make a list of liquids that can exist at 1 atm at densities of less than $0.5\ g/cm^3$. A good source of data is *The Handbook of Chemistry and Physics* (CRC Press, Cleveland).

1.3. A typical mud is 70 weight percent sand and 30 weight percent water. What is its density? The respective densities are

$$\rho_{\text{sand}} = 165\ \text{lbm/ft}^3\quad (2.65\ \text{g/cm}^3)\quad \text{and}\quad \rho_{\text{water}} = 62.3\ \text{lbm/ft}^3\quad (1.0\ \text{g/cm}^3)$$

1.4. Why are specific gravities most often referred to the density of water at 4°C instead of 0°C?

1.5. The American Petroleum Institute (API) gravity (used extensively in the petroleum industry) is defined, in "degrees," by

$$\text{Deg API} = \frac{141.5}{\text{specific gravity}} - 131.5$$

Here the specific gravity is the ratio of the density of the fluid to that of water, both at 60°F. Sketch the relation between degrees API and density in grams per cubic centimeter. What advantages of this scale might have led the petroleum industry to adopt it?

1.6. You are handed a small, heavy-walled glass tube, sealed at both ends, and you are told that it contains a fluid. From observation it is clear that there is no interface in the tube; i.e., only one phase is present. Without opening the tube, can you determine whether the contents are a liquid or a gas? How?

1.7. What are the dimensions of dV/dy? What are the dimensions of shear stress? Shear stress in liquids is often called *momentum flux* [2]. Show that shear stress has the same dimensions as momentum/(area · time). What are the dimensions of viscosity?

1.8. List as many applications as you can of industrial, domestic, or other materials in which nonnewtonian viscosity behavior is desirable. In each case, specify why this behavior is desirable.

1.9. Calculate the surface/volume of a sphere, a cube, and a right cylinder with the height equal to the diameter. Which has the least surface/volume?

1.10. A fluid under tensile stress is unstable [8]; a small disturbance can cause it to boil and thus change to a stable state. Make a list of other unstable situations demonstrable in a chemistry or physics laboratory. The working criterion of instability is that a very small disturbance can cause a large effect.

1.11. Earth may be considered a sphere with a diameter of 8000 mi and an average specific gravity of 5.5. What is its mass? What is its weight? Explain your answer.

1.12. A cubic foot of water at 68°F weighs about 62.3 lbf at sea level.
(*a*) What is its density?
(*b*) What does it weigh on the moon ($g = 6\ \text{ft/s}^2$)?
(*c*) What is its density of the moon?

1.13. How many gallons are there in $1\ \text{mi}^3$? Note that 1 U.S. gal $= 231\ \text{in}^3$, $1\ \text{ft}^3 = 1728\ \text{in}^3$, and 1 mi = 5280 ft. The total proven U.S. oil reserves are roughly 30×10^9 barrels (bbl), where 1 bbl = 42 U.S. gal. How many cubic miles is this?

1.14. In electrochemical equations it is common to write in the symbol $\mathcal{F}$ to remind the user to convert from moles of electrons to coulombs. This is just like the force-mass and thermal energy–mechanical energy conversion factor, namely,

$$\mathcal{F} = 1 = \frac{96{,}500\ \text{C}}{\text{g equiv of electrons}}$$

In 1 g equiv of electrons there are 6.02×10^{23} electrons. How many electrons are there in 1 C?

1.15. As discussed in the text, the slug and the poundal were invented to make the conversion factor (mass · length)/(force · time^2) have a coefficient of 1. A new unit of length or a new unit of time could just as logically have been invented for this. Let us name those units the *toof* and the *dnoces*. What are the values of the toof and the dnoces in terms of the foot and the second?

1.16. In U.S. irrigation practice, water is described in acre-feet, which is the amount required to cover 1 acre of farmland 1 ft deep. What is the mass of 1 acre · ft of water? ($1\ \text{mi}^2 = 640$ acres.) What is the mass of 1 ha · m of water ($1\ \text{km}^2 = 100$ ha)?

1.17. Einstein's equation $E = mc^2$ indicates that the speed of light squared must be expressible in units of energy per unit mass. What is the value of the square of the speed of light in Btu per pound-mass? In joules per kilogram? The speed of light c is 186,000 mi/s $= 2.998 \times 10^8$ m/s.

1.18. A common basis for comparing rocket fuel systems is the *specific impulse*, defined as pounds-force of thrust divided by pounds-mass of fuel and oxidizer per second. The common values are 250 to 400 lbf · s/lbm. One frequently sees the specific impulse referred to simply as "300 s." Is 300 s the same thing as 300 lbf · s/lbm? European engineers regularly express the same quantity in terms of the exhaust velocity of the rocket. If a rocket has a specific impulse of 300 lbf · s/lbm, what is its exhaust velocity?

1.19. Most U.S. engineers work with heat fluxes in units of Btu/(h · ft^2). In the rocket business, the common unit is cal/(s · cm^2). How many Btu/(h · ft^2) equals 1 cal/(s · cm^2)? The proper SI unit is J/(m^2 · s). How many Btu/(h · ft^2) is 1 J/(m^2 · s)?

1.20. The Reynolds number (discussed in Chap. 6) is defined for a pipe as velocity times diameter times density divided by viscosity. What is the Reynolds number (dimensionless) for water flowing at 10 ft/s in a pipe with a diameter of 6 in?

1.21. The flow of fluids through porous media (such as oil sands) is often described by Darcy's equation (see Chap. 12):

$$\frac{\text{Flow}}{\text{Area}} = \frac{\text{permeability}}{\text{viscosity}} \cdot \text{pressure gradient}$$

The unit of permeability is the *darcy*, which is defined as that permeability for which a pressure gradient of 1 atm/cm for a fluid of 1-cP viscosity produces a flow of 1 cm^3/s through an area of 1 cm^2. What are the dimensions of the darcy? What is its numerical value in those dimensions? Give the answer both in English and SI units.

1.22. The surface tension of a fluid exposed to a gas is normally expressed in dynes per centimeter. How many lbf/in correspond to 1 dyne/cm? How many N/m correspond to 1 dyne/cm?

1.23. Determine the value of X in the equation,

$$1.0 \frac{\text{Btu}}{\text{lbm} \cdot {}^\circ\text{F}} = X \frac{\text{cal}}{\text{gm} \cdot {}^\circ\text{C}}$$

1.24. The bar is a common unit of pressure (bar = 10^5 Pa), particularly for high pressure work. What is the relation between the bar and the pressure of the atmosphere at sea level?

1.25. Many European pressure gauges give the pressure in kg/cm^2. Is this kgm or kgf? Why would this be a convenient unit of pressure?

1.26. In the third part of Ex. 1.3, what would have happened if we had taken the force-mass conversion factor as 32.2 lbm · ft/lbf · sec^2 instead of 9.8 kgm · m/kgf · sec^2?

CHAPTER 2

FLUID STATICS

In this chapter we apply Newton's law of motion $F = ma$ to fluids at rest. We will see that this leads to a remarkably simple equation:

$$\frac{dP}{dz} = -\rho g \tag{2.1a}$$

This equation and its applications are almost the whole of fluid statics.

In Chap. 7 we apply Newton's law of motion to moving fluids. What we do in this chapter is really only part of the more general application in Chap. 7. In Chaps. 4, 5, and 6, however, we will need some of the results from this chapter, and the kinds of problem we deal with here are different from (and simpler than) those in Chap. 7; for these reasons a separate chapter on fluid statics is practical at this point. Remember, all we do in this chapter is apply $F = ma$ to a static fluid; the more general application, covering both moving and static fluids, is discussed in Chap. 7.

2.1 THE BASIC EQUATION OF FLUID STATICS

For a simple fluid at rest the pressure is the same in all directions; there are no shear stresses. These facts lead to the basic equation of fluid statics. Consider a small block of fluid which is part of a large mass of fluid at rest in a gravity field; see Fig. 2.1. Since the fluid is at rest, there are no accelerations, and the

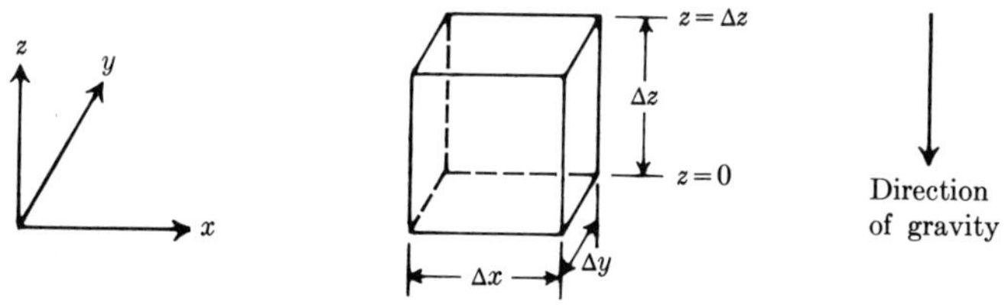

FIGURE 2.1

sum of the forces on any part of the fluid in any direction is zero. Let us consider the z direction, which is opposite to the direction of gravity. The forces which act on the small block of fluid in the z direction are the pressure forces on the top and bottom and the force of gravity acting on the mass of the element. Their sum (positive upward) is

$$(P_{z=0})\,\Delta x\,\Delta y - (P_{z=\Delta z})\,\Delta x\,\Delta y - \rho g\,\Delta x\,\Delta y\,\Delta z = 0 \tag{2.2}$$

Dividing by $\Delta x\,\Delta y\,\Delta z$ and rearranging, we find

$$\frac{P_{z=\Delta z} - P_{z=0}}{\Delta z} = -\rho g \tag{2.3}$$

If we let Δz approach zero, then

$$\operatorname*{limit}_{\Delta z \to 0} \frac{\Delta P}{\Delta z} = \frac{dP}{dz} = -\rho g \tag{2.1b}$$

This is the basic equation of fluid statics, also called the *barometric equation*. It is correct only if there are no shear stresses on the vertical faces of the cube in Fig. 2.1. If there are such shear stresses, then they may have a component in the vertical direction, which must be added to the sum of forces in Eq. 2.1. For simple newtonian fluids, shear stresses in the vertical direction can exist only if the fluid has a different vertical velocity on one side of the cube from that on the other side (see Eq. 1.5). Thus this equation is correct if the fluid is not moving at all, which is the case in fluid statics, or if it is moving but only in the x and y directions, or if it has a uniform velocity in the z direction. In this chapter, we apply it only when a fluid has no motion relative to its container or to some set of fixed coordinates. In later chapters, we apply it to flows in which there is no motion in the z direction or a motion with a uniform z component.

For complicated fluids, such as toothpaste, paints, and jellies, Eq. 2.1 is not correct, because the fluids can sustain small but finite shear stresses without any motion. The equation simply is not applicable. To find its equivalent, it is necessary to make up a sum of forces which includes shear forces on the vertical sides of the cube.

The barometric equation tells the change in pressure with distance upward, where *upward* is opposite to the direction of gravity and is called z. If we want to know the change in pressure with distance in some other, nonvertical direction, call it direction a, then we can write

$$\frac{dP}{da} = \frac{dz}{da}\frac{dP}{dz} = -\rho g \frac{dz}{da} \tag{2.4}$$

But, as shown in Fig. 2.2,

$$\frac{dz}{da} = \frac{\Delta z}{\Delta a} = \cos\theta \tag{2.5}$$

where θ is the angle between the direction a and the z axis. Substituting this equation into Eq. 2.4, we have

$$\frac{dP}{da} = -\rho g \cos\theta \tag{2.6}$$

A particularly interesting direction a is the one at right angles to z, that is, any direction parallel to the x-y plane. For any such direction θ is 90°, $\cos\theta$ is 0, and the pressure does not change with distance. Thus, from Eq. 2.6 we see that for a fluid at rest any surface perpendicular to the direction of gravity is a surface of constant pressure. The most interesting constant-pressure surface of a body of fluid at rest is the one with zero gauge pressure, i.e., the surface in contact with the atmosphere. Since this is a constant-pressure surface, it must be everywhere perpendicular to the direction of gravity. On a global scale, this makes the free surface of the oceans practically a sphere. (The earth is not quite spherical; it is slightly flattened at the poles.) In typical engineering operations, it means that the free surface of a liquid exposed to the atmosphere is practically a horizontal plane (Prob. 2.1).

The product of density and gravity, which appears in Eq. 2.1, is often called the *specific weight*, and given the symbol γ, where

$$\rho g = \gamma = \text{specific weight} \tag{2.7}$$

At any place where the acceleration of gravity is equal to 32.2 ft/s^2 = 9.81 m/s^2 (practically any place on the surface of the earth), the specific weight expressed in pounds-force per cubic foot (or kilograms-force per cubic meter) is numerically equal to the density expressed in pounds-mass per cubic foot (or kilograms-mass per cubic meter).

Example 2.1. Calculate the specific weight of water at a place where the acceleration of gravity is 32.2 ft/s.

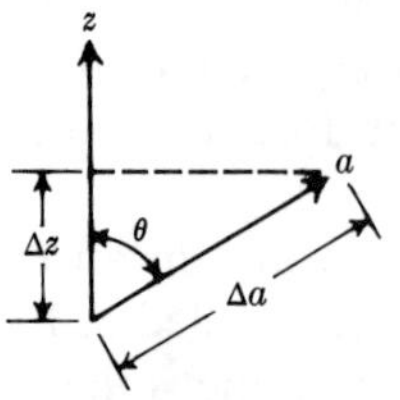

FIGURE 2.2

$$\gamma = \rho g = 62.3\,\frac{\text{lbm}}{\text{ft}^3}\cdot 32.2\,\frac{\text{ft}}{\text{s}^2}\cdot\frac{\text{lbf}\cdot\text{s}^2}{32.2\,\text{lbm}\cdot\text{ft}} = 62.3\,\frac{\text{lbf}}{\text{ft}^3} = 998.2\,\frac{\text{kgf}}{\text{m}^3} \qquad \blacksquare$$

If one deals principally with fluid flows in which the gravity forces are important, then often one can simplify the calculations by replacing the density in all equations by γ/g. In civil engineering hydraulics this is normally the case, and this is common practice. But if one deals mostly with flows whose gravity terms are small compared with the other terms, then it is more convenient to work with ρ than with γ/g. In chemical and mechanical engineering problems the gravity terms are normally small, so γ is seldom used.

2.2 PRESSURE-DEPTH RELATIONSHIPS

Equation 2.1 is a separable, first-order differential equation. It can be separated and integrated as follows:

$$\int dP = -\int \rho g\, dz \tag{2.8}$$

However, to perform the integration, it is necessary to have some relation between ρ, g, and z. In situations on the surface of the earth, g is practically constant (see Sec. 2.6), so we may take it outside the integral sign. Several possible kinds of behavior of ρ in relation to z lead to simple integrations of the equation, as shown below.

No real substances have constant density; the density of every substance increases as the pressure increases. However, for most liquids at temperatures far below their critical temperatures, the effect of pressure on density is very small. For example, raising the pressure of water at 100°F from 1 to 1000 lbf/in^2 causes the density to increase by 0.3 percent. In most engineering calculations, we can neglect such small changes in density. Then we take ρ outside the integral sign in Eq. 2.8 and find that the pressure is

$$P_2 - P_1 = -\rho g(z_2 - z_1) \qquad \text{[constant density]} \tag{2.9}$$

Example 2.2. When the submarine *Thresher* sank in the Atlantic, it was estimated in the newspapers that the accident had occurred at a depth of 1000 ft (304.9 m). What is the pressure of the sea at that depth?

Seawater may be considered incompressible, with density 63.9 lbm/ft^3. The pressure at the surface is atmospheric pressure, which is approximately 14.7 lbf/in^2. The acceleration of gravity is 32.2 ft/s^2. Therefore,

$$P_{1000\text{ ft}} = 14.7\,\frac{\text{lbf}}{\text{in}^2} + 63.9\,\frac{\text{lbm}}{\text{ft}^3}\cdot 32.2\,\frac{\text{ft}}{\text{s}^2}\cdot 1000\,\text{ft}\cdot\frac{\text{ft}^2}{144\,\text{in}^2}\cdot\frac{\text{lbf}\cdot\text{s}^2}{32.2\,\text{lbm}\cdot\text{ft}}$$

$$= 14.7\,\frac{\text{lbf}}{\text{in}^2} + 444\,\frac{\text{lbf}}{\text{in}^2} = 459\,\frac{\text{lbf}}{\text{in}^2}$$

or

$$P_{304.9\text{ m}} = 101.3\text{ kPa} + 1024\ \frac{\text{kg}}{\text{m}^3} \cdot 9.81\ \frac{\text{m}}{\text{s}^2} \cdot 304.9\text{ m} \cdot \frac{\text{Pa}}{\text{N/m}^2} \cdot \frac{\text{N} \cdot \text{s}^2}{\text{kg} \cdot \text{m}}$$

$$= (101.3 + 3062.8)\text{ kPa} = 3.164\text{ MPa}$$ ■

In hydraulics problems and in all problems involving a free surface, we can further simplify Eq. 2.9 by working in gauge pressure. The gauge pressure is zero at the free surface: $P_{1\text{ gauge}} = 0$. We now define the depth as the distance measured downward from the free surface and give it the symbol h,

$$h = z_{\text{free surf}} - z \tag{2.10}$$

in which case Eq. 2.9 simplifies to

$$P = \rho g h \qquad \text{[gauge pressure, constant density]} \tag{2.11}$$

Example 2.3. A cylindrical oil-storage tank is 75 ft deep and contains an oil of density 55 lbm/ft^3. Its top is open to the atmosphere. What is the gauge pressure–depth relation in this tank?

The gauge pressure is zero at the free surface. At the bottom it is

$$P_{\text{bot}} = 55\ \frac{\text{lbm}}{\text{ft}^3} \cdot 32.2\ \frac{\text{ft}}{\text{s}^2} \cdot 75\text{ ft} \cdot \frac{\text{ft}^2}{144\text{ in}^2} \cdot \frac{\text{lbf} \cdot \text{s}^2}{32.2\text{ lbm} \cdot \text{ft}} = 28.7\ \frac{\text{lbf}}{\text{in}^2} = 197.4\text{ kPa}$$

From Eq. 2.11 we know that the pressure-depth relation is linear (see Fig. 2.3). ■

The density of gases changes significantly with the pressure, so we must be cautious about taking the density outside the integral sign in Eq. 2.8. At low pressure the densities of most gases are reasonably well approximated by the perfect-gas law $\rho = PM/(RT)$. Here T is the absolute temperature, in degrees Rankine or in kelvins (T°R = T°F + 459.69 or TK = T°C + 273.15); R is the universal gas constant, whose value in various systems of units is shown in App. A.9; and M is the molecular weight, normally expressed in grams per mole or pounds-mass per pound-mole.[†] Substituting this equation for the density in Eq. 2.1, we find

$$\frac{dP}{dz} = -\frac{PM}{RT}\,g \tag{2.12}$$

[†] This formulation of the perfect-gas law gives the density in pounds-mass per cubic foot. In chemistry we often see the perfect-gas law written as $\rho = P/(RT)$, which gives the density in pound-moles per cubic foot or moles per cubic meter. Multiplying the latter density by the molecular weight (in pounds-mass per pound-mole or grams per mole) gives the density, shown here. Common chemical engineering practice is to retain R as a universal constant and to multiply by the appropriate value of M for the gas being used. Common mechanical engineering practice is to define a new R; in symbols, $R_{\text{mech eng}} = R_{\text{univ}}/M$, which makes each gas have its own unique R. In this book we let R stand for the universal gas constant and multiply by M as needed to get ρ in pounds-mass per cubic foot or kilograms per cubic meter.

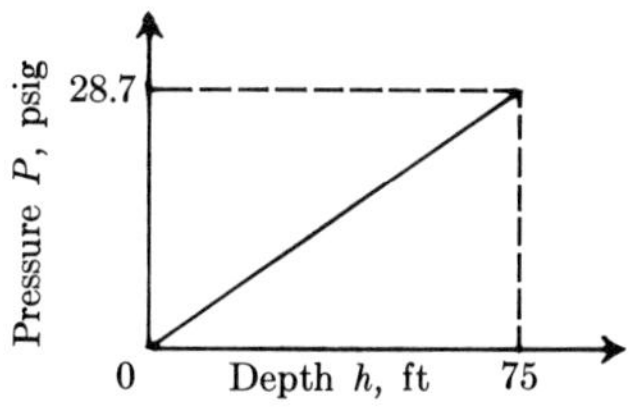

FIGURE 2.3
Pressure-depth relation in Example 2.3.

If the temperature is constant, this can be separated and integrated as follows:

$$\int_1^2 \frac{dP}{P} = \frac{-gM}{RT}\int_1^2 dz \tag{2.13}$$

$$\ln \frac{P_2}{P_1} = \frac{-gM}{RT}(z_2 - z_1) \tag{2.14}$$

$$P_2 = P_1 \exp\left(\frac{-gM\,\Delta z}{RT}\right) \qquad \text{[isothermal, perfect gas]} \tag{2.15}$$

Example 2.4. At sea level the atmospheric pressure is 14.7 psia and the temperature is 70°F. Assuming that the temperature does not change with elevation (a poor assumption, but one that simplifies the mathematics), calculate the pressure at 1000, 10,000, and 100,000 ft. For air the molecular weight M is 29 lbm/lbmol. For $z = 1000$ ft

$$P_2 = P_1 \exp\left\{\frac{-32.2\ \text{ft/s}^2 \cdot 29\ \text{lbm/lbmol} \cdot 1000\ \text{ft}}{[10.73\ \text{lbf/in}^2 \cdot \text{ft}^3/\text{lbmol} \cdot °\text{R}] \cdot 530°\text{R}} \cdot \frac{\text{ft}^2}{144\ \text{in}^2} \cdot \frac{\text{lbf} \cdot \text{s}^2}{32.2\ \text{lbm} \cdot \text{ft}}\right\}$$

$$= P_1 \exp(-0.0354) = \frac{P_1}{\exp 0.0354} = \frac{P_1}{1.036} = 0.965\ \text{atm}$$

We can calculate the pressures at the other two elevations:

Elevation, ft	$gM\,\Delta z/(RT)$	$\exp[gM\,\Delta x/(RT)]$	P, atm
1,000	0.0354	1.036	0.965
10,000	0.354	1.423	0.703
100,000	3.54	34.5	0.0290

■

How much error would we have made if we had used the constant-density formulas instead of taking the change in density into account?

Example 2.5. Rework Example 2.4, assuming that air is a constant-density fluid, which has the same density at all elevations as it has at 14.7 psia at 70°F.

Here we use Eq. 2.9:

$$P_2 - P_1 = -\rho_1 g(z_2 - z_1) = \left(\frac{-P_1 M}{RT}\right) g(z_2 - z_1)$$

$$P_2 = P_1\left[1 - \left(\frac{gM}{RT}\right)(z_2 - z_1)\right]$$

For 1000 ft we find

$$P_2 = P_1\left\{1 - \frac{32.2\ \text{ft/s}^2 \cdot 29\ \text{lbm/lbmol} \cdot 1000\ \text{ft}}{[10.73\ \text{lbf/in}^2 \cdot \text{ft}^3/\text{lbmol} \cdot {}^\circ\text{R}] \cdot 530^\circ\text{R}} \cdot \frac{\text{ft}^2}{144\ \text{in}^2} \cdot \frac{\text{lbf} \cdot \text{s}^2}{32.2\ \text{lbm} \cdot \text{ft}}\right\}$$

$$= P_1(1 - 0.0354) = 0.965\ \text{atm}$$

Again it is easiest to proceed by making a table:

Elevation, ft	$gM\,\Delta z/(RT)$	$1 - gM\,\Delta z/(RT)$	**P, atm**	**P from Eq. 2.15, atm**
1,000	0.0354	0.965	0.965	0.965
10,000	0.354	0.646	0.646	0.703
100,000	3.54	−2.54	−2.54	0.0290

■

From this table we see that up to 1000 ft the assumption of constant density makes a negligible error, at 10,000 ft it makes a 10 percent error, and at 100,000 ft it gives absurd results (negative absolute pressure?). Thus, for ordinary industrial-size equipment (generally less than 1000 ft high) we can accurately calculate changes in gas pressure with elevation as if the gas had a constant density. But in aeronautics and meterological problems, in which the elevations are often from 10,000 to 100,000 ft, we would make disastrous errors by making this simplification.

In Example 2.5 we made the simplifying assumption that the atmosphere was isothermal. Anyone who has gone to the mountains in the summer to get away from the heat did so because the atmosphere is not isothermal. To understand why the air temperature decreases with elevation, consider a mass of air being lifted from one elevation to a higher one (by a wind, e.g., blowing it over a mountain range). The air mass expands because the pressure of the surrounding air decreases as it rises, and the air mass is cooled by doing expansion work on the surrounding air. Air is a fairly poor conductor of heat, so during this process it undergoes an expansion which is practically adiabatic and practically reversible. If it were reversible and adiabatic, then the temperature-pressure-elevation relation would be exactly the isentropic one. For an isentropic atmosphere we can work out the following elevation-temperature and elevation-pressure relationships (Prob. 2.14):

$$P_2 = P_1\left(1 - \frac{k-1}{k} \cdot \frac{gM\,\Delta z}{RT_1}\right)^{k/(k-1)} \qquad \text{[isentropic, perfect gas]} \tag{2.16}$$

$$T_2 = T_1\left(1 - \frac{k-1}{k} \cdot \frac{gM\,\Delta z}{RT_1}\right) \qquad \text{[isentropic, perfect gas]} \tag{2.17}$$

Here k is the ratio of specific heats (discussed in Chap. 8). For air its value is practically constant at 1.4.

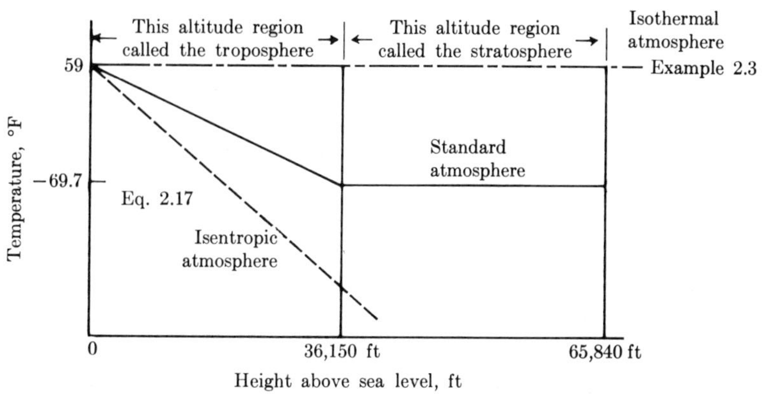

FIGURE 2.4
Comparison of standard atmosphere, isentropic atmosphere, and isothermal atmosphere.

The isothermal atmosphere in Example 2.4 would be observed if the air were a perfect conductor of heat, evening out all temperature differences instantly. The isentropic atmosphere in Eqs. 2.16 and 2.17 would be observed if air were a perfect insulator against heat conduction, transferring no heat at all. Experimental measurements show that the real behavior of the atmosphere is intermediate between these two extremes. Heat is conducted outward from the earth not only by simple conduction in the air (which is fairly slow) but also by winds, which mix cold and warm air layers, and by condensation of water vapor. For purposes of calculation aeronautical engineers have defined a "standard atmosphere," which agrees well with the *average* of many observations. As shown in Fig. 2.4, this standard atmosphere is indeed intermediate between the isothermal and isentropic atmospheres. It is an average; most interesting weather phenomena are caused by deviations from it. For a simple discussion of this see chap. 11 of Prandtl and Tietjens [1]. From the standard-atmosphere temperature we can calculate a "standard" pressure-height curve (Prob. 2.15).

2.3 PRESSURE FORCES ON SURFACES

Static, simple fluids can exert only pressure forces on surfaces adjacent to them. Since pressure is the normal (perpendicular) force per unit area, the pressure forces must act normal to the surface. Moving fluids can exert not only pressure forces, but also shear forces, so that the combined force exerted by a moving fluid on a surface is not necessarily normal to the surface. However, in problems involving moving fluids it is often convenient to treat the pressure and shear forces as separate and thus calculate the pressure force exactly as we do here, but using the pressure distribution on the surface corresponding to the flow situation rather than to the static-fluid one discussed in this chapter.

For an infinitesimal area of surface, the force exerted is

$$dF = P\,dA \tag{2.18}$$

This dF is a vector quantity; it has both direction (perpendicular to the surface) and magnitude. For a plane surface all the differential dF vectors point in the same direction, so that we can find the total force simply by integrating this equation:

$$F = \int P\,dA \tag{2.19}$$

The integration must take into account the curvature of surfaces. Before discussing curved surfaces, we consider the application of the last equation to several kinds of plane surface.

If the pressure over an entire surface is constant, then Eq. 2.19 becomes

$$F = PA \qquad \text{[constant pressure]} \tag{2.20}$$

Because the pressure in gases changes very slowly with position, this is practically true for all moderate-size surfaces exposed to gases.

For *horizontal* plane surfaces exposed to static fluids, the pressure is constant over the entire surface; so Eq. 2.20 gives the required force.

Example 2.6. An oil storage tank has a flat, horizontal, circular roof 150 ft in diameter. The atmospheric pressure is 14.7 psia. What force does the atmosphere exert on the roof?

$$F = PA = 14.7\,\frac{\text{lbf}}{\text{in}^2}\cdot\frac{\pi}{4}\,(150\text{ ft})^2\cdot 144\,\frac{\text{in}^2}{\text{ft}^2} = 3.74\times10^7\text{ lbf} = 166.4\text{ MN}$$

■

The roof of the storage tank can withstand this startingly large force because the gas or air inside it exerts an equal upward force, so the *net* force due to the pressure of the atmosphere and to the pressure of the gas inside the tank is zero. Since these forces ordinarily cancel in force calculations, it is customary to make such calculations in gauge pressure, when both sides of the surface are subjected to the pressure of the atmosphere in addition to the gauge pressure of the fluid.

Example 2.7. A layer of rainwater 4 in deep collects on the roof of the oil storage tank of Example 2.6. What net pressure force does it exert on the roof of the tank?

Here

$$P_{\text{gauge}} = \rho g h = 62.3\,\frac{\text{lbm}}{\text{ft}^3}\cdot 32.2\,\frac{\text{ft}}{\text{s}^2}\cdot\frac{4}{12}\,\text{ft}\cdot\frac{\text{lbf}\cdot\text{s}^2}{32.2\text{ lbm}\cdot\text{ft}}$$

$$= 20.8\,\frac{\text{lbf}}{\text{ft}^2} = 0.144\,\frac{\text{lbf}}{\text{in}^2} = 0.996\text{ kPa}$$

$$F = PA = 20.8\ \frac{\text{lbf}}{\text{ft}^2} \cdot \frac{\pi}{4}\ (150\ \text{ft})^2 = 3.67 \times 10^5\ \text{lbf} = 1.63\ \text{MN}$$

■

We could have found exactly the same answer by asking what the weight W of the fluid on the roof was, i.e.,

$$W = mg = V\rho g = \frac{4}{12}\ \text{ft} \cdot \frac{\pi}{4}\ (150\ \text{ft})^2 \cdot 62.3\ \frac{\text{lbm}}{\text{ft}^3} \cdot 32.2\ \frac{\text{ft}}{\text{s}^2} \cdot \frac{\text{lbf} \cdot \text{s}^2}{32.2\ \text{lbm} \cdot \text{ft}}$$

$$= 3.67 \times 10^5\ \text{lbf} = 1.63\ \text{MN}$$

where V is volume. This is typical of fluid statics problems involving horizontal surfaces. Since we found the basic equation of fluid statics by considering the weight of the fluid, we could work this kind of problem just as well by simply considering the weight of the fluid involved.

For *vertical* plane surfaces the pressure is not constant over the whole surface. Therefore, Eq. 2.19 must be used to find the force, and in general we cannot take the pressure outside the integral sign.

Example 2.8. The lock gate of a canal (Fig. 2.5) is rectangular, 20 m wide and 10 m high. One side is exposed to the atmosphere, the other side to water whose top surface is level with the top of the lock gate. What is the net force on the lock gate?

The net force is the force exerted by the water on the front of the gate minus the force exerted by the atmosphere on the back of the gate. Over the short vertical distance involved, the pressure of the atmosphere may be considered constant = P_{atm}. Thus, the force exerted on the back of the gate by the atmosphere is $P_{\text{atm}}A$, where A is the area of the gate. The pressure at any point in the water is given by Eq. 2.9. Here we define W as the width of the gate and h as the depth below the free surface. Then, substituting Eq. 2.9 in Eq. 2.19, we find

$$F_{\text{water}} = \int P\,dA = \int (P_{\text{atm}} + \rho g h)\,dA = P_{\text{atm}}A + \rho g \int hW\,dh$$

$$= P_{\text{atm}}A + \rho g W\ \frac{h^2}{2}\Bigg]_0^{10\ \text{m}}$$

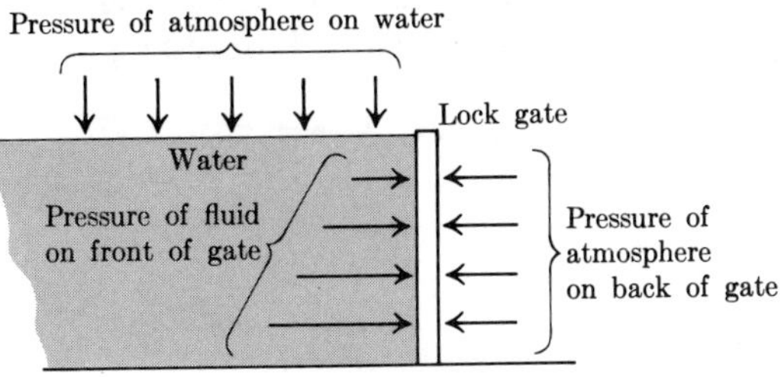

FIGURE 2.5

The net force in the x direction is

$$F_{\text{net}} = F_{\text{water}} - F_{\text{air}} = P_{\text{atm}}A + \rho g W \left.\frac{h^2}{2}\right]_0^{10\text{ m}} - P_{\text{atm}}A$$

The two atmospheric pressure terms cancel, and

$$F_{\text{net}} = 998.2\,\frac{\text{kg}}{\text{m}^3} \cdot 9.81\,\frac{\text{m}}{\text{s}^2} \cdot 20\text{ m} \cdot \left.\frac{h^2}{2}\right]_{h=0}^{h=10\text{ m}} \cdot \frac{\text{N}\cdot\text{s}^2}{\text{kg}\cdot\text{m}}$$

$$= 9.80\text{ MN} = 2.20 \times 10^6\text{ lbf}$$ ■

In this problem—and in all others in which a fluid open to the atmosphere acts on one side of a surface and the atmosphere acts on the opposite side—the effect of the atmospheric pressure cancels. Thus, such problems can be worked most easily by using gauge pressure; if it had been used in this problem, it would have given exactly the result shown above.

In Example 2.8 the integration was quite simple because the surface was rectangular; for more complicated shapes it is more difficult. Such problems frequently can be simplified by using the idea of a centroid. We begin by writing Eq. 2.19 for the gauge pressure of a constant-density fluid:

$$F = \rho g \int h\,dA \tag{2.21}$$

Multiplying by A/A and rearranging, we get

$$F = \rho g A\,\frac{\int h\,dA}{A} \tag{2.22}$$

But $\int h\,dA/A$ is the definition of h_c, the centroid of the depth measured from the free surface, so this equation may be simplified to

$$F = \rho g A h_c \tag{2.23}$$

The centroids of many geometric figures are tabulated in texts on mechanics and strength of materials; so by using those tabulations and Eq. 2.23, often we can avoid performing the integration shown in Eq. 2.21.

Example 2.9. A dam (Fig. 2.6) is a triangle 100 ft across the top and 75 ft deep. Water is up to the top on one side; the other side is exposed to the atmosphere. What is the net horizontal force on the dam?

Here, as in Example 2.8, the atmospheric pressure terms cancel, so we can save effort by working the problem in gauge pressure. Proceeding directly

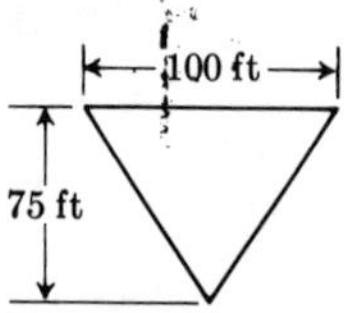

FIGURE 2.6

to integrate Eq. 2.21, we get

$$F = \rho g \int hW \, dh$$

Here

$$W = (100 \text{ ft})\left(1 - \frac{h}{75 \text{ ft}}\right)$$

so that

$$F = \rho g(100 \text{ ft}) \int h\left(1 - \frac{h}{75 \text{ ft}}\right) dh = \rho g(100 \text{ ft})\left[\frac{h^2}{2} - \frac{h^3}{3 \cdot 75 \text{ ft}}\right]_0^{75 \text{ ft}}$$

$$= 62.3 \frac{\text{lbm}}{\text{ft}^3} \cdot 32.2 \frac{\text{ft}}{\text{s}^2} \cdot 100 \text{ ft} \cdot \left[\frac{(75 \text{ ft})^2}{2} - \frac{(75 \text{ ft})^3}{3 \cdot 75 \text{ ft}}\right] \cdot \frac{\text{lbf} \cdot \text{s}^2}{32.2 \text{ lbm} \cdot \text{ft}}$$

$$= 5.84 \times 10^6 \text{ lbf} = 26.0 \text{ MPa}$$

To use Eq. 2.23 we look up the centroid of a triangle about its base and find that it is one-third of its altitude and that its area is its base times one-half of its altitude; so

$$F = 62.3 \frac{\text{lbm}}{\text{ft}^3} \cdot 32.2 \frac{\text{ft}}{\text{s}^2} \cdot \frac{75}{3} \text{ ft} \cdot \frac{100 \text{ ft} \cdot 75 \text{ ft}}{2} \cdot \frac{\text{lbf} \cdot \text{s}^2}{32.2 \text{ lbm} \cdot \text{ft}}$$

$$= 5.84 \times 10^6 \text{ lbf}$$ ■

The simplification afforded by Eq. 2.23 is even greater for circles, ellipses, etc.; the direct integration is very cumbersome, but the calculation via the centroid is as simple as in this example.

The pressure force exerted by a static fluid is always normal to the surface adjacent to the fluid. Thus, the pressure force on a vertical surface is horizontal, and on a horizontal surface it is vertical. For a plane surface which is neither vertical nor horizontal the pressure force is simply perpendicular to the surface and may be computed by Eq. 2.19. In many practical problems we are most interested in the vertical and horizontal components of the pressure force on such a body. Figure 2.7 shows a small section of area dA on such a surface.

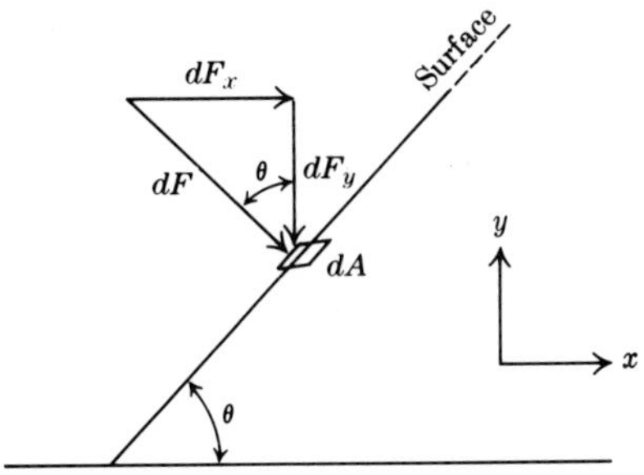

FIGURE 2.7
Force components on a surface.

Here and in the remainder of this chapter, we choose our angle θ as the angle between the normal to the surface on the fluid side and the vertical direction. We could make other choices of θ and find correct results, but the equations shown here are only correct for this choice of θ. The net pressure force is given by Eq. 2.18. The horizontal and vertical components are given by

$$dF_x = P \sin\theta \, dA \tag{2.24}$$

$$dF_y = P \cos\theta \, dA \tag{2.25}$$

Example 2.10. Figure 2.8 shows a weir which is set at an angle of 70° from the horizontal. The weir is 20 ft long and 10 ft wide and has water up to its top. What is the net force due to the water in the direction perpendicular to the weir? What are its x and y components?

Here we use Eq. 2.18 and note that, as in Example 2.8, the atmospheric contributions cancel, so we may use gauge pressure:

$$dF = \int P\, dA = W\rho g \int h\, dl$$

where l is the distance measured downward from the top along the weir. Then

$$h = l \sin\theta$$

$$F = W\rho g\,(\sin\theta)\,\frac{l^2}{2}\bigg]_{l=0}^{l=20\text{ ft}}$$

$$= 10\text{ ft}\cdot 62.3\,\frac{\text{lbm}}{\text{ft}^3}\cdot 32.2\,\frac{\text{ft}}{\text{s}^2}\cdot 0.9397\cdot\frac{(20\text{ ft})^2}{2}\cdot\frac{\text{lbf}\cdot\text{s}}{32.2\text{ lbm}\cdot\text{s}^2}$$

$$= 1.17\times 10^5\text{ lbf} = 520.3\text{ kN}$$

$$F_x = \int P\sin\theta\, dA = Wg\rho\,(\sin\theta)^2\,\frac{l^2}{2}\bigg]_{l=0}^{l=20\text{ ft}} = F\sin\theta$$

$$= 1.17\times 10^5\text{ lbf}\cdot 0.9397 = 1.10\times 10^5\text{ lbf} = 489\text{ kN}$$

$$F_y = \int -P\cos\theta\, dA = -Wg\rho\sin\theta\cos\theta\,\frac{l^2}{2}\bigg]_{l=0}^{l=20\text{ ft}} = -F\cos\theta$$

$$= -1.17\times 10^5\text{ lbf}\cdot 0.342 = -4.0\times 10^4\text{ lbf} = -177.9\text{ kN}$$

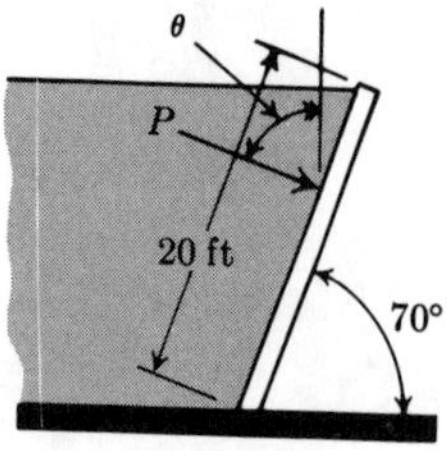

FIGURE 2.8

Here the minus sign indicates a force in the $-y$ direction, i.e., a downward force. ■

Equation 2.18 correctly gives the pressure on any small element of surface, whether curved or plane. For plane surfaces all the force due to various parts of the area is in one direction; for curved surfaces it is not. Equation 2.19, which gives the total force on a body, may also be interpreted as showing the total force acting on a curved surface, but generally this is of no practical use. We generally are interested in the component of the force in some specified direction. Figure 2.9 shows the force on a small element of area of some larger curved surface. To find the total x and y components of the force we write

$$F_x = \int dF_x = \int P \sin\theta \, dA \tag{2.26}$$

$$F_y = \int dF_y = \int -P \cos\theta \, dA \tag{2.27}$$

These are correct for any kind of surface; however, for complicated surfaces their integration may be difficult. They can be simplified by noting that $\sin\theta \, dA$ is exactly equal to the x projection of the element of surface area dA and that $\cos\theta \, dA$ is exactly equal to the y projection of the element of surface area dA.

Hence we can replace these two equations with

$$F_x = \int P \sin\theta \, dA = \int P \, d(x \text{ projection of } A) \tag{2.28}$$

$$F_y = \int -P \cos\theta \, dA = \int -P \, d(y \text{ projection of } A) \tag{2.29}$$

Example 2.11. A dam has been constructed of a large cylindrical pipe 10 ft in diameter and 3 ft long (see Fig. 2.10). Calculate the net vertical and horizontal components of the fluid force on this dam.

Here, as in previous examples, we can work in gauge pressure, because the atmospheric pressure contribution to the fluid force is exactly equal and opposite to the force exerted by the atmosphere on the rear of the dam. Then

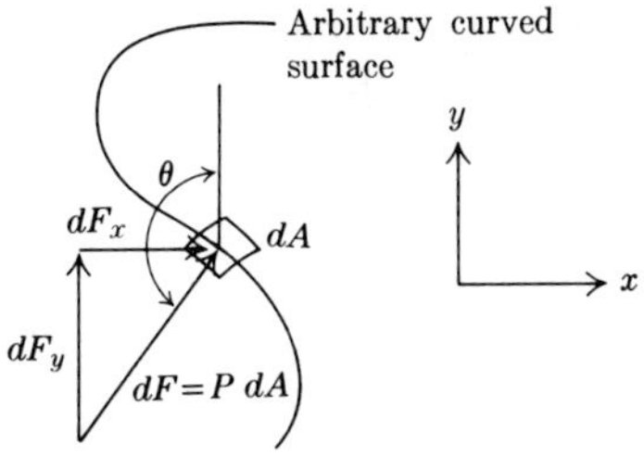

FIGURE 2.9
Force components on a curved surface.

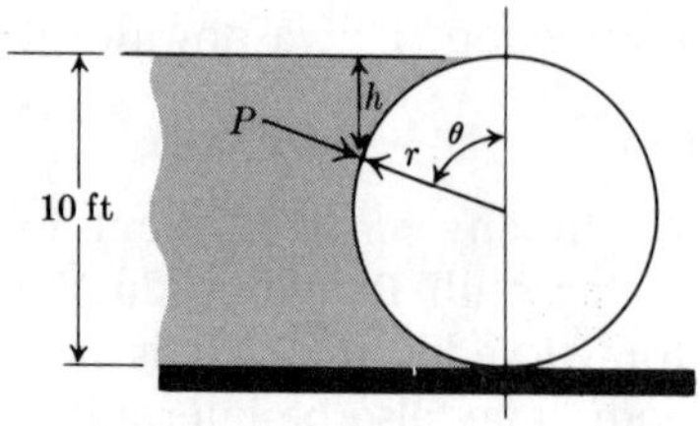

FIGURE 2.10

$$P = \rho g h = \rho g(r - r\cos\theta) \qquad dA = Wr\,d\theta$$

Therefore,

$$F_x = \int P\sin\theta\,dA = W\rho g\int_0^{\pi}(r - r\cos\theta)\sin\theta\,r\,d\theta$$

$$= W\rho g r^2[-\cos\theta + \tfrac{1}{2}\sin^2\theta]_0^{\pi} = 2W\rho g r^2$$

$$= 2\cdot 3\text{ ft}\cdot 62.3\,\frac{\text{lbm}}{\text{ft}^3}\cdot 32.2\,\frac{\text{ft}}{\text{s}^2}\cdot(5\text{ ft})^2\cdot\frac{\text{lbf}\cdot\text{s}^2}{32.2\text{ lbm}\cdot\text{ft}}$$

$$= 9345\text{ lbf} = 41.57\text{ kN}$$

$$F_y = \int -P\cos\theta\,dA = -W\rho g\int_0^{\pi}(r - r\cos\theta)\cos\theta\,r\,d\theta$$

$$= -W\rho g r^2[\sin\theta - \tfrac{1}{2}\sin\theta\cos\theta - \tfrac{1}{2}\theta]_0^{\pi} = W\rho g r^2\,\frac{\pi}{2}$$

$$= F_x\,\frac{\pi}{4} = 7340\text{ lbf} = 32.65\text{ kN}$$ ■

In this case we can greatly simplify the calculation of F_x by noting that the x projection of the cylinder is simply a rectangle 3 ft wide and 10 ft deep. From Example 2.8 we know that for such a rectangle

$$F_x = g\rho W\,\frac{h^2}{2}$$

Here $h = 2r$, so this is identical with the result for F_x found above. For the vertical force the projected-area idea offers no such simplification, because although this shape has a surface facing up and an equal surface facing down, the two surfaces face different pressures and their pressures do not vary in a simple way with the projected surface area. Next we will see a simple way to evaluate the vertical component of the pressure force on such surfaces.

2.4 BUOYANCY

We can calculate the force exerted by static fluids on floating and immersed bodies by integrating the vertical component of the pressure force over the entire surface of the body. This leads to a very simple generalization, called

Archimedes' principle, which is much easier to apply than the integration used in the previous section.

Consider the floating block of wood shown in Fig. 2.11. The block is at rest, so the sum of forces in any direction on it is zero. The only forces acting on it are the gravity force and the total pressure force around its entire surface; these must be equal and opposite. The vertical component of the pressure force integrated around the entire surface of a floating or submerged body is called a *buoyant force*. The buoyant force over the entire surface is then given by

$$F_y = \int -P\cos\theta\, dA$$

However, for the block shown, $\cos\theta$ is zero for the sides, -1 for the bottom, and $+1$ for the top; so

$$F_y = (P_{\text{bot}} - P_{\text{top}})\,\Delta x\,\Delta y \tag{2.30}$$

Here

$$(P_{\text{bot}} - P_{\text{top}}) = \rho g h + \rho_{\text{air}} g (l - h) \tag{2.31}$$

Multiplying by $\Delta x\,\Delta y$ gives

$$F_y = \rho_{\text{liq}} g V_{\text{liq}} + \rho_{\text{air}} g V_{\text{air}} \tag{2.32}$$

where V_{liq} is the volume of liquid displaced and V_{air} is the volume of air displaced. Thus the buoyant force is exactly equal to the weight of both fluids displaced. This is Archimedes' principle. In most cases the term for the weight of air in Eq. 2.32 is negligible compared with that for the weight of the water involved. For floating bodies Archimedes' principle is often restated: "A floating body displaces a volume of fluid whose weight is exactly equal to its own." If a body is completely immersed in a fluid, then there is only one term on the right in Eq. 2.32.

The statements above were worked out for a block with the axis vertical. This was convenient, because the pressure on the vertical sides did not contribute to the buoyant force. However, the result is true for any kind of body because, as shown in Fig. 2.12, any shape at all can be visualized as made up of blocks.

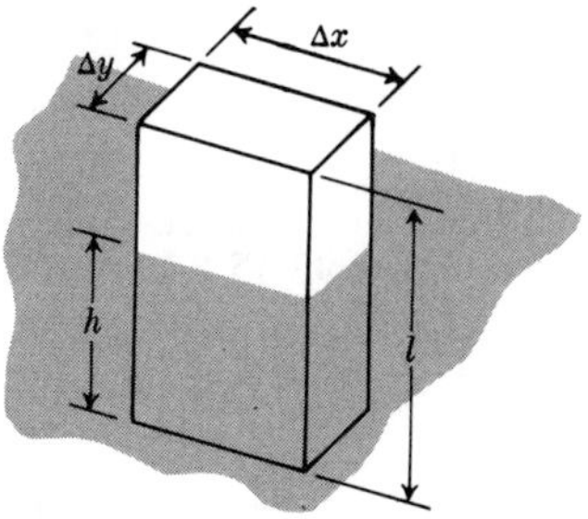

FIGURE 2.11

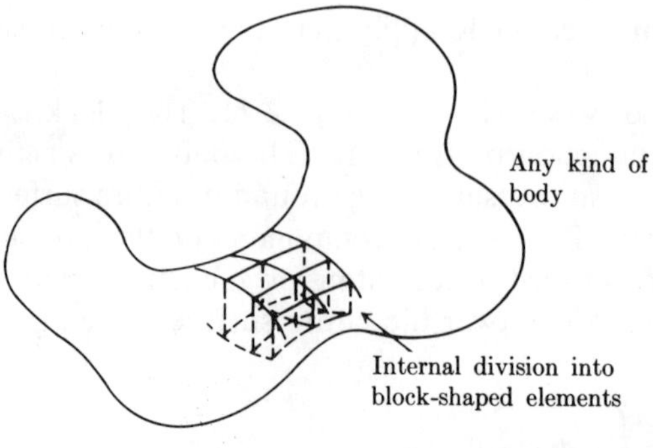

FIGURE 2.12

If the blocks are large, as shown in the figure, then their combined volume will be a rough approximation to the volume of the body. However, as the x and y dimensions of the blocks decrease, the blocks form a steadily improving approximation to the body, becoming identical with it as the x and y dimensions approach zero. Nonetheless, for each block, no matter how small, the foregoing argument holds, and thus Archimedes' principle holds for any shape of body. Thus, although it would be very difficult to perform the indicated pressure integration over a body with a shape like an octopus, if we know its volume (and hence the volume of fluid it displaces), then we can easily calculate the buoyant force by Archimedes' principle.

Example 2.12. A helium balloon is at the same pressure and temperature as the surrounding air (1 atm, 20°C) and has a diameter of 3 m. The weight of the plastic skin of the balloon is negligible. How much payload can the balloon lift?

The buoyant force is the weight of air displaced:

$$F_{\text{buoy}} = \rho_{\text{air}} g V_{\text{balloon}}$$

The weight of helium is

$$W_{\text{hel}} = \rho_{\text{hel}} g V_{\text{balloon}}$$

Therefore, the payload is

$$\text{Payload} = F_{\text{buoy}} - W_{\text{hel}} = V_{\text{balloon}} g(\rho_{\text{air}} - \rho_{\text{hel}})$$

$$= Vg \frac{P}{RT} (M_{\text{air}} - M_{\text{hel}})$$

$$= \frac{\pi}{6} \cdot (3\text{ m})^3 \cdot \frac{9.81\text{ m}}{\text{s}^2} \cdot \frac{1\text{ atm}}{[8.2 \times 10^{-5}\text{ m}^3 \cdot \text{atm}/(\text{mol} \cdot \text{K})](293.15\text{ K})}$$

$$\cdot \left(29 \frac{\text{g}}{\text{mol}} - 4 \frac{\text{g}}{\text{mol}}\right) \cdot \frac{\text{kg}}{1000\text{ g}} \cdot \frac{\text{N} \cdot \text{s}^2}{\text{kg} \cdot \text{m}}$$

$$= 144.2\text{ N} = 32.4\text{ lbf}$$

■

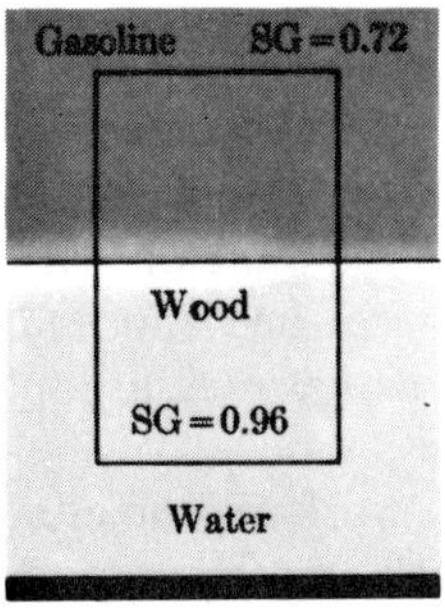

FIGURE 2.13

Example 2.13. A block of wood is floating at the interface between a layer of gasoline and a layer of water (see Fig. 2.13). What fraction of the wood is below the interface?

Here the weight of the wood is equal to the buoyant force, which in turn is equal to the weight of the two fluids displaced:

$$V_{\text{wood}}\rho_{\text{wood}}g = V_{\text{water}}\rho_{\text{water}}g + V_{\text{gas}}\rho_{\text{gas}}g$$

where V_{wood} is the volume of the block and V_{water} and V_{gas} are the volumes of water and gasoline displaced. Dividing by $g\rho_{\text{water}}$, we find

$$V_{\text{wood}}\text{SG}_{\text{wood}} = V_{\text{water}} + V_{\text{gas}}\text{SG}_{\text{gas}}$$

where SG is the specific gravity. But since

$$V_{\text{wood}} = V_{\text{water}} + V_{\text{gas}}$$

We may eliminate V_{gas}

$$V_{\text{wood}}\text{SG}_{\text{wood}} = V_{\text{water}} + (V_{\text{wood}} - V_{\text{water}})\text{SG}_{\text{gas}}$$

and then we find

$$\frac{V_{\text{water}}}{V_{\text{wood}}} = \frac{\text{SG}_{\text{wood}} - \text{SG}_{\text{gas}}}{1 - \text{SG}_{\text{gas}}} = \frac{0.96 - 0.72}{1 - 0.72} = 0.857 \qquad \blacksquare$$

This result appears paradoxical. The gasoline pushes down on the top of the block, not up on it at any point, yet the volume of gasoline displaced enters the buoyant force calculation. However, if we examine the pressure integral around the surface, we see that the pressure difference from top to bottom of the block does indeed involve the gasoline in the way shown.

In leaving this topic we note again that the basic operation is the integration of the vertical components of the pressure force over the entire surface of the body. The convenient result of this integration is Archimedes' principle that the buoyant force is equal to the weight of the fluid displaced.

2.5 PRESSURE MEASUREMENT

Pressures usually are measured by letting them act across some area and opposing them with either a gravity force or the force of a compressed spring.

In the gravity-force method, a device called a *manometer* often is used; its operation is described in Example 2.14.

Example 2.14. Figure 2.14 shows a tank of gas connected to a manometer. The manometer is a U-shaped glass tube open to the atmosphere at one end and containing water. From the elevations shown, calculate the gauge pressure in the vessel.

We want to know the pressure at D. The simple way to work manometer problems is to start with some pressure we know and to work step by step to the pressure we want to find. In this case, we know that the gauge pressure at A is zero, because it is open to the atmosphere. The water is practically a constant-density fluid: therefore, we can use Eq. 2.11 to find the pressure at B:

$$P_B = \rho_{\text{water}} g h_B = (\rho_{\text{water}} g)(3 \text{ ft})$$

To find the pressure at C, we use Eq. 2.9:

$$P_C = P_B - (\rho_{\text{water}} g \cdot \tfrac{1}{2} \text{ ft})$$

To find the pressure at D, we use the same equation (i.e., we assume that the density *change* of the gas is negligible):

$$P_D = P_C - (\rho_{\text{gas}} g \cdot \tfrac{1}{2} \text{ ft})$$

Adding these three equations and canceling like terms, we find

$$\begin{aligned} P_D &= (\rho_{\text{water}} g \cdot 3 \text{ ft}) - (\rho_{\text{water}} g \cdot \tfrac{1}{2} \text{ ft}) - (\rho_{\text{gas}} g \cdot \tfrac{1}{2} \text{ ft}) \\ &= 32.2 \frac{\text{ft}}{\text{s}^2}\left[\left(62.3 \frac{\text{lbm}}{\text{ft}^3} \cdot 2.5 \text{ ft}\right) - \left(0.08 \frac{\text{lbm}}{\text{ft}^3} \cdot 0.5 \text{ ft}\right)\right] \\ &\quad \cdot \frac{\text{lbf} \cdot \text{s}^2}{32.2 \text{ lbm} \cdot \text{ft}} \cdot \frac{\text{ft}^2}{144 \text{ in}^2} \\ &= 1.08 \frac{\text{lbf}}{\text{in}^2} - 0.0003 \frac{\text{lbf}}{\text{in}^2} = 1.08 \frac{\text{lbf}}{\text{in}^2} \text{ gauge} \\ &= 7.46 \text{ kPa} \end{aligned}$$

■

The example illustrates several points.

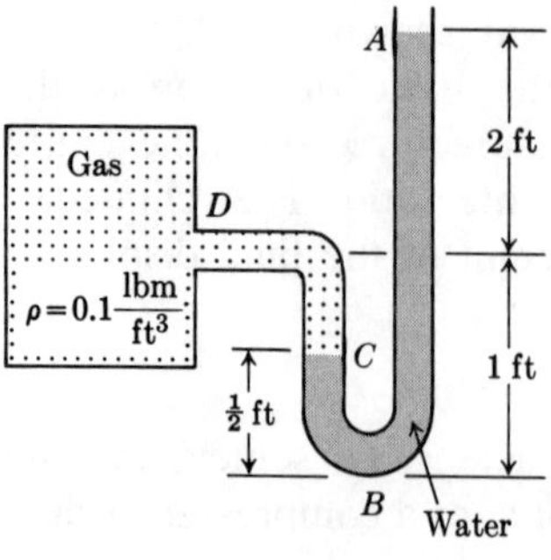

FIGURE 2.14
Simple manometer.

1. The contribution of the section of the manometer full of gas to the answer is very small and generally can be neglected in manometer problems.
2. Manometers which are open to the atmosphere are gauge-pressure devices and should be calculated in gauge pressure.
3. In reading such a device, we normally read an elevation; the actual operational reading in Example 2.14 was 2.5 ft. For many purposes it is convenient to think of pressures and to report them in terms of such manometer reading as heights; the U.S. air conditioning industry, for example, commonly refers to all pressure in air conditioning ducts as "inches of water," and most U.S. vacuum equipment manufacturers refer to vacuums as "inches of mercury."
4. At no place in our last calculation did the cross-sectional area of the manometer tube appear. Therefore, this tube can be any convenient size and need not be made of constant-diameter tubing. The only measurements necessary are the fluid densities, which can be looked up in handbooks, and the differences in elevation, which can be read directly with tape measures and rulers. Thus, manometers require neither calibration nor testing with standards; one simply connects them and takes the reading.
5. It may seem that we went to a lot of trouble for such a simple problem. This is true; working engineers use shorter calculation methods than that shown here. However, with complicated systems, such as two-fluid manometers, the shortcuts are confusing, and the method shown above is always reliable.

Example 2.15. A two-fluid manometer is often used to make it unnecessary to read small differences in elevation. The one shown in Fig. 2.15 is measuring the pressure difference between two tanks. What is that pressure difference?

We want to know $P_A - P_E$. All the fluids have practically constant density, so we can use Eq. 2.9. We begin by calling P_E known. Then

$$P_D = P_E + (\rho_{\text{water}} g)(1\text{ ft})$$
$$P_C = P_D + (\rho_{\text{oil}} g)(2\text{ ft})$$
$$P_B = P_C - (\rho_{\text{oil}} g)(1\text{ ft})$$
$$P_A = P_B - (\rho_{\text{water}} g)(2\text{ ft})$$

Adding and canceling like terms, we find

$$P_A = P_E + (\rho_{\text{water}} g)(1\text{ ft} - 2\text{ ft}) + (\rho_{\text{oil}} g)(2\text{ ft} - 1\text{ ft})$$

$$P_A - P_E = (1\text{ ft})g(\rho_{\text{oil}} - \rho_{\text{water}})$$

$$= 1\text{ ft} \cdot 32.2\,\frac{\text{ft}}{\text{s}^2} \cdot (1.1 - 1.0) \cdot 62.3\,\frac{\text{lbm}}{\text{ft}^2} \cdot \frac{\text{lbf} \cdot \text{s}^2}{32.2\text{ lbm} \cdot \text{ft}} \cdot \frac{\text{ft}^2}{144\text{ in}^2}$$

$$= 0.043\,\frac{\text{lbf}}{\text{in}^2} = 298\text{ Pa}$$ ■

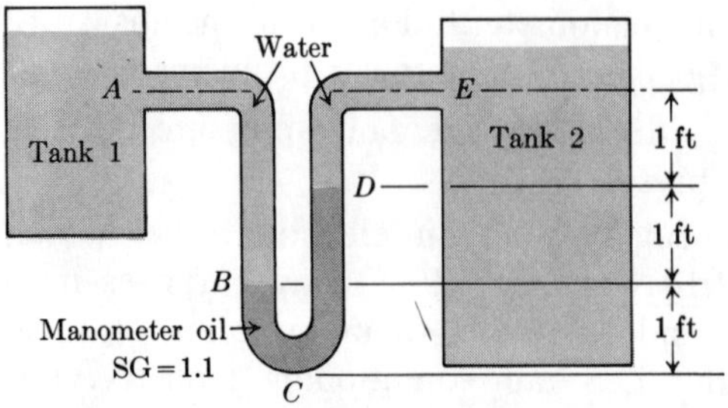

FIGURE 2.15
Two-fluid manometer.

This reading corresponds to a pressure difference of 0.1 ft of water. The actual reading of this two-fluid manometer is 1 ft. If we assume that we can read with an accuracy of ±0.005 ft, then a simple water manometer will have an uncertainty of 5 percent for this difference; the two-fluid manometer shown has an uncertainty of 0.5 percent.

Because a manometer is a device for measuring pressure differences, to use one to measure absolute pressure we must measure the difference between the pressure in question and a perfect vacuum. In principle this is impossible, because there is no such thing as a perfect vacuum, but in practice we may produce vacuums of sufficient quality that the error introduced by calling them perfect is negligible. This idea is used in the mercury barometer shown in Fig. 2.16. This common device is found in most laboratories for measuring the pressure of the atmosphere. The pressure of the atmosphere acts on the mercury in the cup at the bottom and is opposed by the weight of the column of mercury. Calculating this, we find

$$P_A - P_B = \rho_{\mathrm{Hg}} g h$$

where P_B is the pressure in the vapor space above the liquid mercury. In well-built manometers, this will be simply the vapor pressure of mercury, which at 70°F is about 10^{-6} atm; this is so small compared with 1 atm that it can be neglected. Thus, although the barometer, like all manometers, measures pressure differences, it can be used with satisfactory accuracy as an absolute-pressure device in this case.

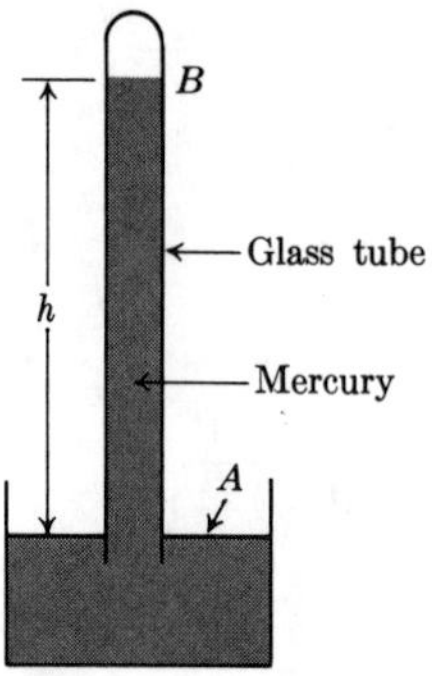

FIGURE 2.16
Mercury barometer.

The second basic way to measure pressure is to let the pressure act on some piston, which compresses a spring, and to measure the displacement. Figure 2.17 shows an impractical but illustrative way of doing this. The fluid whose pressure is to be measured presses on the piston, compressing the spring and moving the pointer along the scale. If we know the area of the piston, the spring constant, and the pointer reading for zero pressure, we can calculate the pressure on the piston from the pointer position.

Example 2.16. The piston in Fig. 2.17 has an area of $100\ \text{cm}^2$, and the spring constant k is 100 N/cm. We set the pointer so that there is a zero reading when both sides of the piston are exposed to the atmosphere. Now we attach the gauge to a tank with an unknown pressure, and the pointer moves to 2.5 cm. What is the pressure in the tank?

Here the net force acting on the piston is

$$F_{\text{net}} = (P_{\text{tank}} - P_{\text{atm}})A = AP_{\text{tank(gauge)}}$$

This must be equal to the force on the spring, which is $k\,\Delta x$, and therefore

$$P_{\text{tank(gauge)}} = \frac{k\,\Delta x}{A} = \frac{(100\ \text{N/cm})(2.5\ \text{cm})}{100\ \text{cm}^2} \cdot \left(\frac{100\ \text{cm}}{\text{m}}\right)^2 = 25\ \text{kPa} = 3.62\ \frac{\text{lbf}}{\text{in}^2}$$ ■

From the last example we observe the following:

1. This device, like the manometer, measures pressure differences. To use it as an absolute-pressure device, we must make it compare pressure with a vacuum. We can do this by placing it in an evacuated chamber.
2. This device, unlike the manometer, requires a precise measurement of its dimensions or a calibration. Spring-type pressure gauges usually are calibrated by comparing their readings with those of manometers like the one shown in Fig. 2.16, or other equivalent devices.

The gauge shown in Fig. 2.17 is impractical because of the problem of leakage around the piston. The most widely used type of spring pressure gauge

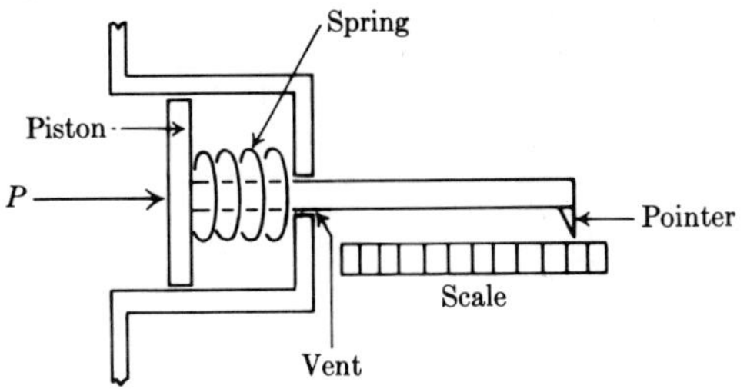

FIGURE 2.17
Piston-and-spring pressure gauge.

is one involving a *bourdon tube*, shown in Fig. 2.18. A bourdon tube is a stiff metal tube bent into a circular shape. The fluid whose pressure is to be measured is inside the tube. One end of the tube is fixed, and the other is free to move inward or outward. The inward or outward movement of the free end moves a pointer through a linkage-and-gear arrangement. Because of the curvature of the tube the side of the tube farthest from the center has more surface area than the side nearest the center. Thus, the pressure inside exerts a net outward force on the tube, causing the tube to straighten as the pressure rises. The tube itself serves as the spring; it is made of metal which is stiff and has a reasonable spring constant. With such a tube the calculation of the movement as a function of the inside and outside pressures would be more difficult than with the piston-and-spring gauge of Fig. 2.17. However, the bourdon tube is a very convenient shape and causes no leakage problems, unlike the piston-and-spring gauge in Fig. 2.17. Since both are calibrated devices, the difficulty in calculating the performance of the bourdon tube is not a real disadvantage. Bourdon-tube pressure gauges are simple, rugged, reliable, and cheap; they are the most widely used type of industrial pressure gauge.

Neither the manometer nor the bourdon-tube gauge is suited to measuring rapidly changing pressures. Both are unsatisfactory for this purpose because of their high inertial mass; this mass makes them move slowly to accommodate a change in pressure, and so their readings lag behind a rapidly changing pressure. For rapidly changing pressures (such as pressure fluctuations in rocket motors), two other types of pressure gauges respond much more quickly. One is the diaphragm gauge, which is similar to that of Fig. 2.17 but has, instead of the piston and spring, a thin metal diaphragm, which acts as both. When the pressure increases, the diaphragm stretches very slightly; the stretch is detected by an electric strain gauge (or other electronic means) and recorded electrically. The advantage of the diaphragm over the bourdon tube is its very low mass, which allows it to move quickly in response to a change in pressure. The second type of rapid-response pressure gauge is the quartz-crystal piezometer, which makes use of the change in electrical properties of quartz crystals with change in pressure.

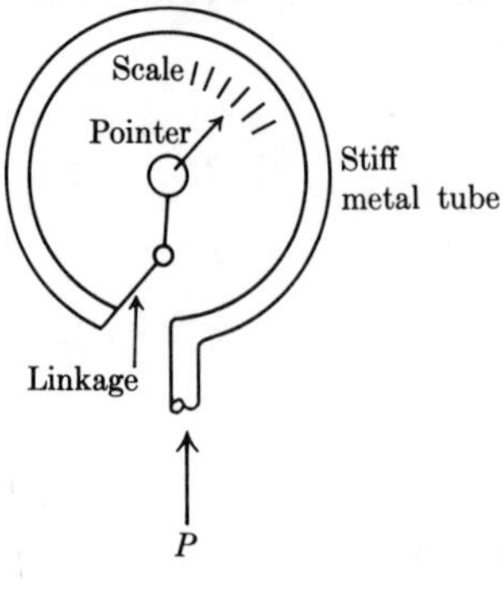

FIGURE 2.18
Bourdon-tube pressure gauge.

2.6 MANOMETER-LIKE SITUATIONS

In Sec. 2.5 we discussed manometers as pressure-measuring devices. There are many other fluid mechanical situations which are understood most easily if we analyze them just as we analyze manometers. Several examples are shown here.

Figure 2.19 shows a schematic cross section of a percolator-type coffee maker. In it, the pot is filled to a height z_1 with water. The basket above the water is filled with ground coffee. The whole assembly is placed on a stove and heated from below. When the water has been warmed, it begins to flow in irregular spurts up the central tube; it is diverted by the cap on the top, falls on the coffee grounds, and percolates through them, extracting the water-soluble constituents of the ground coffee, to make the hot drink many enjoy.

How can the fluid do this? Here we have a fluid flowing from a low elevation to a high one, with no mechanical device lifting it. How can that be? To answer the question, we compute the pressures at B and C. It will be easiest if we do this all in gauge pressure. So the pressure at B is

$$P_B = \rho g z_1$$

and if the fluid in the tube is up to the level where it spills out at D, then the pressure at C is

$$P_C = \rho g z_2$$

and

$$P_B - P_C = g[(\rho z)_1 - (\rho z)_2]$$

When we put the pot on the stove, the density inside and the density outside the riser tube will be the same (that of water) and the liquid in the tube will stand at the same level as the liquid outside, z_1. There will be no flow.

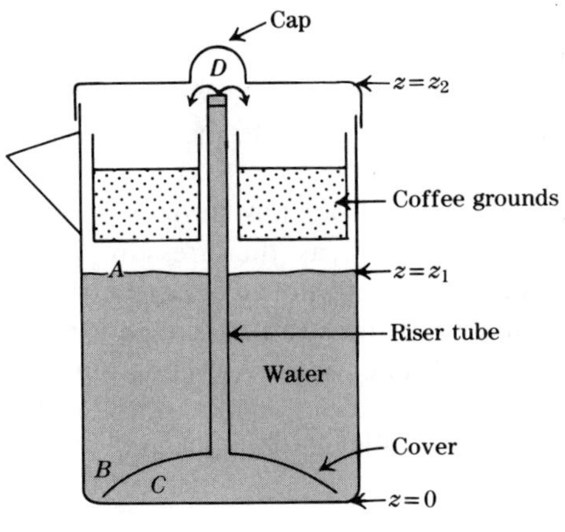

FIGURE 2.19
Coffee percolator.

However, as the water at the bottom is heated by the stove, the loose-fitting cover prevents the water from mixing with the rest of the fluid in the pot, so that a small amount of liquid is heated to its boiling point. When it boils, the bubbles of steam produced in that way flow by buoyancy up through the riser tube. While they do so, the average density of the gas-liquid mixture in the riser tube falls. If there is no net flow under the loose-fitting cover, then the pressure difference from one side of it to the other, $(P_B - P_C)$, must be zero, and the level in the riser tube z_2 must increase to keep this pressure difference equal to zero.

When the generation rate of bubbles becomes high enough that z_2 becomes greater than the height of the top of the riser tube, then a mixture of steam and water will flow out of the top of the tube. If the rate of generation of steam bubbles increases even more, then the average density of the steam-water mixture in the riser tube will fall low enough that $P_B - P_C$ can no longer be zero, but must become a positive number. Then the pressure force due to gravity will force water from the pot under the loose-fitting cover, and the circulation will be established, with flow downward under the cover, up the riser tube, and down through the coffee grounds. For that flow we can no longer use the simple equations of fluid statics, which we have used so far in this discussion; the methods of Chaps. 5, 6, 7, and 14 must be brought in. But this simple discussion shows how the pressure forces that move the fluids in coffee percolators arise. The exact same discussion applies to geysers (where the flow is intermittent instead of the steady flow in the boiling coffee pot) and to the circulation system in most steam boilers, in gas- and propane-fired refrigerators, and in the reboilers of many distillation columns. In all these, the formation of bubbles of steam (or the vapor of some other liquid being boiled) lowers the average density in one leg of the "equivalent manometer," producing the pressure difference which drives the flow.

Such pressure differences can also arise in systems that do not involve boiling liquids. This is illustrated in Example 2.17.

Example 2.17. Figure 2.20 shows a schematic of a home fireplace, with part of the house which surrounds it. The burning logs in the fireplace heat the gases in the chimney to 300°F. If we treat this as a static situation, what will be the difference in pressure between the air in the room adjacent to the fireplace and the air inside the fireplace at the same level?

Here we assume that the house is leaky enough, or has an open window, so that the pressure inside the house is the same as the pressure in the atmosphere outside. (This is true for older houses, but not necessarily true for modern energy-conserving houses which have much less air exchange with the surroundings!) Here, as shown in the figure, we take the elevation datum $z = 0$ at the top of the chimney; we will see that this choice makes the solution simple. We take the pressure at $z = 0$ to be atmospheric pressure and work in gauge pressures:

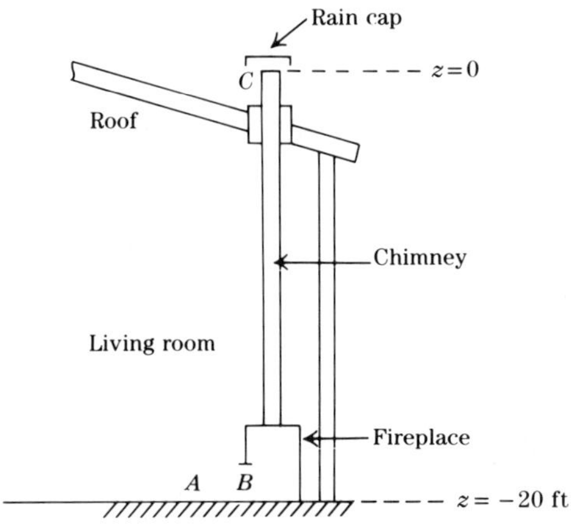

FIGURE 2.20
Home fireplace.

$$P_A = \rho_{\text{air}} g z_1$$

and

$$P_B = \rho_{\text{flue gas}} g z_1$$

so that

$$P_A - P_B = (\rho_{\text{air}} - \rho_{\text{flue gas}}) g z_1$$

Assuming that both the air and the flue gas are ideal gases, we can express each in terms of the perfect gas law:

$$P_A - P_B = g z_1 \frac{P}{R}\left[\left(\frac{M}{T}\right)_{\text{air}} - \left(\frac{M}{T}\right)_{\text{flue gas}}\right]$$

In the most careful work, we need to take into account the small difference in molecular weights of air and flue gas, but here we can assume that the molecular weights are practically equal. Then we multiply and divide by $(M/T)_{\text{air}}$ and substitute ρ_{air} for its perfect-gas equivalent:

$$P_A - P_B = g z_1 \frac{PM_{\text{air}}}{RT_{\text{air}}}\left(1 - \frac{T_{\text{air}}}{T_{\text{flue gas}}}\right) = g z_1 \rho_{\text{air}}\left(1 - \frac{T_{\text{air}}}{T_{\text{flue gas}}}\right)$$

$$= 32.2 \frac{\text{ft}}{\text{s}^2} \cdot 20\,\text{ft} \cdot 0.075 \frac{\text{lbm}}{\text{ft}^3} \cdot \left(1 - \frac{528°\text{R}}{760°\text{R}}\right) \cdot \frac{\text{lbf} \cdot \text{s}^2}{32.2\,\text{lbm} \cdot \text{ft}} \cdot \frac{\text{ft}^2}{144\,\text{in}^2}$$

$$= 0.0032\,\text{lbf/in}^2 = 22\,\text{Pa}$$ ■

This is clearly a very small pressure difference. But, as we will see in Chap. 5, very small pressure differences can produce significant velocities in gas flows; this pressure difference would produce a velocity, in frictionless flow, of about 20 ft/s.

In this example the calculation was made for a static fluid. In the real situation, the fluid will be set in motion by the pressure difference calculated above, and we will need the methods developed in later chapters to compute the velocity. But this calculation shows how pressure differences can arise in both boiling liquids and gases if one side of an "equivalent manometer" is heated to a temperature higher than the other. This explains how chimneys work. For all but the largest furnaces, the airflow is driven through the furnace by the pressure difference computed here. That explains why large furnaces have tall stacks; the available pressure difference is shown above to be proportional to the height of the stack. Many large furnaces now use powered fans to drive the gases through them; small furnaces mostly operate on "natural draft," as shown in this example.

This calculation also explains many meteorological phenomena. Oceans and lakes are heated and cooled slowly by the sun; the ground on the shore heats more rapidly in the daytime and cools more rapidly at night. Thus, during the day the hot ground heats the air above it, and hot air above the ground plays the same role as the flue gas in Example 2.17. The winds blow from the ocean or lake onto the shore. At night the ground cools, and cools the air above it, and the direction of the pressure gradient reverses, causing the wind to blow from the shore out over the body of water. The monsoon rains of India and parts of tropical Africa are the same phenomenon, on a much larger scale. In the summer the heated air over the continents rises, and the cooler air, which is moist, flows in and brings rain.

2.7 VARIABLE GRAVITY

Up until now we have assumed that the acceleration of gravity is constant: $32.2\ \text{ft/s}^2 = 9.81\ \text{m/s}^2$. This is not exactly true in any problem involving two different elevations. However, the change in gravitational acceleration with a change in elevation is quite small. The acceleration of gravity near the surface of the earth is proportional to the reciprocal of the square of the distance from the center of earth. The radius of the earth is about $4000\ \text{mi} = 6440\ \text{km}$, so the acceleration of gravity $1\ \text{mi} = 1.609\ \text{km}$ above the surface is $4000^2/4001^2 = 0.9995$ times the acceleration of gravity at the surface. Few engineering problems include data precise enough to justify making such corrections.

In two types of problem, however, nonconstant gravity is important:

1. In space travel and rocket problems, where the distances from the earth become significant compared with 4000 mi (so the changing value of gravity must be taken into account)
2. Acceleration and centrifugal force problems

Since this chapter is about fluid statics, it seems strange to consider acceleration or centrifugal force problems, in which the fluid is certainly moving. We do so because in these problems the fluid is not moving relative to its container or relative to other parts of the fluid. Really, all problems in terrestrial fluid statics involve moving fluids, because the fluids are on the earth, the earth is rotating about its axis and revolving around the sun, and the sun is moving through space. As long as the individual particles of fluid are not moving relative to each other, we can treat such moving problems by the methods of fluid statics. Such motions of fluids are called *rigid-body motions*.

2.8 PRESSURE IN ACCELERATED RIGID-BODY MOTIONS

We now repeat the derivation of Eq. 2.1 for the case in which an entire mass of fluid is in some kind of accelerated rigid-body motion. Again we use the small, cubical element of fluid shown in Fig. 2.1 and consider it to be part of a larger mass of fluid. In Sec. 2.1 we showed that if the fluid was not being accelerated, then the sum of the forces on it must be zero. If the fluid was being accelerated, then the sum of the forces acting on it, in the direction of the acceleration, must equal the mass times the acceleration. For the cubical element of fluid being accelerated in the vertical direction, we rewrite Eq. 2.1 as

$$(P_{z=0})\,\Delta x\,\Delta y - (P_{z=\Delta z})\,\Delta x\,\Delta y - \rho g\,\Delta x\,\Delta y\,\Delta z = \rho\,\Delta x\,\Delta y\,\Delta z\,\frac{d^2z}{dt^2}$$

Dividing by $\Delta x\ \Delta y\ \Delta z$ and taking the limit as Δz approaches zero, we find

$$\frac{dP}{dz} = \rho\left(g + \frac{d^2z}{dt^2}\right) \tag{2.33}$$

which for constant-density fluids can be integrated to

$$P_2 - P_1 = -\rho\left(g + \frac{d^2z}{dt^2}\right)(z_2 - z_1) \qquad \text{[constant density]} \tag{2.34}$$

For gauge pressure this simplifies further to

$$P = -\rho h\left(g + \frac{d^2z}{dt^2}\right) \qquad \text{[constant density, gauge pressure]} \tag{2.35}$$

Example 2.18. An open tank contains water 5 m deep. It is sitting on an elevator. Calculate the gauge pressure at the bottom of the tank (*a*) when the elevator is standing still, (*b*) when the elevator is accelerating upward at the rate of $5\ \text{m/s}^2$, and (*c*) when the elevator is accelerating downward at the rate of $5\ \text{m/s}^2$.

From Eq. 2.11, part (*a*) is simply

$$P_a = \rho g h = 998.2\,\frac{\text{kg}}{\text{m}^3}\cdot 9.81\,\frac{\text{m}}{\text{s}^2}\cdot 5\,\text{m}\cdot\frac{\text{N}\cdot\text{s}^2}{\text{kg}\cdot\text{m}}\cdot\frac{\text{Pa}}{\text{N/m}^2}$$

$$= 49.0\ \text{kPa} = 7.11\,\frac{\text{lbf}}{\text{in}^2}$$

For (*b*) and (*c*) we use Eq. 2.35:

$$P_b = \rho h\left(g + \frac{d^2z}{dt^2}\right) = 998.2\ \frac{\text{kg}}{\text{m}^3} \cdot 5\ \text{m} \cdot \left(9.81\ \frac{\text{m}}{\text{s}^2} + 5\ \frac{\text{m}}{\text{s}^2}\right) \cdot \frac{\text{N} \cdot \text{s}^2}{\text{kg} \cdot \text{m}} \cdot \frac{\text{Pa}}{\text{N/m}^2}$$

$$= 74.0\ \text{kPa} = 10.73\ \text{lbf/in}^2$$

and

$$P_c = \rho h\left(g + \frac{d^2z}{dt^2}\right)$$

$$= 998.2\ \frac{\text{kg}}{\text{m}^3} \cdot 5\ \text{m} \cdot \left(9.81\ \frac{\text{m}}{\text{s}^2} - 5\ \frac{\text{m}}{\text{s}^2}\right) \cdot \frac{\text{N} \cdot \text{s}^2}{\text{kg} \cdot \text{m}} \cdot \frac{\text{Pa}}{\text{N/m}^2}$$

$$= 24.0\ \text{kPa} = 3.48\ \text{lbf/in}^2$$ ■

If the acceleration is not in the same direction as gravity or in the direction opposite to it, then suppose it is in some direction *a*, as in Fig. 2.2. Then we can take the summation of forces in the *a* direction and substitute $g \cos \theta$ for *g* in Eq. 2.35:

$$\frac{dP}{da} = -\rho\left(g \cos \theta + \frac{d^2a}{dt^2}\right) \tag{2.36}$$

Example 2.19. A rectangular tank of orange juice on a cart is moving in the *x* direction with a steady acceleration of 1 ft/s^2. See Fig. 2.21. What angle does its free surface make with the horizontal?

Here we assume that the tank has been under acceleration so long that the initial sloshing back and forth of the liquid at the start of acceleration has died out and that the fluid is truly in rigid-body motion. In the figure, points *A* and *B* are both on the free surface; neglecting the very slight change in atmospheric pressure over this change in elevation, we may say that the gauge pressure is zero at both points. Then we can calculate the pressure at *C* from the pressure at *A* by using Eq. 2.36. Here we are applying it in the *y* direction ($a = y$), so we have $\cos \theta = 1$ and $d^2y/dt^2 = 0$. Hence, the result is the same as Eq. 2.11:

$$P_c = \rho g\, \Delta y$$

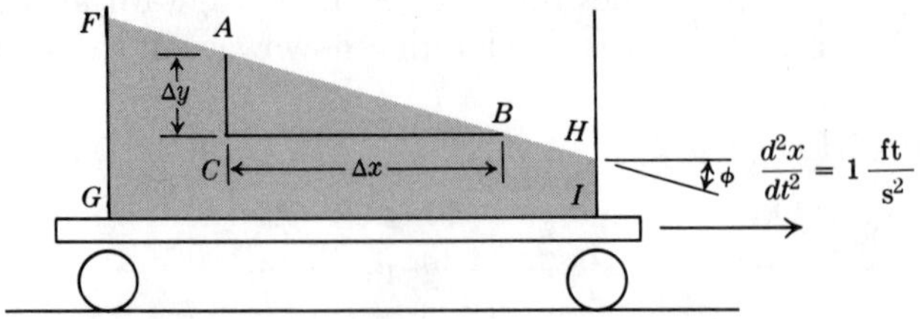

FIGURE 2.21
Accelerating system.

However, we may also calculate the gauge pressure at C by using Eq. 2.36 for the horizontal direction, in which case we have $a = x$, $\cos\theta = 0$, and

$$P_c = -\rho\,\Delta x\,\frac{d^2x}{dt^2}$$

But the pressure at C is the same no matter how we calculate it, so we may eliminate P_c between these two equations and rearrange to obtain

$$\frac{\Delta y}{\Delta x} = \frac{d^2x/dt^2}{g} = \tan\phi$$

where ϕ is the angle shown in Fig. 2.21. For this problem

$$\phi = \arctan\frac{d^2x/dt^2}{g} = \arctan\frac{1\ \text{ft/s}^2}{32.2\ \text{ft/s}^2} = 1.76°$$ ■

To calculate the pressure at any point in the tank, we may now use Eq. 2.11, being careful to measure the depth from the free surface vertically above the point in question. The force on wall *FG*, for example, is exactly the same at every point as it would be if the cart were standing still and filled with liquid up to point *F*. The force on wall *HI* is exactly the same as it would be if the cart were standing still and filled with liquid up to the level of *H*.

Example 2.19, a case of uniform rectilinear acceleration, holds little practical interest because such an acceleration acting for a reasonable period of time (e.g., long enough for the sloshing to die out) would produce enormous velocities. However, it serves as an introduction to the more interesting case of rigid-body *rotation*. Consider an open-topped cylindrical tank of water with a vertical axis. The system is initially at rest; then the tank is set in steady motion, rotating about its vertical axis (say, on a rotating phonograph turntable). At first the fluid in the center will not be affected by the rotation of the walls but will stand still, and only the fluid near the walls will rotate. This sets up motions of parts of the fluid relative to each other, so that this is not a fluid statics problem. Eventually, however, the shear forces due to this relative motion will bring the fluid at the center to the same angular velocity as the tank, and then there is no relative motion within the fluid. Once the fluid in the center reaches the same angular velocity as the wall of the container, the whole of the fluid moves as if it were a rigid body; hence the name "rigid-body rotation." Pressures in rigid-body rotation can be calculated by the method of fluid statics.

Example 2.20. An open-topped can of water with a 30-cm inside diameter is rotating on a phonograph turntable at 78 rpm. It has been rotating a long time and is in rigid-body rotation. What is the shape of the free surface?

A cross section of this system is sketched in Fig. 2.22. Here we use the same procedure as we did for Fig. 2.21, calculating the pressure at C in two directions. To simplify the calculation, we have chosen C to be at exactly the

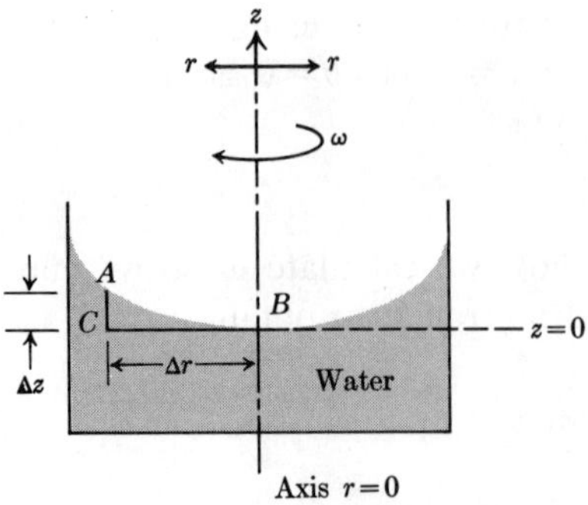

FIGURE 2.22
Rotating system.

same elevation as the lowest point on the free surface. As in Example 2.19, we assume that the pressures at A and B are the same, the local atmospheric pressure. Then, from Eq. 2.36 applied for the z direction, we can write

$$P_{c,\text{ gauge}} = -\rho g\,\Delta z$$

because the rotational acceleration is perpendicular to the z axis. In the radial direction, which is the r direction in Fig. 2.22, the only forces acting on the element of fluid are the pressure forces and the centripetal acceleration, whose magnitude is given by

$$\text{Centripetal acceleration} = -(\text{angular velocity})^2 \cdot \text{radius}$$

$$a_c = -\omega^2 r$$

Substituting this for d^2a/dt^2 in Eq. 2.36 and noting that $\cos\theta$ is zero for the radial direction, we find

$$\frac{dP}{dr} = \rho\omega^2 r$$

We then find the gauge pressure at C:

$$P_c = \int_{r=0}^{r=\Delta r} \rho\omega^2 r\,dr = \rho\omega^2 \frac{r^2}{2}\Bigg]_0^{\Delta r} = \rho\omega^2 \frac{(\Delta r)^2}{2}$$

The pressure at C is the same no matter how we calculate it, so we may eliminate P_c between these two equations and divide by ρg, to find

$$-\Delta z = \frac{\omega^2}{2g}(\Delta r)^2$$

Now if we let the elevation of point B (the lowest point on the free surface) be $z = 0$, then the length Δz is minus the value of z at point A, and Δr is the value of r at point A. So those points on the free surface are described by

$$z = \frac{\omega^2}{2g} r^2 \tag{2.37}$$

The free surface is a parabola with its vertex at the center of the can. The height of the free surface at the wall of the can is

$$z = \frac{(2\pi \cdot 78 \text{ rpm})^2 (15 \text{ cm})^2}{2 \cdot 9.81 \text{ m/s}^2} \cdot \left(\frac{\text{min}}{60 \text{ s}}\right)^2 \cdot \frac{\text{m}}{100 \text{ cm}}$$

$$= 7.65 \text{ cm} = 3.01 \text{ in}$$ ■

To find the pressure at any point in the rotating system (with the axis of rotation vertical) we use Eq. 2.11 and measure the distance down from the free surface directly upward from the point in question. The pressure at any point on the wall of the can in Fig. 2.22 is exactly the same as if the can were not rotating and were filled up to the level at which the rotating free surface would be at the edge.

Example 2.21. An industrial centrifuge has a basket with a 30-in diameter that is 20 in high. Its speed is 1000 rpm; see Fig. 2.23. If the liquid layer against the wall of the centrifuge is 1 in thick at the top, how thick is it at the bottom?

This is really the same problem as Example 2.20, except that only part of the free surface is present. To solve it, we write Eq. 2.37 twice, for points A and B on the figure, and we subtract one from the other:

$$z_A - z_B = \frac{\omega^2}{2g}(r_A^2 - r_B^2) \tag{2.38}$$

The only unknown here is r_B. Solving for it, we find

$$r_B = \left[r_A^2 - (z_A - z_B)\frac{2g}{\omega^2}\right]^{1/2}$$

$$= \left[(14 \text{ in})^2 - \frac{20 \text{ in} \cdot 2 \cdot 32.2 \text{ ft/s}^2}{(2\pi \cdot 1000/\text{min})^2} \cdot \left(\frac{60 \text{ s}}{\text{min}}\right)^2 \cdot \left(\frac{12 \text{ in}}{\text{ft}}\right)\right]^{1/2}$$

$$= (196 \text{ in}^2 - 1.4 \text{ in}^2)^{1/2} = 13.95 \text{ in} = 0.354 \text{ m}$$ ■

Thus, the liquid film is 1.05 in (2.67 cm) thick at the bottom.

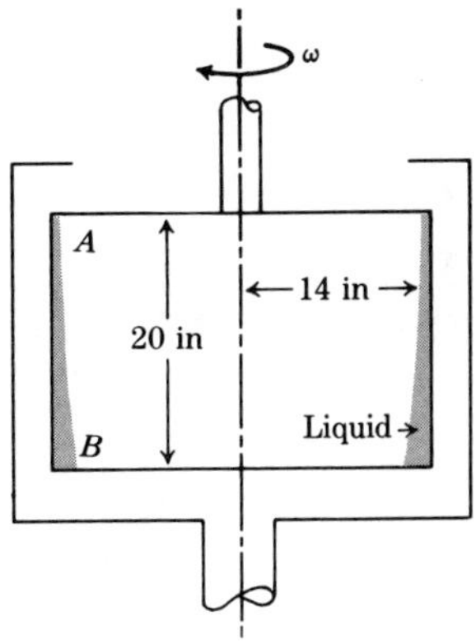

FIGURE 2.23
Centrifuge.

2.9 THIN-WALLED PRESSURE VESSELS

For many vessels, pipes, and other containers, the principal structural requirement is that the walls must withstand the pressure of the fluid in them. The calculation of the required wall thickness of thin-walled vessels is quite simple.

Consider a long, cylindrical pipe, as shown in Fig. 2.24, which contains a gas under some high pressure. A section of the pipe and the forces tending to make it move relative to line AA are shown on the right in the figure. The internal pressure tends to make the pipe move to the right of AA, and the stress in the pipe wall tends to make it move to the left. If it is not in the act of rupturing, it is not moving; so these forces must be equal and opposite. The forces in the two wall sections are

$$F_{\text{wall}} = 2ht\sigma \tag{2.39}$$

Here σ is the stress in the wall material, in psi or Pa, and t is the wall thickness. (Because this stress is resisted by the hoops in barrels, it is commonly referred to as the *hoop stress*.) The component of the pressure force tending to move the section to the right is

$$F_p = \int_{-\pi/2}^{\pi/2} hP\cos\theta\, r\, d\theta = hPr\sin\theta\Big]_{-\pi/2}^{\pi/2} = 2Prh \tag{2.40}$$

Since these forces are equal and opposite, they may be equated and solved for the thickness:

$$t = \frac{rP}{\sigma} \tag{2.41}$$

This equation describes not only the stresses in a pipe wall but also those in a cylindrical pressure vessel.

This approach ignores stresses in the axial direction; for moderate pressures and wall thicknesses this simplification does not cause much error.

Example 2.22. We wish to select a pipe with 1-ft inside diameter that can withstand an internal pressure of 1000 psig. The steel to be used has a safe

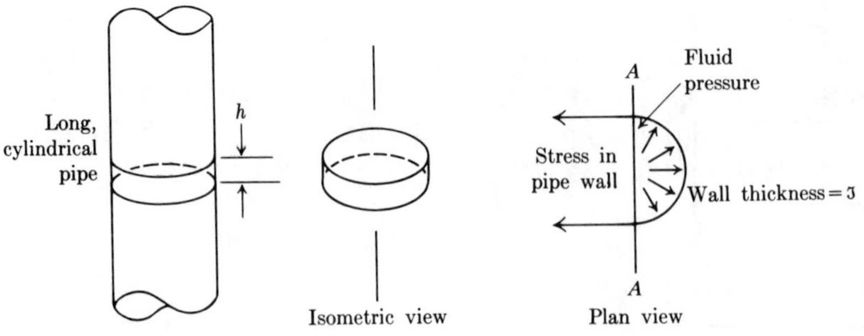

FIGURE 2.24
Stresses in a thin-walled pipe.

tensile stress of 20,000 psi, but to allow for a safety factor of 2 we design for a stress of 10,000 psi. How thick must the pipe walls be?

From Eq. 2.41,

$$t = \frac{Pr}{\sigma} = \frac{(1000\ \text{lbf/in}^2)(6\ \text{in})}{10{,}000\ \text{lbf/in}^2} = 0.6\ \text{in} = 1.52\ \text{cm} \qquad ■$$

This calculation leaves out the extra allowance for corrosion, etc., but illustrates the basic approach used in designing pipelines and pressure vessels for *internal* pressures of as much as 3000 psig (design for external pressures or vacuums is more complicated). For more details see the ASME code [2].

2.10 MORE PROBLEMS IN FLUID STATICS

Having worked out the basic equation and its simplifications for constant density, gauge pressure, isothermal and isentropic ideal gas, centrifugal force fields, etc., we can attack a wide range of problems. In this text we pass over some types of problems that have been widely treated elsewhere. Forces, distribution of forces, overturning moments, etc., on dams, retaining walls, flood gates, etc., are treated in all texts on civil engineering hydraulics, such as that by Streeter [3]. The subject of the buoyancy and stability of ships (why some turn over and others do not) is treated in the same texts. The behavior of lighter-than-air craft is covered in books on aeronautics, such as that by Prandtl and Tietjens [1].

2.11 SUMMARY

1. For simple fluids at rest, the pressure-depth relationship is given by the basic equation of fluid statics $dP/dz = -\rho g$. This equation is found by considering the weight of a small element of fluid and the pressure change with depth necessary to support that weight.
2. For constant-density fluids, the basic equation can be integrated to $P_2 - P_1 = \rho g(z_2 - z_1)$. This equation is an excellent approximation for liquids and a good approximation for gases when the change in elevation is small.
3. For changes in elevation measured in thousands of feet, gases cannot be treated as constant-density fluids. For isothermal, isentropic, or constant-temperature-gradient behavior, the basic equation can be integrated easily.
4. In problems involving liquids with free surfaces, it is generally easiest to work in gauge pressure, in which case the basic equation simplifies further to $P_{\text{gauge}} = \rho g h$.
5. The force exerted by a static fluid on any surface is given by $dF = P\,dA$.
6. The buoyant force exerted by a fluid on a floating or submerged body is equal to the weight of the fluid displaced.
7. Most pressure-measuring devices either balance the pressure against the weight of a column of fluid, in which case the height of the fluid column is

the reading, or let the pressure act on some area, compressing a spring, in which case the deflection of the spring is the reading.

8. Problems involving accelerated motion can be handled by the methods of fluid statics if the particles of fluid do not move relative to each other.

PROBLEMS

See the Common Units and Values for Problems and Examples inside the back cover of the book.

2.1. A large petroleum storage tank is 100 ft in diameter. The free surface is really a very small part of a sphere with radius 4000 mi (the radius of the earth). If one drew an absolutely straight line from the liquid surface at one side of the tank to the liquid surface directly across the diameter on the other side, how deep into the fluid would that line go?

2.2. Calculate the specific weight of water at a place where the acceleration of gravity is $32.2\ \text{ft/s}^2 = 9.81\ \text{m/s}^2$. Express your answer in lbf/ft^3 and in N/m^3. Calculate its specific weight on the moon, where $g = 6\ \text{ft/s}^2 \approx 2\ \text{m/s}^2$.

2.3. Calculate the specific weight of water in SI units.

2.4. Most swimmers find the pressure at a depth of about 10 ft painful to ears. What is the gauge pressure at this depth?

2.5. A new submarine can safely resist an external pressure of 1000 psig. How deep in the ocean can it safely dive?

2.6. The tallest building in the world (excluding TV towers which are not buildings in the common sense) is the Sears Tower in Chicago, which is 1454 ft tall. If the pressure in the supply line to the drinking fountain on the top floor (perhaps 1400 ft high) is 15 psig, what is the required pressure in the supply line at street level? Assume zero flow in the water line.

2.7. The deepest point in the oceans of the world is believed to be in the Marianas Trench, southeast of Japan; there the depth is about 11,000 m. What is the pressure at that point?

2.8. In the deep oil fields of Louisiana, occasionally one encounters a fluid pressure of 10,000 psi at a depth of 15,000 ft. If this pressure is greater than the hydrostatic pressure of the drilling fluid in the well from the surface, the result may be a blowout, which is dangerous to life and property. Assuming that you are responsible for selecting the drilling fluid for an area where such pressures are expected, what is the minimum-density drilling fluid you can use, assuming a surface pressure of 0 psig for the drilling fluid?

2.9. The tank in Fig. 2.25 contains gasoline and water. What is the absolute pressure at the bottom? Sketch the curve of gauge pressure versus depth for this tank.

2.10. An open-ended can 1 ft long is originally full of air at 70°F. The can is now immersed in water, as shown in Fig. 2.26. Assuming that the air stays at 70°F and behaves as a perfect gas, how high will the water rise in the can?

2.11. Normally we assume that liquids are constant-density fluids. To find out how large an error we make in that way, compute the pressure at the deepest point in the oceans (about 11,000 m) in two ways.

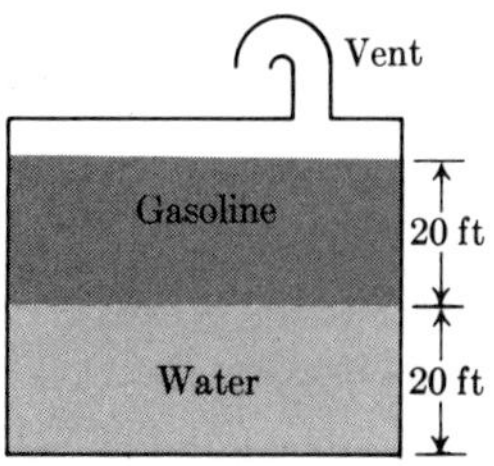

FIGURE 2.25

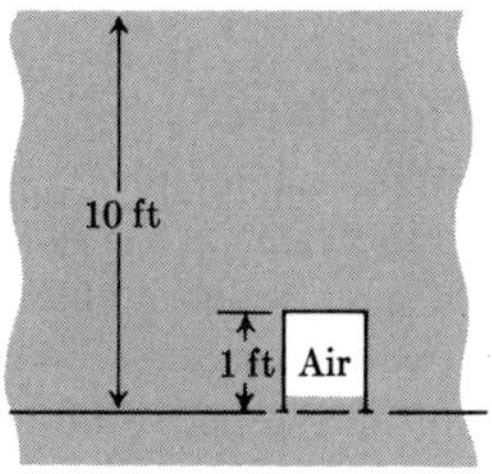

FIGURE 2.26

(*a*) Assume seawater is a constant-density fluid with properties shown on the inside back cover of the book.

(*b*) Assume that the density of water is given by $\rho = \rho_0[1 + \beta(P - P_0)]$. The definitions of the symbols in this equation and the value of β for water are given in App. A.

2.12. On a very cold day at the South Pole, the temperature of the air is −60°F. Assuming that the air remains isothermal up to a 10,000-ft elevation and that the pressure at sea level is 1 atm, calculate the pressure at 10,000 ft.

2.13. An airplane takes off from sea level and is climbing at 2000 ft/min. The plane is not pressurized, so that the pressure of the cabin is falling as the plane rises. At sea level (just after takeoff), how fast is the pressure falling (psi/min or kPa/min)?

2.14. Derive Eqs. 2.16 and 2.17, starting with $P/\rho^k = \text{constant}$ and $\rho = PM/(RT)$.

2.15. For the "standard atmosphere" shown in Fig. 2.4, (*a*) derive the pressure-height relation for the troposphere, (*b*) calculate the pressure at the troposphere-stratosphere interface, and (*c*) derive the pressure-height relation for the stratosphere.

2.16. At what height does the equation for an isentropic atmosphere, Eq. 2.17, indicate that the temperature of the air is 0 K? Assume that the surface temperature is 59°F. What is the physical significance of this prediction? What is the predicted pressure (Eq. 2.16) for this elevation?

2.17. For most problems we assume that $P_{\text{atm}} = 14.7$ psia. This is a reasonable approximation for sea level but not for other elevations. What is the average atmospheric pressure (*a*) at Salt Lake City, whose elevation is 4300 ft; (*b*) at 10,000 ft, the elevation to which the cabins on commercial airliners are pressurized; and (*c*) on the top of Mt. Everest (29,028 ft)? (For simplicity, use the isothermal atmosphere; but see also Prob. 2.19.)

2.18. What is the sea-level temperature gradient in degrees Fahrenheit per foot in (*a*) the standard atmosphere (Fig. 2.4) and (*b*) the isentropic atmosphere (Eq. 2.17) with a surface temperature of 59°F?

2.19. The conditions at sea level are 14.7 lbf/in^2 absolute at 59°F. Calculate the pressure and temperature at 10,000 ft according to (*a*) the isothermal atmosphere, (*b*) the isentropic atmosphere, and (*c*) the standard atmosphere.

2.20. What is the mass of the entire atmosphere of the earth? The earth may be considered a sphere of radius 4000 mi. All the atmosphere is so close to the surface of the earth that all of it may be considered to be subjected to the same acceleration due to gravity.

2.21. The oil storage tank in Examples 2.6 and 2.7 has a vent to the atmosphere, to allow air to move in or out as the tank is filled. This vent is plugged by snow in a blizzard while the oil is being pumped out of the tank, and the gauge pressure in the tank falls to $-1\ \text{lbf/in}^2$ gauge. What is the *net* force on the roof of the tank?

2.22. In the hydraulic lift in Fig. 2.27, the total mass of car, rack, and piston is 1800 kg. The piston has a cross-sectional area of $0.2\ \text{m}^2$. What is the pressure in the hydraulic fluid in the cylinder if the car is not moving?

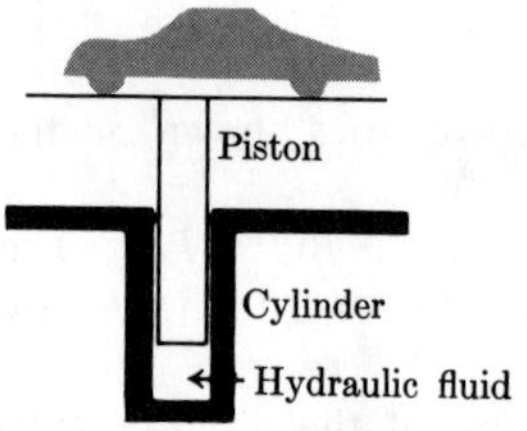

FIGURE 2.27

2.23. The Hoover Dam is approximately 230 m high and 76 m wide. Consider it to be a rectangle. When the water is up to the top, what is the pressure at the bottom? What is the net force tending to move the dam?

2.24. Figure 2.28 is the end-on view of a dam. The top is open to the atmosphere. What is the total force exerted by the water on the dam? What is the force exerted by the atmosphere on the other side? What is the difference between the two?

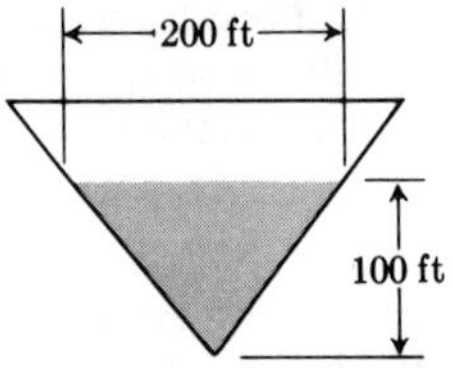

FIGURE 2.28

2.25. A dam has an upstream face which is vertical and has the shape of a semicircle with a diameter of 100 m. Water is up to the top of the dam. The atmosphere presses on the rear of the dam. What is the net horizontal force on the dam? Work this problem in two ways: (*a*) by direct integration of Eq. 2.19 and (*b*) by means of Eq. 2.23. The centroid of a semicircle about its diameter is $2D/(3\pi)$.

2.26. The upstream projection of a dam consists of a quarter-circle and a triangle, as shown in Fig. 2.29. The water is up to the top, as shown. What is the net force on the dam in the downstream direction?

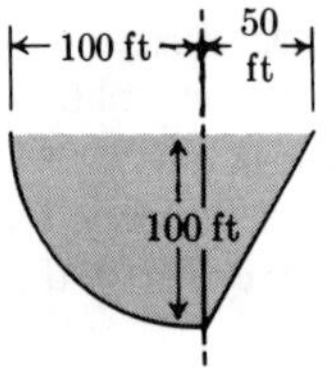

FIGURE 2.29

2.27. A dam has the cross section shown in Fig. 2.30. The ground exerts a force, as shown, on the dam which may be resolved into horizontal and vertical components. Assuming that the dam has a width of 100 ft into the paper and that the weight of the dam is negligible, calculate F_h and F_v.

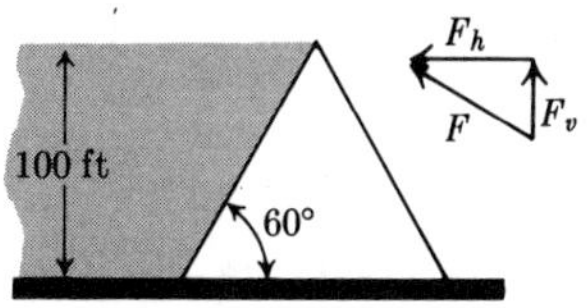

FIGURE 2.30

2.28. A dam is made of a quarter of a cylindrical pipe, as shown in Fig. 2.31. It extends 10 ft into the paper. The water is 10 ft deep. What are the net horizontal and vertical force components on the dam?

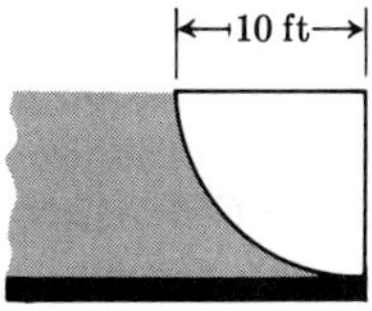

FIGURE 2.31

2.29. A gas storage vessel has a hemispherical end 10 m in diameter, as shown in Fig. 2.32. The gas in the container is at a pressure of 700 kPa gauge. The pressure of the gas is practically independent of position over the inside of the vessel. Calculate the vertical and horizontal components of the net pressure force on the hemispherical end.

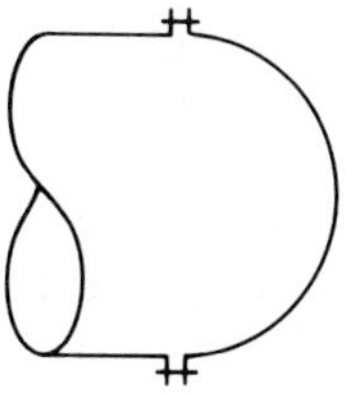

FIGURE 2.32

2.30. Archimedes is said to have discovered the buoyancy rules, which are called Archimedes' principle, when he was asked by the King of Syracuse in Sicily to determine whether a crown was pure gold, as the goldsmith had said it was, or was an alloy. At that time no chemical means were known of settling the question without destroying the crown. Archimedes was struck with the idea of how to do so while taking a bath, and he jumped out of his tub and ran through the streets yelling "Eureka." ("I have found it.") The story goes that he was so excited that he didn't bother to get dressed before doing this.

Suppose that in testing the crown Archimedes found that it had a weight of 5.0 N in air and 4.725 N in water. Assuming that the crown was made of gold or silver or an alloy of both, what percentage by volume was gold? Assume that the density of gold-silver alloys in $\rho_{\text{alloy}} = \rho_{\text{silver}} + (\text{vol \% gold}) \cdot (\rho_{\text{gold}} - \rho_{\text{silver}})/100$.

The densities are as follows: water, 1.0 g/cm^3; gold, 19.3 g/cm^3; silver, 10.5 g/cm^3.

2.31. A helium balloon has a flexible skin of negligible weight and infinite capacity for expansion, so that the helium is always at the same pressure as the surrounding air. If the mass of helium in the balloon is 10 lbm, how much payload can it lift under the following conditions: (*a*) 1 atm and 70°F, (*b*) 0.01 atm and 0°F, and (*c*) 0.001 atm and −100°F? Assume that helium behaves as a perfect gas with $M = 4$ g/mol and that its temperature is always the same as the temperature of the surrounding air.

2.32. Helium is preferred to hydrogen in balloons because it is nonflammable. However, hydrogen has only half the weight of helium ($M_{\text{hydrogen}} = 2$ g/mol, $M_{\text{helium}} = 4$ g/mol). By how much would the payload of the balloon in Example 2.12 have been increased if hydrogen had been used to fill it instead of helium?

2.33. Currently recreational balloons are not filled with hydrogen or helium but with hot air; the pilot has a small propane burner to heat the air in the balloon. If the balloon is a sphere 20 m in diameter, and if the total weight of balloon, pilot, passenger compartment, propane burner, propane tank, ropes, etc. is 200 kg, what average temperature must the air in the balloon have to just barely lift the balloon? Assume that the air inside and the air outside the balloon both are at atmospheric pressure and have equal molecular weights, 29 g/mol. (The latter is slightly inaccurate because of the products of combustion inside the balloon; this inaccuracy is small.)

2.34. A sample of lead is weighed on a pan balance by means of brass weights. It weighs 2.500 lbf.

(*a*) With the same set of brass weights, what would the lead weigh if the entire scale with weights and lead were at the bottom of a tank of water?

(*b*) If they were in a vacuum chamber? Here $SG_{\text{brass}} = 8.5$ and $SG_{\text{lead}} = 11.3$.

2.35. Calculate the vertical component of the force in Example 2.11 by using Archimedes' principle.

2.36. Rework Example 2.13, not by Archimedes' principle, but by assuming that the block has only vertical and horizontal faces and by calculating the difference in pressure between the top and bottom faces.

2.37. A 150-lb drunkard falls in a vat of whiskey. Whiskey has SG = 0.92, while the drunkard has SG = 0.99. The drunkard, who wishes to stay alive long enough to drink his fill of the whiskey, treads water, while keeping his head above water. If his head up to his mouth is 15 percent of his body volume, how much upward force must he exert by treading water to keep his head out of the whiskey?

2.38. It is proposed to build a raft of pine logs to carry a cargo on a river. The cargo will weigh 500 kg, and it must be kept entirely above the water level. How many kilograms of pine logs must we use to make the raft, if the logs may be entirely submerged and they have SG = 0.80?

2.39. A sphere of wood (SG = 0.96) is floating at the interface between water and gasoline. What fraction of the wood is above the dividing surface between the water and the gasoline?

2.40. A sunken battleship weighs 40,000 tons. It may be considered to be all steel; SG = 7.9. We now propose to raise the battleship by sinking steel tanks adjacent to it, attaching them to the battleship, and then blowing the water out of them with compressed air, making them buoyant. Assuming that the compressed-air tanks will have negligible mass, what volume must they have to raise the battleship? Assume that the battleship is in seawater and that the insides of the battleship are completely filled with water.

2.41. The fluid in the manometer of Fig. 2.33 is ethyl iodide with SG = 1.93. The heights are $h_1 = 44$ in and $h_2 = 8$ in.
(*a*) What is the gauge pressure in the tank?
(*b*) What is the absolute pressure in the tank?

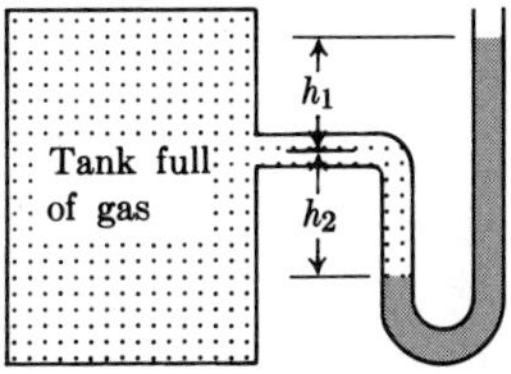

FIGURE 2.33

2.42. The two tanks in Fig. 2.34 are connected through a mercury manometer. What is the relation between Δz and Δh?

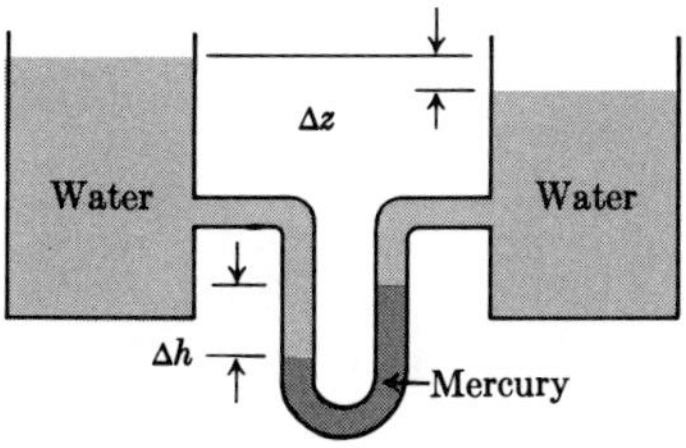

FIGURE 2.34

2.43. Figure 2.35 is a schematic diagram of a general two-fluid manometer. What is $P_A - P_B$ in terms of h, g, ρ_1, and ρ_2? If we want maximum sensitivity—that is, $\Delta h/(P_A - P_B)$ as large as possible—what relation of ρ_1 to ρ_2 should we choose?

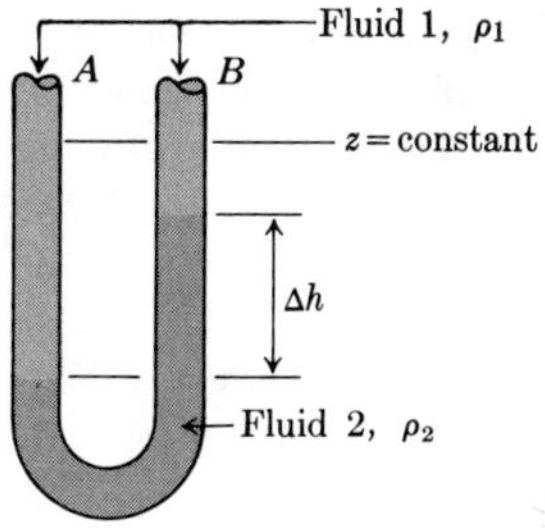

FIGURE 2.35

2.44. For low pressure differences, the inclined manometer shown in Fig. 2.36 is often used (this device is used so often to measure the draft of a furnace that its common name is a *draft tube*). If the scale is set to read zero length at $P_A = P_B$ and the manometer fluid is colored water, what will the reading be at $P_A - P_B = 0.1$ lbf/in^2? What would be the reading of an ordinary manometer with vertical legs for this pressure difference?

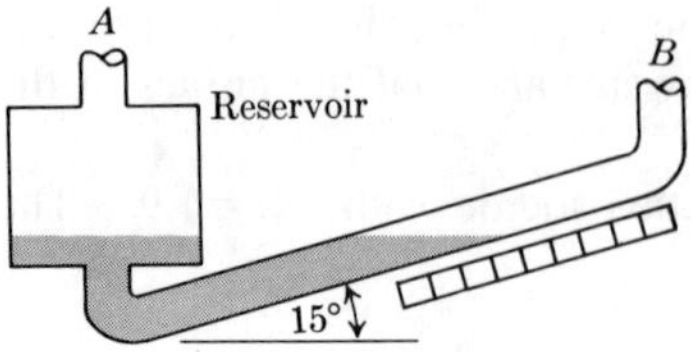

FIGURE 2.36
Draft tube.

2.45. The manometer in Prob. 2.44 has as its reservoir a cylinder with a diameter of 2 in. The tube has a diameter of $\frac{1}{8}$ in. The scale is set to read zero at $P_A = P_B$. When the level is at the 10-in mark, how much has the level in the reservoir fallen?

2.46. A common scheme for measuring the liquid depth in tanks is shown in Fig. 2.37. Compressed air or nitrogen bubbles slowly through a "dip tube" into the liquid. The gas flow rate is so low that the gas may be considered a static fluid. The pressure gauge is 6 ft above the end of the dip tube.

(*a*) If the pressure gauge reads 2 psig, and the dip tube is 6 in from the bottom of the tank, what is the depth of the liquid in the tank? Here $\rho_{\text{liquid}} = 60 \text{ lbm/ft}^3$ and $\rho_{\text{gas}} = 0.08 \text{ lbm/ft}^3$.

(*b*) Customarily engineers read these gauges as if ρ_{gas} were zero. How much error is made by such a simplification?

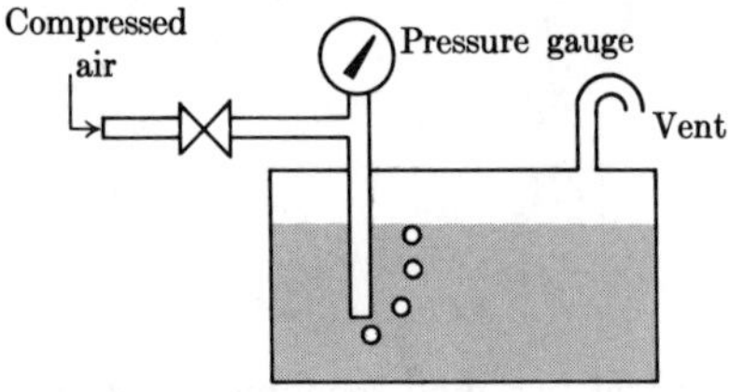

FIGURE 2.37
Compressed-air depth gauge.

2.47. The system shown in Fig. 2.38 is used to measure the density of a fluid in a tank. Compressed air or nitrogen bubbles at a very low rate through two dip tubes, whose ends are vertically 1.00 m apart. The difference in pressure between the two dip tubes is measured by a water manometer, which reads 1.5 m of water. The gas flow rate is so slow that the gas in the dip tubes may be considered a static fluid. The density of the gas is 1.21 kg/m^3. What is the density of the fluid in the tank?

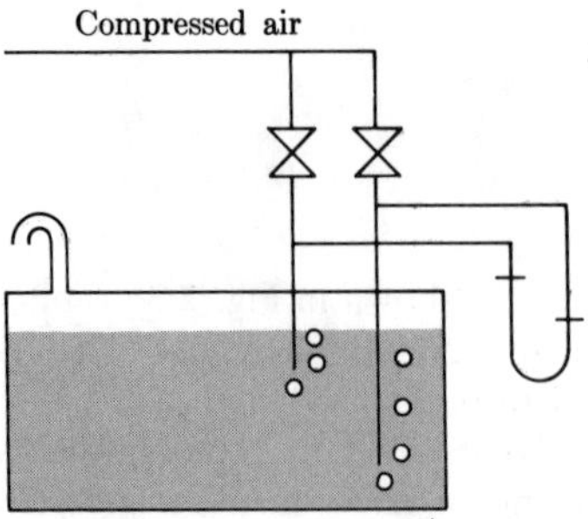

FIGURE 2.38
Compressed-air density gauge.

2.48. A furnace has a stack 100 ft high. The gases in the stack have $M = 28$ g/mol and $T = 300°F$. The outside air has $M = 29$ g/mol and $T = 70°F$. If the pressures of the air and the gas in the stack are equal at the top of the stack, what is the pressure difference at the bottom of the stack?

2.49. An oil well is 10,000 ft deep. The pressure of the oil at the bottom is equal to the pressure of a column of seawater 10,000 ft deep. (This is typical of oil fields; at the time of discovery, most have about the pressure of a hydrostatic column of seawater of equal depth; there are exceptions.) The density of the oil is 55 lbm/ft^3. What is the gauge pressure of the oil at the top of the well (at the surface)?

2.50. A natural gas well contains methane ($M = 16$ g/mol), which is practically a perfect gas. The pressure at the surface is 1000 psig. What is the pressure at a depth of 10,000 ft? How much error would be made by assuming that methane were a constant-density fluid? Assume the temperature is constant at 70°F.

2.51. An oil pipeline was constructed to transport an oil with SG = 0.8 a distance of 10 mi. The country was hilly, so that the line made many ups and downs. These may be considered equivalent to 10 rises of 200 ft, followed by descents of 200 ft. When the pipe was completed, it was tested by pumping water through it. The water flowed satisfactorily with a pressure at the inlet end of 150 psi. Then the oil was slowly fed into the pipe. As the oil flowed, the pressure at the inlet end began to rise, and the flow rate began to fall. Finally, the flow stopped altogether, while the pressure at the inlet side remained at 150 psi. Explain what caused this. (*Hint:* This is a manometer problem.)

2.52. The tank in Fig. 2.39 is completely full of water. Both valves are closed; now we open valve *B* and allow the water to drain out, without opening valve *A*. What is the minimum pressure that will be reached in the tank?

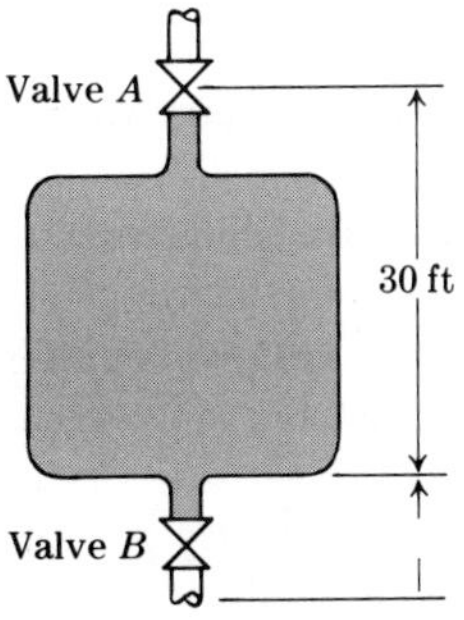

FIGURE 2.39

2.53. In typical suburbs one popular way of making the tops of fence posts level is to use a transparent, plastic garden hose. One partly fills the hose with water, holds the two ends to two different posts, and adjusts the liquid level in one end of the hose (by raising or lowering it) until it is level with the top of the post being used for reference. Then one marks the level of the fluid in the hose on the next post; and so on. Will this system work if there are trapped air bubbles in the hose?

2.54. Rework Example 2.17 for the elevator falling freely, i.e., for downward acceleration of 9.81 m/s^2.

2.55. The device in Fig. 2.40 consists of two pieces of pipe of 1-in inside diameter and is connected to a pressure gauge. The whole apparatus is on a elevator, which moves in the z direction. The pressure gauge reads 5 psig. How fast is the elevator accelerating? Which way?

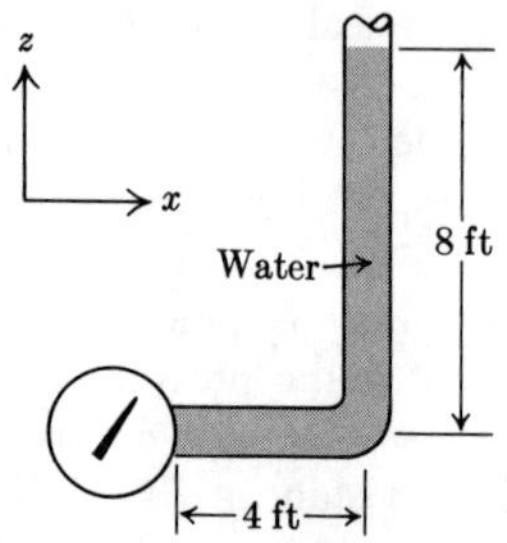

FIGURE 2.40

2.56. The rectangular tank in Fig. 2.41 is sitting on a cart. We slowly accelerate the cart. What is the maximum acceleration we can give the cart without having the fluid spill over the edge of the tank?

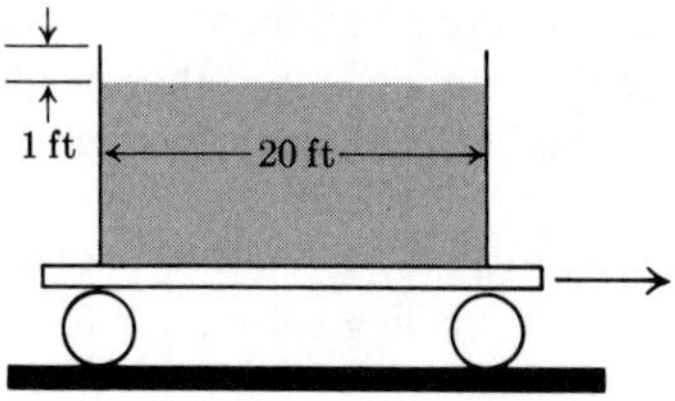

FIGURE 2.41

2.57. A closed tank contains water and heating oil, SG = 0.96, and is completely full of the two liquids, with no air space at the top. The tank is being steadily accelerated in the x direction at 1 ft/s^2. What angle does the water-oil interface make with the vertical?

2.58. If the fluid in the centrifuge in Example 2.21 is water, what is the gauge pressure at the outer wall of the centrifuge (under the layer of water 1 in thick)?

2.59. In the centrifuge in Example 2.21, a solid particle of 0.01-in^3 volume is settling through the fluid. When it is almost at the wall, where the radius is 15 in, what is the buoyant force acting on it? Which way does it act?

2.60. The tank and manometer shown in Fig. 2.42 are mounted on a merry-go-round which is revolving at 10 rpm. The vessel is filled with a gas of negligible density; the manometer fluid is water. What is the pressure in the vessel?

2.61. A cylindrical can contains a layer of gasoline 4 in deep on top of a layer of water 4 in deep. The can is now set on a phonograph turntable and rotated about its vertical axis at 78 rpm. Describe mathematically the shape of the gasoline-air and gasoline-water interfaces.

2.62. Calculate the relation between internal gauge pressure, radius, wall thickness, and wall stress for a thin-walled, spherical pressure vessel.

2.63. Assuming that steel pipes have an allowable wall stress of 6500 psi, calculate the maximum allowable internal pressure for a 5-in schedule 40 pipe. See App. A-3

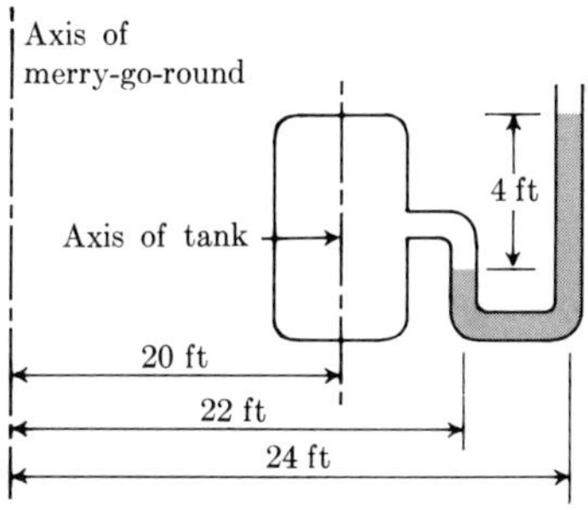

FIGURE 2.42

for its dimensions. In calculating the allowable pressure, subtract 0.100 in from the outside diameter as a *corrosion allowance*.

2.64. From the data given in App. A.3 on the diameter and wall thickness of schedule 40 pipes, show that these roughly correspond to the formula

$$t = A + BD$$

where t is the wall thickness, D is the diameter, A is an arbitrary constant known as the *corrosion allowance*, and B is the value we would compute from Eq. 2.41. Calculate the values of A and B in this equation from the best straight line through a plot of thickness versus diameter.

2.65. Figure 2.43 is a sketch of a fountain arrangement made of two glass jars with rubber stoppers, several lengths of glass tubing, a funnel, and a piece of rubber tubing. The level of the jet and the level of the water in the funnel are exactly the same. The space above the water in each bottle is full of air, as is the rubber tube connecting the two bottles. An inventor has come to us, telling us that with this arrangement the water will squirt high in the air, much higher than the water level in the funnel. Is she right? Explain your answer.

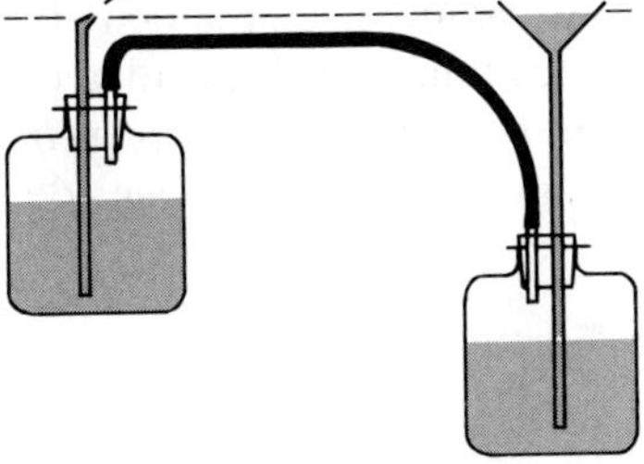

FIGURE 2.43
Gravity fountain.

CHAPTER 3

THE BALANCE EQUATION AND THE MASS BALANCE

Much of engineering is simply careful accounting of things other than money. The accountings are called mass balances, energy balances, component balances, momentum balances, etc. In this chapter we examine the basic idea of a balance and then apply it to mass. The result is the *mass balance*, or *principle of conservation of mass*, or *continuity equation*, one of the four basic ideas listed in Sec. 1.3.

3.1 THE BALANCE EQUATION

Let us illustrate the general balance idea by making a *population balance* around the state of Utah. The population of Utah can change by

1. Births
2. Deaths
3. Immigration
4. Emigration

Adding these with the correct algebraic signs and equating them to the change in population, we get

$$\text{Increase in population} = \text{births} - \text{deaths} + \text{immigration} - \text{emigration} \tag{3.1}$$

This equation is a special case of the general balance equation:

$$\text{Accumulation} = \text{creation} - \text{destruction} + \text{flow in} - \text{flow out} \tag{3.2}$$

We can now make four comments:

1. These equations must apply to some period of time. If we talk simultaneously about the births in 1 year and the deaths in 1 month, the balance will be queer indeed. If in the population balance we are talking about 1 year, then we can divide Eq. 3.1 by 1 year to find

$$\text{Annual increase in population} = \text{annual birth rate} - \text{annual death rate}$$
$$+ \text{annual immigration rate} - \text{annual emigration rate} \tag{3.3}$$

This is a *rate equation*. If someone promises you a million dollars, you will be happy. If that person pays you at the rate of \$0.01 per year, you will be unhappy; we are all normally interested in rates.
2. If we apply the population balance to the state of Utah for a 1-day period, we will find misleading rates. The number of births per day fluctuates widely; the annual rate is practically constant. To get meaningful rates, the period over which measurements are made must be long enough to average out fluctuations. (There are, however, some situations in which we wish to study the short-time fluctuations. An example is the statistical study of turbulence. For such studies it is worthwhile to make balances over time periods short enough for these fluctuations not to "average out.")
3. In the example above, the balance was made over an identifiable set of boundaries (the legal boundaries of the state of Utah; see Prob. 3.1). A general principle of engineering balances is that there can be no meaningful balance without a carefully *defined and stated set of boundaries*. The set of boundaries need not be fixed, but they must be identifiable. For an example of an identifiable set of boundaries that is moving in space and changing shape, suppose a group of people were shipwrecked in the Antarctic and took refuge on a floating iceberg. We could make a population balance around the iceberg, and it would have the same terms in it as our population balance around the state of Utah did. However, the iceberg is not fixed in place, and its size is constantly changing.

   Whatever is inside a set of boundaries is often called *system*. Throughout this text we use the term "system" to describe whatever is inside the boundaries. Everything that is outside the boundaries we call the *surroundings*. Thus the boundaries divide the whole universe into two parts, the system and the surroundings.

For some problems it is convenient to choose as our system the contents of some closed container, which does not allow flow into or out of it. For such a system, the balance equation reduces to

$$\text{Accumulation} = \text{creation} - \text{destruction}$$

Such a system is called a *closed system*. An example might be the population of a sealed space capsule traveling through space, for which the balance equation is

$$\text{Increase in population} = \text{births} - \text{deaths}$$

The closed system is widely used in chemistry, because it is very convenient when a chemical reaction is taking place in a closed container, in which new species may be created by chemical reaction and old ones destroyed but none flow into or out of it.

An *open system* is usually some kind of container or vessel which has flow in and out across its boundaries at some small number of places. This is used much more commonly in engineering than the closed system is and is used extensively in this book.

We consider flows in and out of most open systems only at some small number of places, e.g., a household water heater which has one inlet pipe, one outlet pipe, and one drain pipe. If we choose as a system some arbitrary region of space which can have flow in or out over its entire boundary, then this system is called a *control volume*. When we consider the control volume in this book, we treat it merely as a special kind of open system.

4. The balance equation deals only with *changes* in the thing being accounted for, not with the total amount present. The population balance given above tells the change in the population of Utah but not the numerical value of the population. If we wish to know the numerical value of the population of that state, we may conduct a census. Alternatively, if we could find birth, death, immigration, and emigration data from the time that the first person entered the state to the present, we could compute the change in population, starting with population zero. Mathematically, this is

$$\text{Current population} = \int_{\text{time at population}=0}^{\text{present}} (\text{rate of change of population})\, d(\text{time}) \qquad (3.4)$$

Beginners are often tempted to find a place in their balances for the total amount contained, such as a numerical value of population. Resist this temptation!

To what can the balance equation be applied? It can be applied to any countable set of units or to any extensive property. An *extensive property* is one that doubles when the amount of matter present doubles. Some examples are mass, energy, entropy, mass of any chemical species, momentum, and electric charge. Some examples of countable units are people, apples, pennies, molecules, home runs, electrons, and bacteria.

The balance equation cannot be applied to uncountable individuals (units) or to intensive properties. *Intensive properties* are independent of the amount of matter present. Some examples are temperature, pressure, viscosity, hardness, color, honesty, electric voltage, beauty, and density. An example of uncountable individuals is all the decimal fractions between 0 and 1.

3.2 THE MASS BALANCE

Our example of a balance equation in Sec. 3.1 would be of interest to demographers but not necessarily to engineers. Let us consider the most important engineering balance, the mass balance. Mass obeys the general balance equation: the creation and destruction terms are zero. Thus, the mass balance is

$$\begin{pmatrix}\text{Increase in mass within}\\ \text{the chosen boundaries}\end{pmatrix} = \begin{pmatrix}\text{flow of}\\ \text{mass in}\end{pmatrix} - \begin{pmatrix}\text{flow of}\\ \text{mass out}\end{pmatrix} \tag{3.5}$$

The careful application of this equation is necessary to most fluid-mechanics problems. Naturally we can divide by time and find

$$\begin{pmatrix}\text{Rate of increase of mass}\\ \text{within the chosen boundaries}\end{pmatrix} = \begin{pmatrix}\text{flow rate}\\ \text{of mass in}\end{pmatrix} - \begin{pmatrix}\text{flow rate}\\ \text{of mass out}\end{pmatrix} \tag{3.6}$$

The mass balance cannot be derived from any prior principle. Like all the other basic "laws of nature," it rests on its ability to explain observed facts. Every careful experiment which has been run to test it indicates that it is correct. We will see in Chap. 4 that mass and energy can be converted from one to the other. In most engineering problems, we can neglect this fact and use the simple formulation in Eq. 3.5 (but we cannot neglect it in dealing with atomic bombs or the energy source of the sun).†

Example 3.1. Consider the simple pot-bellied stove, burning natural gas, shown in Fig. 3.1. Applying Eq. 3.6 to this stove, we choose as our system

† There is no experimental evidence on earth that matter is created except by conversion of energy to mass, as enunciated by Einstein. This production of mass from energy is discussed in Chap. 4; the conditions of its occurrence are well known and are demonstrable in the laboratory. A more interesting prospect is the idea of the "steady-state universe," put forward by the British astronomer Fred Hoyle. According to his theories, matter is being created all the time, everywhere; however, the rate is very slow, about 1 hydrogen atom per hour per cubic mile of space [1]. No instruments now exist which could detect such an event, so a confirmation of this theory on an earthbound scale seems impossible at present. Hoyle claims that the experimental observations of the behavior of the farthest galaxies support his theories; other astronomers disagree. Although these theories have no foreseeable application to engineering problems, it is well to keep an open mind on the subject of the *absolute* nature of the mass balance.

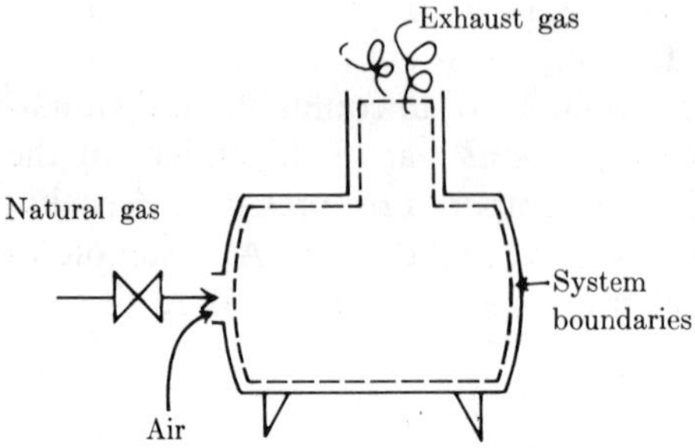

FIGURE 3.1
Pot-bellied stove.

boundaries the walls of the stove. Then Eq. 3.6 becomes

$$\begin{pmatrix}\text{Rate of increase}\\ \text{of mass with the}\\ \text{chosen boundaries}\end{pmatrix} = \begin{pmatrix}\text{mass}\\ \text{flow rate}\\ \text{of gas in}\end{pmatrix} + \begin{pmatrix}\text{mass}\\ \text{flow rate}\\ \text{of air in}\end{pmatrix} - \begin{pmatrix}\text{mass flow}\\ \text{rate of exhaust}\\ \text{gas out}\end{pmatrix} \quad (3.7)$$

■

Here we have two mass-flow-in terms. There is no limit to the number of such terms. Recall our population balance around the state of Utah; we could have immigration by airplane, car, boat, train, etc. We have a term for each and add the terms to get the total immigration term. Similarly, here we add the individual mass-flow-in terms to get the total mass-flow-in term.

The mass balance has several other names, which are in wide use: *principle of conservation of mass*, *continuity equation*, *continuity principle*, and *material balance*. They all mean exactly the same thing as mass balance, namely, that mass obeys the general balance equation, with no creation or destruction.

3.3 STEADY-STATE BALANCES

When the pot-bellied stove in Fig. 3.1 is first lighted after being turned off for a long time, the temperature of its various parts changes rapidly. After a certain time it is warmed up, and thereafter the temperature of the various parts does not change with time. During the warm-up period, the velocities and temperatures of the gases passing through it at some fixed point change with time. For example, a thermometer at some point in the flue registers a continually increasing temperature. After the stove has warmed up, this thermometer will register a constant temperature. When the stove has warmed up and is running steadily, we speak of it as being at steady state.

A steady state does not mean that nothing is changing; it means that nothing is changing with respect to time. Consider a waterfall with a steady flow over it. From the standpoint of a particle of water, there is a rapid increase in velocity as it falls and then a sudden decrease in velocity at the bottom. From the standpoint of an observer watching one specific point in space, the waterfall is always the same: There is always water going by at a fixed velocity. Mathematically, if velocity V is some function of time and position

$$V = f(t, x, y, z) \tag{3.8}$$

then at steady state

$$\left(\frac{\partial V}{\partial t}\right)_{x,\,y,\,z} = 0 \qquad \text{[steady state]} \tag{3.9}$$

Similarly we can write for steady state that

$$\left(\frac{\partial}{\partial t}\right)_{x,\,y,\,z}$$

of *any* measurable property of the system at *any* point is zero. Thus, if we write the balance equation for some measurable quantity such as mass and divide by dt to find the rate form, then the left-hand side (the time rate of mass increase within the system) must be zero, because at every point in the system the mass contained is not changing with time. Entirely analogous arguments indicate that at steady state the accumulation term must be zero for all possible balances, including the energy and momentum balances, which we discuss in Chaps. 4 and 7.

Returning now to the pot-bellied stove of Example 3.1, we see that if we consider the stove's steady-state behavior, the mass balance simplifies to

$$0 = \begin{pmatrix}\text{mass flow rate}\\ \text{of gas in}\end{pmatrix} + \begin{pmatrix}\text{mass flow rate}\\ \text{of air in}\end{pmatrix} - \begin{pmatrix}\text{flow rate of}\\ \text{exhaust gas out}\end{pmatrix} \tag{3.10}$$

This is the familiar "flow in equals flow out" idea, which is true only for steady state, with no creation or destruction.

Example 3.2. For the pot-bellied stove of Example 3.1 we now make a steady-state carbon dioxide balance. By chemical analysis we find that the amount of carbon dioxide in the natural gas and in the air is small enough to ignore; so, omitting the unnecessary terms from Eq. 3.2, we find

$$0 = \begin{pmatrix}\text{creation rate}\\ \text{of carbon}\\ \text{dioxide}\end{pmatrix} + \begin{pmatrix}\text{destruction}\\ \text{rate of}\\ \text{carbon dioxide}\end{pmatrix} - \begin{pmatrix}\text{mass flow rate}\\ \text{of carbon dioxide}\\ \text{out in exhaust gas}\end{pmatrix} \tag{3.11}$$

Chemical analysis of the exhaust gas indicates that it contains about 9% carbon dioxide, so the mass flow rate out is not negligible. Thus, for this equation to be satisfied, there must be significant creation minus destruction of carbon dioxide in the stove; i.e., carbon dioxide is formed by combustion in the stove. In this case, the destruction term is negligible. ■

If we made a similar balance for natural gas, the destruction term would be approximately equal to the mass-flow-in term. In the field of chemical reactions, the creation and destruction terms are very important and cannot be ignored. The momentum balance (Chap. 7) includes creation and destruction terms, as does the entropy balance or the second law of thermodynamics (see any elementary textbook on thermodynamics). Thus, although the two most common balances, the mass and energy balances, have no creation or destruction terms, remember that these terms are very important in some other balances.

3.4 THE STEADY-STATE FLOW, ONE-DIMENSIONAL MASS BALANCE

Consider the steady-state flow of some fluid in a pipe of varying cross section (see Fig. 3.2). If we apply the steady-state mass balance equation to the system shown, we find

$$\text{Mass flow in at point 1} = \text{mass flow out at point 2}$$

In general, the velocity is not the same at every point in a cross section of pipe; it is faster near the center than at the walls. (One may verify this for the analogous open-channel flow by dropping bits of wood or leaves on a flow of water in a ditch or gutter and noting that those in the center go faster than those at the side.) Therefore, to calculate the total flow in across the system boundaries at point 1, we break up the area across which the flow is entering into small subareas A, over each of which the flow is practically uniform:

$$\text{Mass flow in at point 1} = \sum_{\text{many subareas}} \rho A V \tag{3.12}$$

Here the individual elements of area must be taken perpendicular to the local flow velocity. For flow in a pipe or channel this is no problem, because the flow is all in one direction, and the area we normally consider is one perpendicular to the flow. If we take the limit as each subarea becomes infinitely small, the term on the right becomes the integral, over the entire system boundary at point 1, of $\rho V\,dA$. Therefore, the steady-state mass balance for the system in Fig. 3.2 is

$$0 = \int_{\text{area 1}} \rho V\,dA - \int_{\text{area 2}} \rho V\,dA \tag{3.13}$$

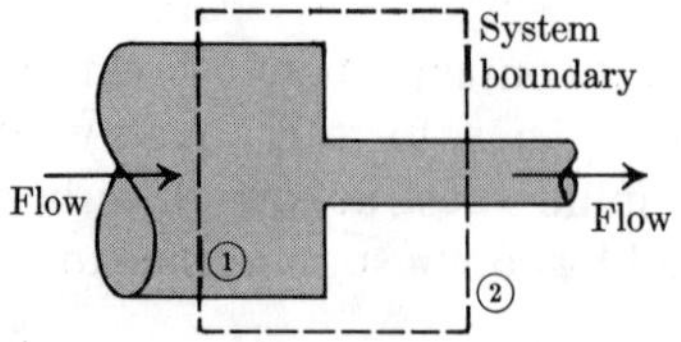

FIGURE 3.2

But we could choose points 1 and 2 to be any locations in the pipe, so for steady-state flow in a pipe or channel this equation becomes

$$\int_{\substack{\text{area at any boundary} \\ \text{perpendicular} \\ \text{to the flow}}} \rho V \, dA = \text{constant} \qquad \text{[steady flow]} \tag{3.14}$$

The velocity of laminar flows (discussed in Chap. 6) and of boundary-layer flows (discussed in Chap. 11) changes rapidly with distance perpendicular to the flow, so that the flow into or out of a system must always be calculated by an integral of the form of Eq. 3.14. However, in most industrial flows in pipes or channels, the velocity is practically constant across the entire cross section of the pipe or channel (except in a very thin layer near the pipe or channel wall). The density and velocity of these flows may be considered constant across the cross section, and then the integrations in Eqs. 3.13 and 3.14 can be easily performed, giving

$$\rho_1 A_1 V_1 = \rho_2 A_2 V_2 = \dot{m} = \text{constant} \tag{3.15}$$

The mass crossing the boundary per unit time is called the *mass flow rate*. Normally it is measured in kilograms per second or pounds-mass per second and given the symbol $\dot{m}$.

Example 3.3. In a natural-gas pipeline at station 1, the pipe diameter is 2 ft, and the flow conditions are 800 psia, 70°F, and 50 ft/s velocity. At station 2, the pipe diameter is 3 ft, and the flow conditions are 500 psia and 70°F. What is the velocity at station 2? What is the mass flow rate?

Solving Eq. 3.15 for V_2, we find

$$V_2 = V_1 \frac{\rho_1}{\rho_2} \frac{A_1}{A_2} = 50 \frac{\text{ft}}{\text{s}} \cdot \frac{\rho_1}{\rho_2} \cdot \frac{(\pi/4)(2\text{ ft})^2}{(\pi/4)(3\text{ ft})^2}$$

The density of natural gas (principally methane) at 800 psia and 70°F is approximately 2.44 lbm/ft^3, and at 500 psia and 70°F it is approximately 1.49 lbm/ft^3. Therefore,

$$V_2 = V_1 \frac{\rho_1}{\rho_2} \frac{A_1}{A_2} = 50 \frac{\text{ft}}{\text{s}} \cdot \frac{2.44\text{ lbm/ft}^3}{1.49\text{ lbm/ft}^3} \cdot \frac{(\pi/4)(2\text{ ft})^2}{(\pi/4)(3\text{ ft})^2} = 36.2 \frac{\text{ft}}{\text{s}} = 11.0 \frac{\text{m}}{\text{s}}$$

$$\dot{m} = \rho_1 V_1 A_1 = 2.44 \frac{\text{lbm}}{\text{ft}^3} \cdot 50 \frac{\text{ft}}{\text{s}} \cdot \frac{\pi}{4} (2\text{ ft})^2 = 384 \frac{\text{lbm}}{\text{s}} = 174.5 \frac{\text{kg}}{\text{s}}$$ ■

For liquids at temperatures well below their critical temperature, the changes in density with moderate temperature and pressure changes are small. Therefore, for liquids we can divide the density out of Eq. 3.15, finding

$$A_1 V_1 = A_2 V_2 = \frac{\dot{m}}{\rho} = \text{constant} \qquad \text{[constant density]} \tag{3.16}$$

Mass divided by density equals volume; therefore the constant in this equation (the mass flow rate divided by the density) is the *volumetric flow rate*, usually measured in cubic feet per second or cubic meters per second and given the symbol Q.†

Example 3.4. Water is flowing in a pipe. At point 1, the inside diameter is 0.25 m, and the velocity is 2 m/s. What are the mass flow rate and the volumetric flow rate? What is the veloicty at point 2, where the inside diameter is 0.125 m?

$$\dot{m} = \rho_1 V_1 A_1 = 998.2 \frac{\text{kg}}{\text{m}^3} \cdot 2 \frac{\text{m}}{\text{s}} \cdot \frac{\pi}{4} (0.25\text{ m})^2 = 98.0 \frac{\text{kg}}{\text{s}} = 216 \frac{\text{lbm}}{\text{s}}$$

$$Q = \frac{\dot{m}}{\rho} = V_1 A_1 = 2 \frac{\text{m}}{\text{s}} \cdot \frac{\pi}{4} (0.25\text{ m})^2 = 0.09817 \frac{\text{m}^3}{\text{s}} = 3.46 \frac{\text{ft}^3}{\text{s}}$$

$$V_2 = V_1 \frac{A_1}{A_2} = 2 \frac{\text{m}}{\text{s}} \frac{(\pi/4)(0.25\text{ m})^2}{(\pi/4)(0.125\text{ m})^2} = 8 \frac{\text{m}}{\text{s}} = 24.2 \frac{\text{ft}}{\text{s}}$$

■

3.5 UNSTEADY-STATE MASS BALANCES

The steady-state behavior of systems, shown in the preceding examples, is very important. Most of the examples and problems in elementary textbooks concern steady-state behavior. However, unsteady-state behavior is also important, perhaps more important. The characteristics of the two types of system are compared in Table 3.1.

A power plant burns coal and produces electricity by means of a boiler, turbine, condenser, generator, etc.; its steady-state behavior is fairly easy to calculate. However, when the power demand on the generator is suddenly increased or decreased, its behavior is much more difficult to calculate. Power

TABLE 3.1
Comparison of steady and unsteady state flow

Property	Steady state	Unsteady state
Calculations	Generally easy	More difficult
Normally requires calculus?	No	Yes
Setup in laboratory	Difficult	Easy
Large-scale industrial use	Desirable	Undesirable
Efficiency	Generally high	Generally lower
Capital cost per unit of production:		
Large volume product (e.g., gasoline)	Low	High
Small-volume product (e.g., pharmaceuticals)	High	Low

† In civil engineering books, the volumetric flow rate Q usually is called the *discharge*.

company engineers would prefer to have a steady load, because then they could always operate the plant at its maximum efficiency. Nonetheless they must plan for and be equipped for sudden load disturbances (e.g., a lightning strike shuts down a major consumer, thus quickly reducing the power demand). Also most industrial disasters, such as explosions and fires, occur not during periods of steady-state operation but during startup or shutdown of some processing unit. Thus we see that the unsteady-state behavior is very important and worthy of our attention.

Unsteady-mass mass balances do not introduce any ideas beyond those seen so far. However, as shown by the following two examples, they generally lead to more complicated mathematics.

Example 3.5. The microchip diffusion furnace in Fig. 3.3 contains air, which may be considered an ideal gas. The vacuum pump is pumping air out prior to beginning the thermal diffusion step. During the pump-out process, the heating coils in the tank hold the temperature in the tank constant at 70°F. The volumetric flow rate at the inlet of the pump, independent of pressure, is $1.0\ \text{ft}^3/\text{min}$. How long does it take the pressure to fall from 1 to 0.0001 atm?

We choose as our system the tank up to the pump inlet. For this system the mass balance gives

$$\left(\frac{dm}{dt}\right)_{\text{sys}} = -\dot{m}_{\text{out}} \tag{3.17}$$

But we know that

$$m_{\text{sys}} = V_{\text{sys}}\rho_{\text{sys}} \tag{3.18}$$

where V_{sys} is the volume of the system, which does not change. Thus,

$$\left(\frac{dm}{dt}\right)_{\text{sys}} = V_{\text{sys}}\frac{d\rho_{\text{sys}}}{dt} \tag{3.19}$$

Furthermore,

$$\dot{m}_{\text{out}} = Q_{\text{out}}\rho_{\text{out}} \tag{3.20}$$

But Q_{out} is constant and

$$\rho_{\text{out}} = \rho_{\text{sys}} \tag{3.21}$$

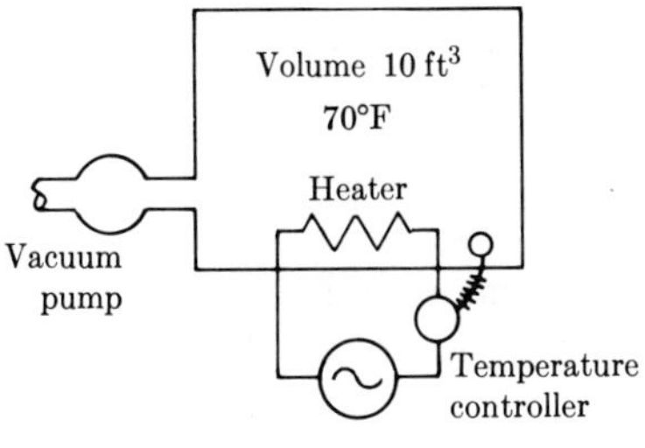

FIGURE 3.3
Microchip diffusion furnace.

so that

$$V_{sys}\frac{d\rho_{sys}}{dt} = -Q_{out}\rho_{sys} \tag{3.22}$$

This is a separable, first-order differential equation, which can be rearranged to

$$\frac{d\rho_{sys}}{\rho_{sys}} = -\frac{Q_{out}}{V_{sys}}\,dt \tag{3.23}$$

and integrated from initial to final states, yielding

$$\ln\frac{\rho_{sys,\ final}}{\rho_{sys,\ init}} = -\frac{Q_{out}}{V_{sys}}\,\Delta t \tag{3.24}$$

For low-pressure gases at constant temperature, the densities are proportional to the pressures, so we can solve for the required time:

$$\Delta t = \frac{V_{sys}}{Q_{out}}\ln\frac{P_{init}}{P_{final}} = \frac{10\ \text{ft}^3}{1\ \text{ft}^3/\text{min}}\ln\frac{1\ \text{atm}}{0.0001\ \text{atm}} = 92.1\ \text{min}$$ ■

In many unsteady-state mass balance problems, it is convenient to take as the system the fluid in some container. Thus, as the mass of fluid increases or decreases, the volume of the system changes.

Example 3.6. A cylindrical tank 3 m in diameter, with vertical axis, has an inflow line of 0.1-m inside diameter and an outflow line of 0.2-m inside diameter. Water is flowing in the inflow line at a velocity of 2 m/s and leaving by the outflow line at a velocity of 1 m/s. Is the level in the tank rising or falling? How fast?

Here we take as our system the instantaneous mass of water in the tank. For this system

$$\left(\frac{dm}{dt}\right)_{sys} = \dot{m}_{in} - \dot{m}_{out} \tag{3.25}$$

For any fluid we have $m = \rho V$ and $\dot{m} = \rho Q$. Substituting these into the last equation and canceling the density, which is constant, we find

$$\left(\frac{dV}{dt}\right)_{sys} = Q_{in} - Q_{out} \tag{3.26}$$

The volumetric flow in or out is equal to VA, so

$$\left(\frac{dV}{dt}\right)_{sys} = 2\ \frac{\text{m}}{\text{s}}\cdot\frac{\pi}{4}\,(0.1\ \text{m})^2 - 1\ \frac{\text{m}}{\text{s}}\cdot\frac{\pi}{4}\,(0.2\ \text{m})^2$$

$$= 0.0157 - 0.0314 = -0.0157\ \frac{\text{m}^3}{\text{s}}$$

The volume of liquid in the tank is decreasing, and the level is falling. The rate of decrease in volume is equal to the cross-sectional area times the rate of fall in the level:

$$\left(\frac{dV}{dt}\right)_{\text{sys}} = A\,\frac{dz_{\text{sur}}}{dt}$$

$$\frac{dz_{\text{sur}}}{dt} = \frac{1}{A}\frac{dV}{dt} = \frac{1}{(\pi/4)(3\text{ m})^2}\left(-0.0157\,\frac{\text{m}^3}{\text{s}}\right)$$

$$= -0.0022\text{ m/s} = -0.00673\text{ ft/s}$$ ■

3.6 MASS BALANCES FOR MIXTURES

In the preceding examples, the flowing materials have been uniform single species, e.g., air or water. In most of the rest of this book we deal with such uniform single species. However, in many problems of great interest two or more components mix inside the system. If we make the simplest possible mixing assumption—perfect mixing of all components—then we can apply the simple balance equation as before and find useful answers. The perfect mixing assumption is obviously a great simplification of what must occur in nature, but it is often used because the results are so simple and useful. Two examples illustrate the idea.

Example 3.7. Figure 3.4 is a sketch of a rectangular city with length L and width W. The wind blows over the city in the x direction with velocity V. Atmospheric turbulence mixes the air over the city up to height H, so we may assume that the air in the "box" with dimensions L times W times H is well mixed and has the same pollutant concentration c everywhere. The air flowing into the upwind side of the city has pollutant concentration b (which stands for background concentration). The city emits pollutants into the atmosphere uniformly over its surface with an emission rate q. [Here q will have units like $\text{kg}/(\text{m}^2 \cdot \text{s})$. This uniform-emission assumption is a fair one for emissions from

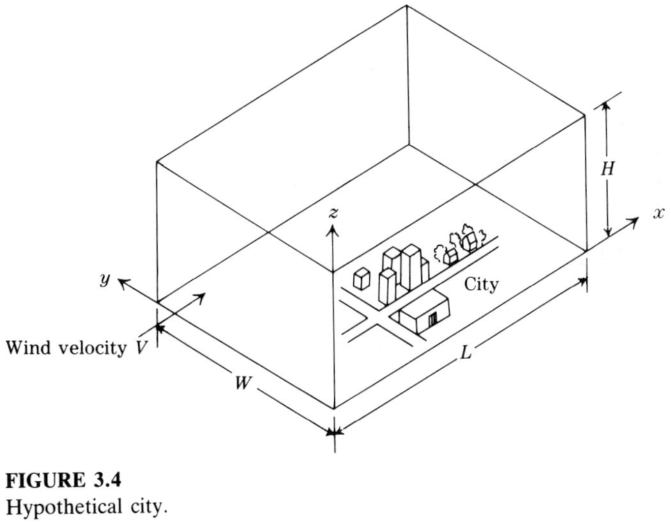

FIGURE 3.4
Hypothetical city.

automobiles or small industry which are more or less uniformly spread over the city, but is a very poor one for emissions from a single large factory or power plant; such emissions are treated a very different way in air pollutant regulation.] What is the concentration of pollutant in the air over the city, in terms of q, V, W, L, and H?

Here we make the steady-state assumption that the concentration is not changing with time, so the algebraic sum of the flows of pollutant in and out must be zero:

$$\begin{aligned} 0 = {} & \text{flow rate of pollutant into city air from upwind} \\ & + \text{flow rate of pollutant into city air from city} \\ & - \text{flow rate of pollutant out of city air at downwind edge of city} \end{aligned} \tag{3.27}$$

The pollutant flow rates are expressed as concentrations (e.g., kilograms per cubic meter) times volumetric flow rates (e.g., cubic meters per second), so

$$0 = bVWH + qLW - cVWH$$

$$c = b + \frac{qL}{VH} \tag{3.28}$$

Equation 3.28 says that the pollutant concentration in the city is equal to that in the air entering the city (the background concentration) plus a term $[qL/(VH)]$ which indicates how much the pollutant concentration has been increased by the emissions from the city itself. This is the "box model" or "proportional" or "rollback" equation, which has played a very important role in the formulation of air pollution regulations in the United States [2]. ■

Example 3.8. Our paint shop will use a special paint that has benzene as a solvent. In the course of an 8-h day, the paint evaporates 200 kg (440 lb) of benzene ($q = 200$ kg/8 h). The shop dimensions are 10 m × 4 m × 4 m. To protect the health of our workers, we must limit the concentration of benzene in the shop air to less than or equal to the industrial hygiene standard for benzene [3], which is 30 mg/m^3. If we wish to keep the concentration c of benzene in the shop at or below this permitted concentration, how large a flow of ventilating air must we supply?

This is very similar to Example 3.7. Here we assume that the benzene is well mixed into the air in the shop and that the air leaving the shop will have the permitted benzene concentration. Making a steady-state benzene balance on the shop, taking the inlet airflow as Q, we get

$$0 = \text{benzene in inlet air} + \text{benzene evaporated by the paint} - \text{benzene in exit air}$$

$$0 = Qb + q - Qc \tag{3.29}$$

Now we observe that there is no benzene in the incoming air ($b = 0$), so we can solve for Q:

$$\begin{aligned} Q = \frac{q}{c} = \frac{200\ \text{kg}/8\ \text{h}}{30\ \text{mg}/\text{m}^3} \cdot 10^6\ \frac{\text{mg}}{\text{kg}} &= 833{,}000\ \frac{\text{m}^3}{\text{h}} \\ &= 13{,}900\ \text{m}^3/\text{min} = 490{,}000\ \text{ft}^3/\text{min} \end{aligned} \tag{3.30}$$

■

This example shows that, for the assumption of perfect mixing of benzene into the shop air, it is quite straightforward to compute the required dilution air to meet the industrial hygiene standard. We also see that this is an impossibly large airflow rate. If we divide the above flow rate by the cross-sectional area of the shop (4 m × 4 m), we find

$$\text{Velocity} = \frac{Q}{A} = 13{,}900\ (\text{m}^3/\text{min})/16\ \text{m}^2$$
$$= 869\ \text{m/min} = 14.5\ \text{m/s} = 47.5\ \text{ft/s}$$

This is a very high velocity, which could hardly be used inside a paint shop. Our practical alternatives are to choose a less toxic solvent, for which the permitted concentration is higher, or to devise some kind of ventilation system, such as a laboratory fume hood, which will prevent the mixing of the benzene with the air that workers breathe. We also need to consider the air pollution consequences of emitting 200 kg/day of benzene to the atmosphere; in most U.S. cities that would require a permit and probably some form of capture or destruction of the benzene.

3.7 MASS BALANCES FOR MULTIDIMENSIONAL FLOWS

The foregoing mass balances were for a small number of flows in and out. Obviously, the same idea can be readily applied to a much larger number of flows in and out. One equation often used in theoretical fluid mechanics (see Chap. 10) is the mass balance equation for an arbitrary point in space. We find this equation by defining the coordinates and components of the local fluid velocity, as shown in Fig. 3.5. Our system is a small open-faced cube.

The mass balance for this system in rate form is

$$\begin{pmatrix}\text{Accumulation rate of}\\ \text{mass in the system}\end{pmatrix} = \begin{pmatrix}\text{all mass flow}\\ \text{rates in}\end{pmatrix} - \begin{pmatrix}\text{all mass flow}\\ \text{rates out}\end{pmatrix} \tag{3.31}$$

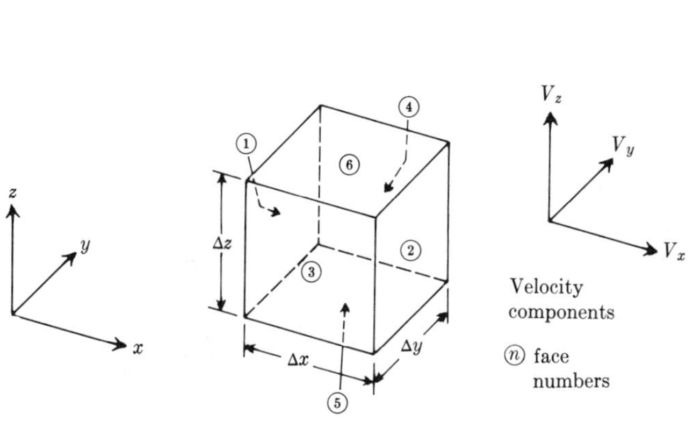

FIGURE 3.5
Notation for three-dimensional mass balance.

The mass in the system at any instant is $\rho \, \Delta x \, \Delta y \, \Delta z$. The flow into the system through face 1 is

$$\dot{m}_1 = \rho_1 V_{x_1} \, \Delta y \, \Delta z \tag{3.32}$$

and the flow out of the system through face 2 is

$$\dot{m}_2 = \rho_2 V_{x_2} \, \Delta y \, \Delta z \tag{3.33}$$

Writing the analogous terms for faces 3, 4, 5, and 6 and inserting all in Eq. 3.31, we find

$$\Delta x \, \Delta y \, \Delta z \, \frac{\partial \rho}{\partial t} = \Delta y \, \Delta z (\rho_1 V_{x_1} - \rho_2 V_{x_2}) + \Delta x \, \Delta z (\rho_3 V_{y_3} - \rho_4 V_{y_4}) + \Delta x \, \Delta y \, (\rho_5 V_{z_5} - \rho_6 V_{z_6}) \tag{3.34}$$

We now divide through by $-\Delta x \, \Delta y \, \Delta z$:

$$-\frac{\partial \rho}{\partial t} = \frac{\rho_2 V_{x_2} - \rho_1 V_{x_1}}{\Delta x} + \frac{\rho_4 V_{y_4} - \rho_3 V_{y_3}}{\Delta y} + \frac{\rho_6 V_{z_6} - \rho_5 V_{z_5}}{\Delta z} \tag{3.35}$$

Now we let Δx, Δy, and Δz each approach zero simultaneously, so that the cube shrinks to a point. Taking the limit of the three ratios on the right-hand side of this equation, we find the partial derivatives

$$-\frac{\partial \rho}{\partial t} = \frac{\partial (\rho V_x)}{\partial x} + \frac{\partial (\rho V_y)}{\partial y} + \frac{\partial (\rho V_z)}{\partial z} \tag{3.36}$$

If the density is constant or the density changes are small enough to be neglected, this simplifies to

$$0 = \frac{\partial V_x}{\partial x} + \frac{\partial V_y}{\partial y} + \frac{\partial V_z}{\partial z} \qquad \text{[constant density]} \tag{3.37}$$

By letting Δx, Δy, and Δz approach zero we have shrunk the system to a single point. Thus Eq. 3.36 is the mass balance for *any point* in space; it is often called the *general continuity equation*. Equation 3.37 is the mass balance for any point in space which contains a constant-density fluid.

3.8 SUMMARY

1. Balances are important in engineering.
2. All balances can be made from the general balance equation (accumulation = creation − destruction + flow in − flow out) by dropping the unnecessary terms.
3. All balances can be divided by time to make rate equations.
4. In any balance it is necessary to choose and state the boundaries over which the balance is made. Whatever is inside the boundaries over which the balance is made is called the *system*. Whatever is outside is called the *surroundings*.

5. The most important engineering balance is the mass balance, in which the creation and destruction terms are zero. This is also called the *continuity equation* or the *principle of conservation of mass*.

PROBLEMS

See the Common Units and Values for Problems and Examples inside the back cover.

3.1. In our balance equation for the population of Utah, we must resolve several questions of definition. For example, are out-of-state students to be counted in the population of Utah? Are tourists driving through the state to be counted while they are there? List several other ambiguous groups for which we must make a definition.

3.2. Write out the balance for the number of one-dollar bills in circulation in the United States.

3.3. Write out the balance for the mass of refined sugar in Idaho (which is a sugar-producing state).

3.4. A toy balloon has just had its neck released. It is zipping through the air and shooting air out its neck. Write out the balance for the mass of air involved. Be careful to state what your boundaries are. Are they fixed in space? Are they fixed in size? Are they identifiable?

3.5. Write a carbon-atom balance for an automobile driving at a constant speed (include the carbon atoms which are in chemical compounds as well as the free carbon atoms).

3.6. Write a mass balance for an exploding firecracker.

3.7. A river has a cross section that is approximately a rectangle 10 ft deep and 50 ft wide. The average velocity is 1 ft/s. How many gallons per minute pass a given point? What is the average velocity (assuming steady flow) at a point downstream, where the channel shape has changed to 7 ft in depth and 150 ft in width?

3.8. There is steady flow in a circular pipe. The average velocity is given by Q/A. Calculate the ratio of the average velocity to the maximum velocity for the following cases:

(*a*) The flow is laminar (to be discussed in Chap. 6), and the local velocity at any point in the pipe is given by

$$V = V_{\max}\left(\frac{r_0^2 - r^2}{r_0^2}\right)$$

where r is the radial distance from the center of the pipe and r_0 is the radius at the wall of the pipe.

(*b*) The flow is turbulent (to be discussed in Chap. 6), and the local velocity is approximately given by

$$V = V_{\max}\left(\frac{r_0 - r}{r_0}\right)^{1/7}$$

3.9. A perfect gas is flowing in a pipe at a constant temperature. The pipe diameter is constant. What is the relation of average velocity to pressure?

3.10. A column of soldiers is marching 12 abreast at a speed of 4 mi/h. To get through a narrow pass, they must crowd in to form a column of 10 soldiers abreast. Assuming steady flow, how fast are the soldiers moving when they are 10 abreast?

3.11. A water tank has an inflow line 1 ft in diameter and two $\frac{1}{2}$-ft-diameter outflow lines. The velocity in the inflow line is 5 ft/s. The velocity out one of the outflow lines is 7 ft/s. The mass of water in the tank is not changing with time. What are the volumetric flow rate, mass flow rate, and velocity in the other outflow line?

3.12. A compressed-air vessel has a volume of 10 ft^3. Cooling coils hold its temperature constant at 70°F. The pressure now in the vessel is 100 lbf/in^2 absolute. Air is flowing in at the rate of 10 lbm/h. How fast is the pressure increasing?

3.13. The tank in Example 3.5 has a leak which admits air at the rate of 10^{-4} lbm/min whenever the pressure is less than atmospheric pressure. Everything else is the same as in Example 3.5. How long does it take the pump to reduce the tank pressure from 1 to 0.01 atm? What is the steady-state pressure in the tank?

3.14. The tank in Example 3.5 has a leak which admits air at the rate of

$$\dot{m} = \frac{10^{-4}\ \text{lbm}}{\text{min} \cdot \text{atm}} (P_{\text{outside}} - P_{\text{tank}})$$

Everything else is the same as in Example 3.5. How long does it take the pump to reduce the pressure in the tank from 1 to 0.01 atm? What is the steady-state pressure in the tank?

3.15. A lake has a surface area of 100 km^2. One river is bringing water into the lake at the rate of 10,000 m^3/s, while another is taking water out at 8000 m^3/s. Evaporation and seepage are negligible. How fast is the level of the lake rising or falling?

3.16. The tank in Fig. 3.6 has an inflow line with a cross-sectional area of 0.5 ft^2 and an outflow line with a cross-sectional area of 0.3 ft^2. Water is flowing in the inflow line at a velocity of 12 ft/s, and gasoline is flowing out the outflow line at a velocity of 16 ft/s. How many pounds-mass per second of air are flowing through the vent? Which way?

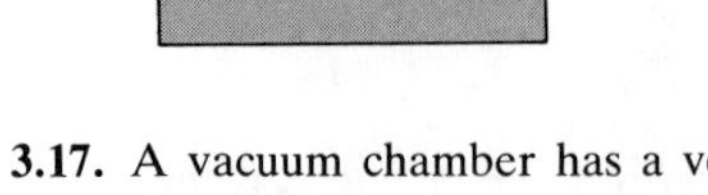

FIGURE 3.6

3.17. A vacuum chamber has a volume of 10 ft^3. When the vacuum pump is running, the steady-state pressure in the chamber is 0.1 lbf/in^2. The pump is shut off, and the following pressure-time data are observed:

Time after shutoff, min	Pressure, psia
0	0.1
10	1.1
20	2.1
30	3.1

Calculate the rate of air leakage into the vacuum chamber when the pump is running. Air may be assumed to be a perfect gas. The air temperature may be assumed constant at 70°F.

3.18. A tank contains 1000 m^3 of salt solution. The salt concentration is 10 kg/m^3. At time zero, salt-free water starts to flow into the tank at a rate of 10 m^3/min. Simultaneously salt solution flows out of the tank at 10 m^3/min, so that the volume of solution in the tank is always 1000 m^3. A mixer in the tank keeps the concentration of salt in the entire tank constant; the concentration in the effluent is the same as the concentration in the tank. What is the concentration in the effluent as a function of time?

3.19. Repeat Prob. 3.18, with the change that there is a layer of solid salt on the bottom of the tank, which is steadily dissolving into the solution at a rate of 5 kg/min.

3.20. Repeat Prob. 3.18, with the change that the outflow is only 9 m^3/min and the total volume of liquid contained in the tank is thus increasing by 1 m^3/min.

3.21. Rework Example 3.5 with the following change. The tank is now somewhat flexible, so that it is being slowly crushed by the surrounding pressure. If it is crushed at such a rate that its volume decreases steadily by 0.1 ft^3/min, and if this rate of volume decrease begins as soon as the vacuum pump starts, how long does it take the pressure to fall from 1 to 0.0001 atm?

3.22. While Moses was crossing the Red Sea, he took up a liter of water, examined it, and then threw it back. The tides, currents, etc., have been steadily mixing the waters of the ocean; so we may assume that the molecules in that liter of water have now been uniformly distributed over all the oceans of the world. If you pick up a liter of water from the ocean and examine it, how many molecules will it contain which were in the liter that Moses examined? State clearly what assumptions and simplifications you make.

3.23. The typical human being breathes about 10 times per minute and takes in about 1 liter per breath. Assuming that the atmosphere has been perfectly mixed since Julius Caesar's time, estimate the number of air molecules that you take in with a single breath which at some time were breathed in and out by Julius Caesar.

CHAPTER 4

THE FIRST LAW OF THERMODYNAMICS

We have seen how the general balance equation applies to people and mass. We now apply it to the abstract quantity energy to find the energy balance, which is also called the *first law of thermodynamics* or the *law of conservation of energy*. This is one of the four basic ideas listed in Sec. 1.3 and one of the few truly fundamental laws of nature. Like the others, it cannot be derived from any more basic principle; rather, it rests on its ability to explain all the pertinent observations of nature ever made and on the fact that all experiments designed to prove or disprove it has indicated that it is true.

4.1 ENERGY

The idea of energy and the first law of thermodynamics arose out of observations of friction heading. By the early 1800s scientists knew that a moving body possessed what we would call kinetic energy. They also knew that if it was allowed to come to rest by sliding across a rough surface, then it lost this kinetic energy, but it and the surface became hotter. Various explanations of this phenomenon were tried, but, principally through the work of Rumford [1], Joule [2], and Mayer [3], the idea was introduced that in any such process there is a quantity called *energy*, which is conserved. This quantity could appear in the form of kinetic energy or heat. We will see that it can also appear in other forms.

Some quantities in engineering have absolute values: temperature, entropy, length, and mass (excluding relativistic effects). For each of these there is defined a standard unit of measurement, and the meaning of a zero amount of the quantity is clear.

Other quantities in engineering have only relative values. The simplest example is elevation. We can speak of elevation relative to mean sea level, relative to some convenient benchmark, relative to the center of the earth. Any of these is useful. However, an "elevation of 23 ft" without mention of the datum is meaningless. Another example of a relative quantity is velocity. Normally we consider velocity relative to the local surface of the earth, and a "velocity of 23 km/h" is perfectly clear. However, this statement has a built-in assumption of a datum, namely, the surface of the earth. If we speak of a star moving at 23 km/s, we probably mean relative to the sun, but we could mean relative to the earth or to the center of our galaxy. We must state our datum, unless the datum is understood by all.

All energy quantities are relative to some arbitrary datum. This statement is made simply so that you will remember it. A more precise statement is that no one has yet found a way to measure or to calculate absolute values of energies. From this it follows that all energy calculations are based on changes in energy or on energies relative to some arbitrary datum. This need not trouble us; all the buildings in the world were designed relative to some arbitrary elevation datum without any particular trouble about the datum.

4.2 FORMS OF ENERGY

As implied above, energy has many forms, and they are interconvertible, subject to some restrictions (the restrictions apply only to the direction of conversion). Let us consider a 1-kg ball of steel. What forms of energy can it possess?

A. Internal Energy

If the steel is at a temperature of 20°C, you can hold it in your hand. If it is at a temperature of 200°C, you cannot hold it in your hand very long. Clearly, the ball at 200°C produces effects which the ball at 20°C cannot. Yet, if we measure the mass of the ball, it is the same at 20°C as it is at 200°C (within the precision of current measuring techniques). If we could label the atoms when the ball was at 20°C and take a census of them when the ball was at 200°C, we would find exactly the same atoms present. Therefore, the difference between what the ball will do at 20°C and what it will do at 200°C is not dependent on changing the mass or identity of the *matter* present. Something else obviously is involved. For now we say that a body which is hot possesses more *internal energy* than the same body does when cold.

Now suppose that instead of an iron ball we have a balloon which contains a mixture of gasoline and oxygen with a total mass of 1 kg at 20°C. Now we can introduce a small spark, and the contents of the balloon become very hot (explosively so). After a moment the contents will be much hotter than at the start, and they will have a different chemical composition; instead of being oxygen and gasoline, they will be carbon dioxide and water vapor. Clearly, the oxygen-gasoline mixture at 20°C can produce effects that the mixture of carbon dioxide and water (when cooled to 20°C) could not. Therefore, there must be a difference in energy. This we classify also as a change of internal energy.

Thus, an *approximate* rule (with exceptions to be seen later) is that internal energy is a measure of hotness plus the ability to cause heat-releasing chemical reactions. A more complete definition is given in Sec. 4.6. We denote internal energy by U and internal energy per unit mass by u.†

B. Kinetic Energy

Let us return to our 1-kg ball of steel. If I hand it to you gently, you can easily hold it. If I deliver it to you at a velocity of 100 m/s and you are foolish enough to get in the way, it will certainly kill you. The fast-moving ball can produce effects different from those produced by the slow-moving ball. This difference we call the difference in kinetic energy. *Kinetic energy* is the energy which a moving body possesses because of its motion. We designate kinetic energy "KE" and kinetic energy per unit mass "ke."

C. Potential Energy

If our 1-kg ball of steel is resting on the floor, it is not likely to damage the floor. If it is resting on a shelf 100 m above the floor and then is gently pushed off, probably it will go right through the floor. When it gets to the floor, its kinetic energy will be very large. However, when it was sitting on a shelf at 100 m above the floor, it did not have that kinetic energy, but obviously it had the potentiality of acquiring it by falling. This potential to do work, or to acquire kinetic energy, we call *potential energy*. In this instance it is the energy which the body possesses by being some distance above "bottom" in a gravity field. We call potential energy "PE" and potential energy per unit mass "pe."

If we fabricate our iron ball into a coil spring, it will have a certain length

† The use of uppercase letters for the total quantity and lowercase letters for the quantity per unit mass is very common in thermodynamics and is seen extensively in this chapter. The quantities per unit mass are often called *specific quantities*; e.g., ke is called the *specific* kinetic energy. This use of two symbols makes sense only for extensive properties—those which double when the mass doubles. It is never used for intensive properties—those which do not depend on the amount of mass present. (What is meant by "the temperature per unit mass"?)

when relaxed. When we compress it, it will have another length. But, given the opportunity, it will return to its original length and in doing so can give some other body kinetic energy. Toy guns work in exactly this way. If we compare the compressed spring with the relaxed one, we see that the compressed spring has the potential of speeding up a projectile or doing some other useful work. This is due to a difference in energy; one might logically call this "spring energy," but most writers have agreed to call it potential energy. This is particularly true in physical chemistry, in which the "springs" are the force fields between atoms or subatomic particles. Throughout the remainder of this chapter we consider potential energy to mean the energy that a body possesses due to its position in a gravitational field; however, we must remember this other meaning.

D. Electrostatic Energy

Suppose we take our ball of iron and some convenient dielectric and fabricate a huge electric condenser out of it. If it is not charged, we can touch the terminals with our fingers without effect. If it is charged, placing our fingers over the terminals will be a shocking experience. The charged condenser can do things that the uncharged one cannot. This difference is due to *electrostatic energy*.

E. Magnetic Energy

If our ball of iron is shaped into a rod or horseshoe and annealed, it will not attract iron filings. If we subject it to a properly designed magnetic field and then remove the field, it will attract the filings; it has become a magnet. A magnet can do things which a bar of unmagnetized steel cannot. The difference is due to *magnetic energy*.

F. Surface Energy

Salad oil, egg yolks, and vinegar do not form a homogeneous mixture. If they are gently shaken together and then allowed to settle, they will separate cleanly. If, however, they are beaten very vigorously, to break up the oil into small droplets, then they will form a stable system called mayonnaise. Under normal conditions mayonnaise will not separate back into salad oil, egg yolks, and vinegar; it is an emulsion. Emulsions possess properties that their unmixed constituents do not. These are due to the *surface energy* of all the microscopic droplets which make up an emulsion.

G. Nuclear Energy

Einstein has shown that matter and energy are interconvertible; their conversion is the basis of nuclear explosives and nuclear power plants and the source of energy in the sun and the stars. We discuss this conversion in Sec. 4.14. For the moment, we restrict ourselves to saying that it is *convenient* to talk of some materials *as if* they possessed another kind of energy, called *nuclear energy*.

Of the kinds of energy which matter can possess, here we deal with only the first three cited above: internal, kinetic, and potential energy. We shall discuss the others briefly in Sec. 4.14.

4.3 ENERGY TRANSFER

If a kilogram of matter can possess energy, how can that energy be transferred from one body to another?

One way to transfer energy from one body to another is to place two bodies at different temperatures in contact with each other. It is a universal experience that in such circumstances the internal energy of the hotter body will decrease and the internal energy of the colder body will increase. Therefore, energy must have flowed from one to the other. The energy which flows directly between two bodies in contact because of a temperature difference we call *heat*.

Our definition of heat is different from the one commonly used. We say, "Heat is energy in transit from one body to another because of a temperature difference." In common English, people use "heat" interchangeably with "temperature." This leads to such common phrases as "It's not the heat; it's the humidity" and "Beat the heat with a brand X air conditioner." Clearly, these rest on the human experience that when the temperature of the air is high, energy will flow into our bodies, uncomfortably. While it is flowing, it is "heat."

The idea of energy "in transit" is also contradicted by common usage. Many people refer to a hot body as containing a large amount of heat rather than a large amount of internal energy. Rain is water in transit from clouds to ground under the influence of gravity. We would scarcely look at a cloud and say, "Look at all that rain," or look at the ocean and refer to "all that rain." Another useful analogy is to electric current. While electrons are flowing from a battery to a condenser, we speak of them as an electric current; but when the flow has stopped, we speak of them as "charge" and speak of a "charged battery" or a "charged condenser." We would scarcely speak of a "currented" condenser or a "currented" battery. These ideas are summarized in Table 4.1.

The second way in which two bodies can exchange energy is by doing *work* upon each other. Again we must distinguish between an engineer's idea

TABLE 4.1
Comparison of three kinds of flows

Name of species at rest	Potential difference causing species to flow	Name of species flowing
Water	Elevation difference	Rain
Electrons or charge	Voltage difference	Current
Energy	Temperature difference	Heat

of work and common English usage. For an engineer,

$$\text{Work} = \int \text{force} \cdot d(\text{distance}) = \int F\,dx \qquad (4.1)$$

Hod carriers lifting plaster up a ladder are doing work in the engineering sense of the word. However, if they are required to stand and hold a load of plaster on their shoulders for an hour, they are not doing work in the engineering sense, although they will certainly feel just as tired as if they had kept moving. Similarly, a baby-sitter is "working" in common usage but not in the engineering sense. If we rub two articles together, such as two pencil erasers, and they resist the rubbing, then we must exert a force and move it through a distance. Thus, rubbing is work in the engineering sense.

We have considered work only as force times a distance: $F\,dx$. There can also be the work of rotating shafts and electrical and magnetic work. We consider these in Sec. 4.14.

The third way in which two bodies can exchange energy is by *radiation*. The sun heats the earth by radiation, and x-rays and gamma rays change the energy of bodies by radiation. Radiation does not fit perfectly into either of the categories *work* and *heat*. However, with a little adjustment of the definitions, it can be made to appear as heat. When a radiation is due to a difference in temperature, such as that between the sun and the earth or between the glowing wires in the toaster and the slice of bread, that radiation fits our definition of heat well, except that the bodies are not in contact. However, if gamma radiation is flowing from a cold piece of radium to a warmer piece of lead, heat appears to flow from a cold body to a hot body. If we focus our attention, not on the average temperature of the bulk of the radium, but on the individual atom emitting the gamma ray, we see that at the instant of emitting the ray it undergoes a nuclear event, which raises its instantaneous "temperature" to a very high value. If we consider this "temperature" instead of the temperature of the large mass of radium, then radiation fits the definition of heat fairly well.

4.4 THE ENERGY BALANCE

Now we are ready to write the general energy balance equation. To write any balance equation we need a well-defined system of boundaries. Let us choose as our system the tank shown in Fig. 4.1.

We begin our balance by excluding from consideration magnetic, electrostatic, surface, and nuclear energies. Thus, the only kinds of energy which 1 lb of matter can contain are internal, kinetic, and potential: $u + \text{ke} + \text{pe}$. Now, the general balance says that

$$\text{Accumulation} = \text{flow in} - \text{flow out} + \text{creation} - \text{destruction} \qquad (4.2)$$

However, observations of nature have led to the conclusion that energy can be neither created nor destroyed (excluding nuclear effects, to be discussed later).

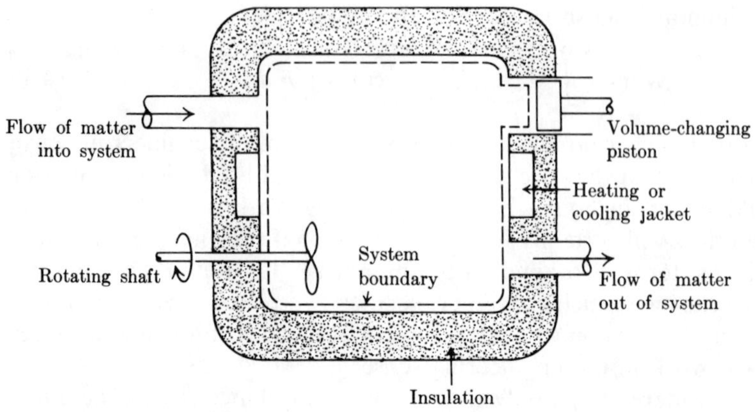

FIGURE 4.1
Tank used as system for energy balance.

So for energy the balance equation is

$$\text{Accumulation} = \text{flow in} - \text{flow out} \tag{4.3}$$

Accumulation is the differential of the energy contained within the system boundary. The only such energy is that associated with the matter within the boundary. If the matter is uniform (if all has the same u, ke, and pe), then accumulation is $d[m(u + \text{pe} + \text{ke})]$ where m is the mass in the system. We use this simplified form for now and consider nonuniform systems in Sec. 4.13.

Energy can enter in three ways. One way is by matter coming in the inlet pipe. For every infinitesimal amount of matter which flows in, the amount of energy which flows in with it is $(u + \text{pe} + \text{ke})_{\text{in}}\, dm_{\text{in}}$. Obviously, for matter flowing out the outlet line, energy can also flow out. The amount of energy flowing via the outlet line is $(u + \text{pe} + \text{ke})_{\text{out}}\, dm_{\text{out}}$. The two other ways energy can flow in or out are via heat through the heating or cooling jacket, which we call dQ, and via mechanical work of various forms, which we call dW.† Substituting these into Eq. 4.3, we find

$$d[m(u + \text{pe} + \text{ke})]_{\text{sys}} = [(u + \text{pe} + \text{ke})_{\text{in}}\, dm_{\text{in}} + dQ_{\text{in}} + dW_{\text{in}}] - [(u + \text{pe} + \text{ke})_{\text{out}}\, dm_{\text{out}} + dQ_{\text{out}} + dW_{\text{out}}] \tag{4.4a}$$

† As discussed in most thermodynamics texts, dQ and dW are inexact differentials. This means that the value of $\int_b^a dQ$ or of $\int_b^a dW$ depends on not only the initial and final states of the system but also on the path followed. Quantities such as dz (an infinitesimal change in elevation) are exact differentials. However, the fact that dQ and dW are inexact differentials has no effect on their role in most fluid mechanics problems, so we do not dwell further on this distinction.

This is a preliminary equation; its final form will appear shortly. We now introduce two sign conventions that are very common in thermodynamics:

$$dQ = dQ_{\text{in}} - dQ_{\text{out}} \tag{4.5}$$

$$dW = -(dW_{\text{in}} - dW_{\text{out}}) \tag{4.6}$$

These stem from the early attempts of thermodynamicists to study the steam engine. For a steady-running or cyclical steam engine, Eq. 4.4*a* reduces to

$$dQ_{\text{in}} - dQ_{\text{out}} = -(dW_{\text{in}} - dW_{\text{out}}) \tag{4.7}$$

Using the definitions in Eqs. 4.5 and 4.6 for the steam engine, we find

$$dQ = dW \tag{4.8}$$

which is so simple and pleasing that most workers in thermodynamics have settled upon it. Others have decided that all flows into the system should be positive and all flows out of the system should be negative (which is the convention in accounting and almost all other balances) so they show $(+dW)$ where this book shows $(-dW)$. You must learn to live with these two different choices of sign for the work term, which appear in different texts. Equation 4.5 says that dQ is the *net* heat coming *into* the system and that dW is the *net* work going *out* of the system. Making the substitution of Eqs. 4.5 and 4.6 into Eq. 4.4*a*, we find this final form of Eq. 4.4*a*:

$$d[m(u + \text{pe} + \text{ke})]_{\text{sys}} = (u + \text{pe} + \text{ke})_{\text{in}}\, dm_{\text{in}} - (u + \text{pe} + \text{ke})_{\text{out}}\, dm_{\text{out}} + dQ - dW \tag{4.4b}$$

4.5 KINETIC AND POTENTIAL ENERGIES

Equation 4.4*b* can be used only if we can find a way to assign numerical values to the various symbols in it. We already have an expression for work, Eq. 4.1. It has the dimension of force times distance; in SI its unit is the joule $(1\ \text{J} = \text{N} \cdot \text{m})$. In the English engineering system of units, its unit is the foot-pound force, abbreviated ft · lbf.

We deduced Eq. 4.4*b* for the system shown in Fig. 4.1, but it applies equally well to many other systems. Let us now choose as our system a 1-kg steel ball. We lift it slowly a distance dz. We insulate it so that during this lifting process there is no heat transfer to or from the surroundings; $dQ = 0$. Moreover, no matter flows into or out of the system: $dm_{\text{in}} = dm_{\text{out}} = 0$. Since no matter flows in or out, we have $d[m(u + \text{pe} + \text{ke})]_{\text{sys}} = m[d(u + \text{pe} + \text{ke})]$. Substituting this in Eq. 4.4*b*, we find

$$m\, d(u + \text{pe} + \text{ke})_{\text{sys}} = -dW \tag{4.9}$$

If we have proceeded without friction heating, the final temperature is the same as the initial temperature; so we conclude that du_{sys} is zero. The final and initial velocities are also zero; so $d(\text{ke})_{\text{sys}}$ is zero. Furthermore, according to

Eq. 4.1, dW equals $-F\,dz$. Here the sign of the work term is negative because work is done on the system. The force needed to lift the ball is the same as the weight of the ball, i.e., its mass times the acceleration of gravity. So

$$m_{\text{sys}}\,d(\text{pe}) = -(-m_{\text{sys}}g\,dz) \tag{4.10}$$

$$d(\text{pe}) = g\,dz \tag{4.11}$$

Here, then, is a convenient equation for the change in potential energy. If the acceleration of gravity is constant (practically true in all earth-bound problems but certainly not true in interplanetary space problems), we may integrate both sides of Eq. 4.11, taking g outside the integral sign, and find

$$\text{pe} = gz + \text{constant} \tag{4.12}$$

Here the constant is chosen so as to make the potential energy zero when the elevation above some arbitrary datum (such as sea level or ground level) is zero. If z is measured above this datum, then the constant is zero, and

$$\text{pe} = gz \tag{4.13}$$

Example 4.1. Determine the change in potential energy of a 10-kg bag of feathers which is raised a vertical distance of 23 m.

Since we are dealing with a *change* in potential energy (as we do in all practical problems), we need not concern ourselves with the datum. We can see this by applying Eq. 4.12 to the initial and final states:

$$\begin{aligned}\Delta\text{pe} &= \text{pe}_{\text{fin}} - \text{pe}_{\text{init}}\\ &= gz_{\text{fin}} + \text{constant} - (gz_{\text{init}} + \text{constant})\\ &= g(z_{\text{fin}} - z_{\text{init}}) = 9.81\,\frac{\text{m}}{\text{s}^2}\cdot 23\text{ m}\\ &= 225.6\text{ m}^2/\text{s}^2 = 2427\text{ ft}^2/\text{s}^2\end{aligned}$$

This is the change in potential energy per unit mass. We calculate the total change in potential energy by multiplying the potential-energy change per unit mass by the mass present:

$$\Delta\text{PE} = m\,\Delta\text{pe} = 10\text{ kg}\cdot 225.6\,\frac{\text{m}^2}{\text{s}^2} = 2256\,\frac{\text{kg}\cdot\text{m}^2}{\text{s}^2} = 53{,}400\,\frac{\text{lbm}\cdot\text{ft}^2}{\text{s}^2}$$

To find the answer in joules or foot-pounds-force, we use the force-mass conversion factor:

$$\Delta\text{PE} = 2256\,\frac{\text{kg}\cdot\text{m}^2}{\text{s}^2}\cdot\frac{\text{J}}{\text{N}\cdot\text{m}}\cdot\frac{\text{N}\cdot\text{s}^2}{\text{kg}\cdot\text{m}} = 2256\text{ J} = 1664\text{ ft}\cdot\text{lbf}$$ ■

Now we take the 1-kg ball of steel and throw it horizontally. Again we take the ball as the system. As before, we can insulate it so that no heat flows in or out ($dQ = 0$). During the throwing process no matter flows into or out of the ball, so $dm_{in} = dm_{out} = 0$. Furthermore, if we proceed without friction heating, the temperature will not change; so $du_{sys} = 0$. If we throw it perfectly horizontally, then there is no change in elevation during the throwing; so $d(gz)_{sys} = 0$. As before, $dW = -F\,dx$. Substituting all these in Eq. 4.4*b* yields

$$m_{sys}\, d(\text{ke}) = -F\, dx \tag{4.14}$$

We may replace F with $m_{sys}a_{sys}$ (according to Newton's law) to get

$$d(\text{ke}) = a_{sys}\, dx \tag{4.15}$$

But $a = dV/dt$, where V is the velocity; so $a\,dx = dV\,dx/dt$. Furthermore, $dx/dt = V$; so $a\,dx = V\,dV$. We can now integrate both sides of Eq. 4.15 to get

$$\text{ke} = \frac{V^2}{2} + \text{constant} \tag{4.16}$$

Here, too, we may choose any value we like for the constant. The logical choice is zero, which makes the kinetic energy zero for a body at rest:

$$\text{ke} = \frac{V^2}{2} \tag{4.17}$$

Example 4.2. What is the kinetic energy of a 0.01-lbm bullet traveling at 2000 ft/s relative to the barrel of the gun it has just left?

Here the velocity is measured relative to the same datum as we wish the kinetic energy to be relative to, so we have no problem with the datum.

$$\text{KE} = m \cdot \text{ke} = m\,\frac{V^2}{2}$$

$$= \frac{(0.01\ \text{lbm})(2000\ \text{ft/s})^2}{2} \cdot \frac{\text{lbf} \cdot \text{s}^2}{32.2\ \text{lbm} \cdot \text{ft}} = 621\ \text{ft} \cdot \text{lbf} = 842\ \text{J}$$ ■

Example 4.3. Suppose that the gun of Example 4.2 were mounted, facing backward, on an airplane which just flew past us at a velocity of 1990 ft/s. What would be the kinetic energy of the bullet relative to the airplane and relative to us?

Obviously, the bullet is moving at 2000 ft/s relative to the airplane, so the kinetic energy relative to the airplane is the same as in Example 4.2. However, relative to us the bullet is moving at 10 ft/s, and its kinetic energy is

$$\text{KE} = m\,\frac{V^2}{2} = \frac{(0.01\ \text{lbm})(10\ \text{ft/s})^2}{2} \cdot \frac{\text{lbf} \cdot \text{s}^2}{32.2\ \text{lbm} \cdot \text{ft}}$$

$$= 0.016\ \text{ft} \cdot \text{lbf} = 0.021\ \text{J}$$ ■

4.6 INTERNAL ENERGY

Now that we have the numerical forms of kinetic energy per unit mass and potential energy per unit mass, we can rewrite Eq. 4.4*b* as

$$d\left[m\left(u + gz + \frac{V^2}{2}\right)\right]_{\text{sys}}$$
$$= \left(u + gz + \frac{V^2}{2}\right)_{\text{in}} dm_{\text{in}} - \left(u + gz + \frac{V^2}{2}\right)_{\text{out}} dm_{\text{out}} + dQ - dW \quad (4.18)$$

This is the semifinal form of the energy balance equation.

At this point we must reconsider our idea of the internal energy. Since the potential energy has to do with elevation and gravity, we should expect its numerical formulation (Eq. 4.13) to involve g and z, as it does. Likewise, the kinetic energy depends on velocity, and its formulation (Eq. 4.17) indicates this. The internal energy, as we suggested before, is related to the "hotness" of a body, so we should expect its formulation to be related to heat in some way. The common unit used for kinetic and potential energies is foot-pounds-force or joules, but this is an inconvenient unit for heat flows or for the internal energy. Instead, we use a "heat" unit, the British thermal unit (Btu) or the calorie (cal). The *British thermal unit* (Btu) is defined as the amount of energy which must be transferred into 1 lbm of water to raise its temperature 1°F starting at 59.5°F. It is the unit used in English-speaking countries to measure most heat flows. (Inquisitive readers can find furnace ratings in Btu per hour on the nameplates of most U.S. household furnaces and water heaters, and they can find the "heating value" of natural gas in Btu per cubic foot on their gas bills.) The *calorie* (cal) is defined as the amount of energy which must be transferred into 1 g of water to raise its temperature 1°C starting at 0°C. The "calorie" that appears in diet and nutrition guides is the kilocalorie (1 kcal = 1000 cal), commonly called a calorie in the United States.

Now suppose we take as our system the tank shown in Fig. 4.1. We close the valves in the inlet and outlet lines, so that $dm_{\text{in}} = dm_{\text{out}} = 0$. We also stop the rotating shaft and do not move the volume-changing piston, so there will be no work done ($dW = 0$). Now we transfer 100 Btu of energy into the tank from the heating jacket:

$$d\left[m\left(u + gz + \frac{V^2}{2}\right)\right]_{\text{sys}} = dQ = 100 \text{ Btu} \quad (4.19)$$

However, in this operation the elevation and velocity of the material in the tank did not change, so that $d(gz)_{\text{sys}} = d(V^2/2)_{\text{sys}} = 0$, and since m_{sys} remained constant, we may take it out of the differential. This leads to

$$du_{\text{sys}} = \frac{dQ}{m_{\text{sys}}} = \frac{100 \text{ Btu}}{m_{\text{sys}}} = \frac{25{,}200 \text{ cal}}{m_{\text{sys}}} = \frac{25.2 \text{ kcal}}{m_{\text{sys}}} \quad (4.20)$$

The potential and kinetic energies per unit mass are expressed in units of foot-pounds-force per pound-mass (ft · lbf/lbm) or joules per kilogram (J/kg). Here we have a change in internal energy expressed in Btu per pound-mass or calories per kilogram. In our balance equation, we obviously need some way to interconvert these units so that the sum $(u + gz + V^2/2)$ is in a consistent set of units. All efforts to calculate this conversion factor from some more basic principle have failed; however, it can be determined experimentally.

Suppose we cool the system in Fig. 4.1 to its initial state by removing 100 Btu of energy via the cooling jacket. Now we start the stirrer and measure the work input required to produce the same temperature rise as was caused by the addition of 100 Btu as heat. In this case Eq. 4.4*b* simplifies to

$$du_{\text{sys}} = \frac{-dW}{m_{\text{sys}}} \tag{4.21}$$

By carefully measuring the temperature changes, we can find the exact number of foot-pounds-force or joules of work that produces the same heating effects as 1 Btu or 1 cal of energy added as heat. This experiment was done by Joule [2] and formed the keystone in constructing the first law of thermodynamics. His experimental result (as corrected by later workers with better equipment) is

$$1\ \text{Btu} = 778\ \text{ft}\cdot\text{lbf} \qquad 1\ \text{cal} = 4.184\ \text{J} \tag{4.22}$$

This is an experimental fact reproducible in any well-equipped laboratory. Using the conversion factors in Eq. 4.22, we can easily convert all the terms in the energy balance to a common basis. In SI the use of the calorie is discouraged; thermal energy quantities are to be expressed only in joules. However, the use of the calorie (or kilocalorie) is quite common in countries using metric units. Today's student will have to be familiar with its use.

We said before that internal energy might be thought of as hotness plus chemical energy. However, there can also be internal-energy changes at constant temperature. Suppose we have some mass of some substance in an absolutely rigid vessel. Now we transfer heat into the vessel. For this process Eq. 4.18 yields

$$m_{\text{sys}}\, du = dQ \tag{4.23}$$

Thus, we see that *for a simple constant-volume heating* we have $du = dQ/m$. What are the possible external signs of such an increase in internal energy?

1. The substance may increase in temperature.
2. The substance may undergo an energy-consuming chemical reaction, such as

$$2NH_3 \rightarrow N_2 + 3H_2$$

3. The substance may undergo a phase change such as ice → water or water → steam.
4. The substance may undergo a crystal-structure change, such as $\alpha_{\text{iron}} \rightarrow \gamma_{\text{iron}}$. This is really a phase change but is not as obvious as those shown above.
5. Any combination of the four items listed above may occur simultaneously.

So we see that any exact definition of an internal-energy change must be based on consideration of all the terms in Eq. 4.18. Thus, Eq. 4.18 *is* the exact definition of the change in internal energy. If we restrict ourselves to a closed system of constant mass, with no changes in kinetic or potential energy, the equation simplifies to

$$m\,du = dQ - dW \tag{4.24}$$

Integrating this, we find

$$U = mu = Q - W + \text{constant} \tag{4.25}$$

Here there is no obvious choice for the constant, as there was in the case of kinetic or potential energy. In making up tabulations of thermodynamic properties, we must arbitrarily select a value for this constant. For refrigerants such as Freon 12 (App. A.2) the constant is chosen so that $u = -Pv$ for the saturated liquid at $-40°F$. This choice is made on the basis of convenience alone. Here v is the *volume per pound-mass or specific volume*. (After we introduce the enthalpy h, we will see that this choice is the same as choosing $h = 0$ for saturated liquid at $-40°F$. This temperature is chosen because $-40°F = -40°C$; thus the common datum temperature for refrigerants is the same in countries using both temperature scales.)

In sum, internal energy may be thought of approximately as hotness plus chemical energy. Its exact formulation is Eq. 4.18, which allows us to calculate changes of internal energy. Using this equation and an arbitrarily selected value in some datum state, we can calculate the numerical value of the internal energy per unit mass for any state of any substance.

4.7 THE WORK TERM

So far we have said little about the work term in Eq. 4.18. Suppose our system is the 1-kg steel ball described previously. The system is practically rigid, and the work done on it generally consists of something, e.g., a hand, pushing it. This work is shown by Eq. 4.1.

Now consider the system shown in Fig. 4.1. Let us assume that the material in the tank is something easily compressed, such as air or steam. In this case we can do work on the system by moving the volume-changing piston; the magnitude of this work is shown by Eq. 4.1. However, the force required to move the piston is equal to the piston's cross-sectional area times the pressure in the tank. Further, the product of the piston's cross-sectional area and the distance traveled is equal to the change in volume of the tank; so

$$dW = F\,dx = PA\,dx = P\,dV \tag{4.26}$$

where V is the volume of the tank. If we close the inlet and outlet lines of the tank, turn off the heating and cooling coils, and then move the piston inward, Eq. 4.18 shows the following:†

† The result is correct for any work done by a moving boundary. However, for boundaries moving at supersonic speeds, the pressure at the boundary may be different from the pressure in the nearby fluid. As long as the P in Eq. 4.27 is the P experienced by the boundary, this result is correct.

$$(m\,du)_{sys} = -dW = -P\,dV \tag{4.27}$$

When we move the piston inward, dV is negative; so du_{sys} is positive. When du_{sys} is positive, the temperature of the system will rise, unless a phase change occurs. This phenomenon can be observed easily in an ordinary bicycle pump. Driving the piston inward causes the air in the pump to become hot. So we see that one form of work we must consider is the work of moving the boundaries of the system. This work is equal to $P\,dV$ and is often simply referred to as "$P\,dV$ work."

4.8 INJECTION WORK

If we considered only such systems as a cannonball or a tank with no flow in or out, we would never need to introduce the idea of injection work. However, it is often advantageous to choose as our system a certain set of boundaries through which mass flows, which we call an *open system*. If, e.g., we wished to analyze the power plant at the Hoover Dam, we would find it easier to choose as our system the power plant from water inlet to water outlet than to choose 1 lbm or 1 kg of water passing through the plant as our system. By choosing the open system with mass flow through it, we will have a much simpler analysis, since we do not have to consider the many changes in pressure, elevation, and velocity along the complex flow path taken by the water through valves, turbines, wicket gates, and so forth. However, we do have to consider the injection work.

Suppose we have as our system the tank shown in Fig. 4.1. We now bring into the tank a mass dm_{in} from the inlet line; nothing flows out, there is no heat transfer, and there is no work due to moving the volume-changing piston or turning the shaft. What will be the energy balance for this operation? This is easiest to see when we do it by a two-step process; see Fig. 4.2.

In the first step we let the mass dm flow in and simultaneously move the volume-changing piston out. We move the piston at a rate such that the fluid originally in the tank is not compressed. This means that all the fluid pushed aside by the fluid coming in is pushed into the space vacated by the volume-changing piston. Thus, no net work is done on the system because, for all the fluid involved, there is no volume change. Therefore, the compression work $P\,dV$ is zero. Thus, the energy balance for step 1 is

$$d\left[m\left(u + gz + \frac{V^2}{2}\right)\right]_{sys} = \left(u + gz + \frac{V^2}{2}\right) dm_{in} \tag{4.28}$$

Now, to get to the desired final state, we must move the volume-changing piston back to its original position. It must move back by a volume exactly equal to the volume of the fluid which moved in, which is $v_{in}\,dm_{in}$; then the work to move it back is $dW = -Pv_{in}\,dm_{in}$. The minus sign indicates that $dV = -v_{in}\,dm_{in}$. The energy balance for this second step is

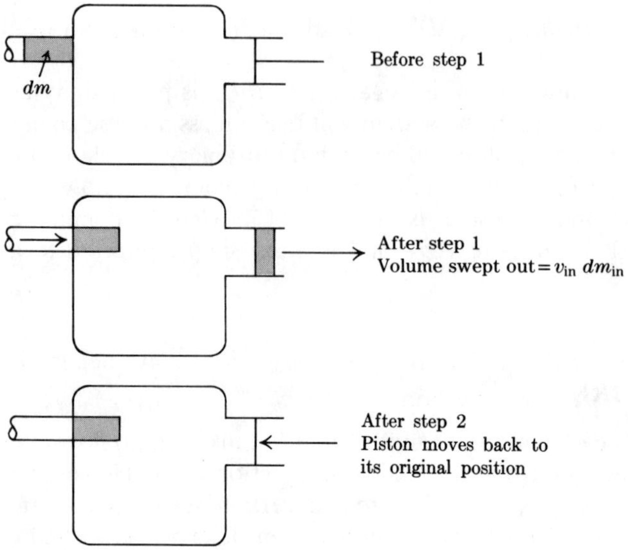

FIGURE 4.2
Two-step process to illustrate injection work.

$$d\left[m\left(u + gz + \frac{V^2}{2}\right)\right]_{sys} = -P\,dV = -(-Pv_{in})\,dm_{in} = (Pv)_{in}\,dm_{in} \tag{4.29}$$

The energy balance for the entire injection process, which is the equivalent of the two steps given above, must be the sum of the energy balances for the two separate steps:

$$\begin{aligned} d\left[m\left(u + gz + \frac{V^2}{2}\right)\right]_{sys} &= \left(u + gz + \frac{V^2}{2}\right)_{in} dm_{in} + (Pv)_{in}\,dm_{in} \\ &= \left(u + Pv + gz + \frac{V^2}{2}\right)_{in} dm_{in} \end{aligned} \tag{4.30}$$

What does this $(Pv)_{in}\ dm_{in}$ term represent? We call it *injection work*, because it is exactly the work that was needed to inject the mass dm_{in} across the system boundaries. It is also sometimes called *intrusion work*, *flow work*, and *flow energy*.

Obviously, we could repeat the calculation for fluid flowing out the outlet line. How, then, will we reconcile this injection work idea with Eq. 4.18? Equation 4.18 is correct as it stands; for the process described above, the $(Pv)_{in}\ dm_{in}$ term is included in the dW term. However, we now break up the dW term as follows:

$$dW = dW_{inj} + dW_{a.o.} = (-Pv)_{in}\,dm_{in} + dW_{a.o.} \tag{4.31}$$

where the subscripts "inj" and "a.o." denote "injection" and "all others," respectively.

We now make this substitution in Eq. 4.18 and factor the injection work terms as shown in Eq. 4.30:

$$d\left[m\left(u + gz + \frac{V^2}{2}\right)\right]_{\text{sys}} = \left(u + Pv + gz + \frac{V^2}{2}\right)_{\text{in}} dm_{\text{in}}$$
$$- \left(u + Pv + gz + \frac{V^2}{2}\right)_{\text{out}} dm_{\text{in}} + dQ - dW_{\text{a.o.}} \quad (4.32)$$

This is the final working form of the energy balance equation. Let us list the restrictions on it:

1. Electrostatic, magnetic, surface, and nuclear energy effects are negligible.
2. The contents of the system are uniform.
3. The inflow and outflow streams are uniform.
4. The acceleration of gravity is constant.

Using this equation we can solve an immense array of problems. Furthermore, by making slight changes we can relax these four restrictions so that the equation will apply to *any* problem.

4.9 ENTHALPY

In Eq. 4.32 the combination $u + Pv$ occurs in the flow-in and flow-out terms. This combination occurs so often in thermodynamics that it has been given a name and a symbol:

$$u + Pv = h = \text{enthalpy per unit mass, or specific enthalpy} \quad (4.33)$$

The enthalpy is also called *total heat*, *inherent heat*, and several other names in older thermodynamics texts. Obviously, it is the combination of the internal energy per unit mass and the injection work per unit mass. You will soon appreciate its convenience in solving practical problems. Substituting Eq. 4.33 in Eq. 4.32, we find its *exact equivalent*:

$$d\left[m\left(u + gz + \frac{V^2}{2}\right)\right]_{\text{sys}}$$
$$= \left(h + gz + \frac{V^2}{2}\right)_{\text{in}} dm_{\text{in}} - \left(h + gz + \frac{V^2}{2}\right)_{\text{out}} dm_{\text{out}} + dQ - dW_{\text{a.o.}} \quad (4.34)$$

Let us summarize how we found this equation:

1. We discussed the general idea of the balance equation.
2. We introduced the abstract quantity energy.
3. We asserted, *without proof*, that this abstract quantity, energy, obeys the balance equation, with the creation and destruction terms set equal to zero. This assertion is unprovable; it rests on its ability to explain all the careful experiments ever run to test it.

4. We chose a fairly general system and listed a set of restrictions that would apply to that system.
5. We wrote out in detail the balance equation for that system, subject to the restrictions and subject to the sign conventions for heat and work, and we found Eq. 4.18.
6. We introduced the idea of injection work, split up the work term in Eq. 4.18, and regrouped terms to find Eq. 4.32.
7. Finally, we introduced the definition of enthalpy to find Eq. 4.34.

This is not a *derivation* of the first law of thermodynamics; that law is underivable. Rather, this is a set of algebraic manipulations and definitions that convert the statement "energy obeys the balance equation without creation or destruction" to a very convenient and useful working equation.

4.10 RESTRICTED FORMS

Equation 4.34 is powerful because it is so general. However, whenever we write it down, we have, in effect, written down the four restrictions listed previously. The procedure recommended for solving all thermodynamics problems is to write Eq. 4.32 or Eq. 4.34, select a system of boundaries, and cancel the terms that appear negligible. Each cancellation represents an assumption. For example, if we assume no heat exchange with the surroundings, then dQ is zero. By crossing out dQ we are making this assumption. When all the unnecessary terms have been canceled, we have not only a working equation but also a list of the assumptions on which that equation is based.

There are several restricted forms of Eq. 4.34 in common use.

Example 4.4. A sample of air and coal is contained in the constant-pressure cylinder shown in Fig. 4.3. This cylinder has a frictionless, weightless piston, so the pressure inside the cylinder is always exactly the same as the pressure of the atmosphere (which we assume is 1 atm). A small spark is now introduced, causing the coal to burn. When the burning is over, the piston has moved so that the volume of the contents is increased by 1 ft^3. The heat transferred to the surroundings was 42 Btu. What is the internal-energy change for this reaction?

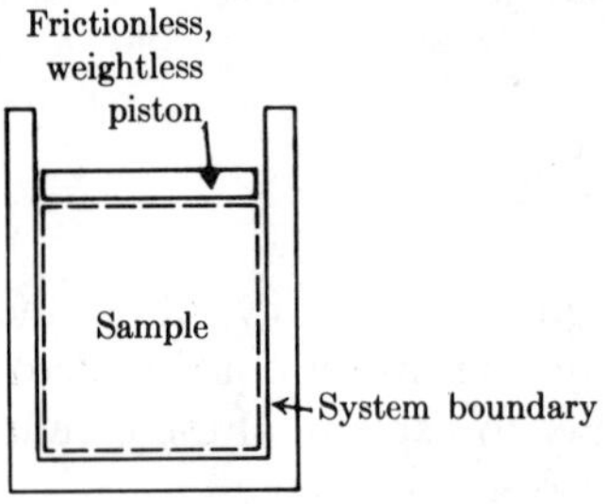

FIGURE 4.3
Simple piston and cylinder.

We choose as our system the contents of the cylinder. In this system there is no flow in or out, so that $dm_{\text{in}} = dm_{\text{out}} = 0$. Furthermore, there is a negligible change in the kinetic or potential energies of the material contained in the system: $d(gz)_{\text{sys}} = d(V^2/2)_{\text{sys}} = 0$. Because there is no mass flow in or out, $d(mu)_{\text{sys}} = m\,du_{\text{sys}} = dU_{\text{sys}}$. Making these substitutions in Eq. 4.34, we find

$$dU_{\text{sys}} = dQ - dW_{\text{a.o.}} \qquad \text{[closed system]} \tag{4.35}$$

This formula appears in most chemistry books as the basic statement of the first law of thermodynamics.† From the foregoing we can see that it is a much more restricted form than the one we chose (Eq. 4.34). We can now substitute for dW from Eq. 4.26:

$$\begin{aligned} dU_{\text{sys}} &= dQ - dW_{\text{a.o.}} = dQ - P\,dV \\ &= -42\ \text{Btu} - 14.7\,\frac{\text{lbf}}{\text{in}^2} \cdot 1\ \text{ft}^3 \cdot \frac{144\ \text{in}^2}{\text{ft}^2} \cdot \frac{\text{Btu}}{778\ \text{ft} \cdot \text{lbf}} \\ &= -44.7\ \text{Btu} = -11{,}270\ \text{cal} = -47{,}153\ \text{J} \end{aligned}$$

■

Here dQ is negative according to our sign convention.

Example 4.5. A steady-flow water power plant has its water inlet 15 m above its water outlet. The water enters the plant with a velocity of 3 m/s and leaves with a velocity of 10 m/s. What is the work done by the plant per kilogram of water passing through it?

We choose as our system the plant from inlet to outlet. If the flow is steady, then, as discussed in Sec. 3.3, we have $d[m(u + gz + V^2/2)]_{\text{sys}} = 0$. Furthermore, for the assumption of only one inlet and one outlet stream dm_{in} equals dm_{out}. We can then divide by dm to find

$$0 = \left(h + gz + \frac{V^2}{2}\right)_{\text{in}} - \left(h + gz + \frac{V^2}{2}\right)_{\text{out}} + \frac{dQ}{dm} - \frac{dW_{\text{a.o.}}}{dm} \qquad \text{[steady flow, open system]} \tag{4.36}$$

This is the steady-flow form of the first law of thermodynamics. It appears in most chemical and mechanical engineering textbooks either as shown or rearranged to

$$dh + g\,dz + d\left(\frac{V^2}{2}\right) = \frac{dQ}{dm} - \frac{dW_{\text{a.o.}}}{dm} \qquad \text{[steady flow, open system]} \tag{4.37}$$

† In this statement the subscript "all others" on the work term is unnecessary, since there can be no injection work or out of a closed system. Thus, this term is usually written simply dW.

Here Eq. 4.36 has been applied to two points negligibly far apart in a steadily flowing stream. Bernoulli's equation, which we discuss in Chap. 5, follows directly from Eq. 4.37.

Returning to the water power plant, we assume that there is no heat transfer to the plant ($dQ = 0$) and that the enthalpy of the outlet water is the same as the enthalpy of the inlet water (this is equivalent to the assumption that the inlet and outlet water streams are at the same temperature and pressures). Then

$$\frac{dW_{\text{a.o.}}}{dm} = g(z_{\text{in}} - z_{\text{out}}) + \frac{V_{\text{in}}^2 - V_{\text{out}}^2}{2}$$

$$= 9.81\ \frac{\text{m}}{\text{s}^2} \cdot 15\ \text{m} + \frac{(3\ \text{m/s})^2 - (10\ \text{m/s})^2}{2}$$

$$= 147.15\ \frac{\text{m}^2}{\text{s}^2} - 45.50\ \frac{\text{m}^2}{\text{s}^2}$$

$$= 101.65\ \frac{\text{m}^2}{\text{s}^2} \cdot \frac{\text{J}}{\text{N}\cdot\text{m}} \cdot \frac{\text{N}\cdot\text{s}^2}{\text{kg}\cdot\text{m}} = 101.65\ \frac{\text{J}}{\text{kg}} = 34.01\ \frac{\text{ft}\cdot\text{lbf}}{\text{lbm}} = 0.044\ \frac{\text{Btu}}{\text{lbm}}$$

■

4.11 SOME COMMON MACHINES AND PROCESSES

Several very common types of machine and process are described by simple forms of the energy balance.

A. Adiabatic Throttle

The *adiabatic throttle* is a reasonable model for any partially open valve, the expansion valve on any refrigerator, a leak from a high-pressure storage container, etc. Such a device is shown in Fig. 4.4. The system boundaries are as shown. We make the following assumptions:

1. The flow is steady.
2. The flow is horizontal.

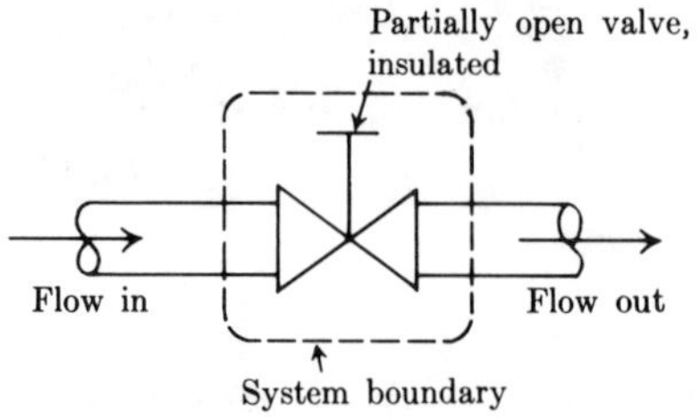

FIGURE 4.4
Adiabatic throttle.

3. The inlet and outlet velocities are negligible.
4. The heat transfer into or out of the system is negligible.

In this system there is no work done except the injection work, which we have excluded from $dW_{a.o.}$; therefore, $dW_{a.o.} = 0$. When we make these substitutions in Eq. 4.34, we find

$$0 = h_{in}\, dm_{in} - h_{out}\, dm_{out} \tag{4.38}$$

Dividing by dm_{in}, which is the same as dm_{out}, and rearranging, we find

$$h_{in} = h_{out} \tag{4.39}$$

Example 4.6. Freon 12 flows steadily through an insulated throttle valve. The upstream conditions are 100 psia and 100°F. The downstream pressure is 20 psia. What is the downstream temperature?

From Eq. 4.39 we have $h_{in} = h_{out}$. Using App. A.2, we find $h_{in} = 88.5$ Btu/lbm. Therefore, the fluid flowing out has a pressure of 20 psia and an enthalpy of 88.5 Btu/lbm. From App. A.2 we can read the outlet temperature as about 75°F. ■

B. Turbine and Compressor

A *turbine* or an *expansion engine* is a device for extracting work from a flowing fluid stream by allowing the stream to decrease in pressure or velocity. *Compressors*, *fans*, *blowers*, and *pumps* are devices for increasing the pressure of a fluid stream by doing work on it. Obviously, the turbine and the compressor are inverses of each other, and some machines are actually designed so that they can serve as expansion engines one moment and compressors the next (the piston-and-cylinder arrangement in automobile engines serves as a compressor in the compression stroke and an expansion engine in the power stroke). The flow diagram of such a device is shown in Fig. 4.5. We use the boundaries shown in the figure and assume:

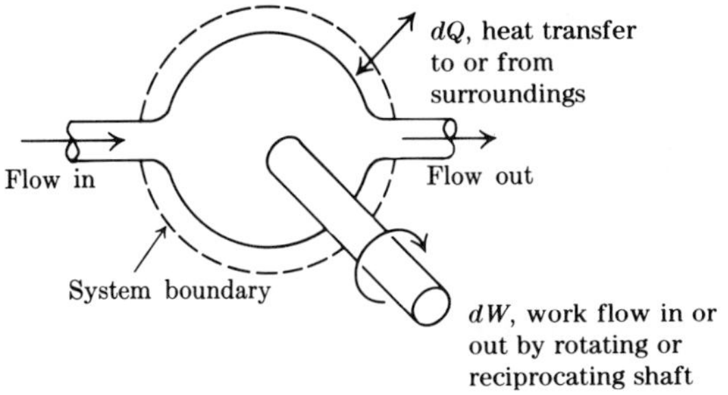

FIGURE 4.5
Turbine, expansion engine, compressor, fan, blower, or pump.

1. Steady flow.
2. Negligible change in elevation from inlet to outlet.

Then Eq. 4.34, rearranged, becomes

$$dW_{\text{a.o.}} = \left(h + \frac{V^2}{2}\right)_{\text{in}} dm_{\text{in}} - \left(h + \frac{V^2}{2}\right)_{\text{out}} dm_{\text{out}} + dQ \tag{4.40}$$

For most such devices we can also assume:

3. Negligible heat transfer to surroundings.
4. Negligible kinetic energies of inlet and outlet streams.

Therefore,

$$\frac{dW_{\text{a.o.}}}{dm} = h_{\text{in}} - h_{\text{out}} \tag{4.41}$$

For a turbine or expansion engine which extracts work from the fluid passing through it, the work per unit mass is positive (because of the sign convention discussed above), and h_{in} must be greater than h_{out}. For a pump, fan, blower, or compressor which does work on the fluid passing through it, the reverse is true: h_{out} must be greater than h_{in}, in which case $dW_{\text{a.o.}}/dm$ is negative, indicating work flowing into the system.

Example 4.7. An insulated, steady-flow, horizontal Freon compressor has the following conditions:

Flow rate:	15,000 lbm/h
Inlet:	10 psia, 20°F
Outlet:	40 psia, 120°F

What power is required to drive this compressor?

From Eq. 4.41 we have $dW_{\text{out}}/dm = h_{\text{in}} - h_{\text{out}}$. From App. A.2 we read $h_{\text{in}} = 81.0$ Btu/lbm and $h_{\text{out}} = 94.2$ Btu/lbm. Therefore,

$$\frac{dW_{\text{a.o.}}}{dm} = 81.0 - 94.2 = -13.2 \text{ Btu/lbm} = -30.67 \text{ kJ/kg}$$

Power (Po) is defined as the rate of doing work, so

$$\text{Po} = \frac{dW}{dt} = \frac{dW}{dm}\dot{m}$$

$$= -13.2\,\frac{\text{Btu}}{\text{lbm}} \cdot 15{,}000\,\frac{\text{lbm}}{\text{h}} \cdot \frac{\text{hp}\cdot\text{h}}{2545\text{ Btu}}$$

$$= -77.8 \text{ hp} = -58.04 \text{ kW}$$

The minus sign here indicates work done on the system. ■

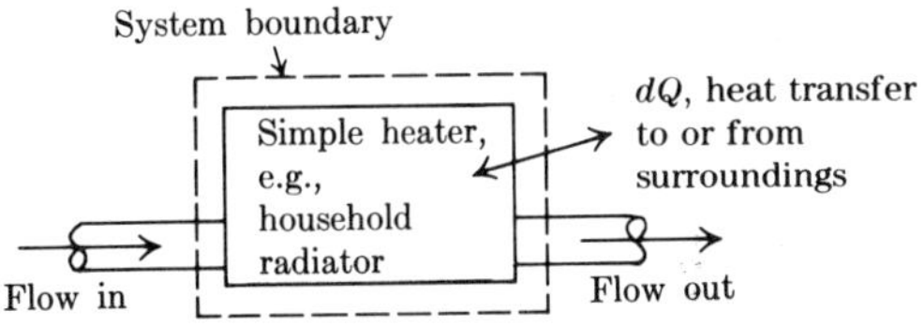

FIGURE 4.6
Simple heater.

C. Simple Heater or Cooler

A *simple heater* is shown in Fig. 4.6. For this system we use the boundaries shown and make the following assumptions:

1. Steady flow
2. Horizontal flow
3. Negligible inlet and outlet velocities

In this system, too, no work is done except the injection work, which we have excluded from $dW_{\text{a.o.}}$; therefore, substituting in Eq. 4.34 gives

$$\frac{dQ}{dm} = h_{\text{out}} - h_{\text{in}} \tag{4.42}$$

Example 4.8. A steady-state Freon cooler is used to remove 10^6 Btu/h from a meat-chilling room. The Freon is to enter as saturated liquid at 20 psia and to leave as saturated vapor at the same pressure. What flow rate of Freon is required?

We use Eq. 4.42 and choose the flowing Freon as our system. In App. A.2 we read $h_{\text{in}} = 6.8$ Btu/lbm and $h_{\text{out}} = 76.4$ Btu/lbm. Then

$$\frac{dQ}{dm} = 76.4 - 6.8 = 69.6 \text{ Btu/lbm}$$

$$\dot{m} = \frac{dQ}{dt}\frac{dm}{dQ} = 10^6 \frac{\text{Btu}}{\text{h}} \cdot \frac{\text{lbm}}{69.6 \text{ Btu}} = 14{,}300 \frac{\text{lbm}}{\text{h}} = 6500 \frac{\text{kg}}{\text{h}} \qquad \blacksquare$$

D. Steady-Flow Chemical Reactor

A steady-flow chemical reactor (e.g., the burner in a household water heater) is shown schematically in Fig. 4.7. At steady, horizontal, low-velocity flow, the energy balance for this reactor is

$$0 = h_{\text{in}}\, dm_{\text{in}} - h_{\text{out}}\, dm_{\text{out}} + dQ \tag{4.43}$$

Here we are dealing with a chemical change, so the material flowing in is not of the same chemical composition as the material flowing out. Consider first the case of a reactor with the same outlet temperature as inlet temperature. Equation 4.43, rearranged, becomes

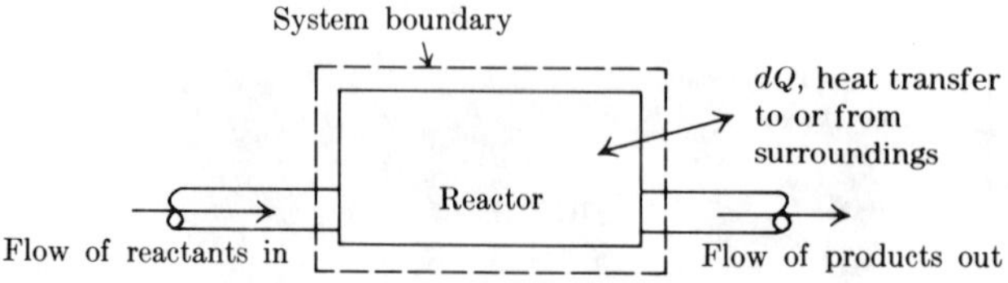

FIGURE 4.7
Steady-flow chemical reactor.

$$(h_{\text{products}} - h_{\text{reactants}})_{T_{\text{constant}}} = \Delta h_{\text{reaction}} = \frac{dQ}{dm_{\text{in}}} \tag{4.44}$$

Thus, the amount of heat added for a steady-flow reactor with equal inlet and outlet temperatures is called the *enthalpy change of the reaction*, denoted $\Delta h_{\text{reaction}}$. This is a relatively easily measured quantity, and it has been tabulated for many reactions [4, p. 449]; e.g., the combustion of methane (the principal ingredient of natural gas) with oxygen is

$$CH_4 + 2O_2 \rightarrow CO_2 + 2H_2O \qquad (\Delta h)_{77°\text{F}} = -380{,}000 \frac{\text{Btu}}{\text{lbmol}} = -211{,}000 \frac{\text{cal}}{\text{mol}}$$

Here $\Delta h_{\text{reaction}}$ is negative because heat flows out of the reactor shown in Fig. 4.7. In all reactions involving the burning of fuels (oil, coal, and wood), heat flows out of the reactor (called a burner in such cases), and in these $\Delta h_{\text{reaction}}$ is negative. In many important industrial reactions, such as the catalytic cracking of petroleum, $\Delta h_{\text{reaction}}$ is positive; heat must be added if the inlet and outlet streams are to have the same temperature.

If the reactor is an adiabatic reactor, then Eq. 4.43 shows us that the inlet and outlet enthalpies are the same. If $\Delta h_{\text{reaction}}$ in the isothermal reaction is negative, then the outlet products from an adiabatic reactor will be hotter than the inlet reactants; if $\Delta h_{\text{reaction}}$ is positive, the products will be cooler. These relations are seen most easily on a plot of enthalpy versus temperature for reactants and products, as shown for the methane-oxygen reaction in Fig. 4.8.

The diagram shows two curves, one for reactants and the other for products. Since any enthalpy is based on an arbitrary datum, we select the enthalpy of the reactants at 77°F (25°C) as zero. Then the enthalpy of the products as 77°F must be $\Delta h_{\text{reaction, 77°F}}$. From these two data the enthalpy of products or of reactants can be constructed with measurements made in simple heaters such as that of Fig. 4.6; thus we may construct the entire curves of Fig. 4.8. For an isothermal reactor the process can be represented by a vertical line, as shown in the figure; the heat removed is equal to the heat of reaction at the temperature of the reactor, which equals the vertical distance between the two curves at that temperature. For an adiabatic reactor, the process can be represented by a horizontal line, as shown in the figure, because the inlet and outlet enthalpies are the same. The outlet temperature of an adiabatic burner is called the *adiabatic flame temperature*. For most fuels these adiabatic flame temperatures are so high that combustion is not complete [4, p. 498].

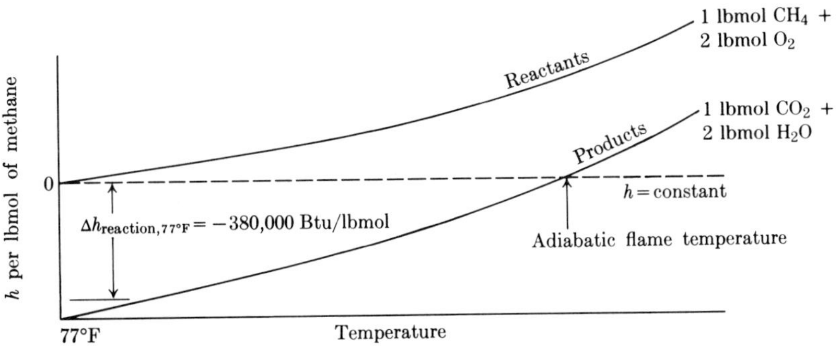

FIGURE 4.8
Enthalpy-temperature diagram for the complete combustion of methane with air.

These four examples should persuade the reader of the great convenience of the enthalpy.

4.12 UNSTEADY-STATE SYSTEMS, ACCUMULATION

Most unsteady-state problems are too difficult and time-consuming for an elementary text. One exception is *bottle-filling problems*; these have some practical significance and are very instructive. Consider a bottle attached to a large supply line with the connecting valve opened (see Fig. 4.9). We use the system boundaries shown and assume:

1. Negligible heat transfer to surroundings during filling process
2. Negligible change in elevation of fluid flowing in
3. Negligible kinetic energies

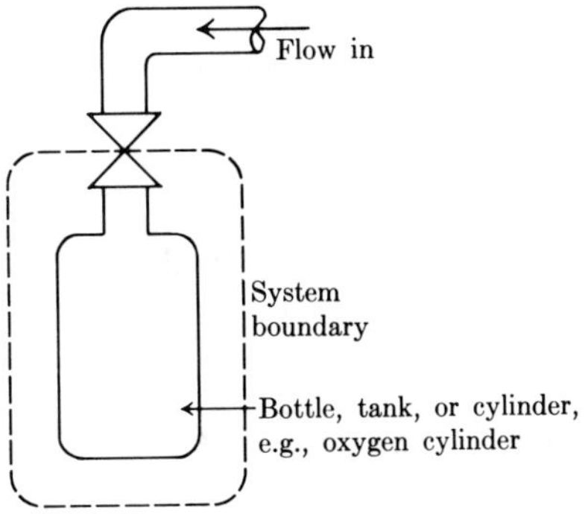

FIGURE 4.9
Constant-volume container being filled.

For the system shown there is no $dW_{a.o.}$, and Eq. 4.34 becomes

$$d(mu)_{sys} = h_{in}\, dm_{in} \tag{4.45}$$

Here we have assumed that all the contents of the bottle are at the same temperature, pressure, etc. If there is adequate mixing due to the inflow, this assumption is a good one. If the assumption is not correct, then $d[(mu)_{sys}]$ must be replaced with $d[\int_{\text{whole system}} u\, dm]$, and the problem is much more complicated. The simplest nonuniform system is considered in Section 4.13.

If we make the assumption of a uniform system, then Eq. 4.45 applies whether there is flow in or flow out, whether h_{in} is constant or variable, and so on. In the simplest possible subcase, originally the bottle is completely empty and the filling line is so large that the conditions in the line are not influenced by the flow rate, so that h_{in} is constant. In such a case, we can integrate both sides of Eq. 4.45:

$$[(mu)_{fin} - (mu)_{init}]_{sys} = h_{in}\, \Delta m_{in} \tag{4.46}$$

From the assumption that the bottle was originally empty we have

$$(mu)_{init,\ sys} = 0$$

because the initial mass is zero. Moreover, Δm_{in} is the same as $m_{fin,\ sys}$. Dividing this out, we find

$$u_{fin,\ sys} = h_{in} \tag{4.47}$$

Example 4.9. A completely evacuated container is attached to a line which contains Freon 12 at 100°F and 20 psia. The valve is opened, and Freon 12 flows in. When the flow stops, the pressure in the container is 20 psia. What is the temperature in this container?

This is exactly the system described above, so Eq. 4.47 applies. From App. A.2 we find that the enthalpy of the material flowing in is 92.0 Btu/lbm. This is equal to the final internal energy per unit mass. We also know the final pressure. If we had a table of internal energy per unit mass for Freon 12, we could simply look up the required temperature. (If we were doing a large number of such problems, it would be worthwhile to make up such a table—modern steam tables contain this information.) Since we do not have one, we must determine the final temperature from App. A.2 and from the definition of the enthalpy (Eq. 4.33).

We require the temperature for which, at a pressure of 20 psia, the internal energy is 92.0 Btu/lbm. First we try 140°F and find $h = 98.2$ Btu/lbm and $v = 2.7\ \text{ft}^3/\text{lbm}$. Then

$$u = h - Pv = 98.2\,\frac{\text{Btu}}{\text{lbm}} - 20\,\frac{\text{lbf}}{\text{in}^2}\cdot 2.7\,\frac{\text{ft}^3}{\text{lbm}}\cdot 144\,\frac{\text{in}}{\text{ft}^2}\cdot\frac{\text{Btu}}{778\ \text{ft}\cdot\text{lbf}}$$

$$= 98.2 - 10.0 = 88.2\ \text{Btu/lbm}$$

Obviously, we have tried too low a temperature. We now continue by trial and error as summarized below:

Trial no.	T_{assumed}, °F	h, Btu/lbm	v, ft³/lbm	Pv, Btu/lbm	$u = h - Pv$, Btu/lbm
1	140	98.2	2.7	10.0	88.2
2	160	101.5	2.8	10.0	91.5
3	180	104.5	2.8	10.0	94.5

From this table we can interpolate and arrive at 161°F (72°C) as the correct temperature. ■

This result is surprising at first glance; the fluid in the bottle is 61 F° hotter than the fluid in the line from which it was withdrawn. If we rewrite Eq. 4.47 in terms of the definition of enthalpy, we see that

$$u_{\text{fin}} - u_{\text{in}} = Pv_{\text{in}} \tag{4.48}$$

The higher temperature of the fluid in the bottle is due to the injection work done by the fluid in the line, pushing it in. It may seem that this work is negligible because the bottle was originally empty, but this is not correct. According to our assumption that the fluid in the line does not change pressure during the injection process, the work of getting it in is really the work of squeezing it through the valve, which is only partially open. If the valve were very large and offered no restriction, then the pressure in the line would have to fall, and we would have a different problem with a different solution.

It is instructive to consider this same problem from a closed-system viewpoint. We choose as our system the mass of gas which is finally contained in the bottle. Originally this gas is in the filling line, as shown in Fig. 4.10.

For the system chosen we make the same assumptions as before. However, in this case, there is no flow into or out of the system chosen ($dm_{\text{in}} = dm_{\text{out}} = 0$). There is, however, $dW_{\text{a.o.}}$ because the fluid behind our system, in the line, pushes it. Applying Eq. 4.34, we find

$$d(mu)_{\text{sys}} = -dW_{\text{a.o.}} \tag{4.49}$$

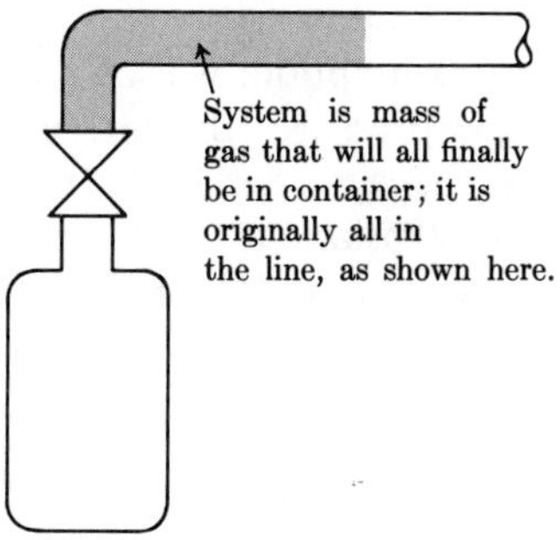

FIGURE 4.10
System chosen for analysis of bottle-filling problem using a closed system.

Here $dW_{a.o.}$ equals $P_{line}(-dV)_{in}$. The minus sign indicates that this is work done on the system. We also have $dV_{in} = v_{in}\, dm_{in}$. If we insert these in Eq. 4.49 and integrate and divide by dm_{in}, we find Eq. 4.48. Some students find this view more intuitively satisfying than the previous one.

This was the simplest bottle-filling case; the bottle was initially evacuated and the line was large enough to have h_{in} constant. If the bottle is not initially evacuated but the line is large enough for h_{in} to be constant, then Eq. 4.45 still applies, but $(mu)_{init}$ cannot be dropped. The problem can still be solved, although not as conveniently as in Example 4.9.

Example 4.10. The vessel in Example 4.9 had a final pressure of 20 psia and a temperature of 161°F. The internal energy per unit mass in the container was 92.0 Btu/lbm; the specific volume, 2.8 ft³/lbm. Now we connect the container to another Freon 12 line, in which the pressure is 40 psia and the temperature is 200°F. We open the valve and let gas flow in until the pressure in the vessel is 40 psia. What is the final temperature in the vessel?

Equation 4.46 applies here. Let us select the size of the vessel as 1 ft³; this is an arbitrary choice for calculation purposes and will not influence the answer. The mass initially in the vessel is

$$m = \frac{V}{v} = \frac{1\ \text{ft}^3}{2.8\ \text{ft}^3 \cdot \text{lbm}} = 0.357\ \text{lbm} = 0.162\ \text{kg}$$

$$[(mu)_{init}]_{sys} = 0.357\ \text{lbm} \cdot 92.0\ \text{Btu/lbm} = 32.8\ \text{Btu}$$

From the chart we find h_{in} as 107.2 Btu/lbm. Now we begin our trial and error by assuming a value of Δm_{in}; let us assume it is 0.2 lbm. This makes the final mass in the container 0.357 + 0.2 = 0.557 lbm and the final specific volume

$$v = \frac{V}{m} = \frac{1\ \text{ft}^3}{0.557\ \text{lbm}} = 1.80\ \frac{\text{ft}^3}{\text{lbm}}$$

Solving Eq. 4.46 for u_{fin}, We find

$$u_{fin} = \frac{h_{in}\,\Delta m_{in} + (mu)_{init}}{m_{fin}}$$

$$= \frac{(102.2\ \text{Btu/lbm} \cdot 0.2\ \text{lbm}) + 32.8\ \text{Btu}}{0.557\ \text{lbm}} = 95.6\ \frac{\text{Btu}}{\text{lbm}}$$

Now, for a check of whether we have made the correct assumption, we see that three properties of the final state are known, i.e., the pressure, the internal energy per unit mass, and the specific volume. If we have made the correct assumption, these must be consistent. In App. A.2 for 40 psia and 1.8 ft³/lbm, we read an enthalpy of 134 Btu/lbm. The corresponding internal energy is

$$u = h - Pv$$

$$= 134\ \frac{\text{Btu}}{\text{lbm}} - 40\ \frac{\text{lbf}}{\text{in}^2} \cdot 1.8\ \frac{\text{ft}^3}{\text{lbm}} \cdot 144\ \frac{\text{in}^2}{\text{ft}^2} \cdot \frac{\text{Btu}}{778\ \text{ft} \cdot \text{lbf}}$$

$$= 134 - 13.3 = 121\ \text{Btu/lbm}$$

This does not agree with the value of 95.6 calculated above, so we now proceed by trial and error.

Trial no.	$\Delta m_{\text{assumed}}$, lbm	m_{fin}, lbm	v_{fin}, ft^3/lbm	u_{fin} (from Eq. 4.46), Btu/lbm	h_{fin} (from P_{fin} and v_{fin}), Btu/lbm	u_{fin} (from P_{fin} and v_{fin}), Btu/lbm
1	0.20	0.557	1.80	95.6	134	121
2	0.30	0.657	1.52	98.6	116	105
3	0.35	0.707	1.41	99.5	106	96

The third trial is approximately correct, and we read from App. A.2 that T_{fin} equals about 190°F (88°C). ■

The preceding two examples are much simpler if the fluid flowing in is a perfect gas. In that case the results can be calculated without trial and error (Prob. 4.22).

The two examples above were done under the assumption that the fluid flowing in has a constant h_{in}. If the fluid is flowing out of a vessel, then h_{out} usually is not constant; however, Eq. 4.45 still holds. In any but the simplest problems, it is necessary to integrate this equation numerically to find a solution.

We discuss one more unsteady-state problem after we have shown how some of the restrictions in Eq. 4.34 can be lifted.

4.13 LESS RESTRICTED SYSTEMS

In the formulation of Eq. 4.34 several restrictions were made. Now we consider under what circumstances we might lift these restrictions.

A. Multiple Flows

We first wrote Eq. 4.34 for one flow in and one flow out. We can generalize the equation to apply to any number of flows in or out by placing summation signs before the flow-in and flow-out terms:

$$d\left[m\left(u + gz + \frac{V^2}{2}\right)\right]_{\text{sys}} = \sum_{\text{all flows in}} \left(h + gz + \frac{V^2}{2}\right)_{\text{in}} dm_{\text{in}} - \sum_{\text{all flows out}} \left(h + gz + \frac{V^2}{2}\right)_{\text{out}} dm_{\text{out}} + dQ - dW_{\text{a.o.}} \tag{4.50}$$

Example 4.11. A fluid mixer has the flow diagram shown in Fig. 4.11. The flow is steady and horizontal. There is a heat leak out of the system of 2.0 Btu/lbm of stream C. Stream C is Freon 12 at 20 psia and 100°F. Stream B is Freon 12 at 20 psia and 140°F. Stream A is saturated liquid Freon 12 at 20 psia. If stream C flows at the rate of 1 lbm/s, what are the flow rates of streams A and B?

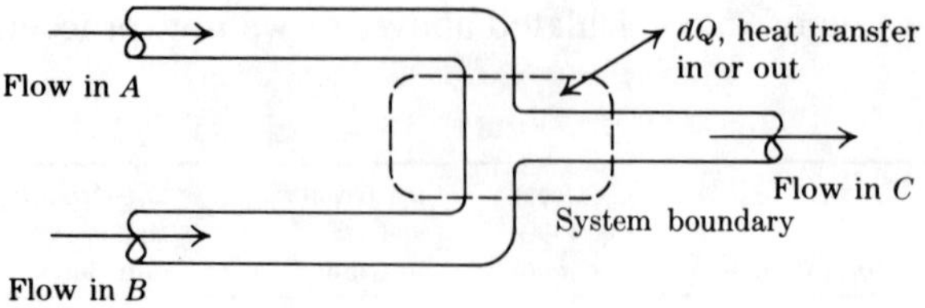

FIGURE 4.11
Simple mixer.

For this problem we assume that the kinetic energies and changes in potential energy are negligible. Thus, Eq. 4.50 reduces to

$$0 = \sum_{\text{all flows in}} h_{\text{in}}\, dm_{\text{in}} - \sum_{\text{all flows out}} h_{\text{out}}\, dm_{\text{out}} + dQ$$

$$= h_A\, dm_A + h_B\, dm_B - h_C\, dm_C + dQ$$

In addition, the mass balance shows that

$$dm_B = dm_C - dm_A$$

We substitute this into the energy balance to eliminate dm_B:

$$0 = h_A\, dm_A + h_B\,(dm_C - dm_A) - h_C\, dm_C + dQ$$

We now solve this for dm_A:

$$dm_A = \frac{dm_C\,(h_B - h_C) + dQ}{h_B - h_A}$$

and finally we divide by dm_C:

$$\frac{dm_A}{dm_C} = \frac{h_B - h_C + dQ/dm_C}{h_B - h_A}$$

From App. A.2 we find that $h_A = 6.5$ Btu/lbm, $h_B = 98.0$ Btu/lbm, and $h_C = 92.0$ Btu/lbm. From the problem statement we know that $dQ/dm_C = -2.0$ Btu/lbm; therefore,

$$\frac{dm_A}{dm_C} = \frac{(98.0 - 92.0) + (-2.0)}{98.0 - 6.5} = \frac{4.0\ \text{Btu/lbm}}{91.5\ \text{Btu/lbm}} = 0.043$$

The problem statement asks for the flow rate of A, $\dot{m}_A$. We can multiply the top and bottom of the last equation by $\dot{m}_C = 1$ lbm/s and rearrange to get

$$\dot{m}_A = \frac{dm_A}{dm_C}\,\dot{m}_C = 0.043 \cdot 1\,\frac{\text{lbm}}{\text{s}} = 0.043\,\frac{\text{lbm}}{\text{s}}$$

$$\dot{m}_B = \dot{m}_A - \dot{m}_C = 1.0 - 0.043 = 0.957\ \text{lbm/s}$$ ■

Note that in this problem only enthalpy and heat-transfer terms appeared. In many such problems there are negligible changes in potential and kinetic energies and negligible $dW_{\text{a.o.}}$, with the result that only "heat terms" appear. For this reason such problems are often referred to as "heat balance" or "enthalpy balance" problems. Although this terminology is commonly used, it is misleading because it implies that heat or enthalpy is conserved. They are conserved only in a very limited class of problems; an "enthalpy balance" around a steam turbine would lead only to confusion.

B. Nonhomogeneous Systems

In Eq. 4.34 it was assumed that the contents of the system were homogeneous, i.e., that each infinitesimal bit of matter in the system had the same values of u, z, and V as all the other infinitesimal bits of matter. In complex systems that is not the case, and the accumulation term should be written as

$$d \int_{\text{all mass}} \left(u + gz + \frac{V^2}{2} \right) dm$$

If these properties vary continuously over the entire mass in the system in some simple way, the resulting problems may be solved analytically, but most often the mathematics are too difficult. However, many complex systems can be conceptually divided into a group of subsystems, each of which is homogeneous. Then for a nonhomogeneous system made up of homogeneous subsystems, we can rewrite Eq. 4.34 as

$$\sum_{\substack{\text{all homogeneous} \\ \text{subsystems}}} d\left[m\left(u + gz + \frac{V^2}{2} \right) \right]_{\text{subsys}}$$

$$= \left(h + gz + \frac{V^2}{2} \right)_{\text{in}} dm_{\text{in}} - \left(h + gz + \frac{V^2}{2} \right)_{\text{out}} dm_{\text{out}} + dQ - dW_{\text{a.o.}} \quad (4.51)$$

Example 4.12. A steel bottle, shown in Fig. 4.12, contains saturated liquid and vapor Freon 12 at 180°F. The bottle is in a tank of water which transfers heat in or out as needed to keep the temperature constant at 180°F. Now 1 lbm is allowed to flow slowly out through the partially opened valve. How much heat is transferred from the water to the bottle, or vice versa?

Here we take as our subsystems the liquid and gas phases in the bottle, and we assume that the changes in kinetic and potential energies are negligible. Then Eq. 4.51 reduces to

$$d(mu)_{\text{liq}} + d(mu)_{\text{gas}} = -h_{\text{out}}\, dm_{\text{out}} + dQ$$

However, here the temperature in the vessel is held constant, so u_{liq} and u_{gas} are constants. Furthermore, since all the gas flows out saturated at a constant pressure, h_{out} is constant and equal to h_{gas}. Thus, we can integrate to

$$u_{\text{liq}}\, \Delta m_{\text{liq}} + u_{\text{gas}}\, \Delta m_{\text{gas}} = -h_{\text{gas}}\, \Delta m_{\text{out}} + \Delta Q$$

This equation contains three unknowns: ΔQ, Δm_{liq}, and Δm_{gas}. To solve it, we need two other relations. One is found from the mass balance:

$$\Delta m_{\text{liq}} + \Delta m_{\text{gas}} = -\Delta m_{\text{out}}$$

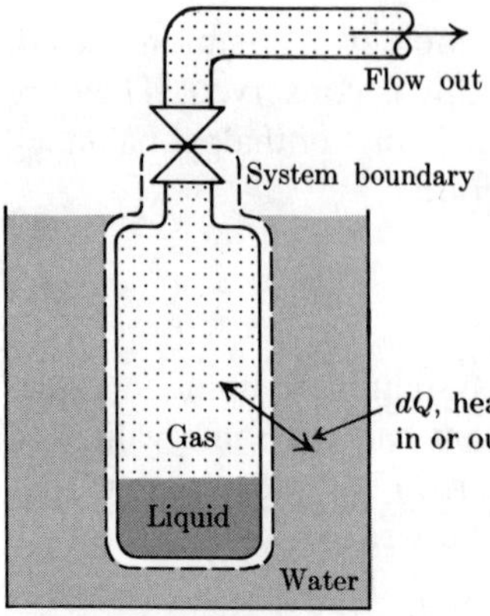

FIGURE 4.12

The minus sign is included because we have chosen Δm_{out} as a positive quantity. The other relation is found from the fact that the total volume of the two phases does not change (it is assumed that the bottle is rigid). This gives

$$\Delta V_{liq} + \Delta V_{gas} = 0 = v_{liq}\,\Delta m_{liq} + v_{gas}\,\Delta m_{gas}$$

Solving this equation for Δm_{gas} and inserting the latter in the mass-balance equations yield

$$\Delta m_{liq}\left(1 - \frac{v_{liq}}{v_{gas}}\right) = -\Delta m_{gas}$$

$$\Delta m_{gas} = -\Delta m_{out} - \Delta m_{liq} = -\Delta m_{out}\left(1 - \frac{v_{gas}}{v_{gas} - v_{liq}}\right)$$

Substituting these into the energy balance yields

$$u_{liq}\left(\frac{v_{gas}}{v_{gas} - v_{liq}}\right)(-\Delta m_{out}) + u_{gas}\left(1 - \frac{v_{gas}}{v_{gas} - v_{liq}}\right)(-\Delta m_{out}) = -h_{gas}\,\Delta m_{gas} + \Delta Q$$

Dividing by Δm_{out} and rearranging, we find

$$\frac{\Delta Q}{\Delta m_{out}} = -u_{liq}\left(\frac{v_{gas}}{v_{gas} - v_{liq}}\right) + u_{gas}\left(\frac{v_{liq}}{v_{gas} - v_{liq}}\right) + h_{gas}$$

From App. A.2 we can find $v_{liq} = 0.015\ \text{ft}^3/\text{lbm}$, $v_{gas} = 0.103\ \text{ft}^3/\text{lbm}$, $h_{gas} = 91.56\ \text{Btu/lbm}$, $h_{liq} = 52.56\ \text{Btu/lbm}$, and $P = 349$ psia. From these we can calculate $u_{gas} = 84.71\ \text{Btu/lbm}$ and $u_{liq} = 51.57\ \text{Btu/lbm}$. Substituting these, we find

$$\frac{\Delta Q}{\Delta m} = \left(-51.57\,\frac{\text{Btu}}{\text{lbm}} \cdot \frac{0.103}{0.103 - 0.015}\right) + \left(84.71\,\frac{\text{Btu}}{\text{lbm}} \cdot \frac{0.015}{0.103 - 0.015}\right)$$
$$+ 91.56\,\frac{\text{Btu}}{\text{lbm}}$$
$$= -60.4 + 14.4 + 91.6 = 45.6\,\frac{\text{Btu}}{\text{lbm}} = 25.3\,\frac{\text{kcal}}{\text{kg}} = 106\,\frac{\text{kJ}}{\text{kg}}$$ ■

C. Variable Gravity

In the development of Eq. 4.34, we assumed that the gravitational field was constant. This made the potential-energy term equal to gz. If the *acceleration of gravity is not constant*, then we must return to Eq. 4.11 and state the potential-energy term as

$$\text{pe} = \int_{z=0}^{z=z} g\,dz \tag{4.52}$$

Making this substitution in Eq. 4.34, we find

$$d\left[m\left(u + \int_{z=0}^{z=z} g\,dz + \frac{V^2}{2}\right)\right]_{\text{sys}} = \left(h + \int_{z=0}^{z=z} g\,dz + \frac{V^2}{2}\right)_{\text{in}} dm_{\text{in}} - \left(h + \int_{z=0}^{z=z} g\,dz + \frac{V^2}{2}\right)_{\text{out}} dm_{\text{out}} + dQ - dW_{\text{a.o.}} \tag{4.53}$$

Example 4.13. If a planet has no atmosphere, a projectile leaving its surface with a high enough velocity will escape its gravitational field. The minimum velocity for such escape is called the *escape velocity*. Calculate the escape velocity for the earth, assuming the earth has no atmosphere. Here

$$\frac{g}{g_0} = \left(\frac{z_0}{z}\right)^2$$

where g = local acceleration of gravity (This equation is correct above the surface of the planet, but not inside it!)
g_0 = acceleration of gravity at earth's surface = 32.2 ft/s^2 = 9.81 m/s^2
z = distance from center of earth
z_0 = radius of the earth = 4000 mi = 6440 km

We choose as our system the projectile. We assume that there is no heat transfer to or from the projectile, the internal energy of the projectile does not change, and the mass of the projectile is constant. Then Eq. 4.53 reduces to

$$m\,d\left(\int_0^z g\,dz + \frac{V^2}{2}\right)_{\text{sys}} = 0$$

We now divide out the mass of the projectile and integrate from the surface of the earth to an infinite distance away from the earth:

$$\int_0^\infty g\,dz - \int_0^{z_0} g\,dz = \int_{z_0}^\infty g\,dz = -\left[\left(\frac{V^2}{2}\right)_\infty - \left(\frac{V^2}{2}\right)_{z_0}\right]$$

The lowest surface velocity at which the projectile escapes is the one for which the velocity goes to zero as the force of gravity goes to zero, or $V = 0$ at $z = \infty$. We now substitute the expression for g and perform the integration:

$$\int_{z_0}^{\infty} g\,dz = g_0 z_0^2 \int_{z_0}^{\infty} \frac{dz}{z^2} = g_0 z_0^2 \left(\frac{-1}{z}\right)\Bigg]_{z_0}^{\infty} = g_0 z_0 = \left(\frac{V^2}{2}\right)_{z_0}$$

Solving for the escape velocity, we find

$$V_{z_0} = \sqrt{2 g_0 z_0} = \sqrt{2 \cdot \frac{32.2\text{ ft}}{\text{s}^2} \cdot 4000\text{ mi} \cdot \frac{5280\text{ ft}}{\text{mi}}} = 36{,}880\ \frac{\text{ft}}{\text{s}} = 11{,}244\ \frac{\text{m}}{\text{s}}$$ ■

This value is high enough that humans were unable to send projectiles out of earth's gravity until the late 1950s. The escape velocity also explains why planets like earth have atmospheres, while our moon does not. The earth's escape velocity is higher than the maximum velocities of gas molecules at the top of the atmosphere, so few molecules escape. For the moon the escape velocity is smaller than the velocities gas molecules regularly have, so that a gas molecule at the top of the atmosphere, if there were one, which was moving outward at its maximum thermal velocity would be able to escape into outer space.

If we calculate the last problem directly from Eq. 4.34 without making this correction for nonconstant gravity, we get

$$V_{z_0} = \sqrt{-2 g_0 z_0}$$

which is obviously incorrect. (What would an imaginary velocity mean?)

Another significant nonconstant-gravity application is the motion of a body in a centrifugal force field, as in a centrifuge. In such a device the change in potential energy of a body is given by

$$d(\text{pe}) = -\omega^2 r\,dr \tag{4.54}$$

where ω is the angular velocity and r is the radius. Comparing Eq. 4.54 with Eq. 4.11, we see that $-\omega^2 r$ plays the same role as g for gravitational potential energy; however, it is not a constant for different radii, so the potential-energy term for a centrifugal force field must enter in the form

$$d\left[m\left(u + gz + \frac{V^2}{2}\right)\right]_{\text{sys}} - d\int_{\text{sys}} \omega^2 r\,dm = \text{etc.} \tag{4.55}$$

4.14 OTHER FORMS OF WORK AND ENERGY

So far we have discussed only $F\,dx$ work and kinetic, potential, and internal energies. In this section we consider some other kinds of work and energy.

In most modern machinery, work is done by rotating shafts. In a simple crank arrangement (Fig. 4.13), a force is being exerted on a crank and is being resisted by the shaft to which the crank is attached; the torque Γ in the shaft is given by

$$\Gamma = FL \tag{4.56}$$

If we now allow the shaft to move, with the force always being applied at right angles to the crank, the distance moved is

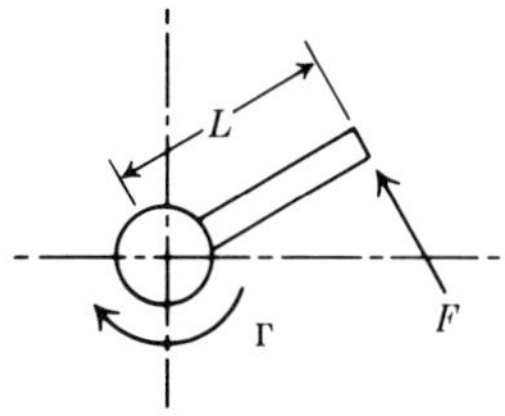

FIGURE 4.13
Relation between force, lever arm, and torque.

$$dx = L\, d\theta \tag{4.57}$$

where $d\theta$ is the angular displacement about the center, in radian measure. Solving Eq. 4.56 for F and Eq. 4.57 for dx, we find

$$dW = F\, dx = \left(\frac{\Gamma}{L}\right) L\, d\theta = \Gamma\, d\theta \tag{4.58}$$

The power is given by

$$\text{Po} = \frac{dW}{dt} = \Gamma \frac{d\theta}{dt} = \Gamma \omega \tag{4.59}$$

Here we see that the power output of a rotating device is the product of its torque Γ and its rotating speed ($\omega = 2\pi \cdot$ revolutions per minute or per second). Generally, the torque that a device can develop is roughly proportional to its size; so to have a given amount of power, we can use a large low-speed device or a small high-speed device. In speedboat, automobile, and airplane power plants, where low weight is important, the trend of the past 150 years has been toward higher and higher rotational speeds, to get lighter and lighter engines. In large stationary engines like those in diesel electric power plants, where weight is of no serious concern, the choice has been to use a large, slow-turning engine (generally less than 1000 rpm), to minimize frictional resistance. Dental drill turbines, which are very small, rotate at 350,000 rpm! Automobile superchargers are also small and rotate at up to 200,000 rpm.

Much of the work of modern societies is done electrically. It is shown in texts on electricity that the force required to move a charge $\mathcal{Q}$ in an electric field is

$$F = \mathcal{Q} \frac{dE}{dx} \tag{4.60}$$

Here dE/dx is the potential (or voltage) gradient. Substituting this in Eq. 4.1, we find

$$W = \int F\, dx = \int \mathcal{Q} \frac{dE}{dx}\, dx \tag{4.61}$$

For any fixed amount of charge, we can integrate this to find

$$W = \mathcal{Q}\, \Delta E \tag{4.62}$$

If we now consider that both the voltage difference and the charge can vary, we can differentiate this equation and find

$$dW = \mathcal{Q}\ d(\Delta E) + \Delta E\ d\mathcal{Q} \tag{4.63}$$

Ordinarily we take either of two views of electric flows. One view is to consider some fixed piece of equipment with a steady flow of electrons through it and a fixed voltage difference across it; this corresponds to a steady-flow open system with respect to electrons (or to "charge," which by the usual conventions is the negative of electrons). In this case, since nothing is changing with time, $d(\Delta E)$ is zero and

$$dW = \Delta E\ d\mathcal{Q} \tag{4.64}$$

Dividing this equation by dt, we have

$$\text{Po} = \frac{dW}{dt} = \frac{d\mathcal{Q}}{dt}\ \Delta E = I\ \Delta E \tag{4.65}$$

here $d\mathcal{Q}/dt = I$ is the current, so this is the familiar statement that the electric power is the product of the voltage difference and the current. This is our usual way of looking at motors, generators, electric cells, and so on.

The other way of looking at electric flows is to consider some fixed amount of charge (or a fixed number of electrons). This corresponds to a closed system for electrons. In this case $d\mathcal{Q}$ is zero, and Eq. 4.63 becomes

$$dW = \mathcal{Q}\ d(\Delta E) \tag{4.66}$$

We normally consider one of the voltages to be a fixed ground voltage, to which we assign the arbitrary value of zero, so that the $d(\Delta E)$ becomes simply dE. Equation 4.66, then, is the usual way of regarding capacitors, cathode-ray tubes, and electron ballistics in general.

So far we have treated energy and mass as two completely distinct entities. In most engineering problems this is a satisfactory approximation, but it is impossible to understand nuclear reactors or atomic explosions or the behavior of the sun without taking into account the conversion of matter to energy. Einstein has shown that this conversion may occur and that when it does, it obeys the rule

$$E = mc^2 \tag{4.67}$$

where c is the speed of light in a vacuum. This equation indicates that c^2 must have the dimensions of energy divided by mass. Simple calculations show that

$$c^2 = 3.85 \times 10^{13}\ \text{Btu/lbm} = 8.95 \times 10^{16}\ \text{J/kg} \tag{4.68}$$

In the statement of Eq. 4.34 we specifically excluded such nuclear effects. We also indicated that mass is conserved. What is really true is that mass and energy together obey a conservation law, with neither creation nor destruction. In this case we could write a mass balance and add it to the energy balance:

$$dm_{\text{sys}} = dm_{\text{in}} - dm_{\text{out}}$$

We now multiply both sides of this equation by c^2 and add it to Eq. 4.34:

$$d\left[m\left(u + gz + \frac{V^2}{2}\right)\right]_{\text{sys}} + c^2\, dm_{\text{sys}} = \left(h + gz + \frac{V^2}{2} + c^2\right)_{\text{in}} dm_{\text{in}}$$
$$- \left(h + gz + \frac{V^2}{2} + c^2\right)_{\text{out}} dm_{\text{out}} + dQ - dW_{\text{a.o.}} \qquad (4.69)$$

Obviously, if there is no conversion of mass to energy, the c^2 terms added here cancel, and we find the same result as that from Eq. 4.34.

Example 4.14. A nuclear power plant is running steadily. It produces 2.5×10^8 W of electric power and rejects 7.5×10^8 W of heat to the cooling water taken from a nearby river. How much matter is being converted to energy each hour?

We choose as our system the complete power plant, excluding the cooling water passing through it. Then in the period between fuel refuelings there is no mass flow into or out of the system (we overlook such things as boiler feedwater makeup due to leaks or water-treating chemicals, since they are insignificant). Furthermore, at steady state the internal, potential, and kinetic energies of the various parts of the plant are not changing (again, this is slightly inaccurate, because chemical changes accompany nuclear fission, but the error is insignificant). Thus, the only terms remaining in Eq. 4.69 are

$$c^2\, dm_{\text{sys}} = dQ - dW_{\text{a.o.}}$$

We divide by dt and solve for dm/dt:

$$\frac{dm}{dt} = \frac{dQ/dt - dW_{\text{a.o.}}/dt}{c^2}$$
$$= \frac{(-7.5 \times 10^8 \text{W}) - (2.5 \times 10^8 \text{ W})}{8.95 \times 10^{16} \text{ J/kg}} \cdot \frac{\text{J} \cdot \text{s}}{\text{W}}$$
$$= -1.12 \times 10^{-8} \text{ kg/s} = -8.9 \times 10^{-5} \text{ lbm/h}$$

This is a large power plant; it converts less than 1/10,000 lbm/h to energy. ■

Einstein's relativity theory shows not only that mass and energy can be converted from one to the other but also that radiant energy (such as light and infrared radiation) has mass. This was demonstrated in classic experiments which showed that light beams passing close to the sun are deflected by the sun's gravity field. The theory shows that the mass of such radiation is proportional to its energy, and the proportionality is given by Eq. 4.68. Based on this and other arguments, physics teaches that in *all* energy transactions there is a change in mass, with the proportionality given by Eq. 4.68.

Example 4.15. A sample of water has a mass of exactly 1.0 lbm. It is heated over a hotplate from 59.5 to 60.5°F. By how much, if any, does its mass increase?

We have chosen the problem so that Eq. 4.34 indicates that

$$m\,du = dQ = 1\ \text{Btu}$$

Dropping the unnecessary terms from Eq. 4.69, we have

$$\Delta m = \frac{\Delta Q}{c^2} = \frac{1\ \text{Btu}}{3.85 \times 10^{13}\ \text{Btu/lbm}} = 2.6 \times 10^{-14}\ \text{lbm} = 5.7 \times 10^{-14}\ \text{kg}$$

Thus we conclude that the final mass of the water is 1.000,000,000,000,026 lbm. ■

Detecting the difference between this mass and a 1-lbm mass would be a serious measurement challenge. However, because Einstein's theory works whenever tests are possible, physicists are confident that such measurements, if made, would show this result. Obviously, for most engineering problems we may neglect this effect.

Other kinds of energy generally considered are surface, electrostatic, and magnetic energies. Although they are easily described, their mathematical formulation is more difficult than that of kinetic or potential energy. The reason is that a change in surface or electrostatic field or magnetic field usually is accompanied by an absorption or rejection of heat or by a change in internal energy, whereas we can represent kinetic energy by a single term, because we can increase a body's kinetic energy without any exchange of heat with the surroundings or any change in internal energy. The usual treatment of the other energies requires simultaneous applications of the first and second laws of thermodynamics [5] and hence is beyond the scope of this chapter. Although electrostatic and magnetic energies play little role in fluid mechanics, and will not be discussed in the rest of this book, they are very important in other areas. Xerographic copying is based on electrostatic energy, and computer memories are based on magnetic energy. An intuitive introduction to surface energy is given in Chap. 17.

4.15 LIMITATIONS OF THE FIRST LAW

The first law of thermodynamics is a conservation law. It *accounts for* the quantity called energy. It says nothing about the direction in which energy changes occur. It is equally well satisfied with water flowing downhill and water flowing uphill. It makes no distinction between gas flowing out of a high-pressure vessel into the atmosphere and gas flowing from the atmosphere into a high-pressure vessel. Recall Example 4.6, in which Freon 12 flowed steadily through a valve with an upstream pressure of 100 psia and a downstream pressure of 20 psia. The first law is satisfied by this process. It would also be satisfied if we interchanged the words "upstream" and "downstream" in that example.

To see whether we can make such an interchange, we must rely on another basic principle of nature, the second law of thermodynamics.

4.16 SUMMARY

1. The first law of thermodynamics resulted from a study of friction heating.
2. This study led to the definition of an abstract quantity called energy and to the statement that (excluding nuclear reactions) energy follows the balance principle, with neither creation nor destruction.
3. All energy quantities are measured relative to some arbitrary datum.
4. The first law of thermodynamics is not derivable or provable; its validity rests solely on its ability to predict the outcome of all the experiments ever run to test it.
5. The first law of thermodynamics can be used to solve an enormous number of problems.
6. The first law of thermodynamics says nothing about the direction of an energy change; that is covered by the second law of thermodynamics.

PROBLEMS

See the Common Units and Values for Problems and Examples inside the back cover. Several problems in this section deal with perfect gases. It may be shown that for a perfect gas the enthalpy and internal energy depend on temperature alone. If a perfect gas has a constant heat capacity (which may be assumed in all the perfect-gas problems in this chapter), it is very convenient to choose an enthalpy datum that leads to $h = C_P T$ and $u = C_V T$, where T is the absolute temperature; these values may be used in the perfect-gas problems in this chapter. For Freon 12 problems, use App. A.2. For steam and CO_2 problems, use any standard table of values.

4.1. The groups $u + gz + V^2/2$ and $h + gz + V^2/2$ occur in most thermodynamics problems. To evaluate the relative magnitude of the individual terms, calculate gz and $V^2/2$ in Btu per pound-mass and joules per kilogram for the following: $z = 10, 100, 1000, 10{,}000$ ft; $V = 10, 100, 1000, 10{,}000$ ft/s. Show these results on a log-log plot.

4.2. A gun fires a bullet vertically upward. The bullet has a mass of 0.1 lbm and leaves the gun at a velocity of 2000 ft/s. Air resistance is negligible.
(*a*) How much kinetic energy does the bullet possess when it leaves the gun?
(*b*) How high will the bullet go?
(*c*) At its highest point, how much potential energy will it have relative to the gun barrel?

4.3. On the moon $g = 6$ ft/s^2. How much work is required to raise a 2.0-lbm ball of steel 10 ft?

4.4. A steel ball with a mass of 3.0 kg is dropped from an airplane and goes 500 m in free fall. Air resistance may be neglected. How much work is done on the ball? What does the work?

4.5. A hydraulic lift is shown in Fig. 4.14. The combined mass of the piston, rack, and car is 4000 lbm. The working fluid is water. There is no heat transfer to or from the water, and the internal energy of the water per unit mass is constant. The water may be considered incompressible.

(*a*) Taking all the water in the reservoir, line, and hydraulic cylinder as the system (i.e., taking the closed-system approach), calculate the work necessary to raise the rack and car 1 ft (neglect the change in potential energy of the water in the system).

(*b*) Taking all the water plus the car and the rack as the system, repeat (*a*).

(*c*) Repeat (*a*), taking an open-system approach; choose as your system the volume of the hydraulic cylinder, excluding the piston, rack, and car. If the pressure is 1000 psia, calculate the volume that must flow in to raise the car 1 ft.

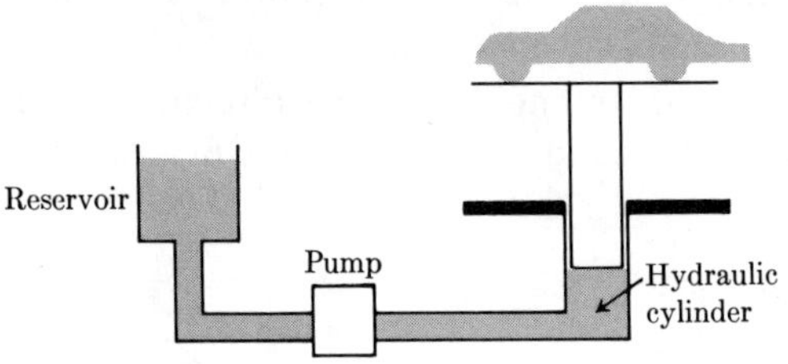

FIGURE 4.14

4.6. A flexible balloon is to be filled with steam. The balloon is constructed of a perfect insulating material, so no heat is transferred from the steam to the balloon. Originally the balloon has zero internal volume. At the end of the process, the balloon has an internal volume of 100 m^3. During the entire process the balloon is surrounded by the atmosphere. The skin of the balloon is flexible, like an accordion, so that no work is needed to stretch it. The steam is taken from a steam line, which always has steam at 300°C and 1000 kPa absolute. What is the temperature of the steam in the balloon at the end of the process?

4.7. Estimate the internal energy per unit mass u in Btu per pound-mass for (*a*) steam at 1600°F and 5000 lbf/in^2, (*b*) water at 32°F and 5000 lbf/in^2, (*c*) Freon 12 at its critical point, and (*d*) carbon dioxide at its critical point. In each case, indicate the datum for your estimate. *Note:* It is satisfactory to give an enthalpy datum.

4.8. One hundred kilograms of steam is contained in a cylinder with a frictionless, zero-mass piston, the other side of which is exposed to the atmosphere. The steam is initially 100 percent quality at 100°C. Enough heat is transferred through the cylinder walls to reduce the quality to 25 percent. How much heat was transferred? Which way?

4.9. An absolutely rigid vessel contains 1.0 lbm of water in the form of a vapor-liquid mixture of 3.1 percent quality. The pressure is 1000 psia. How much heat must be added to or subtracted from this mixture to make it all liquid (zero quality)? Make clear in your answer whether heat is added or subtracted.

4.10. The piston shown in Fig. 4.15 is absolutely frictionless and weightless. The cylinder contains 1 kg of H_2O. Originally the temperature is 101°C. The bottom

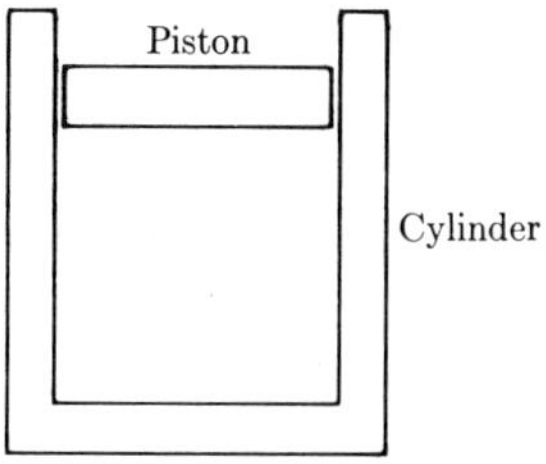

FIGURE 4.15

part of the cylinder is now placed in a constant-temperature bath at 99°C and allowed to come to equilibrium. When the whole system is at 99°C, how much heat has flowed out of it?

4.11. At the Hoover Dam the difference in elevation from lake level to stream below the generators is about 750 ft. With 100 percent efficient machines, how many kilowatt hours can be recovered per kilogram of water passing through the system?

4.12. A gas-liquid mixture of Freon 12 of 90 percent quality at a pressure of 200 psia is passed through a throttling valve. Flow is steady, and the exit pressure is 20 psia. The outlet velocity is very small. What is the outlet temperature?

4.13. A boiler feed pump takes water from a tank at 95°C and 100 kPa absolute and delivers it at 190°C and 2000 kPa absolute. The flow is steady, and the heat loss from the pump is 2 kJ/kg of water passing through it. What is the work input to the pump in kilojoules per kilogram of water passing through it?

4.14. In a power plant the inlet water has a velocity of 100 ft/s and an elevation of 80 ft above the outlet of the plant. The outlet water has a velocity of 5 ft/s. How much work can be extracted by the power plant per unit mass of water passing through the system? The enthalpy of the outlet water may be assumed to be identical to the enthalpy of the inlet water.

4.15. Water flows steadily through a power plant. The enthalpy of the outlet water is the same as that of the inlet water. The water inlet is 40 m above the water outlet. Both inlet and outlet are at atmospheric pressure. The inlet velocity is 9 m/s, and the outlet velocity is 15 m/s. The flow rate is 5000 kg/s. What is the power output of the plant?

4.16. A perfect gas is flowing steadily in a horizontal, adiabatic nozzle. The inlet conditions are $T_1 = 600°F$ and $V_1 = 300$ ft/s. The outlet velocity is 2000 ft/s. The heat capacity C_P is 0.3 Btu/(lbm · °F). What is the temperature of the gas leaving the nozzle?

4.17. A steady-flow steam turbine has the following inlet and outlet streams:

	Inlet	**Outlet**
Temperature, °C	260	140
Pressure, kPa	700	100
Velocity, m/s	Negligible	200

There is heat leakage out of the turbine equal to 20 kJ/kg of steam passing through. How much work does the turbine deliver per kilogram of steam passing through it?

4.18. Steam is available in a large container at 600°F and 250 psia. It is passed through an adiabatic throttling valve into a large-diameter pipe, where its pressure is 14.7 psia and its velocity is negligible. What is its temperature?

4.19. An insulated tank is initially evacuated. It is connected by a valve to a line which is carrying Freon 12 at 50 psia and 100°F. The line is so large that the kinetic energy of the gas in the line is negligible, and the flow into the tank will not affect the temperature and pressure in the line. The valve is now opened and left open until the pressure in the tank is 50 psia. If the contents of the tank are perfectly mixed, what is their temperature?

4.20. An insulated, evacuated 50-m^3 tank is attached to a constant-pressure line containing steam at 500 kPa absolute with a quality of 95 percent. This steam is allowed to flow slowly into the tank. Neglect heat losses and assume that the metal of the tank has negligible heat capacity.

(*a*) How many kilograms of steam will have entered the tank just as its pressure reaches line pressure?

(*b*) What would have been the temperature in the tank if the valve had been closed when the tank pressure reached 300 kPa absolute?

4.21. The rigid vessel shown in Fig. 4.16 has a volume of 1 ft^3 and contains an ideal gas. The temperature is 100°F. The input to the heater is regulated to hold the temperature constant at 100°F. Originally it is at a pressure of 100 psia. The valve is partly opened, and the gas is allowed to escape slowly. When the pressure in the vessel has fallen to 20 psia, the valve is closed. How much heat was added by the heater during this period?

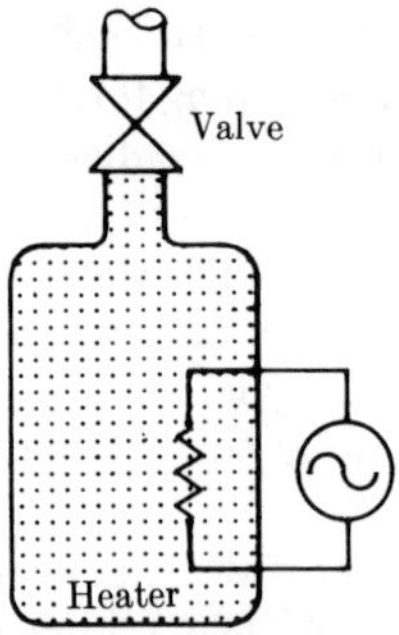

FIGURE 4.16

4.22. Derive the general solution of adiabatic bottle-filling problems (Examples 4.9 and 4.10) for perfect gases. (Here "general" means including the case in which the bottle is not originally empty.) Present your answer as T_f = function of (T_{in}, etc.). Here the initial and final masses may not appear in the solution, but the final and initial pressures may.

4.23. A rigid, adiabatic container contains air at 0.5 atm and 20°C. The surrounding air is at 1.0 atm and 20°C. We now open a valve between the container and the surroundings and allow air to flow in until the inside and outside pressures are equal. What is the final temperature of the air in the container? (The "general solution" to perfect-gas bottle-filling problems, worked out in Prob. 4.22, may be useful here.)

4.24. The rigid, insulated vessel in Fig. 4.17 contains 100 kg of liquid water and a

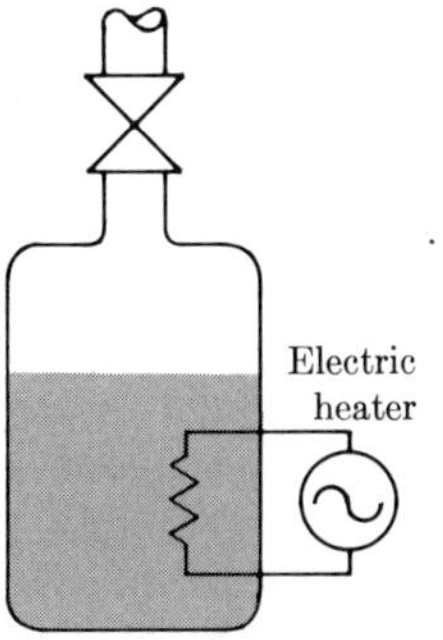

FIGURE 4.17

negligible mass of steam, both in equilibrium at 300 kPa absolute. The valve is opened, and 10 kg of steam is allowed to escape. The electric heater input is regulated to hold the saturation pressure in the vessel constant. When the process is over, the vessel contains 90 kg of water and a negligible mass of steam. How much energy was added to the electric heater?

4.25. Steam of unknown quality and temperature was allowed to pass through a short piece of insulated hose into an open-top barrel of water, where it condensed. During the experiment the following data were collected:

Weight of barrel empty, 25 lbf
Weight of barrel and water initially, 400 lbf
Weight of barrel and water finally, 465 lbf
Temperature of water initially, 50°F
Temperature of water finally, 150°F
Pressure of steam in boiler, 25 psia
Total heat loss from water during process (including heat to warm barrel), 1600 Btu

What was the enthalpy of the steam entering the hose?

4.26. We connect an evacuated, rigid container to a steam line, which contains saturated steam at 150 psia. We allow steam to flow into the container, and at the same time we allow heat to leak out of the container. When the container pressure reaches 150 psia, the steam in the container is exactly at the saturated condition. How much heat was removed from the container per pound of steam admitted?

4.27. Solid carbon dioxide can be manufactured by expanding liquid CO_2 through an adiabatic throttling valve. If CO_2 is available for the process as a saturated liquid at 800 psia and the expansion reduces the pressure to 14.7 psia, how many pounds of solid carbon dioxide are produced per pound of liquid CO_2?

4.28. In rating the energy release of nuclear explosives, the Atomic Energy Commission uses the energy unit "kiloton," where 1 kton $= 10^{12}$ cal. This is roughly the energy release involved in detonating 10^3 tons of TNT. The Hiroshima bomb was reported to be about 20 ktons. How much matter in it was converted to energy?

4.29. A new secret weapon has been proposed for television spies: an assassination pistol that shoots a bullet of solid carbon dioxide. The idea is that this bullet, upon hitting its victim, will convert its kinetic energy to internal energy, thereby vaporizing and leaving no trace at all. The mass of the bullet will be 0.01 lbm. It will hit the victim as a saturated solid at −110°F. How fast must it be moving to totally vaporize, assuming that it expends no energy in deforming the victim and that no heat is transferred between the bullet and the victim?

4.30. Typical high explosives liberate about 8000 Btu/lbm of thermal energy on exploding. It has been suggested that a high-velocity projectile might liberate as much thermal energy on being stopped, by conversion of its kinetic energy to thermal energy. How fast must such a projectile be going in order for its kinetic energy, if all turned to internal energy, to be the same as that of the typical explosive described above? Under what circumstances could a projectile have this kind of velocity?

4.31. Figure 4.18 shows a simple combustion calorimeter. The sample is ignited electrically. After a few minutes the temperature of the water and calorimeter is constant at ΔT higher than the starting temperature. The heat of combustion is defined as

$$\Delta u_{\text{combustion}} = \frac{U_{\text{final products of combustion}} - U_{\text{initial fuel+oxygen}}}{\text{mass of fuel}}$$

Determine the heat of combustion of a sample from the following data:

Sample mass	4 g
Calorimeter mass	500 g
Water mass	5000 g
$C_{P\ \text{calorimeter}}$	0.12 cal/(g · °C)
$C_{P\ \text{water}}$	1.0 cal/(g · °C)
ΔT	5 C°

Ignore the heat capacity of the gases in the calorimeter. Make a list of possible sources of error in this experiment.

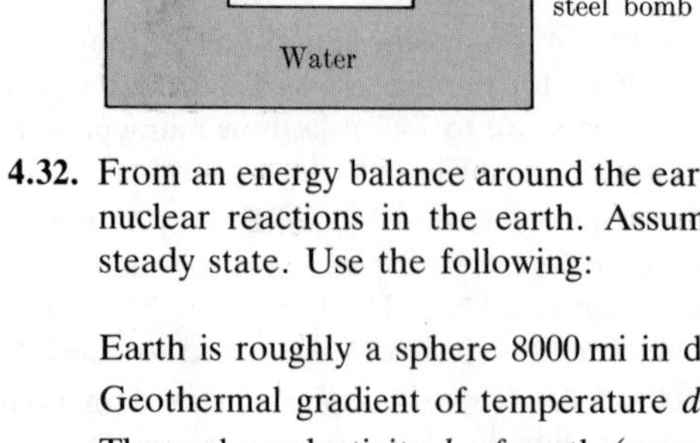

FIGURE 4.18
Combustion calorimeter.

4.32. From an energy balance around the earth, estimate the rate of energy liberated by nuclear reactions in the earth. Assume that heat losses from the earth are at steady state. Use the following:

Earth is roughly a sphere 8000 mi in diameter.
Geothermal gradient of temperature dT/dx is approximately 0.02°F/ft.
Thermal conductivity k of earth (near the surface) is about 1 Btu/(h · °F · ft).

Heat flow is estimated from $dQ/dt = kA\,(dT/dx)$, where A is area.

Also calculate the rate at which matter is being converted to energy in the earth.

4.33. A mixer has the flow diagram shown in Fig. 4.11. The heat leak dQ is 20 kJ removed per kilogram of stream C. There are no other heat losses or additions. The flow is steady; the kinetic and potential energies are negligible. What is the value of x?

	At A	At B	At C
Substance flowing	H_2O	H_2O	H_2O
Flow rate, kg/h	x	y	1.0
Pressure, kPa	1000	101.3	101.3
Condition	Saturated vapor; 100% quality	Saturated liquid; 0% quality	Saturated vapor; 100% quality

4.34. Work out the energy balance for a throttle valve (Eq. 4.39), using the closed-system form of the first law. Choose as your system 1 kg of material flowing down the line.

4.35. Write an energy balance for the sun, indicating which terms are probably important and which are probably negligible.

4.36. A steady-flow water power plant has the following inlet and outlet conditions:

	Inlet	Outlet
Pressure, psig	0	0
Elevation, ft	75	0
Velocity, ft/s	400	50
Temperature, °F	70.0	71.0

The plant is adiabatic. How much work does it deliver per pound-mass of fluid flowing through?

4.37. The container in Fig. 4.19 is initially evacuated. Then the valve is opened, and atmospheric air flows in, compressing the spring. The entire inflow and compression process is adiabatic. The temperature of the atmospheric air is 20°C. When the process is over, the volume of the contained air is 10 m^3, its pressure is 1 atm, and the work done on the piston and spring system is 13 kJ. The air in the container is perfectly mixed. What is the temperature of the air in the container?

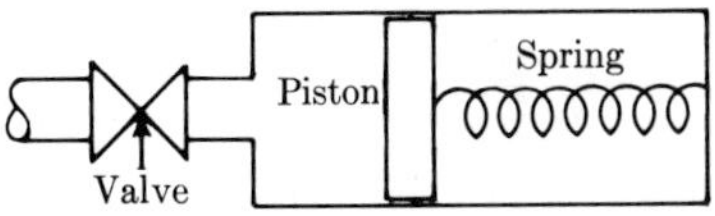

FIGURE 4.19

4.38. The piston in Fig. 4.19 is restrained by a spring. The spring is arranged so that when the absolute pressure in the cylinder is zero, the volume between the piston and the valve is zero, and the pressure volume relationship is

$$P = [10\ \text{lb}/(\text{in}^2 \cdot \text{ft}^3)]V$$

The space between the piston and the valve is now at an absolute pressure of zero. Outside the valve is the atmosphere. We now open the valve and let the air flow in. The process is adiabatic. There is sufficient mixing that the contents of the cylinder are always at a uniform pressure and temperature. What is the temperature of the air in the cylinder when the inflow of air stops?

4.39. In Prob. 4.38, we insert cooling coils in the space between the piston and the valve to cool the incoming gas, so that we maintain the temperature of the gas in the cylinder constant at 20°C = 68°F during the whole inflow process. How much heat will be removed by the time the inflow stops?

4.40. Do any of the following processes violate the first law of thermodynamics?

(*a*) In a rigid, insulated container, 1 lbm of dry saturated steam at 30 psia spontaneously converts to 0.1 lbm of ice at 32°F and to 0.90 lbm of superheated steam at 635°F and 42.5 psia. *Note:* At 635°F and 42.5 psia steam has the properties $v = 15.29\ \text{ft}^3/\text{lbm}$ and $u = 1228\ \text{Btu/lbm}$.

(*b*) A baseball lying on a table spontaneously jumps to another table which is 10 ft higher. When the process is over, the temperature of the ball has fallen sufficiently for du to be -0.01284 Btu/lbm.

(*c*) Freon 12 flows through a throttling valve. The velocities on both sides of the valve are negligible. The conditions before the valve (upstream) are 150°F and 14.7 psia. The conditions after (downstream) are 160°F and 66 psia.

CHAPTER 5

BERNOULLI'S EQUATION

The energy balance for steady, incompressible flow, which is called *Bernoulli's equation*, is probably the most useful single equation in fluid mechanics.

5.1 THE ENERGY BALANCE FOR A STEADY, INCOMPRESSIBLE FLOW

As shown in Chap. 4, the working form of the energy balance is

$$d\left[m\left(u + gz + \frac{V^2}{2}\right)\right]_{\text{sys}}$$
$$= \left(u + \frac{P}{\rho} + gz + \frac{V^2}{2}\right)_{\text{in}} dm_{\text{in}} - \left(u + \frac{P}{\rho} + gz + \frac{V^2}{2}\right)_{\text{out}} dm_{\text{out}} + dQ - dW_{\text{a.o.}} \tag{5.1}$$

This equation has the following restrictions:

1. Electrostatic, magnetic, and surface energies are negligible.
2. The contents of the system are uniform.
3. The inflow and outflow streams are uniform.
4. The acceleration of gravity is constant.

We now add the restriction of a single, steady-state flow through a system. This causes the leftmost term of the equation to drop out, makes dm_{in}

equal dm_{out}, and allows us to divide through by dm_{in} to find

$$\left(u + \frac{P}{\rho} + gz + \frac{V^2}{2}\right)_{\text{in}} - \left(u + \frac{P}{\rho} + gz + \frac{V^2}{2}\right)_{\text{out}} = \frac{dW_{\text{a.o.}}}{dm} - \frac{dQ}{dm} \tag{5.2}$$

Regrouping and multiplying by -1 produces

$$\Delta\left(\frac{P}{\rho} + gz + \frac{V^2}{2}\right) = \frac{-dW_{\text{a.o.}}}{dm} - \left(\Delta u - \frac{dQ}{dm}\right) \tag{5.3}$$

Here ΔP stands for $P_{\text{out}} - P_{\text{in}}$, etc. This equation is the preliminary form of Bernoulli's equation.[†] Before we convert it to the final form, let us see what each term represents physically. The P/ρ terms are injection work terms, representing the work required to inject a unit mass of fluid into or out of the system or both. The gz terms are potential-energy terms, representing the potential energy of a unit mass of fluid above some arbitrary datum plane. Since they appear only as Δgz, it is unnecessary in most problems to know or to state what that datum is. The $V^2/2$ terms show the kinetic energy per unit mass of fluid. The $dW_{\text{a.o.}}/dm$ term represents the amount of work done by the fluid per unit mass of fluid passing through the system (this does not include injection work, which was specifically excluded). This can be work done by turning or reciprocating shafts or by expansion of the boundaries of the system against some restraining force or by several other methods. The minus sign goes with the thermodynamic convention that work is positive when it flows *out* of the system.

5.2 THE FRICTION HEATING TERM

We are all familiar with friction heating, as seen in the smoking brakes and tires of an automobile which has stopped suddenly and in the high temperature of a saw which is cutting wood. We are less familiar with the idea of friction heating in fluids, because the temperature increases produced by friction heating in fluids are generally much less than those produced by rubbing together two solids. These temperature increases are less for the following reasons:

1. The amount of frictional work per unit mass in typical fluid flow problems is generally less than in the examples cited above.

[†] The original form of this equation was developed by Daniel Bernoulli in an entirely different way. By considering momentum balance (Chap. 7) for a frictionless fluid, he found $\Delta(P/\rho + gz + V^2/2) = 0$, the same as Eq. 5.3 but without the two terms to the right of the equals sign. The original equation was not applicable to flows containing pumps or turbines or to flows in which fluid friction was important. Equation 5.3, derived from the energy balance, is applicable to all the flows to which Bernoulli's original equation applies as well as to those which have significant friction and/or pumps. Some writers refer to it as the *extended form of Bernoulli's equation*.

2. The heat capacity of liquids is generally greater than that of solids. For example, the amount of heat required to raise the temperature of 1 lbm of water by 1°F will raise the temperature of 1 lbm of steel by about 8°F.

Friction heating involves the conversion to internal energy of some other kind of energy (kinetic or potential) or of external work (injection, shaft, or expansion). For constant-density materials (gas, liquid, or solid), the only other way† that the internal energy per unit mass can change is through external heating or cooling. Thus,

$$\Delta u = \frac{\text{friction heating}}{\text{lbm}} + \frac{dQ}{dm} \qquad \text{[constant-density materials only]} \tag{5.4}$$

Solving this equation for the friction heating per unit mass, we see that it is given by the $\Delta u - dQ/dm$ term on the right of Eq. 5.3.

This friction heating is not connected with any heating or cooling of the fluid through heat transfer with the surroundings and has the same meaning whether the fluid is being heated or cooled. This may be seen by considering the simple frictionless heater for a constant-density fluid shown in Fig. 5.1. For such a heater, there is no change in elevation or velocity, and because there is no friction, there is no change in pressure. Similarly, there is no pump or compressor work, so Bernoulli's equation simplifies to

$$0 = -\left(\Delta u - \frac{dQ}{dm}\right) \qquad \text{[frictionless heater]} \tag{5.5}$$

If, however, there were friction in the heater, then $\Delta u - dQ/dm$ would be a positive number, whose value would be exactly equal to the amount of friction heating per unit mass.

The increased internal energy produced by friction heating usually is useless for industrial purposes, so that friction heating is often referred to as *friction loss*. Energy does not disappear in this case. Rather, energy of a valuable form is converted to energy of a useless form, hence the "loss" of energy.

As discussed in Sec. 2.2, there is no such thing as an absolutely incompressible fluid. Furthermore, in some situations even a fluid with a very

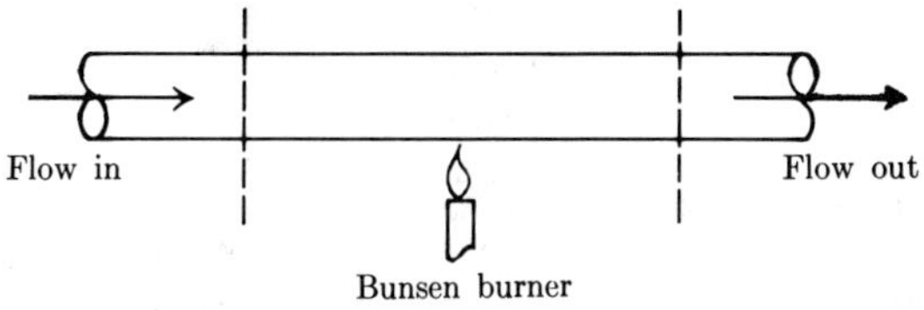

FIGURE 5.1
A simple frictionless heater.

† Here we have excluded the possibility of electric, magnetic, or surface effects.

small compressibility, such as water, behaves in a compressible way. Thus we speak of an *incompressible flow*, by which we mean a flow in which the changes in density are unimportant, rather than of an incompressible fluid. As a general rule, all steady flows of liquids and most steady flows of gases at low velocities (see Sec. 5.6) may be considered incompressible, whereas some unsteady flows of liquids (see Secs. 5.10 and 7.4) and all steady flows of gases at high velocities may not be considered incompressible. We consider the flow of gases at high velocities in Chap. 8, where we will see that the same terms which appear in Bernoulli's equation will reappear in different combinations. Therefore, we apply Bernoulli's equation only to incompressible flows and use only the incompressible-flow meaning of $\Delta u - dQ/dm$, that is, friction heating per unit mass.

To save writing, we now introduce for the friction heating per unit mass a new symbol[†]

$$\Delta u - \frac{dQ}{dm} = \mathscr{F} \tag{5.6}$$

This changes Eq. 5.3 to the final working form of Bernoulli's equation:

$$\Delta\left(\frac{P}{\rho} + gz + \frac{V^2}{2}\right) = \frac{-dW_{\text{a.o.}}}{dm} - \mathscr{F} \tag{5.7}$$

One may show as a consequence of the second law of thermodynamics that $\mathscr{F}$ is zero for frictionless flows and positive for all real flows. One sometimes calculates flows in which it is negative. This indicates that the assumed direction of the flow is incorrect; for the assumed conditions at the in and out locations, the flow is thermodynamically possible only in the opposite direction. However, frictionless flows are reversible, hence any flow described by Bernoulli's equation in which $\mathscr{F}$ is zero could be reversed in direction without any change in magnitude of the velocities, pressures, elevations, etc.

Since for all real flows $\mathscr{F}$ is positive, Eq. 5.7 with a minus sign before $\mathscr{F}$ indicates that friction causes a decrease in pressure or elevation or velocity or of the work that can be extracted by a turbine or an increase in work that must be done by a pump or some combination of these effects.

In Eq. 5.7 we now have only terms which can be measured mechanically; we have eliminated the Q and u terms, which require thermal measurements. Therefore, this equation, the working form of Bernoulli's equation, is often

---

[†] Here we use $\mathscr{F}$ to avoid confusion with F for force. Most civil engineering texts call this quantity gh_f or gh_L, where g is the acceleration of gravity and h_f or h_L stands for *friction head loss* (Sec. 5.4). Some thermodynamics textbooks introduce the idea of the lost work in explaining the second law of thermodynamics. For a constant-density fluid at the heat reservoir temperature, the friction heating per unit mass is exactly equal to the lost work per unit mass, so some texts call this term LW. Other texts call it $(-\Delta P/\rho)_{\text{friction}}$ since for the most common pipe friction problem, steady flow in horizontal, constant-area pipes, $\mathscr{F} = -\Delta P/\rho$.

referred to as the *mechanical energy balance*. Note that "mechanical energy" is conserved only if we include an "energy destruction" term $\mathscr{F}$. This equation has the same restrictions as Eq. 5.1 as well as the restriction that the flow must be steady in and out across the boundaries of the system.

In most applications, we will be dealing with flow in a pipe or channel and will assume that the fluid velocity is constant across a given cross section perpendicular to the flow. This approximation is excellent for most engineering problems; one interesting exception is discussed in Sec. 5.11.

5.3 ZERO FLOW

The basic equation of fluid statics is a limited form of Eq. 5.7. If we apply that equation between any two points in a fluid at rest, there is no external work or friction; so

$$\Delta\left(\frac{P}{\rho} + gz\right) = 0 \tag{5.8}$$

Rearranging, we find

$$\frac{1}{\rho}\,\Delta P = -g\,\Delta z \tag{5.9}$$

or

$$\lim_{\Delta z \to 0} \frac{\Delta P}{\Delta z} = \frac{dP}{dz} = -\rho g \tag{5.10}$$

which is the basic equation of fluid statics. It is found in Chap. 2 by making a force balance around an elemental particle of fluid. The derivation shown here points out only that Eq. 5.7 is general enough to include cases of zero flow.

5.4 THE HEAD FORM OF BERNOULLI'S EQUATION

In many problems, particularly those involving flow of water in dams, canals, and open channels, it is convenient to divide both sides of Eq. 5.7 by g to find

$$\Delta\left(\frac{P}{\rho g} + z + \frac{V^2}{2g}\right) = \frac{-dW_{\text{a.o.}}}{g\,dm} - \frac{\mathscr{F}}{g} \tag{5.11}$$

which is called the *head form of Bernoulli's equation*.

Every term in Eq. 5.11 has the dimension of a length. The lengths are at least conceptually convertible to elevation Δz above some datum plane. These elevations are commonly referred to as *heads*.† Thus, we refer to the various

† "Head" is apparently a variant of "height."

terms in Eq. 5.11 as the *pressure head*, *gravity head*, *velocity head*, *pump or turbine head*, and *friction head loss*.

There is no simple, universal rule for deciding when to use the head form of Bernoulli's equation and when to use the energy form, Eq. 5.7; if correctly applied, both give the same result. Through practice engineers learn which is more convenient for a given problem.

5.5 DIFFUSERS AND SUDDEN EXPANSIONS

In the following sections we will see several examples of flow in which a moving fluid is brought to a halt. Here we consider two ways of slowing down a fluid: a diffuser and a sudden expansion. A diffuser is a gradually expanding pipe or duct, as sketched in Fig. 5.2. Writing Bernoulli's equation for the pipe between locations 1 and 2, we find

$$\frac{P_2 - P_1}{\rho} + \frac{V_2^2 - V_1^2}{2} = -\mathcal{F} \tag{5.12}$$

From the mass balance for a constant-density fluid we have

$$V_2 = \frac{V_1 A_1}{A_2} \tag{5.13}$$

and substituting for V_2 in Eq. 5.12, we find

$$P_2 - P_1 = \rho\,\frac{V_1^2}{2}\left(1 - \frac{A_1^2}{A_2^2}\right) - \rho\mathcal{F} \tag{5.14}$$

This increase in pressure which accompanies the decrease in velocity is often called *pressure recovery*. In such a device, kinetic energy is converted partly to injection work (shown by an increase in pressure) and partly to friction heating.

It is possible to build diffusers so that the friction heating is only about one-tenth of the decrease in kinetic energy, or, as commonly stated, the pressure recovery is about 90 percent of the maximum possible from a frictionless diffuser.

Now consider a fluid flowing through a duct into a large tank of fluid with no net velocity, as shown in Fig. 5.3. This is called a *sudden expansion*. Here point 2 is chosen far away from the fluid inlet, so that the velocity at point 2 is

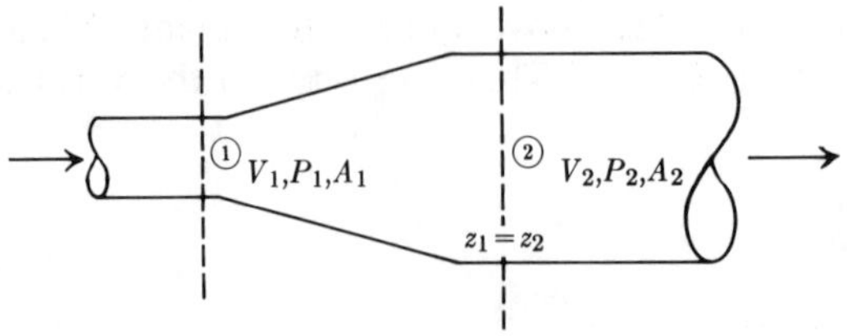

FIGURE 5.2
A diffuser.

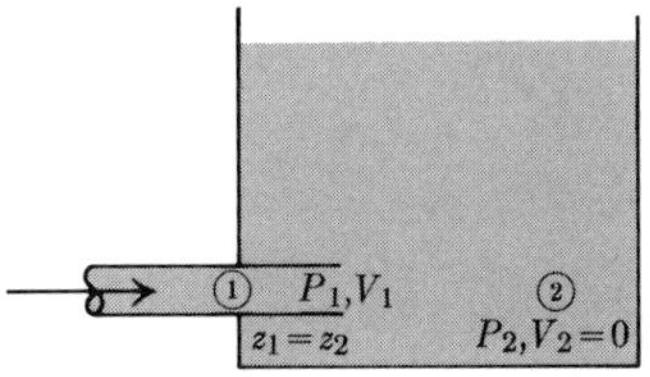

FIGURE 5.3
A sudden expansion.

negligible. Writing Bernoulli's equation for between points 1 and 2, we find

$$P_2 - P_1 = \frac{\rho V_1^2}{2} - \rho \mathscr{F} \tag{5.15}$$

which is quite similar to Eq. 5.14. Here, however, the friction term is much larger than that for the diffuser, because instead of the fluid being brought to rest in an orderly fashion, it is stopped by a chaotic mass of eddies, which converts all its kinetic energy to internal energy. Thus, it is an experimental observation that for such sudden expansions the friction heating per unit mass is exactly equal to the decrease in kinetic energy per unit mass, and there is no pressure recovery at all. Therefore, the pressure of a fluid flowing into such a sudden expansion is the same as the pressure of the fluid into which it flows. This conclusion is limited to flows with velocities less than the speed of sound; it does not apply to supersonic flows, which we discuss in Chap. 8.

These two ways of stopping a fluid are analogous to stopping a fast-moving automobile by letting it run up a hill and thereby converting its kinetic energy to useful potential energy and to stopping it with its brakes and thereby converting its kinetic energy to useless internal energy in the brake shoes and drums. Most students have ridden on roller coasters and hence are comfortable with the idea of converting from potential energy (at the top of the roller coaster) to kinetic energy (at the first valley) and then back to potential energy again (at the top of the next rise). They are less used to the idea of a "pressure hill." But from Bernoulli's equation we see that a rapidly moving fluid stream can convert its kinetic energy to potential energy by climbing a gravity hill, to injection work by climbing a "pressure hill," or to internal energy by friction heating.

5.6 BERNOULLI'S EQUATION FOR GASES

Bernoulli's equation, as we have written it, is exactly correct for constant-density fluids and practically correct for all flows in which the density changes are unimportant. For liquids this includes almost all steady flows. We show here that it also is practically correct for low-velocity gas flows.

Example 5.1. The tank in Fig. 5.4 is full of air at 70°F. The air is flowing out at a steady rate through a smooth, frictionless nozzle to the local atmosphere. What is the flow velocity for various tank pressures?

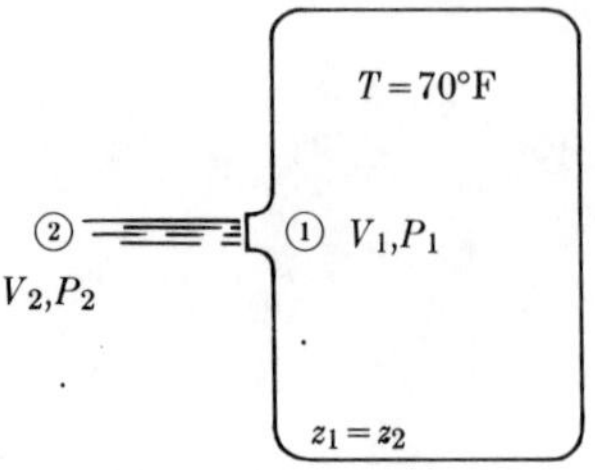

FIGURE 5.4
Bernoulli's equation applied to flow of a gas.

At point 1 the velocity is negligible, and, as discussed in Sec. 5.5, the pressure at point 2 is equal to the local atmospheric pressure, if the flow is subsonic. Making these insertions in Bernoulli's equation, we find

$$V_2 = \left[\frac{2(P_1 - P_{\text{atm}})}{\rho} \right]^{1/2} \tag{5.16}$$

Which value of the density should we use here? It is obviously different at the two states, because the pressure is not the same at the two states. However, if the pressure change is small, the two densities will be practically the same. Let us use the upstream density (but see Prob. 5.5). This is given by substituting the perfect-gas law $\rho = MP_1/(RT_1)$ into Eq. 5.16:

$$V_2 = \left[\frac{2RT_1}{P_1 M} (P_1 - P_{\text{atm}}) \right]^{1/2} \tag{5.17}$$

Using this equation, we can calculate V_2 for various values of P_1. For example, if P_1 is $P_{\text{atm}} + 0.01 \text{ lbf/in}^2$, then

$$V_2 = \left[\frac{2(10.73 \text{ psi} \cdot \text{ft}^3)/(\text{°R} \cdot \text{lbmol}) \cdot 530\text{°R}}{(14.71 \text{ psi})(29 \text{ lbm/lbmol})} \cdot 0.01 \frac{\text{lbf}}{\text{in}^2} \cdot \frac{144 \text{ in}^2}{\text{ft}^2} \cdot \frac{32.2 \text{ lbm} \cdot \text{ft}}{\text{lbf} \cdot \text{s}^2} \right]^{1/2}$$

$$= \left(1235 \frac{\text{ft}^2}{\text{s}^2} \right)^{1/2} = 35 \frac{\text{ft}}{\text{s}} = 10.71 \frac{\text{m}}{\text{s}}$$

Equation 5.17 is based on the assumption of a constant-density fluid, which is not exactly correct here; the correct result for this system, taking gas expansion into account, is developed in Chap. 8. The velocities calculated from Eq. 5.17 and the correct solution are compared in the accompanying table.

$P_1 - P_{\text{atm}}$, psia	Velocity from Eq. 5.17, ft/s	Correct velocity from equations in Chap. 8, ft/s
0.01	35	35
0.1	111	111
0.3	191	191
0.6	267	269
1.0	340	344
2.0	467	477
5.0	679	714

■

From these figures it is clear that to assume that gas flows are incompressible and are described by Bernoulli's equation causes a very small error at low gas velocities. Even at a velocity of 700 ft/s (213 m/s), the error caused by assuming incompressible flow is only about 5 percent.

Most air conditioning and low-speed aircraft problems involve velocities below 200 ft/s (61 m/s), so these problems can be solved with engineering accuracy by Bernoulli's equation. But where there are significant pressure changes for gases in flow, which lead to high velocities, the density changes must be taken into account, as shown in Chap. 8. Observe also the very high velocities caused by very small pressure differences acting on gases. The inverse of this observation is that for ordinary flow velocities, the pressure differences in gases are at least an order of magnitude smaller than those for the corresponding flow velocities in liquids.

Application of Bernoulli's equation to a simple horizontal pump or compressor with equal-sized inlet and outlet pipes leads to

$$\frac{-dW_{\text{a.o.}}}{dm} = \frac{\Delta P}{\rho} + \mathcal{F} \tag{5.18}$$

If we ignore friction, this equation becomes

$$-\left(\frac{dW_{\text{a.o.}}}{dm}\right) = \frac{\Delta P}{\rho} \qquad \text{[frictionless]} \tag{5.19}$$

This is exactly correct for constant-density fluids and correct to better than ± 1 percent for most pumps pumping real liquids. For gases, whose density will change in a compressor, it is not exactly correct. The correct result, taking the change of gas density into account, is developed in Chap. 9. However, if the pressure change ΔP is small compared with the inlet pressure P_{in}, then Eq. 5.19 gives a very good estimate of the required frictionless work. For example, if $\Delta P/P$ is 0.1 or less, then the result from Eq. 5.19 is certain to be within 10 percent of the calculated result which takes density changes into account. This pressure range includes most fans, blowers, air conditioning systems, vacuum cleaners, etc., but does not include air compressors (e.g., those in service stations).

5.7 TORRICELLI'S EQUATION AND ITS VARIANTS

The most interesting applications of Bernoulli's equation include the effects of friction. Before we can solve these, we must learn how to evaluate the $\mathcal{F}$ term, which we do in Chap. 6. However, in many flow problems the friction heating terms are small compared with the other terms and can be neglected. We can solve these by means of Bernoulli's equation without the friction heating term. A good example of this type of problem is the tank-draining problem, which leads to Torricelli's equation.

Example 5.2. The tank in Fig. 5.5 is full of water and open at the top. There is a hole near the bottom, the diameter of which is small compared with the diameter of the tank. What is the velocity of the flow out the hole?

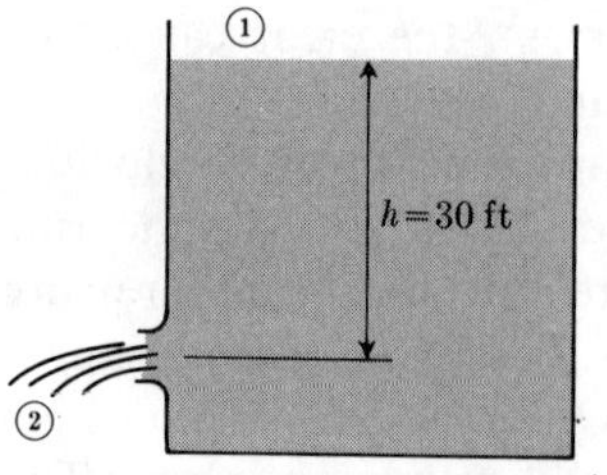

FIGURE 5.5
The flow for Torricelli's equation.

To solve this problem, we apply Eq. 5.7 between the free surface at the top of the tank, location 1, and the jet of fluid just after it has left the tank, location 2. In addition to the assumptions built into Bernoulli's equation, we assume the following:

1. The diameter of the tank is so large that the velocity at the free surface is practically zero, $V_1 = 0$.
2. The pressures at locations 1 and 2 are the local atmospheric pressures. The pressure of the atmosphere is not exactly the same at both points, but it is practically the same; so we assume $\Delta P = 0$.
3. There is no friction or external work.
4. Flow is steady; i.e., the level at the top of the tank is not falling. This means that fluid must be flowing into the tank somewhere exactly as fast as it flows out at location 2.

Subject to these restrictions, we write

$$g(z_2 - z_1) + \frac{V_2^2}{2} = 0$$

Here $z_2 - z_1 = -h$, so

$$V_2 = (2gh)^{1/2} \tag{5.20}$$

This is *Torricelli's equation*, which says that the fluid velocity is exactly the same as the velocity the fluid would attain by falling freely from rest a distance h. Substituting the numerical values, we find

$$V_2 = \left(2 \cdot 32.2 \, \frac{\text{ft}}{\text{s}^2} \cdot 30 \text{ ft}\right)^{1/2} = 43.9 \, \frac{\text{ft}}{\text{s}} = 13.4 \, \frac{\text{m}}{\text{s}} \qquad \blacksquare$$

This is the classic tank-draining solution. It is correct only for situations in which the assumptions made in finding Torricelli's equation apply; in Examples 5.3 and 5.4 we examine some situations in which they may not apply.

Example 5.3. Repeat Example 5.2, making the area of the outlet hole 1 ft^2 and the cross-sectional area of the tank 4 ft^2.

In this case we cannot assume that the velocity at the free surface is zero, as we did in Example 5.2; so

$$g(-h) + \frac{V_2^2 - V_1^2}{2} = 0$$

Using the mass balance for a constant-density fluid, we can solve for V_1 in terms of V_2, A_1, and A_2, and we substitute for V_1:

$$g(-h) + \frac{1}{2}\left[V_2^2 - \left(\frac{V_2 A_2}{A_1}\right)^2\right] = 0$$

$$-gh + \frac{V_2^2}{2}\left[1 - \left(\frac{A_2}{A_1}\right)^2\right] = 0$$

$$V_2 = \left[\frac{2gh}{1 - (A_2/A_1)^2}\right]^{1/2} \tag{5.21}$$

Inserting the numerical values, we find

$$V_2 = \left[\frac{2 \cdot 32.2\ \text{ft/s}^2 \cdot 30\ \text{ft}}{1 - (\frac{1}{4})^2}\right]^{1/2}$$

This is the answer from Example 5.2, divided by $(\frac{15}{16})^{1/2}$:

$$V_2 = \frac{43.9\ \text{ft/s}}{(\frac{15}{16})^{1/2}} = 45.3\ \frac{\text{ft}}{\text{s}} = 13.8\ \frac{\text{m}}{\text{s}} \qquad \blacksquare$$

Why does the water flow faster in this case? All the water in the tank has measurable kinetic energy; it is flowing down at a velocity of 11.4 ft/s. In Example 5.2 the water in the tank has immeasurably small kinetic energy.

What happens in Example 5.3 if the cross-sectional area of the tank is equal to the cross-sectional area of the outlet, i.e., if $A_2 = A_1$? If we substitute this in Eq. 5.21, it predicts an infinite velocity! Therefore, Eq. 5.21 does not describe this situation. Recall the assumptions behind that equation. First, there is the Bernoulli's equation assumption of steady flow. Second, there is the assumption in Eqs. 5.20 and 5.21 that friction is negligible. Suppose we have a vertical pipe of constant cross-sectional area and steady flow downward. Suppose also that the pressure gauges at two different elevations read the same value. Then this situation is analogous to that in Example 5.3, with $A_1 = A_2$. Returning to Eq. 5.7, we see that the only significant terms are

$$g(z_2 - z_1) = -\mathcal{F}$$

This situation, in which the friction forces are dominant, is quite different from the situation shown in Fig. 5.5, from which we found Torricelli's equation, and is not covered by the frictionless assumption of Torricelli's equation.

Example 5.4. Repeat Example 5.2, making the tank contain carbon dioxide gas at the same temperature and pressure as the surrounding atmosphere instead of water.

This looks strange: an open tank full of gas! However, it is easy to demonstrate in the laboratory or kitchen by mixing a little bicarbonate of soda and vinegar in a cup. When the bubbling has stopped, the cup will be filled with carbon dioxide gas. This gas is heavier than air and can be poured from cup to cup, visibly. However, it mixes slowly with the air and ultimately will disperse by diffusion.

To return to the problem, it appears at first to be the same as Example 5.2. However, there is a big difference, namely, that we cannot ignore the difference in atmospheric pressure between locations 1 and 2. The other assumptions for Example 5.2 appear sound, so Bernoulli's equation becomes

$$\frac{P_2 - P_1}{\rho} + g(z_2 - z_1) + \frac{V_2^2}{2} = 0$$

From the basic equation of fluid statics we can calculate

$$P_2 - P_1 = -\rho_{\text{air}} g(z_2 - z_1)$$

Now we must be careful, because there are two densities in our problem: the ρ_{air} shown here and the ρ in Bernoulli's equation. If we follow the derivation of Bernoulli's equation back to its source, we see that the ρ in it is the ρ of the fluid which is flowing; we label it ρ_{CO_2}. Combining these two, we find

$$\frac{-\rho_{\text{air}} g(z_2 - z_1)}{\rho_{CO_2}} + g(z_2 - z_1) + \frac{V_2^2}{2} = 0$$

$$0 = \frac{V_2^2}{2} + g(z_2 - z_1)\left(1 - \frac{\rho_{\text{air}}}{\rho_{CO_2}}\right)$$

Solving for V_2, we find

$$V_2 = \left[2gh\left(1 - \frac{\rho_{\text{air}}}{\rho_{CO_2}}\right)\right]^{1/2} \tag{5.22}$$

Assuming the air and carbon dioxide behave as perfect gases and are at the same temperature and pressure, their densities are proportional to their molecular weights; so

$$V_2 = [2(32.2\ \text{ft/s}^2)(30\ \text{ft})]^{1/2}(1 - \tfrac{29}{44})^{1/2}$$
$$= (43.9\ \text{ft/s})(0.34^{1/2}) = 25.6\ \text{ft/s} = 7.8\ \text{m/s} \qquad ■$$

If the difference in atmospheric pressure is important in Example 5.4, is it important in Example 5.2? Equation 5.22 applies as well to Example 5.2 as it does to Example 5.4. Therefore, if we wish to take the effect of the difference in atmospheric pressure into account in Example 5.2, we should use Eq. 5.22. This is equivalent to multiplying the answer in Example 5.2 by $(1 - \rho_{\text{air}}/\rho_{\text{water}})^{1/2}$. For water and air at normal temperature and pressure, this is about

$$\left(1 - \frac{\rho_{\text{air}}}{\rho_{\text{water}}}\right)^{1/2} = \left(1 - \frac{0.075\ \text{lbm/ft}^3}{62.3\ \text{lbm/ft}^3}\right)^{1/2} = (0.9988)^{1/2} = 0.9994$$

Ignoring the change in atmospheric pressure in Torricelli's equation for air and water causes an error of less than 0.1 percent (much less than the error introduced by some of the other assumptions). We are justified in leaving out this term if the ratio of the density of the surrounding fluid to that of the flowing fluid, ρ_{sf}/ρ_{ff}, is much less than 1. This is true in most hydraulics problems but not in two-liquid problems (Probs. 5.13 and 5.14).

We discuss one more variant of Torricelli's equation in Sec. 5.10.

5.8 BERNOULLI'S EQUATION FOR FLUID FLOW MEASUREMENT

Several important types of fluid-flow-measuring devices are based on the frictionless form of Bernoulli's equation. Where the friction effects in these devices become significant, they are normally accounted for by introducing empirical coefficients and retaining the frictionless form of Bernoulli's equation, rather than by introducing the friction term into Bernoulli's equation. Thus, it seems appropriate to take up these devices before we discuss the friction term in Bernoulli's equation, even though these devices obviously involve some friction.

A. Pitot Tube

The simplest pitot tube (invented by H. Pitot) is sketched in Fig. 5.6. This is sometimes called an *impact tube* or *stagnation tube*. It consists of a bent, transparent tube with one vertical leg projecting out of the flow and another leg pointing directly upstream in the flow.

At location 1 the flow is practically undisturbed by the presence of the tube and hence has the velocity which would exist at location 2 if the tube were not present. At location 2, the flow has been completely stopped by the tube which has been inserted, so $V_2 = 0$. Writing Bernoulli's equation between locations 1 and 2 yields

$$\frac{P_2 - P_1}{\rho} - \frac{V_1^2}{2} = -\mathcal{F} \tag{5.23}$$

But inside the pitot tube the fluid is not moving, so the pressure at location 2 is given by

$$P_2 = P_{atm} + \rho g(h_1 + h_2) \tag{5.24}$$

If all the fluid flow is in the horizontal direction, then the basic equation of

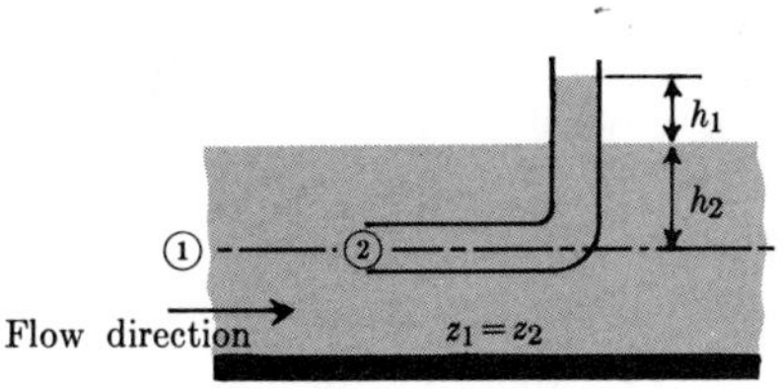

FIGURE 5.6
Pitot tube.

fluid statics can be used to find the vertical change in pressure with depth, so that

$$P_1 = P_{\text{atm}} + \rho g h_2 \tag{5.25}$$

Substituting Eqs. 5.24 and 5.25 in Eq. 5.23 and rearranging, we find

$$V_1 = (2gh_1 + 2\mathscr{F})^{1/2} \tag{5.26}$$

It has been found experimentally that the friction heating term in Eq. 5.26 is normally less than 1 percent of the total; it may be ignored, giving

$$V_1 = (2gh_1)^{1/2} \tag{5.27}$$

The pitot tube allows us to measure a liquid height (a very easy thing to measure) and to calculate a velocity from it by Bernoulli's equation. The device, exactly as shown in Fig. 5.6, is used for finding velocities at various points in open-channel flow and for determining the velocities of boats.

Example 5.5. A pitot tube exactly as shown in Fig. 5.6 is used for measuring the velocity of a sailboat. When the water level in the tube is 1 m above the surface, how fast is the boat going?

$$V_1 = \left(2 \cdot 9.81 \frac{\text{m}}{\text{s}^2} \cdot 1 \text{ m}\right)^{1/2} = \left(19.62 \frac{\text{m}^2}{\text{s}^2}\right)^{1/2} = 4.43 \frac{\text{m}}{\text{s}} = 14.5 \frac{\text{ft}}{\text{s}} \quad \blacksquare$$

B. Pitot-Static Tube

The pitot tube shown in Fig. 5.6 is suitable for open-channel flow but not for flow of the atmosphere or flow in pipes. For the latter two it is combined with a second tube, called a *static tube*, shown in Fig. 5.7. Here some kind of pressure-difference-measuring device is used to indicate the difference between the pressures of the pitot and static tubes. The pressure at point 2 is given by Eq. 5.23. By arguments similar to those used with the pitot tube it can be shown that the pressure difference between point 2 and the inlet of the pressure-difference meter, due to the weight of fluid in the tube connecting them, is exactly balanced by the pressure difference due to gravity from point 1 to the other side of the meter, and hence that the pressure-difference meter

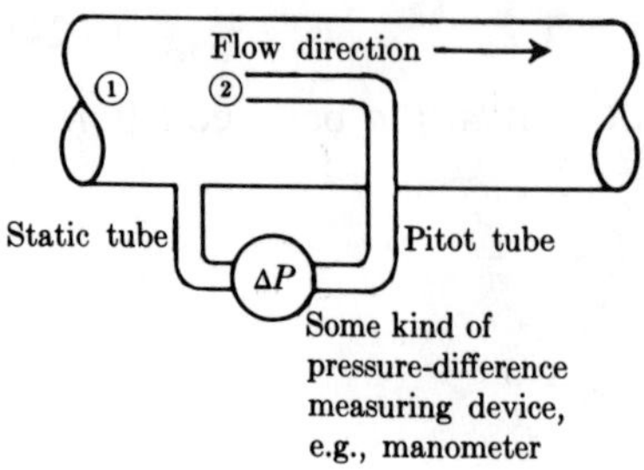

FIGURE 5.7
Pitot-static tube.

really measures $P_2 - P_1$. Furthermore, experimental tests have shown that for a well-designed pitot-static meter the friction effect is negligible, so that we may read the pressure difference from the meter and calculate the velocity from Eq. 5.23 rearranged:

$$V_1 = \left(\frac{2\,\Delta P}{\rho}\right)^{1/2} \tag{5.28}$$

Example 5.6. Air at a density of 0.1 lbm/ft^3 is flowing in the duct in Fig. 5.7. The pressure-difference gauge indicates a difference of 0.5 lbf/in^2. What is the air velocity?

$$V = \left(\frac{2 \cdot 0.5\ \text{lbf/in}^2}{0.1\ \text{lbm/ft}^3} \cdot \frac{144\ \text{in}^2}{\text{ft}^2} \cdot \frac{32.2\ \text{lbm} \cdot \text{ft}}{\text{lbf} \cdot \text{s}^2}\right)^{1/2} = 215\ \frac{\text{ft}}{\text{s}} = 65.6\ \frac{\text{m}}{\text{s}} \quad \blacksquare$$

The pitot-static tube is the standard device for measuring the airspeed of airplanes and is often used for measuring the local velocity in pipes or ducts, particularly in air pollution sampling procedures. One can easily identify the pitot-static probes projecting from the front of modern commercial airplanes; look next time you are at an airport. For measuring flow in enclosed ducts or channels the venturi meter and orifice meters discussed below are more convenient and more frequently used.

C. Venturi Meter

Figure 5.8 shows a horizontal venturi meter. It consists of a truncated cone in which the cross-sectional area perpendicular to flow decreases, a short cylindrical section, and a truncated cone in which the cross-sectional area increases to its original value. There are pressure taps both upstream and in the short cylindrical section (the "throat"); they are connected to some pressure-difference-measuring device, usually a manometer. Applying Bernoulli's equation between locations 1 and 2 and neglecting friction, we find

$$\frac{P_2 - P_1}{\rho} + \frac{V_2^2 - V_1^2}{2} = 0 \tag{5.29}$$

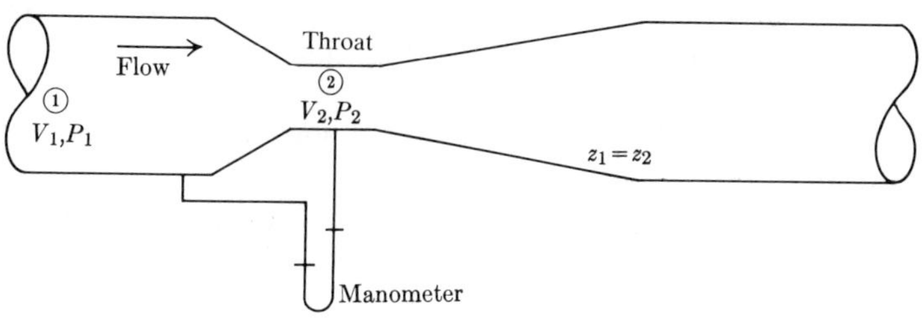

FIGURE 5.8
Venturi meter.

Using the mass balance for a constant-density fluid, we can write V_1 in terms of V_2, A_2, and A_1. Substituting in Eq. 5.29 and rearranging, we find

$$V_2 = \left[\frac{2(P_1 - P_2)/\rho}{1 - A_2^2/A_1^2} \right]^{1/2} \tag{5.30}$$

Example 5.7. The venturi meter in Fig. 5.8 has water flowing through it. The pressure difference $P_1 - P_2$ is 1 lbf/in^3. The diameter at point 1 is 1 ft, and that at point 2 is 0.5 ft. What is the volumetric flow rate through the meter?

From Eq. 5.30

$$V_2 = \frac{\left\{ \dfrac{2 \cdot 1\ \text{lbf/in}^2}{62.3\ \text{lbm/ft}^3} \cdot \dfrac{144\ \text{in}^2}{\text{ft}^2} \cdot \dfrac{32.2\ \text{lbm} \cdot \text{ft}}{\text{lbf} \cdot \text{s}^2} \right\}^{1/2}}{\left\{ 1 - \dfrac{[(\pi/4)(0.5\ \text{ft})^2]^2}{[(\pi/4)(1\ \text{ft})^2]^2} \right\}^{1/2}}$$

$$= 12.7\ \frac{\text{ft}}{\text{s}} = 3.9\ \text{m/s}$$

The flow rate is

$$Q = V_2 A_2 = 12.7\ \frac{\text{ft}}{\text{s}} \cdot \frac{\pi}{4}\ (0.5\ \text{ft})^2 = 2.49\ \frac{\text{ft}^3}{\text{s}} = 0.070\ \frac{\text{m}^3}{\text{s}}$$ ■

It is found experimentally that the flow rate calculated from Eq. 5.30 is slightly higher than that actually observed. This is due partly to the friction heating in the meter, which we have assumed to be zero, and partly to the fact that the flow is not entirely uniform across any cross section of the pipe, as we have tacitly assumed. We could attempt to account for these differences by using a more complicated formula than Eq. 5.30; the more common approach is to introduce an empirical coefficient into Eq. 5.30. This is called the *coefficient of discharge* C_v:

$$V_2 = C_v \left[\frac{2(P_1 - P_2)}{\rho(1 - A_2^2/A_1^2)} \right]^{1/2} \tag{5.31}$$

A large number of experimental tests have shown that C_v depends only on the Reynolds number, a dimensionless group whose significance is discussed in Chaps. 6 and 13; these results are summarized in Fig. 5.9.

Example 5.8. Rework Example 5.7, taking into account the experimental results summarized in Fig. 5.9.

This requires a trial-and-error solution because, to calculate V, we need to know C_v, which is a function of V. The procedure is as follows.

1. Assume $V = V_{\text{Ex. 5.7}} = 12.7$ ft/s.
2. Compute the Reynolds number:

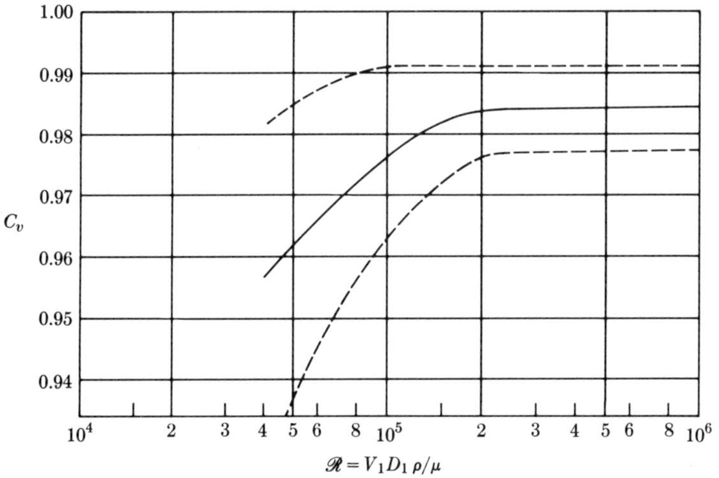

FIGURE 5.9
Discharge coefficients for venturi meters. Here velocities and diameters V_1 and D_1, respectively, are measured at point 1 in Fig. 5.8. The solid line represents the best average of the available data; the dotted lines represent the range of scatter in the experimental data. (From *Fluid Meters, Their Theory and Practice*, 5th ed., ASME, New York, 1959. Reproduced with permission of the publisher.)

$$\mathscr{R} = \frac{V_1 D_1 \rho}{\mu} = \frac{V_2(A_2/A_1)D_1\rho}{\mu}$$

$$= \frac{[(12.7 \text{ ft/s})/4] \cdot 1 \text{ ft} \cdot 62.3 \text{ lbm/ft}^3}{1 \text{ cP} \cdot 6.72 \times 10^{-4} \text{ lbm/(ft} \cdot \text{s} \cdot \text{cP)}} = 2.9 \times 10^5$$

3. On Fig. 5.9 we read $C_v = 0.984$.
4. $V_{\text{rev}} = 0.984 \cdot 12.7 \text{ ft/s} = 12.5 \text{ ft/s}$.
5. We should now repeat steps 2 and 3, using this revised value of V, V_{rev}. However, in comparing these, we ask, "How much C_v would be changed by using $V_{\text{rev}} = 12.5$ ft/s in calculating the Reynolds number (step 2) and then a new value of C_v?" Clearly, because of the shape of Fig. 5.9 this would cause a negligible change; so a revised C_v would be the same, and we accept $V = 12.5$ ft/s as a satisfactory estimate of the velocity. Then

$$Q = 12.5 \frac{\text{ft}}{\text{s}} \cdot \frac{\pi}{4} (0.5 \text{ ft})^2 = 2.45 \frac{\text{ft}^3}{\text{s}} = 0.069 \frac{\text{m}^3}{\text{s}}$$

If the velocity had been much lower, not corresponding to the horizontal part of the curve in Fig. 5.9, this trial-and-error solution probably would have taken several steps; normally the meters are designed to operate at high velocities, on the right-hand side of Fig. 5.9, so that the trial and error is very simple. ■

The foregoing is all based on a horizontal venturi meter. If we use the setup shown in Fig. 5.8 and take the manometer reading as a pressure difference to get our value of $P_1 - P_2$ in Eq. 5.31, then the result is quite independent of the angle to the vertical of the venturi meter. The reason is that the elevation change in the meter is compensated by the elevation change in the manometer legs. Consider the venturi meter in Fig. 5.10.

Applying Bernoulli's equation between points 1 and 2 in Fig. 5.10 and solving for $V_2(1 - A_2^2/A_1^2)^{1/2}$ gives

$$V_2\left(1 - \frac{A_2^2}{A_1^2}\right)^{1/2} = \left[\frac{2(P_1 - P_2)}{\rho} + 2g(z_1 - z_2)\right]^{1/2} \tag{5.32}$$

To solve for $(P_1 - P_2)$, let us call P_2 known and work our way through the manometer step by step:

$$P_3 = P_2 + \rho_1 g(z_2 - z_3)$$

$$P_4 = P_3 + \rho_2 g(z_3 - z_4)$$

$$P_1 = P_4 - \rho_1 g(z_1 - z_4)$$

Adding these equations and canceling like terms, we find

$$P_1 = P_2 + \rho_1 g[(z_2 - z_1) - (z_3 - z_4)] + \rho_2 g(z_3 - z_4)$$

$$P_1 - P_2 = -\rho_1 g(z_1 - z_2) + g(z_3 - z_4)(\rho_2 - \rho_1) \tag{5.33}$$

Substituting this in Eq. 5.32, we see that the elevation $z_1 - z_2$ does indeed cancel, and so

$$V_2 = \left[\frac{2g(z_3 - z_4)(\rho_2 - \rho_1)}{\rho_1(1 - A_2^2/A_1^2)}\right]^{1/2} \tag{5.34}$$

But $g(z_3 - z_4)(\rho_2 - \rho_1)$ is precisely the pressure difference we would have calculated for the manometer reading if we had not taken into account the

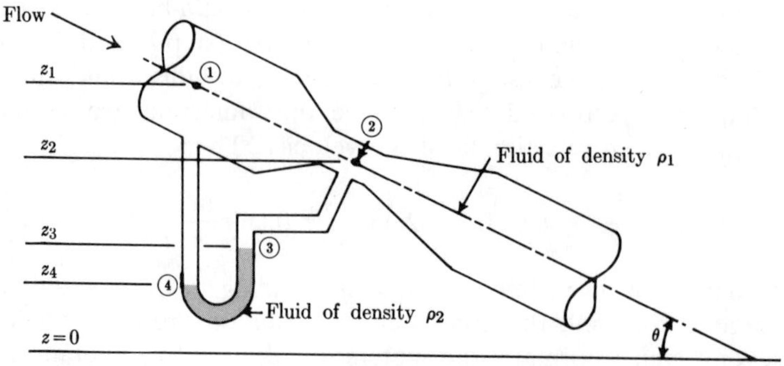

FIGURE 5.10
Inclined venturi meter.

difference in length of the manometer legs. The result found above is true for any angle θ; so we conclude that if the venturi meter is connected as shown in Fig. 5.10, we can neglect the angle to the vertical and simply use Eq. 5.31 (but see Prob. 5.28).

D. Orifice Meter

The venturi meter described above is a reliable flow-measuring device. Furthermore, it causes little pressure loss (i.e., the actual value of $\mathcal{F}$ is small). For these reasons it is widely used, particularly for large-volume liquid and gas flows. However, the meter is relatively complex to construct and hence expensive. Especially for small pipelines, its cost seems prohibitive, so simpler devices have been invented, such as the orifice meter.

As shown in Fig. 5.11, the orifice meter consists of a flat orifice plate with a circular hole drilled in it. There is a pressure tap upstream from the orifice plate and another just downstream. If the flow direction is horizontal and we apply Bernoulli's equation, ignoring friction from point 1 to point 2, we find Eq. 5.30, exactly the same equation we found for a venturi meter. However, in this case we cannot assume frictionless flow and uniform flow across any cross section of the pipe as easily as we can in the case of the venturi meter.

As in the case of the venturi meter, experiments indicate that if we introduce a discharge coefficient and thus form Eq. 5.31, then that coefficient is a fairly simple function of the ratio of the diameter of the orifice hole to the diameter of the pipe, $D_{\text{o.h.}}/D_{\text{p}}$, and the Reynolds number. The relation is shown in Fig. 5.12.

Example 5.9. Water is flowing at a velocity of 1 m/s in a pipe 0.4 m in diameter. In the pipe is an orifice with a hole diameter of 0.2 m. What is the measured pressure drop across the orifice?

Rearranging Eq. 5.31, we find

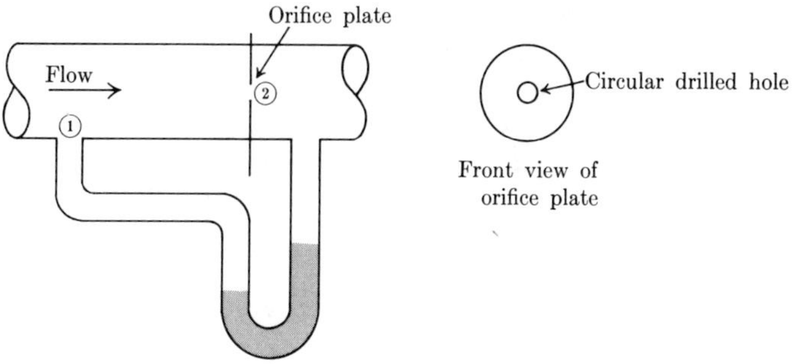

FIGURE 5.11
Orifice meter.

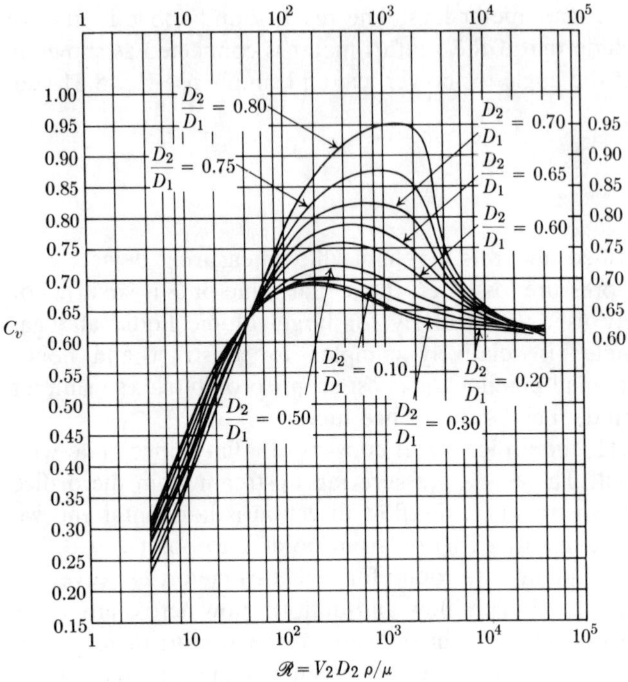

FIGURE 5.12
Discharge coefficients for drilled-plate orifices. [From G. L. Tuve and R. E. Sprenkle, "Orifice discharge coefficients for viscous liquids," *Instruments* 6: 201 (1933). Reproduced by permission of the publisher.]

$$\Delta P = \frac{\rho V_2^2}{2C_v^2}\left(1 - \frac{A_2^2}{A_1^2}\right) = \frac{\rho V_2^2}{2C_v^2}\left(1 - \frac{D_2^4}{D_1^4}\right)$$

From the mass balance for steady flow we know that

$$V_2 = V_1 \frac{A_1}{A_2} = 1\ \frac{\text{m}}{\text{s}} \cdot \frac{(\pi/4)(0.4\ \text{m})^2}{(\pi/4)(0.2\ \text{m})^2} = 4\ \frac{\text{m}}{\text{s}} = \frac{13.1\ \text{ft}}{\text{s}}$$

The Reynolds number $\mathscr{R}$ based on D_2 is calculable and is about 1.6×10^6; so from Fig. 5.12 we have $C_v = 0.62$. Hence

$$P_1 - P_2 = \frac{(998.2\ \text{kg/m}^3)(4\ \text{m/s})^2}{2 \cdot 0.62^2} \cdot (1 - 0.5^4) \cdot \frac{\text{N} \cdot \text{s}^2}{\text{kg} \cdot \text{m}} \cdot \frac{\text{Pa}}{\text{N/m}^2}$$
$$= 19.5\ \text{kPa} = 2.83\ \text{psi}$$

■

Note from Fig. 5.12 that for small orifice holes (D_2/D_1 equal to 0.4 or less) and high flow rates ($\mathscr{R}$ greater than about 1000), C_v is approximately equal to 0.6. These conditions occur in most typical industrial orifice applications, so many practicing engineers automatically write $C_v = 0.6$. In new applications, it is best to check Fig. 5.12 to see whether this simplification applies. By the mathematical methods of potential flow (Chap. 10) one may show that an ideal orifice should have a value $C_v = \pi/(\pi + 2) = 0.611$ [1].

Figure 5.12 is based on a standard location of the upstream and downstream pressure taps. When the taps are in some other location, the value of C_v is different [2]. In comparison with venturi meters, orifice meters have high pressure losses—high $\mathcal{F}$—and correspondingly high pumping costs, but because they are mechanically simple, they are cheap and easy to install. For small-size lines, orifice meters are much more common than venturi meters.

The values of C_v in Fig. 5.12 are applicable only to drilled-plate orifices (sometimes called *square-edge orifices*, because the edges of the hole are not rounded). Some other standard types also are used, and sets of C_v curves for these have been published [3].

E. Rotameters

The three previously discussed devices use a fixed geometry and read a pressure difference which is proportional to the square root of the volumetric flow rate. A rotameter uses a fixed pressure difference and a variable geometry, which is a simple function of the flow rate. Figure 5.13 shows a schematic view of a simple rotameter. It consists of a tapered transparent (glass or plastic) tube, in which the fluid whose flow is to be measured flows upward, and an interior float, which is shown in Fig. 5.13 as a spherical ball.

Suppose the flow is upwards as shown in Fig. 5.13, is steady so that the ball is not moving, and is fast enough to hold the ball steadily suspended in the flow. If we make a force balance around the ball (positive downward), we find

$$0 = F_{\text{gravity}} + F_{\text{pressure above}} - F_{\text{buoyancy}} - F_{\text{pressure below}} \tag{5.35}$$

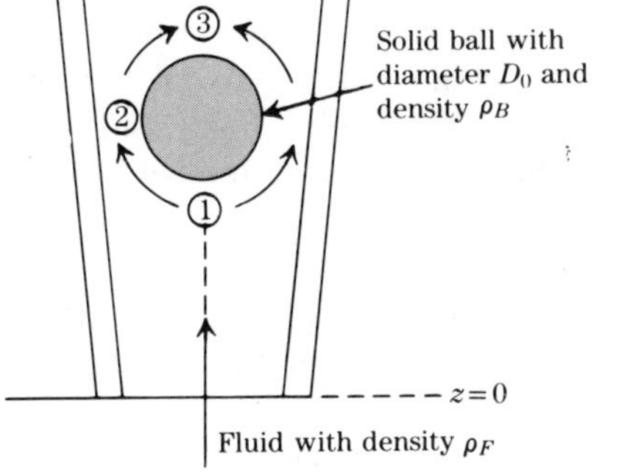

FIGURE 5.13
Rotameter.

If we assume that the pressure below the ball is practically uniform across the ball's lower surface, and similarly for the pressure across the ball's upper surface, and if we remember from Chap. 2 that the z component of that pressure force will be simply the pressure times the projected area of the ball, then we find

$$0 = \frac{\pi}{6} D_0^3 \rho_b g + P_3 \pi D_0^2 - \frac{\pi}{6} D_0^3 \rho_f g - P_1 \pi D_0^2$$

$$\frac{\pi}{6} D_0^3 (\rho_b - \rho_f) g = \pi D_0^2 (P_1 - P_3) \tag{5.36}$$

From Bernoulli's equation

$$P_1 - P_2 = \rho\left(\frac{V_2^2}{2} - \frac{V_1^2}{2}\right) = \rho\,\frac{V_2^2}{2}\left(1 - \frac{A_2^2}{A_1^2}\right)$$

But $(A_2/A_1)^2$ is generally much less than 1, so we can drop the last term above. And as discussed previously, the flow from 2 to 3 is a sudden expansion, so that P_3 is very close to P_2. Making these substitutions in Eq. 5.36 and solving for V_2, we find

$$V_2 = \left(\frac{D_0 g}{3} \cdot \frac{\rho_B - \rho_F}{\rho_F}\right)^{1/2} \tag{5.37}$$

Thus, applying Bernoulli's equation (and some judicious assumptions), we find that for a given diameter of the ball and the given densities of the ball and the fluid, there is only one possible value of V_2 that will keep the ball steadily suspended! Thus for any flow rate Q, the ball must move to that elevation in the tapered tube where $V_2 = Q/A_2$. But

$$A_2 = \frac{\pi}{4}\left[(D_0 + Bz)^2 - D_0^2\right] = \frac{\pi}{4}\left[2Bz + (Bz)^2\right] \tag{5.38}$$

The taper B of the tube is generally small enough that the $(Bz)^2$ term in Eq. 5.38 is small compared to $2Bz$ and can be dropped. Then it follows that the height z at which the ball stands is linearly proportional to the volumetric flow rate Q.

This treatment is simple; more complex treatments [4] lead to similar conclusions. Because of some of the assumptions that went into finding Eq. 5.37, we should not assume that we can compute the true velocities from it; an empirical coefficient like the orifice coefficient would enter. However, most rotameters are treated as calibrated devices; for a given tube, float, and fluid the Qz curve is measured and thereafter one simply reads the float position and looks up the flow rate from that calibration curve.

Example 5.10. Our rotameter has been calibrated for nitrogen at room temperature and atmospheric pressure; the calibration shows that for a reading (float position) of 50 percent of the height of the rotameter tube, the volumetric flow rate is 100 cm^3/min. We now need to measure the flow of helium at room temperature and atmospheric pressure, using the same rotameter. When the reading is 50 percent of full scale, we estimate the helium volumetric flow rate.

From Eq. 5.37 we know that the velocity, and hence the volumetric flow rate, for a given float position is proportional to

$$\left(\frac{\rho_B - \rho_F}{\rho_F}\right)^{1/2}$$

Here the density of the float, if it is made of almost any solid material, is at least 1000 times the density of nitrogen at atmospheric pressure. So we can safely drop the ρ_F in the numerator, from which it follows that the velocity is proportional to $1/(\text{fluid density})^{1/2}$. Thus

$$Q_{\text{hel}} = Q_{\text{nit}}\left(\frac{\rho_{\text{nit}}}{\rho_{\text{hel}}}\right)^{1/2} = Q_{\text{nit}}\left(\frac{M_{\text{nit}}}{M_{\text{hel}}}\right)^{1/2}$$

$$= 100\,\frac{\text{cm}^3}{\text{min}}\left(\frac{28}{4}\right)^{1/2} = 265\,\frac{\text{cm}^3}{\text{min}} \qquad \blacksquare$$

Rotameters are widely used for measuring low flow rates. The simple spherical ball float is used for the smallest flows, and more complex float designs are used for larger flow rates.

5.9 NEGATIVE ABSOLUTE PRESSURES: CAVITATION

In certain flows Bernoulli's equation can predict negative absolute pressures, as shown by the following two examples. In gases negative absolute pressures have no physical meaning at all. When Bernoulli's equation predicts a negative absolute pressure for a gas flow, then the flow probably contains velocities much too high for the assumptions of Bernoulli's equation; the equations developed in Chap. 8 must then be used.

In liquids negative absolute pressures can exist under very rare conditions, but they are unstable. Normally when the absolute pressure on a liquid is reduced to the vapor pressure of the liquid, the liquid boils. This converts the flow to a two-phase flow, which has a much higher value of $\mathscr{F}$ than does the corresponding one-phase flow. Thus, when Bernoulli's equation predicts a pressure less than the vapor pressure of the liquid, the flow as calculated is physically impossible. The actual flow will have a much higher friction effect, and the flow velocity will be less than that assumed in the original calculation.

Example 5.11. Figure 5.14 shows a siphon draining a tank of water. What is the absolute pressure at point 2?

Applying Bernoulli's equation without friction from the free surface, point 1, to the outlet, point 3, we find

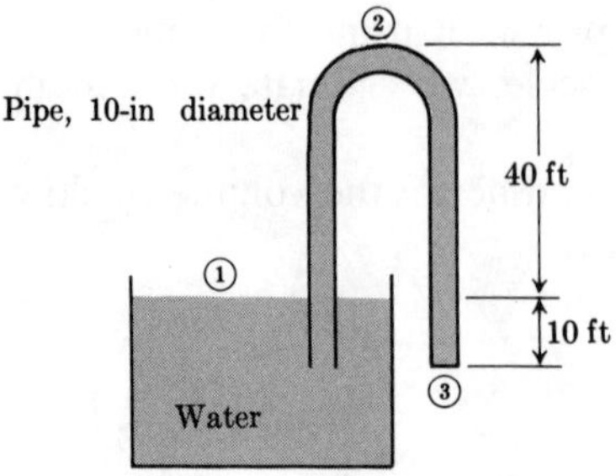

FIGURE 5.14

$$V_3 = [2g(h_1 - h_3)]^{1/2} = [2(32.2\text{ ft/s})(10\text{ ft})]^{1/2}$$
$$= 25.3\text{ ft/s} = 7.71\text{ m/s}$$

Then applying Bernoulli's equation between points 1 and 2, we find

$$P_2 = P_1 - \rho\left[\frac{V_2^2}{2} + g(z_2 - z_1)\right]$$
$$= 14.7\,\frac{\text{lbf}}{\text{in}^2} - 62.3\,\frac{\text{lbm}}{\text{ft}^3}\left[\frac{(25.3\text{ ft/s})^2}{2} + 32.2\,\frac{\text{ft}}{\text{s}^2}\cdot 40\text{ ft}\right]\cdot\frac{\text{lbf}\cdot\text{s}^2}{32.2\text{ lbm}\cdot\text{ft}}\cdot\frac{\text{ft}^2}{144\text{ in}^2}$$
$$= 14.7 - 21.6 = -6.9\text{ lbf/in}^2 = -47.6\text{ kPa} \qquad ?$$ ■

This flow is physically impossible. One may show that when water is open to the atmosphere, such siphons can never lift water more than about 34 ft (10.4 m) above the water surface, even with zero velocity; the siphon shown in Fig. 5.14 will not flow at all. In this example the physically unreal, negative pressure was mostly a result of the gravity term in Bernoulli's equation. Negative absolute pressures can also be predicted by Bernoulli's equation for horizontal flows in which gravity plays no role.

Example 5.12. Water flows from a pressure vessel through a venturi meter to the atmosphere (see Fig. 5.15); $P_1 = 10$ psig and $A_2/A_3 = 0.50$. What is the pressure at location 2?

Applying Bernoulli's equation without friction between locations 1 and 3, we find

$$V_3 = \left[2\,\frac{(P_1 - P_3)}{\rho}\right]^{1/2}$$
$$= \left(2\cdot\frac{10\text{ lbf/in}^2}{62.3\text{ lbm/ft}^3}\cdot\frac{32.2\text{ lbm}\cdot\text{ft}}{\text{lbf}\cdot\text{s}^2}\cdot\frac{144\text{ in}^2}{\text{ft}^2}\right)^{1/2}$$
$$= 38.6\text{ ft/s} = 11.8\text{ m/s}$$

The mass balance gives

$$V_2 = V_3\,\frac{A_3}{A_2} = 38.6\,\frac{\text{ft}}{\text{s}}\cdot\frac{1}{0.50} = 77.2\,\frac{\text{ft}}{\text{s}} = 23.5\,\frac{\text{m}}{\text{s}}$$

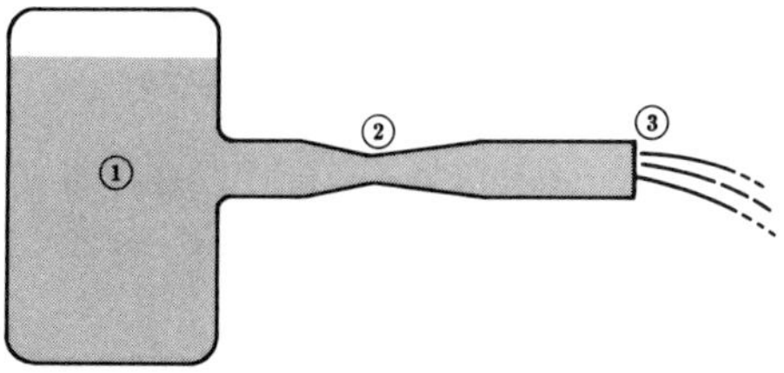

FIGURE 5.15

Applying Bernoulli's equation without friction between locations 1 and 2, we find

$$P_2 = P_1 - \rho \frac{V_2^2}{2}$$

$$= 24.7 \text{ psia} - \frac{(62.3 \text{ lbm/ft}^3)(77.2 \text{ ft/s})^2}{2} \cdot \frac{\text{lbf} \cdot \text{s}^2}{32.2 \text{ lbm} \cdot \text{ft}} \cdot \frac{\text{ft}^2}{144 \text{ in}^2}$$

$$= 24.7 - 40.1 = -15.4 \text{ psia} = -106 \text{ kPa} \quad ?$$ ■

This flow also is physically unreal. At such high velocities, the fictional effects become large; so the frictionless assumption above is a poor one. If the frictional effect were negligible, then the fluid would boil in the venturi and thus convert to a two-phase flow with a much lower flow rate.

Example 5.12 shows that as the velocity increases in horizontal flow, the pressure falls. The pressure decrease can cause the liquid to boil. Dramatic examples of this phenomenon occur in pumps, turbines, and ship's propellers. In these devices, the fluid is often speeded up to a velocity at which it forms a vapor bubble. Then the bubble flows to a region of higher pressure and collapses. The collapse can cause a sudden pressure pulse, and the pulses, occurring at high frequencies, can damage the pump, turbine, etc. The phenomenon of local boiling due to velocity increase is called *cavitation*, the study of which is an important part of modern research in fluid machines [5, 6].

5.10 BERNOULLI'S EQUATION FOR UNSTEADY FLOWS

Bernoulli's equation is a steady-flow equation; however, it can be applied successfully to some unsteady flows, if the changes in flow rate are slow enough to be ignored. To decide how slow the change must be to be ignored, we reason as follows. For a steady flow, $(\partial V/\partial t)_{x,y,z}$ is zero. This means that although an observer riding with the fluid would observe a changing velocity, an observer watching a specific point in the system would observe no change in velocity with respect to time. Then we are safe in ignoring unsteady-flow effects if $(\partial V/\partial t)_{x,y,z}$ for all points in the system is small compared with the acceleration we are considering, i.e., the acceleration of gravity or the acceleration due to pressure forces $(dP/dL)/\rho$. If, however, $(\partial V/\partial t)_{x,y,z}$ at any point in the system is comparable to the largest of the other acceleration terms in the system, then we cannot safely apply Bernoulli's equation to the system.

Example 5.13. If the tank in Fig. 5.5 is cylindrical with a diameter of 10 m, and if the outlet hole is 1 m in diameter, how long does it take the fluid level to drop from 30 m above the tank outlet to 1 m above the tank outlet?

Here we assume that the $(\partial V/\partial t)_{x,\,y,\,z}$ is small; we check that assumption later. Then the instantaneous flow rate is assumed to be given by Bernoulli's equation, which here takes the form of Torricelli's equation:

$$V_2 = (2gh)^{1/2}$$

But by the mass balance for an incompressible fluid,

$$V_2 = V_1 \frac{A_1}{A_2}$$

Where V_1 is the rate at which the free surface of the tank is moving downward, which is equal to $-dh/dt$. So

$$V_2 = \frac{-dh}{dt}\frac{A_1}{A_2} = (2gh)^{1/2}$$

$$\frac{-dh}{h^{1/2}} = \frac{A_2}{A_1}(2g)^{1/2}\,dt$$

$$-\int_{h_1}^{h_2}\frac{dh}{h^{1/2}} = -2h^{1/2}\Big]_{h_1}^{h_2} = \frac{A_2}{A_1}(2g)^{1/2}\int_{t_1}^{t_2}dt = \frac{A_2}{A_1}(2g)^{1/2}t\Big]_{t_1}^{t_2}$$

Therefore,

$$\Delta t = t_2 - t_1 = \frac{-2(h_2^{1/2} - h_1^{1/2})}{(A_2/A_1)(2g)^{1/2}}$$

Inserting numbers, we find

$$\Delta t = \frac{-2[(1\text{ m})^{1/2} - (30\text{ m})^{1/2}]}{\{[(\pi/4)(1\text{ ft})^2]/[(\pi/4)(10\text{ ft})^2]\}(2\cdot 9.81\text{ m/s}^2)^{1/2}}$$

$$= 2.02\times 10^2\text{ s} = 3.37\text{ min}$$

The maximum velocity in this tank is at the outlet, and all other velocities are proportional to it; therefore, the maximum value of $(\partial V/\partial t)_{x,\,y,\,t}$ must occur at the outlet. Differentiating Torricelli's equation with respect to time yields

$$\frac{\partial V_2}{\partial t} = \frac{(2g)^{1/2}}{2h^{1/2}}\frac{dh}{dt}$$

Substituting for dh/dt, we find

$$\frac{\partial V_2}{\partial t} = \left(\frac{g}{2h}\right)^{1/2}\cdot V_2\,\frac{A_2}{A_1} = \left(\frac{g}{2h}\right)^{1/2}(2gh)^{1/2}\,\frac{A_2}{A_1} = g\,\frac{A_2}{A_1}$$

Thus in this example the maximum value of $(\partial V/\partial t)_{x,\,y,\,z}$ is $\frac{1}{100}$ the acceleration of gravity, and the unsteady-flow aspect of the problem can be neglected safely. ■

Most flow problems in which the unsteady flow cannot be neglected and hence Bernoulli's equation cannot be applied involve starting the flow from rest or a sudden stopping of the flow. Consider the pipe and valve shown in Fig. 5.16. Initially the pipe is practically full of fluid, and the valve is closed. Then the valve is suddenly opened. If the friction effect is negligible, the fluid will fall freely, maintaining its cylindrical shape, just as a solid rod would. In this case, the entire outflow process takes place during the flow-starting period; the whole fluid is still accelerating when the last particle of fluid leaves the pipe. Friction and surface tension complicate the picture; but for low-viscosity fluids in short, large-diameter pipes, the result described above is observed experimentally.

In this case, all the fluid has the same velocity, and thus $(\partial V/\partial t)_{x,\,y,\,z}$ is the same at all points where there is fluid. Here it is equal to g, so it is the same size as the largest accelerations in Bernoulli's equation, and the test indicates that we cannot safely apply Bernoulli's equation to this problem.

The other general type of unsteady-flow problem which cannot be solved by Bernoulli's equation is the problem with sudden valve closing, which leads to a phenomenon called *water hammer*.

Figure 5.17 shows a tank from which a liquid flows through a pipe, at the end of which is a quick-closing valve. If the liquid is flowing steadily and the valve is suddenly closed, the flow during the closing process cannot be described by Bernoulli's equation. Bernoulli's equation would indicate that once the valve closed, the pressure throughout the system would be the pressure given by the basic equation of fluid statics. Actually, at the time the valve is being closed, the fluid in the pipe has significant kinetic energy, and the sudden shutting of the valve requires that kinetic energy be converted either to internal energy, with a rise in temperature, or to injection work, with a rise in pressure.

In this chapter we concentrated on problems most easily solved by the energy balance (of which Bernoulli's equation is a restricted form). The

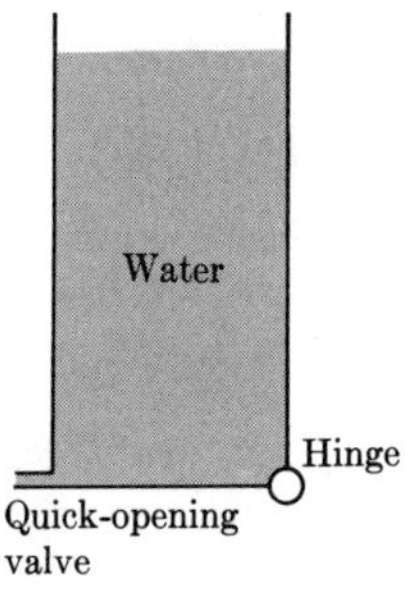

FIGURE 5.16

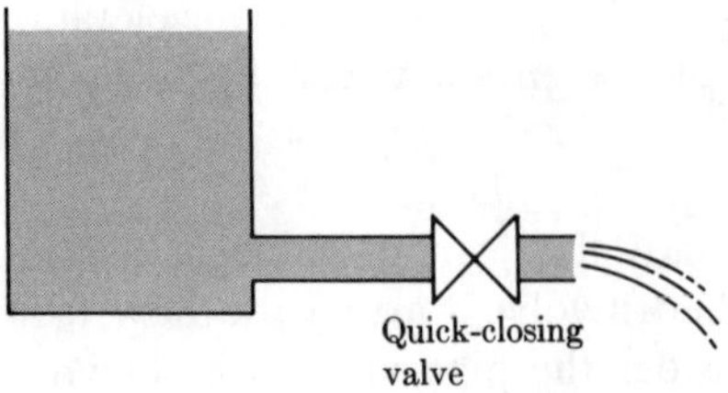

FIGURE 5.17

problem of suddenly stopping the fluid in Fig. 5.17 and the problem of starting it from rest are both solved more easily by the momentum balance (Chap. 7). We return to this problem and the problem of what happens when the valve in Fig. 5.17 is suddenly opened in Sec. 7.4. For now we simply note that while Bernoulli's equation is immensely useful, there are some problems for which it is not useful; the starting and stopping of the flow in Fig. 5.17 is one such problem.

5.11 NONUNIFORM FLOWS

So far in this chapter and in the vast majority of problems in pipes, channels, ducts, etc., we assume that the velocity is practically uniform across the pipe, duct, or channel so that we may associate one velocity with the entire flow at one area perpendicular to the flow. In most flows of practical interest to chemical engineers, this simplification introduces negligible errors. However, there are some very simple and common flows for which this is not the case. The simplest and most illustrative example is the flow over a sharp-edged weir.

Figure 5.18 shows schematically the flow in an open channel, which passes over a sharp-edged weir. You can study a very similar flow in the kitchen sink by pouring water out of a pot or a cup, at a high enough velocity that the flow does not dribble down the side of the pot or cup, but rather flows freely away from the edge, as shown in Fig. 5.18. The flow over the weir in Fig. 5.18 is much simpler than the flow out of a cup, because the weir is assumed to be straight and to extend a long way into and out of the page, so the complications where it meets the walls of the channel and the complications due to the curvature of the cup or pot can be ignored.

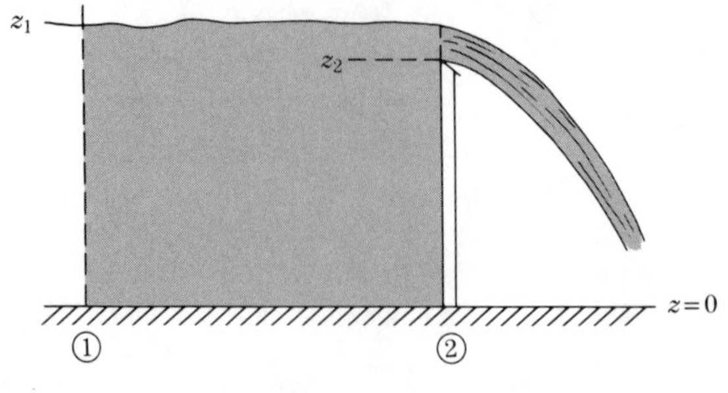

FIGURE 5.18
Flow over a weir.

Here at 1, far upstream from the weir, the velocity is presumably uniform (ignoring the effects of friction), equal to V_1. Clearly at 1 we do not have a single value of z, but have elevations ranging from $z = 0$ to $z = z_1$. Similarly we do not have a single pressure, but have gauge pressures ranging from zero at the free surface to $P = \rho g z_1$ at the bottom. Fortunately this causes little problem because the sum $P/\rho + gz + V^2/2$ is constant at 1, independent of z, because the pressure rises as z declines, according to the basic equation of fluid statics, to keep the sum of the $P/\rho + gz$ terms constant.

The same is not true at 2. There the flow is open to the air at both sides, so that the gauge pressure at 2 must be zero, at all elevations above the weir. We can thus write Bernoulli's equation between an arbitrary upstream point at 1 and some elevation z (above z_2) at point 2. Because the sum $P/\rho + gz$ is constant at 1, we choose $z = z_1$, for which $P_1 = 0$, and we write

$$gz_1 + \frac{V_1^2}{2} = gz + \frac{V_2^2}{2}$$

$$V_2 = [2g(z_1 - z) + V_1^2]^{1/2} \tag{5.39}$$

This says that (1) at the free surface the velocity should be the same as the velocity at 1, which in most cases is negligibly small, and (2) at the bottom of the overflowing stream, the velocity should be a maximum, with the value given by the above equation, with $z = z_2$. You can try this by pouring water from a cup in the sink, and you will see that this is not exactly the case. The internal friction in the flow does not allow the fluid at the surface to go that slowly, adjacent to the much faster-flowing fluid just below it. Instead the faster-flowing fluid drags the surface fluid along, faster than Eq. 5.39 predicts. Also as the surface fluid speeds up, its elevation falls, so that instead of being completely level up to the weir, as Fig. 5.18 shows, the free surface actually falls slightly just before the weir, to satisfy the energy balance.

Ignoring this disagreement between Eq. 5.39 and what we can see in the sink, we can compute the expected flow rate by considering a length W into the page in Fig. 5.18. To simplify the integration, we now measure elevations downward from the surface, defining $h = z_1 - z$, and write

$$Q = \int V\,dA = W \int_0^{h_1} (2gh + V_1^2)^{1/2}\,dh$$

Normally V_1 is small enough that we can drop it from the right side of the above equation, and we integrate to find

$$Q = W\sqrt{2g}\,\frac{h_1^{3/2}}{\frac{3}{2}} \tag{5.40}$$

Experimental results show that the $\frac{3}{2}$ dependence of Q on h_1 is correct, but that the flow rate is less than predicted by Eq. 5.40, typically about 67 percent of what that equation predicts [7].

The same differences in velocity from top to bottom of the flow that we calculate here are certainly present in all the horizontal-flow examples in this chapter. However, in a flow like that shown in Fig. 5.5, the difference in elevation from top to bottom of the exit flow is so small compared to the elevation change from the free surface in the tank to the centerline of the exit that we make a negligible error in ignoring the minor differences in velocity from top to bottom of the exit flow. The same is true of most flows of practical interest to chemical engineers. But for shallow gravity-driven flows, e.g., the flow over weirs in distillation columns, clarifiers, etc., we must take them into account.

For inherently two- or three-dimensional flows, such as the flow around an airplane, the simple application of Bernoulli's equation from one point to another in the flow, which we have used here, is applicable only if the two points chosen are on a single streamline, as discussed in Chap. 10.

5.12 SUMMARY

1. Bernoulli's equation is the energy balance equation for steady flow of constant-density fluids.
2. For constant-density fluids, the term $\Delta u - dQ/dm$ in Bernoulli's equation represents the friction heating per unit mass.
3. Although no fluid has exactly constant density, Bernoulli's equation can be applied with negligible error to almost all steady flows of liquids and to steady flows of gases at low velocities.
4. A large number of fluid-measuring devices are based on the frictionless form of Bernoulli's equation. Friction is present in these devices. The commonly used equations for these devices retain the frictionless Bernoulli's equation form and add empirical correction factors to deal with the effects of friction.
5. Bernoulli's equation can predict negative absolute pressures for some impossible flows. For gas flows, this prediction normally means that the velocities are too high for Bernoulli's equation to apply. For liquids, it normally means that the fluid will boil, leading to a two-phase flow and a much lower velocity than predicted.
6. Although Bernoulli's equation is a steady-flow equation, it can be used for unsteady flows if the time rate of change of velocity at every point in the system is small compared with the accelerating forces (e.g., the acceleration of gravity). It is not useful for problems involving sudden starting and stopping of flows; those are best solved with the momentum balance.
7. Normally we ignore the differences in velocity perpendicular to the flow in applying Bernoulli's equation to the flow in pipes and channels. This causes negligible errors, except in shallow gravity-driven flows, such as the flows over weirs.

PROBLEMS

See the Common Units and Values for Problems and Examples inside the back cover of the book.

5.1. If a body falls 1000 ft in free fall and then is stopped by friction in such a way that all its kinetic energy is converted to internal energy, by how much will its internal energy per unit mass increase? How much will the temperature of the body increase if (*a*) it is steel, $C_V = du/dT = 0.12\ \text{Btu}/(\text{lbm} \cdot {}^\circ\text{F})$, and (*b*) it is water, $C_V = du/dT = 1.0\ \text{Btu}/(\text{lbm} \cdot {}^\circ\text{F})$? Here C_V is the heat capacity at constant volume.

5.2. In the head form of Bernoulli's equation, show that each term has the dimension of a length.

5.3. Water is flowing in a pipe at a velocity of 8 m/s. Calculate the pressure increase and the increase in internal energy per unit mass for the following ways of bringing it to rest: (*a*) a completely frictionless diffuser with infinitely large A_2, (*b*) a diffuser which has 90 percent of the pressure recovery of a frictionless diffuser, with infinitely large A_2, and (*c*) a sudden expansion.

5.4. A fluid is flowing in a frictionless diffuser in which $A_2/A_1 = 3$ and $V_1 = 10$ ft/s. Calculate the pressure recovery $(P_2 - P_1)$ (*a*) when the fluid is water and (*b*) when the fluid is air.

5.5. Rework Example 5.1, calculating the density from the formula $\rho = MP_{av}/(RT)$, where P_{av} is $0.5(P_1 + P_{atm})$. Compare the results with those shown in the table in that example.

5.6. Torricelli's equation can be reduced to a simple plot of V as a function of h. Prepare such a plot for heights up to 1000 ft.

5.7. The tank shown in Fig. 5.19 has an outflow area of 2 ft². The diameter of the tank is so large that it may be considered infinite. The height h is 12 ft. How many cubic feet per second are flowing out? Assume frictionless flow.

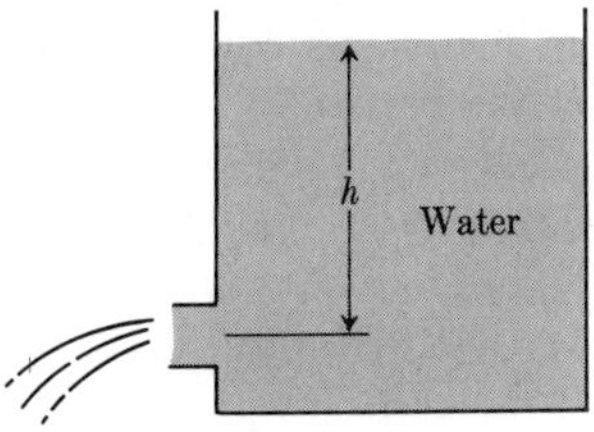

FIGURE 5.19

5.8. Repeat Example 5.2 when the fluid is gasoline.

5.9. Repeat Prob. 5.7 except that now the horizontal cross-sectional area of the tank is 5 ft².

5.10. An ocean liner strikes an iceberg, which tears a 5-m² hole in its side. The center of the hole is 10 m below the ocean surface. The emergency safety doors of the ocean liner have not yet closed, so that the part of the ship which is open to the hole is also open to the atmosphere. Estimate the volumetric flow rate of water into the ship.

5.11. Rework Example 5.4, making the fluid in the tank be air at the same temperature and pressure as the air of the atmosphere. Is the answer from Eq. 5.20 plausible? Would the answer from Eq. 5.22 be plausible?

5.12. The tank in Fig. 5.20 is full of helium, of molecular weight 4 g/mol, at the same temperature and pressure as the surrounding atmosphere. Assuming steady, frictionless flow, how fast is the helium flowing through the hole?

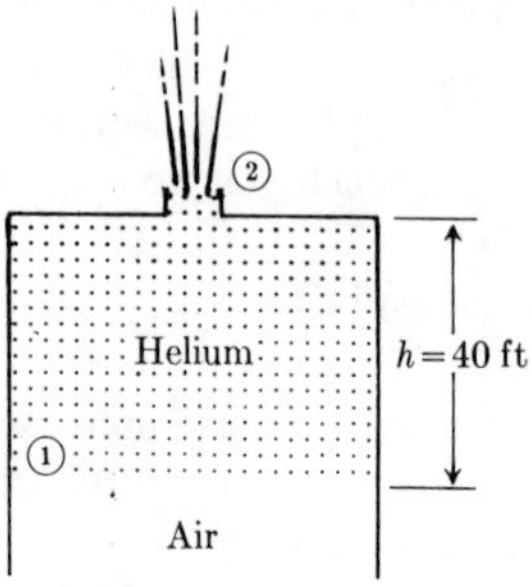

FIGURE 5.20

5.13. In Fig. 5.21 a tank of water is immersed in a larger tank of gasoline, and the water is flowing through a hole in the bottom. What is the velocity of this flow?

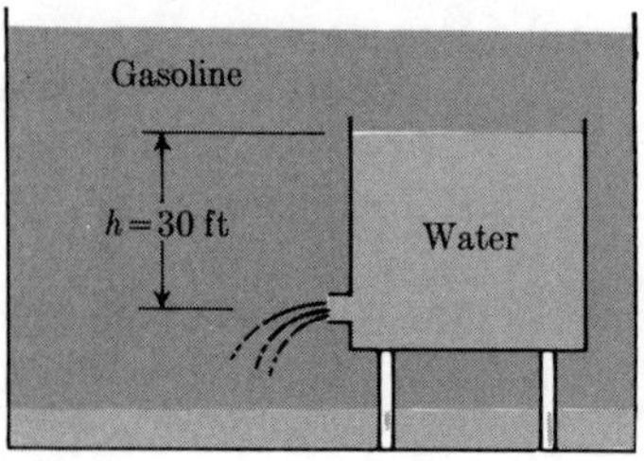

FIGURE 5.21

5.14. In the vessel in Fig. 5.22 water is flowing steadily in frictionless flow under the barrier. What is the velocity of the water flow under the barrier?

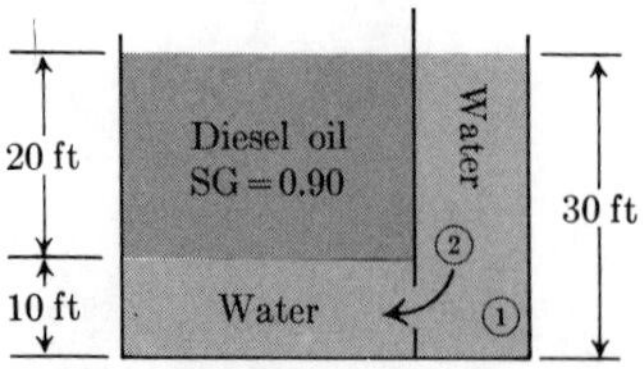

FIGURE 5.22

5.15. In the tank and standpipe in Fig. 5.23 which way is the fluid flowing? *Hint:* Write Bernoulli's equation, taking the two free surfaces as points 1 and 2. Compute the magnitude and sign of $\mathscr{F}$ for flow in each direction.

5.16. In the tank in Fig. 5.24 water is under a layer of compressed air, which is at a pressure of 20 psig. The water is flowing out through a frictionless nozzle, which is 5 ft below the water surface. What is the velocity of the water?

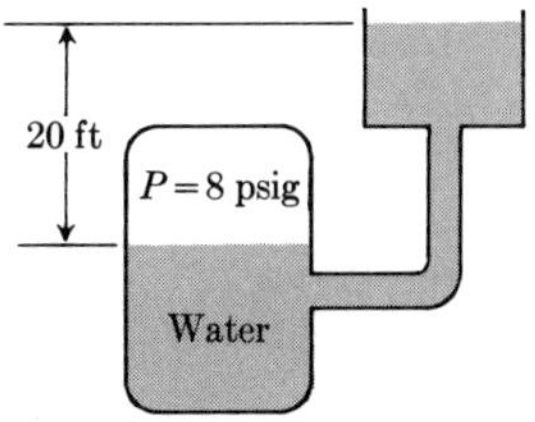

FIGURE 5.23

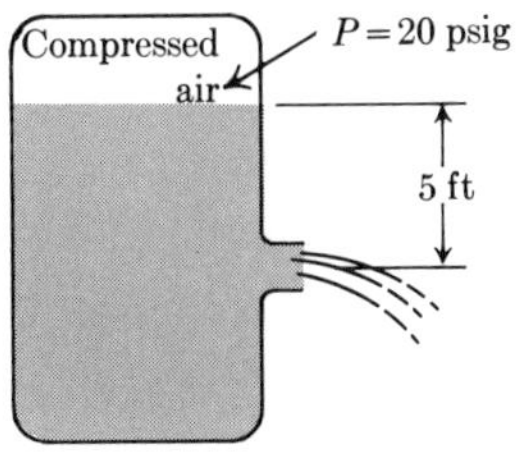

FIGURE 5.24

5.17. The system in Fig. 5.25 consists of a water reservoir with a layer of compressed air above the water and a large pipe and nozzle. The pressure of the air is 50 psig, and the effects of friction can be neglected. What is the velocity of the water flowing out through the nozzle?

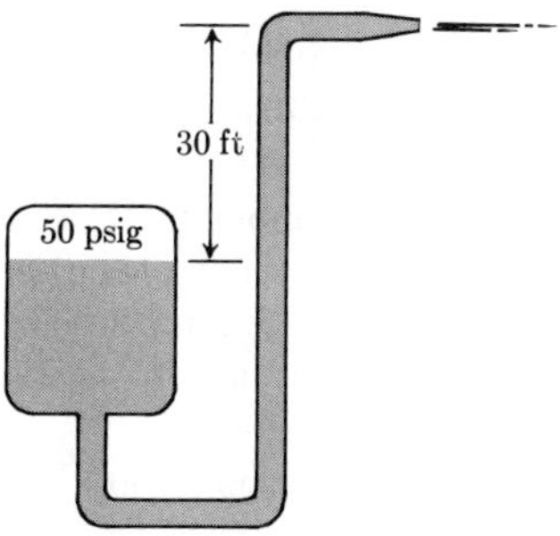

FIGURE 5.25

5.18. The tank in Fig. 5.26 has a layer of mercury under a layer of water. The mercury is flowing out through a frictionless nozzle. The heights are $h_1 = 1$ m and $h_2 = 8$ m. What is the velocity of the fluid leaving the nozzle?

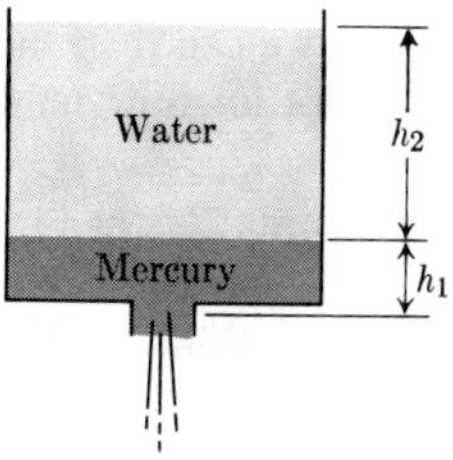

FIGURE 5.26

5.19. The compressed-air-driven water rocket shown in Fig. 5.27 is ejecting water vertically downward through a frictionless nozzle. When the pressure and elevation are as shown, what is the velocity of the fluid leaving the nozzle?

5.20. An industrial centrifuge is sketched in Fig. 5.28. The fluid in the basket is water. The radii are $r_1 = 21$ in and $r_2 = 20$ in. The basket is revolving at 2000 rpm. There is a small hole in the outer wall of the centrifuge, through which the fluid is flowing in frictionless flow. What is the velocity of flow through this hole?

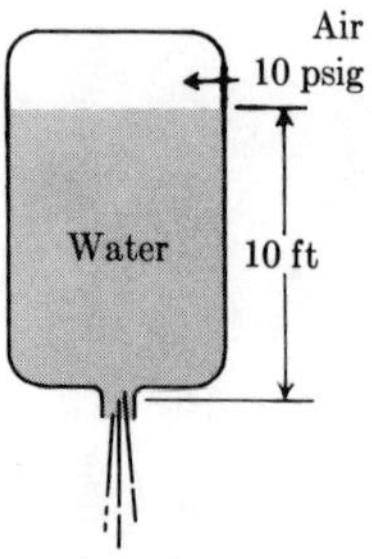

FIGURE 5.27
Compressed-air water rocket.

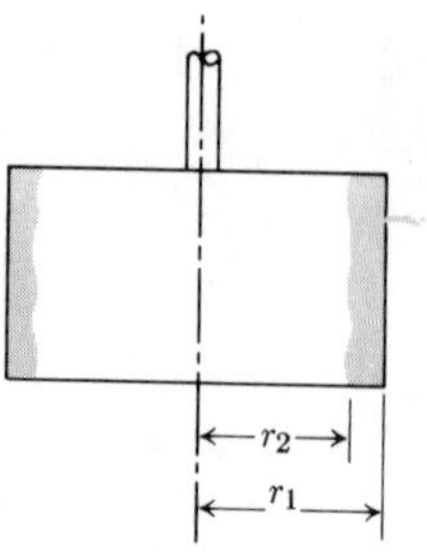

FIGURE 5.28
Centrifuge basket.

5.21. Flow-recorder charts frequently have a square scale rather than a linear one. Why?

5.22. Equipping speedboats with pitot-tube speedometers has been proposed. For ease of construction it is decided that the tubes shall not extend more than 10 ft above the water. What is the maximum speed at which they can be used?

5.23. A pitot-static tube is to be used to measure a flow of air. The manometer fluid is water. We decide not to use the tube for flows so slow that the elevation difference in the manometer is less than 0.1 in because smaller differences are hard to read. What is the smallest air velocity at which we can use this pitot-static tube?

5.24. Repeat Prob. 5.23 except that now we measure the flow of gasoline and the manometer fluid is water.

5.25. A pitot-static tube is used to measure an airplane's airspeed. When the pressure-difference gauge reads 0.3 psig, how fast is the plane going (*a*) at sea level where the air density is about 0.075 lbm/ft^3 and (*b*) at an altitude of 10,000 ft where the air density is about 0.057 lbm/ft^3?

5.26. If the venturi meter in Example 5.7 is to be used on a day-to-day basis, then it will be useful to have a plot of flow rate versus pressure drop, so that we can read the pressure drop and simply look up the flow rate. Sketch such a plot for flow rates of 1 to 10 ft^3/s.

5.27. The venturi meter in Example 5.7 is now set at 30° to the horizontal, as in Fig. 5.10. The flowing fluid is gasoline. The fluid in the bottom of the manometer is colored water. The reading of the manometer is $z_3 - z_4 = 1$ ft. What is the volumetric flow rate of the gasoline?

5.28. Repeat Prob. 5.27 except that the two pressure taps have been replaced with pressure gauges. These are placed on the side of the pipe, so that they indicate pressures on the pipe centerline. The gauge at point 1 reads 7 psig, and the gauge at point 2 reads 5 psig. The difference in elevation between the gauges $z_1 - z_2$ is 2 ft. What is the volumetric flow rate of the gasoline?

5.29. In the apparatus in Fig. 5.29, what is the volumetric flow rate?

5.30. The venturi meter in Fig. 5.30 has air flowing through it. The manometer, as shown, contains both mercury and water. The cross-sectional areas at the upstream location and at the throat are 10 and 1 ft^2, respectively. What is the volumetric flow rate of the air? The discharge coefficient C_v equals 1.0.

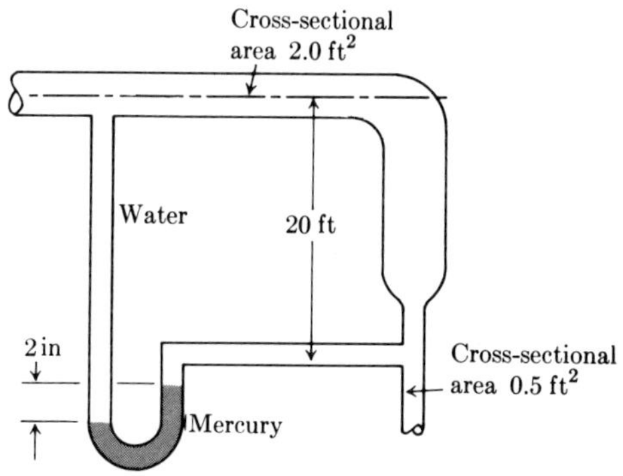

FIGURE 5.29

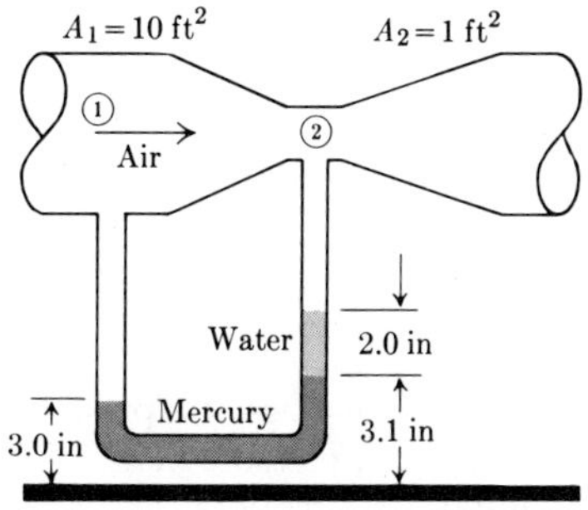

FIGURE 5.30

5.31. The carburetors in automobiles are much more complicated versions of the carburetor shown in Fig. 5.31, but they operate basically the same as that simple one would. The cross-sectional areas at points 1 and 3 are large enough that the velocities there can be considered negligible compared to the velocity at point 2 and the pressures at points 1 and 3 considered equal to atmospheric pressure. The gasoline enters from a constant-liquid-level, atmospheric-pressure reservoir through a large-diameter tube and a small jet, which may be considered a frictionless nozzle with diameter D_j. The diameter at the throat of the venturi, point 2, is D_2.

(*a*) Write the equation for the *air-fuel ratio*, which is the mass flow rate of air divided by the mass flow rate of fuel, in terms of the diameters of the throat, jet, etc.

(*b*) How does this air-fuel ratio change with changes in airflow rate to the engine? (The airflow rate to the engine is governed by the settling of the throttle plate, which is connected to the driver's accelerator pedal and located between the part of the carburetor shown here and the engine.)

(*c*) If we want an air-fuel ratio of 15 lb/lb (which is typical of gasoline engines), what ratio of D_j/D_2 should we choose?

(*d*) If the carburetor shown here gives an air-fuel ratio of 15 at sea level, will it give the same, a higher, or a lower air-fuel ratio in Denver, elevation 5280 ft above sea level?

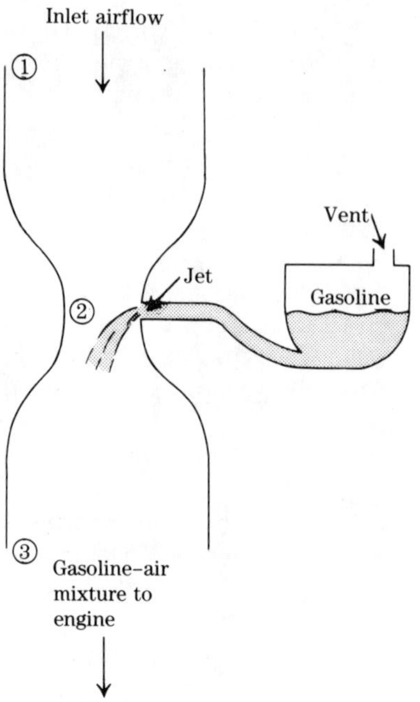

FIGURE 5.31
Elementary carburetor.

5.32. In the United States, natural gas is normally piped inside buildings at a pressure of 4 in of water, while propane is piped inside buildings at a pressure of 11 in of water. Why? *Hint:* The molecular weights are: natural gas, 16 g/mol; propane, 44 g/mol.

5.33. An oil with density 55 lbm/ft^3 is flowing through the orifice in Fig. 5.32. The oil velocity is 1 ft/s in the pipe, and $C_v = 0.6$. What is the indicated value of $P_1 - P_2$?

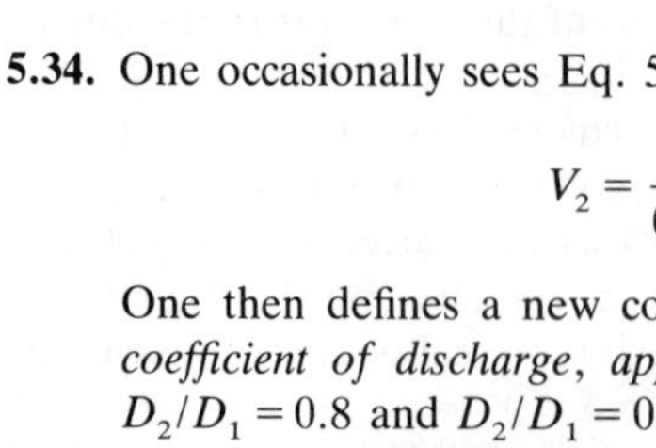

FIGURE 5.32

5.34. One occasionally sees Eq. 5.31 written

$$V_2 = \frac{C_V}{(1 - A_2^2/A_1^2)^{1/2}}\left(\frac{2(P_1 - P_2)}{\rho}\right)^{1/2}$$

One then defines a new coefficient $C = C_V/(1 - A_2^2/A_1^2)^{1/2}$, which is called the *coefficient of discharge, approach-velocity-corrected*. Sketch this coefficient for $D_2/D_1 = 0.8$ and $D_2/D_1 = 0.2$ on a graph like Fig. 5.12.

5.35. We have a flow of mercury in a 1-in-diameter pipe. The velocity of the mercury is 1 ft/s. We want to select a drilled plate to insert in the pipe, so that the pressure-drop signal across it will be 3 lbf/in^2 gauge. What diameter should we select for the orifice hole?

5.36. A venturi meter (Fig. 5.8) has $A_2/A_1 = 0.5$. The fluid flowing is water. The pressure at point 1 is 20 psia. What is the velocity at point 2 which corresponds to a pressure of 0.0 psia at point 2? If the water is at 200°F, its vapor pressure is 11.5 psia. What is the highest velocity possible at point 2 at which water at 200°F will not boil?

5.37. For a siphon similar to that sketched in Fig. 5.14, we wish the fluid to have a velocity of 10 ft/s in the siphon pipe. Assuming frictionless flow and that the minimum pressure allowable is 1 psia, what is the maximum height of the siphon above the liquid surface level?

5.38. A ship's propeller has an outside diameter of 15 ft. When the ship is loaded, the uppermost part of the propeller is submerged 4 ft. If the water is at 60°F (with a vapor pressure of 0.26 psia), what is the maximum speed of the propeller, in revolutions per minute, at which cavitation cannot be expected to occur at the tip of the propeller?

5.39. Is cavitation likely to be as severe a problem with the propellers of submarines as with the propellers of surface ships? Why?

5.40. The tank in Fig. 5.33 is cylindrical and has a vertical axis. Its horizontal cross-sectional area is 100 ft^2. The hole in the bottom has a cross-sectional area of 1 ft^2. The interface between the gasoline and the water remains perfectly horizontal at all times. That interface is now 10 ft above the bottom. How soon will gasoline start to flow out the bottom? Assume frictionless flow.

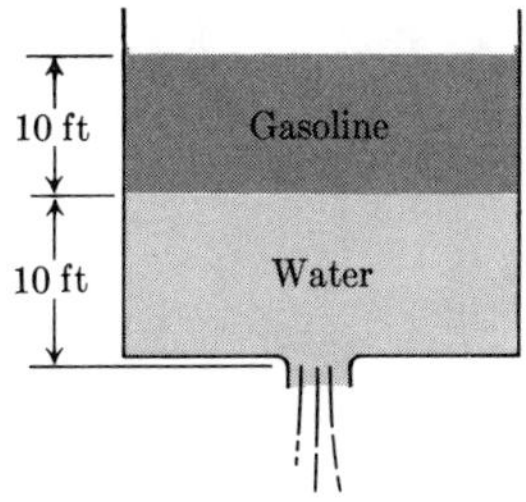

FIGURE 5.33

5.41. The open-topped tank shown in Fig. 5.34 is full to the top with water. The bottom opening is uncovered, so that the water runs out into the air. The cross-sectional

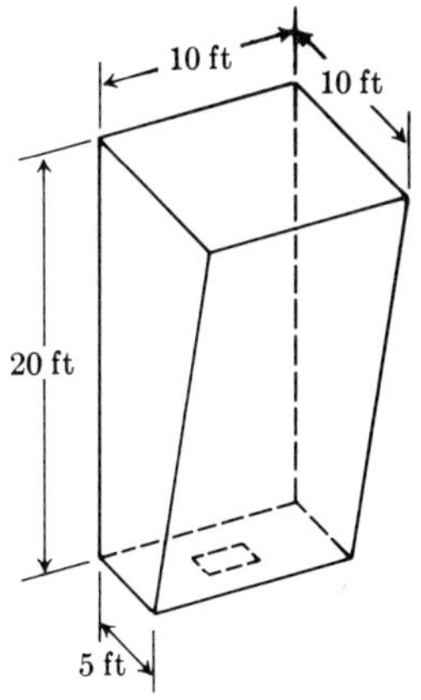

FIGURE 5.34

area of the bottom opening is 1 ft^2. How long does it take the tank to empty? Assume frictionless flow.

5.42. An open-ended tin can (Fig. 5.35) has a hole punched in its bottom. The can is empty and is suddenly immersed in water to depth h_1 and then held steady. The area of the hole is 0.5 in^2, and the horizontal cross-sectional area of the can is 20 in^2. Assuming that the flow through the hole in the bottom of the can is frictionless, how long does it take the can to fill up to the level of the surrounding water?

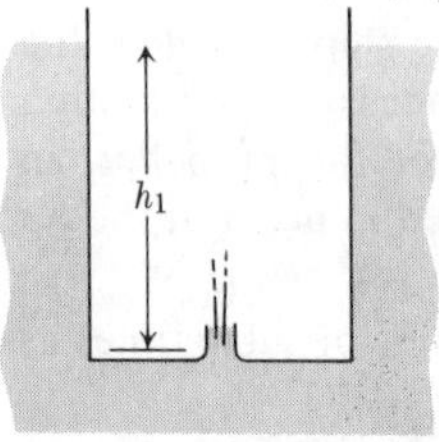

FIGURE 5.35

5.43. Figure 5.36 shows a toy fluid mechanics demonstrator, which consists of a wooden spool, a piece of cardboard, and a thumbtack. When you blow hard enough into the spool, the cardboard is held firmly against the spool; when you stop blowing, the cardboard falls away by gravity. Sketch a pressure-radius plot for pressure along line AA in the sketch. Use the axes shown in the lower part of the figure.

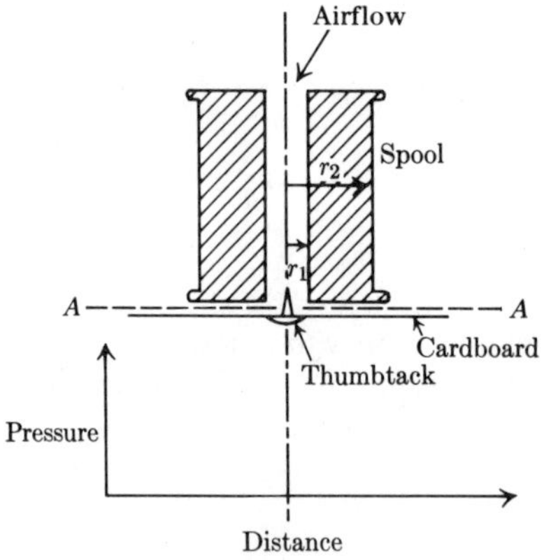

FIGURE 5.36
Toy fluid mechanics demonstrator.

5.44. For frictionless pumps and compressors pumping constant-density fluids, the required work is given by Eq. 5.19. If the fluid is an ideal gas, then that equation becomes $-dW_{\text{a.o.}}/dm = (RT/M)(\Delta P/P)$. Almost all real compressors are intermediate between adiabatic (no heat transfer to the surroundings) and isothermal (complete thermal equilibrium with the surroundings). For those two cases, the required work for ideal gases is shown in Chap. 9 to be

$$\frac{-dW_{\text{a.o.}}}{dm} = \frac{RT}{M} \ln \frac{P_2}{P_1} \qquad \text{[isothermal, frictionless]}$$

$$\frac{-dW_{\text{a.o.}}}{dm} = \frac{RT_1}{M} \cdot \frac{k}{k-1}\left[\left(\frac{P_2}{P_1}\right)^{(k-1)/k} - 1\right] \qquad \text{[adiabatic, frictionless]}$$

Here T_1 is the inlet temperature; k is the ratio of specific heats (discussed in Chap. 8), which is a practically constant value for any gas (=1.40 for air); and P_1 and P_2 are the inlet and outlet pressures, respectively. To show how these formulas compare with the constant-density formula, prepare a plot of $[-M/(RT_1)](dW_{\text{a.o.}}/dm)$ versus P_2/P_1 for air, showing curves for each of the three equations for $1.0 < P_2/P_1 < 1.3$.

5.45. Figure 5.37 shows an air-cushion car, the type widely used to slide heavy loads over relatively smooth surfaces. In it, a fan or blower forces air under pressure into the confined space under the car. This air supports the car and its load. Some of the air continually leaks out through the gap between the skirt of the car and the ground; the fan must supply enough air to make up for this leakage. Assuming that the car and its payload have a total mass of 5000 lbm, that the car is circular with a diameter of 10 ft, and that the clearance between the skirt of the car and the floor is 0.01 in, calculate the airflow rate. Then, assuming that the blower is 100 percent efficient and isothermal (Prob. 5.44), calculate the required blower horsepower.

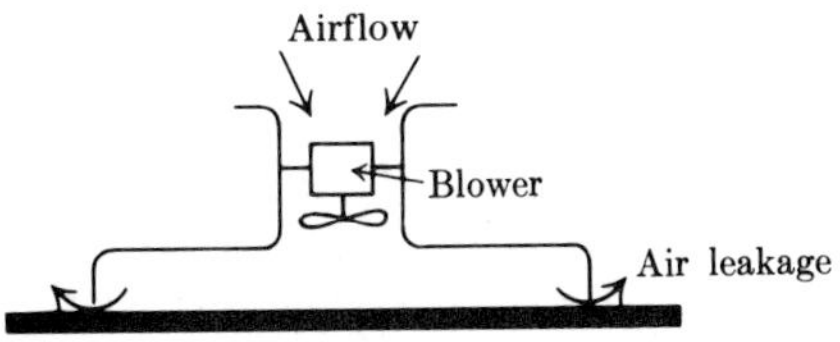

FIGURE 5.37
Air-cushion car.

5.46. Water flows in a channel 50 m wide at 100 m^3/s, and it spills over a sharp-edged weir.
(*a*) Estimate the difference in elevation between the upstream flow and the top of the weir.
(*b*) If the upstream channel is 5 m deep, what is the upstream velocity?
(*c*) How large a percentage error are we likely to have made in neglecting this velocity in formulating Eq. 5.40?

5.47. In Example 5.2 we computed the exit velocity by Torricelli's equation, which does not take into account the fact that at the bottom of the jet the velocity will be higher than at the top, as discussed in Sec. 5.11. How large an error are we likely to have made? If the jet is passing through a perfectly rounded entrance with an outlet diameter of 0.5 ft, and if the centerline of the jet is 30 ft below the fluid surface, how much difference should there be between the velocities at the top and bottom of the jet?

5.48. A slow-moving stream of water flows from a faucet into a sink. It is observed that the width of the stream decreases with distance from the faucet. If the flow leaves the faucet, vertically downward, in the form of a cylindrical jet with diameter 0.25 in and a velocity of 1 ft/s, what will be its diameter 1 ft below the faucet?

5.49. A meteorologist discussing a record-breaking hurricane said, "It had a pressure of 850 mbar in the center, so it had winds of 250 mi/h!" Explain this statement in terms of Bernoulli's equation.

CHAPTER 6

FLUID FRICTION IN STEADY, ONE-DIMENSIONAL FLOW

In Chap. 5 we found the working form of Bernoulli's equation (Eq. 5.7)

$$\Delta\left(\frac{P}{\rho} + gz + \frac{V^2}{2}\right) = \frac{-dW_{\text{a.o.}}}{dm} - \mathscr{F}$$

and applied it to problems in which we could set the friction term $\mathscr{F}$ equal to zero. In this chapter we show how to evaluate the $\mathscr{F}$ term for the very important and practical case of steady flow in one dimension, as in a pipe, duct, or channel. Using the $\mathscr{F}$'s we evaluate here, we can use the above equation for a much wider range of problems than those considered so far, including many problems of great practical interest to chemical engineers. The student should keep in mind that our main reason for evaluating $\mathscr{F}$ is to put the proper relation for $\mathscr{F}$ into Eq. 5.7 and then to solve the resulting equation for the appropriate pressures, velocities, elevations, pipe diameters, etc.

The form of the friction-loss term is very dependent on the geometry of the system. The problem is much simpler if the flow is all in one direction, as in a pipe, rather than in two or three dimensions, as is flow around an airplane. Therefore, first we consider fluid friction in long, constant-diameter pipes in steady flow. This case is of great practical significance and is the easiest case to treat mathematically. The starting and stopping of flow in pipes is discussed in Sec. 7.4.

In Sec. 6.14 we consider the frictional drag on particles in steady, rectilinear motion which, although it is two-dimensional, gives results quite similar to those found in long, straight pipes.

In Chaps. 10 and 11 we investigate two- and three-dimensional flows by using some of the ideas from this chapter and introducing several others.

6.1 THE PRESSURE-DROP EXPERIMENT

The classic pressure-drop experiment is performed on an apparatus like that shown in Fig. 6.1. In this experiment we set the flow of the fluid with the flow-regulating valve. We measure the flow rate with the bucket and stop watch. At steady state we read pressure gauges P_1 and P_2 and record their difference. Usually we are interested in the pressure drop per unit length, so we divide the pressure drop by distance Δx (the length of the test section) and plot $(P_1 - P_2)/\Delta x$ versus volumetric flow rate Q.

Regardless of what newtonian liquid is flowing or of what kind of pipe we use, the result is always of the form shown in Fig. 6.2, and for all gases *at low velocities* the result is the same as that shown. The salient features of Fig. 6.2 are that *for one specific fluid flowing in one specific pipe*:

1. At very low flow rates, the pressure drop per unit length is proportional to the volumetric flow rate to the 1.0 power.
2. At intermediate flow rates, there is a region where the experimental results are not easily reproduced.
3. At very high flow rates, the pressure drop per unit length is proportional to the volumetric flow rate raised to a power which varies from 1.8 (for very smooth pipe) to 2.0 (for very rough pipes).

The experiment is relatively easy to run, and the curves have been found for many combinations of pipe and fluid. However, since not all possible combinations have been tested, it would be convenient to have some way of

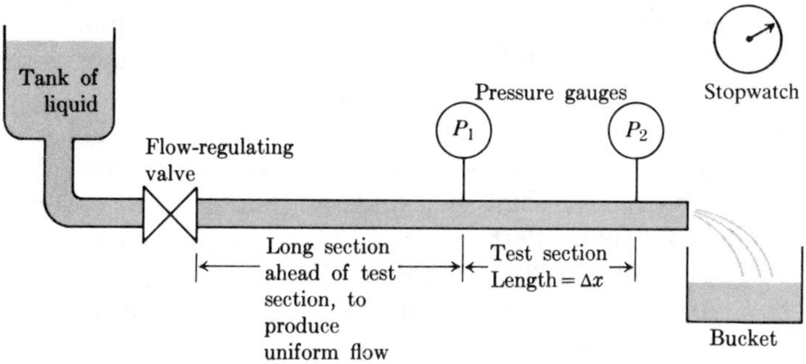

FIGURE 6.1
Apparatus for pressure-drop experiment.

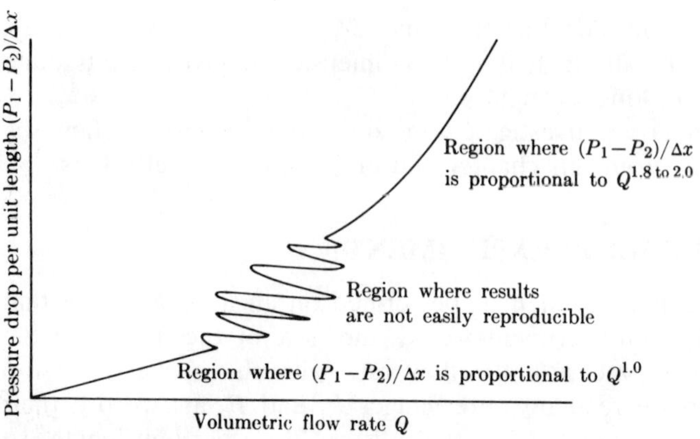

FIGURE 6.2
Typical pressure-drop curve for a specific fluid in a specific pipe.

calculating the results of a new combination from the results of an old one. Furthermore, no inquisitive mind will be satisfied with Fig. 6.2 without asking why it has such a peculiar shape.

6.2 REYNOLDS' EXPERIMENT

Osborne Reynolds [1] explained the strange shape of Fig. 6.2. In an apparatus similar to that of Fig. 6.1 but made of glass, he arranged to introduce a liquid dye into the flowing stream at various points. In the low-flow-rate region (in which the pressure drop per unit length is proportional to the flow rate), the dye he introduced formed a smooth, thin, straight streak down the pipe; there was no mixing perpendicular to the axis of the pipe. This type of flow, in which all the motion is in the axial direction, is now called *laminar flow* (the fluid appears to move in thin shells or layers, or lamina).

In contrast to the result in laminar flow, in the high-flow-rate region, where the pressure drop is proportional to the flow rate to the 1.8 to 2.0 power, no matter where he introduced the dye it rapidly dispersed throughout the entire pipe. A rapid, chaotic motion in all directions in the pipe was superimposed on the overall axial motion and caused the rapid, crosswise mixing of the dye. This type of flow is called *turbulent flow*.

The two types of curve (linear and approximately parabolic) in Fig. 6.2 thus were shown to represent two radically different kinds of flow. The distinction is very important, as we shall see; students should try to observe both types in the world about them. Perhaps the easiest example to see is the smoke from a cigarette rising in a still room (Fig. 6.3). The smoke rises in a smooth, laminar flow for some distance, and then the flow converts to turbulent flow, with random chaotic motion perpendicular to the major, upward flow direction. This case, although easy to demonstrate in the labora-

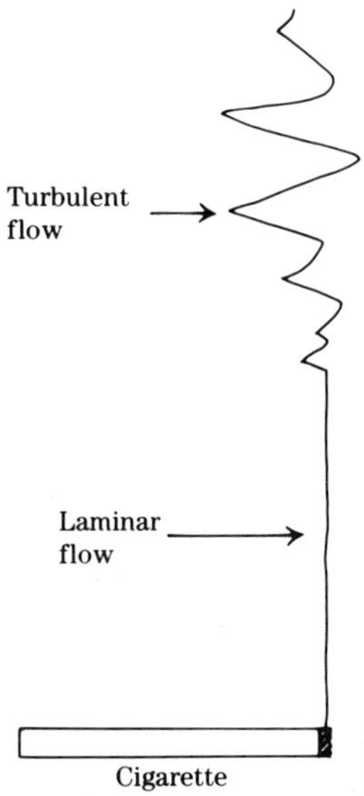

FIGURE 6.3
Laminar and turbulent flows.

tory or in the living room, is much harder to analyze mathematically than Reynolds' pipe flow experiment, so we return to the latter.

Reynolds showed further that the region of unreproducible results between the regions of laminar and of turbulent flow is the region of transition from one type of flow to the other; it is called the *transition region*. The reason for the poor reproducibility here is that laminar flow can exist in conditions in which it is not the stable flow form, but it fails to switch to turbulent flow unless some outside disturbance, such as microscopic roughness on the pipe wall or very small vibrations in the equipment, triggers the transition. Thus, in the transition region, the flow can be laminar or turbulent, and the pressure drop can vary by a factor of 2. Under some circumstances the flow can alternate between being laminar and turbulent, causing the pressure drop to oscillate between a higher and a lower value.

Besides clarifying the strange shape of Fig. 6.2, Reynolds made the most celebrated application of dimensional analysis (Chap. 13) in the history of fluid mechanics. He showed that for smooth, circular pipes, for all newtonian fluids, and for all pipe diameters, the transition from laminar to turbulent flow occurs when the dimensionless group $DV\rho/\mu$ has a value of about 2000. Here D is the pipe diameter, V is the average fluid velocity in the pipe, ρ is the fluid density, and μ is the fluid viscosity. This dimensionless group is now called the *Reynolds number* $\mathcal{R}$. For flows other than pipe flow, some other appropriate length is substituted for the pipe diameter in the Reynolds number, as discussed later.

The results of Reynolds' experiments are summarized in Table 6.1.

TABLE 6.1
Comparison of laminar, transition, and turbulent flows

	Type of flow		
	Laminar	**Transition**	**Turbulent**
Behavior of dye streak	Dye in 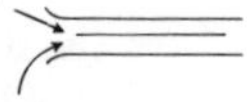 Flow	Oscillates between laminar and turbulent	Dye in 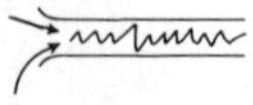 Flow
Pressure drop proportional to	$Q^{1.0}$	Oscillates from one value to another; very difficult to measure	$Q^{1.8}$ (very smooth pipes) to $Q^{2.0}$ (very rough pipes)
Reynolds number	<2000	≈ 2000 to 4000	>4000

6.3 LAMINAR FLOW

Laminar flow is the simplest of the three flows, so we discuss it first. Consider a steady laminar flow of an incompressible newtonian fluid in a horizontal circular tube or pipe. A section of the tube Δx long with inside radius r_0 is shown in Fig. 6.4. We arbitrarily select a rod-shaped element of the fluid, symmetrical about the center, with radius r, and we compute the forces acting on it. Here it is assumed that location 1 is well downstream from the place where the fluid enters the tube. This analysis is not correct for the tube entrance. The flow is steady and all in the axial direction. There is no acceleration in the x direction, so the sum of the forces acting in the x direction on the rod-shaped element we have chosen must be zero. There is a pressure force acting on each end, equal to the pressure times the cross-sectional area of the end. These act in opposite directions; their sum in the positive x direction is

$$\text{Pressure force} = P_1(\pi r^2) - P_2(\pi r^2) = \pi r^2(P_1 - P_2) \tag{6.1}$$

Along the cylindrical surface of our rodlike element, the pressure forces have no component in the x direction and can be ignored, but there is a shear force

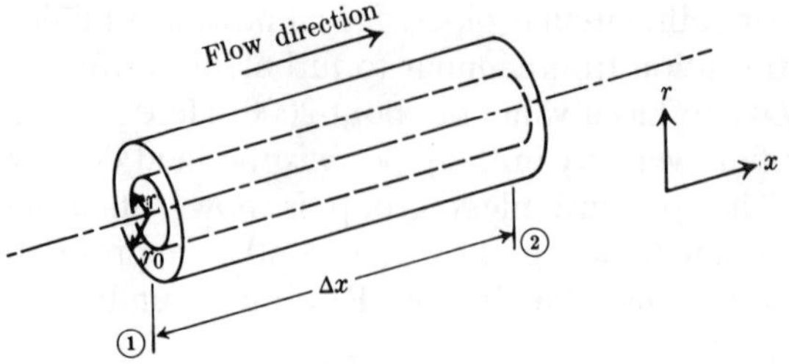

FIGURE 6.4
Force balance system in pipe flow.

resisting the flow. The shear force acts in the direction opposite to the flow, and its magnitude is

$$\text{Shear force} = 2\pi r\,\Delta x\,(\text{shear stress at } r) \tag{6.2}$$

Since the pressure force and shear force are the only forces acting in the x direction, and since the sum of the forces is zero, these must be equal and opposite. Equating their sum to zero and solving for the shear stress at r, we find

$$\tau = \text{shear stress at } r = \frac{-r\,(P_1 - P_2)}{2\,\Delta x} \tag{6.3}$$

This equation applies to steady laminar or turbulent flow of any kind of fluid in a circular pipe or tube.

Here we have applied Newton's law $F = ma$ to the particularly simple case in which there is no acceleration and the sum of the forces is therefore zero; in Chap. 7 we will see how to apply it to more complicated cases.

We saw in Chap. 1 that for newtonian fluids in laminar motion the shear stress is equal to the product of the viscosity and the velocity gradient. Substituting in Eq. 6.3, we find

$$\mu\,\frac{dV}{dr} = -r\,\frac{P_1 - P_2}{2\,\Delta x} \tag{6.4}$$

For steady laminary flow, the pressure gradient $(P_1 - P_2)/\Delta x$ does not depend on radial position in the pipe; so we may integrate this to

$$V = \frac{-r^2}{4\mu} \cdot \frac{P_1 - P_2}{\Delta x} + \text{constant} \tag{6.5}$$

To solve for the value of the constant, we need one observational fact: For the flow of everything except rarefied gases, the fluid at the solid surface clings to the surface. This is not intuitively obvious, and we cannot derive it from some prior principle. The behavior is quite different from that of solids, whose sliding surfaces do slip over one another so that there is a sharp discontinuity in velocity at the sliding boundary. However, we may observe that it is so by watching the behavior of bits of wood or leaves on the surface of a stream: those at the center move rapidly, those near the bank slowly, and those right at the bank not at all. (This condition is often referred to as the *no-slip condition*; one kind of rarefied gas flow is called, logically enough, *slip flow*.) From this observational fact it follows that at $r = r_0$ (at the pipe wall) $V = 0$; so

$$0 = \frac{-r_0^2}{4\mu} \cdot \frac{P_1 - P_2}{\Delta x} + \text{constant} \tag{6.6}$$

Substituting this value of the constant in Eq. 6.5 and factoring, we find

$$V = \frac{r_0^2 - r^2}{4\mu} \cdot \frac{P_1 - P_2}{\Delta x} \tag{6.7}$$

This equation says that for steady, laminar flow of newtonian fluids in circular pipes:

1. The velociy is zero at the tube wall ($r = r_0$).
2. The velocity is a maximum at the center of the pipe ($r = 0$).
3. The magnitude of this maximum velocity is

$$V_{\text{max}} = \frac{r_0^2}{4\mu} \frac{P_1 - P_2}{\Delta x}$$

4. The pressure drop per unit length is independent of the fluid density and is proportional to the first power of the local velocity and the first power of the viscosity.
5. The velocity-radius plot is a parabola; see Fig. 6.5.

In engineering generally we are more interested in the total volumetric flow rate Q than in the local velocity V. To find the Q of a uniform-velocity flow, we multiply the velocity by the cross-sectional area perpendicular to the flow. The velocity of the laminar flow described above is not uniform, so we must integrate velocity times area over the whole pipe cross section:

$$Q = \int_{\text{tube}} V\,dA = \int_{r=0}^{r=r_0} \frac{r_0^2 - r^2}{4\mu} \cdot \frac{P_1 - P_2}{\Delta x} \cdot 2\pi r\,dr$$

$$= \frac{P_1 - P_2}{\Delta x} \cdot \frac{\pi}{2\mu} \left[\frac{r_0^2 r^2}{2} - \frac{r^4}{4} \right]_{r=0}^{r=r_0}$$

$$= \frac{P_1 - P_2}{\Delta x} \cdot \frac{\pi}{\mu} \cdot \frac{r_0^4}{8} = \frac{P_1 - P_2}{\Delta x} \cdot \frac{\pi}{\mu} \cdot \frac{D_0^4}{128} \tag{6.8}$$

This equation was developed by Hagen and also, independently, by Poiseuille [2, p. 160]. In the United States it is most commonly called the *Poiseuille equation*.[†] It shows that the pressure drop $(P_1 - P_2)/\Delta x$ is proportional to the

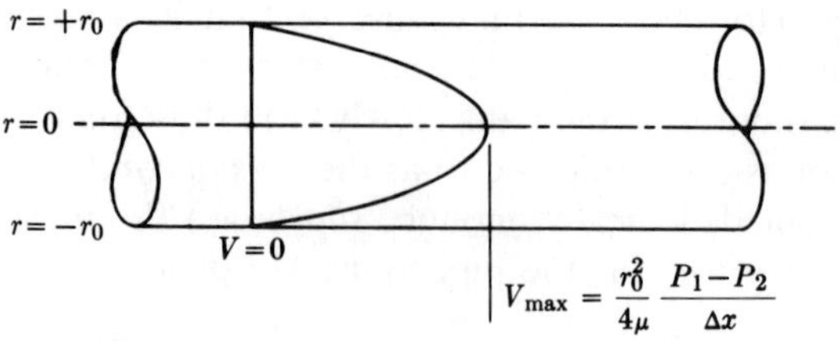

FIGURE 6.5
Velocity distribution in steady, laminar flow of a newtonian fluid in a circular pipe.

[†] Pronounced pwah-zoo-y.

first power of the volumetric flow rate Q, as shown in Fig. 6.2. The solution is immensely satisfying; using only very simple mathematics, we find a complete description of the flow. The description has been experimentally verified so well that when laminar-flow experiments disagree with it, the experiments are probably in error.

From Eq. 6.8 it can also be shown (Prob. 6.4) that

$$V_{\text{av}} = \frac{Q}{\pi r_0^2} = \frac{V_{\text{max}}}{2} \tag{6.9}$$

$$\text{Average ke} = \frac{\int (V^2/2)\,dQ}{\int dQ} = \frac{\int (V^2/2)V(2\pi r)\,dr}{\int V(2\pi r)\,dr} = V_{\text{av}}^2 \tag{6.10}$$

To see how this fits in with Bernoulli's equation (Eq. 5.11), we apply Bernoulli's equation from point 1 to point 2 in Fig. 6.4 and substitute for $-\Delta P$ from Eq. 6.8:

$$\mathscr{F} = \frac{-\Delta P}{\rho} = Q\,\Delta x\,\frac{\mu}{\rho}\,\frac{128}{\pi D_0^4} \tag{6.11}$$

Here we have found an equation which relates the $\mathscr{F}$ in Bernoulli's equation to the flow rate, diameter, length, and (viscosity/density) of a horizontal flow in which gravity plays no role. If we repeat the entire derivation for a vertical flow in a pipe in which the pressure is constant throughout (Prob. 6.1), we find that the ΔP term is replaced with a $\rho g\,\Delta z$ term. When we substitute this in Eq. 6.11 and then calculate $\mathscr{F}$ from Bernoulli's equation, we find

$$\mathscr{F} = -g\,\Delta z = Q\,\Delta x\,\frac{\mu}{\rho}\,\frac{128}{\pi D_0^4} \tag{6.12}$$

Thus, for either horizontal or vertical laminar flow

$$\mathscr{F} = Q\,\Delta x\,\frac{\mu}{\rho}\,\frac{128}{\pi D_0^4} \tag{6.13}$$

We may readily extend the argument to show that this equation applies also to flows at any angle to the vertical, so that it is the general description of friction heating in laminar flow of newtonian fluids in circular pipes.

Example 6.1. Oil at a rate of 100 gal/min is flowing steadily from tank A to tank B through 3000 ft of 3-in schedule 40 pipe (actual inside diameter is 3.068 in); see Fig. 6.6. The oil has a density of 58 lbm/ft^3 and a viscosity of 100 cP. The levels of the free surfaces are the same in both tanks. Tank B is vented to the atmosphere. What is the gauge pressure in tank A?

Applying Bernoulli's equation between the free surface in tank A (point 1) and the free surface in tank B (point 2), we see that the velocities are negligible. Since there is no change in elevation and no pump or compressor work, we have

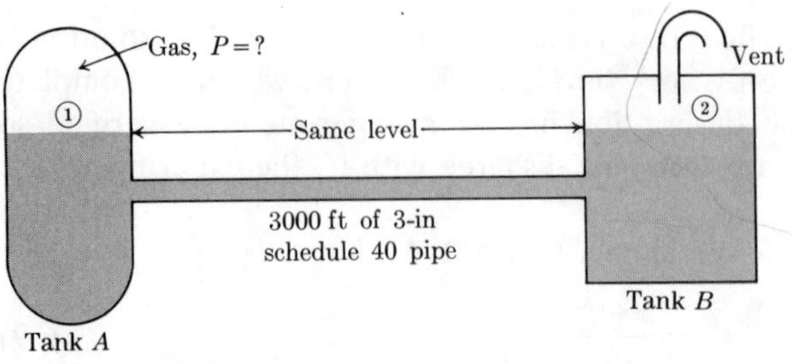

FIGURE 6.6

$$\Delta \frac{P}{\rho} = -\mathscr{F}$$

The density is constant, so in tank A the gauge pressure $P_1 - P_2$ is $(-\rho\mathscr{F})$. *If the flow in the pipe is laminar*, then we can solve for this pressure from Eq. 6.13. The average velocity is

$$V_{\text{av}} = \frac{Q}{A} = \frac{100\ \text{gal/min}}{(\pi/4)(3.068\ \text{in})^2} \cdot \frac{144\ \text{in}^2}{\text{ft}^2} \cdot \frac{\text{min}}{60\ \text{s}} \cdot \frac{\text{ft}^3}{7.48\ \text{gal}}$$
$$= 4.37\ \text{ft/s} = 1.33\ \text{m/s}$$

Therefore, the Reynolds number is

$$\mathscr{R} = \frac{[(3.068/12)\ \text{ft}](4.37\ \text{ft/s})(58\ \text{lbm/ft}^3)}{100\ \text{cP} \cdot 6.72 \times 10^{-4}\ \text{lbm/(ft} \cdot \text{s} \cdot \text{cP)}} = 97$$

As shown before, steady pipe flow is laminar if $\mathscr{R} < 2000$; so we have laminar flow here and are safe in substituting for $\mathscr{F}$ from Eq. 6.13. Multiplying through by $-\rho$, we find

$$-\Delta P = P_1 - P_2 = Q\,\frac{128}{\pi}\,\frac{\mu}{D_0^4}\,\Delta x$$
$$= 100\,\frac{\text{gal}}{\text{min}} \cdot \frac{128}{\pi} \cdot \frac{100\ \text{cP}}{(3.068\ \text{in})^4} \cdot 3000\ \text{ft} \cdot 231\,\frac{\text{in}^3}{\text{gal}}$$
$$\cdot 2.09 \times 10^{-5}\,\frac{\text{lbf} \cdot \text{s}}{\text{cP} \cdot \text{ft}^2} \cdot \frac{\text{min}}{60\ \text{s}} \cdot \frac{\text{ft}}{12\ \text{in}} = 93\,\frac{\text{lbf}}{\text{in}^2} = 641\ \text{kPa}$$

This is the gauge pressure in tank A. ■

Example 6.2. A typical capillary viscometer has the flow diagram shown in Fig. 6.7. It consists of a large-diameter reservoir and a long, small-diameter, vertical tube. The sample is placed in the reservoir, and the flow rate due to gravity is determined. The tube is 0.1 m long and has 1-mm inside diameter. The height of the fluid in the reservoir above the inlet to the tube is 0.02 m. The fluid being tested has a density of 1050 kg/m^3. The flow rate is 10^{-8} m^3/s. What is the viscosity of the fluid?

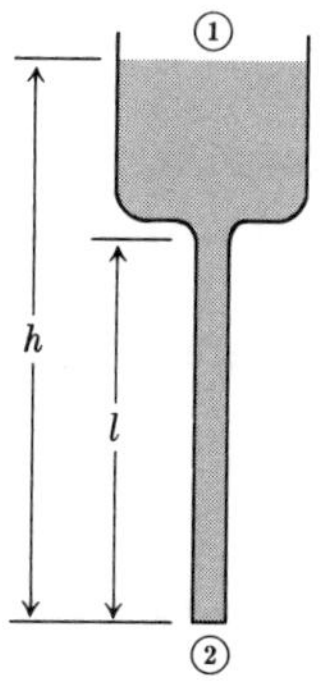

FIGURE 6.7
Typical capillary viscometer.

Applying Bernoulli's equation between the free surface in the reservoir (point 1) and the fluid leaving the bottom of the viscometer (point 2), we see that the pressure at each point is atmospheric and that there is no pump or compressor work. We can neglect the velocity in the reservoir, so Bernoulli's equation becomes

$$g(z_2 - z_1) + \frac{V_2^2}{2} = -\mathcal{F}$$

It can be shown that the kinetic-energy term here is negligible compared with the other two terms. This is found in most laminar-flow problems, so we may drop the kinetic-energy term and find

$$\mathcal{F} = -g\,\Delta z$$

Substituting for $\mathcal{F}$ from Eq. 6.13 and solving for μ, we find

$$\mu = \frac{\rho g(-\Delta z)\pi D_0^4}{128 Q\,\Delta x}$$

$$= \frac{1050\ \text{kg/m}^3 \cdot 9.81\ \text{m/s}^2 \cdot 0.12\ \text{m} \cdot \pi \cdot (0.001\ \text{m})^4}{(128 \times 10^{-8}\ \text{m}^3/\text{s})(0.1\ \text{m})} \cdot \frac{10^3\ \text{cP} \cdot \text{m} \cdot \text{s}}{\text{kg}}$$

$$= 30.3\ \text{cP} = 0.0303\ \text{Pa} \cdot \text{s}$$

■

Viscometers slightly more complicated than the one described above are very widely used. In using them we recognize that:

1. They must not be used at flow rates so high that flow is turbulent (Prob. 6.8).
2. They must be long enough that the error introduced by applying the Poiseuille equation (which applies only well downstream from the entrance) to the whole tube is small. For a brief introduction to the problem of *entrance flow*, to which Poiseuille's equation does not apply, see Sec. 11.6 and Prob. 11.16.

3. The Poiseuille equation applies only to newtonian fluids, so this type of apparatus can be used only for such fluids.
4. Because the viscosity measured by such a device is proportional to D^4, a small error in the diameter measurement leads to a large one in the viscosity measurement. For this reason, these devices ordinarily are calibrated by using a fluid of known viscosity and by determining the appropriate average diameter from the calibration.
5. As discussed in Chap. 5, the $\mathcal{F}$ term represents the conversion of mechanical energy to internal energy. Normally this conversion results in an increase of temperature. It can be shown (Prob. 6.6) that in this case the temperature change is negligible. However, in more viscous liquids, which are pumped through capillary viscometers, there can be a significant temperature rise. In most fluids, a small temperature rise can cause a large viscosity change, so the temperature rise must be minimized.

6.4 TURBULENT FLOW

Why does the preceding analysis not work for turbulent flow? Equation 6.3 is correct for steady laminar or turbulent flow of any kind of fluid, but the substitution of $\mu \, dV/dy$ for the shear stress is correct only for laminar flow of newtonian fluids. In laminar flow in a tube, there is no motion perpendicular to the tube axis. In turbulent flow, there is no *net* motion perpendicular to the tube axis, but there does exist an intense, local, oscillating motion perpendicular to the tube axis. The transfer of fluid perpendicular to the net axial motion causes an increase in shear stress over the value given above for laminar flow of newtonian fluids. This is seen most easily in an analogy. Consider two students playing catch with baseballs. One is standing on the ground. The other is on a railroad car (see Fig. 6.8).

In Fig. 6.8*a*, the railroad car is not moving, and both students throw the ball back and forth in the $+y$ and $-y$ directions. Each time one catches the

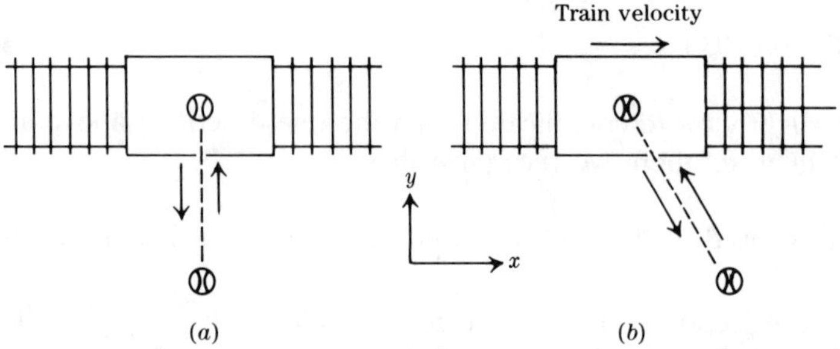

FIGURE 6.8
Illustration by analogy of shear forces due to turbulence. (*a*) Top view of students playing catch, neither moving; (*b*) top view of students playing catch, one moving perpendicular to the direction of throwing.

ball, the student experiences a force, and if the student throws it back at the same speed, the student exerts an equal force in the opposite direction. Therefore, the net effect of throwing the ball back and forth is that a force is exerted on each one, tending to move them apart in the $+y$ and $-y$ directions. There is no force in the x direction.

In Fig. 6.8*b*, the train is moving at constant speed in the x direction. Each student still throws the ball in the $+y$ and $-y$ directions. However, because of their relative motion, each one receives the ball moving, not in a y direction, but at an angle between the x and y directions; the directions of the balls (relative to the two students) are shown by the arrows. Since each one receives the balls in these directions, the force exerted by a student in stopping the ball consists of both the y component, which the other student put into the ball by throwing it, and the x component due to their relative motion. When the train is moving, in addition to the y-directed force there is a force tending to retard the train and to drag the stationary student along in the x direction.

Exactly the same thing happens in turbulent flow. The exchange of fluid between the faster-moving fluid in the center of the tube and the slower-moving fluid near the wall increases the shear stress over that which would exist in laminar flow. This extra stress is called a *Reynolds stress* after Reynolds, who first explained it. Thus, the actual $\mathscr{F}$ in turbulent flow is greater than that predicted by Poiseuille's equation.

In the case of the students throwing the balls, the extra stress is proportional to the velocity of the train and the number of times per second which they throw the balls back and forth. In the case of the corresponding stress in a fluid in turbulent flow, the stress is proportional to the velocity gradient dV/dy times the average mass of fluid passing back and forth across a surface of constant y (across which there is no *net* flow). Since the velocity goes from zero at the pipe wall to the average velocity near the center, the velocity gradient should be some function of V_{av}/D. If we now assume that this is a linear proportion and that the magnitude of the flow of mass back and forth across a surface of constant y is proportional to the average velocity, then it follows that

$$\mathscr{F} \propto \frac{V^2}{D} \tag{6.14}$$

Furthermore, the friction heating should be proportional to the length of the pipe. Including this idea, we find

$$\mathscr{F} \propto \frac{\Delta x\, V^2}{D} \tag{6.15}$$

This equation rests on the plausible assumption that the friction heating is proportional to the length of the pipe and on the more questionable assumption that it is proportional to V^2/D. Do these assumptions agree with the experimental data? Yes and no. To save writing, we now define a new term, the *friction factor f*, which is equal to twice the proportionality constant in Eq. 6.15:

$$\mathscr{F} = 2f\frac{\Delta x\, V^2}{D} = 4f\frac{\Delta x}{D}\cdot\frac{V^2}{2} \tag{6.16}$$

$$f = \text{friction factor} = \frac{\mathscr{F}}{4(\Delta x/D)(V^2/2)} \tag{6.17}$$

To test these assumptions, Blasius and Stanton [2, p. 323] calculated the friction factor for a large variety of pipe flow experiments with smooth pipes. They found that the friction factor was not a constant, as predicted by the analysis above, but decreased slowly with increasing Reynolds number. However, all the data for smooth pipes of various diameters at all velocities for a large range of fluids formed a single curve on a plot of friction factor versus Reynolds number (see Fig. 6.9).

Once plots like this came into common use, it became apparent that they were very good for smooth pipes, such as glass pipes or drawn metal tubing, but that the pressure drops they predicted were too low for rough pipes, such as those made from cast iron or concrete. It appeared that the roughness of the pipe surface influenced f. To resolve the question, Nikuradse [3] measured the pressure drop in various smooth pipes to the inside of which he had glued sand grains. For a given value of ε/D, where ε is the size of the sand particle and D is the pipe diameter, he could plot all his results on one curve on Fig. 6.9, but there were different curves for different values of ε/D. This ratio ε/D is called the *relative roughness*. Figure 6.10 is currently the most commonly used friction factor plot; it was prepared by Moody [4, pp. 671–684], who based it on Nikuradse's data and on all the other available data on flow in pipes.[†] Moody [4] also gave the working values for the absolute roughness; they are shown in Table 6.2.

Figure 6.10 shows that as the relative roughness becomes greater and greater, the assumptions which went into Eq. 6.15 become better and better;

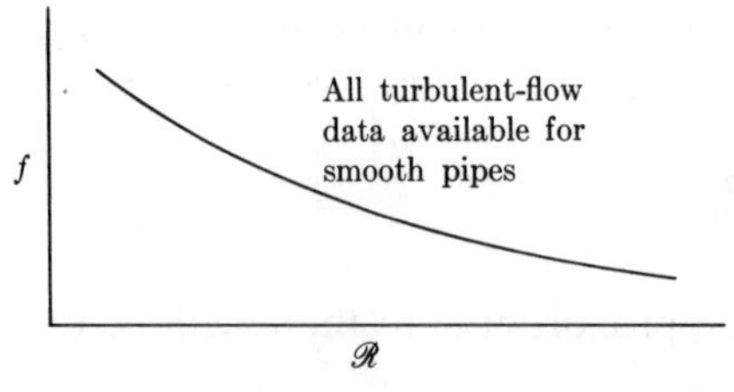

FIGURE 6.9
Blasius and Stanton's plot.

[†] The friction behavior of a pipe with sand grains glued to the wall is somewhat different from that of a commercial pipe. This is believed to be due to the wide range of sizes and shapes of the rough spots in a commercial pipe as compared with the uniform size and shape of the sand grains used by Nikuradse. Moody [4] made Fig. 6.9 according to the Colebrook equation [5] which agrees with the data on commercial pipes. The differences between the two kinds of roughness are discussed by Schlichting [6].

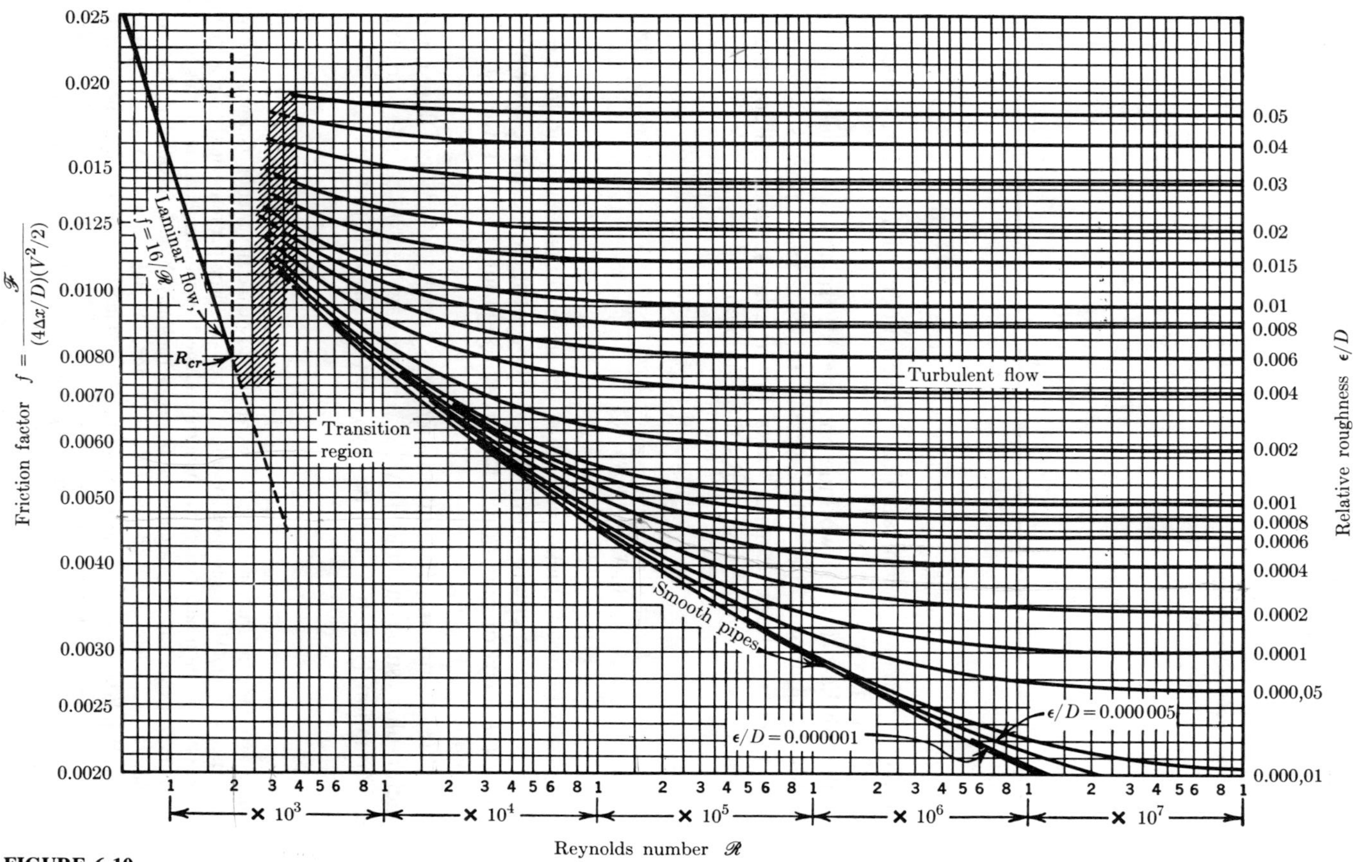

FIGURE 6.10
Friction factor plot for pipes. [From L. W. Moody, "Friction factors for pipeflow," *Trans. ASME 66:* 672 (1944). Reproduced by permission of the publisher.]

TABLE 6.2
Values of surface roughness for various materials†

Material	Surface roughness	
	ε, ft	ε, in
Drawn tubing (brass, lead, glass, etc.)	0.000005	0.00006
Commercial steel or wrought iron	0.00015	0.0018
Asphalted cast iron	0.0004	0.0048
Galvanized iron	0.0005	0.006
Cast iron	0.00085	0.010
Wood stave	0.0006–0.003	0.0072–0.036
Concrete	0.001–0.01	0.012–0.12
Riveted steel	0.003–0.03	0.036–0.36

† From Moody [4].

the friction factor becomes a constant that is independent of diameter, velocity, density, and fluid viscosity.

To make life hard for the working engineer, there are two values of the friction factor in common use. That shown in Eq. 6.17 appears in most chemical and mechanical engineering books, but in other mechanical engineering books and in civil engineering books there appears

$$f_{\text{civ}} = \frac{\mathcal{F}}{(\Delta x/D)(V^2/2)} = 4f_{\text{chem, mech}} \tag{6.18}$$

The existence of the two values means that whenever engineers plan to use a chart like Fig. 6.10 or an equation with f in it, they must check to see on which of the two f's the chart or equation is based. Throughout this book we use the value of f defined by Eq. 6.17.†

There is really not much point in having a curve for laminar flow on a friction factor plot, since laminar flow in a pipe can be solved analytically. Poiseuille's equation (Eq. 6.8) can be rewritten (Prob. 6.12) as

† Why are there two different friction factors? If we decide to define the friction factor as shear stress at wall of pipe $= \tau = f\rho(V^2/2)$ and then solve for the pressure drop per unit length, we are led to the definition of the friction factor used by chemical and mechanical engineers and in this book (Eq. 6.17). However, as we will see, in many problems, particularly those involving both friction and kinetic energy, we eventually find the friction factor entering in a term of the form $(V^2/2)[1 + 4f(\Delta x/D)]$. When it was seen that this group occurred in problems, it was argued that it would be easier to make up a new friction factor, one that was 4 times as large as the other, and that we should not have to bother with the 4 in this group. The larger friction factor is the f_{civ} in Eq. 6.18. Occasionally we see even a third friction factor, which is $\frac{1}{2} f_{\text{civ}} = 2f_{\text{chem,mech}}$. Throughout this book we use the f that leads to the equations with $4f\ \Delta x/D$ in them.

$$f = \frac{16}{\mathcal{R}} \tag{6.19}$$

Plotting any equation this simple is unnecessary. However, the laminar-flow line usually is included in friction factor plots, as it is in Fig. 6.10.

6.5 THE THREE FRICTION FACTOR PROBLEMS

The friction factor plot (Fig. 6.10) relates six parameters of the flow:

1. Pipe diameter
2. Average velocity
3. Fluid density
4. Fluid viscosity
5. Pipe roughness
6. The friction heating per unit mass $\mathcal{F}$

Therefore, given any five of these, we can use Fig. 6.10 to find the sixth.

Most often, instead of being interested in the average velocity V_{av}, we are interested in the volumetric flow rate

$$Q = \frac{\pi}{4} D^2 V_{av}$$

The three most common types of problem are the following:

Type	Given	To find
1	$D, \varepsilon, \rho, \mu, Q$	$\mathcal{F}$
2	$D, \varepsilon, \rho, \mu, \mathcal{F}$	Q
3	$\varepsilon, \rho, \mu, \mathcal{F}, Q$	D

Generally, type 1 can be solved directly, whereas types 2 and 3 require simple trial and error.

Example 6.3. Two reservoirs are connected by 2000 ft of 3-in schedule 40 pipe (actual inside diameter is 3.068 in). We wish to pump 200 gal/min of water from one to the other. The levels in the reservoirs are the same, and both are open to the atmosphere; see Fig. 6.11. What are the pump work per unit mass, the required pump power, and the required pump pressure rise?

Applying Bernoulli's equation from the free surface of the first reservoir, point 1, to the free surface of the second, point 2, we see that all terms are zero except

$$0 = -\frac{dW_{a.o.}}{dm} - \mathcal{F}$$

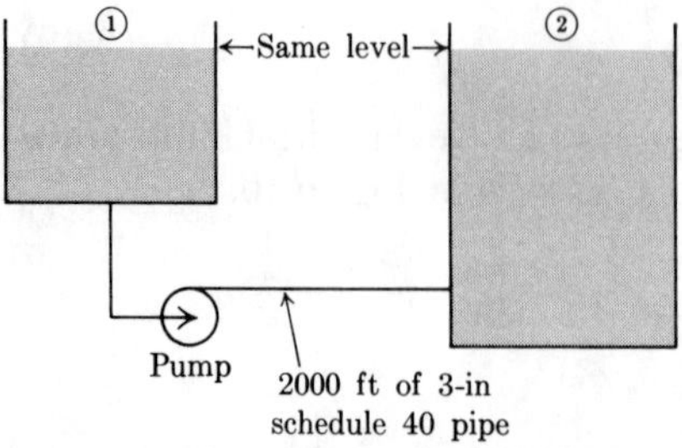

FIGURE 6.11

The pump work (negative because of the thermodynamic sign convention) is equal to minus the friction loss. This is the type 1 problem from the previous list.

From Table 6.2 we see that for commercial steel pipe the roughness ε is 0.0018 in, so

$$\frac{\varepsilon}{D} = \frac{0.0018 \text{ in}}{3.068 \text{ in}} = 0.0006$$

The velocity of the water is given by

$$V = \frac{Q}{A} = \frac{200 \text{ gal/min}}{(\pi/4)(3.068 \text{ in})^2} \cdot \frac{231 \text{ in}^3}{\text{gal}} \cdot \frac{\text{ft}}{12 \text{ in}} \cdot \frac{\text{min}}{60 \text{ s}} = 8.68 \frac{\text{ft}}{\text{s}} = 2.09 \frac{\text{m}}{\text{s}}$$

Therefore, the Reynolds number is

$$\mathscr{R} = \frac{[(3.068/12) \text{ ft}](8.68 \text{ ft/s})(62.3 \text{ lbm/ft}^3)}{(1.0 \text{ cP})[6.72 \times 10^{-4} \text{ lbm/(ft} \cdot \text{s} \cdot \text{cP)}]} = 2.05 \times 10^5$$

From Fig. 6.10 for this $\mathscr{R}$ and for $\varepsilon/D = 0.0006$ we read $f = 0.0048$. Therefore,

$$\begin{aligned}\frac{\Delta P}{\rho} &= -\mathscr{F} = -4f \frac{\Delta x}{D} \frac{V^2}{2} \\ &= \frac{-4 \cdot 0.0048 \cdot 2000 \text{ ft} \cdot (8.68 \text{ ft/s})^2}{2[(3.068/12) \text{ ft}]} \cdot \frac{\text{lbf} \cdot \text{s}^2}{32.2 \text{ lbm} \cdot \text{ft}} \\ &= -176 \text{ ft} \cdot \text{lbf/lbm} = -524 \text{ J/kg}\end{aligned}$$

This is negative because it is work done on the fluid. The pump power required is

$$\text{Po} = \frac{dW_{\text{a.o.}}}{dt} = \frac{dW_{\text{a.o.}}}{dm} \dot{m}$$

where

$$\dot{m} = 200 \frac{\text{gal}}{\text{min}} \cdot \frac{\text{min}}{60 \text{ s}} \cdot 62.3 \frac{\text{lbm}}{\text{ft}^3} \cdot \frac{\text{ft}^3}{7.48 \text{ gal}} = 27.8 \frac{\text{lbm}}{\text{s}}$$

$$\text{Po} = -176 \frac{\text{ft} \cdot \text{lbf}}{\text{lbm}} \cdot 27.8 \frac{\text{lbm}}{\text{s}} \cdot \frac{\text{hp} \cdot \text{s}}{550 \text{ ft} \cdot \text{lbf}} = -8.90 \text{ hp} = -6.63 \text{ kW}$$

To find the pressure rise across the pump, we apply Bernoulli's equation from the pump inlet to the pump outlet. There is no change in elevation or velocity, and the friction in the pump may be neglected; so

$$\frac{\Delta P}{\rho} = \frac{-dW_{\text{a.o.}}}{dm}$$

$$\Delta P = \frac{-\rho\, dW_{\text{a.o.}}}{dm} = \left(-62.3\ \frac{\text{lbm}}{\text{ft}^3}\right)\left(-176\ \frac{\text{ft}\cdot\text{lbf}}{\text{lbm}} \cdot \frac{\text{ft}^2}{144\ \text{in}^2}\right)$$

$$= 76.1\ \text{lbf/in}^2 = 524\ \text{kPa}$$ ■

Although the solution of this example is long, it involves no trial and error and no complicated ideas. The procedure is problem statement → fluid properties → Reynolds number → friction factor → $\mathscr{F}$ → $dW_{\text{a.o.}}/dm$ → pump power → pump ΔP. The pump power computed here is the amount of mechanical power delivered to the fluid. For a 100 percent efficient pump, this would also be the power input to the pump. Real pumps are never 100 percent efficient; their actual behavior is discussed in Chap. 9.

In all such problems (and the ones which follow) it is necessary to convert the volumetric flow rate (gallons per minute or cubic feet or meters per second) to linear velocity (in this case, 200 gal/min in a 3-in schedule 40 pipe = 8.68 ft/s). This routine calculation can be simplified by the use of App. A.3 which shows the volumetric flow rate in gallons per minute corresponding to a velocity of 1 ft/s for all standard U.S. pipe sizes. In Example 6.3 we could have looked up the value of 23.00 (gal/min)/(ft/s) for 3-in schedule 40 pipe and computed

$$V = \frac{200\ \text{gal/min}}{23.0(\text{gal/min})/(\text{ft/s})} = 8.68\ \frac{\text{ft}}{\text{s}}$$

Example 6.4. A gasoline storage tank drains by gravity to a tank truck; see Fig. 6.12. The pipeline between the tank and the truck is 200 ft of 1-in schedule 40 commercial steel pipe. The properties of gasoline are given in the Common Units and Values for Problems and Examples. Both tank and truck are open to the atmosphere, and the level in the tank is 20 ft above the level in the truck. What is the flow rate of the gasoline?

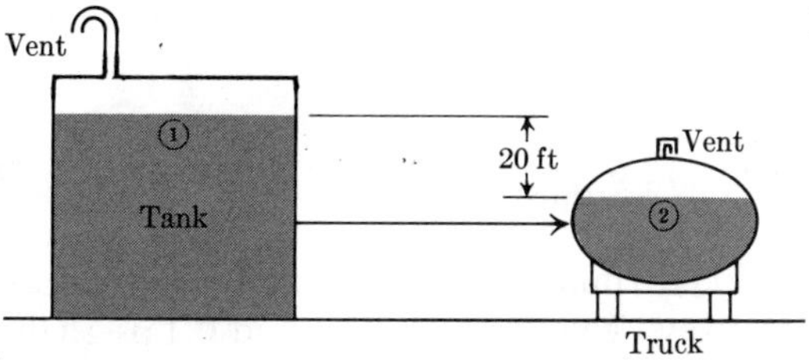

FIGURE 6.12

Applying Bernoulli's equation between the free surface in the tank, point 1, and the free surface in the truck, point 2, we see that all terms cancel except

$$\Delta(gz) = -\mathcal{F} = -4f\,\frac{\Delta x}{D}\,\frac{V^2}{2} \tag{6.20}$$

This is a type 2 problem. The equation contains two unknowns, V and f; therefore, to solve it, we need an additional equation or relationship among the variables listed. The second relationship is provided by Fig. 6.10, which relates f and V. If we could represent Fig. 6.10 by some *simple* equation, we could use it to eliminate f or V from Eq. 6.20 and thus proceed to an analytic solution. However, the data on Fig. 6.10 have not been successfully represented by any such *simple* equation, so it is most convenient to proceed by trial and error. Later we will see that this trial and error and the next one are fairly simple computer problems.

Here we know the fluid properties and the pipe diameter (1.049-in inside diameter) from App. A.3. From Table 6.2 we have $\varepsilon = 0.0018$ in, so

$$\frac{\varepsilon}{D} = \frac{0.0018}{1.049\text{ in}} = 0.0017$$

From Fig. 6.10 we see that for this value of the relative roughness the possible range of f for turbulent flow is 0.0056 to about 0.008. As our first guess, let us try $f = 0.0057$. Then from Eq. 6.20, rearranged to solve for V, we have

$$V = \left[\frac{2g(-\Delta z)}{4f}\cdot\frac{D}{\Delta x}\right]^{1/2} = \left[\frac{2\cdot 32.2\text{ ft/s}^2\cdot 20\text{ ft}}{2\cdot 0.0057}\cdot\frac{(1.049/12)\text{ ft}}{200\text{ ft}}\right]^{1/2} = 4.96\,\frac{\text{ft}}{\text{s}}$$

$$\mathcal{R} = \frac{[(1.049/12)\text{ ft}](4.96\text{ ft/s})(45\text{ lbm/ft}^3)}{(0.6\text{ cP})[6.72\times 10^{-4}\text{ lbm/(ft}\cdot\text{s}\cdot\text{cP)}]} = 4.84\times 10^4$$

Now we can check our assumed value of f; from Fig. 6.10 for $\mathcal{R} = 4.84\times 10^4$ and $\varepsilon/D = 0.0017$, the friction factor f is 0.0065 (as closely as we can read it). For the second trial we let f be 0.0065. Then

$$V = 4.96\,\frac{\text{ft}}{\text{s}}\left(\frac{0.0057}{0.0065}\right)^{1/2} = 4.65\,\frac{\text{ft}}{\text{s}}$$

$$\mathcal{R} = (4.84\times 10^4)\,\frac{4.65\text{ ft/s}}{4.96\text{ ft/s}} = 4.54\times 10^4$$

For this value of $\mathcal{R}$, f (as closely as we can read it from Fig. 6.10) is 0.0065, which agrees with our assumption. To find the flow rate in gallons per minute, we use App. A.3:

$$Q = 4.65\,\frac{\text{ft}}{\text{s}}\cdot\frac{2.69\text{ gal/min}}{\text{ft/s}} = 12.6\,\frac{\text{gal}}{\text{min}} = 7.95\times 10^{-4}\,\frac{\text{m}^3}{\text{s}}$$ ■

Example 6.5. We wish to transport 500 ft^3/min of air horizontally from an air conditioner to an outbuilding 800 ft away. The air is at 40°F and 0.1 psig (the atmospheric pressure is 14.7 psia). At the outbuilding the pressure is to be 0.0 psig. We will use a circular sheet-metal duct, which has a roughness of 0.00006 in. Find the required duct diameter.

Here we are applying Bernoulli's equation to a compressible fluid. However, as discussed in Sec. 5.6, for low fluid velocities Bernoulli's equation gives the same result as the analysis which takes compressibility into account. Applying Bernoulli's equation from the inlet of the duct, point 1, to its outlet, point 2, we find

$$\frac{\Delta P}{\rho} = -\mathscr{F} = -4\mathscr{F}\frac{\Delta x}{D}\cdot\frac{V^2}{2} \tag{6.21}$$

This is the type 3 problem, in which we know everything but the pipe diameter. The equation contains the three unknowns f, D, and V; therefore, we need two additional relations. One is supplied by Fig. 6.10; the other is supplied by the continuity equation, which shows that Q (given in the problem statement) is equal to $V(\pi/4)D^2$. We could use this relation to eliminate V or D from Eq. 6.21, but this is not particularly convenient. Rather, we proceed by trial and error. First we guess a pipe diameter and calculate the pressure drop from Eq. 6.21. Then we compare the calculated pressure drop with the known value, 0.1 psi, and readjust the guessed pipe diameter until we find the diameter for which the pressure drop is 0.1 psi.

For the density of air, we use the average pressure between inlet and outlet and the perfect-gas law:

$$\rho_{\text{air}} = \frac{PM}{RT}$$

$$= \frac{14.75\ \text{lbf/in}^2 \cdot 29\ \text{lbm/(lbmol)}}{[10.73\ \text{lbf}\cdot\text{ft}^3/(\text{in}^2\cdot\text{lbmol}\cdot{}^\circ\text{R})]\cdot(500^\circ\text{R})} = 0.080\ \frac{\text{lbm}}{\text{ft}^3}$$

For air at 40°F we have $\mu = 0.017$ cP (see App. A.1).

For our first trial we select a diameter of 1 ft. Then

$$\frac{\varepsilon}{D} = \frac{0.00006\ \text{in}}{12\ \text{in}} = 0.000005$$

$$V = \frac{(500\ \text{ft}^3/\text{min})(\text{min}/60\ \text{s})}{(\pi/4)(1\ \text{ft})^2} = 10.6\ \frac{\text{ft}}{\text{s}} = 3.23\ \frac{\text{m}}{\text{s}}$$

$$\mathscr{R} = \frac{1\ \text{ft}\cdot 10.6\ \text{ft/s}\cdot 0.080\ \text{lbm/ft}^3}{0.017\ \text{cP}\cdot 6.72\times10^{-4}\ \text{lbm/(ft}\cdot\text{s}\cdot\text{cP)}} = 7.41\times10^4$$

From Fig. 6.10 we read $f = 0.0049$, and from Eq. 6.21

$$\Delta P = \rho_{\text{air}}\left(-4f\,\frac{\Delta x}{D}\cdot\frac{V^2}{2}\right)$$

$$\Delta P = -0.080\,\frac{\text{lbm}}{\text{ft}^3}\cdot\frac{4}{2}\cdot 0.0049\cdot\frac{800\text{ ft}}{1\text{ ft}}\cdot\left(10.6\,\frac{\text{ft}}{\text{s}}\right)^2\cdot\frac{\text{lbf}\cdot\text{s}^2}{32.2\text{ lbm}\cdot\text{ft}}\cdot\frac{\text{ft}^2}{144\text{ in}^2}$$

$$= -0.015\text{ lbf/in}^2 = -103\text{ Pa}$$

Our first guess of the diameter is too large, because it would result in a pressure drop due to friction which is only about one-seventh of that available; i.e., a pipe 1 ft in diameter will do, but we can use a smaller one and still get the required flow with the available pressure difference. To make our second guess, we observe that if we hold the total volumetric flow rate, fluid density, friction factor, and pipe length constant in Eq. 6.21, then the pressure drop will be proportional to $(\text{velocity})^2/\text{diameter}$. But the velocity is proportional to $(\text{diameter})^{-2}$, so the pressure drop is proportional to $1/(\text{diameter})^5$. (This is only approximate, because the friction factor will change, but it is a very good approximation.) Therefore, for a second guess,

$$\frac{\text{New guess of diameter}}{\text{Old guess of diameter}} = \left(\frac{\text{calculated pressure drop for first guess}}{\text{desired pressure drop}}\right)^{1/5}$$

$$= \left(\frac{0.015\text{ lbf/in}^2}{0.1\text{ lbf/in}^2}\right)^{1/5} = 0.68$$

$$\text{New guess of } D = 0.68\text{ ft} = 8.15\text{ in} = 0.207\text{ m}$$

Then we have

$$V = \frac{500\text{ ft}^3/\text{min}}{(\pi/4)(0.68\text{ ft})^2}\cdot\frac{\text{min}}{60\text{ s}} = 22.9\,\frac{\text{ft}}{\text{s}}$$

$$\mathcal{R} = 7.4\times10^4\cdot\frac{22.9\text{ ft/s}}{10.6\text{ ft/s}}\cdot\frac{0.68\text{ ft}}{1.0\text{ ft}} = 1.09\times10^5$$

From Fig. 6.10 we read $f = 0.0045$; so

$$-\Delta P = \frac{4}{2}\cdot 0.0045\cdot\frac{800\text{ ft}}{0.68\text{ ft}}\cdot 0.08\,\frac{\text{lbm}}{\text{ft}^3}\cdot\left(22.9\,\frac{\text{ft}}{\text{s}}\right)^2\cdot\frac{\text{lbf}\cdot\text{s}^2}{32.2\text{ lbm}\cdot\text{ft}}\cdot\frac{\text{ft}^2}{144\text{ in}^2}$$

$$= 0.096\text{ lbf/in}^2 = 662\text{ Pa}$$

Thus, a diameter a little less than 8.15 in is required. The exact choice would probably be made from some readily available diameter, such as $8\frac{1}{4}$ or $8\frac{1}{2}$ in. ■

6.6 SOME COMMENTS ABOUT THE FRICTION FACTOR METHOD AND TURBULENT FLOW

1. The three preceding examples show how we use the friction factor plot (Fig. 6.10). However, the calculations are not reliable to better than ±10 percent, because the exact values of the roughnesses are seldom known to

better than that accuracy. Furthermore, roughnesses of pipes change over time as they corrode or collect deposits. The values in Table 6.2 are only estimates.

2. The plot is made up for sections of pipe which contain no valves, elbows, sudden contractions, sudden expansions, etc. These are probably present in all the actual situations described in Examples 6.3, 6.4, and 6.5. In Secs. 6.8 and 6.9 we discuss how to account for them.
3. The data on which the plot is based are all taken well downstream of the entrance to the pipe. We discuss the entrance region briefly in Chap. 11.
4. The friction factor plot is a generalization of experimental data. One should not attach too much theoretical significance to it. So far no one has been able to calculate friction factors for turbulent flow without starting with experimental data.
5. It can be easily shown that for turbulent flows the heat-transfer coefficient and mass-transfer coefficient are related fairly simply to the friction factor f. This is so because the eddy which transports momentum (and thus increases the shear stress) also can transport heat and mass and thus increase the heat and mass transfer. This subject is discussed in heat-transfer and mass-transfer texts as the *Reynolds analogy*.
6. In Chap. 11 we discuss briefly the measured velocity distributions in turbulent pipe flow. Now we simply note that the velocity profile of turbulent pipe flow is much flatter than that of laminar pipe flows. Most of the fluid flows in a central core, which moves almost as a unit at nearly the same velocity throughout. There is a small layer near the pipe wall in which the velocity drops rapidly from the high velocity of the central core to the zero velocity at the wall. Thus, it is quite reasonable to treat the average velocity of a turbulent pipe flow as the velocity representing the entire flow.

6.7 MORE CONVENIENT METHODS

The friction factor plot (Fig. 6.10) is a very great generalization; all the pressure-drop data for all newtonian fluids, pipe diameters, and flow rates are put on a single graph. However, as shown in Examples 6.3, 6.4, and 6.5, the plot is tedious to use. Therefore, working engineers have rearranged the same experimental data in numerous forms which are more convenient. The resulting methods are more convenient but less compact than the friction factor methods; for instance, one might be using 20 charts instead of only one.

Suppose that we are going to build an oil refinery, a city water supply system, or an aircraft carrier. We will have to deal with a very large number of fluid flows. We could calculate the friction effect for each from Fig. 6.10. However, in any of these projects, we would probably use U.S. standard pipe sizes for practically all the flows. From App. A.3 we see that they constitute a fairly small number of sizes. For pipes of any size and of the same material, the diameters and relative roughness are constant. Therefore, for a given size of pipe there are only four variables; $\mathcal{F}$ per foot, volumetric flow rate, fluid density, and fluid viscosity. These can be plotted (for one pipe size) in a way that makes calculations of friction loss very convenient. Thus, if we spend the time to make about 10 such plots for the common U.S. pipe sizes, we can save considerable work in designing our refinery, water system, or aircraft carrier. Naturally, oil companies, water supply companies, and the Navy have done just that. (Currently computer design programs make most pipe sizing decisions. But they have not eliminated the need for these convenient methods, because experienced engineers always use the convenient methods for quick preliminary estimates and for plausibility checks on computer design program outputs.)

In making such plots and tables it is customary to set them up for the most common problem, which is the long, horizontal, constant-diameter pipe. For such a pipe Bernoulli's equation, rearranged, is

$$\frac{-\Delta P}{\Delta x} = \frac{\rho}{\Delta x}\mathcal{F} = \rho\left(\frac{4f}{D}\cdot\frac{V^2}{2}\right) \tag{6.22}$$

so the charts customarily can be read directly in $-\Delta P/\Delta x$. If we must use such a chart for some other type of problem, we may read the appropriate $-\Delta P/\Delta x$ and then use Eq. 6.22 to find $\mathcal{F}$.

Figure 6.13 is an example of such a plot. This figure shows, for a 3-in schedule 40 pipe, the pressure drop per 1000 ft as a function of volumetric flow rate, kinematic viscosity (viscosity/density), and density. The plot is logarithmic on both axes, but the log scale is different for each. A plot like this can be made up directly from Fig. 6.10 (Prob. 6.25).

Example 6.6. Rework Example 6.3, using Fig. 6.13.

Here the Bernoulli's equation analysis is the same as in Example 6.3. We start on the chart at the right at 200 gal/min and read horizontally to the left to 1.0 cSt. Here the 1.0-cSt line has merged into the zero-viscosity boundary. (This boundary corresponds to the horizontal lines on the right side of Fig. 6.10 for rough pipes. Once the Reynolds number is high enough to reach such a section, increasing it further by lowering the viscosity will not change the friction factor or the friction heating per unit length.) The specific gravity of the fluid is 1; so, reading vertically downward from the zero-viscosity boundary to the bottom of the figure (at SG = 1.0), we find that pressure drop per 1000 ft is 34 lbf/in^2. Thus, the total pressure drop is $(34\ \text{lbf/in}^2)/(1000\ \text{ft})\cdot 2000\ \text{ft} = 68\ \text{lbf/in}^2$. The other parts of Example 6.3 can be calculated for this value of ΔP. ■

The pressure drop in Example 6.3 is 76 psi, and in Example 6.6 it is 68 psi. These disagree, but the disagreement is within the 10 percent uncer-

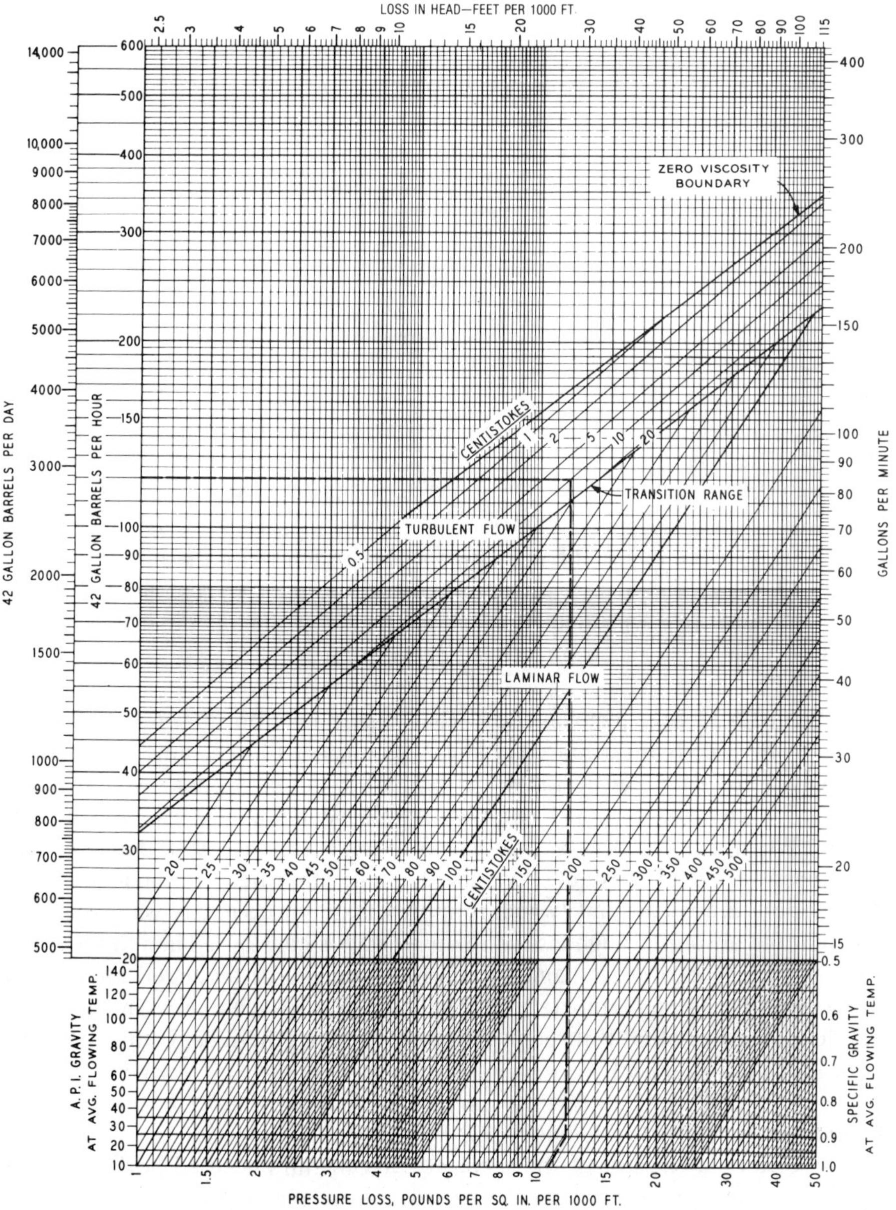

FIGURE 6.13
Pressure drop in a 3-in schedule 40 pipe, with 3.068-in inside diameter. For example shown: flow rate = 120 barrels per hour (BPH); kinematic viscosity = 10 cSt, specific gravity = 0.9, pressure loss (follow dashed line) = 10.7 psi/1000 ft. (Courtesy of the Board of Engineers, Standard Oil Company of California.)

tainty which exists in all such calculations. The cause of the disagreement is the slightly lower value of the relative roughness chosen in Fig. 6.13 in contrast with the value in Table 6.2.

Example 6.7. Rework Example 6.1, using Fig. 6.13. The specific gravity of the oil is

$$\frac{58\ \text{lbm}/\text{ft}^3}{62.4\ \text{lbm}/\text{ft}^3} = 0.93$$

The kinetic viscosity of the oil is

$$\frac{100\ \text{cP}}{0.93\ \text{g}/\text{cm}^3} \cdot \frac{\text{cSt} \cdot \text{g}/\text{cm}^3}{\text{cP}} = 107.6\ \text{cSt}$$

Now we enter Fig. 6.13 on the right at 100 gal/min and scan horizontally to 107.6 cSt (this requires interpolating between 100 and 150 cSt). Then we read vertically downward to SG = 0.93 and then along the diagonal to the pressure drop of 31 lbf/in^2 per 1000 ft. Thus, the pressure drop is 31 $\text{lbf}/(\text{in}^2 \cdot 1000\ \text{ft}) \cdot 3000\ \text{ft} = 93\ \text{lbf}/\text{in}^2$, which agrees exactly with the result of Example 6.1. The agreement is exact here because the relative roughness does not enter the pressure-drop calculation for laminar flow. The laminar-flow section of Fig. 6.13 is constructed directly from Eq. 6.11. ■

Figure 6.13 is a "convenience" chart made up from Fig. 6.10. It is well suited to the needs of an oil company, which spends large sums of money in pumping fluids with a wide range of viscosities; but it is poorly suited to the needs of a city water supply company, which deals almost exclusively with water. When Fig. 6.13 was made from Fig. 6.10, the pipe diameter and roughness were held constant. If we are dealing with water, we can assume that the temperature is constant (which is approximately true in city water systems) and that the absolute roughness of the pipe wall is constant. Then the pressure drop as a function of pipe diameter and flow rate can be tabulated for all flows of water at the chosen temperature. Appendix A.4 is such a table, made up for the flow of water at 60°F through schedule 40 pipe (the most common size in U.S. industrial practice).

Example 6.8. Rework Example 6.3, using App. A.4.

We start at the left of the table at 200 gal/min and read horizontally to the column for 3-in pipe, where the pressure drop is indicated as $(3.87\ \text{lbf}/\text{in}^2)/(100\ \text{ft})$. The pipe is 2000 ft long, so the pressure drop is $[(3.87\ \text{lbf}/\text{in}^2)/(100\ \text{ft})] \cdot (2000\ \text{ft}) = 77.4\ \text{lbf}/\text{in}^2$. This does not agree exactly with the 76 lbf/in^2 calculated in Example 6.3, but the agreement is within the uncertainty in reading Fig. 6.10. Having found the pressure drop by using App. A.4, we can readily calculate the remaining parts of Example 6.3. ■

In using App. A.4, remember where it comes from; each calculation such as that in Example 6.3 produced one table entry. By making such calculations for a large number of flow rates and pipe diameters, we can make up App. A.4. Thus, that appendix is simply Fig. 6.10, rearranged for the special case of 60°F water flowing in schedule 40 pipes.

Just as oil refinery engineers have made charts convenient to their class of problems, so also air conditioning engineers have prepared Fig. 6.14, on which they can quickly solve most pressure-drop problems in common air conditioning use [7]. Here they have recognized that almost all flow in air conditioning ducts is of air at about 70°F and 14.7 psia, for which the viscosity and density are known, and that most of the ducts are made of galvanized steel or aluminum, for which ε is known. Thus they have one fewer variable than the oil refinery engineers (who treat fluids with a variety of viscosities); so instead of having to have a separate plot for each size pipe, like Fig. 6.13, they can have one plot which covers all sizes.

Example 6.9. Repeat Example 6.5, using Fig. 6.14. Here the figure was made for air at 70°F, which we can tell from the assumed density of 0.075 lbm/ft^3, while our problem is for 40°F and a density of 0.080 lbm/ft^3. We know that the viscosities do not match perfectly either. However, we ignore these differences for the moment and simply use Fig. 6.14. We know the flow rate (500 ft^3/min), the pressure drop (0.1 psi), and the pipe length (800 ft). We need to convert the pressure drop to pressure drop per unit length which we can do as either

$$\frac{\Delta P}{\Delta x} = \frac{0.1\ \text{psi}}{800\ \text{ft}} \cdot \frac{27.71\ \text{in H}_2\text{O}}{\text{psi}} = 0.00346\ \frac{\text{in H}_2\text{O}}{\text{ft}}$$

or

$$\frac{\Delta P}{\Delta x} = \frac{0.1\ \text{psi}}{800\ \text{ft}} \cdot \frac{6.895 \times 10^3\ \text{Pa}}{\text{psi}} \cdot \frac{3.28\ \text{ft}}{\text{m}} = 2.83\ \frac{\text{Pa}}{\text{m}}$$

You may verify that these give the same entry point on the abscissa. Reading at the intersection of this pressure drop and 500 $\text{ft}^3/\text{min(cfm)}$, we find that the required pipe diameter is about 8.2 in and the velocity is about 1350 ft/min = 22.5 ft/s. The practically perfect agreement with Example 6.5 simply shows that Fig. 6.14 was made up by using the standard friction factor plot. The small differences between the density and viscosity of air at 70 and 40°F do not affect the answer to chart-reading accuracy. ■

Such convenient charts as Fig. 6.13, Fig. 6.14, and App. A.4 are widely used in industry for routine calculations. When engineers leave the university and join industrial firms, they find that their colleagues have a large supply of them. It is worthwhile for young engineers to trace them to their sources. Not only will engineers discover how the convenient methods fit in with the ideas learned in the university, but also they will see more clearly the limitations of the convenient methods. Then they can use them for the routine parts of

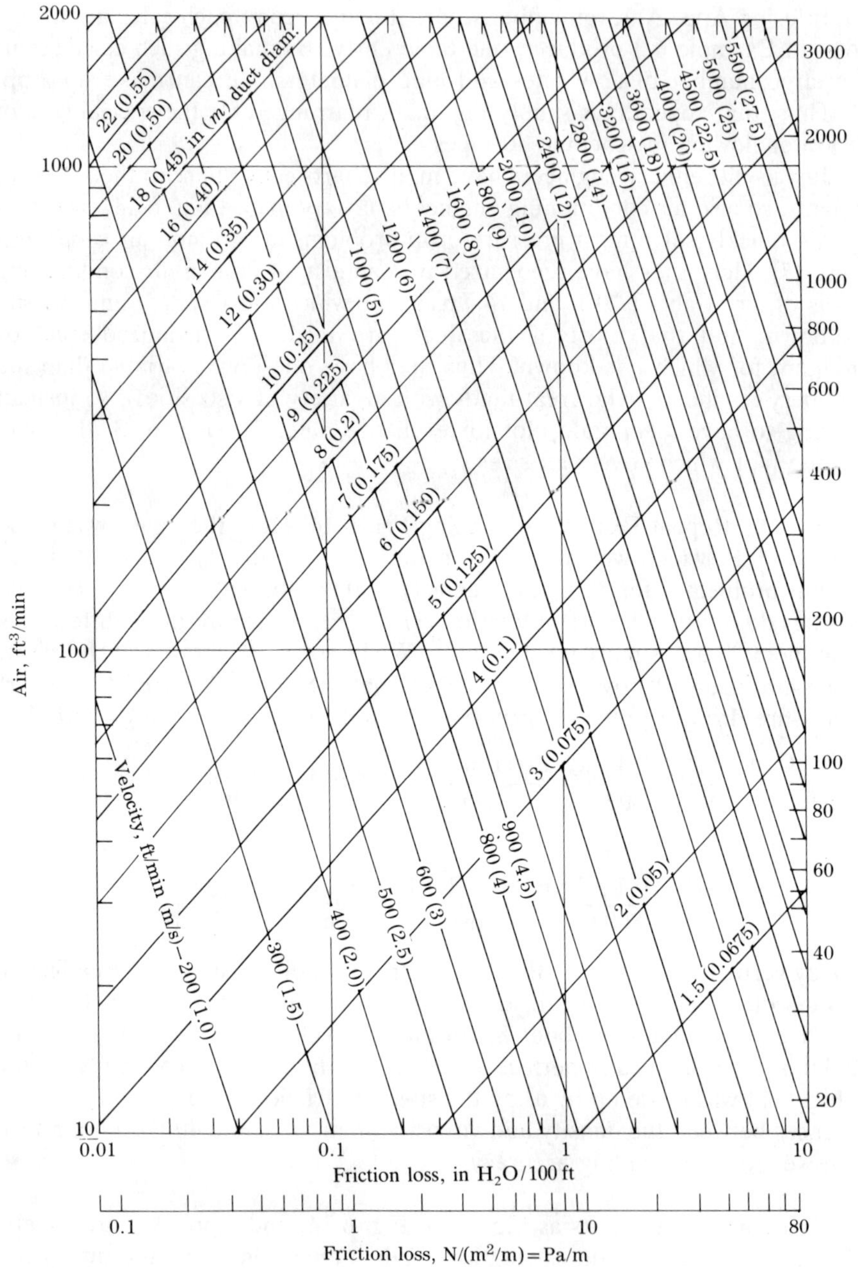

FIGURE 6.14
Friction of air in straight ducts for volumetric flow rates of 10 to 2000 ft^3/min (20 to 3000 m^3/h), based on standard air of 0.075 lb/ft^3 (1.2 kg/m^3) density flowing through average, clean, round galvanized metal ducts having approximately 40 joints per 100 ft (30 m). Do not extrapolate below chart. (Reprinted from the *1972 ASHRAE Handbook—Fundamentals*, with permission.)

complex jobs, saving their creative efforts for the nonroutine parts that will test their talents and education.

6.8 COMPUTER METHODS

Before the computer age, the friction factor plot and the convenience plots made from it were the only means that engineers had of solving fluid friction problems in pipes. For an occasional such calculation or a problem outside the normal range of the engineer's experience, they are probably still the best way. However, for routine pipe friction calculations, engineers use computers. To do so, they need equations equivalent to the friction factor plot.

For laminar flow that is very simple because (as we discussed before) the laminar-flow part of the friction factor plot is simply $f = 16/\mathcal{R}$. For turbulent and transition flow, it is more difficult. The common friction factor plot (Fig. 6.10) is based on the Colebrook equation [5]

$$\frac{1}{\sqrt{f}} = -4\log\left(\frac{\varepsilon/D}{3.7} + \frac{1.255}{\mathcal{R}\sqrt{f}}\right) \tag{6.23}$$

This equation is difficult to use, because f appears on both sides, once as the argument of a logarithm. Given values of $\mathcal{R}$ and f, we can find ε/D; or given values of f and ε/D, we can find $\mathcal{R}$. But given values of $\mathcal{R}$ and ε/D, we must use a trial-and-error solution to find f. Although computers are good at such trial-and-error solutions, they are slow and require extensive programming.

A more convenient approach uses Wood's approximation [8]. He showed that for the range of interest of the friction factor plot, Eq. 6.23 could be replaced with

$$\begin{aligned} f &= a + b\mathcal{R}^{-c} \\ a &= 0.0235\left(\frac{\varepsilon}{D}\right)^{0.225} + 0.1325\left(\frac{\varepsilon}{D}\right) \\ b &= 22\left(\frac{\varepsilon}{D}\right)^{0.44} \\ c &= 1.62\left(\frac{\varepsilon}{D}\right)^{0.134} \end{aligned} \tag{6.24}$$

He reports that for the range of values on the friction factor plot Eq. 6.24 differs from Eq. 6.23 by less than the uncertainty in the data upon which both equations are based.

Example 6.10. Repeat Example 6.3, using Eq. 6.24. From that example we know that $\varepsilon/D = 0.0006$ and $\mathcal{R} = 2.05 \times 10^5$. Then

$$a = 0.0235(0.0006)^{0.225} + 0.1325(0.0006) = 0.004507$$

$$b = 22(0.0006)^{0.44} = 0.8410$$

$$c = 1.62(0.0006)^{0.134} = 0.5995$$

$$f = 0.004507 + 0.8410(2.05 \times 10^5)^{-0.5995} = 0.00506$$

This is 5 percent larger than the $f = 0.0048$ found in Example 6.3, but again it is well within the 10 percent uncertainty of all transition region and turbulent friction factors. The rest of that example is the same as before, with f increased from 0.0048 to 0.00506. ■

If we incorporated Eq. 6.24 directly into Eq. 6.16, we would find

$$\mathcal{F} = 4(a + b\mathcal{R}^{-c})\frac{\Delta x}{D} \cdot \frac{V^2}{2} \tag{6.25}$$

with the values of a, b, and c taken from Eq. 6.24. For the type 2 problem where V is unknown, this results in an equation with V to a fractional power, which can be solved directly without trial and error. For the type 3 problem, where D is unknown, the problem remains one of trial and error, because D appears in so many places. Most programmers have preferred the trial-and-error procedures in Examples 6.3, 6.4, and 6.5, using Eq. 6.24 or one of its many published equivalents to replace the chart lookup on Fig. 6.10.

6.9 FITTING LOSSES

So far all the discussion has been about steady flow well downstream of the pipe entrance in a straight circular pipe. This is the simplest and one of the most important cases of fluid friction. However, in many fluid systems we must take into account the effect of valves, elbows, etc. They are much more complex to analyze than the one-dimensional flows we have considered so far. (The student is advised to take apart an ordinary household faucet and study its flow path; it is much more complicated than that of a straight pipe.) Efforts have been made to calculate the friction losses in such fittings, and the results have been correlated in several convenient ways which allow us to treat them as if they were one-dimensional problems.

One popular way to correlate these results is to assume that for a given flow

$\mathcal{F}$ through valve or fitting

$$= \text{constant} \cdot \mathcal{F} \text{ through length of pipe equal to 1 pipe diameter} \tag{6.26}$$

If for a given kind of valve or fitting this constant turns out to be independent of the kind of flow, then this correlation will be very easy to use. For practical purposes, in turbulent flow the constant in this equation is independent of pipe size, flow rate, and nature of the fluid flowing. If we know the value of the constant for a particular kind of fitting, we can calculate an "equivalent length of pipe," which would have the same friction effect as the fitting, and we can add this length to the actual length of pipe to find an *adjusted length*, which

TABLE 6.3
Equivalent lengths for various kinds of fittings[†]

Type of fitting	Equivalent length L/D (dimensionless)
Globe valve, wide open	340
Angle valve, wide open	145
Gate valve, wide open	13
Check valve (swing type)	135
90° standard elbow	30
45° standard elbow	16
90° long-radius elbow	20

[†] From Crane Technical Paper No. 410. Reproduced by permission of the Crane Company.

gives the same friction effect as does the actual pipe including fittings. The constant in the equation is dimensionless. Typical values [9] are shown in Table 6.3; they are referred to as *equivalent lengths*.

Example 6.11. Rework Example 6.3 on the assumption that, in addition to the 2000 ft of 3-in schedule 40 pipe, the line contains two globe valves, a swing check valve, and nine 90° elbows.

Using the constants in Table 6.3, we can calculate the equivalent length of 3-in schedule 40 pipe which would have the same friction effect as these fittings. This is

$$\begin{aligned} 2 \times 340 &= 680 \\ 1 \times 135 &= 135 \\ 9 \times 30 &= \underline{270} \\ & 1085 \end{aligned}$$

From Eq. 6.26 we see that this is the number of pipe diameters needed to have the same friction loss as the fittings. Thus, the equivalent length is 1085[(3.068/12) ft] = 277 ft. Therefore, the adjusted length of the pipe is

$$\begin{aligned} \text{Adjusted length} &= \text{actual length of pipe} \\ &\quad + \text{equivalent length to account for fittings} \\ &= 2000 \text{ ft} + 277 \text{ ft} \end{aligned} \tag{6.27}$$

Then the pump work, pump power, and pressure drop must be the values found in Example 6.3 multiplied by 2277 ft/2000 ft. ■

Laminar flow has yielded few experimental data on which to base pressure-drop correlations for valves and fittings. Generally the adjusted length calculated by the method given above is correct for turbulent flow but too large for laminar flow. Empirical guides to estimating the adjusted length for laminar flow have been published [9].

We should not attach much theoretical significance to these empirical relations for fitting losses. They are simply the results of careful tests of specific cases, arranged in a way that is useful in predicting the behavior of new systems.

6.10 ENLARGEMENTS AND CONTRACTIONS

The first part of this chapter was devoted to the steady flow of a fluid in a part of a circular pipe, well downstream from the pipe entrance. However, in each of the examples there were places where the flow entered and left a pipe. In Examples 6.1, 6.3, and 6.4, the fluid flowed from the first reservoir into a pipe and left the pipe to enter a second reservoir. Friction losses are associated with these transitions; see Fig. 6.15.

For the enlargements we can calculate the friction effect on the basis of some simple assumptions; the calculation, made by means of the momentum balance, is given in Sec. 7.3. The results of that calculation can be put in the form

$$\mathcal{F} = \frac{KV^2}{2} \tag{6.28}$$

where K is an empirical constant, called the *resistance coefficient*, which is dependent on the ratio of the two pipe diameters involved, and V is the larger of the two velocities involved. The experimental data agree with it reasonably well [10]. No one has successfully calculated the friction effect of sudden contractions without having resort to experimental data. However, it can be shown [10] that the experimental data can also be represented by the same equation. The experimental values of K for contractions are shown in Fig. 6.16, as are the calculated values of K for sudden expansions.

From Fig. 6.16 it is clear that the larger the change of diameter, the greater the pressure losses. The reasons for the losses are as follows:

1. In a sudden expansion, the fluid is slowed down from relatively high velocity and high kinetic energy in the small pipe to relatively low velocity and low kinetic energy in the large pipe. If this process took place without friction, the kinetic energy would be converted to injection work with a resulting pressure increase. In a sudden expansion, the process takes place as a fluid mixes and eddies around the enlargement. The kinetic energy of the fluid is

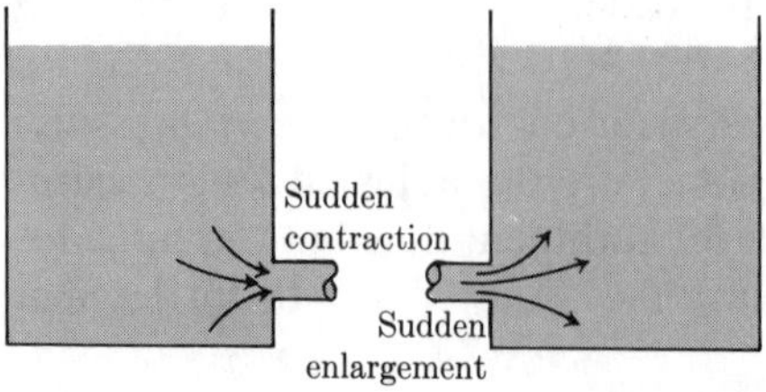

FIGURE 6.15
Sudden contractions and enlargements.

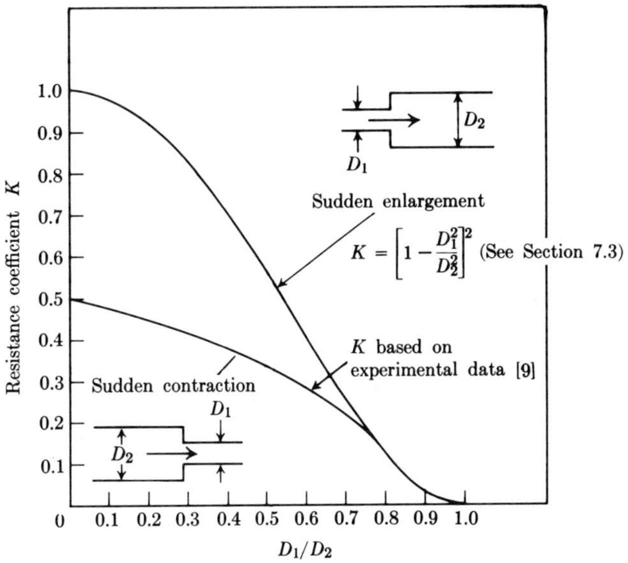

FIGURE 6.16
Resistance due to sudden enlargements and contractions. The resistance coefficient K is defined in Eq. 6.28. (From Crane Technical Paper No. 410. Reproduced by permission of the Crane Company.)

converted to internal energy. Therefore, when the downstream velocity is zero, the friction loss is equal to the upstream kinetic energy. This is shown by Eq. 6.28 with $K = 1$, which is the value for zero downstream velocity in Fig. 6.16 (see the discussion in Sec. 5.5).

2. In a sudden contraction, the flow does not come into the pipe entirely in the axial direction. Rather, it comes from all directions, as sketched in Fig. 6.15. On entering the pipe, the flow follows the pattern shown in Fig. 6.17.

The fluid forms a neck, called the *vena contracta*, just downstream of the tube entrance. The flow into the neck is caused by the radial inward velocity of the fluid approaching the tube. Because it is coming inward, the fluid overshoots the tube wall and goes into the neck. This neck is surrounded by a collar

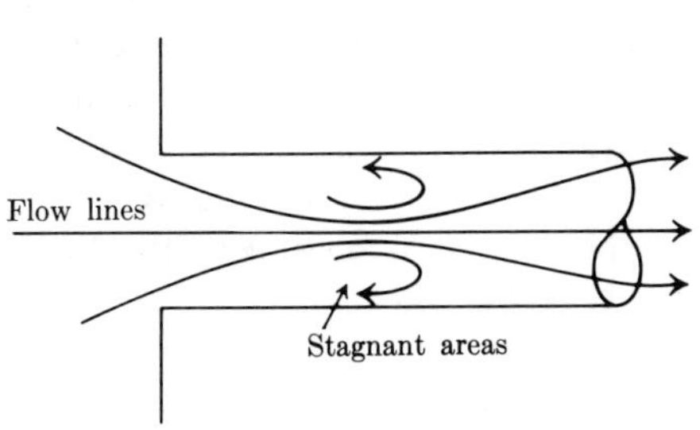

FIGURE 6.17
Flow pattern (turbulent flow) in a sudden contraction.

of stagnant fluid. In the neck the velocity is greater than the velocity farther downstream. Thus, the kinetic energy decreases from the neck to some point downstream, where the velocity is practically uniform over the cross section of the pipe. Not all this kinetic energy is recovered as increased pressure; it leads to the friction loss shown in Fig. 6.16.

Our discussion of entrance and exit losses has concerned turbulent flow only. In laminar flow, these effects generally are negligible, because the kinetic energies usually are negligible compared with the viscous effects.

Example 6.12. Calculate the error made in Example 6.3 by neglecting the expansion and contraction losses.

From Fig. 6.16, for flow from a tank to a pipe, the coefficient K is 0.5, and for flow from a pipe to a tank, it is 1.0. Thus, the friction loss due to the expansion and contraction should be $1.0 + 0.5 = 1.5$ times the kinetic energy of the fluid in the pipe. In Example 6.3 the velocity was 8.68 ft/s; therefore, it can be shown that

$$-\Delta P_{\substack{\text{expansion}\\ \text{and contraction}}} = \rho\mathcal{F} = \frac{\rho K V^2}{2}$$

$$= 62.3\,\frac{\text{lbm}}{\text{ft}^3}\cdot\frac{1.5}{2}\cdot\left(8.68\,\frac{\text{ft}}{\text{s}}\right)^2\cdot\frac{\text{ft}^2}{144\,\text{in}^2}\cdot\frac{\text{lbf}\cdot\text{s}^2}{32.2\,\text{lbm}\cdot\text{ft}}$$

$$= 0.76\,\text{lbf/in}^2 = 5.24\,\text{kPa}$$ ■

This is 1 percent of the 76 lbf/in^2 previously calculated. In this example the pipe was long (several thousand feet). If the pipe were short, the contraction and expansion losses would be just as large, but the percentage error in neglecting them would be much greater.

6.11 FLUID FRICTION IN ONE-DIRECTIONAL FLOW AND OTHER GEOMETRIES

Steady flow in a circular pipe is one of the simplest flow problems; in laminar flow, the velocity is one-directional, independent of time, and dependent on only one dimension, the radius.

A somewhat harder problem is steady flow of an incompressible newtonian fluid in some duct or pipe which is of constant cross section but not circular, such as a rectangular duct or an open channel. The problem of laminar flow of a newtonian fluid can be solved analytically for several shapes. Generally the velocity depends on two dimensions. In several cases of interest, the problems can be solved by the same method we used to find Eq. 6.8, i.e., setting up a force balance around some properly chosen section of the flow, solving for the shear stress, introducing the newtonian law of viscosity for the shear stress, and integrating to find the velocity distribution. From the velocity distribution the flow rate–pressure-drop relation is found.

The fact that these are all similar to the solution for laminar flow in a horizontal circular tube may be seen by comparing the horizontal, steady flow solutions with that for a circular tube. For a circular tube

$$Q = \frac{P_1 - P_2}{\Delta x} \cdot \frac{1}{\mu} \cdot \frac{\pi}{128} D_0^4 \tag{6.8}$$

For a slit between two parallel plates (Prob. 6.43),

$$Q = \frac{P_1 - P_2}{\Delta x} \cdot \frac{1}{\mu} \cdot \frac{1}{12} lh^3 \tag{6.29}$$

where h is the distance between plates and l is the width of the slit. If both sides are divided by l, the left-hand side becomes the volumetric flow rate per unit width. For an annulus (Prob. 6.44),

$$Q = \frac{P_1 - P_2}{\Delta x} \cdot \frac{1}{\mu} \cdot \frac{\pi}{128} (D_o^2 - D_i^2)\left[D_o^2 + D_i^2 - \frac{D_o^2 - D_i^2}{\ln (D_o/D_i)}\right] \tag{6.30}$$

where D_o is the outer diameter and D_i is the inner diameter. These equations differ only by the terms at the far right, which account for the different geometries. Most of the cases which can be worked out by simple mathematics have been summarized by Bird et al. [11, Chap. 2 and problems at end of chapter] and Sakiadis [12].

We are no more able to calculate the pressure drop in steady, *turbulent* flow in a noncircular conduit than we are in a circular one. However, it seems reasonable to expect that we could use the friction-loss results for circular pipes to estimate the results for other shapes. Let us *assume* that the shear stress at the wall of any conduit is the same for a given average fluid flow velocity independent of the shape of the conduit. Then, from a force balance on a horizontal section like that leading to Eq. 6.3, we conclude that in steady flow

$$\Delta P(\text{area perpendicular to flow}) = \text{wall shear stress} \cdot \text{wetted perimeter} \cdot \text{length} \tag{6.31}$$

Rearranging this, we find

$$\frac{\Delta P}{\Delta x} = \text{wall shear stress} \cdot \frac{\text{wetted perimeter}}{\text{area perpendicular to flow}} \tag{6.32}$$

We now define a new term:

$$\text{Hydraulic radius} = \frac{\text{cross-sectional area perpendicular to flow}}{\text{wetted perimeter}} \tag{6.33}$$

For a circular pipe, the hydraulic radius (HR) is

$$\text{HR} = \frac{\pi r^2}{2\pi r} = \frac{r}{2} = \frac{D}{4} \tag{6.34}$$

If the assumptions which went into Eq. 6.32 are correct, then we can construct the ratio of the pressure drop per unit length in a noncircular conduit to that in a circular one:

$$\frac{(\Delta P/\Delta x)_{\text{noncirc}}}{(\Delta P/\Delta x)_{\text{circ}}} = \frac{1/\text{HR}}{4/D} \tag{6.35}$$

$$\left(\frac{\Delta P}{\Delta x}\right)_{\text{noncirc}} = \left(\frac{\Delta P}{\Delta x}\right)_{\text{circ}} \left(\frac{D}{4\text{HR}}\right) \tag{6.36}$$

But for turbulent flow $(\Delta P/\Delta x)_{\text{circ}}$ is given by the friction factor equation, Eq. 6.22. Substituting, we get

$$\left(\frac{\Delta P}{\Delta x}\right)_{\text{noncirc}} = \frac{-4f\rho V^2}{2D} \cdot \frac{D}{4\text{HR}} = \frac{f\rho V^2}{2\text{HR}} \tag{6.37}$$

Alternatively, we may write

$$\mathscr{F}_{\text{noncirc}} = \frac{-f\,\Delta x}{\text{HR}} \cdot \frac{V^2}{2} \tag{6.38}$$

What value of f should we use in Eqs. 6.37 and 6.38? Experimental results indicate that those equations work *fairly well* if one uses the ordinary friction factor plot (Fig. 6.10) but replaces the diameter in the Reynolds number and in ε/D with 4HR. The equations do not work well for shapes that depart radically from circles, such as long, narrow slits.

Example 6.13. Air at 1 atm and 68°F is flowing in a long, rectangular duct whose cross section is 1 ft by 1.5 ft. The average velocity is 40 ft/s. The roughness of the duct is 0.00006 in. What is the pressure drop per unit length?

First, we calculate the hydraulic radius:

$$\text{HR} = \frac{1.5\ \text{ft}^2}{2 \cdot 1\ \text{ft} + 2 \cdot 1.5\ \text{ft}} = 0.3\ \text{ft}$$

Then

$$\mathscr{R} = \frac{V\rho(4\text{HR})}{\mu} = \frac{40\ \text{ft/s} \cdot 0.075\ \text{lbm/ft}^3 \cdot 1.2\ \text{ft}}{0.018\ \text{cP} \cdot 6.72 \times 10^{-4}\ \text{lbm/(ft} \cdot \text{s} \cdot \text{cP)}} = 2.98 \times 10^5$$

$$\frac{\varepsilon}{D} = \frac{\varepsilon}{4\text{HR}} = \frac{0.00006\ \text{in}}{(1.2\ \text{ft})(12\ \text{in/ft})} = 0.000004$$

From Fig. 6.10 (for this low an ε/D we use the "smooth-tubes" curve) we find $f = 0.0036$. So, using Eq. 6.37, we find

$$-\frac{\Delta P}{\Delta x} = \frac{0.0036(0.075\ \text{lbm/ft}^3)(40\ \text{ft/s})^2}{0.6\ \text{ft}} \cdot \frac{\text{ft}^2}{144\ \text{in}^2} \cdot \frac{\text{lbf} \cdot \text{s}^2}{32.2\ \text{lbm} \cdot \text{ft}}$$

$$= 1.55 \times 10^{-4}\ \frac{\text{lbf/in}^2}{\text{ft}} = 3.5\ \frac{\text{Pa}}{\text{m}}$$

We may check this by using Fig. 6.14. Here we enter at 2400 ft/min and 14-in diameter, finding (by short extrapolation) about 4.0 Pa/m. ■

Example 6.14. We wish to transport 1000 ft³/s of water steadily in a concrete-lined irrigation canal, which is rectangular, 10 ft by 10 ft, and open at the top. How much must we slope the canal?

First we apply Bernoulli's equation from some upstream point in the canal to some downstream point in the canal. Since both points are open to the atmosphere, the pressures are the same. For steady flow of a constant-density fluid in a canal of constant cross-sectional area, the velocities at the two points are the same. There is no pump or turbine work. Therefore, the remaining terms are

$$g\,\Delta z = -\mathcal{F}$$

This says that the decrease in potential energy is exactly equal to the energy "loss" due to friction, i.e., the mechanical energy converted to internal energy. Substituting for $\mathcal{F}$ from Eq. 6.38, we find

$$-g\,\Delta z = \frac{f\,\Delta x\,V^2}{2\text{HR}} \qquad \frac{-\Delta z}{\Delta x} = \frac{fV^2}{2\text{HR}(g)}$$

In Example 6.13 the wetted perimeter was the entire perimeter of the duct. Here we do not include the perimeter facing the air, because the air exerts little resistance to the flow, compared with the walls of the canal. You can verify this by watching the flow of leaves or bits of wood on any open stream or irrigation ditch; those at the center move much faster than those at the edges. If the air restrained the flow as much as the solid walls, then the whole top surface of the flow would not move at all, just as the fluid right at the solid boundaries does not move. Therefore, the hydraulic radius is

$$\text{HR} = \frac{\text{flow area}}{\text{wetted perimeter}} = \frac{10\text{ ft}\cdot 10\text{ ft}}{3\cdot 10\text{ ft}} = 3.33\text{ ft}$$

The velocity is

$$\frac{1000\text{ ft}^3/\text{s}}{100\text{ ft}^2} = 10\,\frac{\text{ft}}{\text{s}}$$

The Reynolds number is

$$\mathcal{R} = \frac{13.3\text{ ft}\cdot 10\text{ ft/s}\cdot 62.3\text{ lbm/ft}^3}{1.0\text{ cP}\cdot 6.72\times 10^{-4}\text{ lbm/(ft}\cdot\text{s}\cdot\text{cP)}} = 1.23\times 10^7$$

The absolute roughness of a concrete-lined irrigation ditch is estimated from Table 6.2 to be 0.01 ft, so the relative roughness is

$$\frac{\varepsilon}{4\text{HR}} = \frac{0.01\text{ ft}}{13.3\text{ ft}} = 0.00075$$

Thus, from Fig. 6.10 we read $f = 0.0046$; therefore,

$$\frac{-\Delta z}{\Delta x} = \frac{0.0046(10\text{ ft/s})^2}{2\cdot 3.33\text{ ft}\cdot 32.2\text{ ft/s}^2} = 0.0021$$

This slope is dimensionless, or it can be thought of as being feet of drop per foot of length. The canal must slope downward about 11 ft/mi (2.1 m/km) to carry the required 1000 ft^3/s of water. ■

6.12 MORE COMPLEX PROBLEMS INVOLVING BERNOULLI'S EQUATION

Now that we can evaluate all the terms in Bernoulli's equation, we may consider some of the more interesting types of problem which this equation can be used to solve.

Example 6.15. A large, high-pressure chemical reactor contains water at a pressure of 2000 lbf/in^2. A 3-in schedule 80 line connecting to it ruptures at a point 10 ft from the reactor. What is the flow rate through this break?

This is an unsteady-state problem because the reactor pressure will fall during the outflow. However, if the reactor is large, the unsteady-state contribution can be neglected, and we do so here.

Applying Bernoulli's equation from the free liquid surface in the reactor to the exit of the pipe, we neglect the potential-energy terms, which are negligible, and the small velocity at the free surface. The remaining terms are

$$\frac{P_2 - P_1}{\rho} + \frac{V_2^2}{2} = -\mathscr{F}$$

The flow rate in this case will be much higher than that used in common industrial practice, so Fig. 6.13 and App. A.4 will be of no use to us. Here the friction loss consists of two parts: the entrance loss into the pipe and the loss due to the flow through the 10 ft of pipe. Substituting from Eqs. 6.28 and 6.16, we find

$$\mathscr{F} = \underbrace{K\frac{V^2}{2}}_{\textit{entrance}} + \underbrace{4f\frac{\Delta x}{D}\cdot\frac{V^2}{2}}_{\textit{straight pipe}}$$

Substituting these in Bernoulli's equation as given above, we find

$$\frac{P_2 - P_1}{\rho} = \frac{-V_2^2}{2} - \frac{KV_2^2}{2} - 4f\frac{\Delta x}{D}\cdot\frac{V_2^2}{2} = \frac{-V_2^2}{2}\left(1 + K + 4f\frac{\Delta x}{D}\right)$$

$$V_2 = \left[\frac{2(P_1 - P_2)/\rho}{1 + K + 4f(\Delta x/D)}\right]^{1/2}$$

From Fig. 6.16 we can read $K = 0.5$ (the diameter of the line is much smaller than the tank diameter). From App. A.3 we see that for 3-in schedule 80 pipe, the inside diameter is 2.900 in. Then from Table 6.2

$$\frac{\varepsilon}{D} = \frac{0.0018\text{ in}}{2.900\text{ in}} = 0.00062$$

It is safe to assume that the Reynolds number here will be very high; so on Fig. 6.10 we select as our first guess a friction factor at the far right of the diagram, which for an ε/D of 0.00062 gives us $f = 0.0043$. Then

$$V_2 = \left\{ \frac{2(2000 - 15)\ \text{lbf/in}^2}{62.3\ \dfrac{\text{lbm}}{\text{ft}^3} \cdot \left[1 + 0.5 + \dfrac{4 \cdot 0.0043 \cdot 10\ \text{ft}}{(2.900/12)\ \text{ft}} \right]} \cdot \frac{32.2\ \text{lbm} \cdot \text{ft}}{\text{lbf} \cdot \text{s}^2} \cdot \frac{144\ \text{in}^2}{\text{ft}^2} \right\}^{1/2}$$

$$= \left[\frac{2 \cdot 1985 \cdot 32.2 \cdot 144}{62.3(1 + 0.5 + 0.71)} \cdot \frac{\text{ft}^2}{\text{s}^2} \right]^{1/2} = 365\ \frac{\text{ft}}{\text{s}} = 111\ \frac{\text{m}}{\text{s}}$$

We then check to see whether our assumed friction factor is correct:

$$\mathcal{R} = \frac{[(2.9/12)\ \text{ft}](62.3\ \text{lbm/ft}^3)(365\ \text{ft/s})}{1.0\ \text{cP} \cdot 6.72 \times 10^{-4}\ \text{lbm/(ft} \cdot \text{s} \cdot \text{cP)}} = 8.2 \times 10^6$$

From Fig. 6.10 we see that our assumed f was correct. From App. A.3 we see that for 1 ft/s the flow rate is 20.55 gal/min; so the flow rate is

$$Q = 365\ \frac{\text{ft}}{\text{s}} \cdot \frac{20.55\ \text{gal/min}}{\text{ft/s}} = 7500\ \frac{\text{gal}}{\text{min}} = 0.47\ \frac{\text{m}^3}{\text{s}}$$

This is the instantaneous flow rate. As the flow continues, the pressure and flow rate both decrease. ■

The rapid methods, Fig. 6.13 and App. A.4, are of no use in this case, because the flow velocity is much larger than normal pipeline velocities. Ignoring the kinetic energy of the exit fluid, or the friction loss in the pipe, or the entrance loss would have given a significantly incorrect answer.

Example 6.16. A fire truck (see Fig. 6.18) is sucking water from a river and delivering it through a long hose to a nozzle, from which water issues at a velocity of 100 ft/s. The total flow rate is 500 gal/min. The hoses have a diameter equivalent to that of a 4-in schedule 40 pipe and may be assumed to have the same relative roughness. The total length of hose, corrected for valves, fittings, entrance, etc., is 300 ft. What power is required of the fire truck's pump?

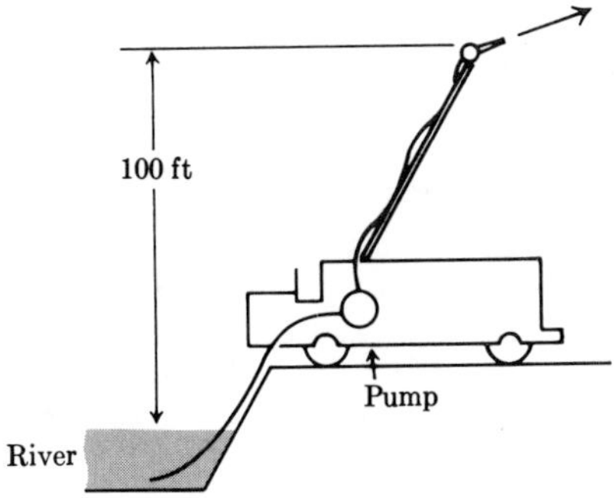

FIGURE 6.18

Applying Bernoulli's equation from the surface of the river, point 1, to the outlet of the nozzle, point 2, we find

$$g(z_2 - z_1) + \frac{V_2^2}{2} = \frac{-dW_{\text{a.o.}}}{dm} - \mathscr{F}$$

Here we may find the friction-loss term from App. A.4 and Eq. 6.22:

$$\mathscr{F} = \frac{-\Delta P}{\Delta x}\frac{\Delta x}{\rho} = \frac{5.65\ \text{lbf/in}^2}{100\ \text{ft}} \cdot \frac{300\ \text{ft}}{62.3\ \text{lbm/ft}} \cdot \frac{32.2\ \text{lbm}\cdot\text{ft}}{\text{lbf}\cdot\text{s}^2} \cdot \frac{144\ \text{in}^2}{\text{ft}^2}$$

$$= 1260\ \text{ft}^2/\text{s}^2$$

Then we have

$$\frac{-dW_{\text{a.o.}}}{dm} = 32.2\ \frac{\text{ft}}{\text{s}^2} \cdot 100\ \text{ft} + \frac{(100\ \text{ft/s})^2}{2} + 1260\ \frac{\text{ft}^2}{\text{s}^2}$$

$$= 9480\ \text{ft}^2/\text{s}^2$$

$$\dot{m} = 500\ \frac{\text{gal}}{\text{min}} \cdot 8.33\ \frac{\text{lbm}}{\text{gal}} \cdot \frac{\text{min}}{60\ \text{s}} = 69.5\ \frac{\text{lbm}}{\text{s}} = 31.6\ \frac{\text{kg}}{\text{s}}$$

Therefore,

$$\text{Po} = \frac{dW_{\text{a.o.}}}{dt} = \frac{dW_{\text{a.o.}}}{dm}\,\dot{m}$$

$$= -9480\ \frac{\text{ft}^2}{\text{s}^2} \cdot 69.5\ \frac{\text{lbm}}{\text{s}} \cdot \frac{\text{lbf}\cdot\text{s}^2}{32.2\ \text{lbm}\cdot\text{ft}} \cdot \frac{\text{hp}\cdot\text{s}}{550\ \text{ft}\cdot\text{lbf}}$$

$$= -37\ \text{hp} = -27.6\ \text{kW}$$

■

A very important class of problems is illustrated by example in Fig. 6.19. Water flows from one reservoir through a pipe to a division point (called a *node*), from which it flows to two other reservoirs via separate pipes. The elevations of the reservoirs are shown. The task is to compute the flow through each pipe branch, assuming steady flow. This problem arises very often in water supply networks, where multiple reservoirs are connected to multiple users. The three-reservoir example illustrates the idea, but not the complexity

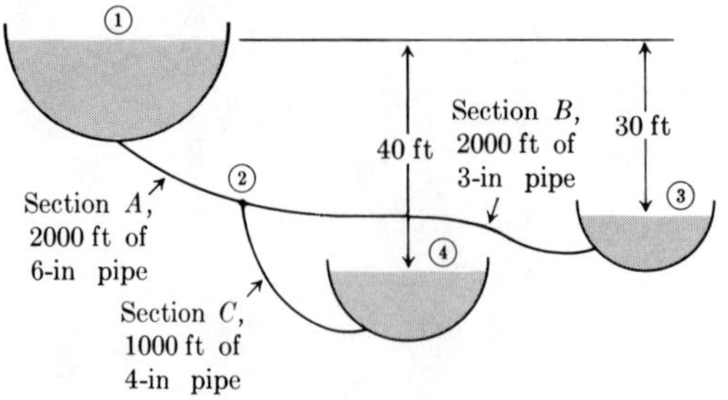

FIGURE 6.19

that can exist in water supply networks which have grown and been modified over time.

For such long pipes, the kinetic-energy terms in Bernoulli's equation will be negligible, and the gauge pressures at the free surfaces are all zero. If these two simplifications are not appropriate, the problems can still be solved, but not as simply as shown here. Writing Bernoulli's equation for the three sections of the pipe, we find

$$\frac{P_2}{\rho} + g(z_2 - z_1) = -\mathcal{F}_A = -\left(4f\,\frac{\Delta x}{D}\cdot\frac{V^2}{2}\right)_A$$

$$\frac{-P_2}{\rho} + g(z_3 - z_2) = -\mathcal{F}_B = -\left(4f\,\frac{\Delta x}{D}\cdot\frac{V^2}{2}\right)_B$$

$$\frac{-P_2}{\rho} + g(z_4 - z_2) = -\mathcal{F}_C = -\left(4f\,\frac{\Delta x}{D}\cdot\frac{V^2}{2}\right)_C$$

From the mass balance we find

$$Q_A = Q_B + Q_C$$

$$V_A = V_B\,\frac{A_B}{A_A} + V_C\,\frac{A_C}{A_A}$$

Here we have four equations relating four unknowns (the three V's and P_2). However, to solve the problem, we must use the correct values of the three f's which are related to the pipe diameters and the V's by the friction factor chart (Fig. 6.10) or by Eq. 6.23 or 6.24. Thus, we could also think of this as a system with seven unknowns and seven equations (taking the friction factor chart or Eq. 6.23 or 6.24 three times). Because of the forms of Eqs. 6.23 and 6.24, there is no possibility of solving these equations analytically. The solution must be by trial and error.

In the problem statement, the elevation at point 2 is not given, because we do not need to know it to solve the problem. We can see this by adding the first two Bernoulli equations, finding

$$g(z_3 - z_1) = -(\mathcal{F}_A + \mathcal{F}_B)$$

$$= -\left[\left(4f\,\frac{\Delta x}{D}\cdot\frac{V^2}{2}\right)_A + \left(4f\,\frac{\Delta x}{D}\cdot\frac{V^2}{2}\right)_B\right]$$

and finding a similar equation by combining the first and third Bernoulli equations. Both the elevation and the pressure at point 2 are thereby eliminated from the problem. The procedure to solve the problem, either by hand or in a computer, is to guess a set of flows in the three pipe branches, subject to the constraint that the algebraic sum of the flows into any node (e.g., point 2 in Fig. 6.19) must be zero. With this assumed set of flow rates, one then computes the $\mathcal{F}$'s for each of the pipe branches. Then one compares the sum of the $\mathscr{F}$'s from one reservoir to another with the $g\,\Delta z$ term between the same two reservoirs. When the proper set of flows has been chosen, these will all agree. For this three-branch, one-node example, the trial-and-error method is quite easy (Prob. 6.60). For more complex examples, it is not. A widely used systematic procedure for solving this type of system was developed by Cross [13]. Computer programs are available to carry out that solution [14].

6.13 ECONOMIC PIPE DIAMETER

From the foregoing we can easily calculate the flow rate, given the pipe diameter and pressure drop, or calculate the pipe diameter, given the flow rate and pressure drop, and so forth. A much more interesting question is, Given the flow rate, what size of pipe should we select? It is possible that the choice is dictated by aesthetics; e.g., the pipe goes through a lobby, and we want it to be the same size as other exposed pipes in the lobby. Or the choice may be dictated by the supply; e.g., we have on hand a large amount of surplus 4-in pipe which we want to use. Most often the choice is based on economics; the engineer is asked to make the most economical selections, all things considered.

For economic analysis we must consider two possibilities:

1. The fluid is available at a high pressure and eventually will be throttled to a low pressure, so the energy needed to overcome friction losses may come from the available pressure drop.
2. The fluid is not available at a high pressure, so a pump or compressor is needed to overcome the effects of fluid friction.

The first is simple: We select the smallest size of pipe which will carry the required flow with the available pressure drop. Example 6.5 is that case.

If the effects of friction must be overcome by a pump or compressor, then the total annual costs of the pump pipeline system are the following:

1. Power to run the pump
2. Maintenance charges on pump and line
3. Capital-cost charges for both line and pump

How these change with increasing line size is sketched in Fig. 6.20. The figure indicates the following:

1. The larger the pipe diameter, the greater the capital charges. The cost of pipeline is roughly proportional to the pipe diameter; bigger pipes cost more to buy, require more expensive supports, take longer to install, etc. The cost of the pump is proportional to the cost of the pipe and is included in it.
2. The maintenance cost is not affected much by pipe size.

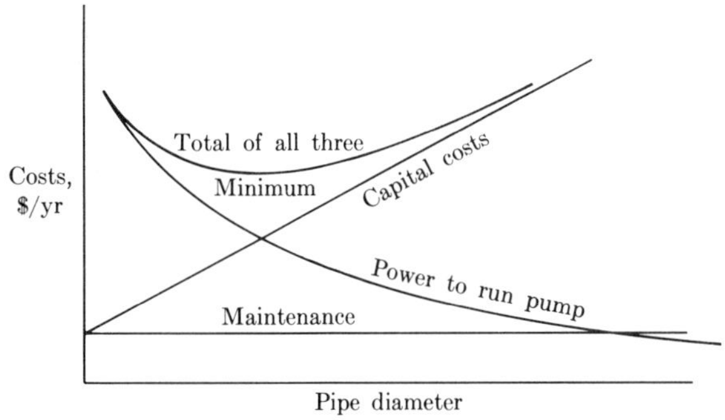

FIGURE 6.20

3. The pumping cost goes down rapidly as the pipe size goes up. The pumping cost is proportional to the pressure drop (see Example 6.3), which for turbulent flow is proportional to the velocity to the 1.8 to 2.0 power divided by the diameter. The velocity (for constant flow rate) is proportional to the reciprocal of the square of the diameter, so the pumping cost is proportional to the reciprocal of the diameter to the 4.6 to 5 power.

As Fig. 6.20 shows, the sum of these has a rather sharp minimum. This minimum occurs at the economic pipe diameter. Recognize here that we are taking the sum of a power cost during some finite period, e.g., a year, and the annual charge for owning the pipeline and the pump, whose lifetime will be many years. There are a variety of sophisticated ways of doing this, treated in books on plant design [15]. Here we consider the *simplest possible* kind of economic analysis:

$$\text{Purchase price} = \text{PP} \cdot \text{pipe diameter} \cdot \text{pipe length} \tag{6.39}$$

where the purchase price is what we would have to pay a contractor for both supplies and labor to build the complete pipeline and pump for us and PP is a constant with dimensions $/[inch (of diameter) · ft (of length)].

$$\text{Annual capital charge} = \text{CC} \cdot \text{purchase price} \tag{6.40}$$

where capital charge (CC) is a constant, with dimension (1/year) and

$$\text{Annual pumping cost} = \text{PC} \cdot \text{pump power} \tag{6.41}$$

where pumping cost (PC) is a constant with dimensions $/[hp · year].

As shown in Fig. 6.20, the maintenance cost is practically independent of the pipe diameter, so we do not include it in the analysis. We then wish to find the minimum of

$$\text{Total annual cost} = \text{PC} \cdot \text{Po} + \text{CC} \cdot \text{PP} \cdot \text{diameter} \cdot \text{length} \tag{6.42}$$

Assuming that the pipe is horizontal, we may apply Bernoulli's equation from the pump inlet, point 1, to the pipe outlet, point 2, and see that there is no change in elevation or velocity. We assume that the pressure at the pump inlet is the same as the pressure at the pipe outlet; i.e., the pump has to overcome only the effects of friction. Then from Eq. 6.16 we have

$$\frac{-dW_{\text{a.o.}}}{dm} = \mathcal{F} = 2f\,\frac{\Delta x}{D}\,V^2 \tag{6.43}$$

$$\text{Po} = \frac{-dW_{\text{a.o.}}}{dm}\,\dot{m} = 2f\,\frac{\Delta x}{D}\,V^2\dot{m} \tag{6.44}$$

but we have

$$V = \frac{\dot{m}}{\rho(\pi/4)D^2} \tag{6.45}$$

and therefore

$$\text{Po} = \frac{\dot{m}^3 2f\,\Delta x\,(4/\pi)^2}{\rho^2 D^5} \tag{6.46}$$

Substituting Eq. 6.43 and the cost of the pipe in Eq. 6.39, we find

$$\text{Total annual cost} = \text{PC}\cdot\dot{m}^3 2f\,\Delta x\left(\frac{4}{\pi}\right)^2\frac{1}{\rho^2}\cdot\frac{1}{D^5} + \text{CC}\cdot\Delta x\cdot\text{PP}\cdot D \tag{6.47}$$

We now differentiate the total annual cost with respect to diameter D and set the derivative equal to zero:

$$0 = \frac{d(\text{cost})}{dD} = \text{PC}\cdot\dot{m}^3 2f\,\Delta x\left(\frac{4}{\pi}\right)^2\frac{1}{\rho^2}\cdot\frac{-5}{D^6} + \text{CC}\cdot\Delta x\cdot\text{PP} \tag{6.48}$$

Solving for D_{econ}, we find

$$D_{\text{econ}} = \left[\frac{10\cdot\text{PC}\cdot\dot{m}^3 f(4/\pi)^2(1/\rho^2)}{\text{CC}\cdot\text{PP}}\right]^{1/6} \tag{6.49}$$

This equation shows that the economic pipe diameter is independent of how long the pipe is. This should be no surprise: Both the pumping and capital costs are proportional to the pipe length. The equation also shows that the economic diameter is proportional to the friction factor to the one-sixth power; hence, we can use a rough estimate of the friction factor and make very little error.

Example 6.17. We wish to transport 200 gal/min of water 5000 ft in a horizontal, schedule 40, carbon-steel pipe. We will install a pump to overcome the friction loss. Given the economic data shown below, what is the economic pipe diameter?

$$\text{PC} = \frac{\$270}{\text{hp}\cdot\text{yr}} \qquad \text{PP} = \frac{\$2}{\text{in of diameter}\cdot\text{ft of length}} \qquad \text{CC} = \frac{0.40}{\text{yr}}$$

First we guess that the pipe will have an inside diameter of 3 in. Then from Table 6.2 we have $\varepsilon/D = 0.0018/3 = 0.0006$. The friction factor will probably be about 0.0042. The mass flow rate is 200 gal/min · 8.33 lbm/gal = 1666 lbm/min. Substituting these and the values of PC, CC, and PP in Eq.

6.49 produces

$$D_{\text{econ}} = \left[\frac{\dfrac{\$270}{\text{hp}\cdot\text{yr}} \cdot \left(\dfrac{1666\ \text{lbm}}{\text{min}}\right)^3 \cdot 10 \cdot 0.0042 \cdot \left(\dfrac{4}{\pi}\right)^2 \cdot \left(\dfrac{\text{ft}^3}{62.3\ \text{lbm}}\right)^2}{0.4/\text{yr}\cdot\$2/(\text{in}\cdot\text{ft})} \right]^{1/6}$$

$$\cdot \left(\frac{\text{hp}\cdot\text{min}}{3.3\times10^4\ \text{ft}\cdot\text{lbf}} \cdot \frac{\text{lbf}\cdot\text{s}^2}{32.2\ \text{lbm}\cdot\text{ft}} \cdot \frac{\text{min}^2}{3600\ \text{s}^2} \cdot \frac{\text{ft}}{12\ \text{in}} \right)^{1/6}$$

$$= (5.95\times10^{-4}\ \text{ft}^6)^{1/6} = 0.290\ \text{ft} = 3.48\ \text{in} = 0.088\ \text{m}$$

Because of the approximate nature of the economic data used, a 4-in pipe would probably be selected. It would be appropriate to check the assumed friction factor (Prob. 6.62). ■

Because calculations such as these are long and tedious, companies that install many pipelines have solved the problem for a large number of cases and have summarized the results in convenient form. The most popular method is to calculate the economic velocity:

$$\text{Economic velocity} = \frac{\text{volumetric flow rate}}{(\pi/4)(\text{economic diameter})^2} \tag{6.50}$$

Substituting for the economic diameter from Eq. 6.49, we find

$$V_{\text{econ}} = \frac{\dot{m}/\rho}{\dot{m}(1/\rho^{2/3})f^{1/3}\cdot\text{constant}} = \text{constant}\cdot\frac{1}{f^{1/3}\rho^{1/3}} \tag{6.51}$$

This equation says that for a given set of cost data the economic velocity is independent of the mass flow handled and dependent on only the fluid density and the friction factor. More thorough analyses and far more complicated cost equations lead to substantially the same conclusion. For example, for schedule 40 carbon-steel pipe, Boucher and Alves [16] give the data shown in Table 6.4.

The table refers to turbulent flow only. For laminar flow, the value of f goes up quite rapidly as the viscosity increases, making the economic velocity go down. Oil companies spend more money pumping viscous liquids (crude oils, asphalt, heating oils, etc.) than do any other companies; therefore they have made up the most convenient economic-velocity plots for laminar flow.

TABLE 6.4
Economic velocity for schedule 40, carbon-steel pipe

Fluid density, lbm/ft	Economic velocity, ft/s
100	5.1
50	6.2
10	10.1
1	19.5
0.1	39.0
0.01	78.0

Figure 6.21 shows such a plot. It can be used to rapidly select the economic pipe diameter for laminar flow, subject to the restriction that the economic data on the line to be installed must be the same as those shown on the plot. Figure 6.21 has nomenclature similar to that of Fig. 6.13, and the comments on the latter are applicable here. Figure 6.21 also shows the economic diameter for turbulent flow.

Why does App. A.4 show the velocity in feet per second for all the water flows given? From Table 6.4 and Fig. 6.21 we can see that for water (which is almost always in turbulent flow in industrial equipment) an economic velocity is almost always about 6 ft/s. Thus, working engineers often simply select pipe sizes for water or similar fluids by looking at App. A.4 for the pipe size which gives a velocity of about 6 ft/s (2 m/s).

Table 6.4 and Fig. 6.21 are for one set of costs; for other costs the results are different. However, because of the $\frac{1}{6}$ factor in Eq. 6.49, the different costs change the economic diameter very little (see Prob. 6.66).

6.14 FLOW AROUND SUBMERGED OBJECTS

The flow around a submerged object is generally more complicated than the flow in a straight pipe or channel, because it is two- or three-dimensional. To understand the *details* of the flow around any submerged object, we must first take up the subjects of potential flow and the boundary layer, which we do in Chaps. 10 and 11.

Frequently we are not interested in the details of the flow but only in the practical problem of predicting the force on a body due to the flow of fluid around it. For example, the airplane designer wants to know the "air resistance" of the plane to select the right engine, the submarine designer wants to know the "water resistance" to determine how fast the submarine can go, and the designer of a chimney wants to know the maximum wind force on it to decide how much bracing is needed. These forces are now all called *drag forces*, following aeronautical engineering terminology. By using experimental data on such flows we can treat the problems as if they were one-dimensional.

Probably the first systematic investigation of drag forces was undertaken by Isaac Newton [17], who dropped hollow spheres from the inside of the dome of St. Paul's Cathedral in London and measured their rate of fall. He calculated that the drag force on a sphere should be given by

$$\text{Drag force} = F = \pi r^2 \rho_{\text{air}} \frac{V^2}{2} \tag{6.52}$$

Subsequent workers found that this equation had to be modified by introducing a coefficient, which we call the *drag coefficient* C_d. This coefficient is not a constant equal to 1, as Newton believed, but varies with varying conditions, as we will see. Introducing it and dividing both sides of Eq. 6.52 by the cross-sectional area of the sphere, we find

$$\frac{F}{A} = C_d \rho \frac{V^2}{2} \tag{6.53}$$

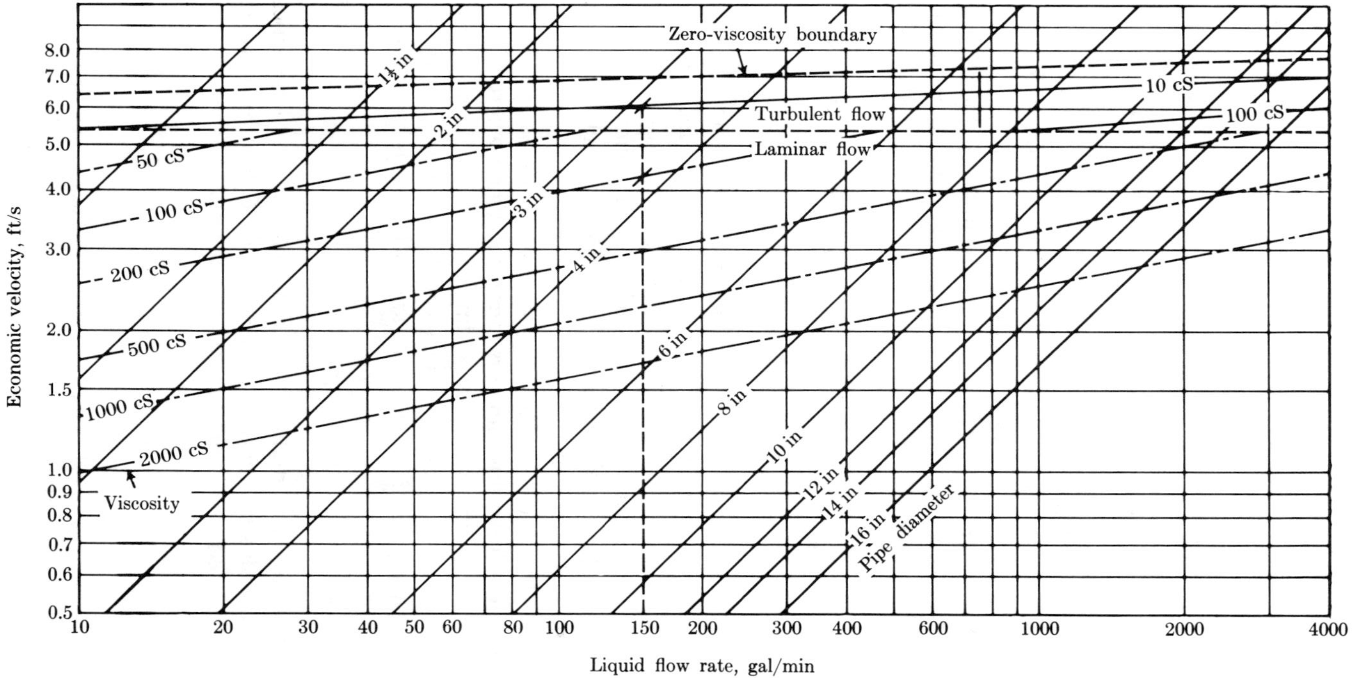

FIGURE 6.21
Economic pipe size for pumped liquids (carbon-steel pipe). Assumptions: Pumping cost = \$135 per horsepower-year; line cost = \$1 per inch of diameter per foot; fixed charges per year on line = 0.40 times line cost; liquid specific gravity = 0.80 (not very critical). Examples: for 150 gal/min and 200 cSt, use the 4-in line; for 150 gal/min and 10 cSt, use the 3-in line. (These prices are from the early 1960s. Current prices are higher, but they have risen more or less together, so that the economic pipe sizes have not changed much.) (Courtesy of the Board of Engineers, Standard Oil Company of California.)

Compare this with the equation for the pressure drop in a long, straight, horizontal pipe:

$$-\Delta P = 4f \frac{\Delta x}{D} \rho \frac{V^2}{2} \tag{6.54}$$

From these equations we see that C_d plays the same role as f. Equation 6.54 contains the factor $\Delta x/D$, which describes the geometry of the system (long, thin pipes have more pressure drop than short, thick ones), but since all spheres have the same shape, there was no need to include such a factor in Eq. 6.53.

In steady pipe flow it was found experimentally that f depends on only the Reynolds number and the relative roughness. It has been found similarly that the drag coefficient for smooth spheres in steady motion depends on only the Reynolds number. Here we must redefine the Reynolds number, which previously included the pipe diameter. The common practice is to define a *particle Reynolds number*, in which the particle diameter takes the place of the pipe diameter:

$$\text{Particle Reynolds number } \mathcal{R}_p = \frac{\text{particle diameter} \cdot \text{velocity} \cdot \text{fluid density}}{\text{fluid viscosity}} \tag{6.55}$$

With this definition all the steady-state drag data on single, smooth spheres moving in infinite, quiescent, newtonian fluids at moderate velocities can be represented by a single curve on Fig. 6.22. This figure shows also drag coefficients for disks and cylinders, to be discussed later. It is limited to steady velocities of less than about one half the local speed of sound; velocities higher than this are discussed elsewhere [18].

Figure 6.22 and the friction factor plot (Fig. 6.10) show marked similarities. In both at low Reynolds numbers there is a region where f or C_d is proportional to $1/\mathcal{R}$ or $1/\mathcal{R}_p$, that is, a straight line of slope $-45°$ on log paper. For pipes this is Poiseuille's equation, which can be written $f = 16/\mathcal{R}$; for spheres (straight line on Fig. 6.22) it is *Stokes law*,† which can be written

$$C_d = \frac{24}{\mathcal{R}_p} \tag{6.56}$$

Both figures have a horizontal section in which f or C_d is practically independent of the Reynolds number. The value of f or C_d on this horizontal section cannot be calculated; it is found from experimental data.

† Stokes law, like the Poiseuille equation, can be derived mathematically without the aid of experimental data. In doing so we must assume that the flow is laminar and Newton's law of viscosity holds and that the resulting terms in the equations involving velocities squared are negligible. The latter condition is called *creeping flow*. Even with these assumptions the derivation takes several pages [19]. The greater complexity, compared with the Poiseuille equation, is due to the three-dimensional nature of the flow around a sphere.

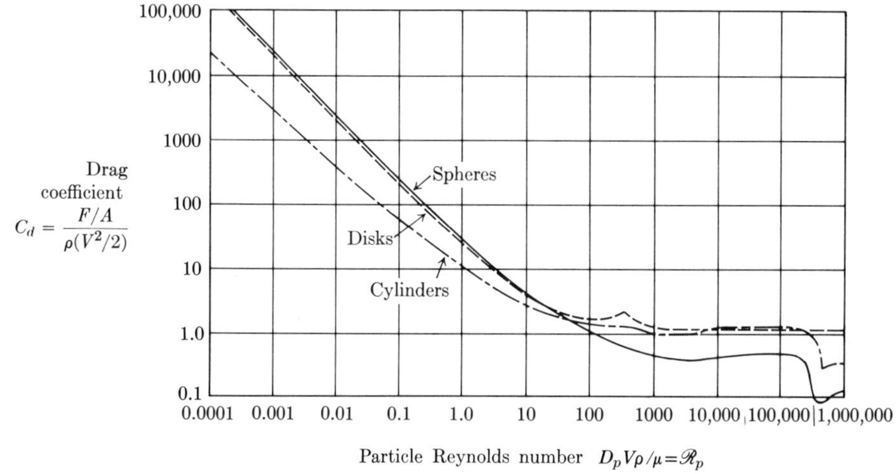

FIGURE 6.22
Drag coefficients for spheres, disks, and cylinders. [From C. E. Lapple and C. B. Shepherd, "Calculation of particle trajectories," *Ind. Eng. Chem. 32:* 605–617 (1940). Copyright © by the American Chemical Society. Reproduced by permission.]

However, the curve of the sphere drag coefficient has some marked differences from the friction factor plot. It does not continue smoothly to higher and higher Reynolds numbers, as does the f curve; instead, it takes a sharp drop at an $\mathscr{R}_p$ of about 300,000. Also it does not show the upward jump that characterizes the laminar-turbulent transition in pipe flow. Both differences are due to the different shapes of the two systems. In a pipe all the fluid is in a confined area, and the change from laminar to turbulent flow affects all the fluid (except for a very thin film at the wall). Around a sphere the fluid extends in all directions to infinity (actually the fluid is not infinite, but if the distance to the nearest obstruction is 100 sphere diameters, we may consider it so), and no matter how fast the sphere is moving relative to the fluid, the entire fluid cannot be set in turbulent flow by the sphere. Thus, there cannot be the sudden laminar-turbulent transition for the entire flow, which causes the jump in Fig. 6.10. The flow very near the sphere, however, can make the sudden switch, and the switch is the cause of the sudden drop in C_d at $\mathscr{R}_p = 300{,}000$. This sudden drop in drag coefficient is discussed in Sec. 11.6. Leaving until Chaps. 10 and 11 the reasons *why* the curves in Fig. 6.22 have the shapes they do, for now we simply accept the curves as correct representations of experimental facts and show how to use them to solve various problems.

Consider a spherical particle settling through a fluid under the influence of gravity. Figure 6.23 shows the forces acting on a particle falling through a fluid. Writing Newton's law for the particle, we find

$$ma = \rho_{\text{part}} \frac{\pi}{6} D^3 g - \rho_{\text{fluid}} \frac{\pi}{6} D^3 g - C_d \frac{\pi}{4} D^2 \rho_{\text{fluid}} \frac{V^2}{2} \tag{6.57}$$

If the particle starts from rest, its initial velocity is zero, so the drag force in

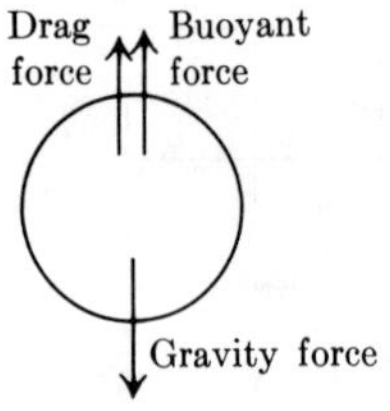

FIGURE 6.23
Forces on a settling particle.

this equation is initially zero. The particle accelerates rapidly; as it accelerates, the drag force increases as the square of the velocity increases, until it equals the gravity force minus the buoyant force. This is the state of *terminal velocity*; the sum of the forces acting is zero, so the particle continues to move at a constant velocity. To find this velocity, we set the acceleration to zero in Eq. 6.57 and solve for V:

$$V^2 = \frac{4Dg(\rho_{\text{part}} - \rho_{\text{fluid}})}{3C_d\rho_{\text{fluid}}} \tag{6.58}$$

This equation is correct for any value of $\mathscr{R}_p$. If the particle is very small, it probably obeys Stokes law, $C_d = 24/\mathscr{R}_p$. Substituting this in Eq. 6.58 and rearranging, we find

$$V = \frac{D^2g(\rho_{\text{part}} - \rho_{\text{fluid}})}{18\mu} \qquad \text{[Stokes law]} \tag{6.59}$$

Example 6.18. A small particle of dust may be considered a sphere. It has the density of 100 lbm/ft and a diameter of 0.0001 in. It is settling in still air at 68°F and has been doing so for some time. How fast is it falling?

Let us assume that Stokes law applies. Then

$$V = \frac{(0.0001\ \text{in})^2(32.2\ \text{ft/s}^2)(100\ \text{lbm/ft}^3 - 0.075\ \text{lbm/ft}^3)}{18 \cdot 0.018\ \text{cP}}$$
$$\cdot \frac{\text{cP} \cdot \text{ft} \cdot \text{s}}{6.72 \times 10^{-4}\ \text{lbm}} \cdot \frac{\text{ft}^2}{144\ \text{in}^2} = 0.0010\ \frac{\text{ft}}{\text{s}} = 3.7\ \frac{\text{ft}}{\text{h}} = 1.7\ \frac{\text{m}}{\text{h}}$$

Now we must check our assumption that Stokes law applies here. The particle Reynolds number is

$$\mathscr{R}_p = \frac{0.001\ \text{in} \cdot 0.075\ \text{lbm/ft}^3 \cdot 0.0010\ \text{ft/s}}{0.018\ \text{cP}} \cdot \frac{\text{cP} \cdot \text{ft} \cdot \text{s}}{6.72 \times 10^{-4}\ \text{lbm}} \cdot \frac{\text{ft}}{12\ \text{in}}$$
$$= 5.2 \times 10^{-4}$$

From Fig. 6.22 it appears that Stokes law (the straight-line part of the figure) holds to an $\mathscr{R}_p$ of about 0.3; so the assumption was good here (see Prob. 6.80). ■

Example 6.19. A solid steel sphere with SG = 7.85 and a diameter of 0.02 m is falling at its terminal velocity through water. What is its velocity?

As a first trial we assume that Stokes law (Eq. 6.59) applies; then

$$V = \frac{(0.02\ \text{m})^2(9.81\ \text{m/s}^2)(7.85-1)(998.2\ \text{kg/m}^3)}{18 \cdot 1.002 \times 10^{-3}\ \text{Pa} \cdot \text{s}} \cdot \frac{\text{N} \cdot \text{s}^2}{\text{kg} \cdot \text{m}} \cdot \frac{\text{Pa}}{\text{N/m}^2}$$

$$= 1488\ \text{m/s} = 4880\ \text{ft/s} \qquad ?$$

The corresponding particle Reynolds number is

$$\mathcal{R}_p = \frac{(0.02\ \text{m})(998.2\ \text{kg/m}^3)(1488\ \text{m/s})}{1.002 \times 10^{-3}\ \text{Pa} \cdot \text{s}} \cdot \frac{\text{Pa}}{\text{N/m}^2} \cdot \frac{\text{N} \cdot \text{s}^2}{\text{kg} \cdot \text{m}}$$

$$= 2.96 \times 10^7 \qquad ?$$

Clearly, Stokes law *does not apply* here. However, Eq. 6.58 applies for any value of the Reynolds number. To solve it, we must assume a value of the drag coefficient, calculate the corresponding velocity, and check our assumed C_d. This is a fairly simple matter of trial and error. On our first trial we let C_d equal 0.4; then

$$V = \left[\frac{4(0.02\ \text{m})(9.81\ \text{m/s}^2)(7.85-1)(998.2\ \text{kg/m}^3)}{3 \cdot (998.2\ \text{kg/m}^3)(0.4)} \right]^{1/2}$$

$$= 2.12\ \text{m/s} = 6.94\ \text{ft/s}$$

The corresponding particle Reynolds number is

$$\mathcal{R}_p = \frac{(0.02\ \text{m})(998.2\ \text{kg/m}^3)(2.12\ \text{m/s})}{1.002 \times 10^{-3}\ \text{Pa} \cdot \text{s}} \cdot \frac{\text{Pa}}{\text{N/m}^2} \cdot \frac{\text{N} \cdot \text{s}^2}{\text{kg} \cdot \text{m}}$$

$$= 4.2 \times 10^4$$

From Fig. 6.22 we see that the drag coefficient corresponding to this Reynolds number is about 0.5. On our second trial we use a C_d of 0.5; then

$$V = \left(2.12\ \frac{\text{m}}{\text{s}}\right)\left(\frac{0.4}{0.5}\right)^{1/2} = 1.90\ \frac{\text{m}}{\text{s}}$$

$$\mathcal{R}_p = \left(4.2 \times 10^{-4}\right) \cdot \left(\frac{1.90}{2.12}\right) = 3.8 \times 10^4$$

From Fig. 6.22 we see that the assumed C_d and $\mathcal{R}_p$ agree; so this velocity is the desired solution. ■

The foregoing pertains entirely to spheres. We can use Eq. 6.53 for other shapes, if we agree on what area A represents. Generally, in drag measurements it refers to the "frontal" area perpendicular to the flow; that is the definition on which the coefficients in Fig. 6.22 are based. Moreover, we must decide on which dimensions to base the Reynolds number in our correlation of C_d versus $\mathcal{R}_p$: in Fig. 6.22 the Reynolds number for cylinders takes the cylinder diameter as D, and that for disks takes the disk diameter.

Example 6.20. A cylindrical chimney is 5 ft in diameter. The wind is blowing horizontally at a velocity of 20 mi/h. What is the wind force per foot of height on the chimney?

For this case the Reynolds number is

$$\mathcal{R} = \frac{5\text{ ft} \cdot 20\text{ mi/h} \cdot 0.075\text{ lbm/ft}^3}{0.018\text{ cP}} \cdot \frac{5280\text{ ft}}{\text{mi}} \cdot \frac{\text{h}}{3600\text{ s}} \cdot \frac{\text{cP} \cdot \text{s} \cdot \text{ft}}{6.72 \times 10^{-4}\text{ lbm}}$$

$$= 0.91 \times 10^6$$

From the curve for cylinders on Fig. 6.22 we can read $C_d = 0.35$; therefore,

$$F = AC_d\rho \frac{V^2}{2}$$

$$= \frac{5\text{ ft}^2/(\text{ft of height}) \cdot 0.35 \cdot 0.075\text{ lbm/ft}^3 \cdot (20\text{ mi/h})^2}{2}$$

$$\cdot \left(\frac{5280\text{ ft}}{\text{mi}}\right)^2 \cdot \left(\frac{\text{h}}{3600\text{ s}}\right)^2 \cdot \frac{\text{lbf} \cdot \text{s}^2}{32.2\text{ lbm} \cdot \text{ft}} = 1.75 \frac{\text{lbf}}{\text{ft of height}} = 2.61 \frac{\text{kgf}}{\text{m}} \quad \blacksquare$$

This is not a very large force. However, we see from Eq. 6.53 that the force is proportional to the square of the wind velocity; so for a 100 mi/h wind the force is 25 times as great. To make matters worse, the wind force on long, thin objects can be oscillatory. The oscillatory motion is caused by the formation of vortices, which break away rhythmically. If the frequency of the shedding of these vortices is close to the natural frequency of oscillation of the system, then the wind force can drive that natural frequency disastrously. The most famous case of this was the Tacoma Narrows Bridge, which was destroyed by such oscillations [20].

In aircraft calculations the drag coefficient of a wing usually is based on the wing's horizontal surface rather than on the area perpendicular to the flow. In addition, aeronautical engineers define a lift coefficient C_l with exactly the same form as Eq. 6.53. In the equation F stands for the upward force of the air exerted on an airplane's wings, A stands for the horizontal wing surface, and C_d is replaced with C_l.

Example 6.21. An airplane has wings 15 m long (tip to tip) and 1.5 m wide (front to rear). The lift coefficient in level flight is $C_l = 0.8$, and the drag coefficient is $C_d = 0.04$. The drag and lift on parts other than the wing may be neglected. How much force must be exerted by the propeller to keep the plane moving 150 km/h? What is the maximum weight of the loaded airplane in this condition?

The drag force is

$$F = AC_d\rho \frac{V^2}{2}$$

$$= \frac{15\text{ m} \cdot 1.5\text{ m} \cdot 0.04 \cdot (1.21\text{ kg/m}^3)(150\text{ km/h})^2}{2} \cdot \left(\frac{1000\text{ m}}{\text{km}} \cdot \frac{\text{h}}{3600\text{ s}}\right)^2 \cdot \frac{\text{N} \cdot \text{s}^2}{\text{kg} \cdot \text{m}}$$

$$= 945\text{ N} = 212\text{ lbf}$$

This drag force is equal and opposite to the "thrust" force which must be supplied by the engine and propeller to keep the plane moving at this speed. The lift force is given by

$$F = AC_l\rho \frac{V^2}{2}$$

which is the same as the drag force multiplied by C_l/C_d:

$$945\text{ N} \cdot \frac{0.8}{0.04} = 18{,}900\text{ N} = 4249\text{ lbf}$$

The lift is equal to the maximum gross, loaded weight of the aircraft. ■

The last example shows why the lift and drag coefficients are so useful to aeronautical engineers. Their ratio C_l/C_d is equal to the allowable ratio of total aircraft weight to thrust of power plant. Normally both C_l and C_l/C_d are functions of aircraft speed and of the angle between the oncoming airstream and the wing surface [21]. This also shows why commercial aircraft fly as high as they can. To maintain level flight, they need a lift equal to their weight and a thrust equal to their drag. The lift and drag are both proportional to ρV^2. For a given weight, the required speed goes up as the square root of the air density goes down. The drag has the same relationship. So the higher they go, the lower the air density and the faster they can go for a given hourly fuel input. Thus their fuel cost per hour remains constant as they go up, but their fuel cost per mile goes down. (They also deliver the customers to their destination sooner, which the customers like, and pay for fewer hours of work to pilots and flight attendants.)

6.15 SUMMARY

1. The steady flow of fluids in constant-cross-section conduits can be of two radically different kinds: laminar, in which all the motion is in the flow direction, and turbulent, in which there is a choatic crosswise motion perpendicular to the net flow direction.
2. In laminar flow, the pressure drop per unit length is proportional to the first power of the volumetric flow rate. The entire flow behavior can be calculated simply. The calculation requires the observational fact that the fluid clings to solid surfaces; i.e., the velocity at the surface is zero.
3. In turbulent flow, the pressure drop per unit length is proportional to the flow rate to the 1.8 to 2.0 power. The behavior cannot be calculated without experimental data.
4. All experimental data on the turbulent flow of fluids in circular pipes can be represented on one plot, the friction factor plot.
5. The friction factor plot can be rearranged into other forms which are less general but more convenient. It can be represented by equations, but these are not convenient for hand use.
6. All data on turbulent flow through valves and fittings can be correlated by assuming that each kind of fitting contributes as much friction as a certain

number of pipe diameters of straight pipe. That number is about the same for one kind of fitting and is independent of pipe size, fluid properties, etc.

7. Laminar flow in a few kinds of noncircular conduits can be analyzed by the same technique as used for circular pipes.
8. Turbulent flow in many kinds of noncircular conduits can be estimated by substituting 4 times the hydraulic radius for the diameter in the friction factor plot and the friction factor equation.
9. The economic size for a pipe is the size with the lowest sum of annual charges for the purchased cost of the pipe and pump and the annual power cost of running the pump or compressor needed to overcome friction. For turbulent flow, this results in an economic velocity which is practically independent of everything but fluid density; it is about 6 ft/s for most liquids and about 40 ft/s for air under normal conditions.
10. The forces of fluids flowing over bodies are ordinarily correlated by the drag equation, in which the drag coefficient plays the same role as does the friction factor in pipe flow.

PROBLEMS

See the Common Units and Values for Problems and Examples inside the back cover. In all problems in this chapter, unless a statement is made to the contrary, assume that all pipes are schedule 40 commercial steel (see App. A.3).

6.1. Derive the equivalents of Eqs. 6.3 and 6.8 for fluid flow in the vertical direction, taking gravity into account. Then generalize them for fluid flow at any angle, taking gravity into account.

6.2. Air is flowing through a horizontal tube with 1.00-in inside diameter. What is the maximum average velocity at which laminar flow will be of the stable flow pattern? What is the pressure drop per unit length at this velocity?

6.3. Repeat Prob. 6.2 for water.

6.4. Show the derivations of Eqs. 6.9 and 6.10.

6.5. Show the effect on the calculated viscosity, in the viscometer in Example 6.2, of the 10 percent error in the measurement of (*a*) flow rate, (*b*) fluid density, and (*c*) tube diameter.

6.6. In Example 6.2 how much does the internal energy per unit mass of the fluid increase as it passes through the viscometer? Assume that there is no heat transfer from the fluid to the wall of the viscometer. If the heat capacity of the fluid is 0.5 Btu/(lbm · °F), how much does the fluid's temperature rise?

6.7. A circular, horizontal tube contains asphalt. Its viscosity is 100,000 cP(= 1000 P), and its density is 70 lbm/ft^3. The tube radius is 1 in. Asphalt may be considered a newtonian fluid for the purposes of this problem, although it is not always one. We now apply a pressure gradient of 1.0 $(\text{lbf/in}^2)/\text{ft}$. What is the steady-state flow rate?

6.8. What is the Reynolds number in Example 6.2? What is the lowest fluid viscosity for which this viscometer should be used?

6.9. If the students in Fig. 6.8(*a*) throw a standard 5-oz baseball back and forth, if its velocity in flight is 40 mi/h, and if each one throws it an average of once every 10 s, what is the average force in the y directions tending to separate them?

6.10. Show that in Fig. 6.8(b) the force in the x direction is independent of how fast the balls move in the y directions.

6.11. Show that if we define the shear stress at the pipe wall as $\tau = f\rho V^2/2$ and then calculate the pressure gradient for horizontal flow, we find Eq. 6.16.

6.12. Show that Poiseuille's equation may be rewritten as $f = 16/\mathcal{R}$.

6.13. A fluid is flowing in a pipe. The pressure drop is 10 lbf/(in^2 · 1000 ft). We now double the flow rate, holding the diameter and fluid properties constant. What is the pressure drop if the new Reynolds number (a) is 10 and (b) is 10^8?

6.14. Water is flowing at an average velocity of 7 ft/s in a 6-in pipe. What is the pressure drop per unit length?

6.15. An oil with a kinematic viscosity of 4.3 cSt and a specific gravity of 0.80 is flowing in a 3-in pipe. The pressure drop per 1000 ft is 30 psig. What is the flow rate in gallons per minute?

6.16. We wish to transport 200 gal/min of fluid through a 3-in pipe. The available pressure drop is 28 psi per 1000 ft. The fluid properties are SG = 0.75 and $\mu = 0.1$ cP. Is a 3-in pipe big enough?

6.17. Oil is flowing at a rate of 150 gal/min with $\mu = 1.5$ cP and SG = 0.87 in a 3-in pipe 1000 ft long. What is the pressure drop? Calculate (a) by Fig. 6.10 and (b) by Fig. 6.13.

6.18. In Example 6.4 how much does the temperature of the gasoline rise as it flows through the pipe? Assume that there is no heat transfer from the gasoline and that its heat capacity is 0.6 Btu/(lbm · °F).

6.19. Two large water reservoirs are connected by 5000 ft of 8-in pipe. The level in one reservoir is 200 ft above the level in the other, and water is flowing steadily through the pipe from one reservoir to the other. Both reservoirs are open to the atmosphere. How many gallons per minute are flowing?

6.20. In Prob. 6.19 we wish to replace the existing pipe with a new one that will transmit 10,000 gal/min under the same conditions. What size pipe should we choose?

6.21. Two tanks are connected by 500 ft of 3-in pipe. The tanks contain an oil with $\mu = 100$ cP and SG = 0.85. The level in the first tank is 20 ft above the level in the second, and the pressure in the second is 10 psi greater than the pressure in the first. How much oil is flowing through the pipe? Which way is it flowing?

6.22. We are offered some pipes made of a new kind of plastic. To test their roughness, we pumped water through a 3-in schedule 40 pipe made of this material at an average velocity of 40 ft/s. The observed friction factor was 0.0070. Estimate the absolute roughness of this plastic.

6.23. As discussed in Sec. 6.5, the friction factor plot (Fig. 6.10) relates six variables and therefore can be used for finding any of the six if the other five are known. Examples 6.3, 6.4, and 6.5 show how to find three of these quantities, given all the others. Problem 6.22 shows how to find a fourth, given all the others. The remaining two variables are the density and viscosity of the flowing fluid. Turbulent-flow pressure drops are practically never used for determining fluid viscosities or densities. Discuss why this is so.

6.24. When the plot of friction versus Reynolds number (Fig. 6.10) was introduced, there was some controversy over whether this was the best way to represent the experimental data. Some preferred a plot of $1/f^{1/2}$ versus $\mathcal{R}f^{1/2}$. What would be the advantage of this plot over Fig. 6.10? The interested reader can see this discussed elsewhere [4; see Hunter Rouse's comment at the end of the article] or worked out in an example [22].

6.25. Draw the equivalent of Fig. 6.13 for a 2-in schedule 40 pipe, on a piece of 2 cycle by 2 cycle log paper. Show (*a*) the zero-viscosity boundary, (*b*) the laminar-flow region, (*c*) the turbulent-flow region, and (*d*) the transition region.

6.26. Calculate the pressure drop per unit length for the flow of 100 gal/min of air in a 3-in schedule 40 pipe, using Fig. 6.13.

6.27. Oil with a kinematic viscosity of 20 cSt is flowing in 3-in pipe. According to Fig. 6.13, what is the highest flow rate at which the flow is certain to be laminar? What Reynolds number does this correspond to? What is the lowest flow rate at which the flow is certain to be turbulent? What Reynolds number does this correspond to?

6.28. Do any of the values in App. A.4 correspond to laminar flow?

6.29. For 24-in pipe, check whether the values in App. A.4 correspond to a constant friction factor or whether the pipe size is so large that the friction factor corresponds to the "smooth tubes" curve in Fig. 6.10.

6.30. Rework Example 6.4, using App. A.4. Show what corrections, if any, are needed in the solution if gasoline is assumed to have the same flow properties as water.

6.31. Estimate the pressure loss for 1000 ft^3/min of air flowing in a 12-in-diameter pipe 1000 ft long.

6.32. Estimate the required pipe diameter to transport 100 m^3/h of air with a friction loss of 1 Pa/m.

6.33. Estimate the volumetric flow rate of air for a pressure drop of 5 Pa/m in a duct with 0.125-m diameter.

6.34. Estimate the pressure drop for 500 ft^3/min of helium ($M = 4$ g/mol) flowing in a 6-in-diameter pipe 500 ft long (*a*) by using Fig. 6.10 and (*b*) by using Fig. 6.14 and suitable corrections for its much lower density than that of air.

6.35. Find the friction factor from the Colebrook equation (Eq. 6.23) for $\mathcal{R} = 2.05 \times 10^5$ *and* $\varepsilon/D = 0.0006$.

6.36. What happens to Eq. 6.24 as $\mathcal{R}$ approaches infinity? Compare the resulting values with the comparable ones on Fig. 6.10 for $\varepsilon/D = 0.05$, 0.01, 0.001, and 0.0001.

6.37. Water is flowing at a rate of 500 gal/min in a horizontal 10-in pipe. The pipe is 50 ft long and contains two standard 90° elbows and a swing-type check valve. What is the pressure drop?

6.38. A piping system consists of 100 ft of 2-in pipe, a sudden expansion to 3-in pipe, and then 50 ft of 3-in pipe. Water is flowing at 100 gal/min through the system. What is the pressure difference from one end of the pipe to the other?

6.39. Two large water tanks are connected by a 10-ft piece of 3-in pipe. The levels in the tanks are equal. When the pressure difference between the tanks is 30 lbf/in^2, what is the flow rate through the pipe?

6.40. The water in Fig. 6.24 is flowing steadily. What is the flow rate?

6.41. We are going to lay a length of 6-in steel pipe for a long distance and allow water to flow through it by gravity. If we want a flow rate of 500 gal/min, how much must we slope the pipe (i.e., by how many feet of drop per foot of pipe length)?

6.42. A 1-gal can has the dimensions shown in Fig. 6.25. A piece of $\frac{1}{4}$-in, schedule 40, galvanized pipe is inserted in the bottom. The pipe is horizontal. The tank is full of water. The end of the pipe is unplugged, and the water is allowed to flow out of the tank.

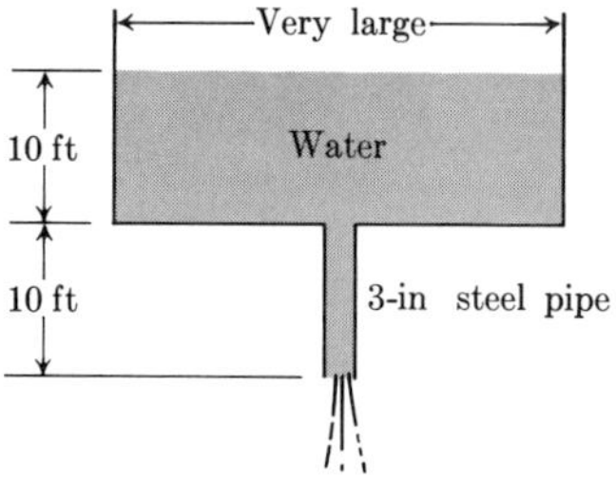

FIGURE 6.24

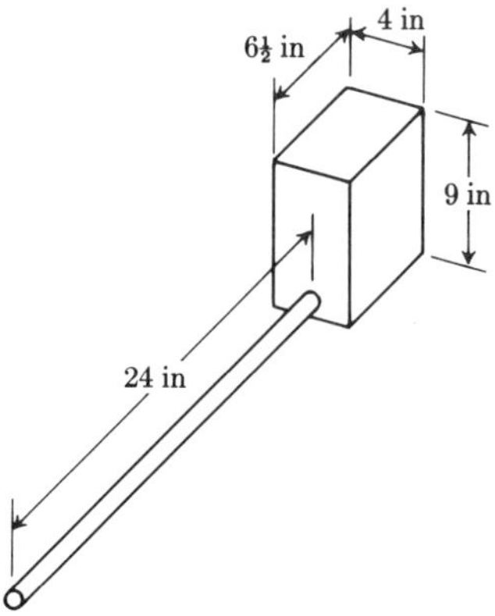

FIGURE 6.25

(*a*) How long will it take the level in the tank to fall from 7 in above the centerline of the pipe to 1 in above the centerline of the pipe? Make whatever assumptions seem plausible.

(*b*) As the level falls, the flow slows down, until it finally converts from turbulent to laminar. How far will the level be above the centerline of the pipe when this transition occurs?

6.43. Derive Eq. 6.29. It is suggested that you use the coordinates shown in Fig. 6.26. Here the flow is in the x direction from left to right, and the slit extends a distance l in the z direction. Choose as your element for the force balance a piece symmetric about the y axis (other choices are possible, but lead to more difficult mathematics).

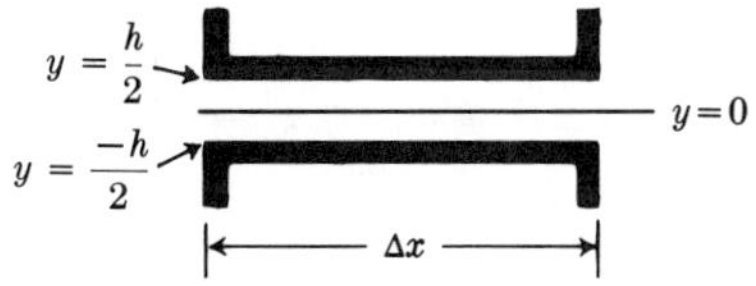

FIGURE 6.26

6.44. Derive Eq. 6.30. This equation is derived in detail in Bird et al. [11, section 2.4].

6.45. The wooden frame of a window is 2 in thick (see Fig. 6.27). The bottom of the window closes against the sill with a space between the frame and sill of 0.001 in. The width of the window (distance perpendicular to the paper in the sketch) is 2 ft. When the wind is blowing toward the window and creating a pressure difference of 0.01 psi across the window, what is the volumetric flow rate of air through the space between the frame and the sill?

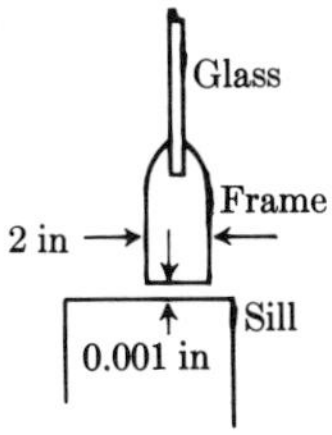

FIGURE 6.27

6.46. The cylindrical vessel in Fig. 6.28 is full of water at a pressure of 1000 psi. The top is held on by a flanged joint, which has been ground smooth and flat, with a clearance of 10^{-5} in, as shown. The diameter of the vessel is 10 ft. Estimate the leakage rate through this joint.

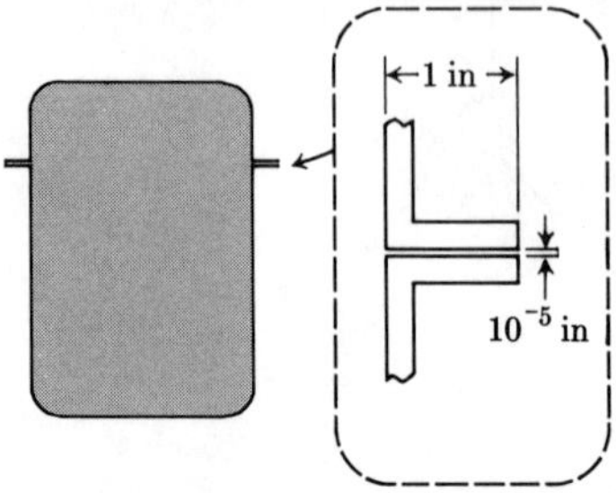

FIGURE 6.28

6.47. A thin film of fluid flows down a flat, vertical wall in steady laminar flow under the influence of gravity. Derive the relation between film thickness, viscosity, density, and flow rate per unit width of the wall. Assume that the shear stress between the film and the surrounding air is zero. (In the real situation surface waves may form; ignore this possibility.)

6.48. Calculate the hydraulic radius for (*a*) a semicircle with the top closed, (*b*) a semicircle with the top open, (*c*) a closed square, and (*d*) an annulus.

6.49. Rework Example 6.5, assuming that a square duct is to be used.

6.50. If the canal in Example 6.14 has a slope of 5 ft/mi, how many cubic feet per second will it transport?

6.51. In Fig. 6.29, the 3-in pipe is joined to the tank by a well-designed adapter, in which there is no entrance loss. What is the instantaneous velocity in the pipe?

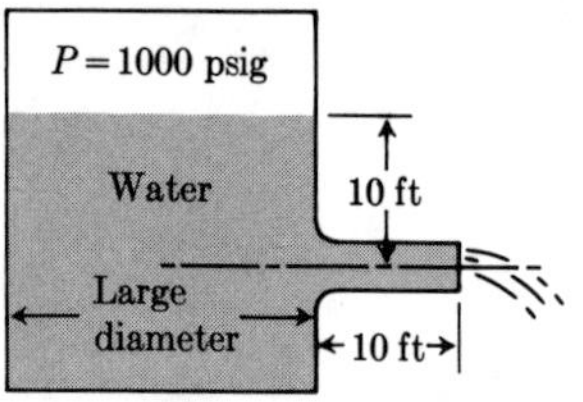

FIGURE 6.29

6.52. In Fig. 6.30 water is being pumped through a 3-in pipe. The length of the pipe plus the equivalent length for fittings is 2300 ft. The design flow rate is 150 gal/min.

(*a*) At this flow rate, what pressure rise across the pump is required?

(*b*) If there are no losses in pump, motor, coupling, etc., how many horsepower must the pump's motor deliver?

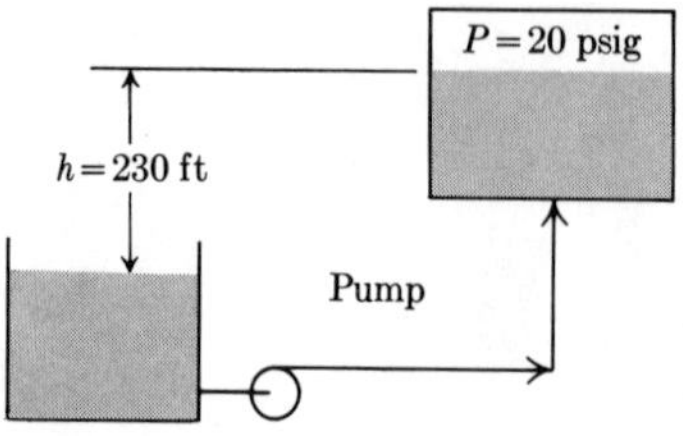

FIGURE 6.30

6.53. The tank in Fig. 6.31 is attached to 10 ft of 5-in pipe. The losses at the entrance from the reservoir to the pipe are negligible. What is the velocity at the exit of the pipe?

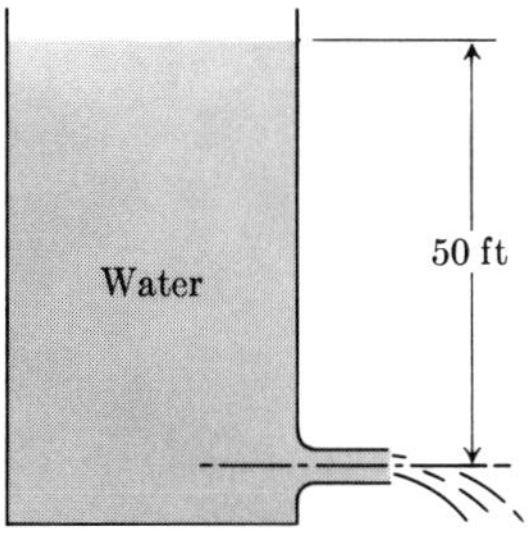

FIGURE 6.31

6.54. The flue gas in the stack in Fig. 6.32 is at 350°F and has a molecular weight equal to that of air. The air outside is at 68°F. The stack diameter is 5 ft, and the friction factor in the stack is 0.005. In passing through the furnace, the air changes significantly in density, because it is heated by the combustion and then cooled in giving up heat to the working parts of the furnace. Thus, we cannot rigorously apply Bernoulli's equation in the form we use in this problem (we could do so by integrating from point of point, over points so close together that the density change is negligible, but that would be very difficult in such a complex flow through a furnace). However, experimental data on the friction effects of furnaces indicate that if we treat them as constant-density devices with flowing fluids having the density of the inlet air, then $\mathcal{F}$ is approximately $KV_2^2/2$, where V_2 is the velocity in the stack and K is an empirical constant that here is assumed equal to 3.0. Thus, in applying Bernoulli's equation to this furnace and stack, assume that the air passes through the furnace into the base of the stack at constant density and then at a point changes to the density of the gases in the stack. On that basis, estimate the velocity of the gases in the stack.

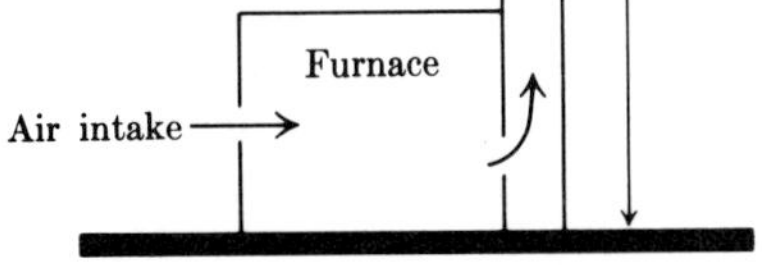

FIGURE 6.32

6.55. The vessels in Fig. 6.33 are connected by 1000 ft of 3-in pipe (neglect fitting and entrance and exit losses). In each vessel the diameter is so large that V is negligible. The fluid is an oil with $\nu = \mu/\rho$ of 100 cSt, and $\rho = 60\ \text{lbm/ft}^3$. How many gallons per minute are flowing? Which way?

6.56. We have two vessels, shown in Fig. 6.34, and are given the following information.

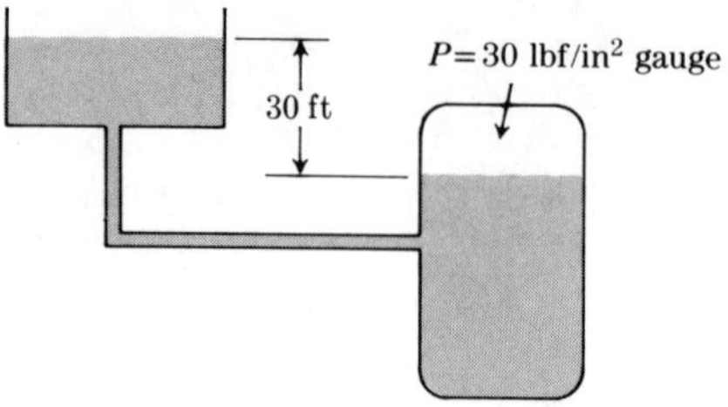

FIGURE 6.33

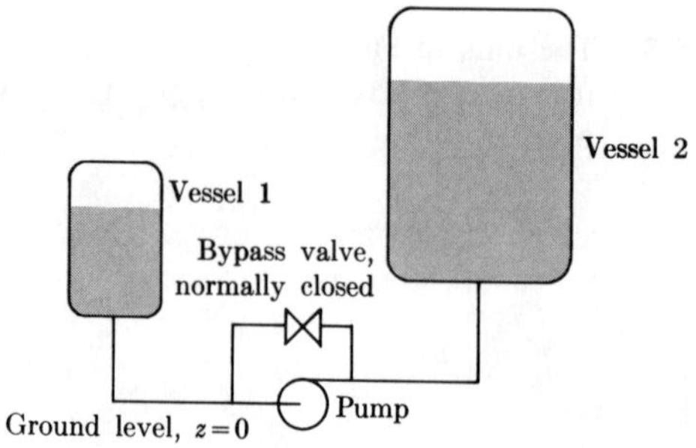

FIGURE 6.34

	Vessel 1	Vessel 2
P_{max}	20 psig	81 psig
P_{min}	8 psig	47 psig
Maximum liquid level, above $z = 0$	43 ft	127 ft
Minimum liquid level, above $z = 0$	21 ft	100 ft

The connecting line between the vessels is a 3-in pipe 627 ft long. It contains six elbows, four gate valves, and one globe valve. The fluid to be pumped has a specific-gravity range of 0.80 to 0.85 and a kinematic viscosity range of 2 to 5 cSt. The flow rate is 150 to 200 gal/min. We are ordering the pump. What values should we specify of (*a*) the flow rate and (*b*) the pump head $\Delta P/(\rho g)$, in feet? *Note:* For this problem the head form of Bernoulli's equation is more convenient.

6.57. If we shut off the pump in the system in Prob. 6.56 and open the bypass around it, what are the maximum and minimum values of the flow? Which way does it go? Neglect the friction losses in the pump bypass line. Assume that the globe valve is always wide open.

6.58. Figure 6.35 shows a siphon which will be used to empty water out of a tank. The siphon is made of 10-in pipe 60 ft long. When the water is at its minimum level, as shown, what are the flow rate and the pressure at the top (point *A*)? The bend at the top of the siphon is equivalent to two 90° long-radius elbows.

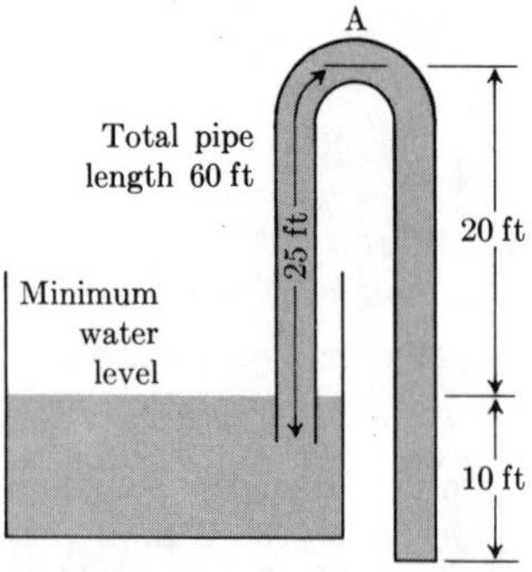

FIGURE 6.35

6.59. The National Park Service has recently decided to construct a pipeline to carry water across the Grand Canyon from the relatively water-rich north rim to the arid south rim. A cross section of the system is shown in Fig. 6.36. The length of pipeline between the springs and the river crossing is 10 mi, and between the river crossing and the pumping station it is 4 mi. The desired flow rate is 1000 gal/min. The pressure at the springs and at the pumping station may be assumed atmospheric. Because all the material must be brought into place on muleback, which is costly, there is a considerable incentive to make the pipe as lightweight as possible. Recommend what material the pipe should be made of, what its inside diameter would be, and what its thickness should be.

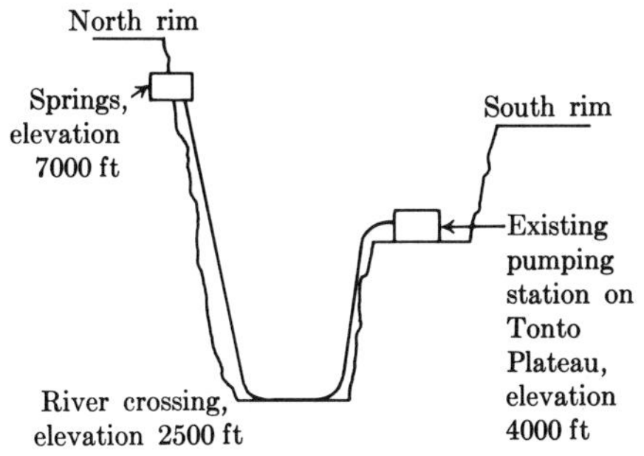

FIGURE 6.36
Grand Canyon.

6.60. Solve for the flows in Fig. 6.19.

6.61. Repeat Prob. 6.60 for the case where $z_3 - z_1 = -5$ ft and $z_4 - z_1 = -40$ ft. *Hint:* In this case the flows will not all be in the same direction as in Prob. 6.60.

6.62. Check the assumed friction factor in Example 6.17. For the value of relative roughness shown, the range of possible friction factors in turbulent flow is 0.004 to 0.01. How much would the economic diameter differ at $f = 0.01$?

6.63. The type of calculation of economic diameter of pipes shown in Sec. 6.13 was apparently first performed by Lord Kelvin [23] in connection with the problem of selecting the economic diameter for long-distance electric wires. To see how he got his result, derive the formula for the economic diameter of an electric conductor (analogous to Eq. 6.49), using the following information: The purchased cost of the whole transmission line (including poles, insulators, land, construction labor, etc.) is A times the mass of metal in the wire, where A has dimensions of dollars per pound-mass. The annual cost of owning the transmission line (which includes interest on the capital investment in the line, taxes, and maintenance) is B times the purchased cost of the whole transmission line, where B has dimensions of 1/year. The electric energy which is lost due to resistive heating in the wire costs C, where C has dimensions of dollars per kilowatthour. The resistive heating is given by $Q = I^2R$, where R is the resistance of the wire and I is the current. The resistance of the wire is given by $R = r\,\Delta x/[(\pi/4)D^2]$, where r is the resistivity (in ohm-feet), Δx is the length of the wire, and D is the wire diameter.

Your formula for the economic diameter should be written in terms of the current to be carried (*not* in terms of the voltage) and in terms of the other variables listed above, plus any others you consider necessary.

6.64. It has been proposed to solve Los Angeles' air pollution problem by pumping out the contaminated air mass every day. The area of the Los Angeles basin is 4083 mi^2. The contaminated air layer is roughly 2000 ft thick. Suppose we plan to pump it out every day a distance of 50 mi to Palm Springs. (It is assumed that the residents of Palm Springs will not object, which is not a very good assumption.)

(*a*) Estimate the economic velocity in the pipe.
(*b*) Estimate the required pipe diameter.
(*c*) Estimate the pressure drop.
(*d*) Estimate the pumping power requirement.
(*e*) Comment on the feasibility of this proposal.

6.65. It has been proposed to solve the water problem in Los Angeles by importing water from the mouth of the Columbia River, where vast amounts flow into the sea. One way to do this would be with a pipeline and pumping station. Both ends of the pipe would be at sea level, so the only pumping cost would be the cost of overcoming the friction loss. The pipe length would be about 1500 mi. Assuming that we wish to move 10^7 acre · ft/year (1 acre · ft = 43,560 ft^3), estimate the horsepower of pumps required. State your assumptions clearly.

6.66. If in Example 6.17 the fluid were water contaminated with hydrofluoric acid, we would have to use a special corrosion-resistant pipe. Suppose that this pipe had a purchased cost PP exactly 10 times that of carbon-steel pipe. What would be the economic pipe diameter, pressure drop, and pump horsepower?

6.67. If Table 6.4 were based on Eq. 6.51 and the friction factor held constant, then the product of the economic velocity and the cube root of the density would be a constant. How much does it vary from being a constant? What is the cause of this variation?

6.68. Does the result in Example 6.17 agree exactly with the data in Table 6.4 and Fig. 6.21? If not, how much does it disagree, and what is the most probable cause of the discrepancy?

6.69. You are to design the fuel line for a lunar-landing rocket. Money is unimportant; low mass is the main goal. Decide what the significant "economic" factors in this problem are, and write in general form the equivalent of Eq. 6.49 for this problem.

6.70. How much does inflation affect the economic pipe diameter? For Example 6.17, if we double PP and PC, how much does that change the economic diameter? Suppose we double PC but do not change PP; what does that do to the economic diameter?

6.71. In Examples 6.18 and 6.19 it was assumed that the particles had reached terminal velocity. We can make an order-of-magnitude estimate of the time required to reach terminal velocity, in cases in which the terminal velocity is in the range of Stokes law, by assuming that the instantaneous value of C_d during the acceleration period is given by Stokes law. For this assumption derive the formula for V/V_{term} as a function of time, and calculate the time required to reach 99 percent of terminal velocity. The actual time required before this percentage of terminal velocity is reached is greater than will be given, because the drag force for accelerated particles at low $\mathcal{R}_p$ is more than the steady-state value [24].

6.72. Rework Example 6.18 for the particle settling in water at 68°F instead of in air.

6.73. Rework Example 6.19 for the ball falling in glycerin instead of in water, where $\mu_{\text{glyc}} = 800$ cP and $\rho_{\text{glyc}} = 78.5\ \text{lbm/ft}^3$.

6.74. A spherical balloon is 10 ft in diameter and has a buoyant force 0.1 lbf greater than its weight. What is its terminal velocity when rising through air?

6.75. A standard baseball has a diameter of 2.9 in and a mass of 0.32 lbm. Good fast-ball pitchers can throw one at about 100 mi/h.

(*a*) Neglecting the effect of the stitching on the ball and the spin of the ball, calculate the drag force of the air on the ball.

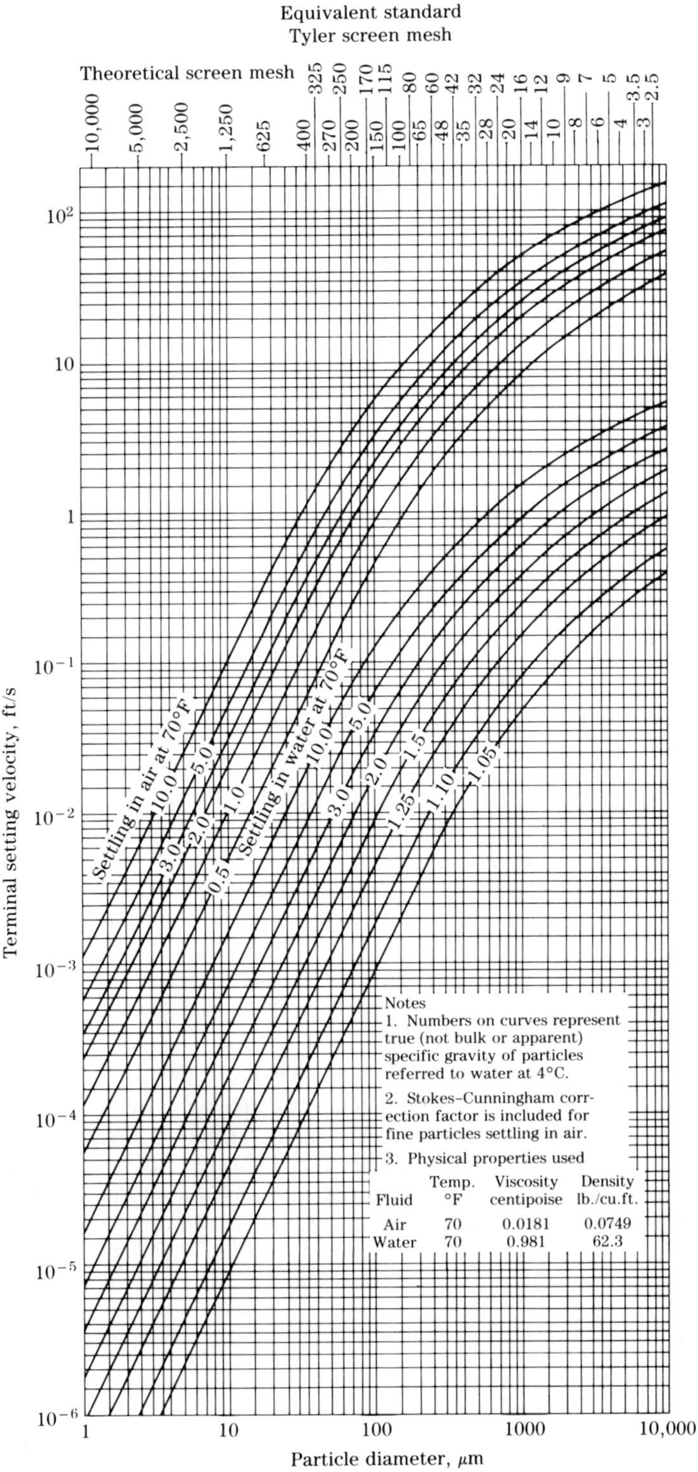

FIGURE 6.37
Terminal velocities of spherical particles of different densities settling in air and water at 70°F under the influence of gravity. (From C. E. Lapple et al., *Fluid and Particle Mechanics*, University of Delaware, Newark, 1951, p. 292.)

(*b*) The distance from the pitcher's mound to home plate is 60 ft. If the ball left the pitcher's hand at 100 mi/h, how fast will it be going when it reaches home plate?

6.76. A parachutist jumps from a plane and falls in free fall for a while before opening her chute. If her projected area perpendicular to the direction of fall is 1 ft^2, she weighs 120 lbf, and C_d is 0.7, what is her terminal velocity? How many seconds must she fall to reach 99 percent of her terminal velocity, assuming that the drag coefficient is independent of velocity? How far does she fall in this time?

6.77. We wish to design a parachute. The requirement is that at terminal velocity the rider must have a velocity equal to the maximum velocity the rider would reach by jumping to the ground from a 10-ft-high roof. The rider weighs 150 lb. The parachute will be circular, and its drag coefficient $C_d = 1.5$. What diameter must the parachute have?

6.78. Occasionally sporty automobiles advertise that they have very low drag coefficients, typically about 0.3 for teardrop-shaped cars. That drag coefficient is based on the frontal area. Suppose that a car has that drag coefficient and a width of 6 ft and a height of 5 ft and is going 70 mi/h.

(*a*) What is the air resistance of the car?

(*b*) How much power must be expended to overcome this air resistance?

6.79. In Examples 6.18 and 6.19, we assumed Stokes law, calculated the corresponding velocity, and then checked the Reynolds number to see whether the assumption of Stokes law was a good one. If our assumption was not a good one, then the velocity calculated in the first step was a wrong velocity, and the calculated Reynolds number was wrong, too. Is there any chance that this procedure can lead to a combination of velocity and Reynolds number which indicated that Stokes law should be obeyed when actually the Reynolds number based on the correct solution is outside the range of Stokes law?

6.80. The problem of the settling of small particles through air and water is so common that for convenience certain charts summarize the results of calculations such as those in Example 6.18. One of the best is Lapple's, shown as Fig. 6.37. Check the results of Example 6.18 and Prob. 6.72 on this chart.

6.81. A spherical raindrop with a diameter of 0.001 in is falling in still air. It has been falling long enough to be at its terminal velocity. How fast is it falling?

6.82. In James Bond movies, the hero is often swimming and has to dive deep into the water to escape the bullets from the enemy helicopter flying above him. How deep should he dive? Assume the bullet is a sphere of diameter 0.5 in and mass 0.27 lbm. It hits the surface of the water vertically at a velocity of 1000 ft/s or less and will not inflict serious injury if it is slowed down to a velocity of 100 ft/s or less. For the purposes of this problem, assume that the drag coefficient is constant, independent of velocity, and equal to 0.1.

CHAPTER 7

THE MOMENTUM BALANCE

Newton's second law of motion, often called Newton's *equation of motion*, is commonly written

$$F = ma \tag{7.1}$$

It is easily applied in this form to the motion of rigid bodies like the falling bodies of elementary physics. It can also be easily applied in this form to the motion of fluids which are moving in rigid-body motion, as discussed in Chap. 2. However, for fluids which are moving in more complicated motions, e.g., in pipes or around airplanes, it is difficult to use Eq. 7.1 in the form shown. Therefore, in this chapter we shall rewrite the equation in the form of a momentum balance. The momentum balances given in this chapter are rearrangements of Eq. 7.1, and they, too, are often referred to in the engineering literature as *equations of motion*.

The momentum balance form will prove very convenient for solving fluid-flow problems. In particular, it will allow us to find out something about complicated flows through a system without having to know in detail what goes on inside the system. In this way the momentum balance is similar to the mass and energy balances. For example, by using the mass and energy balances we can find out some things about a turbine or compressor from the inlet and outlet streams only, without knowing in detail what goes on inside. We shall frequently apply also the momentum balance "from the outside."

Remember that all of this chapter is simply the manipulation and application of Eq. 7.1.

7.1 MOMENTUM

Momentum, like energy, is an abstract quantity. Unlike energy, it is defined in terms of simpler quantities, mass and velocity. The definition of momentum is given in terms of the momentum of a body:

$$\text{Momentum of body} = \text{mass of body} \times \text{velocity of body}$$

$$\text{Momentum} = mV \tag{7.2}$$

It makes no sense to speak of momentum as separate from bodies because, as we see from this equation, if there is no mass, there is no momentum. Only bodies—of solid, liquid, or gas—have mass, so momentum can exist only in connection with some body.†

Furthermore, momentum is a vector. We have applied the balance equations to mass and energy, which are scalar. Here we apply it to a vector and get similar results. Most often in dealing algebraically with vectors, one uses the scalar components of the vector rather than the vector itself. For example, Eq. 7.1 may be written in vector form:

$$\mathbf{F} = m\mathbf{a} \tag{7.3}$$

However, any vector can be resolved into the vector sum of three scalar components multiplied by unit vectors, both in three mutually perpendicular directions.‡ For example,

$$\mathbf{F} = F_x\mathbf{i} + F_y\mathbf{j} + F_z\mathbf{k} \tag{7.4}$$

where F_x, F_y, and F_z are the scalar components of vector $\mathbf{F}$ in the x, y, and z directions and $\mathbf{i}$, $\mathbf{j}$, and $\mathbf{k}$ are unit vectors in the x, y, and z directions, respectively. Similarly, we can resolve the acceleration vector $\mathbf{a}$ and rewrite Eq. 7.3 in the following forms:

$$F_x\mathbf{i} + F_y\mathbf{j} + F_z\mathbf{k} = m(a_x\mathbf{i} + a_y\mathbf{j} + a_z\mathbf{k})$$

$$(F_x - ma_x)\mathbf{i} + (F_y - ma_y)\mathbf{j} + (F_z - ma_z)\mathbf{k} = 0 \tag{7.5}$$

But this equation is the equation of a new vector, the $\mathbf{F} - m\mathbf{a}$ vector, which is seen to be zero. For a vector to be zero, each of its scalar components must be zero, so this equation is exactly equivalent to

$$F_x - ma_x = 0 \qquad F_y - ma_y = 0 \qquad F_z - ma_z = 0 \tag{7.6}$$

---

† Neutrinos and light quanta (photons) apparently possess momentum but not rest mass, and thus they might be considered exceptions to this statement. However, they are observable only when moving at high velocities, at which time they have considerable energy and hence relativistic mass, so that this statement is correct even for them.

‡ As far as we know, we live in a three-dimensional universe, so we speak of three mutually perpendicular directions. If we lived in an n-dimensional universe, there would be n mutually perpendicular directions. In some problems it is convenient to consider n-dimensional "spaces," in which a vector is resolved into n "perpendicular" components.

(*Note:* Boldface type indicates a vector quantity, e.g., **F**)

This shows us that we may consider any vector equation as a shorthand form of writing three scalar equations. Vector calculus is a powerful tool for deriving equations describing multidimensional problems. In electromagnetic problems and problems involving moving coordinate axes (e.g., gyroscopes), it is easiest to work directly with the vector quantities. However, for solving practical fluid mechanics problems, it is almost always more convenient to use the three scalar (component) equations, which are the exact equivalent of the vector equation. In this chapter we show the momentum balance both as a vector equation and as its more useful scalar equivalents.

One distinct complication with the momentum balance as compared with the mass and energy balances concerns the algebraic signs of the momentum terms. In the energy and mass balances, we have little trouble with signs, because we seldom consider negative energies and never consider negative masses.[†] But if we wish to represent a velocity in the $-x$ direction, we write it as

$$\mathbf{V} = V_x\mathbf{i} + V_y\mathbf{j} + V_z\mathbf{k}$$

where V_z and V_y are zero, and V_x is a negative number. Therefore, in scalar momentum equations we must be more careful of algebraic signs than we were with the mass and energy balances.

7.2 THE MOMENTUM BALANCE

In Chap. 3 we saw that the general balance equation (Eq. 3.2) can be applied to any extensive property—any property which is proportional to the amount of matter present. Since momentum is proportional to the amount of matter present, it is an extensive property and must obey a balance equation. Here, as in all other balance equations, we must be careful to choose and define our system.

Figure 7.1 shows the system used to state the momentum balance; it consists of some tank or vessel with flow of matter in or out and system boundaries as shown. The momentum contained in the system boundaries is

$$\text{Momentum inside system boundaries} = \int_{\substack{\text{all mass}\\ \text{in system}}} \mathbf{V}\, dm \tag{7.7}$$

We simplify this by assuming that all the mass inside the system has the same velocity, so that this integral simplifies to $(m\mathbf{V})_{\text{sys}}$. The momentum-accumulation term becomes

$$\text{Momentum accumulation} = d(m\mathbf{V})_{\text{sys}} \tag{7.8}$$

[†] Although in an absolute sense energies can never be negative, energies relative to an arbitrary datum can be negative. In one-component systems, such as steam power plants or refrigeration systems, the datum usually is chosen so that none of the energy terms is negative. However, the common datum for combustion work and chemical reaction problems results in negative energy terms.

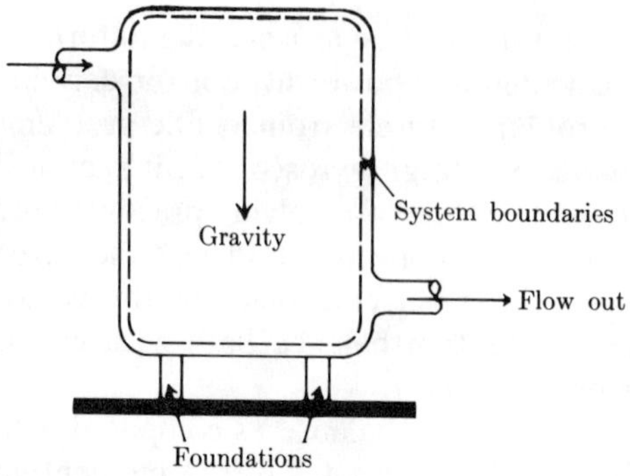

FIGURE 7.1
System used for stating the momentum balance.

For one flow in and one flow out, as in the figure, the momentum flow in minus momentum flow out is

$$\text{Momentum flow in} - \text{momentum flow out} = \mathbf{V}_{\text{in}}\, dm_{\text{in}} - \mathbf{V}_{\text{out}}\, dm_{\text{out}} \tag{7.9}$$

If there is more than one flow in or out, there will be summation terms for momentum flows in and out, just as there are summation terms for mass and energy flows into and out of a system in the mass and energy balances.

Now to account for the creation or destruction of momentum, we invoke Eq. 7.1, which can be rewritten as

$$\mathbf{F} = m\mathbf{a} = m\,\frac{d\mathbf{V}}{dt} \tag{7.10}$$

The possible changes of mass of the system are accounted for by the flow-in or flow-out terms, and the creation or destruction terms must apply equally well to constant-mass and variable-mass systems. For a constant-mass system, we can take the m inside the differential sign in the last equation and rearrange, to show that

$$d(m\mathbf{V})_{\text{sys}} = \mathbf{F}\, dt \tag{7.11}$$

so that the momentum creation or destruction term is $\mathbf{F}\,dt$.[†]

[†] Equation 7.11 implies that momentum is creatable, which can be misleading. If a person standing on the earth throws a ball, the momentum of the ball is increased in one direction and the momentum of the earth is increased in the opposite direction by an equal amount. Thus, the momentum of the earth-ball system is unchanged. Because the mass of the earth is so much larger than that of the ball, we do not perceive this change in the earth's velocity (Prob. 7.1). However, Eq. 7.11 is correct, because creating momentum in one system results in the creation of equal and opposite momentum in some other system (usually the earth); so the net change of momentum of the universe is zero.

Modern physicists prefer Eq. 7.11 to Eq. 7.1 as the basic statement of Newton's law of motion. The reason is that as the velocity of a body approaches the speed of light, the force exerted on it results mostly in an increase in mass rather than an increase in velocity. Thus, Eq. 7.1 is limited to constant-mass systems and excludes any system that is being accelerated to a speed near that of light; and Eq. 7.11 applies not only to constant-mass systems but also to bodies being accelerated to speeds near that of light, such as the particles in linear accelerators and cyclotrons.

If more than one force is acting, the **F** in Eq. 7.11 must be replaced with a sum of forces. Usually several forces act in fluid flow problems, so we write the momentum balance with a $\Sigma\,\mathbf{F}$. The forces acting on the system shown in Fig. 7.1 are the external pressure on all parts of its exterior and the force of gravity. Other forces which we might consider are electrostatic or magnetic forces. If we had chosen our system such that the boundary passed through the foundations in Fig. 7.1, then there would be a compressive force in the structural members of the foundation, which would have to be taken into account.

Writing all the terms together, we find the vector form of the momentum balance:

$$d(m\mathbf{V})_{\text{sys}} = \mathbf{V}_{\text{in}}\,dm_{\text{in}} - \mathbf{V}_{\text{out}}\,dm_{\text{out}} + \sum \mathbf{F}\,dt \tag{7.12}$$

Here we have not included a destruction term, because the $\Sigma\,\mathbf{F}$ in the equation is the vector sum of all the forces acting on the system. If this sum is in the opposite direction of the velocities, then the $\Sigma\,\mathbf{F}\,dt$ term is a momentum destruction term; it will enter with a minus sign. Most often we divide Eq. 7.12 by dt to find the *rate form* of the momentum balance:

$$\frac{d(m\mathbf{V})_{\text{sys}}}{dt} = \mathbf{V}_{\text{in}}\dot{m}_{\text{in}} - \mathbf{V}_{\text{out}}\dot{m}_{\text{out}} + \sum \mathbf{F} \tag{7.13}$$

This is not a *derivation* of the momentum balance, but simply a restatement of Newton's second law in a convenient form. Furthermore, Newton's laws, like the laws of thermodynamics and the law of conservation of mass, are underivable; they cannot be demonstrated from any prior principle but rest solely on their ability to predict correctly the outcome of any experiment ever run to test them.

Equations 7.12 and 7.13 are balance equations entirely analogous to the mass and energy balances discussed in Chaps. 3 and 4. They have the same basic restriction of those balances, namely, that they may be applied only to a carefully defined system. With the mass balance, we can choose any system in which we can account for all flows of matter across the boundaries. With the energy balance, we must choose a system in which we can account for both flows of matter across the boundaries and heat flows across the boundaries and for work due to electric current, changing magnetic fields, moving boundaries, and rotating and reciprocating shafts. In applying the momentum balance, it is necessary to choose a system in which it is possible to account for all flows of matter across the boundaries and all external forces acting on the system. In most fluid flow problems, this means that it must be possible to calculate the pressure on every part of the outside of the system. Notice, however, that there is no term in Eq. 7.12 or 7.13 for heat flows, rotating shafts, or electric current flows, so we may choose systems for the momentum balance without necessarily being able to calculate those quantities over the boundaries of the system (we must account for electrostatic or magnetic fields, if they are significant). As we do with mass and energy balances, we may consider closed systems, in which the $\dot{m}$ terms are zero, or steady-flow systems, in which the accumulation is zero. Skill in applying the momentum balance is largely a matter of choosing a system in which one can conveniently calculate all the terms in the balance.

Equations 7.12 and 7.13 are vector equations; each can be represented by three scalar equations, showing the components of the vectors in the x, y, and z directions or the r, θ, and z directions or in spherical coordinates. The x-component scalar equations equivalent to them are

$$d(mV_x)_{\text{sys}} = V_{x_{\text{in}}}\, dm_{\text{in}} - V_{x_{\text{out}}}\, dm_{\text{out}} + \sum F_x\, dt \tag{7.14}$$

$$\frac{d(mV_x)_{\text{sys}}}{dt} = V_{x_{\text{in}}}\, \dot{m}_{\text{in}} - V_{x_{\text{out}}}\, \dot{m}_{\text{out}} + \sum F_x \tag{7.15}$$

The corresponding y and z equations can be found from these simply by replacing all the x subscripts with y or z subscripts.

To illustrate the application of the momentum balance, we consider first two very simple examples not involving fluids.

Example 7.1. A baseball is thrown in a horizontal direction. What terms of the momentum balance apply?

Taking the ball as our system and using the x component of the momentum balance, we see that there is no flow of matter in or out; therefore,

$$d(mV_x)_{\text{sys}} = (m\, dV_x)_{\text{sys}} = F_x\, dt$$

This is clearly a simply restatement of $F = ma$ for a constant-mass system.

■

Example 7.2. A duck has a mass of 3 lbm and is flying due west at 15 ft/s. The duck is struck by a bullet with a mass of 0.05 lbm, which is moving due east with a velocity of 1000 ft/s. The bullet comes to rest in the duck's gizzard. What is the final velocity of the duck-bullet system?

Here the problem is one-dimensional, so we work with the x-directed scalar equation, Eq. 7.14, and choose east as the $+x$ direction. First we work the problem by taking as our system the combined bullet and duck. No matter is flowing in or out of this system, nor does any external force act on it (we ignore the wind force, if a wind is blowing). So Eq. 7.14 becomes

$$d(mV_x)_{\text{sys}} = 0$$

$$(mV_x)_{\text{sys, fin}} = (mV_x)_{\text{duck, init}} + (mV_x)_{\text{bullet, init}}$$

When we solve for $V_{x_{\text{sys, fin}}}$, we find

$$V_{x_{\text{sys, fin}}} = \frac{(mV_x)_{\text{duck, init}} + (mV_x)_{\text{bullet, init}}}{m_{\text{duck-bullet, fin}}}$$

$$= \frac{3\ \text{lbm}(-15\ \text{ft/s}) + 0.05\ \text{lbm} \cdot 1000\ \text{ft/s}}{3.05\ \text{lbm}}$$

$$= +1.6\ \text{ft/s} = +0.73\ \text{m/s}$$

Now we do the problem over, taking the duck as our system. In this case there is mass flow into the system, so we have

$$d(mV_x)_{\text{sys}} = V_{x_{\text{in}}}\, dm_{\text{in}}$$

Integrating, we find

$$(mV_x)_{\text{sys, fin}} - (mV_x)_{\text{sys, init}} = V_{x_{\text{in}}} m_{\text{in}}$$

When we solve for $V_{x_{\text{sys, fin}}}$, we find

$$V_{x_{\text{sys, fin}}} = \frac{(V_{x_{\text{in}}} m_{\text{in}}) + (mV_x)_{\text{sys, init}}}{m_{\text{sys, fin}}}$$

This is exactly the same as the result we got by taking the combined system. ■

This example shows the great advantage of the momentum balance. The details of the collision are very complicated when we wish to know the exact distance-time-shape history of the bullet in traversing the various feathers, bones, muscles, and internal organs of the duck. But from the momentum balance alone we can write the final velocity of the bullet-duck system without knowing those details.

Example 7.2 also illustrates the fact that some problems can be solved by the momentum balance but cannot be solved by the energy balance. If we write the energy balance for Example 7.2, taking the combined duck-bullet as our system and neglecting the small change in volume of the system as the bullet enters the duck, we find it reduces to

$$\left[m\left(u + \frac{V^2}{2}\right)\right]_{\text{sys, fin}} = \left[m\left(u + \frac{V^2}{2}\right)\right]_{\text{duck, init}} + \left[m\left(u + \frac{V^2}{2}\right)\right]_{\text{bullet, init}}$$

Here we know the mass of the system and the initial kinetic and internal energies of its two parts, but these alone do not allow us to solve this equation for either the final internal energy or the final kinetic energy. However, from the momentum balance we were able to find the final velocity, and we can then use it in this equation to find the final internal energy. In this chapter in several other examples the momentum balance must be applied before the energy balance can be used (see Prob. 7.3).

7.3 SOME STEADY-FLOW APPLICATIONS OF THE MOMENTUM BALANCE

If we choose as our system some pipe, duct, or channel with steady flow through it in one direction, e.g., the x direction, then Eq. 7.15 becomes

$$0 = \dot{m}(V_{x_{\text{in}}} - V_{x_{\text{out}}}) + \sum F_x \qquad \text{[steady flow]} \tag{7.16}$$

The application of this steady-flow one-dimensional momentum balance will be illustrated by several examples.

Many applications of Eq. 7.16 involve jets. A jet is a stream of fluid which is not confined within a pipe, duct, or channel; examples are the stream of water issuing from a garden hose and the exhaust gas stream from a jet engine. If a jet is flowing at a subsonic velocity, its pressure will be the same as the pressure of the surrounding fluid. If a jet enters or leaves a system or device at subsonic speed, it will enter and leave at the pressure of the surrounding fluid, although its pressure may be different inside the device.

Example 7.3. The police are using fire hoses to disperse an unruly crowd. The fire hoses deliver 0.01 m^3/s of water at a velocity of 30 m/s. One member of the crowd has picked up a garbage can lid and is using it as a shield to deflect the flow. She is holding it vertically, so the jet splits into a series of jets going off in the y and z directions, with no x component of the velocity; see Fig. 7.2. What force must she exert to hold the garbage can lid?

By applying Eq. 7.16 in the x direction and taking the lid and the adjacent fluid as our system, we find

$$F_x = -0.01\,\frac{m^3}{s} \cdot 998.2\,\frac{kg}{m^3} \cdot \left(30\,\frac{m}{s} - 0\right) \cdot \frac{N \cdot s^2}{kg \cdot m} = -299.5\text{ N} = -67.3\text{ lbf} \quad \blacksquare$$

Here the pressure around the external boundary of the system is all atmospheric, so this force is simply the force exerted by the arms of the woman holding the lid. It is negative, because she is exerting this force in a direction opposite to the x axis. Here we could also have chosen as our system the fluid alone. Then to solve for the force, we would have had to calculate the pressure exerted on it by the garbage can lid at every point of the system boundary. To do this, we would need a detailed description of the flow. From such a detailed description, if it were available, we could calculate

$$F_x = \int_{\text{all area}} P\,dA$$

for the lid, finding the same answer. Thus, by a proper choice of system we can

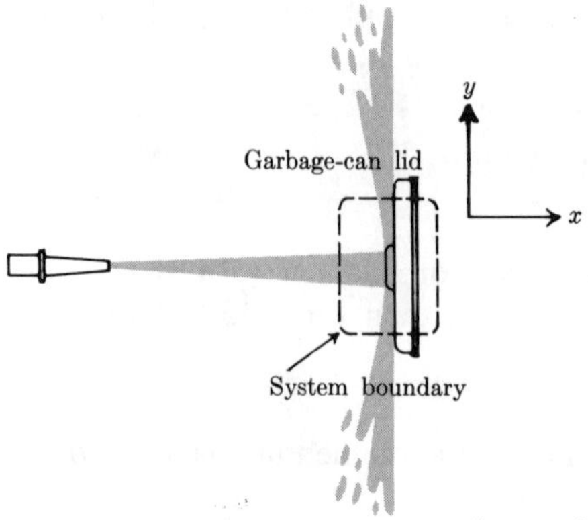

FIGURE 7.2

find the desired force without a detailed description of the flow; we can apply the momentum balance "from the outside."

Example 7.4. The member of the crowd in Example 7.3 now turns the lid around so that she can hold it by the handle. However, because of the shape of the lid, the flow goes off as shown in Fig. 7.3, with an average x component of the velocity of -15 m/s. Now what force must she exert?

By applying Eq. 7.16 exactly as before we find

$$F_x = -0.01\ \frac{\text{m}^3}{\text{s}} \cdot 998.2\ \frac{\text{kg}}{\text{m}^3} \cdot \left[30\ \frac{\text{m}}{\text{s}} - \left(-15\ \frac{\text{m}}{\text{s}}\right)\right] \cdot \frac{\text{N} \cdot \text{s}^2}{\text{kg} \cdot \text{m}}$$
$$= -449.3\ \text{N} = -101\ \text{lbf} \qquad \blacksquare$$

Example 7.5. Now the woman in Examples 7.3 and 7.4 turns the lid so that it deflects the stream off at a 45° angle to the vertical without changing its velocity, as shown in Fig. 7.4. Now what force must she exert?

In this case the exit stream has a velocity of 30 m/s. Its velocity components are

$$V_x = V \cos 45° = 30\ \frac{\text{m}}{\text{s}} \cdot 0.707 = 21.2\ \frac{\text{m}}{\text{s}}$$
$$V_y = V \sin 45° = 30\ \frac{\text{m}}{\text{s}} \cdot 0.707 = 21.2\ \frac{\text{m}}{\text{s}}$$

So, applying Eq. 7.16 twice, we find

$$F_x = -0.01\ \frac{\text{m}^3}{\text{s}} \cdot 998.2\ \frac{\text{kg}}{\text{m}^3} \cdot \left(30\ \frac{\text{m}}{\text{s}} - 21.2\ \frac{\text{m}}{\text{s}}\right) \cdot \frac{\text{N} \cdot \text{s}^2}{\text{kg} \cdot \text{m}} = -87.7\ \text{N} = -19.7\ \text{lbf}$$
$$F_y = -0.01\ \frac{\text{m}^3}{\text{s}} \cdot 998.2\ \frac{\text{kg}}{\text{m}^3} \cdot \left(0 - 21.2\ \frac{\text{m}}{\text{s}}\right) \cdot \frac{\text{N} \cdot \text{s}^2}{\text{kg} \cdot \text{m}} = 211.6\ \text{N} = 47.6\ \text{lbf}$$

By the pythagorean theorem we have

$$F = \sqrt{F_x^2 + F_y^2} = \sqrt{(-87.7\ \text{N})^2 + (211.6\ \text{N})^2} = 229.1\ \text{N} = 51.5\ \text{lbf} \qquad \blacksquare$$

FIGURE 7.3

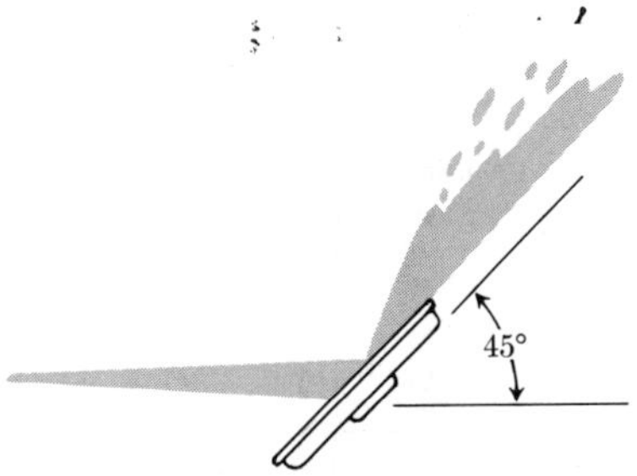

FIGURE 7.4

In the previous examples, the jets were open to the atmosphere, so their gauge pressure was always zero. Thus we had no difficulty deciding on the sign of pressure forces. The next two examples involve pressure forces inside pipes; they require us to consider the sign of the pressure forces. The easiest way to decide on the proper sign of a pressure force is to take the system boundary perpendicular to the axis in which we are applying the momentum balance, e.g., perpendicular to the x direction for an x-directed momentum balance. If we do that, we see that the pressure force acts *inward* on our system and simultaneously *outward* on the surroundings. If you think of a spherical balloon and imagine it divided into two systems with the system boundary passing vertically through its middle, then the internal pressure, which stretches the skin of the balloon, acts equally on each half of the balloon. It acts to the left on the leftmost system and to the right on the rightmost system. If we had chosen the x axis to point to the right, then the pressure force would be negative (act in the negative x direction) for the system consisting of the left half of the balloon and the pressure force would be positive (act in the positive x direction) for the system consisting of the right half of the balloon.

Example 7.6. A nozzle is attached to a fire hose by a bolted flange; see Fig. 7.5. What is the force tending to tear apart that flange when the valve in it is closed?

We take as our system the mass of fluid which is enclosed in the nozzle from the plane of the flange to the valve. Applying Eq. 7.16, we see that $\dot{m}$ is zero, so the sum of the x components of the forces on this body of fluid must be zero. In this case the sum of the forces is the sum of the pressure forces in the x direction. In the plane of the flange, the fluid outside the system exerts a pressure force on the system equal to PA. This is all in the x direction, because this surface is normal to the x axis. The x component of the pressure force exerted by the nozzle must be equal and opposite to this pressure force. The magnitude of these forces is

$$F_x = PA = (P_g + P_{\text{atm}})A = 100\ \text{lbf} \cdot 10\ \text{in}^2 + P_{\text{atm}}A = 1000\ \text{lbf} + P_{\text{atm}}A$$

where P_g is the gauge pressure and P_{atm} is the atmospheric pressure.

Now we choose as a second system the nozzle itself. From Eq. 7.16 we see that for it, too, the sum of the x components of the forces must be zero.

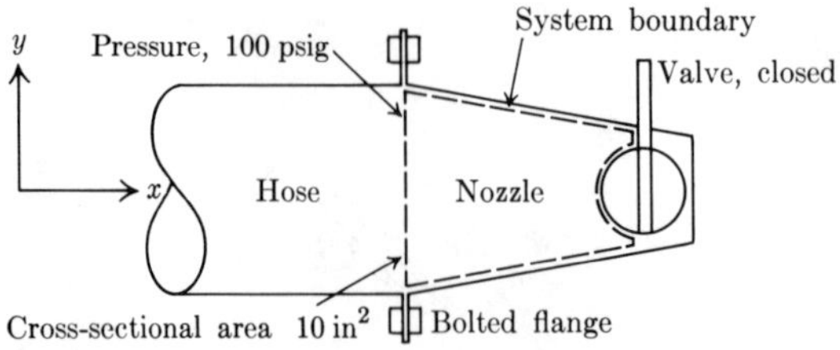

FIGURE 7.5
Closed nozzle.

The forces acting on it are diagramed in Fig. 7.6. We have previously calculated the force which it exerts on the fluid; by Newton's third law, we know that the fluid exerts an equal and opposite force on it, which is given as F_{fluid} in the figure. The bolts also exert a force, as shown, and the atmosphere exerts a pressure on all those parts not exposed to the fluid. The atmospheric pressure force is not all in the x direction, but as we showed in Chap. 2, we could compute the x component of the atmospheric pressure force, which would be $-PA_x$, where A_x is the x projection of the net area exposed to the atmosphere. (The force is negative because it acts opposite to the x direction.) Therefore, summing the forces shown in Fig. 7.6, setting the sum equal to zero, and solving for F_{bolts}, we find

$$\begin{aligned} F_{\text{bolts}} &= -F_{\text{liq}} - F_{\text{atm}} \\ &= -(1000\ \text{lbf} + P_{\text{atm}}A) - (-P_{\text{atm}}A) \\ &= -1000\ \text{lbf} = -4.448\ \text{kN} \end{aligned}$$

The bolt force shown is negative because it acts in the negative x direction. ■

In this example the atmospheric pressure terms canceled. Because this is a common occurrence, engineers ordinarily work such problems in gauge pressures and so need to show pressure forces on only those parts of the boundary of the system where the pressure is different from atmospheric. If we had done that here, we would have found exactly the same answer.

We could have solved this problem more easily by not using the momentum balance, but it illustrates the method, which will be useful now.

Example 7.7. The valve on the end of the fire hose in Example 7.6 is opened. The fluid flows steadily out the end of the nozzle at a velocity of 100 ft/s. The area of the outlet nozzle is 1 in^2. The pressure at the flanged joint is still 100 lbf/in^2 gauge. Now what is the force tending to tear apart the flange?

Again, we choose as our system the fluid enclosed in the nozzle. From Eq. 7.16 we have

$$0 = \dot{m}(V_{x_{\text{in}}} - V_{x_{\text{out}}}) + \sum F_x$$

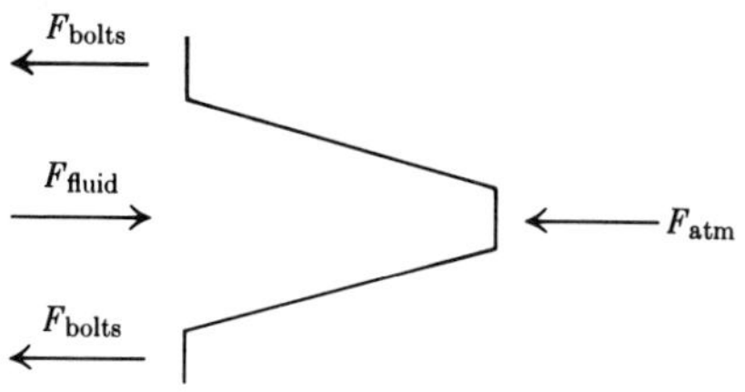

FIGURE 7.6
Forces acting on nozzle.

From the mass balance for steady flow we have

$$\dot{m} = \rho A_{\text{out}} V_{\text{out}}$$

$$= 62.3 \, \frac{\text{lbm}}{\text{ft}^3} \cdot 1 \, \text{in}^2 \cdot 100 \, \frac{\text{ft}}{\text{s}} \cdot \frac{\text{ft}^2}{144 \, \text{in}^2} = 43.3 \, \frac{\text{lbm}}{\text{s}} = 19.7 \, \frac{\text{kg}}{\text{s}}$$

We find the inlet velocity from the mass balance for a constant-density fluid:

$$V_{x_{\text{in}}} = V_{x_{\text{out}}} \frac{A_{\text{out}}}{A_{\text{in}}} = 100 \, \frac{\text{ft}}{\text{s}} \cdot \frac{1 \, \text{in}^2}{10 \, \text{in}^2} = 10 \, \frac{\text{ft}}{\text{s}} = 3.05 \, \frac{\text{m}}{\text{s}}$$

So the net force on the fluid in the x direction is

$$\sum F_x = -\dot{m}(V_{x_{\text{in}}} - V_{x_{\text{out}}})$$

$$= -43.3 \, \frac{\text{lbm}}{\text{s}} \cdot \left(\frac{10 \, \text{ft}}{\text{s}} - \frac{100 \, \text{ft}}{\text{s}} \right) \cdot \frac{\text{lbf} \cdot \text{s}^2}{32.2 \, \text{lbm} \cdot \text{ft}} = 121 \, \text{lbf} = 538 \, \text{N}$$

This is the sum of the forces on the fluid, as sketched in Fig. 7.7.

As discussed in Chap. 5, the fluid leaving such a nozzle will be at the same pressure as the surrounding atmosphere if the flow is subsonic, which it is here. Therefore, if we use gauge pressure, then the pressure force on the system, as the stream leaves the nozzle, is zero, and the 121 lbf given is the algebraic sum of the pressure force exerted on the system at its left boundary and the x component of the force exerted by the nozzle. Thus,

$$121 \, \text{lbf} = PA - F_{x_{\text{noz}}}$$

and the force exerted by the nozzle on the fluid is -879 lbf. By comparing this result with the one in Example 7.6, we can readily deduce that the force on the nozzle bolts in this case is -879 lbf. ■

Why is the force acting on the bolts less in this case? The pressure force acting on the system at its left boundary is the same as in the no-flow case. However, when there is this flow, some of that force is being used to accelerate the fluid and so is not being resisted by the nozzle bolts.

This example also illustrates how the momentum balance helps us solve problems from the outside without looking inside. We also could have found the force on the nozzle by determining the pressure and shear stress at every point on the internal surface of the nozzle; see Fig. 7.8. The x components of these pressure forces and shear forces are equal to the x component of the force on the nozzle:

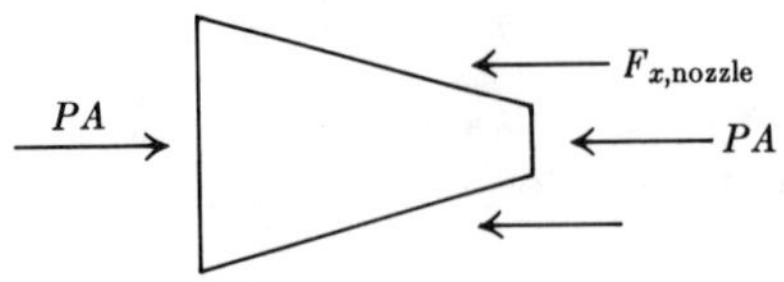

FIGURE 7.7
Forces acting on fluid.

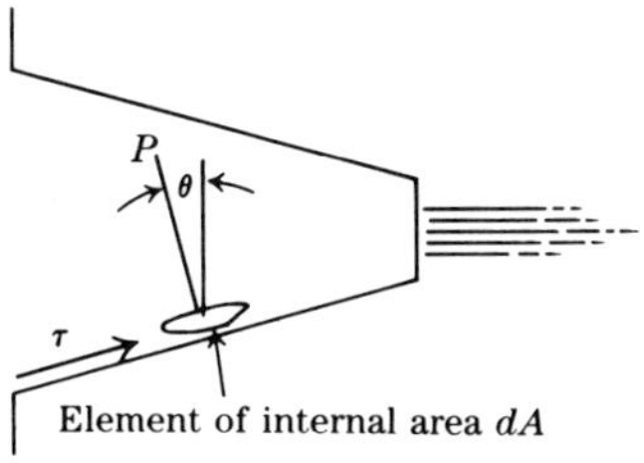

FIGURE 7.8
Alternative way of computing the force on nozzle.

$$-F_{x_{\text{noz}}} = \int (P \sin \theta + \tau \cos \theta)\, dA \tag{7.17}$$

Determining the local values of P and τ for all the internal surface of even a relatively simple device like this nozzle is a formidable task. For a complicated shape, it is beyond our current ability. Nevertheless, we computed the overall force (which we were seeking) by the momentum balance fairly easily.

Rockets are easy to analyze by means of the steady-flow one-dimensional momentum balance. Figure 7.9 shows a cutaway view of a liquid-fuel rocket being fired while rigidly attached to a test stand. What goes on inside the rocket is fairly complicated. We could, in principle, determine the force that the rocket exerts on the test stand by choosing as our system the solid parts of the rocket and by excluding the fluids inside the tanks, pumps, combustion chamber, and nozzle. Then if we could determine the pressure and shear stress at every point in the system, we could find the total force by taking the integral

$$F = \int_{\text{int.ext.surf}} (P \sin \theta + \tau \cos \theta)\, dA$$

over the whole internal and external surfaces; that would be a giant task.

But if we want only to know what this force is, we can take the outside of the rocket as our system boundary and apply the momentum balance. With this

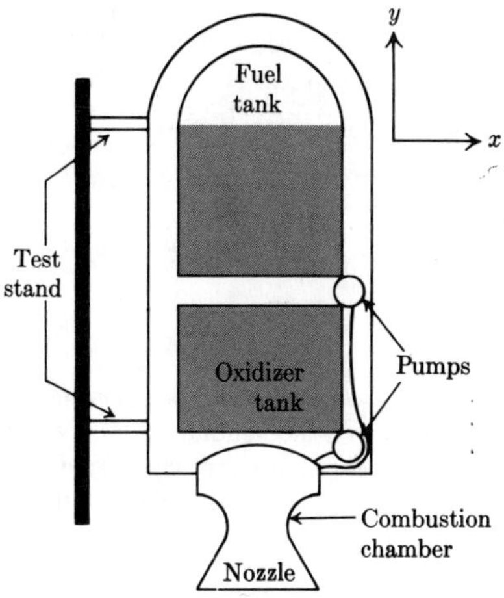

FIGURE 7.9
Liquid-fuel rocket.

boundary $[d(mV_y)/dt]_{sys}$ is zero, because the system momentum is not changing with time.[†] There is no flow into the system, so Eq. 7.16 reduces to

$$F_y = V_{y_{out}} \dot{m}_{out} \tag{7.18}$$

The external forces acting on the system are the pressure forces around the entire boundary of the system as chosen and the force exerted by the test-stand support structure. The latter force is the one we are seeking, so we can split the F_y in Eq. 7.18 into two parts and rearrange to

$$F_{y_{t.stand}} = V_{y_{out}} \dot{m}_{out} - (y \text{ component of pressure force on system}) \tag{7.19}$$

In Sec. 5.5 we noted that for flows moving slower than the velocity of sound we can safely assume that a flow leaving an enclosed system and flowing into the atmosphere is at the same pressure as the atmosphere. We have made this assumption in the previous examples of this chapter. However, in the case of the rocket, the flow leaving the system is generally supersonic, so we can no longer make this assumption. From Fig. 7.9 it is obvious that the pressure on the outside of the system is atmospheric everywhere except across the exit of the nozzle. Thus, the net y component of the pressure force on the system is

$$\begin{aligned} y \text{ component of pressure force} &= \text{exit area} \cdot (P_{exit} - P_{atm}) \\ &= \text{exit area} \cdot \text{gauge pressure of exhaust} \end{aligned} \tag{7.20}$$

Since we have chosen the y direction as positive upward, both $F_{y_{t\,stand}}$ and $V_{y_{out}}$ are negative. Multiplying Eq. 7.19 by -1, we find

$$-F_{y_{t.stand}} = +F_{y_{rocket}} = -V_{y_{out}} \dot{m}_{out} + A_{exit} P_{g\,exh} \tag{7.21}$$

Here we show the force exerted by the rocket as the negative force exerted by the test stand, because they are equal and opposite. The force exerted by the rocket is referred to as its *thrust*.

Example 7.8. A rocket on a test stand is sending out 1000 kg/s of exhaust gas at a velocity of -3000 m/s (negative because it is in the $-y$ direction). The exit area of the nozzle is 7 m^2, and the pressure at the exhaust nozzle exit is 35 kPa gauge. What is the thrust of the rocket?

From Eq. 7.21 we find

[†] Although the system as a whole is not moving in the y direction, some parts of it are, because of the internal fuel flows. Thus, the overall system has some y-directed momentum, but since this presumably is not changing with time, $[d(mV_y)/dt]_{sys} = 0$. During the motor-starting period, this simplification is not correct, but for a rocket standing still it is always small compared with the other terms in the momentum balance.

$$F_{\text{rocket}} = \text{thrust} = -\left(-3000\ \frac{\text{m}}{\text{s}} \cdot 1000\ \frac{\text{kg}}{\text{s}}\right) + 7\ \text{m}^2 \cdot 35\ \text{kPa}$$

$$= 3\ \text{MN} + 0.245\ \text{MN} = 3.25\ \text{MN} = 0.73 \times 10^6\ \text{lbf} \qquad \blacksquare$$

From this it is clear that the higher the exhaust velocity, the greater the thrust per unit mass of fuel consumed. If the exit pressure were exactly atmospheric pressure $P_{\text{gauge}} = 0$, then the thrust would be directly proportional to the exhaust velocity. For horizontally firing rockets, such as artillery rockets and airplane takeoff assist rockets, the atmospheric pressure remains constant, and one could, in principle, design the rocket nozzle for $P_{\text{exh}} = P_{\text{atm}}$. However, this generally results in an impractically large nozzle (too much air resistance or too difficult to fabricate), so the nozzle is usually designed for a P_{exh} significantly greater than the P_{atm} at the exit of the nozzle.

Vertically firing rockets, such as ballistic missiles and satellite launchers, must operate over a wide range of atmospheric pressures, from those at sea level to those in outer space. The nozzle is designed for some average pressure; therefore P_{exh} will equal P_{atm} only at one particular altitude in the flight.

Ignoring this complication for the moment, we can say that for $P_{\text{exh}} = P_{\text{atm}}$ the exhaust velocity is a simple, reliable, direct method of comparing the efficiency of various rocket engines. The early German workers in rocketry used it as a comparison basis. But U.S. workers have preferred to use the *specific impulse* I_{sp} as a comparison basis:

$$I_{\text{sp}} = \frac{\text{thrust produced, lbf}}{\text{fuel flow, lbm/s}} \tag{7.22}$$

If $P_{\text{exh}} = P_{\text{atm}}$, then

$$\text{Thrust} = I_{\text{sp}}\dot{m} = -V_{y_{\text{out}}}\dot{m} \tag{7.23}$$

indicating that I_{sp} must be exactly the same as $-V_{y_{\text{out}}}$ except for a conversion of units. By inspection this must be the conversion involving force, mass, length, and time.

Example 7.9. For a rocket with an exhaust velocity of -9000 ft/s and with $P_{\text{exh}} = P_{\text{atm}}$, what is the value of I_{sp}?

$$I_{\text{sp}} = -(V_{y_{\text{exh}}}) = \frac{-(-9000\ \text{ft/s})}{32.2\ \text{lbm} \cdot \text{ft/(lbf} \cdot \text{s}^2)} = 279.5\ \frac{\text{lbf} \cdot \text{s}}{\text{lbm}} = 1.243\ \frac{\text{kN} \cdot \text{s}}{\text{kg}} \qquad \blacksquare$$

It is common practice in the rocket industry simply to write this as 279.5 s. This is incorrect, because 1 lbf does not equal 1 lbm and they should not be canceled. Nonetheless, the cancellation is common in rocket publications.

Our discussion has concerned only rockets fixed to a test stand; we consider moving rockets in Sec. 7.5.

As shown in Chap. 6, when there is a sudden expansion in turbulent flow in a pipe, there is a resulting "friction loss." Such an expansion is shown in Fig. 7.10. Assuming that there is a uniform velocity at points 1 and 2, we may write the continuity equation for an incompressible fluid (Chap. 3) as

$$V_1 A_1 = V_2 A_2 \tag{7.24}$$

and Bernoulli's equation (Chap. 5) as

$$\frac{P_2 - P_1}{\rho} + \left(\frac{V_2^2}{2} - \frac{V_1^2}{2}\right) = -\mathcal{F} \tag{7.25}$$

Given V_1, A_1, and A_2, we can calculate V_2, but we cannot calculate $P_2 - P_1$ unless we know $-\mathcal{F}$. However, we can apply the steady-flow momentum balance (Eq. 7.16) to this flow. When we take the system as the fluid from point 1 to point 2, we find

$$\sum F_x = -\dot{m}(V_{x_1} - V_{x_2}) \tag{7.26}$$

Here

$$\sum F_x = P_1 A_1 + P_{1a} A_{1a} - P_2 A_2 - \int \tau \, dA_w \tag{7.27}$$

where P_{1a} is the average pressure over the annular area and the integral $\int \tau \, dA_w$ is the total shear force at the walls of the pipe, due to viscous friction. For a sudden expansion, the other terms in the momentum balance are large compared with this one, so we drop it.

Because of our previous discussions of the pressure of a fluid leaving a vessel, it is not implausible that P_{1a} is the same as P_1, that is, the pressure is the same for the entire cross section at point 1. Making this assumption and substituting in Eq. 7.27 and then in Eq. 7.26, we find

$$P_2 A_2 - P_1 A_2 = \dot{m}(V_{x_1} - V_{x_2})$$

But $\dot{m} = \rho V_{x_2} A_2$ and $V_{x_2} = V_{x_1}(A_1/A_2)$, so that

$$P_2 - P_1 = \rho V_{x_1}^2 \frac{A_1}{A_2}\left(1 - \frac{A_1}{A_2}\right) \tag{7.28}$$

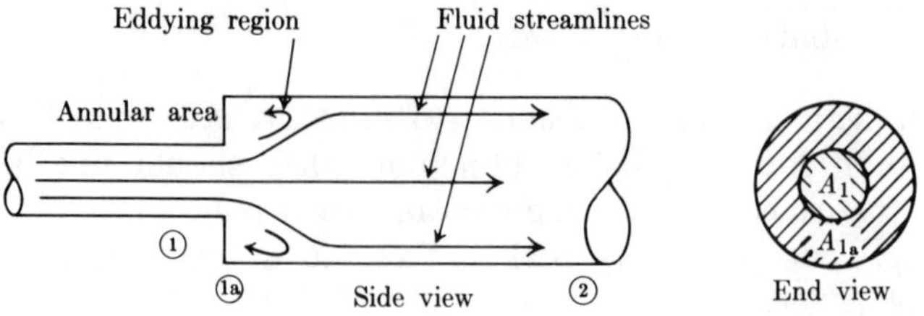

FIGURE 7.10
Sudden expansion.

Substituting this equation in Eq. 7.25 and using Eq. 7.24 to eliminate V_{x_2}, we find

$$-\mathscr{F} = V_{x_1}^2 \frac{A_1}{A_2}\left(1 - \frac{A_1}{A_2}\right) - \frac{V_{x_1}^2}{2}\left(1 - \frac{A_1^2}{A_2^2}\right) \tag{7.29}$$

which may be regrouped and factored to give

$$\mathscr{F} = \frac{V_{x_1}^2}{2}\left(1 - \frac{A_1}{A_2}\right)^2 \tag{7.30}$$

Comparing this equation with Eq. 6.28, which describes the same situation, we see that the two equations are the same if

$$K = \left(1 - \frac{A_1}{A_2}\right)^2 \tag{7.31}$$

This is the function plotted in Fig. 6.14. Experimental tests indicate that Eq. 7.31 is indeed a good predictor of experimental results, so the assumption of constant pressure across the cross section at point 1 seems a very good one.

It is interesting to compare what we did here with what we did in Sec. 6.3, where we applied a force balance to find Poiseuille's equation. That kind of force balance was usable in that case because there was no change in velocity of any part of the fluid, so that the sum of the forces acting on the fluid was zero. Here the fluid is decelerated, so the sum of the forces acting on it is not zero. The simple force balance used in Sec. 6.3 is a strongly restricted form of the momentum balance; with the complete momentum balance we can deal with much more complex flows, like the one examined here and in the next section.

One interesting phenomenon in open-channel flow is *hydraulic jump*, which is the rapid conversion of a fast, shallow flow to a slow, deep flow. This is easily observed in gutters during heavy rainstorms, at the bottom of chutes and spillways, and in any sink (Prob. 7.29). Figure 7.11 shows a cross section through such a jump.

The cross section through the jump extends into and out of the paper. Consider a section 1 ft thick into the paper, and assume that the velocity across any section perpendicular to the flow is uniform. Then for steady flow of an incompressible fluid such as water, the mass balance gives

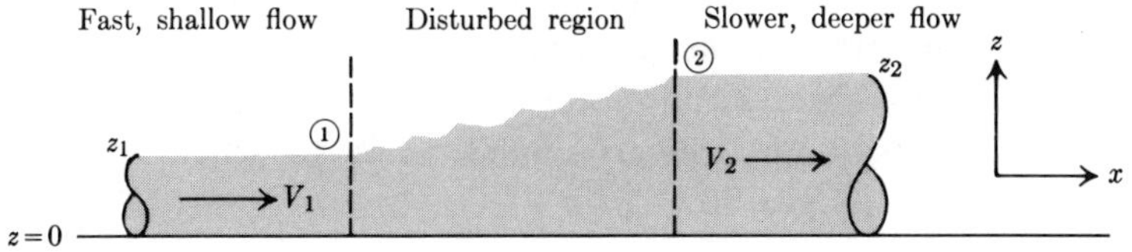

FIGURE 7.11
Hydraulic jump.

$$V_1 z_1 = V_2 z_2 \tag{7.32}$$

In a typical problem we might know two of the four quantities here. This equation provides a relation for finding a third; one more is needed. Bernoulli's equation written between states 1 and 2 shows

$$\frac{P_2 - P_1}{\rho} + g(z_2 - z_1) + \frac{V_2^2 - V_1^2}{2} = -\mathscr{F} \tag{7.33}$$

Here, as in Sec. 5.11, not all the fluid enters or leaves at the same z or the same P, so we must use appropriate average values for the z's and P's. If the friction term were negligible, this equation would supply the needed extra relation, but mathematical analysis and experimental tests indicate that it is quite large; so although it supplies an additional relation between the unknowns in Eq. 7.32, it does us no good because it introduces another unknown, $\mathscr{F}$.

Equation 7.16, however, can be used to supply the missing relationship. Taking as our system the section of fluid between points 1 and 2, we see that the only forces acting in the x direction are the shear force on the bottom, which is negligibly small and will be ignored, and the pressure forces on each side of the liquid in the system, each of which is of the form

$$F = \int P\,dA = l \int_{z=0}^{z_{\text{surf}}} g\rho(z_{\text{surf}} - z)\,dz = lg\rho\,\frac{z_{\text{surf}}^2}{2} \tag{7.34}$$

Since the flow at points 1 and 2 is all in the x direction, we may write Eq. 7.16 and drop the x subscripts to find

$$0 = l\rho z_1 V_1(V_1 - V_2) + \frac{l\rho g}{2}(z_1^2 - z_2^2) \tag{7.35}$$

Equations 7.35 and 7.32 can be solved for z, which gives us Eq. C.5 of Appendix C:

$$z_2 = \frac{-z_1}{2} \pm \sqrt{\left(\frac{z_1}{2}\right)^2 + \frac{2V_1^2 z_1}{g}} \tag{C.5}$$

Here the minus sign before the radical has no physical meaning. From Eqs. C.5 and 7.32 we can calculate the value of $\mathscr{F}$ (see Prob. 7.26).

This topic is traditionally included in fluid mechanics books for the following reasons:

1. Hydraulic jump is an interesting example of a problem that cannot be solved without using the momentum balance.
2. Hydraulic jump is readily observed in nature.
3. Shock waves and hydraulic jumps are very similar, as we see when we study shock waves in high-velocity gas flow.

Hydraulic jumps are easily demonstrated in any kitchen sink and easily studied in any well-equipped laboratory. Shock waves are much harder to demonstrate and study. Therefore, from visual observation and mathematical analysis of hydraulic jumps we can gain an intuitive understanding of shock waves. We return to their similarity in Chap. 8.

Equations C.5 and 7.32 are equally well satisfied, whether a flow is from left to right or from right to left in Fig. 7.11. However, if we calculate $\mathcal{F}$ for both, we see that right-to-left flow in the figure (deep, slow flow to shallow, fast flow) results in a negative value of $\mathcal{F}$. This is forbidden by the second law of thermodynamics, so the flow can be only in the sense indicated in the figure. We see here a strong parallel with what we will see concerning shock waves, in which the continuity, energy, and momentum equations are also satisfied by flow in either direction; but the second law of thermodynamics shows that only one direction is possible.

Equation C.5 may be put into an interesting form by dividing through by z_1:

$$\frac{z_2}{z_1} = -\frac{1}{2} + \sqrt{\frac{1}{4} + 2\,\frac{V_1^2}{gz_1}} \tag{7.36}$$

The group $V_1^2/(gz_1)$ is dimensionless and is called the *Froude number* after William Froude (1810–1879). Its significance is discussed in Chap. 13.

It is clear that z_2/z_1 must always be greater than or equal to 1 (a value of 1 would correspond to a jump of negligible height, i.e., one that was vanishingly small). You may verify from Eq. 7.36 that $z_2/z_1 = 1$ for a Froude number of 1 and that z_2/z_1 is greater than 1 for a Froude number greater than 1. In studying normal shock waves, we see that another dimensionless group, the Mach number, plays a similar role.

7.4 STARTING AND STOPPING FLOWS

The previous examples have been for steady flows. The momentum balance is powerful enough to deal with unsteady flows as well. Two very simple examples illustrate this power.

Example 7.10. Figure 7.12 shows a large water reservoir which discharges through a long, horizontal pipe, at the end of which is a valve. What is the velocity-time behavior of this system when the valve is suddenly opened?

Here the steady-state behavior can be found by Bernoulli's equation from point 1 to point 3 (see Prob. 7.31)

$$V_\infty = \left[\frac{(P_2 - P_3)}{\rho} \cdot \frac{D}{2fL}\right]^{1/2}$$

where V_∞ is the steady-state velocity. Using the values in Table A.4, we find that the steady-state velocity in the pipe is 2.45 m/s (8.03 ft/s) and the steady-state friction factor is 0.0042.

To estimate the starting behavior, we take as our system the pipe from its entrance, point 2, to its exit, point 3. Here we can assume that the pressure at

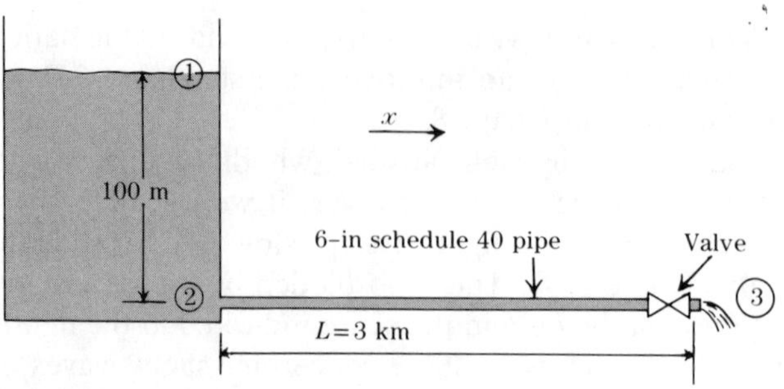

FIGURE 7.12
Long pipe with quick opening and closing valve.

point 2 does not change during the starting of the flow and is given by $P_2 = \rho g(z_1 - z_2)$. Applying the x-directed momentum balance (Eq. 7.14), we assume that the density of the fluid does not change, so that the mass of fluid in the system is constant and the mass flow rates and velocities in and out at any instant are equal. Then

$$m_{\text{sys}}\, dV_{\text{sys}} = \sum F\, dt = \left[(P_2 - P_3)\frac{\pi}{4}D^2 - \tau\pi DL\right] dt \tag{7.37}$$

Here the shear force acts in the direction opposite to the pressure force; at steady state they will be equal. Replacing τ by its expression in terms of the friction factor and expressing the mass of the system in terms of its volume and density, we find

$$\rho\frac{\pi}{4}D^2L\, dV = \sum F\, dt = \left[(P_2 - P_3)\frac{\pi}{4}D^2 - f\rho\frac{V^2}{2}\pi DL\right] dt$$

$$dV = \left(\frac{P_2 - P_3}{\rho L} - \frac{4f}{D}\cdot\frac{V^2}{2}\right) dt = \frac{2f}{D}(V_\infty^2 - V^2)\, dt$$

$$\frac{dV}{V_\infty^2 - V^2} = \frac{2f}{D}\, dt \tag{7.38}$$

Here f is not constant because the flow starts in the laminar region, so that f is initially large, then declines, increases sharply during the transition, and declines slowly in the turbulent region. But the term involving f is significant only near the end of the starting transient, so we can treat f as a constant and perform the indicated integration:

$$t = \frac{D}{4fV_\infty}\ln\frac{V_\infty + V}{V_\infty - V} + C \tag{7.39}$$

Here at $t = 0$, $V = 0$, so the ln term on the right is $\ln 1 = 0$, from which it follows that the constant of integration $C = 0$. We may also check to see that Eq. 7.39 gives the correct steady-state solution by setting $t = \infty$. The only way the right-hand side can be infinite is for the denominator of the ln term to be zero, which requires that $V = V_\infty$. To find the velocity-time relation, we evaluate

$$\frac{D}{4fV_\infty} = \frac{[(6.065/12)\text{ ft}]\cdot \text{m}/3.28\text{ ft}}{4\cdot 0.0042\cdot 2.45\text{ m/s}} = 3.74\text{ s}$$

and then make up the following table:

Velocity, m/s	Time, s
0.1	0.31
1	3.24
2	8.57
2.4	17.11
2.44	23.2
2.449	31.8
2.45	Infinite

We see that the velocity increases quickly at first and then asymptotically increases to the steady-state value. ■

Example 7.11. Repeat Example 7.10 for the case in which the fluid is flowing steadily and the valve at the end of the pipe is instantaneously closed. Here we begin by rearranging Eq. 7.37 to get

$$\frac{dV_{\text{sys}}}{dt} = \frac{(P_2 - P_3)(\pi/4)D^2 - \tau\pi DL}{m_{\text{sys}}} \tag{7.40}$$

If, as the problem suggests, we stop the fluid instantaneously, then the left-hand side of this equation must be minus infinity! The only way the right-hand side can be minus infinity is for P_3 to become infinite! If it were possible to stop the fluid instantaneously, and if the fluid did not increase in density nor the pipe wall stretch, then that is exactly what would happen. One might compare this situation to dropping an egg off a tall building. The egg increases steadily in velocity as it falls, but the forces acting on it are gentle enough that it is unharmed. When it reaches the pavement, its deceleration is very rapid, practically infinite; the egg responds by splattering. The observational fact is that we generally cannot stop the flow instantaneously, but with readily available valves, closed as quickly as possible, we can stop the fluid quickly enough to generate very large pressures adjacent to the valve.

To solve the problem, we must take into account the fact that the liquid will compress, slightly but significantly. In real problems the expansion of the pipe due to the increased pressure must also be taken into account; it makes the pressure less than the value we compute here. If we are able to stop the flow by shutting the valve at point 3 instantaneously, then the layer of fluid adjacent to the valve will be stopped. It will stop the next layer, and so the region of stopped fluid will propagate backward up the pipe to the reservoir. (This is analogous to the big freeway pileups that occur during heavy fogs.

Someone slows down and is hit by a faster-moving car from behind. The first crash produces a pile of stopped, wrecked cars. This pile then enlarges in the upstream direction as more and more cars pile into the stopped wreckage.) The rate of propagation of the boundary between stopped and moving fluid (assuming rigid pipe walls) is the local speed of sound. That is not proved here, but may seem clearer after we have discussed the speed of sound in Chap. 8. From Chap. 8 we can borrow the fact that for water the speed of sound is about $c = 5000$ ft/s (1520 m/s), so that the stopped layer of water will reach the reservoir in $t = L/c = (3000\text{ m})/(1520\text{ m/s})$, or about 2 s after the valve is closed.

To compute the pressure in the stopped fluid, we take the viewpoint of the person riding on the interface between the moving fluid and the stopped fluid. This interface is moving at the speed of sound in the liquid c. From the observer's viewpoint, this is a steady process, so the momentum balance becomes

$$0 = \dot{m}_{\text{in}}(V_{\text{in}} - V_{\text{out}}) + \sum F$$

$$0 = c\rho A\,\Delta V - A\,\Delta P$$

$$\Delta P = c\rho\,\Delta V \tag{7.41}$$

where ΔP is the pressure change across the moving boundary and ΔV is the velocity change across it, which in this case is the velocity of the fluid which has not been stopped yet minus zero. Inserting numerical values, we find

$$\Delta P = c\rho\,\Delta V = 1520\,\frac{\text{m}}{\text{s}} \cdot 998.2\,\frac{\text{kg}}{\text{m}^3} \cdot 2.45\,\frac{\text{m}}{\text{s}} \cdot \frac{\text{N}\cdot\text{s}^2}{\text{kg}\cdot\text{m}} \cdot \frac{\text{Pa}}{\text{N/m}^2}$$

$$= 3.72\text{ kPa} = 539\text{ psi}$$ ■

This is a very large pressure, and it explains why this phenomenon, called *water hammer*, can pose a serious problem, particularly in large hydroelectric structures. One often produces the same result at home by closing a faucet quickly; a pounding sound in the plumbing indicates that a high pressure has been generated. The treatment here is the simplest possible; many more interesting details and complications are shown in the books on this subject [1].

Some students find it more intuitively satisfying to arrive at Eq. 7.41 by asking how long ($t = L/c$) the pressure force ($A\,\Delta P$) must act to decelerate the mass of fluid in the pipe ($AL\rho$) from its initial velocity to zero (ΔV). Substituting those values in $F = ma$ leads directly to Eq. 7.41.

7.5 RELATIVE VELOCITIES

All the examples in Sec. 7.4 concern systems fixed in space. When a system is moving, the momentum balance still applies, but it is often convenient to introduce the idea of a relative velocity. Figure 7.13 shows a student on the ground throwing a ball to a student on a moving cart. The velocity of the ball V

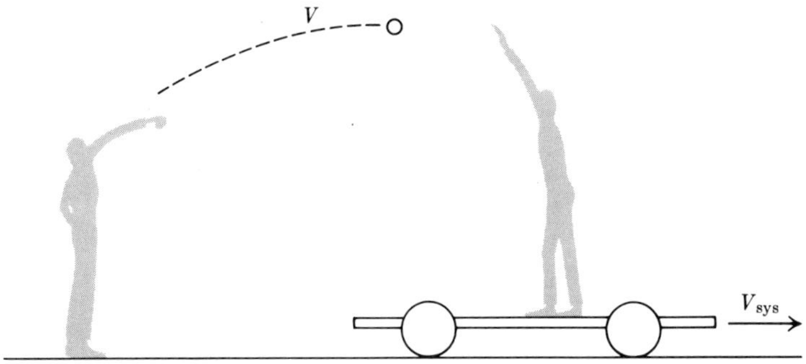

FIGURE 7.13
Relative velocities.

is 10 m/s. The cart is moving with a velocity V_{sys} of 5 m/s. As seen by the student who threw it, the ball is moving at 10 m/s. As seen by the student who catches it, the ball is moving at 5 m/s, because that is the velocity with which it is overtaking the cart. In general,

$$\mathbf{V} = \mathbf{V}_{sys} + \mathbf{V}_{rel} \tag{7.42}$$

where **V** is the velocity of a body or a stream of fluid relative to some set of fixed coordinates, $\mathbf{V}_{sys}$ is the velocity of the system (the cart in this case) relative to the same set of fixed coordinates, and $\mathbf{V}_{rel}$ is the velocity of the body or stream of fluid as seen by an observer riding on the moving system. When velocities are near the speed of light, this becomes more complicated, but such velocities seldom occur in fluid mechanics.

Equation 7.42 is a vector equation; like all other vector equations, it is simply a shorthand way of writing three scalar equations. In this text we use only its scalar equivalents, such as

$$V_x = V_{x_{sys}} + V_{x_{rel}} \tag{7.43}$$

To illustrate the utility of this equation, consider a rocket in horizontal flight with no air resistance. We choose the rocket as our system and simplify by letting $P_{exh} = P_{atm}$. Then since there is no flow into the system or any external force acting in the x direction, Eq. 7.14 becomes

$$d(mV_x)_{sys} = -V_{x_{out}}\, dm_{out} \tag{7.44}$$

Expanding the left side and substituting for $V_{x_{out}}$ from Eq. 7.43, we have

$$m_{sys}\, dV_{x_{sys}} + V_{x_{sys}}\, dm_{sys} = (V_{x_{sys}} + V_{x\ \text{rel, out}})\, dm_{out} \tag{7.45}$$

Because all the velocities are in the x direction, we can drop the x subscript. Now we note that $dm_{out} = -dm_{sys}$. Making this substitution and canceling like terms, we can divide by m_{sys} to find

$$dV_{sys} = V_{\text{rel, out}} \frac{dm_{sys}}{m_{sys}} \tag{7.46}$$

If the exhaust velocity relative to the rocket is constant (which is practically true of most rockets), then we can readily integrate this to

$$(V_{\text{fin}} - V_{\text{init}})_{\text{sys}} = V_{\text{rel, out}} \ln \frac{m_{\text{fin}}}{m_{\text{init}}} \tag{7.47}$$

This equation, often referred to as the *rocket equation* or the *burnout velocity equation*, indicates the limitation on possible speeds of various kinds of rockets.

Example 7.12. A single-stage rocket is to start from rest, so $V_{\text{init}} = 0$. The mass of fuel is 0.9 of the total mass of the loaded rocket; $m_{\text{fin}}/m_{\text{init}} = 0.1$. The specific impulse of the fuel is 430 lbf · s/lbm, and the pressures are $P_{\text{exh}} = P_{\text{atm}}$. What speed will this rocket reach in horizontal flight if there is no air resistance?

From Eq. 7.23 we have

$$V_{\text{rel, out}} = -I_{\text{sp}} = -430 \frac{\text{lbf} \cdot \text{s}}{\text{lbm}} \cdot 32.2 \frac{\text{lbm} \cdot \text{ft}}{\text{lbf} \cdot \text{s}^2} = -13{,}850 \frac{\text{ft}}{\text{s}} = -4220 \frac{\text{m}}{\text{s}}$$

Therefore,

$$V_{\text{fin}} = -13{,}850 \frac{\text{ft}}{\text{s}} (\ln 0.1) = 31{,}900 \frac{\text{ft}}{\text{s}} = 9730 \frac{\text{m}}{\text{s}}$$ ■

This example shows the probable maximum speed attainable with single-stage rockets using chemical fuels. It appears that the maximum I_{sp} for chemical fuels is about 430 lbf · s/lbm. Better structural design may reduce the value of $m_{\text{fin}}/m_{\text{init}}$, but it is unlikely to go much under 0.1 if there is any significant payload involved. For higher velocities staged rockets are needed. Equation 7.47 does not include the effects of gravity or air resistance. These can be included but make the equation somewhat more complicated (Prob. 7.38). For more on rockets see Ley [2] and Sutton [3].

Another example of the utility of the relative-velocity concept concerns the interaction of a jet of fluid and a moving blade. Such interactions are the basis of turbines and rotating compresssors as used in turbojet and gas-turbine engines and in the steam and water turbines that produce most of the world's electricity. In Examples 7.3, 7.4, and 7.5 we saw how the interaction of a jet and solid surface produced a force on the surface. For this force to do work, it must move through a distance. The work is given by $dW = F\,dx$, and the power (or rate of doing work) is given by $\text{Po} = dW/dt = F\,dx/dt$. The latter is equal to the force times the velocity of the system: $\text{Po} = FV_{\text{sys}}$.

A curved blade is moving in the x direction and deflecting a stream of fluid; see Fig. 7.14. Consider this first from the viewpoint of an observer riding with the blade. As far as the observer can tell, the blade is standing still; no work is being done. Therefore, with no change in pressure and elevation, Bernoulli's equation tells the observer that

$$\frac{V_{\text{out}}^2}{2} - \frac{V_{\text{in}}^2}{2} = -\mathscr{F} \tag{7.48}$$

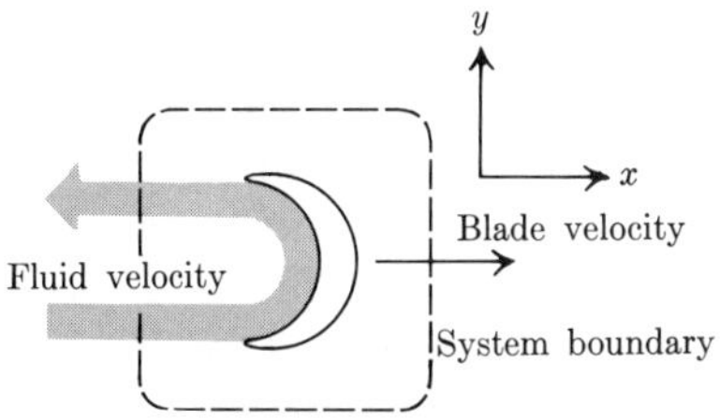

FIGURE 7.14
Jet-blade interaction.

and that if there is no friction, the outlet velocity is equal in magnitude to the inlet velocity but in a different direction. If there is friction, the outlet velocity will be less, but it cannot possibly be more. Here in the energy balance V^2 is a scalar, so we have no concern about the signs of V_{in} and V_{out}.

Now assume frictionless flow and that the blade in the figure is shaped so that the inlet and outlet streams have only x velocity and no y velocity. Then by applying Eq. 7.16 and dropping the x subscript (because all the velocities are in the x direction) we can write

$$F = \dot{m}(V_{\text{out}} - V_{\text{in}}) \tag{7.49}$$

This is the F exerted by the blade. The fluid exerts an equal and opposite force on the blade. Substituting Eq. 7.43 twice, we find

$$\begin{aligned} F &= \dot{m}(V_{\text{rel, out}} + V_{\text{blade}} - V_{\text{rel, in}} - V_{\text{blade}}) \\ &= \dot{m}(V_{\text{rel, out}} - V_{\text{rel, in}}) \end{aligned} \tag{7.50}$$

The velocity of the blade cancels out of the force equation, so the force is the same whether viewed by an observer riding on the blade or by an observer standing still.

The work[†] done by the fluid per unit time (the power) is

$$\text{Po} = \frac{dW}{dt} = -F\,\frac{dx}{dt} = -FV_{\text{blade}} = -\dot{m}(V_{\text{rel, out}} - V_{\text{rel, in}})V_{\text{blade}} \tag{7.51}$$

and therefore the work done per unit mass of fluid is

$$\frac{dW}{dm} = (V_{\text{rel, in}} - V_{\text{rel, out}})V_{\text{blade}} \tag{7.52}$$

As shown above, from Bernoulli's equation for frictionless flow we know that $V_{\text{rel, out}} = -V_{\text{rel, in}}$; therefore,

$$\frac{dW}{dm} = 2V_{\text{rel, in}} \cdot V_{\text{blade}} \tag{7.53}$$

Now suppose that the velocity of the jet is fixed. This would occur if it were a jet of water entering the power plant at the base of a dam with constant

[†] The work terms in this chapter are all exclusive of injection work and would have the symbol $W_{\text{a.o.}}$ in Chaps. 4, 5, and 6. Here we drop the subscript because it causes no confusion to do so.

upstream water level (in which case we could calculate the jet velocity by Bernoulli's equation) or if it were a jet of steam from a boiler with constant steam temperature and pressure (in which case we could calculate the jet velocity by the methods developed in Chap. 8). For a fixed jet velocity, at what velocity should we run the blade to get the maximum amount of useful work from it? Here we want to maximize dW/dm; from calculus we know that this will be a maximum when $d(dW/dm)/dV_{\text{blade}}$ is zero. First we replace $V_{\text{rel, in}}$, using Eq. 7.43:

$$\frac{dW}{dm} = 2(V_{\text{in}} - V_{\text{blade}})V_{\text{blade}} \tag{7.54}$$

Then we differentiate both sides of this equation with respect to V_{blade} and set the derivative equal to zero, finding

$$\frac{d(dW/dm)}{dV_{\text{blade}}} = 0 = 2(V_{\text{in}} - 2V_{\text{blade}}) \tag{7.55}$$

$$V_{\text{blade}} = \frac{V_{\text{in}}}{2} \tag{7.56}$$

The blade speed must be one-half the jet speed for maximum efficiency. The same result for the most efficient blade speed can be found from the energy viewpoint; this is also the velocity at which the fluid after leaving the blades has exactly zero velocity and, therefore, zero kinetic energy relative to a stationary observer.

7.6 A VERY BRIEF INTRODUCTION TO AERONAUTICAL ENGINEERING

Chapters 5 and 6 are devoted to problems which could be understood most easily by applying the energy balance, and this chapter is devoted to other problems, which could be understood most easily by applying the momentum balance. Some problems are most easily understood by applying *both* the energy and the momentum balances. A very interesting example is the elementary analysis of flight, which helps explain the behavior of airplanes and helicopters and birds and insects.

An airplane (or a bird or a flying insect) is a fluid mechanical device; it flies by making a fluid—the air—move. Consider an airplane in constant-velocity, level flight; see Fig. 7.15. The airplane has no acceleration in either the x or the y direction. Therefore, the sum of the forces acting on the airplane in each of these directions is zero. These forces are shown in the figure and given their common aeronautical engineering names. The force which gravity exerts, the *weight*, acts downward. To counteract this, the air must exert an equal and opposite upward force, called the *lift*. It is the function of the airplane's wings to make the air exert this force.

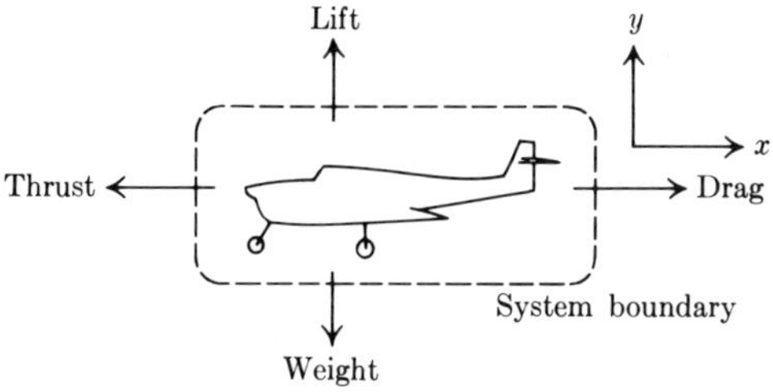

FIGURE 7.15
Airplane in constant-velocity, level flight.

To see how the wing does this, we make a momentum balance around the airplane and take as our system the airplane plus an envelope of air around it, large enough for the pressure on the outside of the envelope to be constant. This system boundary is also shown in the figure. We base our coordinate system on the airplane, so the airplane appears to stand still and the air to flow toward it. Applying the y component of the momentum balance in constant-velocity flight, we see that there is no accumulation: we have $d(mV)_{sys} = 0$.† We have assumed that the pressure around the outside of the system is uniform; then the only external forces acting on the plane are the force of gravity and the force exerted by the air. Thus,

$$F = \text{weight of plane} = \dot{m}(V_{y_{out}} - V_{y_{in}}) \tag{7.57}$$

Since not all the air comes in or goes out at the same velocity, the two V_y's in this equation must be some appropriate average velocities, obtained by an integral of the flow per unit surface area over the entire surface of the system. However, we need not worry about this integration, if we merely think of these velocities as some appropriate average.

In the direction of the $+y$ axis, F is negative. The flow through $\dot{m}$ is positive, so $V_{y_{out}} - V_{y_{in}}$ must be negative; the air must be accelerated in the $-y$ direction, downward. Thus, we see that, to stay in level flight, the airplane must accelerate the surrounding air downward. This is precisely what a swimmer does in "treading water"; by accelerating the water downward, the swimmer stays up.

Example 7.13. An airplane with a mass of 1000 kg (and thus a weight of 9810 N) is flying in constant-velocity, horizontal flight at 50 m/s. Its wingspread is 15 m, and we assume that it influences a stream of air as wide as its wingspread and 3 m thick. How much average vertical downward velocity must it give this air? Assume that the air comes in at zero vertical velocity.

---

† In the most exact work we would have to consider the decrease in mass due to the burning of fuel, but that is small enough to neglect here.

$$\dot{m} = \rho A V_{x,\text{ in}} = 1.21\,\frac{\text{kg}}{\text{m}^3}\cdot(15\text{ m}\cdot 3\text{ m})\cdot 50\,\frac{\text{m}}{\text{s}} = 2723\,\frac{\text{kg}}{\text{s}} = 6000\,\frac{\text{lbm}}{\text{s}}$$

$$V_{y_{\text{out, av}}} = \frac{F_g}{\dot{m}} = \frac{9810\text{ N}}{2723\text{ kg/s}}\cdot\frac{\text{kg}\cdot\text{m}}{\text{N}\cdot\text{s}^2} = 3.60\,\frac{\text{m}}{\text{s}} = 11.8\,\frac{\text{ft}}{\text{s}}$$ ■

This is the approach "from the outside." Without knowing any details of the flow around the various parts of the airplane, we can find the average downward velocity it must give the air that it influences to stay in level flight. If we chose as our system the airplane itself, we would see that it has no flow in or out, excluding the negligible engine intake and exhaust, and therefore the sum of the forces on it must be zero. For the airplane as the system, these forces are the gravity force and the pressure force integrated over its entire surface. To find the latter, it is necessary to analyze the flow in detail around the entire airplane. It can be done in the case of some simple structures, such as certain types of wings, by means of Bernoulli's equation; this type of analysis is introduced in Chaps. 10 and 11. Briefly, the result of the detailed calculation is that the wing is shaped so that the air must move faster over its top surface than over its bottom surface. Therefore, according to Bernoulli's equation, the pressure on the top is less, so there is a net upward pressure force on the wing. Thus, we may regard the wing as a Bernoulli-equation device which speeds up the flow on one side relative to the other and develops a pressure force, or we may regard it as a momentum balance device which turns the net flow downward and thus produces an upward force. Both views are correct.

From the balance of forces in the x direction we see that, in order to fly, the plane must overcome the air resistance, which is called *drag*. In constant-velocity level flight, the drag is equal and opposite to the forward force, or *thrust*, developed by the power plant.

If, as described, the purpose of the wings is to turn the air downward, then it would appear that any surface which was placed at an angle to the oncoming airstream would do. This is true. The first surfaces tried were flat ones, such as paper gliders. Those gliders will fly, and so will a flat-winged airplane. However, a flat wing has much more drag than a properly curved one producing an equivalent lift. This was discovered by birds in their evolution and later by the pioneers in aviation. By careful analysis and much experimentation, wings have been built which have ratios to lift to drag as high as 20.

Example 7.14. A light plane is being designed with an overall aircraft lift/drag ratio of 10. The available power plants have thrust/weight ratios of 2. What percentage of the total loaded weight of the aircraft will be the power plant? Assume that the plane will be used only in constant-velocity, level flight.

Under these circumstances, lift equals gross weight and drag equals thrust; therefore lift/drag is weight/thrust; then

$$\frac{\text{Weight}}{\text{Engine weight}} = \frac{\text{weight}}{\text{thrust}}\cdot\frac{\text{thrust}}{\text{engine weight}} = 10\cdot 2 = 20$$ ■

Suppose that we wish to design a helicopter, using a power plant with the same thrust/weight ratio. For a helicopter in hovering flight, thrust is vertically upward and equals the weight. Thus, with the given engine, 50 percent of the gross weight of the helicopter must be engine. This illustrates the fact that horizontal flight, with a wing, is much more efficient than hovering flight. But why is this so? From Eq. 7.52, we see that the upward force is equal to the velocity change of the air times the mass flow rate of the air. Thus, we may lift a given load by making a small velocity change in a large airflow rate or a large velocity change in a small airflow rate.

But now let us consider the work which must be done to accelerate this quantity of air. We apply Bernoulli's equation to the system shown in Fig. 7.15. Here again there is negligible change in pressure in the air passing through the system and negligible change in elevation. Solving for the external work gives

$$\frac{dW}{dm} = -\mathcal{F} - \frac{\Delta V^2}{2} \tag{7.58}$$

This is the negative work which must be done on the air by the airplane's power plant. Here the ΔV^2 is the change in the square of the average value of the velocity, which is given by $V = (V_x^2 + V_y^2)^{1/2}$. The power is

$$\text{Po} = \frac{dW}{dm}\dot{m} = \dot{m}\left(-\mathcal{F} - \frac{\Delta V^2}{2}\right) \tag{7.59}$$

If we neglect the friction term, the power required is proportional to the change in the square of the velocity.

Thus from the momentum balance (Eq. 7.57), we see that an airplane with a given weight can be lifted by any flow which has the proper combination of $\dot{m}\,\Delta V_y$ but that the power to be supplied by the engine is proportional to $\dot{m}\,\Delta V^2$. So, to lift the maximum weight with the minimum power, we should make $\dot{m}$ as large as possible and thereby make ΔV_y and ΔV^2 as small as possible. This is the same as saying that the wings should be as long and thin as possible. However, long, thin wings are difficult to build and are not very satisfactory for high-speed flight. The fliers most interested in efficiency are soaring birds and human glider fliers; both have settled on the longest, thinnest wings which seem structurally feasible. Commercial aircraft designers have sacrificed some of this efficiency to achieve better high-speed performance and sturdier wings. The first airplane to fly around the world without refueling had even longer and narrower wings than any glider or any bird; this feat became possible only when the development of fiber-reinforced plastics made it possible to build extremely long, thin wings.

This same consideration of the energy and momentum effects of an aircraft shows why the helicopter is an inefficient weight-lifting device. In hovering flight, it can move only as much air as it can suck into its blades, so it must give it a very large velocity change. This leads to a higher power requirement. It also explains why helicopters hover as little as possible; as soon as they can, they move forward, so that the amount of air influenced by the rotors is increased by their forward motion. The same arguments explain why bees and very small birds such as hummingbirds can hover, but larger birds cannot. Insects and hummingbirds have small values of mass per unit wing area, so they can stay up in inefficient hovering flight. Big birds have higher values of mass per unit wing area and cannot hover [4].

Finally this same consideration explains why the commercial aircraft industry is replacing simple jet engines with fan-jet or high-bypass engines, which move more air than a simple jet engine, making a smaller change in its velocity and hence getting better fuel efficiency.

7.7 THE ANGULAR MOMENTUM BALANCE; ROTATING SYSTEMS

In the study of rotating systems, it is convenient to define a quantity called the *angular momentum* of a body:

Angular momentum of a small body = mass · radius · tangential velocity

$$L = mrV_\theta \tag{7.60}$$

The geometric significance of these terms is seen most easily by examining a body in motion and using polar coordinates; see Fig. 7.16.[†] We see from Eq. 7.60 that the angular momentum of a body depends not only on the body's velocity and mass but also on the point chosen for the origin of the coordinate system. This causes no confusion if we always make clear our choice of origin. Since the idea of angular momentum is used most often in rotating systems,

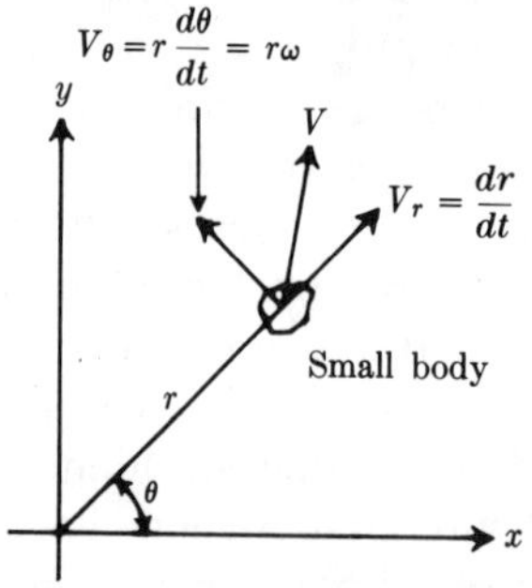

FIGURE 7.16
Velocity components in polar coordinates. Here **V** is the velocity vector, V_θ is the tangential component of the velocity (or simply the tangential velocity), and V_r is the radial component of the velocity (or simply the radial velocity).

[†] This equation is really one component of a vector equation, which is written $\mathbf{L} = m\mathbf{r} \times \mathbf{V}$. The other components of this vector equation refer to motions of the axis of rotation, which are significant in systems such as gyroscopes but are seldom important in fluid mechanics. Therefore, we refer the reader to texts on mechanics for the three-dimensional vector form of this equation.

generally it is easiest to choose the origin so that it coincides with the axis of rotation.

The r of a large body is not constant over the entire mass, so we must find the angular momentum by integrating over the entire mass:

$$L = \int_{\text{mass}} rV_\theta \, dm \tag{7.61}$$

As shown in Fig. 7.16, the tangential velocity V_θ is equal to $r\omega$; therefore, for constant angular velocity over an entire body (i.e., a rotating rigid body) this simplifies to

$$L = \omega \int_{\text{mass}} r^2 \, dm = \omega I \tag{7.62}$$

where I is the angular moment of inertia.

It can be readily shown that angular momentum, like linear momentum, obeys the balance equation, but with the difference that in place of force acting we have torque acting, where

$$\begin{aligned} \text{Torque} &= \text{tangential force} \cdot \text{radius} \\ \Gamma &= F_\theta r \end{aligned} \tag{7.63}$$

So the angular momentum balance (for a fixed axis of rotation) becomes

$$dL = (rV_\theta)_{\text{in}} \, dm_{\text{in}} - (rV_\theta)_{\text{out}} \, dm_{\text{out}} + \Gamma \, dt \tag{7.64}$$

Again it is convenient to divide by dt to find the rate form:

$$\left(\frac{dL}{dt}\right)_{\text{sys}} = (rV_\theta)_{\text{in}} \dot{m}_{\text{in}} - (rV_\theta)_{\text{out}} \dot{m}_{\text{out}} + \Gamma \tag{7.65}$$

This equation, often referred to as the *moment-of-momentum equation*, is one of the basic tools in the analysis of rotating fluid machines, turbines, pumps, and other devices [5]. In the steady-state flow $(dL/dt)_{\text{sys}}$ is zero and $\dot{m}_{\text{in}}$ equals $\dot{m}_{\text{out}}$; so we have

$$\Gamma = \dot{m}[(rV_\theta)_{\text{out}} - (rV_\theta)_{\text{in}}] \tag{7.66}$$

which is known as *Euler's turbine equation*.

Example 7.15. A centrifugal water pump impeller rotates at 1800 r/min; see Fig. 7.17. The water enters the blades at a radius of 1 in and leaves the blades at a radius of 6 in. The total flow rate is 100 gal/min. The tangential velocities in and out may be assumed equal to the tangential velocity of the rotor at those radii. What is the steady-state torque exerted on the rotor?

From Eq. 7.66

$$\dot{m} = 100 \, \frac{\text{gal}}{\text{min}} \cdot 8.33 \, \frac{\text{lbm}}{\text{gal}} = 833 \, \frac{\text{lbm}}{\text{min}} = 378 \, \frac{\text{kg}}{\text{min}}$$

$$(V_\theta)_{\text{in}} = r_{\text{in}} \omega \qquad (V_\theta)_{\text{out}} = r_{\text{out}} \omega$$

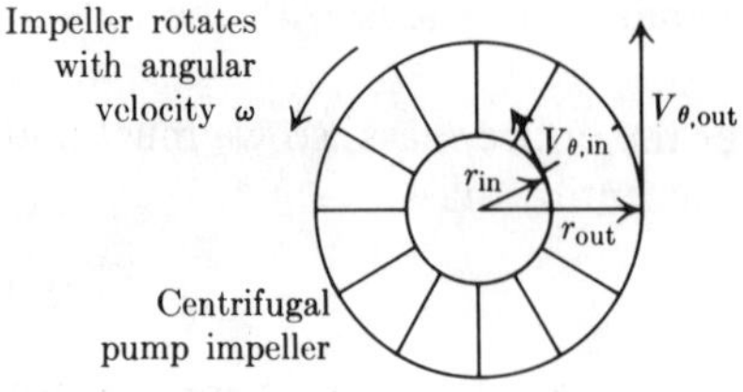

FIGURE 7.17
Centrifugal pump impeller.

$$\Gamma = \dot{m}\omega(r_{\text{out}}^2 - r_{\text{in}}^2)$$

$$\Gamma = 833\,\frac{\text{lbm}}{\text{min}} \cdot \frac{2\pi(1800)}{\text{min}} \cdot \left[\left(\frac{1}{2}\,\text{ft}\right)^2 - \left(\frac{1}{12}\,\text{ft}\right)^2\right] \cdot \frac{\text{lbf}\cdot\text{s}^2}{32.2\,\text{lbm}\cdot\text{ft}} \cdot \frac{\text{min}^2}{3600\,\text{s}^2}$$
$$= 19.8\,\text{ft}\cdot\text{lbf} = 26.8\,\text{N}\cdot\text{m}$$ ■

This is the net torque acting on the rotor; it is the algebraic sum of the positive torque exerted by the shaft driving the rotor and the negative torque exerted by friction between the rotor and the surrounding fluid.

7.8 THE MOMENTUM BALANCE FOR THREE-DIMENSIONAL FLOW

In the preceding material of this chapter, the flow considered was one-dimensional; even in Example 7.5, all the fluid leaving the system had the same velocity and direction. Most practical fluid mechanics problems (flow in pipes, channels, jets, etc.) are one-dimensional. However, there are some important two- or three-dimensional flows, particularly in flows of very viscous materials and of fluids around airplanes and ships. In such cases it is useful to have the momentum balance set up for a three-dimensional flow. The notation to be used is shown in Fig. 7.18.

Our system is the small cube shown. We may think of it as being a wire frame with flow in or out through all six faces. The frame itself is fixed in space

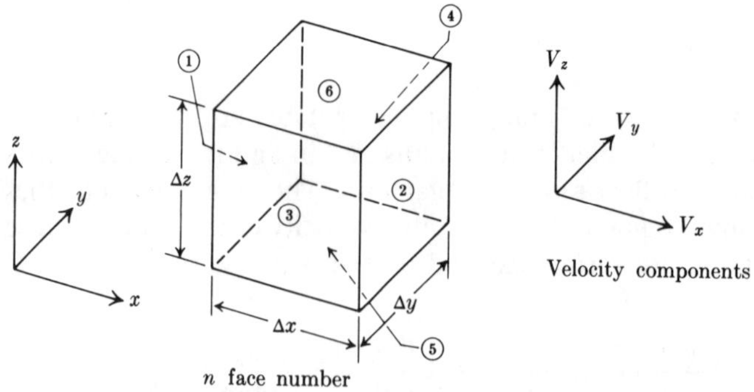

FIGURE 7.18
Notation for three-dimensional momentum balance.

and does not move. For this system we can write three separate, independent momentum balances, one for each of the x, y, and z directions. The rate form of the x-directed momentum balance for this sytem (see Eq. 7.15) is

$$\frac{\partial}{\partial t}(mV_x)_{\text{sys}} = \sum_{1-6}(V_{x_{\text{in}}} - V_{x_{\text{out}}})\dot{m} + \sum(x \text{ component of all forces acting}) \tag{7.67}$$

where the summation includes the faces numbered 1 to 6.

Previously we have written the first term as d/dt; here we write it as $\partial/\partial t$ because V_x is a function of both time and position; that is, $V_x = V_x(x, y, z, t)$. Thus, our momentum balance is a partial differential equation, and $\partial/\partial t$ implies d/dt at some fixed location.

It is common practice in fluid mechanics not to include in the partial differentials the subscripts indicating which variables are being held constant. This is so because with very few exceptions the independent variables are x, y, z, and t or r, θ, z, and t in polar coordinates. Thus, the symbol $\partial(\rho V_x)/\partial x$ really means $[\partial(\rho V_x)/\partial x]_{y,z,t}$. Throughout this text any partial derivative that does not have a subscript indicating what is being held constant is assumed to have as independent variables x, y, z, and t or r, θ, z, and t in polar coordinates.

The mass of fluid in the system is equal to the system volume times the density of the fluid in the system; so we may write

$$m_{\text{sys}} = \rho\,\Delta x\,\Delta y\,\Delta z \tag{7.68}$$

The flow in through face 1 is

$$\dot{m}_1 = \rho_1 A V_{x_1} = \rho_1\,\Delta y\,\Delta z\,V_{x_1} \tag{7.69}$$

and the flow out through face 2 is

$$\dot{m}_2 = \rho_2 A V_{x_2} = \rho_2\,\Delta y\,\Delta z\,V_{x_2} \tag{7.70}$$

Thus, the contribution of faces 1 and 2 to the first term on the right of the equals sign is

$$\Delta z\,\Delta y\,(\rho_1 V_{x_1}^2 - \rho_2 V_{x_2}^2)$$

The flow in through face 3 is

$$\dot{m}_3 = \rho_3 A V_{y_3} = \rho_3\,\Delta x\,\Delta z\,V_{y_3} \tag{7.71}$$

and the flow in through face 4 is similar. The contributions from those faces to the first term on the right of the equals sign are

$$\Delta z\,\Delta y\,(\rho_3 V_{x_3} V_{y_3} - \rho_4 V_{x_4} V_{y_4})$$

Faces 5 and 6 contribute similarly:

$$\Delta x\,\Delta y\,(\rho_5 V_{x_5} V_{z_5} - \rho_6 V_{x_6} V_{z_6})$$

The forces acting on the cube are the force of gravity on the entire body of the fluid and the normal and shear forces on the faces of the cube. The gravity force in the x direction is

$$(\text{Gravity force})_x = mg \cos\theta = \rho g \,\Delta x\,\Delta y\,\Delta z \cos\theta \tag{7.72}$$

where θ is the angle between the gravity vector and the x axis.

The only normal forces on the cube which have components in the x direction are the normal forces on faces 1 and 2. We denote a normal force in the x direction by symbol σ_{xx}; then the normal-force contribution may be written[†]

$$\text{Normal force} = \Delta y\,\Delta z\,(\sigma_{xx_1} - \sigma_{xx_2}) \tag{7.73}$$

The shear forces on faces 1 and 2 have no component in the x direction and do not enter the x-directed momentum balance. The shear forces on faces 3 and 4 are given the symbol τ_{xy} and those on faces 5 and 6 the symbol τ_{xz}, so the shear-force contribution is

$$\text{Shear force} = \Delta x\,\Delta z\,(\tau_{xy_4} - \tau_{xy_3}) + \Delta x\,\Delta y\,(\tau_{xz_6} - \tau_{xz_5}) \tag{7.74}$$

Making all these substitutions in Eq. 7.67, we divide by $\Delta x\,\Delta y\,\Delta z$ and find

$$\frac{\partial}{\partial t}(\rho V_x) = \left(\frac{\rho_1 V_{x_1}^2 - \rho_2 V_{x_2}^2}{\Delta x} + \frac{\rho_3 V_{x_3} V_{y_3} - \rho_4 V_{x_4} V_{y_4}}{\Delta y} + \frac{\rho_5 V_{x_5} V_{z_5} - \rho_6 V_{x_6} V_{y_6}}{\Delta z}\right)$$
$$+ \rho g \cos\theta + \frac{\sigma_{xx_1} - \sigma_{xx_2}}{\Delta x} + \frac{\tau_{xy_4} - \tau_{xy_3}}{\Delta y} + \frac{\tau_{xz_5} - \tau_{xz_5}}{\Delta z} \tag{7.75}$$

Now we let Δx, Δy, and Δz simultaneously approach zero and take the limit, which makes the terms on the right become partial derivatives,

$$\frac{\partial}{\partial t}(\rho V_x)$$
$$= -\left[\frac{\partial(\rho V_x^2)}{\partial x} + \frac{\partial(\rho V_x V_y)}{\partial y} + \frac{\partial(\rho V_x V_z)}{\partial z}\right] + \rho g \cos\theta + \frac{-\partial\sigma_{xx}}{\partial x} + \frac{\partial\tau_{xy}}{\partial y} + \frac{\partial\tau_{xz}}{\partial z} \tag{7.76}$$

This may be simplified by using the mass balance for a three-dimensional flow (Eq. 3.36) to eliminate several terms. First we expand the left side of the equation and the first term in brackets on the right, choosing our terms on the right in a special way:

$$\rho\frac{\partial V_x}{\partial t} + \underline{V_x \frac{\partial \rho}{\partial t}} = -\left[\underline{V_x \frac{\partial(\rho V_x)}{\partial x}} + \rho V_x \frac{\partial(V_x)}{\partial x} + \underline{V_x \frac{\partial(\rho V_y)}{\partial y}} + \rho V_y \frac{\partial V_x}{\partial y}\right.$$
$$\left. + \underline{V_x \frac{\partial(\rho V_z)}{\partial z}} + \rho V_z \frac{\partial V_x}{\partial z}\right] + \text{remaining terms} \tag{7.77}$$

[†] In most circumstances the normal force is practically the same as the pressure. But in some circumstances it is not, so we keep the general term. This is discussed further in Sec. 7.9.

The underlined terms in this equation are exactly V_x times the terms in Eq. 3.36, and therefore the underlined term on the left is exactly equal to the underlined terms on the right. Dropping the underlined terms on both sides of this equation and rearranging, we find

$$\rho\left(\frac{\partial V_x}{\partial t} + V_x \frac{\partial V_x}{\partial x} + V_y \frac{\partial V_x}{\partial y} + V_z \frac{\partial V_x}{\partial z}\right) = \rho g \cos \theta + \frac{-\partial \sigma_{xx}}{\partial x} + \frac{\partial \tau_{xy}}{\partial y} + \frac{\partial \tau_{yz}}{\partial z} \tag{7.78}$$

By letting Δx, Δy, and Δz simultaneously go to zero, we have shrunk our system to a point, so that the last equation is the momentum balance for some point in space. As shown here, it applies to any point in any kind of flow which does not include magnetic or electrostatic forces. If the latter are significant, they will enter in forms similar to that of the gravity term.

Equation 7.78, as it stands, is of little practical use, because the right-hand term is written in a form which is not readily evaluated. In Sec. 7.9 we examine one way of evaluating it.

Equations 7.76 and 7.78, which are alternative forms of the x-directed momentum balance, represent two ways of regarding fluid mechanics problems. In Eq. 7.76 the left-hand side is the time rate of change of momentum of the fluid contained in an infinitesimal volume at some fixed point in space. The right-hand terms, in order, are the increase of momentum due to flow of matter into and out of this volume and the net forces on the system due to gravity, normal, and shear forces. This viewpoint, which is the viewpoint of an observer fixed in space, is called the *eulerian viewpoint*.

Equation 7.78, which is Eq. 7.76 rearranged and simplified, has on its left side the time rate of change of the x momentum of an infinitesimal element of fluid, as seen by an observer who is not fixed in space but who rides with the fluid. The entire left side of Eq. 7.78 is often abbreviated DV_x/Dt and is called the *Stokes derivative*, the *substantive derivative*, or the *derivative following the motion*. In this case attention is focused on a specific piece of fluid rather than a region of space. In this approach to the subject, there can be no flow of matter into or out of the piece of matter we are watching. Thus, the left side is the increase in momentum of this piece of matter as it moves along, and the terms on the right are the gravity, normal, and shear forces acting on this fluid particle. This viewpoint is called the *lagrangian viewpoint*. Both the eulerian and the lagrangian viewpoints are used in fluid mechanics, the choice between them depending on the problem at hand.

7.9 THE NAVIER-STOKES EQUATIONS

To use Eq. 7.78, it is normally necessary to replace the normal-stress and shear-stress terms with terms involving measurable properties, such as viscosities and velocities. No one has yet found a way to do this without introducing very severe restrictive assumptions. The set of assumptions commonly used is as follows:

1. The fluid has constant density.
2. The flow is laminar throughout.
3. The fluid is newtonian (Eq. 1.5).
4. The three-dimensional stresses in a flowing, constant-density newtonian fluid have the same form as the three-dimensional stress in a solid body that obeys Hooke's law (i.e., a perfectly elastic, isotropic solid).

If we make these assumption, then it can be shown that

$$\frac{-\partial \sigma_{xx}}{\partial x} + \frac{\partial \tau_{xy}}{\partial y} + \frac{\partial \tau_{xz}}{\partial z} = \frac{-\partial P}{\partial x} + \mu\left(\frac{\partial^2 V_x}{\partial x^2} + \frac{\partial^2 V_x}{\partial y^2} + \frac{\partial^2 V_x}{\partial z^2}\right) \tag{7.79}$$

The derivation of this equation is shown in numerous texts [6, p. 66; 7]. The intuitive meaning of three of the terms on the right is obvious. The fourth is a bit harder to see. The $\partial P/\partial x$ term is the result of pressure force on the infinitesimal cube. The $\mu(\partial^2 V_x/\partial y^2)$ and $\mu(\partial^2 V_x/\partial z^2)$ terms represent the net shear force on the cube due to changes of the x velocity in the y and z directions. One can see how these arise by assuming that the shear forces are independent of each other and by substituting Newton's law of viscosity, Eq. 1.5, in the $\partial \tau_{xy}/\partial y$ and $\partial \tau_{xz}/\partial z$ terms in Eq. 7.78.

The $\mu(\partial^2 V_x/\partial x^2)$ term is more subtle; it arises because, according to the fourth assumption listed above, the normal force is not the same as the pressure. For a fluid at rest, the normal force is the same in all directions and is equal to the pressure. For a moving viscous fluid, it is not the same in all directions, because of the interactions of the various perpendicular shear forces. The pressure is defined as the average of the normal forces in three perpendicular directions; this is the pressure which appears in Eq. 7.79. The $\mu(\partial^2 V_x/\partial x^2)$ term results from this difference between the pressure and the normal force. It may be visualized by considering a piece of taffy being pulled while one end is held fixed: as the taffy stretches in one direction, it contracts in the two perpendicular directions. Thus, although the applied force is in one direction, there are resulting normal forces in the directions perpendicular to the direction of pulling. The normal force is tensile in the direction of pulling and compressive in the perpendicular directions; in any one direction it is not equal to the pressure. When taffy is pulled, with one end fixed, there is an increase in velocity with distance from the fixed end; therefore $(\partial^2 V_x/\partial x^2)$ is positive, leading to a force of the type shown here. In most applications of Eq. 7.79, the latter term drops out, either because of the geometry of the system or because it is assumed negligible compared with the other terms.

Substituting Eq. 7.79 in Eq. 7.78 yields

$$\rho\left(\frac{\partial V_x}{\partial t} + V_x \frac{\partial V_x}{\partial x} + V_y \frac{\partial V_x}{\partial y} + V_z \frac{\partial V_x}{\partial z}\right)$$

$$= \rho g \cos\theta - \frac{\partial P}{\partial x} + \mu\left(\frac{\partial^2 V_x}{\partial x^2} + \frac{\partial^2 V_x}{\partial y^2} + \frac{\partial^2 V_x}{\partial z^2}\right) \tag{7.80}$$

This is the differential momentum balance for the x direction, subject to the list of assumptions given above. Analogous balances can be made up for the y and z directions; the three together make up the *Navier-Stokes equations*.

The three Navier-Stokes equations can be put in very compact form by using the shorthand notation of vector calculus [6, p. 66; 7; 8, p. 80]. Furthermore, it is often convenient to use these equations in polar or spherical coordinates; their transformations to those coordinate systems are shown in many texts [6, p. 66; 8, p. 80]. The corresponding equations for fluids with variable density are also shown in numerous texts [6, p. 66; 7; 8, p. 80]. If we set $\mu = 0$ in the Navier-Stokes equations, thus dropping the rightmost term, we find the *Euler equation* which is often used for three-dimensional flow where viscous effects are negligible.

Example 7.16. To illustrate the application of the Navier-Stokes equations, we investigate the flow between two stationary, infinite, parallel plates a distance h apart (see Fig. 7.19).

Before we can begin with the Navier-Stokes equations, we must make the assumptions listed previously, namely, that we are considering only laminar flow of a constant-density, newtonian fluid. Then, from the geometry of the problem we can make the following assumptions:

1. There is no flow in the z or y directions: $V_z = V_y = 0$.
2. The velocity at any z is not a function of x or y; that is, the fluid appears to flow in sheets which are parallel to the plates. This is equivalent to $\partial V_x / \partial y = \partial V_x / \partial x = 0$.
3. The direction of gravity is perpendicular to the plates, so that $\cos \theta = 0$. Making these simplifications in Eq. 7.80, we find

$$\rho\left(\frac{\partial V_x}{\partial t}\right) = -\frac{\partial P}{\partial x} + \mu \frac{\partial^2 V_x}{\partial z^2} \tag{7.81}$$

This equation can be solved for various kinds of time-dependent flows. In this example we limit ourselves to finding the steady-flow solution, for which the term to the left of the equals sign is zero. For steady-state flow, subject to

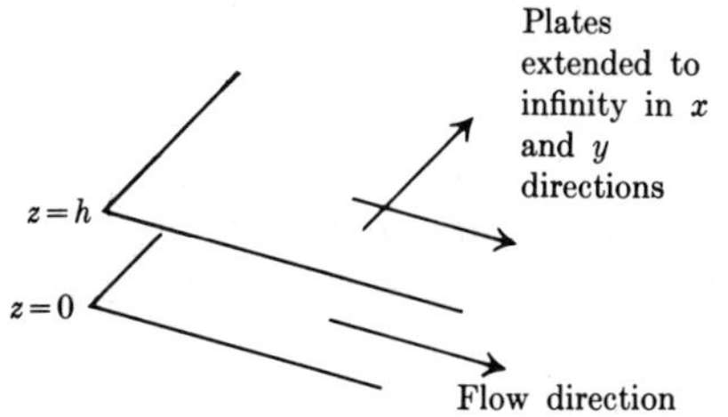

FIGURE 7.19
Flow between two parallel plates.

the assumptions given above, the pressure depends on x alone, and V_x depends on z alone, so we may replace all the partial derivatives with ordinary derivatives. We can further assume that the pressure gradient dP/dx is a constant and then separate the variables and integrate twice:

$$\frac{dP}{dx} = \mu \frac{d}{dz}\left(\frac{dV_x}{dz}\right) \qquad \frac{dP}{dx}\int dz = \mu \int d\,\frac{dV_x}{dz}$$

$$\frac{dP}{dx} z = \mu \frac{dV_x}{dz} + C_1 \qquad \frac{dP}{dx}\int z\,dz = \mu \int dV_x + C_1 \int dz$$

$$\frac{dP}{dx}\frac{z^2}{2} = \mu V_x + C_1 z + C_2$$

where C_1 and C_2 are constants of integration. To solve for the constants of integration we use the boundary conditions that at the surfaces of the plates, where $z = 0$ and $z = h$, the velocity V_x is zero. Substituting these in the last equation above leads to

$$C_2 = 0 \qquad C_1 = \frac{dP}{dx}\frac{h}{2}$$

$$V_x = \frac{1}{2\mu}\left(\frac{dP}{dx}\right)(z^2 - zh) = \frac{-(dP/dx)}{2\mu}(zh - z^2)$$

This gives the velocity distribution. To find the volumetric flow rate for a section of distance l in the y direction, we must integrate:

$$Q = \int V\,dA = \frac{-(dP/dx)l}{2\mu}\int_{z=0}^{z=h}(zh - z^2)\,dz$$

$$= \frac{-(dP/dx)l}{2\mu}\left[\frac{z^2h}{2} - \frac{z^3}{3}\right]_{z=0}^{z=h} = \frac{-(dP/dx)lh^3}{12\mu}$$

which is Eq. 6.29. ■

This example illustrates the advantages and disadvantages of the Navier-Stokes equations. We could have found Eq. 7.81 just as easily from the force balance shown in Sec. 6.3 or from a one-dimensional momentum balance, as we did from the Navier-Stokes equations. However, in more complicated problems in spherical or cylindrical geometry, it is often quite difficult to set up the proper force balance or momentum balance. So it is convenient to start with the Navier-Stokes equations as a list of all the terms and then to drop out terms when necessary to find a solution, as was done here. This approach in several other problems is shown by Bird et al. [8].

There are very few flows for which the Navier-Stokes equations can be solved without the introduction of strong, additional simplifying assumptions. Thus, their great utility is not that they allow us to solve problems which could not be solved some other way but that they require us to enumerate our assumptions in reducing the real problem to one that can be handled mathematically. In that way we obtain a better grasp of the limitations of our solutions. For complex flows, for which analytical solutions are impossible, we can now use large computers to generate numerical solutions. For laminar flow of constant-density newtonian fluids, the Navier-Stokes equations form the normal starting point for the numerical approximations in such solutions.

Finally, the Navier-Stokes equations, as written above, do not apply to turbulent flows or to nonnewtonian fluids. The analogous form for turbulent flow can be derived [6, p. 561], but it is so complex that it is of little, if any, use.

7.10 SUMMARY

1. Momentum is the product of mass and velocity.
2. The momentum balance is simply the restatement of Newton's second law $F = ma$ in a form which is convenient for fluid flow problems.
3. The momentum balance is useful in allowing us to solve some fluid flow problems from the outside, without having to know in detail what goes on inside.
4. For rotating systems it is convenient to introduce an additional defined quantity called the angular momentum. This quantity can also be shown to obey a simple balance equation.
5. The Navier-Stokes equations are the differential momentum balances for a three-dimensional flow, subject to the assumptions that the flow is laminar and of a constant-density newtonian fluid and that the stress deformation behavior of such a fluid is analogous to the stress deformation behavior of a perfectly elastic isotropic solid. These equations are useful in setting up momentum balances for three-dimensional flows, particularly in cylindrical or spherical geometries.

PROBLEMS

See the Common Units and Values for Problems and Examples inside the back cover.

7.1. The earth has a mass of roughly 10^{25} lbm. A person standing on the earth throws a 1-lbm rock vertically upward at a velocity of 20 ft/s. How much does the velocity of the earth increase in the opposite direction? When the rock has fallen back to earth, what is the velocity of the earth compared with its velocity before the rock was thrown?

7.2. A 5-lbm gun fires a 0.05-lbm bullet. The bullet leaves the gun at a velocity of 1500 ft/s in the x direction. If the gun is not restrained, what is its velocity just after the bullet leaves? Work this problem in two ways: (*a*) by taking the gun as the system and (*b*) by taking the combined gun and bullet as the system.

7.3. In Example 7.2, what fractions of the initial kinetic energies of the duck and of the bullet are converted to internal energy?

7.4. A fire hose directs a stream of water against a vertical wall. The flow rate of the water is 50 kg/s, and its incoming flow velocity is 80 m/s. The flow away from the impact point has zero velocity in the x direction. What is the force exerted by this stream on the wall?

7.5. A jet engine of the type used on commercial aircraft has a thrust of 10,000 lbf. All the air enters and leaves at atmospheric pressure. All the air comes into the engine at 800 ft/s and leaves at 1500 ft/s. The mass of fuel added to the exhaust gas may be neglected. How many pounds of air per second flow through the engine?

7.6. In a steady-state methane-air flame at approximately atmospheric pressure, the temperature is raised from 70 to 3200°F. The incoming air-gas mixture and the products of combustion may both be considered ideal gases with a molecular weight of 28 g/mol. The flame is a thin, flat region perpendicular to the gas flow. If the flow comes into the flame at a velocity of 2 ft/s, what is the pressure difference from one side of the flame to the other? This problem and its consequences are discussed elsewhere [9].

7.7. A pipe is delivering a liquid in steady laminar flow, as described in Chap. 6. What is the ratio of the momentum in the outflowing stream to the momentum which would exist if the flow were of uniform velocity over the entire cross section of the pipe?

7.8. The pipe bend in Fig. 7.20 is attached to the rest of the piping system by two flexible hoses which transmit no forces. The fluid enters in the x direction and leaves in the $-y$ direction. The flow rate is 500 lbm/s, and the inlet and outlet velocities are each 100 ft/s. The cross-sectional areas at points 1 and 2 are both 0.5 ft^2. The pressure at point 1 is 50 psig and at point 2 is 40 psig. Calculate F_x and F_y, the x and y components, respectively, of the force in the pipe support.

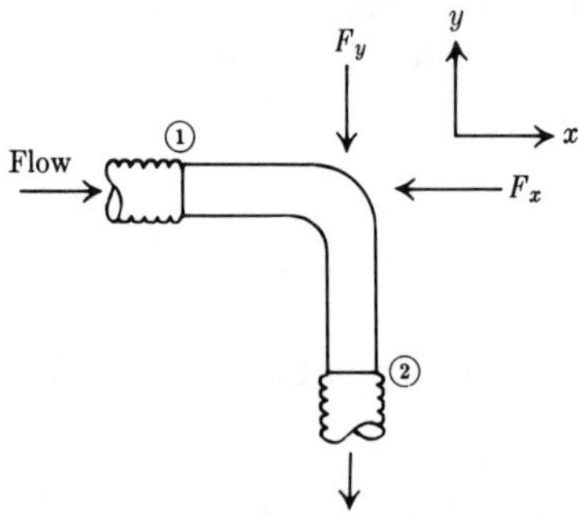

FIGURE 7.20

7.9. A new type of elevator is sketched in Fig. 7.21. The stream of water from a geyser will be regulated to hold the elevator at whatever height is required. Assuming that the maximum flow of the jet is 500 lbm/s at a velocity of 200 ft/s, what is the relation between the weight of the elevator and the maximum height to which the jet can lift it?

7.10. The pipe U-bend in Fig. 7.22 is connected to a flow system by flexible hoses which transmit no force. The pipe has an inside diameter of 3 in. Water is flowing through the pipe at a rate of 600 gal/min. The pressure at point 1 is 5 psig and at point 2 is 3 psig. What is the vertical component of the force in the support? Neglect the weight of the pipe and fluid.

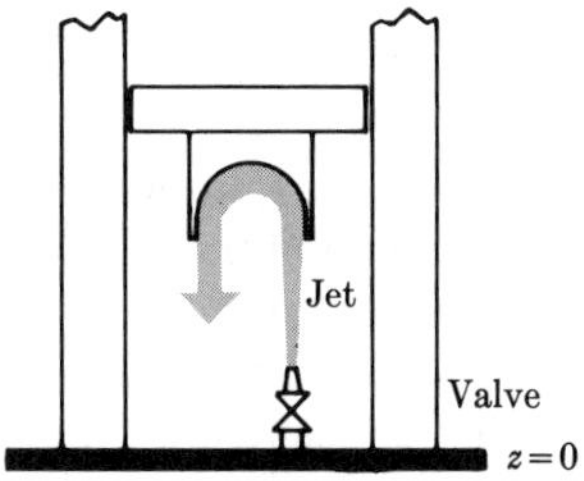

FIGURE 7.21

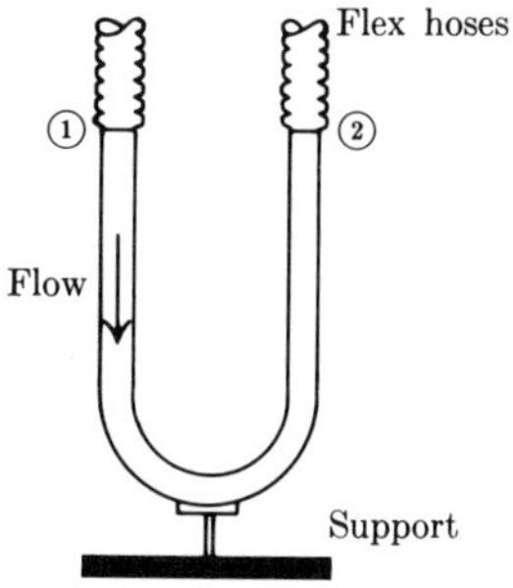

FIGURE 7.22

7.11. The U-bend shown in Fig. 7.23 is connected to the rest of the piping system by flexible hoses. The inside diameter of the pipe is 3 in. The fluid flowing is water, and its average velocity in the pipe is 50 ft/s. The gauge pressure at point 1 is 30 psig and at point 2 is 20 psig. What is the horizontal component of the force in the support?

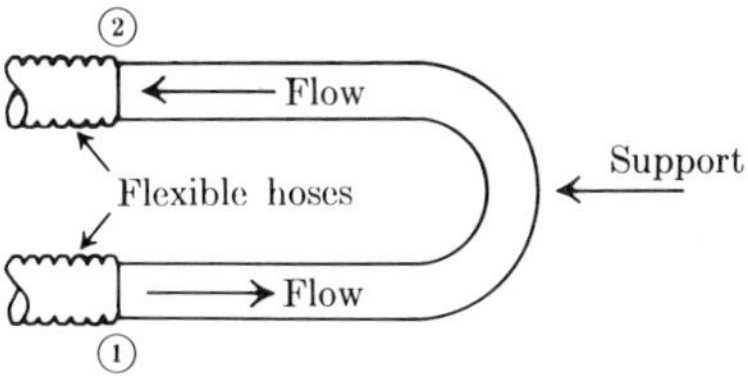

FIGURE 7.23

7.12. The short length of pipe in Fig. 7.24 is used on a sprayer. The pipe and hose both have cross-sectional areas of 1 in^2. The flow velocity is 200 ft/s. The pressure as measured at the flanged joint is 30 psig.

(*a*) What is the force tending to tear apart the flange?

(*b*) How is this force transmitted by the fluid to the pipe?

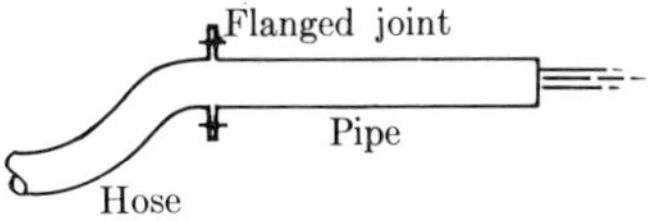

FIGURE 7.24

7.13. A nozzle is bolted onto a pipe by the flanged joint shown in Fig. 7.25. The cross-sectional area perpendicular to the flow at point 1 is 12 in^2 and at point 2 is 3 in^2. At point 2 the flow is open to the atmosphere. The pressure at point 1, measured by a pressure gauge, is found to be 40 psig. The flow rate is 1200 in^3/s. What is the force tending to tear the nozzle off the pipe?

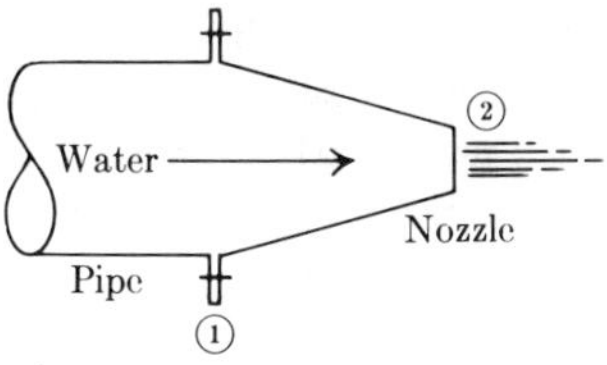

FIGURE 7.25

7.14. A sailboat is moving in the y direction. The wind approaches the boat at an angle of 45° to the y direction and is turned by the sails such that it leaves in exactly the $-y$ direction. Assuming that the average velocity of the incoming and outgoing wind is 10 m/s and that the mass flow rate of air being turned by the boat's sails is 200 kg/s, what are the x and y components of the force exerted by the boat's sails on the air? These are the opposite of the forces exerted by the air on the boat. The y component of the wind force drives the boat in its direction of travel. What does the x component do?

7.15. A pump together with the electric motor which drives it is mounted on a wheeled cart, so that it can be rolled about to various places in the plant for various pumping tasks. It is connected to the vessels to be pumped by flexible hoses that transmit no forces and is connected electrically by a flexible cord that transmits no forces.

The pump inlet and outlet are both parallel to each other and parallel to the x axis. The inlet pipe diameter is 4 in, and the outlet pipe diameter is 3 in. The flow rate through the pump is 230 gal/min of water. The pressure at the inlet is 10 psig, and the pressure at the outlet is 50 psig.

(*a*) How much force must we exert on the pump-motor-cart assembly to keep it from moving?

(*b*) In which direction must we exert this force?

7.16. A large rocket engine ejects 200 kg/s of exhaust gases at a velocity of 4000 m/s. The pressure of the exhaust gas is equal to the atmospheric pressure. What thrust does the engine produce?

7.17. Calculate I_{sp} for the rocket in Example 7.8. Why is this different from the result in Example 7.9?

7.18. The compressed-air-driven water rocket shown in Fig. 7.26 is ejecting water vertically downward through a frictionless nozzle. The exit area of the nozzle is 1 in^2. When the pressure and elevation are as shown, how much thrust does the rocket produce?

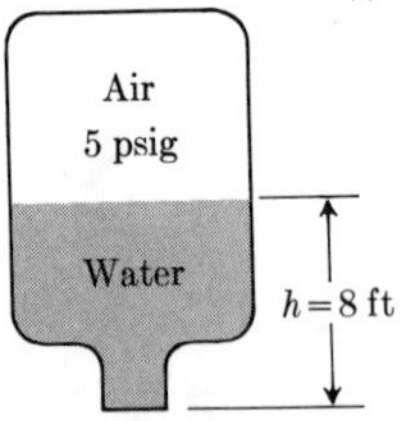

FIGURE 7.26

7.19. A typical high-pressure oxygen cylinder of the type commonly found in welding shops and laboratories falls over. The valve at the top is broken off in the fall, making a hole with a cross-sectional area of 1 in^2. At the time of the accident, the cylinder is full, so that the internal pressure is 2000 psia. The internal temperature is 70°F. The flow through the nozzle cannot be described by Bernoulli's equation, because it is a high-velocity gas flow. By using the methods of high-velocity gas flow (to be developed in Chap. 8) one may estimate that the outlet velocity is 975 ft/s, that the outlet density is 7.0 lbm/ft^2, and that the pressure in the plane of the outlet is 1060 psia.

(*a*) How much thrust does the oxygen cylinder exert?

(*b*) Is it a worthwhile safety practice to fasten these cylinders so they cannot fall over?

7.20. A typical garden hose has an inside diameter of $\frac{3}{4}$ in. The water stream flowing from it has a velocity of 10 ft/s.

(*a*) If such a hose is left loose, will the end move about?

(*b*) Would the hose move about if it were perfectly straight, or must it be curved?

(*c*) What is the maximum plausible value of the force involved in any such motion?

7.21. The 3-ft-diameter, horizontal main cooling line from a nuclear reactor breaks. The pressure inside the reactor is 1000 psia, and the water surface inside the reactor is 20 ft above the broken line. The exiting fluid (a steam-water mixture) has a density of 50 lbm/ft^3. Estimate the horizontal force on the pipe-reactor system due to the flow through the broken pipe. Assume frictionless flow.

7.22. The rocket motor sketched in Fig. 7.27 has the nozzle bolted to the combustion chamber. From the data below calculate (*a*) the thrust of the rocket motor and (*b*) the force (compressive or tensile) at the joint between the nozzle and combustion chamber. The flow rate is 200 lbm/s.

	Cross-sectional area, ft^2	Pressure, psia	Velocity, ft/s
Point 1	5	300	300
Point 2	30	14.7	3000

Note: The fuel is stored in the combustion chamber.

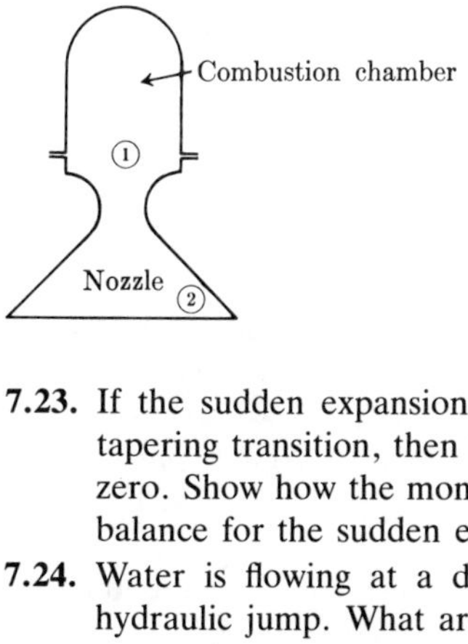

FIGURE 7.27

7.23. If the sudden expansion in Fig. 7.10 were replaced with a gradually outward-tapering transition, then the friction losses would be very small, often practically zero. Show how the momentum balance of that flow differs from the momentum balance for the sudden expansion.

7.24. Water is flowing at a depth of 2 ft and a velocity of 50 ft/s. It undergoes a hydraulic jump. What are the depth and velocity after the jump?

7.25. Water is flowing at a depth of z_1. What is the minimum velocity at which this water could undergo a hydraulic jump? Why?

7.26. Show that for a hydraulic jump, $\mathcal{F}/g = (z_2 - z_1)^3/4z_1z_2$.

7.27. Water is flowing in a horizontal gutter with velocity 10 ft/s and depth 0.1 ft. We now place a large brick in the gutter, which stops the flow. The flow is stopped by a hydraulic jump, which then moves upstream from the brick. The brick is large enough that there is no flow over it. The fluid between the brick and the jump has zero velocity. The jump moves upstream, with a velocity V_j. What is the numerical value of V_j?

This is messy problem analytically. It is fairly easy if you take the viewpoint of someone riding on the jump (the lagrangian viewpoint) and solve by trial and error for the jump velocity that satisfies the hydraulic jump equation in the moving frame of reference.

7.28. Water flows through a hydraulic jump, entering at a velocity of 50 ft/s and a depth of 10 ft. How much does the temperature of the water increase in this jump? For water $C_V = du/dt = 1.0$ Btu/(lbm · °F).

7.29. Hydraulic jump in a radial geometry is easily demonstrated in any sink, as shown in Fig. 7.28. Derive an equation analogous to Eq. C.5 for this radial hydraulic jump.

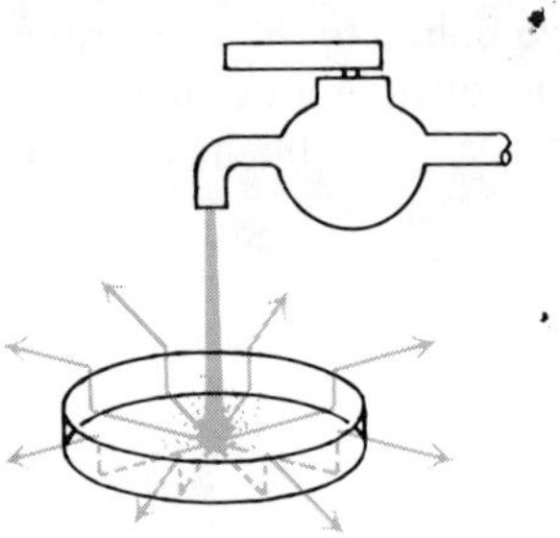

FIGURE 7.28

7.30. Figure 7.28 shows a hydraulic jump, in radial geometry, easily demonstrated in a sink. If the sink drain is open, then the flow is steady and the depths and velocities in the sink are not changing with time. If we now close the drain, without changing the flow from the faucet, what will happen? Describe the situation in terms of the mathematical description of hydraulic jumps.

7.31. In Example 7.10 the pipe was long enough that we ignored the kinetic energy in the fluid leaving the downstream end of the pipe.

(*a*) Rework Example 7.10, taking that kinetic energy into account, and show that the change is negligible.

(*b*) Rework Example 7.10 for a 1-ft-long pipe for which the term involving f will be negligible compared to the term involving the kinetic energy.

7.32. If a fluid were absolutely incompressible (which no materials known to humans are), then the speed of sound in that fluid would be infinite. What happens to the pressure rise in Example 7.11 as we replace the water with fluids which are less and less compressible?

7.33. Repeat Example 7.11 when the flowing liquid is propane, for which (at 70°F) the density is 31.1 lbm/ft^3 and the speed of sound is 2150 ft/s.

7.34. Example 7.11 makes clear that rapid valve closure can cause very high pressures at the valve. That raises the obvious question of how slowly one must close a valve to avoid water hammer. The approximate answer is that if the time to close is longer than the time for a sound wave to make a round trip from the valve to the reservoir, then no significant water hammer will occur. Again, see Parmakian [1] for more details.

(*a*) In Example 7.11, how long is this?

(*b*) Why is the time required that for a round trip, rather than the time for a one-way trip? *Hint:* For instantaneous closure, all the fluid is brought to rest in time $t = L/c$. When it has all been brought to rest, it is at a pressure much higher than the pressure in the reservoir. What will happen then?

7.35. Problem 7.34 shows that the time required for a valve to close without causing water hammer is linearly proportional to the length of the pipe, while Example 7.11 shows that the expected pressure rise is independent of the length of the pipe. Why?

7.36. When a rocket is moving in the $+x$ direction with velocity V_1 and this velocity is equal and opposite to the velocity of the exhaust gas relative to the rocket, then the exhaust velocity relative to fixed surroundings is zero. Thus, according to Eq. 7.44, $d(mV)/dt = 0$. Does this mean that the rocket is not accelerating? Explain.

7.37. Show the equivalent of Eq. 7.47 for vertical flight with a constant value of the acceleration of gravity and zero air resistance.

7.38. The rocket in Example 7.12 is now fired vertically. The specific impulse and mass ratio are the same as in that example. The rocket consumes all its fuel in 1 min. Calculate its velocity at burnout, taking gravity into account.

7.39. A rocket starts from rest on the ground and fires vertically upward. During the entire upward firing, the velocity of the exhaust gas, measured relative to the rocket, is 4000 m/s. The pressure in the exit plane of the rocket nozzle is always exactly equal to the surrounding atmospheric pressure. The mass of the rocket and fuel before launching is 100,000 kg. The mass of the burned-out rocket is 20,000 kg. The entire burning process takes 50 s. What is the velocity of the rocket at burnout? Ignore air resistance.

7.40. A cylindrical tank, shown in Fig. 7.29, is sitting on a platform with absolutely frictionless wheels on a horizontal plane. There is no air resistance. At time 0, the level in the tank is 10 ft above the outlet, and the whole system is not moving. Then the outlet is opened, and the system is allowed to accelerate to the left. The flow through the outlet nozzle is frictionless. What is the final velocity, assuming that (*a*) the mass of the tank and cart is zero and (*b*) the mass of the tank and cart is 3000 lbm?

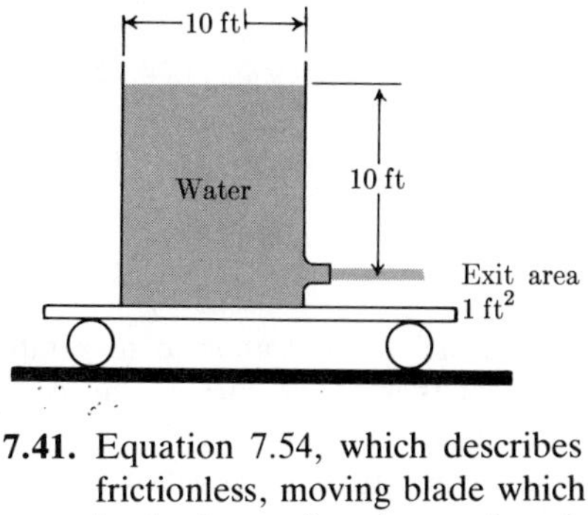

FIGURE 7.29

7.41. Equation 7.54, which describes the work per unit mass to be obtained from a frictionless, moving blade which changes the velocity of a jet of fluid, can be put in the form of an instructive plot by dividing both sides by $V_{\text{in}}^2/2$, to find

$$\frac{dW/dm}{V_{\text{in}}^2/2} = 4\left(1 - \frac{V_b}{V_{\text{in}}}\right)\left(\frac{V_b}{V_{\text{in}}}\right)$$

and then plotting the left side of this equation against V_b/V_{in}.

(*a*) Prepare such a plot, covering the V_b/V_{in} range from −0.5 to 1.5.
(*b*) Explain the meaning of the maximum in the curve, the zero values, and the negative values. Explain the physical significance of the left side of this equation.

7.42. For the turbine blade and fluid jet system shown in Fig. 7.14,
(*a*) Which blade velocities cause the work to be zero?
(*b*) Explain physically why these velocities lead to zero work.
(*c*) For which velocities is the work negative? Why?

7.43. The cart in Fig. 7.30 has a mass of 2000 kg. It is resting on frictionless wheels on a solid, level surface and encounters no air resistance. At time 0 it is standing still, and a jet from a fire hose is used to start it moving. The mass flow rate of the fluid from the fire hose is 100 kg/s, and its velocity relative to fixed coordinates is 50 m/s. The cup on the rear of the cart turns the jet around so that it leaves in the $-x$ direction with the same velocity relative to the cart with which it entered. Calculate the velocity-time behavior of the cart. (Assume the jet is unaffected by gravity.)

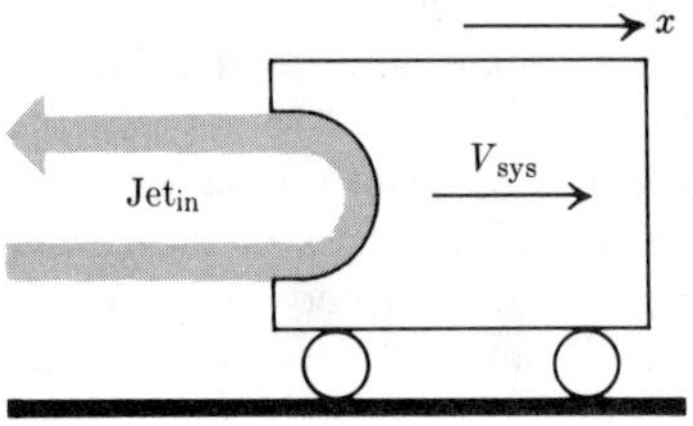

FIGURE 7.30

7.44. All other things being equal, is it easier for an airplane to take off on a short runway on a hot day or on a cold day? Why?

7.45. An ordinary garden hose sprinkler is sketched in Fig. 7.31. All the fluid enters at the axis ($r = 0$) and leaves through the nozzles ($r = 6$ in). If the total flow rate is 5 gal/min and the rotor is held in place by someone's hand, how much torque will the rotor exert? Each of the jets leaving the sprinkler has a diameter of 0.25 in.

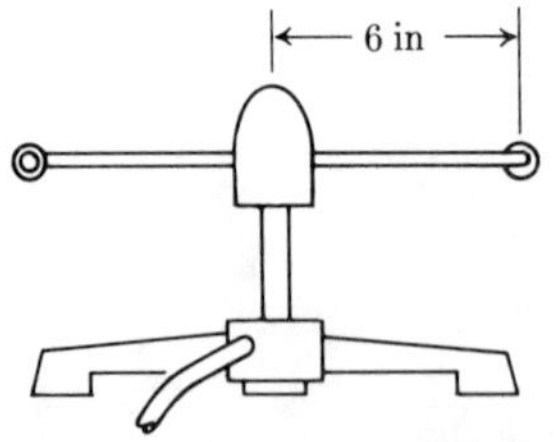

FIGURE 7.31

7.46. If the garden hose sprinkler in Prob. 7.45 is turned on and allowed to rotate freely, at what speed will it rotate? Assume that there is no air resistance or friction in the bearing of the sprinkler.

7.47. If we put the sprinkler in Prob. 7.45 under water and turn it on, will it rotate? Which way?

7.48. Why do helicopters have either one main propeller and a small tail propeller or two main propellers rotating in opposite directions? How is the corresponding problem solved in propeller-driven airplanes?

7.49. The left-hand term of Eq. 7.78 is often called the Stokes derivative or the derivative following the motion. Show that if $V_x = V_x(x, y, z, t)$, then

$$\frac{dV_x}{dt} = \frac{\partial V_x}{\partial t} + V_x \frac{\partial V_x}{\partial x} + V_y \frac{\partial V_x}{\partial y} + V_z \frac{\partial V_x}{\partial z}$$

i.e., that the Stokes derivative is indeed the derivative of the velocity for a small element of fluid.

7.50. Repeat Example 7.16 for the case in which one plate at $z = h$ is moving steadily in the x direction with velocity A. This type of flow with the plate moving is called *Couette flow*; the solution is shown in graphical form in Schlichting [6, p. 85].

7.51. A constant-density newtonian fluid is flowing as a thin film down a vertical wall in laminar flow; see Fig. 7.32. Find the velocity distribution and the volumetric flow rate per unit width of wall by using the Navier-Stokes equations (z component) on the assumptions that there is no flow in the x or y directions, that the z component of the velocity is zero at the solid wall, that there is no shear stress at the liquid-air surface, and that the flow is steady-state. (Waves may appear on the fluid surface in this situation; ignore that possibility for this problem.)

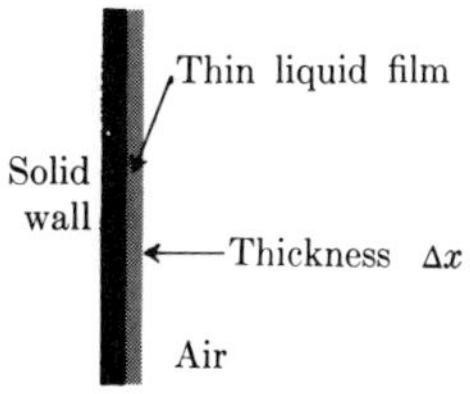

FIGURE 7.32

7.52. In Prob. 7.51 we are now blowing air upward next to the fluid film, at a high enough velocity that the assumption of no shear stress between the liquid and the air is no longer a good one. Instead we now assume that there is a shear stress at the gas-liquid interface in the upward direction as seen by the liquid with magnitude A (where A has dimensions of force divided by area). Repeat Prob. 7.51 for this modified circumstance.

7.53. The cylindrical coordinate form of the Navier-Stokes equations is shown in many textbooks, e.g., Bird et al. [8]. Starting with that form, derive the Poiseuille equation.

7.54. If one sets $\mu = 0$ in the Navier-Stokes equations (Eq. 7.80), then these become Euler's equations of motion. Show that if one considers steady flow in one direction only (say, the x direction), Euler's equation of motion reduces to the classical form of Bernoulli's equation (i.e., Bernoulli's equation with the $dW_{\text{a.o.}}/dm$ and $\mathcal{F}$ terms set equal to zero).

7.55. For nonnewtonian fluids, the viscosity is not a constant, but depends on the shear rate (see Chap. 15). For many such fluids, the relation between shear stress and shear rate can be satisfactorily represented by the "power law" $\tau = \mu(dV/dy)^n$, where n is an empirical constant. Solve the problem posed in Example 7.16 for a power-law fluid. Is this problem best approached by modifying the Navier-Stokes equations, which apply only to newtonian fluids, or by making a force balance, as was done in Sec. 6.3? Could you work the problem starting with Eq. 7.78 and taking $\sigma_{xx} = P$?

7.56. A tank of fluid has a long, narrow rectangular slot in its bottom. A fluid flows steadily out of the slot, in the form of a sheet. From the slot the sheet of fluid falls through the air, eventually falling on a solid surface. This is the arrangement used to put coatings on various products, which move on a conveyer belt under the falling sheet of fluid.

For this situation, write out the z component (i.e., vertical component) of the Navier-Stokes equation. Indicate which terms are zero or negligibly small. Indicate what additional information would be needed to solve for the velocity as a function of z and x.

7.57. A constant-density newtonian fluid of infinite extent is adjacent to a solid wall. At time 0 the wall is suddenly set in motion, with velocity V_0. (This is roughly what would happen in the water near the side of a speedboat that started from rest at full throttle.) Write the differential equation for V_x as a function of y and t for this problem by starting with Eq. 7.80 and canceling the terms which will be zero. List the boundary conditions. The solution of the resulting differential equation is shown in Bird et al. [8, p. 125].

CHAPTER 8

ONE-DIMENSIONAL HIGH-VELOCITY GAS FLOW

In this chapter we apply the ideas of the preceding chapters to the flow of gases at high velocities. No further principles are introduced in this chapter; those previously introduced will meet all our needs. Nevertheless, this is a separate chapter because in high-velocity gas flow several phenomena occur which either are not present at all or are present in negligible amounts in the flow of liquids at their ordinary velocities and in the flow of gases at low velocities. For working purposes we define *high velocity* as a velocity in excess of about 200 ft/s (61 m/s).

The principal differences between high-velocity gas flows and the flows we have studied so far are the following:

1. In an expanding high-velocity gas flow, the gas can convert significant amounts of internal energy to kinetic energy. This results in large decreases in gas temperature and in velocities much higher than would be predicted by Bernoulli's equation, which has a constant-density assumption.
2. The changes in density which accompany high-velocity gas flows will complicate the mathematics. In typical situations we have one more unknown and one more equation to deal with than in the corresponding constant-density flow. As the student sees that the equations in this chapter are longer and more complex than those in the preceding chapters, the student should remember that this is the reason for the added complexity.

3. The velocities in high-velocity gas flows are frequently equal to or greater than the local speed of sound. Information about small disturbances in fluid flow propagates through fluids at the speed of sound, so when the fluid is moving at the speed of sound or faster, certain kinds of information will be unable to travel upstream against the flow. This leads to some special phenomena in high-velocity gas flow, the most important of which are choking and shock waves. Choking is very common in chemical engineering practice.

8.1 THE SPEED OF SOUND

In high-velocity gas flow, velocities are often reached that are comparable to the speed of sound, so the speed of sound plays an important part in what follows. The speed of sound is the speed at which a *small* pressure disturbance moves through a continuous medium. Sound, as our ears perceive it, is a series of small air-pressure disturbances oscillating in a sinusoidal fashion in the frequency range from 20 to 20,000 cycles per second, or hertz (Hz). The magnitude of the pressure disturbances is generally less than 10^{-3} lbf/in^2 absolute (7 Pa) [1].

Suppose that we have a bar of steel 1 mi long. We tap the steel sharply on one end; our tap causes the near end of the bar to move 0.001 in. If the steel was *absolutely incompressible*, the far end of the bar would also move 0.001 in *instantly*. It does not; it moves about one-third of a second after we tap the near end. Nothing in this world is absolutely incompressible.

Consider a pipe full of some fluid, with pistons at each end. We tap one of the pistons. This causes the pressure adjacent to the piston to rise. This moves the next layer of fluid, whose pressure rises, and so on, causing a small pressure pulse to pass down the pipe. This is shown schematically in Fig. 8.1. It is easier to analyze this problem if we ride along with the pressure pulse. We appear to be standing still, and the walls of the pipe seem to be rushing past us. The fluid in the pipe also is rushing toward us and rushing away behind us. We measure the velocity, pressure, and density of the fluid ahead of us and behind us; the values of those ahead are slightly different from the values of those behind. This situation is represented in Fig. 8.2.

The pressure pulse is assumed to have the small volume shown in the figure. The mass flow into it is the same as the mass flow out, so we can apply the steady-flow mass balance equation

$$\rho AV = (\rho + d\rho)A(V + dV) \tag{8.1}$$

Dividing by A, we expand the right-hand side and cancel the ρV term to get

$$0 = V\,d\rho + \rho\,dV + d\rho\,dV \tag{8.2}$$

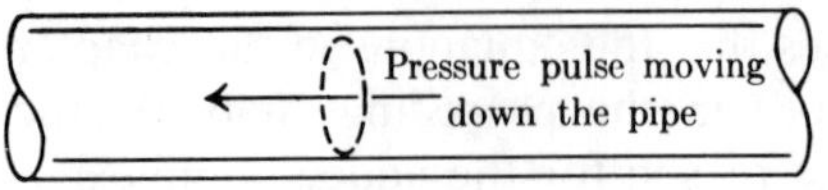

FIGURE 8.1

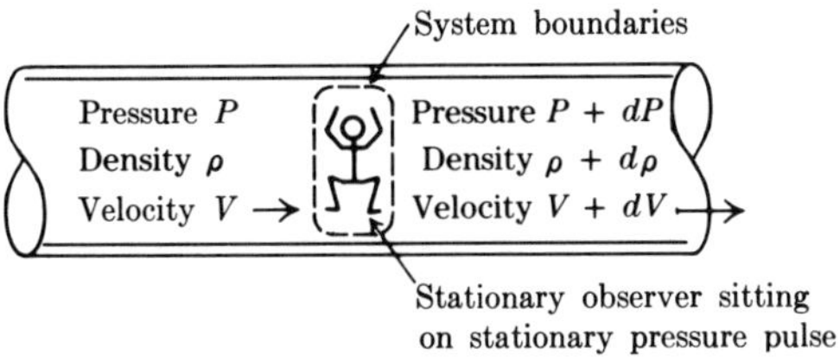

FIGURE 8.2

The far right-hand term here is the product of two differentials and may be ignored, so that

$$0 = V\,d\rho + \rho\,dV \tag{8.3}$$

Applying the momentum balance (Eq. 7.13) to the system shown in Fig. 8.2, we see that there is no accumulation, that there is a steady flow in and out, and that the only forces acting are the pressures on either side; so

$$0 = \dot{m}[V - (V + dV)] + A[P - (P + dP)] \tag{8.4}$$

Replacing $\dot{m}$ with ρAV and simplifying, we find

$$\rho\,dV = \frac{-dP}{V} \tag{8.5}$$

Substituting this value of $\rho\,dV$ in Eq. 8.3 and solving for V produces

$$V = \left(\frac{dP}{d\rho}\right)^{1/2} \tag{8.6}$$

This equation tells us how fast the fluid flows toward the observer riding on the small pressure pulse, which is the same as the velocity at which the pressure pulse moves past a stationary observer. Note that in deriving this equation we have never relied on our initial assumption that the material through which the pressure wave is passing is a fluid; the derivation is equally valid for a liquid, a solid, and a gas.

The equation was worked out for a single step change in pressure. Sound, as we experience it, is a sinusoidally varying pressure wave. However, it may be thought of as a combination of small step changes in pressure following each other. Thus, this equation, which we found for a single step change, is applicable to any shape of pressure change, such as a sound wave.

Equation 8.6, as it stands, is ambiguous, because the derivative $dP/d\rho$ is ambiguous. The pressure P generally is not only a function of ρ but also depends on the temperature. Newton [2] derived its equivalent and assumed that the derivative referred to a constant-temperature process, i.e., it was $(\partial P/\partial\rho)_T$. On this assumption he calculated the speed of sound in air and obtained an answer which was about 80 percent of the experimentally measured velocity. His assumption was plausible—we do not see the air being heated by sound passing through it—but it was wrong. Later workers decided that what really happens is that a layer of gas is heated by being compressed and then cooled by expanding against the adjacent layer of gas. The net result is that the gas undergoes what is practically a reversible, adiabatic compression-expansion. For it to be adiabatic, it must occur so fast that there is little opportunity for the warmer gas in the middle of the pressure wave to transfer heat to the cooler surrounding gas. The success of this assumption in predicting the measured velocities of sound waves leads us to believe that the process is indeed this fast. The temperature rise in a sound wave is very small (see Prob. 8.2). In a reversible adiabatic process, the temperature is not constant, but the entropy is. Therefore, the final equation for the speed of sound is

$$V = \left(\frac{\partial P}{\partial \rho}\right)_S^{1/2} \tag{8.7a}$$

The speed of sound is a special quantity, which we want to keep separate from the local velocity of fluid flow. Therefore, we introduce the symbol c for it. This makes Eq. 8.7*a*

$$c = \left(\frac{\partial P}{\partial \rho}\right)_S^{1/2} \tag{8.7b}$$

This equation is correct for solids, liquids, and gases. For solids and liquids the easily measured $(\partial P/\partial \rho)_T$ is practically the same as the $(\partial P/\partial \rho)_S$ here [3]; so we may write

$$c = \left(\frac{\partial P}{\partial \rho}\right)_S^{1/2} \approx \left(\frac{\partial P}{\partial \rho}\right)_T^{1/2} \qquad \text{[solids and liquids]} \tag{8.8}$$

with satisfactory accuracy. Handbooks [4] usually tabulate, not $(\partial P/\partial \rho)_T$ of solids and liquids, but the bulk modulus K

$$K = \rho\left(\frac{\partial P}{\partial \rho}\right)_T \tag{8.9}$$

or the isothermal compressibility, which is the reciprocal of the bulk modulus. In terms of the bulk modulus

$$c \approx \left(\frac{\partial P}{\partial \rho}\right)_T^{1/2} = \left(\frac{K}{\rho}\right)^{1/2} \qquad \text{[solids and liquids]} \tag{8.10}$$

Example 8.1. Calculate the speed of sound in steel and in water at 20°C.

For steel at that temperature $K = 1.94 \times 10^{11}$ Pa and $\rho = 7800 \text{ kg/m}^3$.

$$c = \left(\frac{1.94 \times 10^{11} \text{ Pa}}{7800 \text{ kg/m}^3} \cdot \frac{\text{N/m}^2}{\text{Pa}} \cdot \frac{\text{kg}\cdot\text{m}}{\text{N}\cdot\text{s}^2}\right)^{1/2}$$

$$= 4.99 \text{ km/s} = 16.4 \times 10^3 \text{ ft/s}$$

For water at that temperature $K = 3.14 \times 10^5 \text{ lbf/in}^2$ and $\rho = 62.3 \text{ lbm/ft}^3$,

$$c = \left(\frac{3.14 \times 10^5 \text{ lbf/in}^2}{62.3 \text{ lbm/ft}^2} \cdot \frac{144 \text{ in}^2}{\text{ft}^2} \cdot 32.2 \frac{\text{lbm} \cdot \text{ft}}{\text{lbf} \cdot \text{s}^2} \right)^{1/2}$$

$$= 4.83 \times 10^3 \text{ ft/s} = 1.6 \text{ km/s}$$ ■

For real gases $(\partial P/\partial \rho)_S$ is a complicated function of pressure and temperature. However, the perfect-gas law is a reasonable approximation of the behavior of most gases at low pressures and of low-boiling gases, such as air, up to reasonably high pressures. For a perfect gas it is shown in App. D that

$$\left(\frac{\partial P}{\partial \rho} \right)_S = \frac{kP}{\rho} \tag{D.26}$$

Here k is the ratio of specific heats C_P/C_V as defined in App. D (this ratio is called γ in many texts). It is dimensionless; its values are given in Table 8.1.

In most engineering calculations, we consider k a constant in a given problem even though for most gases it decreases slightly with increasing temperature. If we substitute for P/ρ in Eq. D.26 from the perfect-gas law and insert the result in Eq. 8.7b, we find

$$c = \left(\frac{\partial P}{\partial \rho} \right)_S^{1/2} = \left(\frac{kP}{\rho} \right)^{1/2} = \left(\frac{kRT}{M} \right)^{1/2} \quad \text{[perfect gas]} \tag{8.11}$$

where M is the molecular weight and R is the universal gas constant.

This equation shows why Newton's calculated value of the speed of sound was only 80 percent of the observed value. If we substitute $(\partial P/\partial \rho)_T$ for $(\partial P/\partial \rho)_S$ in this equation, we find the same result, except that k is replaced with 1. Since k for air is 1.4, this incorrect substitution lowers the calculated value to 80 percent of the value shown in this equation.

Example 8.2. What is the speed of sound in air at 68°F?

We will need $R^{1/2}$. In speed-of-sound calculations, it is convenient to convert the most commonly used form of R as shown:

TABLE 8.1
Values of the ratio of specific heats

Gas	k	Comment
Monatonic gases: He, Ar, Ne, Kr, etc.	1.6667	Exactly
Diatomic gases: N_2, O_2, H_2, CO, NO, air	1.40	Not quite as exact and somewhat temperature-dependent
Triatomic gases: H_2O, CO_2, etc.	1.30–1.33	
More complex gases	1.3 or less	

$$R^{1/2} = \left(10.73 \frac{\text{lbf}}{\text{in}^2} \frac{\text{ft}^3}{\text{lbmol} \cdot {}^\circ\text{R}} \cdot \frac{144 \text{ in}^2}{\text{ft}^2} \cdot \frac{32.2 \text{ lbm} \cdot \text{ft}}{\text{lbf} \cdot \text{s}^2}\right)^{1/2}$$

$$= 223 \frac{\text{ft}}{\text{s}} \cdot \left(\frac{\text{lbm}}{\text{lbmol} \cdot {}^\circ\text{R}}\right)^{1/2} = 91.2 \frac{\text{m}}{\text{s}} \cdot \left(\frac{\text{g}}{\text{mol} \cdot \text{K}}\right)^{1/2} \tag{8.12}$$

From this we have

$$c = \left(\frac{kRT}{M}\right)^{1/2} = R^{1/2}\left(\frac{kT}{M}\right)^{1/2} = 223 \frac{\text{ft}}{\text{s}} \cdot \left(\frac{\text{lbm}}{\text{lbmol} \cdot {}^\circ\text{R}}\right)^{1/2} \cdot \left(\frac{1.4 \cdot 528^\circ\text{R}}{29 \text{ lbm/lbmol}}\right)^{1/2}$$

$$= 1126 \text{ ft/s} = 344 \text{ m/s}$$ ■

The speed of sound of a perfect gas, as shown above, is a function of the temperature and not of the velocity. Keep in mind that the speed of sound is a property of the matter, not a property of the flow. If the temperature changes as the fluid flows, then the speed of sound will change from point to point; but at any point it is the same in a flowing gas as it would be in the same gas standing still at the same temperature. This argument applies equally well to solids and to liquids. Observe also that the magnitudes are $c \approx 3$ mi/s for steel, 1 mi/s for water, and $\frac{1}{5}$ mi/s for air.

Our discussion of the speed of sound has the built-in assumption that a sound wave is a *small* pressure pulse. The difference in velocity and pressure of the medium before and after the sound wave is very small. Another type of pressure pulse, in which these differences are large, called a shock wave, is discussed in Sec. 8.5.

8.2 STEADY, FRICTIONLESS, ADIABATIC, ONE-DIMENSIONAL FLOW OF A PERFECT GAS

Many of the most interesting features of high-velocity gas flow can be seen in the simplest of all cases, the steady, frictionless, adiabatic, one-dimensional flow of a perfect gas. We study this type of flow in detail; other types are treated more briefly, because they have so much in common with this one.

The flow is assumed to be in some kind of a duct or pipe or closed channel of varying cross-sectional area A. As long as this cross-sectional area changes slowly with distance down the duct, the velocities in the direction perpendicular to the main flow will be small enough to neglect and we can treat the flow as one-dimensional. The gas is assumed to be perfect and to have a constant heat capacity C_P.

The open-system energy balance (Chap. 4) between any two points R and 1 in such a duct, for steady flow without heat transfer or turbines or compressors, is

$$\left(h + gz + \frac{V^2}{2}\right)_R = \left(h + gz + \frac{V^2}{2}\right)_1 \tag{8.13}$$

It can be readily shown (Prob. 8.8) that the potential-energy changes Δgz are negligible for most high-velocity gas flows, so we drop the gz terms from this equation. Next we assume that state R is some upstream reservoir, where the cross-sectional area perpendicular to the flow is very large; therefore, V_R is negligible. This condition is referred to in various texts as the *reservoir*, *stagnation*, or *total* condition. We call it the reservoir condition and use the subscript R.

Substituting $V_R = 0$ in Eq. 8.13, we find

$$V_1^2 = 2(h_R - h_1) = 2C_P(T_R - T_1) = \frac{2Rk}{M(k-1)}(T_R - T_1) \tag{8.14}$$

Here we have substituted $C_P\,\Delta T$ for Δh and then substituted $Rk/[M(k-1)]$ for C_P. Both substitutions are justified in App. D, Eqs. D.2 and D.10.

If we now divide both sides of this equation by RkT/M, we find

$$\frac{MV_1^2}{RkT_1} = \frac{2}{k-1}\left(\frac{T_R}{T_1} - 1\right) \tag{8.15}$$

But, as shown previously, RkT_1/M is the square of the speed of sound at state 1, or c_1^2, so the left side is $(V_1/c_1)^2$. The ratio V/c is called the *Mach number* $\mathcal{M}$ in honor of the Austrian physicist Ernst Mach. This ratio plays a crucial role in the study of high-velocity gas flows (and is widely reported in the press describing the speed of supersonic aircraft). It is the ratio of the *local* flow velocity to the *local* speed of sound. For subsonic flows $\mathcal{M}$ is less than 1; for sonic flows it equals 1; for supersonic flows it is greater than 1. Making this definition, we can rearrange Eq. 8.15 to

$$\frac{T_R}{T_1} = \mathcal{M}_1^2\,\frac{k-1}{2} + 1 \tag{8.16}$$

Example 8.3. Air flows from a reservoir in which its velocity is negligible and its temperature is 68°F (20°C). The flow is steady and adiabatic. What is the temperature of the gas at the point where the Mach number is 2.0?

Air is a diatomic gas, so $k = 1.4$, as shown in Table 8.1. From Eq. 8.16 we have

$$\frac{T_R}{T_1} = 2.0^2\left(\frac{1.4-1}{2}\right) + 1 = 1.80$$

$$T_R = 68°\text{F} = 528°\text{R} = 293.15\ \text{K}$$

$$T_1 = \frac{T_R}{1.80} = \frac{528°\text{R}}{1.8} = 293°\text{R} = -167°\text{F} = 163\ \text{K} = -110°\text{C}$$ ■

The startlingly low temperature indicated above shows clearly that the expanding gas is converting its internal energy to kinetic energy; the decrease in internal energy is indicated by the large decrease in temperature. The ratio T_R/T_1 obtained depends only on k and $\mathcal{M}_1$, not on the identity of the gas or on the reservoir temperature.

Example 8.4. What is the velocity of the air in Example 8.3 at the point where the Mach number is 2.0?

We need to know the speed of sound at state 1. From Eq. 8.11 we have

$$c = 223\ \frac{\text{ft}}{\text{s}} \cdot \left(\frac{\text{lbm}}{\text{lbmol} \cdot {}^\circ\text{R}} \right)^{1/2} \cdot \left(\frac{1.4 \cdot 293^\circ\text{R}}{29\ \text{lbm/lbmol}} \right)^{1/2} = 839\ \frac{\text{ft}}{\text{s}}$$

$$V_1 = c_1 \mathcal{M}_1 = 839\ \frac{\text{ft}}{\text{s}} \cdot 2.0 = 1678\ \frac{\text{ft}}{\text{s}} = 511\ \frac{\text{m}}{\text{s}}$$ ■

Equation 8.16 applies to any adiabatic, steady flow of a perfect gas, with or without friction (we look at its application to adiabatic flow with friction in Sec. 8.4). Now we add the assumption that the flow is frictionless. Frictionless, adiabatic flow of any nonreacting gas is isentropic; so we may use the relations among temperature, pressure, and density for an isentropic change of a perfect gas, as developed in App. D. Observe that in this chapter pressures are always absolute pressure, never gauge pressure! The temperatures in the equations are always in degrees Rankine or kelvins, never in degrees Celsius or Fahrenheit!

$$\frac{P_R}{P_1} = \left(\frac{T_R}{T_1} \right)^{k/(k-1)} \tag{D.17}$$

$$\frac{\rho_R}{\rho_1} = \left(\frac{T_R}{T_1} \right)^{1/(k-1)} \tag{D.19}$$

Substituting the value of T_R/T_1 from Eq. 8.16, we find

$$\frac{P_R}{P_1} = \left(\mathcal{M}_1^2 \frac{k-1}{2} + 1 \right)^{k/(k-1)} \tag{8.17}$$

$$\frac{\rho_R}{\rho_1} = \left(\mathcal{M}_1^2 \frac{k-1}{2} + 1 \right)^{1/(k-1)} \tag{8.18}$$

Example 8.5. For the air in Example 8.4 the reservoir pressure is 2 bar, and the reservoir density is 2.39 kg/m^3. What are the pressure and density at the point in the flow where $\mathcal{M}$ is 2.0?

The term in the parentheses in Eqs. 8.17 and 8.18 is precisely T_R/T_1, found in Example 8.3 to be 1.80; therefore,

$$\frac{P_R}{P_1} = 1.80^{1.4/(1.4-1)} = 1.80^{3.5} = 7.82$$

$$P_1 = \frac{P_R}{7.82} = \frac{2\ \text{bar}}{7.82} = 0.256\ \text{bar} = 3.71\ \frac{\text{lbf}}{\text{in}^2}$$

$$\frac{\rho_R}{\rho_1} = 1.80^{2.5} = 4.35$$

$$\rho_1 = \frac{\rho_R}{4.35} = \frac{2.39\ \text{kg/m}^3}{4.35} = 0.549\ \frac{\text{kg}}{\text{m}^3} = 0.034\ \frac{\text{lbm}}{\text{ft}^3}$$ ■

Comparing these results with the temperature ratio T_R/T_1, we see that the pressure and the density change much more rapidly in frictionless, adiabatic flow than does the temperature.

Equations 8.16, 8.17, and 8.18 allow us to calculate the change of temperature, pressure, and density with a change in Mach number for isentropic, steady flow of a perfect gas. From the Mach number and the temperature we can calculate the velocity. The other item of interest is the cross-sectional area perpendicular to flow. By applying the mass balance equation for steady flow between states R and 1 and solving for A_R/A_1, we find

$$\frac{A_R}{A_1} = \frac{\rho_1 V_1}{\rho_R V_R} \tag{8.19}$$

We have defined the reservoir conditions such that A_R is infinite and V_R is zero. If we insert these values in Eq. 8.19, we see that both sides are large without bound (i.e., infinite). The reservoir condition, which is the most convenient reference condition for temperature, pressure, and density, is therefore a very poor reference condition for the cross-sectional area; we will choose a better one. In any such flow there is or could be a state at which the Mach number is exactly 1. Even if such a state does not exist for the flow in question, pretending that it exists will help us solve the problem. Let us refer to this state as the *critical state* and denote it by an asterisk. The mass balance equation between some arbitrary state and the critical state is

$$\frac{A_1}{A^*} = \frac{\rho^* V^*}{\rho_1 V_1} \tag{8.20}$$

Substituting the values of the density ratio in terms of the temperature for isentropic flow and of the velocity in terms of the Mach number and the speed of sound and then eliminating the temperature ratio, we find

$$\frac{A_1}{A^*} = \frac{1}{\mathcal{M}_1}\left[\frac{\mathcal{M}_1^2(k-1)/2+1}{(k-1)/2+1}\right]^{(k+1)/2(k-1)} \tag{8.21}$$

which is plotted in Fig. 8.3 (see App. E for details of this calculation).

Figure 8.3 leads to the commonplace conclusion that, for Mach numbers less than 1, to get the fluid to go faster, we must reduce the cross-sectional area perpendicular to flow. This is how garden hose nozzles work; the flow area

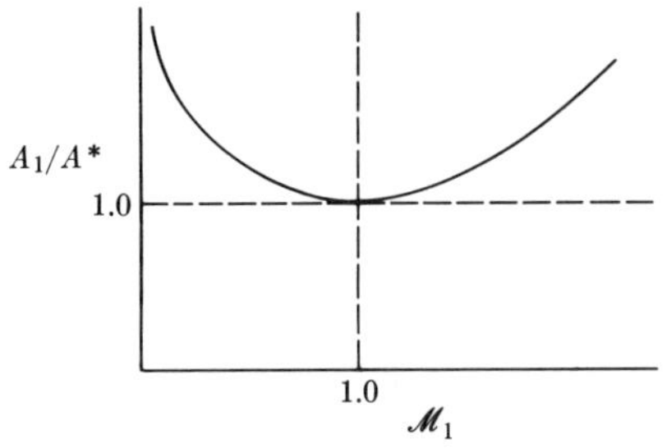

FIGURE 8.3
Plot of Eq. 8.21.

decreases, so the velocity increases for a constant mass flow rate. However, when the Mach number is greater than 1, we are led to the startling conclusion that to get the fluid to go faster, we must increase the area! If this is intuitively obvious to the reader, he has better intuition than the author does. Let us simultaneously plot ρ, A, and V against distance through such a nozzle; see Fig. 8.4.

Here we have let V be some small value at the inlet and increase linearly with distance. Because this is an expanding flow, the density decreases with distance. In the subsonic range, V goes up faster than ρ goes down; so A must decrease to keep ρAV constant. However, as the fluid goes faster and faster, ρ drops more and more rapidly, until at $\mathcal{M} = 1$ it is decreasing just as rapidly as V is increasing. In supersonic flow, $\mathcal{M} > 1$, the density falls very rapidly, more rapidly than the velocity is increasing, and therefore A must increase. This is shown in the figure.

From the preceding information we can also calculate the mass flow rate through our duct. Since the flow is steady, the mass flow rate must be the same at all points:

$$\dot{m} = \rho AV \tag{8.22}$$

Here the product ρAV is the same for every point in the duct, including the point where the flow is sonic (the critical point). Writing it for that point, we divide both sides by A^* and note that V^* equals c^*, which we can write in terms of k, R, M, using Eqs. 8.11 and 8.16. Similarly, we can write ρ^* in terms of ρ_R by using Eq. 8.18. Making these substitutions in Eq. 8.22, we find

$$\frac{\dot{m}}{A^*} = \frac{\rho_R(kRT_R/M)^{1/2}}{[(k-1)/2+1]^{(k+1)/2(k-1)}} \tag{8.23}$$

An alternative form of this equation which is sometimes useful is found by substituting for ρ_R its equivalent from the perfect-gas law, $MP_R/(RT_R)$:

$$\frac{\dot{m}}{A^*} = \frac{P_R}{T_R^{1/2}}\left(\frac{Mk}{R}\right)^{1/2} \frac{1}{[(k-1)/2+1]^{(k+1)/2(k-1)}} \tag{8.24}$$

We finally have a complete set of equations for describing the frictionless, adiabatic, one-dimensional, steady flow of a perfect gas in a duct. If we know

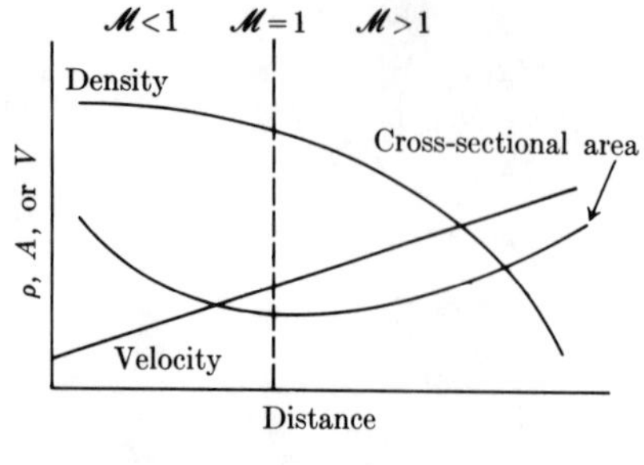

FIGURE 8.4
Density, velocity, and area variations through a nozzle.

the identity of the gas (and hence k and M) and the temperature and pressure in the reservoir, we can calculate the temperature, pressure, velocity, density, Mach number, and mass flow rate per unit area perpendicular to flow at any downstream location where any one of these variables has a known value.

Example 8.6. Air at 30 lbf/in^2 absolute and 200°F flows from a reservoir into a duct. The flow is steady, adiabatic, and frictionless. The flow rate is 10 lbm/s. What are the cross-sectional area, temperature, pressure, and Mach number at the point in the duct where the velocity is 1400 ft/s?

We do not know the Mach number at the point in question; to use Eqs. 8.16, 8.17, and 8.18, we must find it first. From Eq. 8.14, rearranged, we get

$$T_1 = T_R - V_1^2\left(\frac{k-1}{k}\right)\frac{M}{2R}$$

$$= 660°\text{R} - \frac{(1400\ \text{ft/s})^2(1.4-1)[29\ \text{lbm/(lbmol)}]}{2\cdot 1.4\cdot 4.98\times 10^4\ \text{ft}^2/\text{s}^2[\text{lbm/(lbmol}\cdot°\text{R)}]}$$

$$= 660°\text{R} - 163°\text{R} = 497°\text{R} = 276\ \text{K}$$

Therefore,

$$c = 223\ \frac{\text{ft}}{\text{s}}\cdot\left(\frac{\text{lbm}}{\text{lbmol}\cdot°\text{R}}\right)^{1/2}\left[\frac{1.4\cdot 497°\text{R}}{29\ \text{lbm/(lbmol)}}\right]^{1/2} = 1092\ \frac{\text{ft}}{\text{s}} = 333\ \frac{\text{m}}{\text{s}}$$

$$\mathcal{M}_1 = \frac{1400\ \text{ft/s}}{1092\ \text{ft/s}} = 1.282$$

Now we can use Eqs. 8.17 and 8.18:

$$\frac{P_R}{P_1} = \left(\frac{T_R}{T_1}\right)^{k/(k-1)} = \left(\frac{660°\text{R}}{497°\text{R}}\right)^{3.5} = 2.70$$

$$P_1 = \frac{P_R}{2.70} = \frac{30\ \text{lbf/in}^2\ \text{abs}}{2.70} = 11.1\ \frac{\text{lbf}}{\text{in}^2} = 76.5\ \text{kPa}$$

For a flow rate of 10 lbm/s, we can calculate the area at the critical condition from Eq. 8.24:

$$\frac{\dot m}{A^*} = \frac{(30\ \text{lbf/in}^2)\{[29\ \text{lbm/(lbmol)}]\cdot 1.4\}^{1/2}[32.2(\text{lbm}\cdot\text{ft})/(\text{lbf}\cdot\text{s}^2)]}{223\ \text{ft/s}\cdot[\text{lbm/(lbmol}\cdot°\text{R)}]^{1/2}\cdot(660°\text{R})^{1/2}\cdot(0.4/2+1)^{2.4/0.8}}$$

$$= 0.62\ \frac{\text{lbm}}{\text{s}\cdot\text{in}^2} = 437\ \frac{\text{kg}}{\text{s}\cdot\text{m}^2}$$

Therefore,

$$A^* = \frac{\dot m}{0.62\ \text{lbm/(in}^2\cdot\text{s)}} = \frac{10\ \text{lbm/s}}{0.62\ \text{lbm/(in}^2\cdot\text{s)}} = 16.1\ \text{in}^2 = 0.030\ \text{m}^2$$

Now that we know A^*, we can calculate the area at which $V = 1400$ ft/s from Eq. 8.21:

$$\frac{A}{A^*} = \frac{1}{1.282}\left(\frac{1.282^2 \cdot 0.4/2 + 1}{0.4/2 + 1}\right)^{2.4/2(0.4)}$$

$$= \frac{1}{1.282}\left(\frac{1.329}{1.20}\right)^3 = 1.059$$

Therefore, $A = 1.059 A^* = 1.059 \cdot 16.1\ \text{in}^2 = 17.0\ \text{in}^2 = 1.1 \times 10^{-2}\ \text{m}^2$. ■

Certainly by now the reader has observed that this calculation involves a lot of algebra, much of it concerning computation of the quantity

$$\frac{\mathcal{M}_1^2(k-1)}{2} + 1$$

to various powers. For constant k this function obviously can be tabulated to the powers of interest for various values of $\mathcal{M}$, saving us much of the algebra. This is done for $k = 1.4$ in App. A.5.

Several columns in the second part of App. A.5 apply to normal shock waves, which we discuss in Sec. 8.5. The table is based on Eqs. 8.16, 8.17, 8.18, and 8.21. In addition, there is a column labeled V/c^* (the derivation of the equation for this ratio is in App. F, Eq. F.12). This ratio is useful in a problem in which we know the velocity and the reservoir conditions and wish to find $\mathcal{M}$. We could solve for the local temperature and local speed of sound, as we did in Example 8.6, but that is tedious. We would like some velocity ratio like V/V_R to appear in the table, but that is an impossible choice, because V_R is zero. The logical choice is $V/V^* = V/c^*$; its use is illustrated below.

Example 8.7. Rework Example 8.6, using App. A.5.

First, we calculate c^*. To do this, we need T^*, which we find by looking in the table for T/T_R at $\mathcal{M} = 1.0$, finding $T^*/T_R = 0.83333$; then

$$T^* = 660°\text{R} \cdot 0.83333 = 550°\text{R}$$

$$c^* = \left(\frac{kRT^*}{M}\right)^{1/2} = 223\ \frac{\text{ft}}{\text{s}} \cdot \left(\frac{\text{lbm}}{\text{lbmol} \cdot °\text{R}}\right)^{1/2} \cdot \left[\frac{1.4 \cdot 550°\text{R}}{29\ \text{lbm}/(\text{lbmol})}\right]^{1/2}$$

$$= 1149\ \text{ft/s} = 350\ \text{m/s}$$

Now we compute

$$\frac{V}{c^*} = \frac{1400\ \text{ft/s}}{1149\ \text{ft/s}} = 1.218$$

We go to the V/c^* column in the table and read down until we find $V/c^* = 1.218$. This corresponds to a Mach number of about 1.28. In the same row on the table we read

$$\frac{T}{T_R} = 0.7532 \qquad \frac{P}{P_R} = 0.3708 \qquad \frac{A}{A^*} = 1.058$$

Thus,

$$T = 0.7532T_R = 0.7532 \cdot 660°\text{R} = 497°\text{R}$$

$$\text{P} = 0.3708P_R = 0.3708 \cdot 30\ \text{lbf/in}^2\ \text{abs} = 11.1\ \text{lbf/in}^2\ \text{abs}$$ ■

This is clearly a gigantic saving in an effort over solving this kind of problem longhand, as we did in Example 8.6. The homework problems should convince the student of the utility of App. A.5. Appendix A.5 makes clear why the calculations have all been done in terms of the reservoir and critical conditions. How else could one make up such a table?

It is also instructive to consider the following variant of Example 8.6.

Example 8.8. Rework Example 8.6 for the point where the velocity is 4000 ft/s.

As in Example 8.7, $c^* = 1149$ ft/s, so $V/c^* = 4000/1149 = 3.48$. We now look in the V/c^* column of App. A.5 for 3.48. In that table the highest value of V/c^* is 1.69. Looking in the original document from which App. A.5 was extracted, we find that for Mach numbers up to 100 the value of V/c^* approaches 2.4489 as a limit; so 4000 ft/s must not be possible in this flow! If we return to our method of solution in Example 8.6, we see that substituting $V = 4000$ ft/s leads to

$$T_1 = 660°\text{R} - 1340°\text{R} = -680°\text{R} \qquad ?$$ ■

From this we see that the energy balance sets a maximum possible velocity for an expanding flow. Once all the internal energy has been turned into kinetic energy, the gas can no longer accelerate. Obviously, there are other limits to the velocity. Air will turn into a liquid long before it reaches 0°R, and the assumption of constant C_P becomes false at very low temperatures. Furthermore, unless the air were very dry, it would be expected to form a fog at these extremely low temperatures. This explains why high-speed wind tunnels have either big air dryers or big air preheaters.

How is the maximum velocity to be reconciled with Eq. 8.16, which indicates that one can calculate a value of T_R/T_1 for any Mach number, no matter how large? The answer is that as the Mach number goes higher and higher, it does so by driving the temperature lower and lower. Since we have $c = (kRT/M)^{1/2}$, we can raise the Mach number by raising the velocity or by lowering the temperature or both. At high Mach numbers the calculated temperature is very low.

This may be visualized by plotting Eq. 8.14 as T versus V; see Fig. 8.5. The figure is one-half of a parabola with a maximum value V_{max} equal to $(2C_PT_R)^{1/2}$. Again we must remember that the assumptions made in deriving the equation for steady, adiabatic, frictionless flow of an ideal gas in a duct become inaccurate as T approaches zero.

Our discussion has entirely concerned flow from a reservoir. If the initial conditions are stated, not for a reservoir where $V = 0$, but for some other point, we can develop the equivalents of all the equations shown so far, with

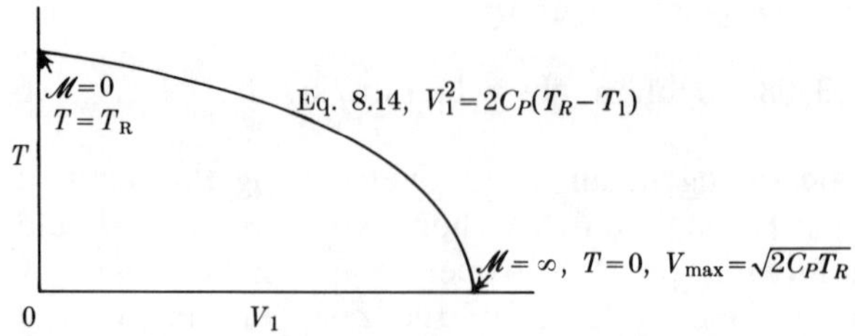

FIGURE 8.5
Variation of temperature with velocity for steady, adiabatic flow of a perfect gas.

the initial velocity not taken equal to zero. That is even messier algebraically than what we have done so far! A much more convenient approach is to utilize App. A.5 to solve for the reservoir condition which corresponds to the given starting condition.

Example 8.9. Air is flowing in a duct in steady, frictionless, adiabatic flow. At the place in the duct where the Mach number is 0.5, the temperature is 20°C. What is the temperature at the point in the duct where the Mach number is 2.0? See Fig. 8.6.

From App. A.5 we get for $\mathcal{M}_1 = 0.5$ the value $T/T_R = 0.9524$; therefore,

$$T_R = \frac{T_1}{0.9524} = \frac{293.15\text{ K}}{0.9524} = 307.8\text{ K} = 554°\text{R}$$

Now, to find T_2, we look in App. A.5 for $\mathcal{M}_2 = 2.0$ and find $T/T_R = 0.5556$; therefore,

$$T_2 = 0.5556\,T_R = 0.5556 \cdot 307.8\text{ K} = 171\text{ K} = 308°\text{R} = -152°\text{F}$$ ■

Clearly, we could use an analogous procedure to find the density, pressure, etc., at point 2 if they were given for point 1. This procedure allows us to use the tables based on flow from a reservoir for computing flow from any state to any other state.

Thus, we ought to look on the reservoir state, not as a state which necessarily exists in a real reservoir, but rather as a convenient reference condition which allows us to solve many common problems by simple application of tabulated values.

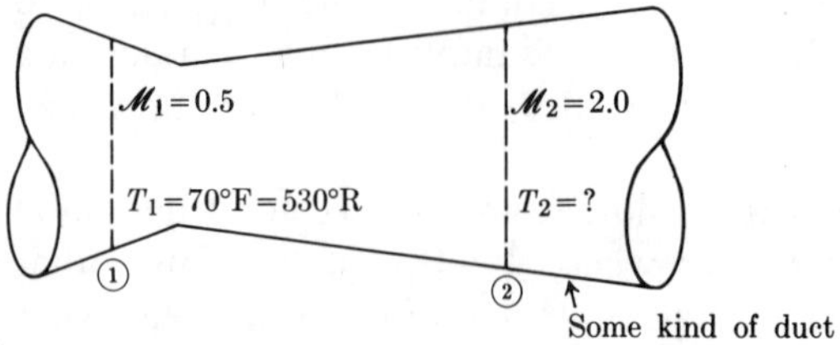

FIGURE 8.6

The existence of two reference states, the reservoir and the critical, is a source of some confusion for students. The states are visualized most easily by means of Fig. 8.7. We assume that at any point in the flow we could introduce a frictionless three-way valve. The flow is actually going through the valve in the direction of the outgoing flow, so that the introduction of this imaginary valve does not change the flow. Now, if we switched the valve so that the flow was diverted frictionlessly to the reservoir, then the temperature, pressure, density, etc., that we measured in the reservoir would constitute the reservoir conditions corresponding to the incoming flow. Similarly, if we diverted the flow through the sonic nozzle, then the conditions measured at the throat of the nozzle would be the critical conditions corresponding to the incoming flow. Thus the imaginary valve allows us to visualize what we would do experimentally to find the reservoir or critical conditions corresponding to the incoming flow. However, since the two nozzles which we should use in the figure are assumed frictionless, we need not actually perform the experiments; we can calculate what would happen in them by using the equations given in this section.

For a reversible, adiabatic flow, neither the reservoir conditions nor the critical conditions change from point to point in the system. Thus, a measurement or calculation of them for any point in the flow gives the values for the entire flow. However, in flow with friction (Sec. 8.4) or flow with heating or cooling (not treated in this text) or for normal shock waves (Sec. 8.5), the reservoir and critical conditions change from point to point in the flow. For such flows we must be conceptually prepared to use the apparatus shown in Fig. 8.7 at many points in the flow, and we must not assume that the values we find (or calculate) at one point would be the same as those at another.

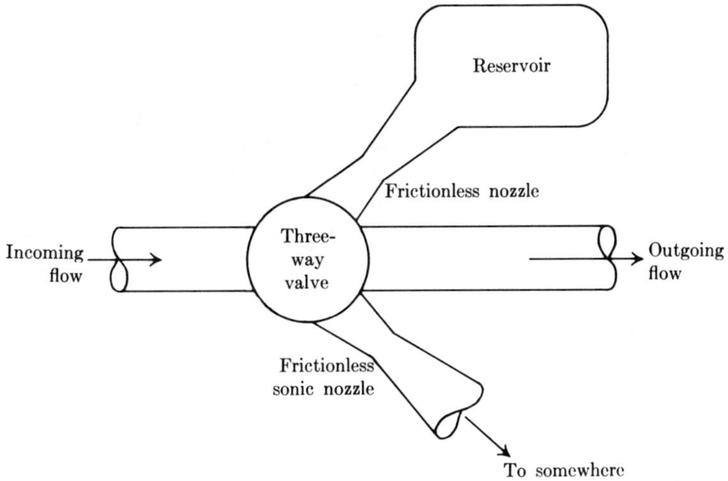

FIGURE 8.7
Visualization of reservoir and critical conditions.

8.3 NOZZLE CHOKING

Suppose that we connect a high-pressure reservoir full of air to a low-pressure reservoir full of air by means of a converging nozzle. We assume that by suitable pumps, etc., we can maintain the pressure in each reservoir at any value we select and that we have some method of measuring the mass flow rate of gas passing through the nozzle. See Fig. 8.8.

At first the pressure in both reservoirs is the same high pressure P_1. Since there is no pressure gradient across the nozzle, there is no flow. Then holding the pressure in the high-pressure reservoir constant at P_1, we begin to lower pressure P_2 in the low-pressure reservoir. For each value of P_2 we measure the mass flow rate $\dot{m}$, and we plot $\dot{m}$ versus P_2/P_1. The results are shown in Fig. 8.9.

We observe that the mass flow rate increases steadily as we lower P_2, until P_2/P_1 equals 0.5283, and then further lowering of P_2 does not increase the mass flow rate. If we refer to App. A.5, we see that if the assumptions of isentropic, one-dimensional, steady flow apply, then this pressure ratio corresponds exactly to the sonic velocity ($\mathcal{M} = 1$) at the throat of the nozzle. Lowering the downstream pressure more does not increase the mass flow rate, because the flow at the narrowest point in the flow is sonic. We also observe that the value of $\dot{m}/A$ at the throat is exactly the value predicted by Eq. 8.24 if we take state 1 as the reservoir state.

Why does lowering the downstream pressure not cause the fluid to flow faster in the nozzle? Suppose we attach an observer to a balloon and let her ride along with the fluid through the nozzle. When she gets to the nozzle throat, she observes that the downstream pressure is lower than she had anticipated from the isentropic-flow equations, and she shouts back to those of us who are behind her to come faster. Figure 8.10 shows the fate of her shout.

The shout never leaves the spot where she made it (in the upstream direction). The sound signal, that there is a sharp pressure drop downstream of the nozzle, can never be communicated to the gas upstream of the nozzle. Thus, once the flow becomes sonic at the throat, nothing we can do downstream will increase the mass flow rate at that point. This situation, in which the flow at the throat is sonic, is called *choking*. One speaks of the nozzle as being choked because no more mass can get through it without a change in upstream conditions. The adjustment of the lower pressure takes place downstream of the throat by a *rarefaction* which is neither an isentropic nor a one-dimensional process and is not covered by the one-dimensional equations we develop in this chapter.

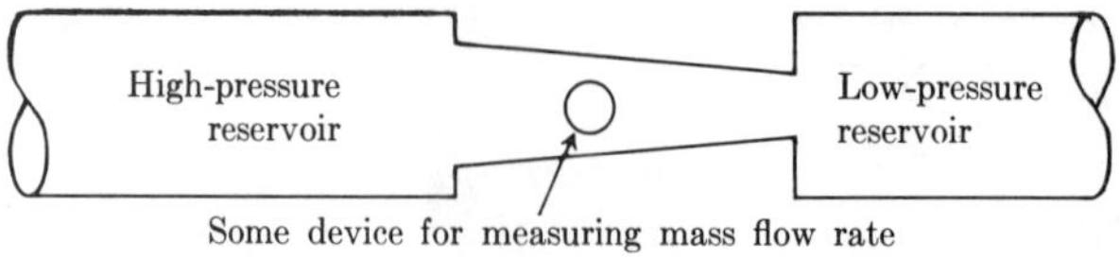

FIGURE 8.8

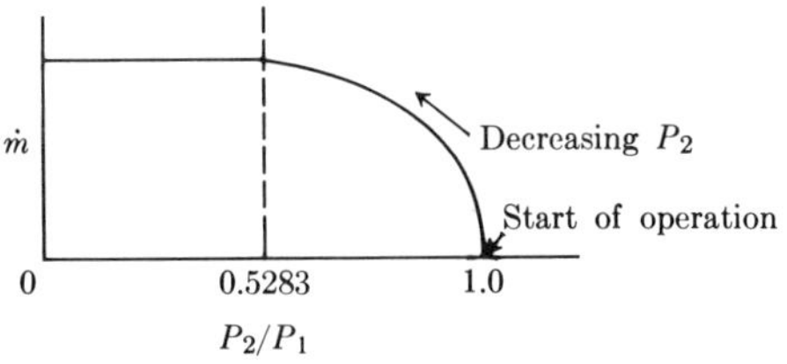

FIGURE 8.9

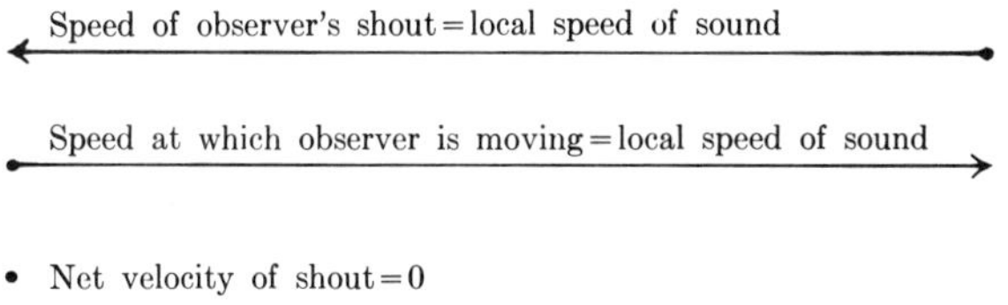

FIGURE 8.10

This type of choking is very common in valves, orifices, and vacuum systems. For air, any time that the pressure ratio across a valve, orifice, or leak into a vacuum system is 0.5283 or less, choked flow is occurring. When one lets air out of bicycle tire, the flow is normally choked. Most control valves for gas flow operate in the choked condition. Most safety relief valves on gaseous systems, when they open, operate in the choked condition. Choked orifices are a common way of providing a small flow of a gas, at a flow rate independent of the downstream pressure. Although these systems are not frictionless or one-dimensional, their behavior can be described reasonably well by the set of equations for one-dimensional, isentropic flow derived in Sec. 8.2. Of all the ideas developed in this chapter, the one most likely to help a practicing engineer understand the behavior of a system which nonengineers do not understand is the idea of choking in gas flows.

Here we can explain the fact, merely stated in Sec. 5.5, that when a fluid, liquid, or gas flows as a jet into another fluid, the pressure of the jet will be the same as that of the surrounding fluid, if the flow is subsonic, but not if the flow is sonic or supersonic. If the flow is subsonic and the pressure of the surrounding fluid is less than that of the jet, that information will propagate back along the flow, causing the flow to speed up until the pressures match. If the flow is sonic or supersonic, that information cannot propagate upstream, and the jet can have a pressure different from the surrounding pressure.

8.4 HIGH-VELOCITY GAS FLOW WITH FRICTION, HEATING, OR BOTH

In previous sections we omitted the possible influence of friction, heat transfer from the surroundings to the flowing gas, and chemical reaction (such as combustion) in the flowing gas. For the one-dimensional flow of perfect gases, it is possible to deduce mathematical solutions for flow with friction in constant-area ducts, flow with heating in constant-area ducts, flow with combustion in constant-area ducts, and so forth. For subsonic flow these solutions agree satisfactorily with experimental results, but for supersonic flow the one-dimensional assumption becomes unreliable, because friction at the tube wall causes oblique shock waves, which are not covered by one-dimensional flow theory.

Rather than develop the complete mathematics of these flows here, we simply indicate some of the salient results, referring the reader to other sources, where the complete mathematics are shown. We concentrate on the two types of flow of most practical interest to chemical engineers.

A. Adiabatic Flow with Friction

This type of flow occurs when a gas flows through a length of pipe at high velocity. If the pipe is insulated or the flow is rapid, the heat transfer to the fluid will be negligible and the flow will be practically adiabatic.

For this type of flow, the mass balance, energy, and perfect-gas equations take the same form as for steady, isentropic flow of a perfect gas. However, in Sec. 8.2 we used the isentropic relations from App. D; here we cannot, because the effect of friction is to increase the entropy of the flowing gas. In their place we use the momentum balance (Eq. 7.13), written for two points dx apart in the flow direction. For steady flow this becomes

$$0 = \rho AV\, dV - A\, dP - \tau_{\text{wall}} \pi D\, dx \tag{8.25}$$

For the flow of incompressible fluids, the shear stress at the wall, τ_{wall}, could be represented in terms of the friction factor f by

$$\tau_{\text{wall}} = f\rho \frac{V^2}{2} \tag{8.26}$$

Experimental data [5] indicate that the same relation holds for flow of gases at high velocities and that the values of f determined from the friction factor plot (Fig. 6.10) for various Reynolds numbers and pipe roughnesses apply equally well to compressible and incompressible fluids. Equation 8.26 may be substituted in Eq. 8.25 and, together with the energy, continuity, and perfect-gas equations, solved to show the change in pressure, temperature, etc., with distance down the pipe. The results are Eq. 8.16, which shows the relation of temperature to Mach number and is the same with or without friction, and

$$\frac{4f\,\Delta x}{D} - \frac{1}{k}\left(\frac{1}{\mathcal{M}_1^2} - \frac{1}{\mathcal{M}_2^2}\right) + \frac{k+1}{2k}\ln\left\{\frac{\mathcal{M}_2^2}{\mathcal{M}_1^2} \cdot \frac{1+[(k-1)/2]\mathcal{M}_1^2}{1+[(k-1)/2]\mathcal{M}_2^2}\right\} = 0 \tag{8.27}$$

The derivation of this equation takes a full page of text; it is shown in detail in Streeter and Wylie [6].

The most common and interesting problem of this type is sketched in Fig. 8.11, in which gas flows through a converging nozzle, assumed isentropic, and then through a length of straight pipe where the friction is significant. This is the situation of a high-pressure relief valve or bursting disk discharging through a pipe to a flare or stack and the situation of a high-pressure vessel discharging through a pipe which has broken some distance from it.

In the type of apparatus shown in this figure, as the fluid flows down the pipe, friction causes its pressure to decrease. This pressure decrease makes the density fall, and hence the velocity must increase. Since the friction effect is proportional to the velocity squared, the pressure drop $-dP/dx$ is not the same for every foot of pipe, as it is for incompressible flows, but rather increases with distance down the pipe.

For the apparatus shown in the figure, if pressures P_0 and P_3 are the same, there will be no flow. If P_3 is then lowered while P_0 is held constant, the flow rate will increase and the Mach number at the outlet of the pipe will steadily increase until the outlet flow is at $\mathcal{M}_2 = 1$. Then further reduction of P_3 will not cause the flow rate to increase, because the flow at the end of the pipe will be choked, just as converging-diverging nozzles become choked (Sec. 8.3). When the flow at the outlet is subsonic, we have $P_2 = P_3$. Once the flow at the outlet becomes sonic, the choked situation exists and P_3 may be lowered further without any change in P_2.

Normally in such a system the flow through the nozzle is practically isentropic, because the effect of friction in the nozzle is small compared with the effect of area change. Therefore, to calculate the flow rate for a given pressure ratio (or the pressure ratio for a given flow rate, etc.), one uses the equations for a converging, isentropic nozzle to find the change in properties from the reservoir to point 1 and Eq. 8.27 to find the change in properties from point 1 to point 2. To solve for the flow rate for a given set of values of the P_0 and P_3, one guesses the Mach number at the outlet of the nozzle $\mathcal{M}_1$. Then from Eq. 8.24 one can compute $(\dot{m}/A)_1$. One computes the outlet Mach number from Eq. 8.27, the outlet temperature from Eq. 8.16, and the outlet speed of sound and hence the outlet velocity from Eq. 8.11. Then because $\dot{m}/A$ is the same at point 2 as at point 1, one can compute the pressure at 2 by material balance. If this pressure matches the assumed value of P_3, then the

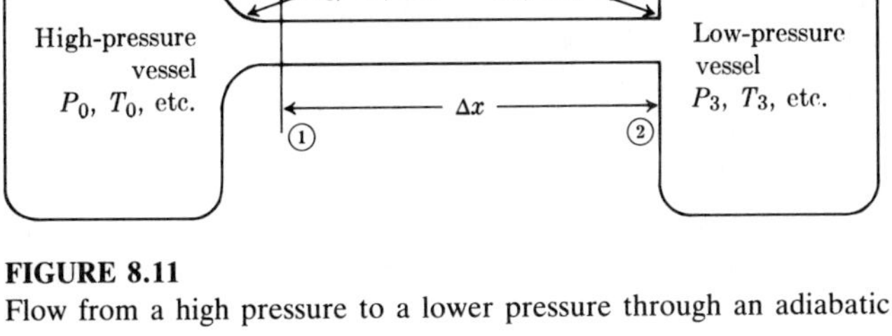

FIGURE 8.11
Flow from a high pressure to a lower pressure through an adiabatic pipe with friction.

guessed value of $\mathcal{M}_1$ is correct; if not, one proceeds by trial and error to find the right value of $\mathcal{M}_1$ and then the other corresponding values.

This is clearly a tedious procedure. Several authors have performed this trial and error for a large variety of cases and presented the results in the form of convenient plots. The plot due to Levenspiel [7] is shown in Fig. 8.12. It is principally for gases with $k = 1.4$ but also shows lines for $k = 1.2$ and 1.6, thus covering the range of practical interest in values of k. On it the ordinate is P_3/P_0 while the abcissa is $(\dot{m}/A)/(\dot{m}/A)^*$, the ratio of the mass flow rate per unit area to the maximum possible mass flow rate per unit area for the nozzle alone, computed from Eq. 8.24. To save writing on the figure, $N = 4f\,\Delta x/D$. (One may quickly check to see that the $N = 0$ curve, which corresponds to the nozzle alone with no pipe, is made up directly from the isentropic flow equations summarized in the tables in App. A.5.) The use of Fig. 8.12 is illustrated in the following example.

Example 8.10. In Fig. 8.11, P_0 is 30 lbf/in^2 and T_0 is 200°F. The pipe is 1-in, schedule 40 steel pipe 8 ft long. Find the flow rate for various values of P_3.

For 1-in schedule 40 pipe, the relative roughness is about 0.0018, and from Fig. 6.10 for high Reynolds numbers the friction factor is about 0.0055.

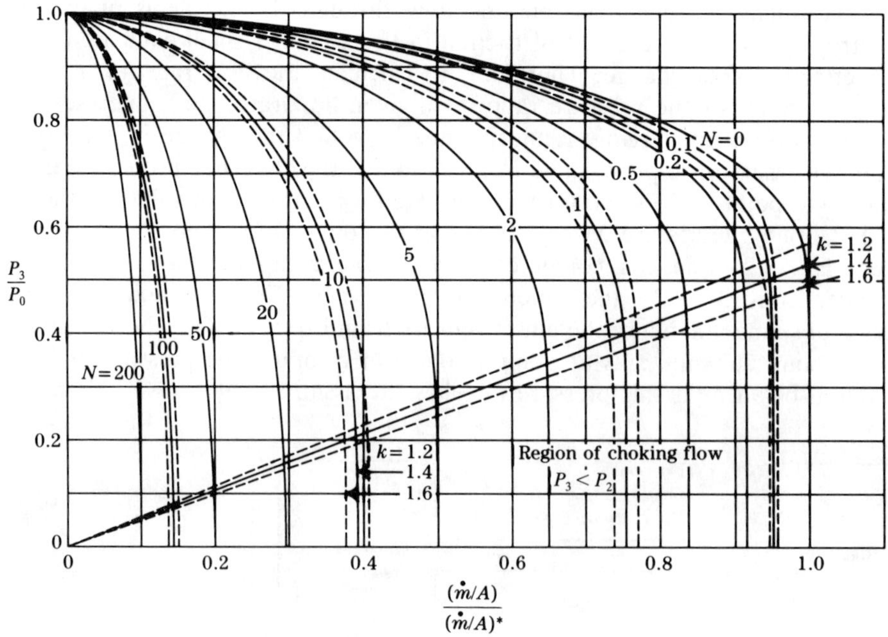

FIGURE 8.12
Pressure–mass flow rate relationship for the apparatus shown in Fig. 8.11. [O. Levenspiel, "The discharge of gases from a reservoir through a pipe," *AIChEJ 23:* 402–403 (1977). Reproduced by permission of the publisher.]

Thus,

$$N = \frac{4f\,\Delta x}{D} = \frac{4 \cdot 0.0055 \cdot 8\text{ ft}}{(1.049/12)\text{ ft}} = 2.01$$

From Example 8.6 we know that for these values of P_0 and T_0, $(\dot{m}/A)^* = 0.62\text{ lbm}/(\text{s} \cdot \text{in}^2)$. For $P_3 = 27\text{ lbf/in}^2$, P_3/P_0 is 0.9; so from Fig. 8.12, reading the curve for $N = 2$,

$$\frac{\dot{m}/A}{(\dot{m}/A)^*} = 0.36$$

so that

$$\left(\frac{\dot{m}}{A}\right)^* = 0.36 \cdot 0.62\,\frac{\text{lbm}}{\text{s} \cdot \text{in}^2} = 0.22\,\frac{\text{lbm}}{\text{s} \cdot \text{in}^2} = 152\,\frac{\text{kg}}{\text{s} \cdot \text{m}^2}$$

and for a 1-in schedule 40 pipe

$$\dot{m} = 0.22\,\frac{\text{lbm}}{\text{s} \cdot \text{in}^2} \cdot 0.864\text{ in}^2 = 0.19\,\frac{\text{lbm}}{\text{s}} = 0.086\,\frac{\text{kg}}{\text{s}}$$

The corresponding Reynolds number is about 7×10^5; so for $P_3 = 27\text{ lbf/in}^2$ or less, the flows correspond to the flat part of the friction factor curve, and we can use $f = 0.0055$. Now, to find the subsequent flow rates, we make a table:

P_3, lbf/in²	$\frac{P_3}{P_0}$	$\frac{\dot{m}/A}{(\dot{m}/A)^*}$	$\dot{m}$, lbm/s
30	1.0	0.00	0.00
27	0.9	0.36	0.19
24	0.8	0.48	0.30
21	0.7	0.56	0.35
18	0.6	0.61	0.38
15	0.5	0.64	0.397
≤ 10	0.34	0.65	0.403

■

For $P_3 = 10.2\text{ lbf/in}^2$ the flow is choked; subsequent lowering of the downstream pressure will not increase the mass flow rate.

Lapple [8] showed that these calculations agree very well with experimental tests run with this type of apparatus. If the nozzle is not a smoothly rounded one like that shown in Fig. 8.11, but rather is a square-cornered one, Lapple suggests using the same contraction coefficient for it as introduced in Sec. 6.9, i.e., adding to the friction effect of the pipeline $4f(\Delta x/D) = 0.5$ for the sudden entrance. In this type of problem, in which flow is from an upstream reservoir, the calculation method shown in Fig. 8.12 is convenient. For adiabatic flows with friction without an upstream reservoir, the same mathematical solution can be put in another form, using the choked outlet state as a reference condition; this form is shown by Shapiro [5].

All the foregoing applies to subsonic flows. To obtain a steady supersonic flow in a duct, the duct must be attached to a converging-diverging nozzle. The resulting flow is described by Eqs. 8.16, and 8.27, but the resulting behavior is somewhat different, as discussed by Shapiro [5].

B. Isothermal Flow

The general case of isothermal flow with friction is discussed by Shapiro [5]. However, as he points out, at Mach numbers approaching the choking condition, an infinite heat-transfer rate would be required to keep the flow isothermal. Thus, for common pipe sizes and lengths, high-velocity gas flow is always much closer to adiabatic than to isothermal. The one exception of interest is the flow of natural gas and similar materials (e.g., ethylene) through long-distance pipelines. These may be 100 mi long between pumping stations and are normally buried in the ground, which supplies heat as needed to keep the flow isothermal.

The commonly used formulas for calculating flow in these lines are based on the momentum equation (Eq. 8.25), the mass balance equation, and the perfect-gas law. Substituting Eq. 8.26 in Eq. 8.25 and dividing by A, we find

$$\rho V\,dV + dP = -4f\rho\,\frac{V^2}{2}\,\frac{dx}{D} \tag{8.28}$$

In the previous section we used all three terms in this equation. Here we can greatly simplify the calculations by noting that for long pipelines the first term, $\rho V\,dV$, is negligible compared with the others (Prob. 8.40) and can be dropped. Then from the mass balance equation, V is replaced with $\dot{m}/(\rho A)$, and Eq. 8.28 is simplified to

$$dP = \frac{-4f}{2}\left(\frac{\dot{m}}{A}\right)^2 \frac{1}{\rho}\,\frac{dx}{D} \tag{8.29}$$

But for a perfect gas we have

$$\rho = \frac{PM}{RT}$$

so that

$$P\,dP = \frac{-4f}{2}\cdot\frac{RT}{DM}\left(\frac{\dot{m}}{A}\right)^2 dx$$

which may be integrated and rearranged to

$$\dot{m} = \left[\frac{(P_1^2 - P_2^2)D^5 M(\pi/4)^2}{4f\,\Delta x\,RT}\right]^{1/2} \tag{8.30}$$

If we substitute $f = 0.0080/(\text{pipe diameter in inches})^{1/3}$ in this equation, we obtain the *Weymouth equation*, which was widely used in the design of early natural-gas pipelines. The historical trend of the gas pipeline industry is to use higher and higher gas pressures; as the pressure increases, the gas departs further and further from the ideal-gas state. Later workers have corrected the Weymouth equation to take this departure into account [9].

As discussed in Sec. 6.13, the economic velocity in a pipeline is primarily dependent on the density of the fluid flowing. For long-distance natural-gas pipelines, the pressures are normally in the range of 500 to 1000 psia, so densities are of the order of 1 to 2 lbm/ft^3. From Table 6.4 we can estimate the economic velocity at about 20 ft/s, which is typical of these pipelines. Thus, this kind of flow does not really correspond to the subject of this section—high-velocity gas flow; however, it fits in naturally here, after we have developed the equations for high-velocity gas flow.

8.5 NORMAL SHOCK WAVES

Sections 8.5, 8.6, and 8.7 concern supersonic flows. They are of great practical interest to aeronautical engineers and to chemical engineers in the rocket propulsion industry. They are of less practical interest to most chemical engineers, who will find them an interesting sidelight but who rarely encounter supersonic flows in engineering practice.

Suppose that a nozzle is steadily discharging a gas stream at $\mathcal{M} = 2.0$ into a low-pressure reservoir and that a valve in the end of this nozzle is suddenly shut. This will stop the flow. In Sec. 7.4 we discussed the analogous problem for the flow of a liquid. There the flow was subsonic, and the boundary between the stopped and moving fluid propagated against the flow at the local speed of sound relative to the fluid. Here the fluid is moving faster than the local speed of sound, so how can the information that the valve is closed propagate upstream against the flow? This information cannot move upstream against a supersonic flow, if it moves only at the speed of sound; therefore, there must be some way of conveying this information as a pressure signal which moves faster than the local speed of sound. We said before that a sound wave was a *small* pressure disturbance which moved at the local speed of sound. Now we consider shock waves, which are *large* pressure disturbances that can move faster than the local speed of sound.

Shock waves occur in nature in the air surrounding explosions (the shock wave causes much of the destruction of buildings, etc., in an atomic bomb blast) and in the sudden closing of a valve in a duct with high-velocity flow. Sonic booms are shock waves. Shock waves can be produced in the laboratory in a duct or nozzle with supersonic flow. In such cases the shock wave will stand still in one place while the fluid flows through it. The latter is the easier to analyze mathematically, so we use it as a basis for calculations. The nomenclature for a shock wave is shown in Fig 8.13.

The situation is quite similar to that shown in Fig. 8.2, as is much of the following mathematical development. Writing the equivalent of Eq. 8.1, the continuity or mass balance equation, for this flow, we find

$$\rho_x V_x = \rho_y V_y \tag{8.31}$$

Equation 8.4, the momentum balance, becomes

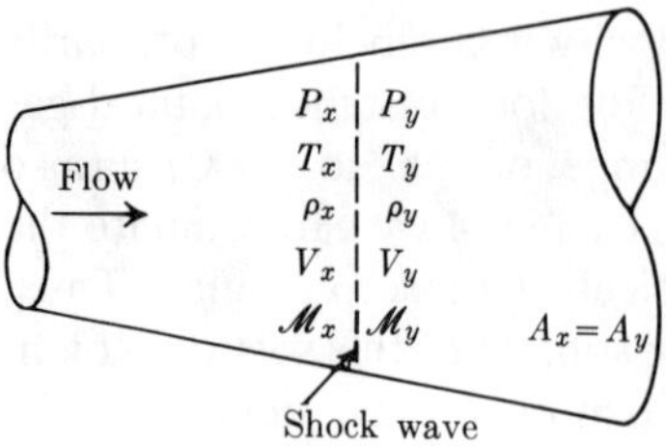

FIGURE 8.13
Nomenclature of a shock wave. Subscripts x and y refer to conditions upstream and downstream, respectively, of the shock wave.

$$V_x - V_y = \frac{P_y}{\rho_y V_y} - \frac{P_x}{\rho_x V_x} \tag{8.32}$$

These two equations alone were enough to solve for the speed of a sound wave, because we were able to neglect one term, $d\rho\, dV$, which was small in Eq. 8.2. It is not small here, so we need another relation, the energy balance (Eq. 8.16).

Equations 8.16, 8.31, and 8.32 can be solved to show the changes in temperature, pressure, and Mach number in a shock wave. The algebra involved is shown in App. F. The results of that algebra are

$$\mathcal{M}_y^2 = \frac{\mathcal{M}_x^2 + 2/(k-1)}{[2k/(k-1)]\mathcal{M}_x^2 - 1} \tag{F.16}$$

$$\frac{P_y}{P_x} = \frac{2k}{k+1}\mathcal{M}_x^2 - \frac{k-1}{k+1} = \frac{2k\mathcal{M}_x^2 - (k-1)}{k+1} \tag{F.22}$$

$$\frac{T_y}{T_x} = \frac{\mathcal{M}_x^2\left[\dfrac{k-1}{2}\right] + 1}{\dfrac{k-1}{2} \cdot \left\{\dfrac{1 + \mathcal{M}_x^2[(k-1)/2]}{k\mathcal{M}_x^2 - [(k-1)/2]}\right\} + 1} \tag{F.18}$$

These three functions are somewhat messy, but for a gas with a constant k they can be tabulated as a function of $\mathcal{M}_x$. Their tabulated values for $k = 1.4$ are shown in App. A.5 with some useful combinations.

If a shock wave is a large pressure disturbance and a sound wave is a small pressure disturbance, what is the dividing line between them? There is none; as the pressure disturbance of shock waves becomes smaller and smaller, these equations become closer and closer to those for a sound wave, until they are identical. In the case of a sound wave, we had the flow in Fig. 8.2 coming toward the wave at $\mathcal{M}_x = 1.0$. The patient student may verify that substituting $\mathcal{M}_x = 1.0$ into Eqs. F.16, F.18, and F.22 shows that there is no change in temperature, pressure, or velocity through the shock waves. Thus, an $\mathcal{M}_x = 1.0$ shock wave has the same properties as a sound wave, and we may consider a sound wave to be a shock wave that moves at the lowest possible velocity for any pressure disturbance ($\mathcal{M}_x = 1.0$).

Notice the form of flow in a shock wave, shown in Fig. 8.13. The fluid enters at a supersonic velocity, a low pressure, and a low temperature and leaves at a subsonic velocity, a higher pressure, and a higher temperature. All the foregoing and all the derivations in App. F place no restriction on the direction of the shock. In the derivations we never made use of the fact that the flow was from supersonic to subsonic or vice versa. Considering these equations alone, we could conclude that the flow could be either from high velocity and low pressure to low velocity and high pressure or the reverse. The first might logically be called a *compression shock* and the second a *rarefaction shock*.

The second law of the thermodynamics, however, shows that only compression shocks are possible (see Prob. 8.44). A compression shock results in an increase in entropy whereas a rarefaction shock, if it existed, would result in a decrease in entropy, which is impossible in an adiabatic, steady-flow system. Thus, we conclude that this type of shock waves always requires a supersonic flow upstream and a subsonic flow downstream.

In Sec. 7.3 we considered hydraulic jumps, which occur in open-channel flow. They are compared with shock waves in Table 8.2. From this comparison we see that there is a strong similarity between the two. The principal difference is that because a shock wave occurs in a fluid which is compressible, we have an additional variable, the density, and we must add the energy equation to the others needed to solve for hydraulic jump, with a resulting increase in mathematical complexity.

Our discussion has been restricted to shock waves in which the flow is perpendicular to the wave; that is the only kind which can occur in one-dimensional flow. Such waves are called *normal shock waves* or *normal shocks* because of the perpendicular relationship. In two-dimensional flow, another kind, called an *oblique shock*, occurs, in which the flow is not perpendicular to the shock wave. Oblique shock waves form at the leading edge of the wings of supersonic aircraft and cause sonic booms; for a discussion of them see Shapiro [5].

TABLE 8.2
Comparison of hydraulic jumps with shock waves

	Hydraulic jumps	**Shock waves**
Flowing material	Liquid	Gas
Type of flow	Open channel	Closed duct
Inlet flow	High velocity, low depth	High velocity, low pressure
Outlet flow	Low velocity, high depth	Low velocity, high pressure
Equations needed to solve	Mass balance, momentum	Mass balance, momentum, energy
Permissible direction of occurrence determined by:	Second law of thermodynamics	Second law of thermodynamics
Upstream	Froude number > 1	Mach number > 1
Downstream	Froude number < 1	Mach number < 1

8.6 RELATIVE VELOCITIES

Thus far we have discussed flows which move past a stationary observer. They are quite important and are found in wind tunnels, turbine nozzles, high-pressure valves, etc. Equally important are systems in which the gas stands still and the system moves; examples are airplanes and missiles. In principle, we could develop a separate set of equations for these systems, but it is much simpler to learn to apply the equations which we already have applied to stationary systems and to use the tables in App. A.5.

The application of the equations in preceding sections to moving coordinate systems is quite simple, once we grasp the idea that the reservoir conditions are a function of the frame of reference. This appears startling at first, but it must be so.

Example 8.11. An airplane is moving at $\mathcal{M} = 2$ in air at $0°C = 273.15$ K and 50 kPa. What are the reservoir temperature and pressure of this air?

If we choose the ground as our frame of reference, then the velocity of the air relative to the ground is zero, so we conclude that the reservoir condition is $0°C = 273.15$ K and 50 kPa. However, this would be a very impractical choice if we wished to analyze the performance of the airplane. We would rather choose a coordinate system based on the airplane. In such a case, the air is moving at $\mathcal{M} = 2$ toward the plane, so it is certainly not standing still and hence not in its reservoir state. To find the reservoir temperature, we may slow the air down to zero velocity relative to the airplane by an adiabatic device and then measure its temperature. From Eq. 8.16 we see that the reservoir temperature is

$$T_R = (273.15)\left(2.0^2 \frac{1.4-1}{2} + 1\right) = 491.7 \text{ K} = 885°\text{R}$$

If our device for slowing down the air were frictionless (e.g., a perfect diffuser, Sec. 8.6), then the pressure at its outlet would be the reservoir pressure. From Eq. 8.17 we can compute the result:

$$P_R = 50 \text{ kPa} \cdot 1.80^{3.5} = 391 \text{ kPa} = 56.8 \text{ lbf/in}^2$$ ■

Once we have made this shift of frame of reference and the corresponding shift in reservoir conditions, we can solve the problem by using the relations developed previously for steady flow in a duct.

We might also think about this problem in another way. Suppose we want to test the airplane standing still in a wild tunnel. We want the air to come in at $\mathcal{M} = 2$ at 0°C and 50 kPa. What must be the conditions in the reservoir of the wind tunnel? Obviously, they are the values computed above. From a fluid mechanics standpoint, it makes no difference whether the plane is moving relative to the air or the air is moving relative to the plane.

From this example we see clearly that the reservoir condition is a function of the coordinate system chosen. Notice also the high reservoir temperature we compute. This is the temperature of the air in contact with the outside of the plane, where the velocity relative to the plane is zero. The high value indicates why supersonic aircraft must be refrigerated to keep the interior at a temperature suitable for human occupancy.

From the moving airplane the air appears to have the same temperature as it appears to have from a stationary position. Changing frames of reference changes only reservoir conditions, not local conditions. The temperature at any point, such as T_1, T_x, and T_y in the preceding equations, is the same for an observer riding with the fluid or an observer standing still. The observer moving relative to the fluid may have some difficulty measuring the temperature (see Prob. 8.49), but with suitable instruments one can get the same reading as the stationary observer.

The equations that we developed for normal shock waves standing still in a nozzle also apply to moving systems.

Example 8.12. An atom bomb blast in still air raises the pressure around the bomb to a high value. This high pressure causes a shock wave which flows outward. The geometry is spherical; the shock wave expands like a balloon. The pressure inside the system steadily falls as the shock wave moves out, but at any particular instant the behavior of the shock wave is given to a fair approximation by the steady-flow equations developed in Sec. 8.5. At the instant when the pressure inside the shock is 2.0 atm, how fast does the shock wave move, and what are the pressure, temperature, and velocity behind it? Outside the shock wave, the air conditions are 14.7 psia and 530°R.

We use the notation of Fig. 8.13, with the subscript x standing for the still air into which the shock wave is advancing and the subscript y applying to the gas inside the expanding high-pressure region. Then we have $P_y/P_x = 2$, and from App. A.5 $\mathcal{M}_x = 1.36$. The speed of sound in the air (Example 8.2) is 1128 ft/s, so the shock wave moves into the still air at $V = \mathcal{M}_x c_x = 1.36 \cdot 1128$ ft/s $= 1534$ ft/s $= 468$ m/s. Behind the shock the pressure is $2 \cdot 14.7 = 29.4$ psia as stated. From App. A.5 we find $T_y/T_x = 1.229$; therefore,

$$T_y = 530°\text{R} \cdot 1.229 = 651.4°\text{R} = 362\ \text{K}$$

and so we can compute the speed of sound at y as 1251 ft/s, and from App. A.5 we find $\mathcal{M}_y = 0.757$. Then

$$V_y = \mathcal{M}_y c_y = 0.757 \cdot 1251\ \text{ft/s} = 947\ \text{ft/s} = 289\ \text{m/s}$$

Thus, in the region just inside the wave, the air is moving away at 947 ft/s, as seen by an observer riding on the shock wave. Switching to fixed coordinates, we see that the shock wave is moving at 1534 ft/s, so the air adjacent to it on the high-pressure side is moving at $1534 - 947 = 587$ ft/s outward from the center.

The simple picture here hides much of the complexity of the fluid mechanics of blast waves. It is implied here that the pressure inside the expanding spherical high-pressure region is constant. That is not correct, for the pressure is lowest at the center and highest at the edge of the expanding wave [10]. ■

8.7 NOZZLES AND DIFFUSERS

Figure 8.3 is a plot of A/A^* versus $\mathcal{M}$ for steady, one-dimensional, frictionless, adiabatic flow of a perfect gas. It is, in effect, a design guide for a supersonic nozzle. If we wish the Mach number to increase linearly with distance in steady, isentropic flow of a perfect gas, then the cross-sectional area-distance relation must be exactly the curve in Fig. 8.3. This is the diagram for a *converging-diverging nozzle*, a type of nozzle commonly referred to as a *de Laval* nozzle after its inventor, who used it in the first practical steam turbine [11].

We have already discussed (see Fig. 8.4) the intuitive explanation of the necessity of the converging-diverging shape. In all our derivations for frictionless, adiabatic, steady flow of a perfect gas, we never said whether the gas was speeding up or slowing down; we said only that the flow was steady, frictionless, and adiabatic. Therefore, all the equations work equally well for an accelerating flow and a decelerating flow. In Fig. 8.3 the flow could be from left to right, as we have tacitly assumed before, or from right to left. If the latter, the gas would enter the nozzle in supersonic flow and emerge in subsonic flow. When a nozzle is used this way, it is called a *diffuser*. There is no known way to obtain a steady, supersonic flow other than by means of a converging-diverging nozzle (there are several ways to produce unsteady supersonic flows, e.g., explosives). There is no known *isentropic* way to slow down a steady, supersonic flow to subsonic speeds other than by means of a *converging-diverging diffuser*. We can steadily convert a supersonic flow to a subsonic flow by means of a normal shock, which is not isentropic.

From Fig. 8.3 and the equations on which it is based, we would assume that we could change the Mach number by any amount over as short a distance as we wished by changing the area rapidly, because there is no restriction on the equations as to dA/dl, where l is the length of the nozzle. However, if this rate of area change becomes too great, then our one-dimensional, frictionless flow assumptions become unreliable, and the observed flow no longer follows our isentropic equations. In engineering practice, the converging section has a wide angle (that is, dA/dl has a large negative value), and the diverging section has a small angle. Such a nozzle is shown in Fig. 8.14.

We are now able to calculate the flow characteristics of such a nozzle for frictionless, adiabatic flow, using the equations of isentropic flow and the equations of normal shock waves. The results we calculate here are a *reasonable approximation* of the behavior of real nozzles, in which friction is always significant. We assume that the cross-sectional area perpendicular to flow at each point in the nozzle is known, and we prepare a plot of P/P_{R_1} versus distance for various flows. We begin by making the pressures in the high-

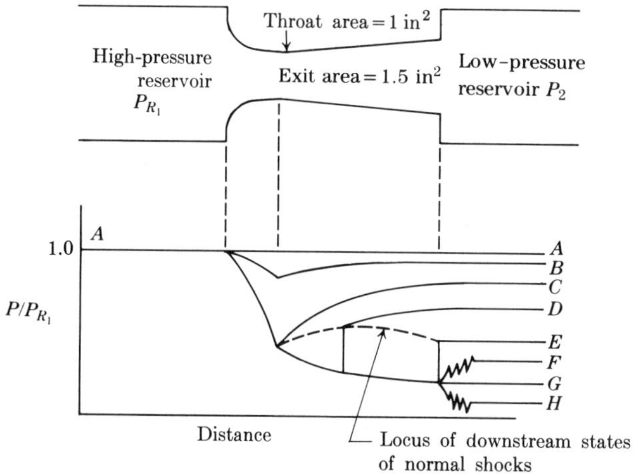

FIGURE 8.14
Pressure-distance plot for various flow rates in a converging-diverging nozzle.

pressure and low-pressure reservoirs the same. Then there is no flow, so the pressure is the same as the reservoir pressure throughout the nozzle; see line AA in Fig. 8.14.

Throughout the calculation we hold the upstream reservoir pressure P_{R_1} constant. We begin to lower the downstream reservoir pressure P_2. As long as the exiting gas is in subsonic flow, the exit pressure will be the same as the downstream pressure; so if we set the downstream pressure, we have set the pressure at the exit of the nozzle. From this pressure and the area of the nozzle, we can calculate the pressure at the throat or at any other point whose cross-sectional area we know, as shown in the following example.

Example 8.13. We now set the downstream reservoir pressure at 0.9506 times the upstream reservoir pressure (point B in Fig. 8.14). What is the pressure at the throat of the nozzle?

We look in App. A.5 for $P/P_R = 0.9506$ and find that the exit Mach number is 0.27. We also observe that A_{exit}/A^* is 2.2385. Then we can find

$$\frac{A_{\text{throat}}}{A^*} = \frac{A_{\text{exit}}}{A^*} \cdot \frac{A_{\text{throat}}}{A_{\text{exit}}} = 2.2385 \cdot \frac{1 \text{ in}^2}{1.5 \text{ in}^2} = 1.4923$$

We see in App. A.5 that this corresponds to a Mach number of about 0.43 at the throat, and therefore we have $P/P_R = 0.88$ at the throat. In the same way, we could calculate P/P_{R_1} at any point in the nozzle at which A was known, thereby completing the entire curve AB in Fig. 8.14. ■

Notice that in this calculation there is no sonic flow anywhere in the nozzle. The A^* in the calculations does not represent any area which really exists in the nozzle for this flow; it is the area which the throat would have if the flow there were sonic with the given exit area and exit P/P_R. By choosing successively lower values of P_{exit} and repeating this example, we can make a family of curves like curve AB for which the flow is entirely subsonic. The lowest pressure for which we can do this is the one for which the flow is exactly sonic at the throat.

Example 8.14. What must P_{exit} be for sonic flow at the throat in Fig. 8.14?
If the flow at the throat is sonic, then

$$A_{\text{throat}} = A^* \qquad \text{and} \qquad \frac{A_{\text{exit}}}{A_{\text{throat}}} = \frac{A_{\text{exit}}}{A^*} = 1.5$$

From App. A.5 we see that this corresponds to $\mathcal{M}_{\text{exit}} = 0.43$ and $P_{\text{exit}}/P_{R_1} = 0.88$. This is point C on Fig. 8.14. For the various cross-sectional areas existing in the nozzle, we can look up the corresponding P/P_{R_1} for this exit pressure and thus complete curve AC. ■

What happens if we lower P even more, say, to $P_2/P_{R_1} = 0.8022$? If we use exactly the same procedure as in Example 8.14, we find that $A_{\text{throat}}/A^* = 0.8175$. This is physically impossible, because in such a nozzle the minimum area occurs at the point of $\mathcal{M} = 1$ (see Fig. 8.3), and this calculation indicates that the throat area is less than this minimum area. The impossible answer results from an incorrect assumption. We assumed isentropic flow through the whole nozzle in Examples 8.13 and 8.14; for subsonic flow throughout this is correct, but here it is not. If we start with the conditions shown by curve AC in Fig. 8.14 and if we lower P_2, the flow tends to go faster. However, the nozzle is choked, because the flow is sonic at the throat, so the mass flow rate $\dot{m}$ cannot increase. With the same mass flow rate at a lower pressure, the flow immediately downstream of the throat becomes supersonic. However, the downstream pressure along the whole nozzle is not low enough for the flow to be supersonic throughout, so the flow will convert back to subsonic somewhere in the nozzle via a normal shock.

Consider the situation in which we take the subsonic flow after a normal shock wave and slow it isentropically to zero velocity, i.e., to the downstream reservoir condition. We previously showed by an energy balance that for abnormal shock wave $T_{R_x} = T_{R_y}$. What about P? By rearranging Eq. 8.17 twice, we find

$$P_x = \frac{P_{R_x}}{\{\mathcal{M}_x^2[(k-1)/2]+1\}^{k/(k-1)}} \qquad P_y = \frac{P_{R_y}}{\{\mathcal{M}_y^2[(k-1)/2]+1\}^{k/(k-1)}}$$

Substituting these in Eq. F.22 and rearranging produces

$$\frac{P_{R_y}}{P_{R_x}} = \left\{\frac{\mathcal{M}_y^2[(k-1)/2]+1}{\mathcal{M}_x^2[(k-1)/2)+1}\right\}^{k/(k-1)} \frac{2k\mathcal{M}_x^2-(k-1)}{k+1} \tag{8.33}$$

This is another messy equation, but fortunately it, too, is tabulated in App. A.5 for gases with $k = 1.4$.

The decrease in reservoir pressure has an interesting consequence. In Eq. 8.24 we showed that $\dot{m}/A^*$ was proportional to P_R. Thus, when P_R decreases, as it does in a normal shock wave, $\dot{m}/A^*$ also decreases. Equation 8.24 shows that

$$\frac{(\dot{m}/A^*)_x}{(\dot{m}/A^*)_y} = \frac{P_{R_x}}{P_{R_y}} \tag{8.34}$$

For steady flow $\dot{m}_x = \dot{m}_y$, and therefore $A_y^*/A_x^* = P_{R_x}/P_{R_y}$. We may look at the same phenomenon from the viewpoint of reversibility. Isentropic flow is reversible; the whole flow could run backward. A normal shock wave is irreversible; once it has occurred, the whole flow could not run backward, the same flow could not fit back through the same nozzle.

Thus, in this case, when the normal shock is present, although we have $A_{\text{throat}}/A^* = 1$, we cannot directly calculate the value of A_{exit}/A^*, because the value of A^* is not the same at these two points. At this point you might do well to reread what we said about A^* at the end of Sec. 8.2.

Example 8.15. Construct the curve analogous to curve AC on Fig. 8.14 for $P_2/P_{R_1} = 0.8022$. In this case we do not know the downstream reservoir pressure P_{R_2}, so we cannot yet calculate the outlet Mach number. The key unknown here is the upstream Mach number at which the normal shock occurs. Since the exit flow is subsonic, it must exit at $P_{\text{exit}} = 0.8022P_{R_1}$ and must have the A/A^* which corresponds to its Mach number. There is only one upstream Mach number at the shock which will satisfy these conditions; we find it by trial and error, using App. A.5. On our first guess we assume that $\mathcal{M}_x = 1.10$, for which we can read $P_{R_y}/P_{R_x} = 0.9989$. Here, because all the flow is isentropic except for the flow through the shock wave, we have $P_{R_1} = P_{R_x}$ and $P_{R_2} = P_{R_y}$. Then

$$\frac{P_2}{P_{R_2}} = \frac{P_2}{P_{R_1}} \cdot \frac{P_{R_x}}{P_{R_y}} = \frac{0.8022}{0.9989} = 0.8031$$

and

$$\frac{A_x^*}{A_y^*} = \frac{P_{R_y}}{P_{R_x}} = 0.9989$$

therefore,

$$\frac{A_{\text{throat}}}{A_1^*} \cdot \frac{A_x^*}{A_y^*} = 1.5 \cdot 1.0 \cdot 0.9989 = 1.4984$$

From the value of P_2/P_{R_2} given above (0.8031) we can read $\mathcal{M}_{\text{exit}} = 0.568$ and $A/A^* = 1.23$. This does not agree with the value calculated, so we must try a different Mach number. The results of the trial and error are shown in Table 8.3.

For an assumed upstream Mach number of 1.49, the two calculated values of A_{exit}/A^* practically agree; this is the correct solution. Now we can

TABLE 8.3
Results of trial-and-error solution to Example 8.14

	Trial number			
	1	2	3	4
$\mathcal{M}_1$ assumed	1.10	1.25	1.50	1.49
P_{R_y}/P_{R_x}	0.9989	0.9871	0.9298	0.9329
P_2/P_{r_2}	0.8031	0.8126	0.8628	0.8599
$\mathcal{M}_{\text{exit}}$	0.568	0.553	0.465	0.470
A_{exit}/A^* calculated from $\mathcal{M}_{\text{exit}}$	1.23	1.255	1.41	1.402
A_{exit}/A^* calculated from $A_{\text{exit}}/A_{\text{throat}}$ and P_{R_2}/P_{R_1}	1.4984	1.4806	1.394	1.399

locate the shock in the duct: It occurs at the place where the upstream Mach number is 1.49 and thus where $A/A_{\text{throat}} = 1.169$. We calculate pressures to the left of this point by the supersonic isentropic flow equations, using $A^* = A_{\text{throat}}$. We calculate pressures to the right of this point by the subsonic, isentropic flow equations, using $A^* = A_{\text{throat}}/0.9329$. The entire curve is sketched in Fig. 8.14 as curve AD. ■

If we continue to lower P_2, the shock wave will move to higher and higher values of M_x, to the right on Fig. 8.14. The farthest to the right that is possible (with the shock inside the nozzle) is at the nozzle's very exit; we can calculate the upstream Mach number there from the area ratio (by linear interpolation in App. A.5), $\mathcal{M}_x = 1.854$, and $P_x/P_{R_1} = 0.1601$, so that $P_2/P_{R_1} = 0.1601 P_y/P_x = 0.1601 \cdot 3.844 = 0.6154$. This condition is shown as curve AE on the figure. If we continue to lower the pressure in the downstream reservoir below that shown at E, then the shock wave cannot occur in the nozzle at all and the flow must exit at $\mathcal{M} = 1.854$, the pressure shown at G. If the pressure in the reservoir is more than that at G (e.g., that at F), then there will be a shock wave outside the nozzle, which will bring the flow up to the pressure of the reservoir. This will be not a normal shock but rather a two- or three-dimensional shock. If the pressure is exactly equal to that at G, the flow will be isentropic throughout and there will be no shock wave at all. If the pressure in the downstream reservoir is less than that at G (e.g., that at H), then the pressure adjustment outside the nozzle will take place via a two- or three-dimensional rarefaction.

8.8 PITOT TUBES FOR HIGH-VELOCITY GAS FLOW

An interesting application of the one-dimensional, isentropic flow theory and normal shock theory we have developed is in the calculation of the behavior of pitot-static tubes. In Sec. 5.8 we showed that for a pitot-static tube the velocity of the oncoming gas stream was given by

$$V_2 = \left[\frac{2(P_1 - P_2)}{\rho} \right]^{1/2} \tag{5.28}$$

This formula gives good results for low-velocity gas flow (less than about 200 ft/s), but for high-velocity gas flows Bernoulli's equation is no longer applicable. For all velocities between zero and sonic velocities, we can assume that the part of the mainstream which is stopped by the impact tube is stopped practically isentropically. If that is correct, then the pressure measured at P_1 is the reservoir pressure P_R for the flow. Thus, we can use Eq. 8.17, solved for V_2:

$$\frac{V_2}{(kRT_2/M)^{1/2}} = \mathcal{M}_2 = \left[\frac{(P_1/P_2)^{(k-1)/k}}{(k-1)/2} \right]^{1/2} \tag{8.35}$$

$$V_2 = \left\{ \frac{2kRT_2}{M(k-1)} \left[\left(\frac{P_1}{P_2} \right)^{(k-1)/k} - 1 \right] \right\}^{1/2} \tag{8.36}$$

In Prob. 8.58 it is shown that as P_1/P_2 approaches 1.0, this equation approaches Eq. 5.28. Thus, at low velocities the two equations give the same answer.

If the velocity in the duct is supersonic, then the fluid cannot be brought isentropically to rest at the inlet of the impact tube. This is a corollary of the previously stated fact that the only known isentropic path from a supersonic to a subsonic flow is through a converging-diverging diffuser, and none is present here. Therefore, there must be a nonisentropic transition from supersonic to subsonic flow from the mainstream to the mouth of the impact tube, by means of a normal shock wave. At which Mach number does this shock wave occur? We may hypothesize that it occurs at the Mach number of the undisturbed flow, because a converging nozzle would be needed to bring the stream to a lower Mach number—if it occurred at a lower one—and a diverging nozzle to bring it to a higher Mach number, and no such nozzle is available. Thus, we can guess that the flow to the mouth of the impact tube is isentropic, except for a normal shock wave at the Mach number of the free stream.

Example 8.16. A pitot-static tube immersed in an airstream shows an impact pressure of 15 lbf/in^2 absolute and a static pressure of 10 lbf/in^2 absolute. What is the Mach number of the flow?

On the basis of the assumptions given above,

$$\frac{P}{P_R} = \frac{10}{15} = 0.6666$$

From App. A.5 the Mach number is 0.78. If we were interested in the velocity, we also would need to know the absolute temperature. ■

Example 8.17. Repeat Example 8.16 for the static pressure $P_{\text{stat}} = 5$ psia.

Assuming isentropic flow, we find that

$$\frac{P}{P_R} = \frac{5}{15} = 0.3333$$

From App. A.5 this would correspond to a free-stream Mach number of 1.36. However, this is not correct, because, as discussed above, the transition from a supersonic flow in a pitot tube takes place by means of a normal shock. Thus, in this case the impact pressure is not the reservoir pressure of the flow but the reservoir pressure which exists downstream of a normal shock at the free-stream Mach number. The reading of the manometer $P_{\text{stat}}/P_{\text{imp}}$ is therefore P_x/P_{R_y}. This function is tabulated in App. A.5 for various values of the free-stream Mach number. In that table the free-stream Mach number is 1.285.

■

The last two examples were based on *assumed* behavior of the flow into the impact tube. Shapiro [5] reports that these assumptions have been shown experimentally to be reasonably accurate.

8.9 SUMMARY

1. The speed of sound is the speed of propagation of a *small* pressure disturbance. For perfect gases the speed of sound is given by $(kRT/M)^{1/2}$.
2. From an energy balance alone we can relate temperature to the Mach number for steady, adiabatic flow (isentropic or nonisentropic). By using the pressure-temperature relation for an isentropic change in an ideal gas, we can complete the mathematical description of steady, frictionless, adiabatic, perfect-gas flow. The mass balance equation is used to solve for the cross-sectional area perpendicular to the flow.
3. When flow in a nozzle or duct becomes sonic, further lowering of the downstream pressure will not cause the flow upstream of the sonic point to increase. This condition is called choking.
4. In high-velocity subsonic flow with friction, the effect of the friction is to increase the velocity, ultimately leading to sonic flow and choking.
5. A normal shock wave is a large pressure disturbance which travels faster than the local speed of sound. Normal shock waves are irreversible, causing an increase in entropy of the fluid flowing through them.
6. The only known way of producing a steady supersonic flow is via a converging-diverging nozzle. Such a nozzle is also the only known isentropic way of converting a steady, supersonic flow to a subsonic flow. A supersonic flow may also be converted to a subsonic one by a shock wave, which is not isentropic. Unsteady, supersonic flows can be produced in several ways, including explosions.
7. High-velocity gas flow has great practical significance in aerodynamics, rocket and turbine design, high-speed combustion, ballistics, etc. This

chapter treats only the simplest cases, showing why this kind of flow is different from the flow of liquids or the flow of gases at low velocities. Those who wish to find out more about this fascinating subject should consult Shapiro [5] or Liepmann and Roshko [12].

PROBLEMS

See the Common Units and Values for Problems and Examples inside the back cover. In all problems in this section, unless stated to the contrary, assume that gases are perfect gases and have the following properties:

Gas	M, lbm/(lbmol) = g/mol	k
Air	29	1.4
Helium	4	1.667
Hydrogen	2	1.4
Steam	18	1.33

8.1. Assuming a friction factor $f = 0.005$, calculate the pressure drop per foot due to friction for flow in a 2-in inside-diameter (ID) circular pipe at a velocity of 1000 ft/s for (*a*) water and (*b*) air. Use the constant-density formulas developed in Chap. 6.

8.2. Assuming that a sound wave causes an isentropic pressure rise of 10^{-3} psia in air, calculate the temperature rise caused by such a wave. Isentropic relations for perfect gases (e.g., air) are shown in App. D.

8.3. Calculate the speed of sound in wood. The bulk modulus of most woods is about 1.5×10^6 lbf/in^2. The density of most woods is about 60 lbm/ft^2. Is the speed of sound likely to be the same with the grain as across the grain of the wood?

8.4. Calculate the speed of sound in acetic acid at 20°C. The bulk modulus of acetic acid is 0.110×10^{10} Pa. The density of acetic acid at 20°C is about 1.05 g/cm^3.

8.5. Calculate the bulk modulus of water at 212°F from the Keenan et al. steam tables [13] or their equivalent.

8.6. Calculate the speed of sound in helium gas at 500°F.

8.7. Uranium hexafluoride (the gas used in the gaseous diffusion separation of uranium isotopes) has a molecular weight of 352. Assuming that for it $k = 1.2$, calculate its speed of sound at 200°F.

8.8. Air is flowing in a vertical nozzle 1 ft high. In the nozzle its velocity changes from 1 to 2000 ft/s. What is the ratio of the change in potential energy to the change in kinetic energy?

8.9. Rework Example 8.3 for helium gas.

8.10. Rework Example 8.4 for hydrogen gas.

8.11. Rework Example 8.5 for helium gas. Assume that the reservoir pressure is 30 psia and the reservoir temperature 70°F.

8.12. Air flows from a reservoir into a nozzle. The flow is isentropic. In the reservoir the pressure is 60 psia and the temperature 100°F. At some point in the nozzle the Mach number is 0.60. At that point what are the pressure, temperature, and velocity?

8.13. Repeat Prob. 8.12 for helium.

8.14. Air is flowing from a reservoir into a nozzle in isentropic flow. In the reservoir the pressure is 60 psia and the temperature 40°F. At the point where the velocity is 1300 ft/s, what are the temperature and pressure?

8.15. Helium is flowing from a reservoir through a nozzle in isentropic flow. The reservoir temperature and pressure are 100°F and 14.7 psia. At the point in the nozzle where the Mach number is 1.00, what are the pressure, temperature, velocity, and density?

8.16. The compressed-air line at a service station contains air at a pressure of 50 psia and a temperature of 70°F. We open the valve and let some of this air flow into the atmosphere, through a valve which may be considered an isentropic nozzle. What is the temperature of the compressed air when it flows into the atmosphere? This is cold enough that it should cause condensation of the water in the atmosphere, forming a cloud. Do we normally see such a cloud? If not, why not?

8.17. Air flows from a reservoir into a nozzle in isentropic flow. In the reservoir the velocity is negligible, T_0 is 100°F, and P_0 is 14.7 psia. At the point in the nozzle where V is 1200 ft/s, what are P, T, and $\mathcal{M}$?

8.18. Hydrogen is flowing steadily and isentropically through a nozzle. The reservoir conditions are 60°F and 20 psia. What is the Mach number at the point where the velocity is 6000 ft/s?

8.19. Steam is flowing steadily and isentropically from a reservoir through a converging-diverging nozzle. In the reservoir the pressure is 50 psia and the temperature 600°F. What are the temperature, pressure, and velocity at the point where the Mach number is 2.0?

8.20. Helium flows from a reservoir through a converging-diverging nozzle. The flow is steady and isentropic. At the point where the Mach number is 1.0, the temperature is 500°R and the pressure 20 psia. What are the reservoir pressure and temperature?

8.21. Air at 70°F and 30 psia flows from a reservoir through a nozzle with a throat area of 1 in^2. What is the maximum mass flow rate which can be passed through this nozzle?

8.22. A vacuum tank is connected to a vacuum pump through a valve. The tank has a volume of 100 ft^3; the pump has a volumetric flow rate of 10 ft^3/min, independent of the density of the material flowing through it. The valve (when wide open) is the equivalent of a reversible nozzle with cross-sectional area 10^{-4} ft^2. All of the piping of the system is very large compared to the dimensions of the valve. The vacuum tank is full of air at a pressure of 1 atm. Heating coils hold the temperature of the air in the vacuum tank at 70°F. The valve is opened wide and the pump started. How long does it take the tank pressure to fall to 0.1 atm?

8.23. Air in a reservoir at 100°F and 100 psia passes through an isentropic, converging nozzle into a second vessel, whose pressure is 80 psia. The area of the nozzle at its minimum is 2 in^2. What is the mass flow rate?

8.24. A laboratory wants to design a steady-state, supersonic wind tunnel. This tunnel is to exhaust a steady flow of gas at $\mathcal{M} = 2$ to the atmosphere. The exit area is to be 1 in^2. The upstream reservoir is to be at 70°F. Calculate the required upstream pressure, mass flow rate, and compressor horsepower needed to supply this reservoir. *Note:* For a perfect gas in an isothermal compressor, the work per unit mass is $-W/m = (RT/M)\ln(P_2/P_1)$.

8.25. Check the results of Examples 8.3 and 8.5, using App. A.5.

8.26. You have been commissioned to make up for helium a table analogous to App. A.5. Show that you know how to do this by calculating all the table entries corresponding to App. A.5 for $\mathcal{M} = 0.8$.

8.27. Section 5.6 shows a comparison of the velocities for flow through a converging nozzle, calculated by means of (*a*) Bernoulli's equation, assuming that air was a constant-density fluid, and (*b*) the equations for isentropic flow. Show the calculations in the latter case, and check your answers against those shown in Sec. 5.6.

8.28. Steam at 100 psia and 600°F is contained in a reservoir. It is expanded through a nozzle in steady, isentropic flow to a velocity of 2000 ft/s. Calculate the temperature, pressure, and Mach number at the point where the velocity is 2000 ft/s from (*a*) the Keenan et al. and Keyes steam tables [13] or their equivalent and (*b*) the equations for steady, isentropic flow of a perfect gas.

8.29. In Example 8.9 assume that at the point where $\mathcal{M} = 0.5$ the pressure is 20 psia and the density 0.102 lbm/ft^3. What are the pressure and density at the point where $\mathcal{M} = 2.0$?

8.30. Rework Example 8.9, not by making the assumption of an upstream reservoir but rather by deriving the equivalent of Eq. 8.16 for two arbitrary states, state 1 and state 2, at each of which the velocity is not negligible.

8.31. A rocket has the throat area and exit area shown in Fig. 8.15. It is being fired while held down by the test stand. In the combustion chamber the velocity is negligible, the pressure is 200 psia, the temperature is 2000°R, the molecular weight is 20 lbm/(lbmol), and k is 1.4. The flow through the nozzle is steady and isentropic. What is the thrust of this rocket?

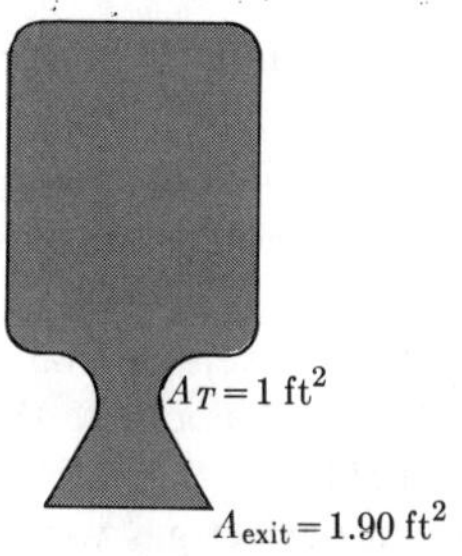

FIGURE 8.15

8.32. In the vacuum system shown in Fig. 8.16, it has been decided that the pressure in the chamber, 0.01 mmHg, is not low enough. Someone proposes that if we use a bigger vacuum pump, we can get the pressure lower. Is this true? What do you recommend?

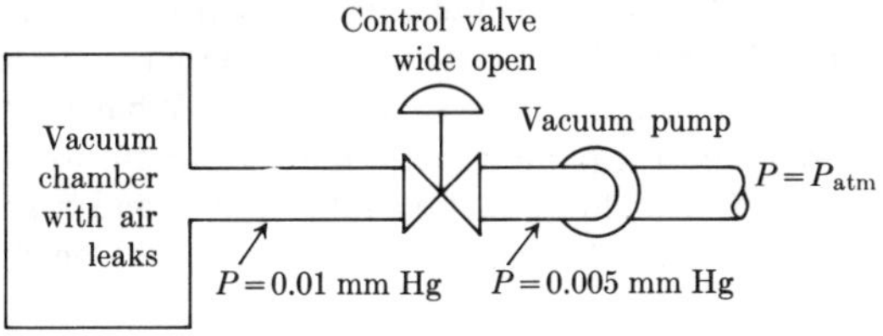

FIGURE 8.16

8.33. A process vessel has a fluctuating pressure. The range of the fluctuations is from 15 to 35 psia. The pressure never exceeds 35 psia. A system is to be designed to admit air into this vessel at a steady rate of 1 lbm/h. The compressed-air supply main is at 100 psia and 70°F. This temperature and this pressure do not fluctuate. Your coworker who is competing for a forthcoming promotion has proposed to install a conventional flow rate control system consisting of an orifice meter, differential-pressure transducer, controller, and control valve in a line between the compressed-air main and the vessel. Have you a suggestion that is likely to win you the promotion?

8.34. For the data shown in Example 8.10, when the flow is choked at the pipe outlet, what are the pressure, temperature, and Mach number at the outlet of the nozzle (station 1 on Fig. 8.11)?

8.35. Make a sketch to indicate what the lines of constant outlet Mach number would look like in Fig. 8.12.

8.36. For the case shown in Example 8.10, prepare a plot of pressure versus length from one reservoir to the other for a P_3/P_0 of (*a*) 0.8, (*b*) 0.4, and (*c*) 0.2.

8.37. The $N = 0$ line in Fig. 8.12 corresponds to an isentropic nozzle only. Check to see whether this line corresponds to the equations in Sec 8.2 by calculating the mass flow rate per unit area for air corresponding to $\mathscr{M}_1 = 0.5$ and $\mathscr{M}_1 = 1.0$ from the equations in that section and by comparing the results with the $N = 0$ line in Fig. 8.12.

8.38. A pressure vessel contains air at 30 psia and 100°F. The air is to be vented to the atmosphere through a frictionless nozzle and a length of 1-in, schedule 40 steel pipe of undetermined length.

(*a*) Prepare a sketch of the flow rate in pounds-mass per second versus the length of this pipe.

(*b*) Repeat for a square-cornered pipe entrance instead of a frictionless nozzle.

8.39. Demonstrate the stepwise procedure for making up Fig. 8.12. For $N = 2$, choose $\mathscr{M}_1 = 0.3$ and calculate the value of the mass flow rate and outlet pressure, using the stepwise procedure shown in the text. Compare the value to the one shown on Fig. 8.12.

8.40. In developing Eq. 8.30 we dropped the $\rho V\, dV$ term in Eq. 8.28, asserting that it was much smaller than the others. For a typical long-distance pipeline, the pressure is 750 psia at the outlet of a compressor station and 500 psia at the inlet of the next compressor station. The fluid may be considered a perfect gas with a constant temperature of 70°F and a molecular weight of 18 lbm/(lbmol). If the velocity at the outlet of the first pumping station is 20 ft/s, what is the ratio of the first two terms in Eq. 8.28?

8.41. Sketch a plot of P_2 versus distance, as predicted by Eq. 8.30, for constant mass flow rate, friction factor, etc.

8.42. A natural-gas line has an inside diameter of 36 in. The compressor stations are located 60 mi apart; the pressure at the outlet of the first is 750 psia, and the pressure at the inlet of the second is 500 psia. The gas may be considered a perfect gas with molecular weight 18. The temperature is constant at 70°F. What is the mass flow rate, according to the Weymouth equation? How much different would the answer have been if we had used Bernoulli's equation with friction and based the friction calculation on (*a*) the upstream density and velocity and (*b*) on the average density and velocity?

8.43. For $\mathcal{M}_x = 1.50$ calculate $\mathcal{M}_y$, T_y/T_x, P_y/P_x, and ρ_y/ρ_x for a normal shock in a gas with $k = 1.4$. Compare your results with those in App. A.5.

8.44. Show that compression shocks are thermodynamically possible and that rarefaction shocks are not. The procedure is as follows. Substitute Eqs. F.18 and F.22 into Eq. D.30 to find

$$M\frac{s_y - s_x}{R} = \ln\left\{\left[\frac{\left(\dfrac{2k}{k-1}\mathcal{M}_x^2 - 1\right)\left(1 + \dfrac{k-1}{2}\mathcal{M}_x^2\right)}{\mathcal{M}_x^2\dfrac{(k+1)^2}{2(k-1)}}\right]^{k/(k-1)}\left[\frac{k+1}{2k\mathcal{M}_x^2 - (k-1)}\right]\right\}$$

Then show that for $k < 1.67$ (i.e., for all gases) $s_y - s_x$ is positive for $\mathcal{M}_x > 1$ and negative for $\mathcal{M}_x < 1$.

8.45. Why does App. A.5 not contain a column of figures for A_2^*/A_1^* for normal shock waves?

8.46. Appendix A.5 contains a column labeled P_x/P_{R_y}. Can this column be constructed directly from any combination of other columns in the table? Check your results by calculating P_x/P_{R_y} from those columns and comparing it with the tabulated result for an $\mathcal{M}_x$ of 1.5 and 2.0. For what kind of problem would such a ratio be useful?

8.47. An earth satellite is entering the upper atmosphere at 30,000 km/h. The air it is entering has a temperature of 225 K. Estimate the temperature of the gas in contact with the surface of the satellite.

8.48. A jet fighter is flying at $\mathcal{M} = 2$ in still air which has a temperature of 0°F and a pressure of 4.0 psia. The air inlet to the jet engine is a converging-diverging duct. Inside the engine the flow is subsonic. The cross-sectional area at the throat is 2.0 ft^2. What is the mass flow rate through this diffuser, assuming isentropic flow?

8.49. A wind tunnel has a flow with $\mathcal{M} = 2$ at $T = 300$°R. If we insert a thermometer at this point, what will it read? *Hint:* If we assume that the air has zero thermal conductivity, the solution is quite simple. If the air can conduct heat (as it actually does), then the problem is more complex, and we cannot give more than an approximate answer without experimental heat-transfer data.

8.50. Air at 70°F and 10 psia is flowing at 500 ft/s in a pipe. A valve at the end of the pipe is suddenly closed. This causes a shock wave to form at the closed end of the duct and to move up it. Calculate the speed at which this wave moves up the duct and the temperature and pressure in the closed end of the duct. *Hint:* The gas downstream of the shock is standing still. Regardless of which coordinate system is chosen, we have $V_x - V_y = 500$ ft/s. If we take the coordinate system to be based on the moving shock wave, then we can use the tabulated values in App. A.5 to solve (by trial and error) for the upstream Mach number which has this relation between the two velocities. (An analytical solution of this type of problem is shown by Zucker [14].)

8.51. At some point in the nozzle shown in Fig. 8.14, the cross-sectional area is 1.5 times the throat area. Assuming that the flow is isentropic, steady flow of a perfect gas with no shock waves and $k = 1.4$, what are the possible Mach numbers at that point, when the Mach number at the throat is 1.0? What is the Mach number at that point, when the Mach numbers at the throat are 0.1, 0.5, and 0.9?

8.52. A converging-diverging nozzle has an outlet area equal to 1.9 times the throat area. Using App. A.5, prepare a plot of outlet Mach number versus P_{out}/P_R for all possible outlet conditions, subject to the assumptions of isentropic flow, with or without normal shock waves.

8.53. In a converging-diverging nozzle, the cross-sectional area at point x (downstream of the throat) is 1.5 times the cross-sectional area at the throat. Air is flowing steadily through the nozzle. The reservoir temperature is 530°R.
(*a*) For isentropic flow list all possible values of T at point x.
(*b*) For flow that is isentropic except for possible shock waves, list all possible values for T at point x.

8.54. In a converging-diverging nozzle, the exit area is 1.50 times the throat area. The flow of air is isentropic except for the possibility of shock waves. What is the Mach number at the throat when the Mach number at the exit is (*a*) 0.30 and (*b*) 0.50?

8.55. Air flows through a supersonic wind tunnel; see Fig. 8.17. The flow is steady and isentropic, except that somewhere in the system there is a normal shock wave. If $A_2/A_1 = 1.01$ and $\mathcal{M} = 1.0$ at both A_1 and A_2, what is the upstream Mach number at the place where the normal shock occurs?

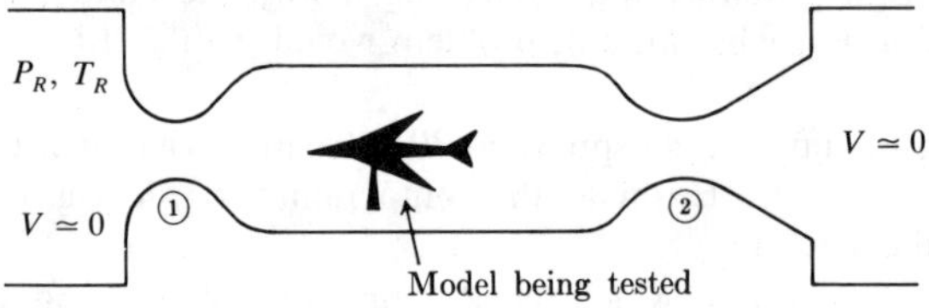

FIGURE 8.17

8.56. Repeat Example 8.15 for an outlet pressure of $0.70P_{R_1}$. What is the lowest outlet pressure for which you can perform this calculation?

8.57. In the nozzle in Fig. 8.18 air is flowing, and there is a normal shock wave at A. The remainder of the flow is isentropic. For this shock wave we have $\mathcal{M}_x = 3.0$. What is P_{R_y}/P_{R_x}? Some of the following values from the NACA tables [15] may be useful. For $\mathcal{M}_x = 3$ we have $\mathcal{M}_y = 0.4752$, $P_y/P_x = 10.33$, $T_y/T_x = 2.679$, and $\rho_y/\rho_x = 3.857$.

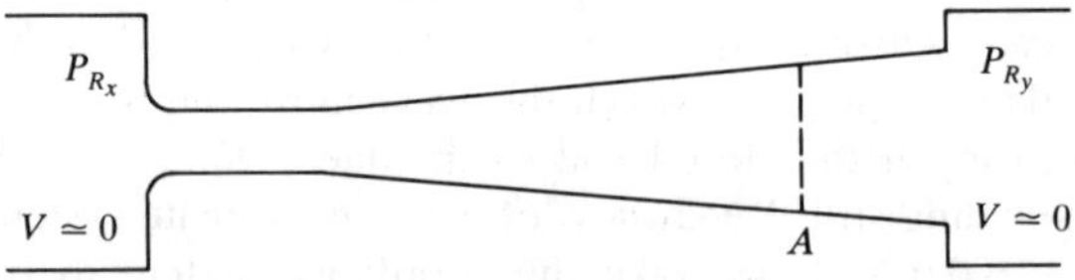

FIGURE 8.18

8.58. Show that as P_1/P_2 approaches 1, Eq. 8.36 approaches Eq. 5.28. *Hint:* Represent $(P_1/P_2)^{(k-1)/k}$ by its series $y^x = 1 + x \ln y + (x^2 \ln^2 y)/2! + (x^3 \ln^3 y)/3! \ldots \ln(1+x) = x - x^2/2 + x^3/3 \ldots$

8.59. Air is flowing steadily in a nozzle. At some point in the nozzle, the Mach number is 2.0 and the absolute pressure 4.0 psia. We now insert a pitot-static tube which has a differential-pressure gauge. What will the gauge read?

CHAPTER 9

PUMPS, COMPRESSORS, AND TURBINES

In Chapters 4, 5, and 6, we have written energy balance equations which involve a $dW_{\text{a.o.}}$ term (see Sec. 4.8 for a definition of $dW_{\text{a.o.}}$). For steady-flow problems this term generally represents the action of a pump, fan, blower, compressor, turbine, etc. Here we discuss the fluid mechanics of the devices which actually perform that $dW_{\text{a.o.}}$.

9.1 POSITIVE-DISPLACEMENT PUMPS

A pump is a device which does work *on* a liquid; $dW_{\text{a.o.}}/dm$ is negative. The distinction between pumps and compressors which do work on a gas is arbitrary, but it is in common use. Most mechanical pumps are one of these:

1. Positive-displacement
2. Centrifugal
3. Special designs with characteristics intermediate between the two

In addition, there are nonmechanical pumps (i.e., electromagnetic, ion, diffusion, jet, etc.), which are not considered here.

Positive-displacement (PD) pumps work by allowing a fluid to flow into some enclosed cavity from a low-pressure source, trapping the fluid, and then forcing it out into a high-pressure receiver by decreasing the volume of the cavity. These are extremely common; examples are the fuel and oil pumps on most automobiles, the pumps on most hydraulic systems, and the hearts of most animals.

Figure 9.1 shows the cross-sectional view of a simple PD pump. The operating cycle of such a pump is as follows, starting with the piston at the top:

1. The piston starts downward, creating a slight vacuum in the cylinder.
2. The pressure of the fluid in the inlet line is high enough relative to this vacuum to force open the left-hand valve, whose spring has been designed to let the valve open under this slight pressure difference.
3. Fluid flows in during the entire downward movement of the piston.
4. The piston reaches the bottom of its stroke and starts upward. This raises the pressure in the cylinder higher than the pressure in the inlet line, so the inlet valve is pulled shut by its spring.
5. The pressure continues to rise until it is higher than the pressure in the outlet line.
6. When the pressure in the cylinder is higher than the pressure in the outlet line, the outlet valve is forced open.
7. The piston pushes the fluid out into the outlet line.
8. The piston starts downward again; the spring closes the outlet valve, because the pressure in the cylinder has fallen, and the cycle begins again.

Suppose that we test such a pump, using a pump test stand, as shown in Fig. 9.2. With the pump discharge pressure regulator (a control valve) we can regulate the discharge pressure and, using the bucket, scale, and clock, determine the flow rate corresponding to that pressure. For a given speed of the pump's motor, the results for various discharge pressures are shown in Fig. 9.3.[†]

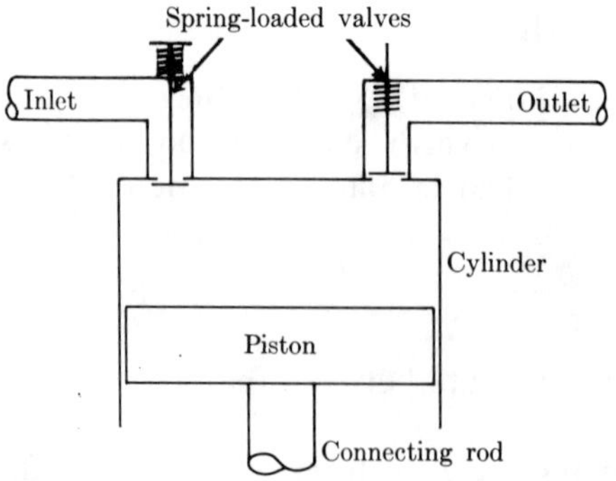

FIGURE 9.1
Positive-displacement pump.

[†] The input and output pressures and velocities of a PD pump vary cyclically. All the values shown in this section are average pressures or average velocities.

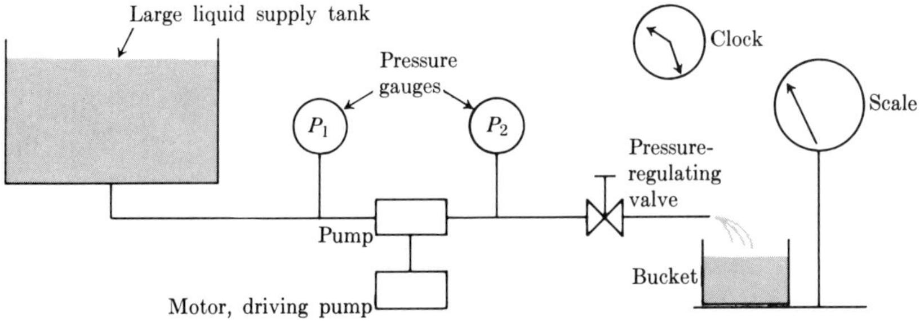

FIGURE 9.2
Pump test stand.

From Fig. 9.3 we see that PD pumps are practically constant-volumetric-flow-rate devices (at a fixed motor speed) and that they can generate large pressures. The danger that these large pressures will break something is so severe that these pumps must always have some kind of safety valve to relieve the pressure if a line is accidentally blocked.

For a perfect PD pump and an absolutely incompressible fluid, the volumetric flow rate equals the volume swept out per unit time by the piston, or

$$\text{Volumetric flow rate} = \text{piston area} \cdot \text{piston travel} \cdot \frac{\text{cycles}}{\text{time}} \tag{9.1}$$

For an actual pump the flow rate will be slightly less because of various fluid leakages.

If we write Bernoulli's equation (Eq. 5.7) from the inlet of this pump to the outlet and solve for the work input to the pump, we find

$$\frac{-dW_{\text{a.o.}}}{dm} = \Delta\left(\frac{P}{\rho} + gz + \frac{V^2}{2}\right) + \mathscr{F} \tag{9.2}$$

Here the first term on the right of the equals sign represents the "useful" work done by the pump: increasing the pressure, elevation, or velocity of the fluid. The second represents the "useless" work done in heating either the fluid or the surroundings. The normal definition of pump efficiency is

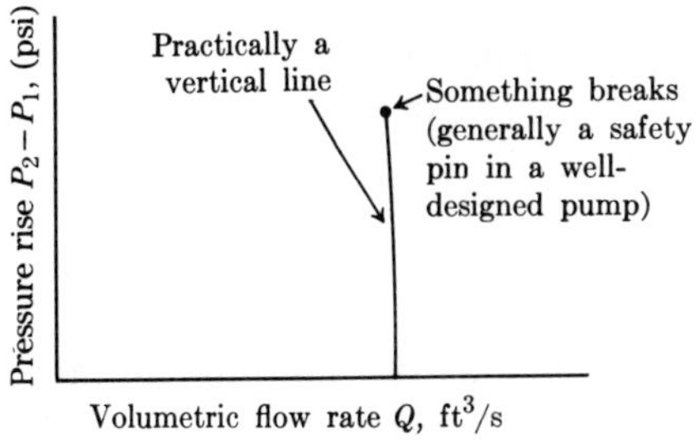

FIGURE 9.3
Pump performance curve for a PD pump.

$$\text{Pump efficiency } \eta = \frac{\text{useful work}}{\text{total work}} = \frac{\Delta(P/\rho + gz + V^2/2)}{-dW_{\text{a.o.}}/dm} \tag{9.3}$$

This gives the pump efficiency in terms of a unit mass of fluid passing through the pump. It is often convenient to multiply the top and bottom of this equation by the mass flow rate $\dot{m}$, which makes the denominator exactly equal to the power supplied to the pump:

$$\eta = \frac{\dot{m} \cdot \Delta(P/\rho + gz + V^2/2)}{\text{power supplied}} \tag{9.4}$$

Example 9.1. A pump is pumping 50 gal/min of water from a pressure of 30 psia to a pressure of 100 psia. The changes in elevation and velocity are negligible. The motor which drives the pump is supplying 2.80 hp. What is the efficiency of the pump?

The mass flow rate through the pump is

$$\dot{m} = \frac{50\text{ gal}}{\text{min}} \cdot 8.33\,\frac{\text{lbm}}{\text{gal}} = 417\,\frac{\text{lbm}}{\text{min}} = 189\,\frac{\text{kg}}{\text{min}}$$

so, from Eq. 9.4,

$$\eta = \frac{417\,\dfrac{\text{lbm}}{\text{min}} \cdot \dfrac{100\text{ lbf/in}^2 - 30\text{ lbf/in}^2}{62.3\text{ lbm/ft}^3} \cdot \dfrac{144\text{ in}^2}{\text{ft}^2} \cdot \dfrac{\text{hp}\cdot\text{min}}{33{,}000\text{ ft}\cdot\text{lbf}}}{2.80\text{ hp}} = 0.73 \qquad \blacksquare$$

From this calculation we see that the numerator in Eq. 9.4 has the dimension of horsepower. We may think of it as the useful work in horsepower. This numerator is often referred to as the *hydraulic horsepower* of the pump, in which case the pump efficiency becomes equal to the hydraulic horsepower divided by the total horsepower supplied to the pump. In Example 9.1 we can calculate that the hydraulic horsepower is 2.04 hp.

For large PD pumps, the efficiency can be as high as 0.90; for small pumps it is less. One may show (Prob. 9.4) that for the pump in this example the energy which was converted to friction heating and thereby heated the fluid would cause a negligible temperature rise. The same is not true of gas compressors, as discussed in Sec. 9.3.

If we connect our PD pump to a sump, as shown in Fig. 9.4, and start the motor, what will happen? A PD pump is generally operable as a vacuum pump. Therefore, the pump will create a vacuum in the inlet line. This will make the fluid rise in the inlet line.

If we write the head form of Bernoulli's equation, Eq. 5.11, between the free surface of the fluid (point 1) and the inside of the pump cylinder, there is no pump work over this section; so

$$z_2 - z_1 = h = \frac{P_1 - P_2}{\rho g} - \frac{V_2^2}{2g} - \frac{\mathscr{F}}{g} \tag{9.5}$$

If, as shown in Fig. 9.4, the fluid tank is open to the atmosphere, then $P_1 = P_{\text{atm}}$. The maximum value of h possible corresponds to $P_2 = 0$. If there is

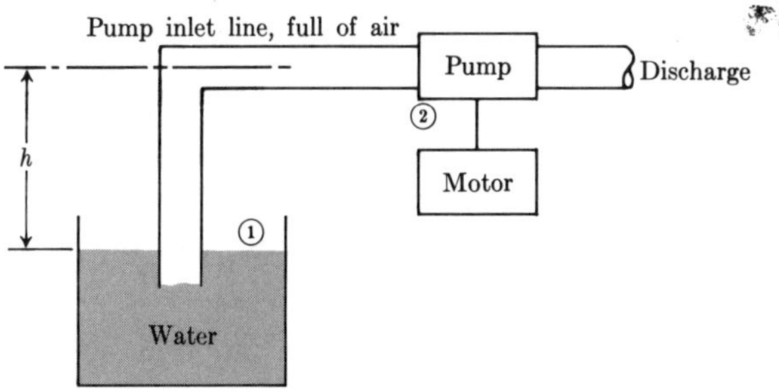

FIGURE 9.4
Suction lift.

no friction and the velocity at 2 is negligible, then

$$h_{max} = \frac{P_{atm}}{\rho g} \tag{9.6}$$

For water under normal atmospheric pressure and room temperature, this is about 34 ft = 10 m. This height is called the *suction lift*.

The actual suction lift obtainable with a PD pump is less than that shown by Eq. 9.6 because

1. There is always some line friction, some friction effect through the pump inlet valve, and some inlet velocity.
2. The pressure on the liquid cannot be reduced to zero with causing the liquid to boil. All liquids have some finite vapor pressure. For water at room temperature, it is about 0.3 psia or 0.02 atm. If the pressure is lowered below this value, the liquid will boil.

Example 9.2. We wish to pump 200 gal/min of water at 150°F from a sump. We have a PD pump which can reduce the absolute pressure in its cylinder to 1 psia. We have an $\mathscr{F}/g$ (for the pipe only) of 4 ft. The friction effect in the inlet valve may be considered the same as that of a sudden expansion (see Sec. 5.5) with the inlet velocity equal to the fluid flow velocity through the valve, which here is 10 ft/s. The atmospheric pressure at this location is never less than 14.5 psia. What is the maximum elevation above the water level in the sump at which we can place the pump inlet?

The lowest pressure we can allow in the cylinder P_2 is 3.72 psia, the vapor pressure of water at 150°F. If the pressure were lower than this, the water would boil, interrupting the flow. The density of water at 150°F is 61.3 lbm/ft^3. Thus,

$$h_{max} = \frac{14.5\ \text{lbf/in}^2 - 3.7\ \text{lbf/in}^2}{61.3\ \text{lbm/ft}^3 \cdot 32.2\ \text{ft/s}^2} \cdot \frac{144\ \text{in}^2}{\text{ft}^2} \cdot 32.2\ \frac{\text{lbm}\cdot\text{ft}}{\text{lbf}\cdot\text{s}^2} - \frac{(10\ \text{ft/s})^2}{2\cdot 32.2\ \text{ft/s}^2} - 4\ \text{ft}$$
$$= 25.4\ \text{ft} - 1.6\ \text{ft} - 4\ \text{ft} = 19.8\ \text{ft} = 6.04\ \text{m}$$

An experienced engineer would select a lower suction lift if possible. ■

We have discussed only the piston-and-cylinder types of PD pump. These have a large number of moving, wearing parts and hence are expensive to buy and maintain. Several other types of PD pumps have been developed which are simpler and cheaper than this type, in small sizes, e.g., gear, sliding vane, peristaltic, and screw pumps. Although the mechanical arrangements are different, the operating principle and the performance of these are similar to those of piston-and-cylinder types.

9.2 CENTRIFUGAL PUMPS

A centrifugal pump raises the pressure of a liquid by giving it a high kinetic energy and then converting that kinetic energy to injection work. The water pump on most automobiles is a typical centrifugal pump. As shown in Fig. 9.5, it consists of an impeller (i.e., a wheel with blades) and some form of housing with a central inlet and a peripheral outlet.

In such a pump, the fluid flows in the central inlet into the "eye" of the impeller, is spun outward by the rotating impeller, and flows out through the peripheral outlet. To analyze such a pump on a *very simplified basis*, we write Bernoulli's equation (Eq. 5.7) between the inlet pipe (point 1) and the outer tips of the impeller blades (point 2). The elevation change and inlet velocity are negligible. We assume that the friction losses also are negligible. Although P_2 is greater than P_1, this term is small compared with the change in kinetic energy. So we can write

$$\frac{V_2^2}{2} = \frac{-dW_{\text{a.o.}}}{dm}, \quad \text{[approximately]} \tag{9.7}$$

This equation indicates that the pump work, which enters through the rotating shaft, principally goes to increase the kinetic energy of the fluid as it flows across the impeller from the eye to the tips of the blades. The tangential velocity at any point is

$$\text{Tangential velocity} = \text{radius} \cdot \text{angular velocity} \tag{9.8}$$

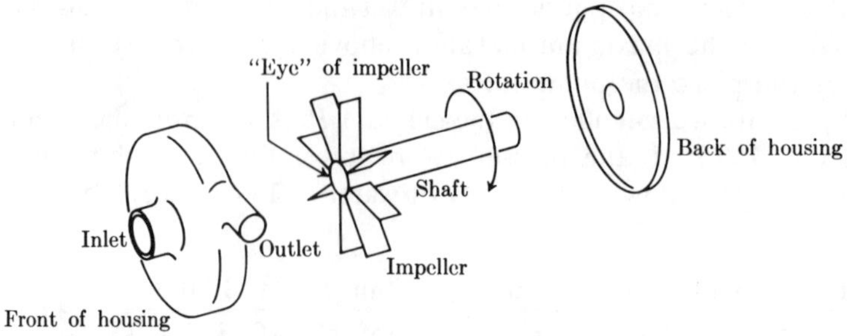

FIGURE 9.5
Exploded view of a centrifugal pump, such as an automobile water pump.

The angular velocity (2π rpm) is constant for the whole impeller, but the radius increases from zero at the eye to a significant value at the tip of the blades; so the tangential velocity increases markedly for the fluid from the point where it enters the pump to the tip of the blades.

Now we apply Bernoulli's equation from the tip of the blades (point 2) to the outlet pipe (point 3). Again the change in elevation is negligible, and we neglect friction. The pump does no work on the fluid after it leaves the tip of the blades, so $dW_{\text{a.o.}}/dm$ is zero. The outlet velocity is small and may be neglected. Thus

$$\frac{P_3 - P_2}{\rho} = \frac{V_2^2}{2} \qquad \text{[approximately]} \tag{9.9}$$

This equation indicates that the section of the pump from the tip of the rotor blades to the outlet pipe converts the kinetic energy of the fluid to increased pressure (i.e., to injection work). This whole section is called the *diffuser*. Thus, the centrifugal pump may be considered a two-stage device:

1. The impeller increases the kinetic energy of the fluid at practically constant pressure.
2. The diffuser converts this kinetic energy to increased pressure.

Equations 9.7 and 9.9 suggest that for a given pump size and speed, $\Delta P/(\rho g)$ should be constant; so for a pump test run with an apparatus like that in Fig. 9.2, a plot of $\Delta P/(\rho g)$ should be a horizontal line. Figure 9.6 shows the results of such a test for a large, high-efficiency pump. As predicted by the simple model, for low flow rates $\Delta P/(\rho g)$ is indeed independent of the flow rate. Equations 9.7 and 9.9 indicate that $\Delta P/(\rho g)$ depends on the square of the tip speed of the rotor; so we can change it at will by changing either the rotor diameter or the angular velocity. Over a wide range of values this is also experimentally observed.

Example 9.3. A typical centrifugal pump runs at 1800 rpm (mainly because of the convenient availability of 1800 rpm electric motors). If the fluid being pumped is water, what is the maximum pressure difference across the pump for impeller diameters of 1, 3, and 10 in?

Using Eq. 9.9, we find that

$$P_2 - P_1 = \frac{\rho V_1^2}{2} = \rho\,\frac{(r\omega)^2}{2} = \frac{\rho}{2}\left(\frac{D}{2}\,2\pi\,\text{rpm}\right)^2$$

For an impeller of 1-in diameter,

$$P_2 - P_1 = \frac{62.3\ \text{lbm/ft}^3}{2}\cdot\left(\frac{\text{ft}}{12}\,\pi\,\frac{1800}{\text{min}}\right)^2\cdot\frac{\text{min}^2}{3600\ \text{s}^2}\cdot\frac{\text{lbf}\cdot\text{s}^2}{32.2\ \text{lbm}\cdot\text{ft}}\cdot\frac{\text{ft}^2}{144\ \text{in}^2}$$

$$= 0.41\ \text{psi} = 2.86\ \text{kPa}$$

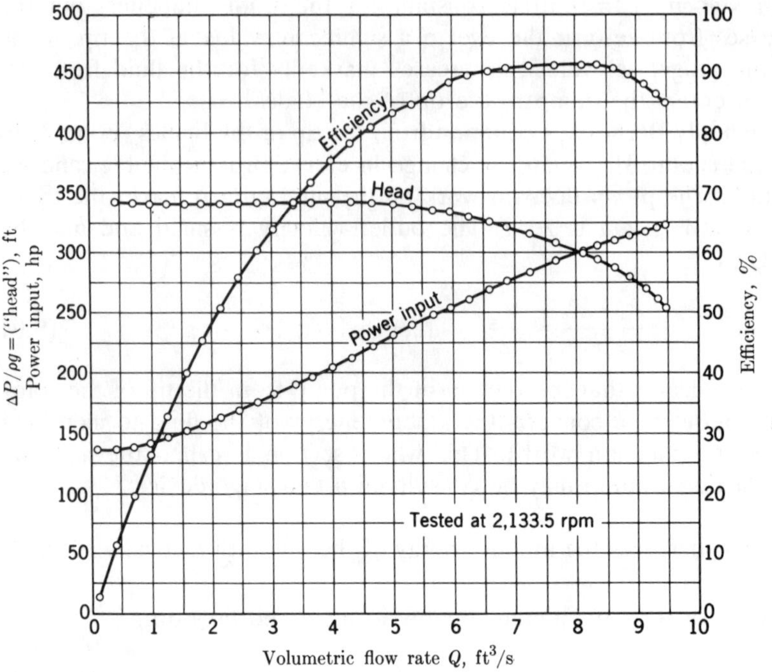

FIGURE 9.6
Pump test results on a centrifugal water pump with impeller diameter of 14.62 in. (From R. L. Daugherty and J. B. Franzini, *Fluid Mechanics with Engineering Applications*, 6th ed., McGraw-Hill, New York, 1965. Reproduced by permission of the publisher.)

For a 3-in-diameter impeller,

$$P_2 - P_1 = 0.41 \text{ psi} \left(\frac{3 \text{ in}}{1 \text{ in}}\right)^2 = 3.7 \text{ psi}$$

and for a 10-in diameter impeller,

$$P_2 - P_1 = 0.41 \text{ psi} \left(\frac{10 \text{ in}}{1 \text{ in}}\right)^2 = 41 \text{ psi} \qquad \blacksquare$$

This example illustrates the fact that centrifugal pumps supply small pressure rises with small impellers and reasonable pressure rises with large impellers. When a large pressure rise is required, we can obtain it by hooking several centrifugal pumps in series (head to tail). The normal practice is to put several impellers on a common shaft and to design a casing so that the outlet from one impeller is fed through a diffuser directly to the inlet of the next impeller. This is particularly true of deep-well centrifugal irrigation pumps, which have many impellers on a common shaft. The impeller size is limited by the diameter of the well.

The performance curve of a real pump, shown on Fig. 9.6, indicates some of the limitations of our simple model:

1. As the flow rate becomes large, the pressure rise decreases, which is not predicted by the model.
2. The model would indicate 100 percent efficiency for all flow rates, whereas the actual efficiency varies over a wide range, peaking near 90 percent at the design operating range.
3. The value of $\Delta P/(\rho g)$ predicted by the model is 286 ft, about 84 percent of the observed value. By more complicated theoretical models [1, 2], one can predict pump performance with better accuracy.

Note the striking difference between the outlet pressure behavior of PD pumps and that of centrifugal pumps. The discharge pressure of a PD pump is determined by the pressure in the outlet line. The piston will continue to raise the pressure of the fluid in its cylinder until one of two things happens:

1. This pressure exceeds the pressure in the outlet line, and the outlet valve opens.
2. Something breaks, or a safety valve opens.

For a centrifugal pump, however, there is a maximum pressure rise across the pump set by the tip speed of the impeller and the fluid density. If the difference between the inlet and outlet pressures becomes greater than this, fluid will flow backward through the pump—in the outlet and out the inlet.

Suppose that we repeat the pump-starting experiment shown in Fig. 9.4 for a centrifugal pump. As in Example 9.3, we let the pump turn at 1800 rpm and have an impeller of 10-in diameter. If the pump is full of water, the pump has a pressure rise of 42 psi; so there is no problem with the suction lift other than those discussed before in connection with PD pumps. However, when we start the pump, the fluid around the impeller is air. Therefore, from Eq. 9.9 we know that the pressure rise across the pump is

$$P_2 - P_1 = \rho \frac{V_1^2}{2} = 41\ \text{psi} \cdot \frac{\rho_{\text{air}}}{\rho_w} = 41\ \text{psi} \cdot \frac{0.075\ \text{lbm/ft}^3}{62.3\ \text{lbm/ft}^3} = 0.05\ \text{psi} \qquad !$$

Thus, if this pump is discharging into the atmosphere and the sump is open to the atmosphere, then the pump, when full of air, can lift the water only a height of

$$h = \frac{\Delta P}{\rho g} = \frac{0.05\ \text{lbf/in}^2}{62.3\ \text{lbm/ft}^3 \cdot 32.2\ \text{ft/s}^2} \cdot \frac{32.2\ \text{lbm} \cdot \text{ft}}{\text{lbf} \cdot \text{s}^2} \cdot \frac{144\ \text{in}^2}{\text{ft}^2} = 0.1\ \text{ft} = 0.03\ \text{m}$$

Thus, to get a centrifugal pump going, it is not enough to start the motor. One must also replace the air in the system with liquid. This is called *priming*. Numerous schemes for performing this function are available; in addition, special self-priming pumps have been patented and manufactured.

In Example 9.2 we saw that in calculating the permissible suction lift for a PD pump it was necessary to account for the pressure drop across the inlet valve. The difficulty to be avoided there was boiling the liquid as its pressure was reduced through the valve. The analogous problem with centrifugal pumps is boiling in the eye of the impeller. As the fluid enters the eye, it has a small velocity in the axial direction and negligible rotational velocity. To be picked up by the blades of the impeller, the fluid must be brought up to the rotational speed of the impeller blades. We again apply Bernoulli's equation between the inlet pipe (point 1) and the point on the blades of the impeller where the pump work starts to increase the pressure of the fluid (point $1a$). If we assume that the friction effects are negligible, ignore changes in elevation, and ignore the velocity of the inlet fluid, we conclude that

$$P_{1a} = P_1 - \rho \frac{V_{1a}^2}{2} \tag{9.10}$$

Thus, the pressure falls and boiling may occur, as discussed in Sec. 5.9. Centrifugal pumps are often used to pump boiling liquids, e.g., at the bottom of many distillation columns, flash drums, evaporators, reflux drums, reboilers, condensers, etc. (see Fig. 9.7).

In this case the pump must be located far enough *below* the boiling surface so that the pressure rise due to gravity from the boiling surface to the pump eye is larger than the pressure drop in the pump eye. This elevation is shown as h in Fig. 9.7. This distance required below the boiling surface is called the *net positive suction head* (NPSH). Many pump manufacturers measure the required NPSH for their pumps and report it, as in Fig. 9.8. The pressure in Fig. 9.8 is the pressure measured at the pump inlet. If there is significant frictional pressure drop between the vessel and the pump, then the height h in Fig. 9.7 must be increased to overcome this additional pressure drop.

As shown in Fig. 9.6, a large, high-efficiency, centrifugal pump may have an efficiency greater than 0.90. However, such high-efficiency pumps are expensive and are justified only for very high-capacity applications (e.g., the

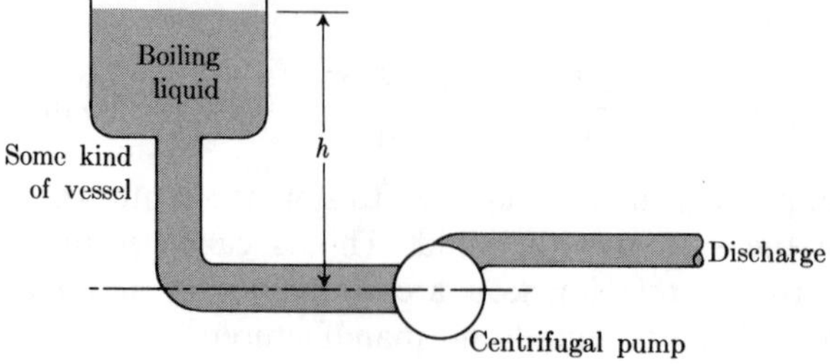

FIGURE 9.7
Centrifugal pump used to pump a boiling liquid.

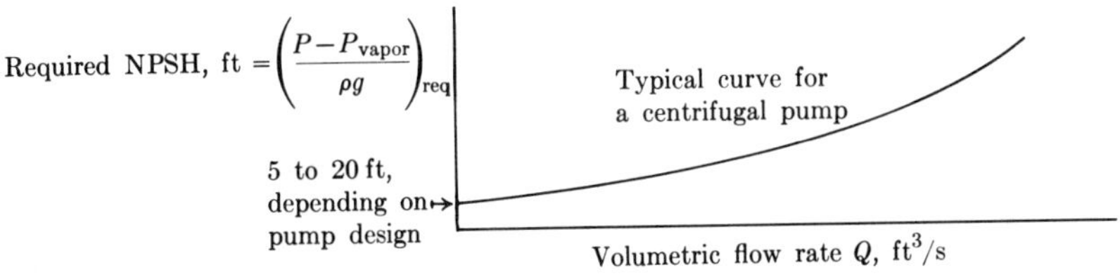

FIGURE 9.8
NPSH requirement of a typical centrifugal pump.

TABLE 9.1
Characteristic of positive-displacement and centrifugal pumps

Characteristic	Positive-displacement	Centrifugal
Normal flow rate	Low	High
Normal pressure rise per stage	High	Low
Variable held constant over normal operating range	Flow rate	Pressure rise
Self-priming	Yes	No
Number of moving and wearing parts	Many	Few
Outlet stream	Pulsing	Steady
Works well on high-viscosity fluids	Yes	No

aqueduct from the Colorado River to Los Angeles). For most applications pumps are designed for simpler, cheaper construction and efficiencies of 0.50 to 0.80. The efficiency of centrifugal pumps decreases rapidly as the viscosity of the pumped fluid increases, so they are seldom used for fluids more viscous than 1000 cP.

The characteristics of PD and centrifugal pumps are compared in Table 9.1. The great majority of the world's pumps are of the two types described in the table. However, many other types fulfill some special need better than do the two described in the table. Among these are the following:

1. Propeller pumps, which pump large quantities of fluid at low pressure differences more efficiently than a centrifugal pump
2. Turbine or regenerative pumps, which resemble centrifugal pumps but because of a quite different impeller design can produce much larger pressure rises than centrifugal pumps of the same size and speed

9.3 POSITIVE-DISPLACEMENT COMPRESSORS

A compressor is a pump which pumps a gas, with P_{outlet}/P_{inlet} significantly greater than 1.0. If the pressure change for a gas in passing through a pump is small, that pump is called a *blower* or *fan*. Blowers and fans work practically the same way as centrifugal pumps, and their behavior can be readily predicted by the equations developed for centrifugal pumps. As shown in Sec. 9.2, an impeller with a 10-in diameter rotating at 1800 rpm in atmospheric air can produce a maximum pressure rise of about 0.05 psia. For many air conditioning applications this is adequate. Normally the impellers of blowers are larger than 10 in, producing pressure rises as great as 1 psi. To compress a gas to a final (absolute) pressure of more than about 1.1 times its inlet pressure requires a compressor, and the change in density of the gas must be taken into account.

A PD compressor has the same general form as a PD pump; see Fig. 9.1. The operating sequence is the same as that described in Sec. 9.1. The differences are in the size and speed of the various parts. The pressure-volume history of the gas in the cylinder of such a compressor is shown in Fig. 9.9.

Here we have simplified the behavior of real compressors by assuming that when the piston reaches the top of its travel, there is no volume left between the piston and the top of the cylinder. This is a *zero-clearance compressor*. Later we examine the consequences of the fact that real compressors generally leave a little gas in the cylinder at the top of the stroke; those consequences are minor.

For a zero-clearance compressor, at the top of the stroke (point A) the pressure is equal to the outlet pressure P_{outlet}, and the volume enclosed in the cylinder is zero. As soon as the piston begins to descend, the pressure falls instantaneously. When the pressure reaches P_{inlet}, the inlet valve opens, still at volume $V = 0$ for a zero-clearance compressor, point B. Then the gas from the inlet line flows in at a constant pressure until the piston reaches the bottom of its stroke at V_2 (point C). As soon as the piston starts up, the inlet valve closes, and then both valves remain closed while the gas in the cylinder is compressed from its inlet pressure to its outlet pressure. When the gas reaches that pressure, the outlet valve opens ($P = P_{\text{outlet}}$, $V = V_1$), at point D. Then as the piston continues to rise, the gas is forced out into the outlet line, completing the cycle at point A.

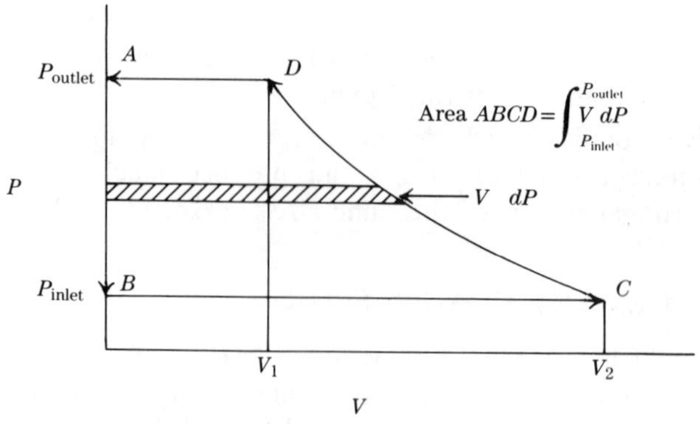

FIGURE 9.9
Pressure-volume behavior of a PD compressor.

The work of any single-piston process is given by

$$W = \int F\,dx = \int PA\,dx = \int P\,dV \tag{9.11}$$

The work done by the compressor on the gas is the gross work done on the gas (area under curve *CDA*) minus the work done by the gas on the piston as the gas flowed in (area under curve *BC*); thus, the net work is the area enclosed by curve *ABCD*. This is the work done on the gas. It is equal to the gross work required to drive the compressor only if there is no friction or gas leakage, etc. (that is, 100 percent efficient operation).

This area on a *PV* diagram is the algebraic sum of three areas: the area $P_{\text{inlet}}V_2$, which is the work done by the gas in the inlet line driving back the piston, the area $\int_C^D P\,dV$ which is the work input of the compression step, and the area $P_{\text{outlet}}V_1$, which is the work to drive the gas out of the cylinder into the outlet line. We can see that although for any one of these three steps the work is given by a $P\,dV$ integral, the algebraic sum of three such $P\,dV$ integrals is a $V\,dP$ integral. This integral is the work done *by* the compressor *on* the gas. In normal thermodynamics convention (and Bernoulli's equation), this would be negative work, because work done by the system is taken as positive, and

$$-W_{\text{per cycle}} = \int_{\text{inlet}}^{\text{outlet}} V\,dP \qquad \text{[zero-clearance compressor]} \tag{9.12}$$

Compressors are often used to compress gases which can be reasonably well represented by the perfect-gas law $PV = nRT$. If a compressor works slowly enough and has good cooling facilities, then the gas in the cylinder will be at practically a constant temperature throughout the entire compression process. Then we may substitute nRT/P for V in Eq. 9.12 and integrate:

$$-\Delta W = \int_{P_1}^{P_2} V\,dP = nRT \int_{P_1}^{P_2} \frac{dP}{P} = nRT \ln \frac{P_2}{P_1} \qquad \text{[isothermal]} \tag{9.13}$$

However, in most compressors the piston moves too rapidly for the gas to be cooled much by the cylinder walls. If it moves very rapidly, the gas will undergo what is practically a reversible, adiabatic process, i.e., an isentropic process. In that case, we may rearrange App. Eq. D.24 to

$$PV^k = \text{constant} = P_1V_1^k \qquad \text{[adiabatic]} \tag{9.14}$$

Inserting this in Eq. 9.12 and making algebraic manipulations, as shown in App. G, we find

$$-\Delta W = \int_{P_1}^{P_2} V\,dP = \frac{nRT_1k}{k-1}\left[\left(\frac{P_2}{P_1}\right)^{(k-1)/k} - 1\right] \qquad \text{[adiabatic]} \tag{9.15}$$

Most often we divide both sides of Eqs. 9.13 and 9.15 by the number of moles compressed n to find the work per mole.

Example 9.4. A 100 percent efficient compressor is required to compress air from 1 to 3 atm. The inlet temperature is 68°F. Calculate the work per pound-mole for an isothermal compressor and an adiabatic compressor.

For an isothermal compressor,

$$\frac{-\Delta W}{n} = RT \ln \frac{P_2}{P_1} = 1.987 \frac{\text{Btu}}{\text{lbmol} \cdot {}^\circ\text{R}} \cdot 528^\circ\text{R} \cdot \ln 3 = 1153 \frac{\text{Btu}}{\text{lbmol}} = 2.68 \frac{\text{kJ}}{\text{mol}}$$

For an adiabatic compressor,

$$\frac{-\Delta W}{n} = 1.987 \frac{\text{Btu}}{\text{lbmol} \cdot {}^\circ\text{R}} \cdot 528^\circ\text{R} \cdot \left(\frac{1.4}{0.4}\right) \cdot (3^{0.4/1.4} - 1) = 1355 \frac{\text{Btu}}{\text{lbmol}}$$

$$= 3.15 \frac{\text{kJ}}{\text{mol}}$$

The difference between the two answers is due to the rise in temperature of the gas in the adiabatic compressor. From Eq. D.17 we can calculate that the adiabatic compressor's outlet temperature was 1.369 times its inlet temperature, or 723°R. Thus, by the perfect-gas law we know that its exit volume is 1.369 times the exit volume for the isothermal compressor, and the integral of $V\,dP$ must be larger. ■

Equations 9.13 and 9.15 indicate that the result in Example 9.4 is a general result; i.e., the required work for an adiabatic compressor is always greater than that for an isothermal compressor with the same inlet and outlet pressures. Therefore, it is advantageous to try to make real compressors as nearly isothermal as possible. One way to do this is to cool the cylinders of the compressor, and this is generally done with cooling jackets or cooling fins on compressors. Another way is by staging and intercooling; see Example 9.5.

Example 9.5. Air is to be compressed from 1 to 10 atm. The inlet temperature is 68°F. What is the work per mole for (*a*) an isothermal compressor, (*b*) an adiabatic compressor, and (*c*) a two-stage adiabatic compressor in which the gas is compressed adiabatically to 3 atm, then cooled to 68°F, and then compressed from 3 to 10 atm?

(*a*) For an isothermal compressor,

$$\frac{-\Delta W}{n} = 1.987 \frac{\text{Btu}}{\text{lbmol} \cdot {}^\circ\text{R}} \cdot 528^\circ\text{R} \cdot \ln 10 = 2416 \frac{\text{Btu}}{\text{lbmol}} = 5.62 \frac{\text{kJ}}{\text{mol}}$$

(*b*) For an adiabatic compressor,

$$\frac{-\Delta W}{n} = 1.987 \frac{\text{Btu}}{\text{lbmol} \cdot {}^\circ\text{R}} \cdot 528^\circ\text{R} \cdot \frac{1.4}{0.4} \cdot (10^{0.4/1.4} - 1) = 3418 \frac{\text{Btu}}{\text{lbmol}} = 7.96 \frac{\text{kJ}}{\text{mol}}$$

(c) For a two-stage adiabatic compressor,

$$\frac{-\Delta W}{n} = 1.987 \frac{\text{Btu}}{\text{lbmol} \cdot {}^\circ\text{R}} \cdot 528^\circ\text{R} \cdot \frac{1.4}{0.4} \cdot \left[3^{0.4/1.4} - 1 + \left(\frac{10}{3}\right)^{0.4/1.4} - 1\right]$$

$$= 2862 \frac{\text{Btu}}{\text{lbmol}} = 6.66 \frac{\text{kJ}}{\text{mol}} \quad ■$$

This example illustrates the advantage of staging and intercooling. There is a limit to the amount of work savings possible; with an infinite number of stages with intercooling an adiabatic compressor would have the same performance as an isothermal compressor (Prob. 9.21). Thus, the behavior of an isothermal compressor represents the best performance obtainable by staging. The optimum number of stages is found by an economic balance between the extra cost of each additional stage and the improved performance as the number of stages is increased.

In Example 9.5 the interstage pressure was arbitrarily selected as 3 atm. It can be shown (Prob. 9.17) that for a two-stage compressor the optimum interstage pressure (that which requires the least amount of total work) is given by

$$P_{\text{interstage}} = (P_{\text{discharge}} \cdot P_{\text{inlet}})^{1/2} \tag{9.16}$$

This is the interstage pressure which makes the pressure ratio $P_{\text{out}}/P_{\text{in}}$ the same for each stage. By similar calculations it can be shown (Prob. 9.18) that for more than two stages the optimum interstage pressures are those which have the same pressure ratio for each stage.

In calculating the work requirement for adiabatic compressors, we must evaluate the factor $(P_2/P_1)^{(k-1)/k} - 1$ in every problem. To save time, specialists in this field have prepared tables of this factor, which they call the *X factor*, for various pressure ratios and $k = 1.395$; see Perry [3].

All the foregoing concerned zero-clearance compressors, ones in which no gas is left in the cylinder at the end of the discharge stroke. For mechanical reasons it is impractical to build a compressor with zero clearance. So in real compressors there is always a small amount of gas in the top of the cylinder, which is repeatedly compressed and expanded. If the compression and expansion are reversible, either adiabatic or isothermal, then they contribute as much work on the expansion step as they require on the compression step, and thus they contribute nothing to the net work requirement of the compressor. For real compressors the compression and the expansion of the gas in the clearance volume contribute to the inefficiency of the compressor; compressor designers make the clearance volume as small as practical.

Equations 9.13 and 9.15 were derived with the assumption of a zero-clearance compressor. They apply equally well to real compressors if we understand the n in them to represent the net number of moles passing through, not the total number of moles present in the cylinder at the start of the compression stroke. Because we normally analyze the work of compressors on the basis of work per mole or unit mass processed, this causes no difficulty. In these derivations and in Fig. 9.9, we assumed no pressure drop through the inlet and outlet valves. All real compressors of this type have some pressure drop through these valves; it is a significant part of the friction heating that causes the efficiency to be less than 100 percent.

Large PD compressors typically have stage pressure ratios P_{out}/P_{in} of 3 to 5 and efficiencies (Sec. 9.5) of 75 to 85 percent.

9.4 ROTARY COMPRESSORS

The PD compressor has been a common industrial tool for a century. However, it is a complicated, heavy, expensive, low-flow-rate device. The need to supercharge aircraft reciprocating engines and the development of turbojet and gas-turbine engines demanded the development of lightweight, efficient, low-cost, high-flow-rate compressors. The resulting compressors, which were developed for aircraft service, are now being applied industrially in high-capacity applications, for example, in ammonia plants.

The two types of compressor developed for aircraft engines are centrifugal and axial-flow. The centrifugal compressor is a centrifugal pump with a very high-speed (for example, 20,000 rpm) and large-diameter rotor. To give high-pressure ratios, centrifugal compressors are normally staged without intercooling; the pressure rise per stage is small.

Axial-flow compressors pass the gas between numerous (for example, 20) rows of blades arranged in an annulus; see Fig. 9.10. The gas is successively accelerated by a moving row of rotor blades and then slowed by a stator blade, which converts the kinetic energy imparted by the rotor blades to pressure (i.e., injection work). The advantages of the axial-flow compressor over centrifugal compressors are the small cross-sectional area perpendicular to the gas flow, which makes it easy to build into a streamlined airplane, and the lower velocities, which lead to lower friction losses and slightly higher efficiency.

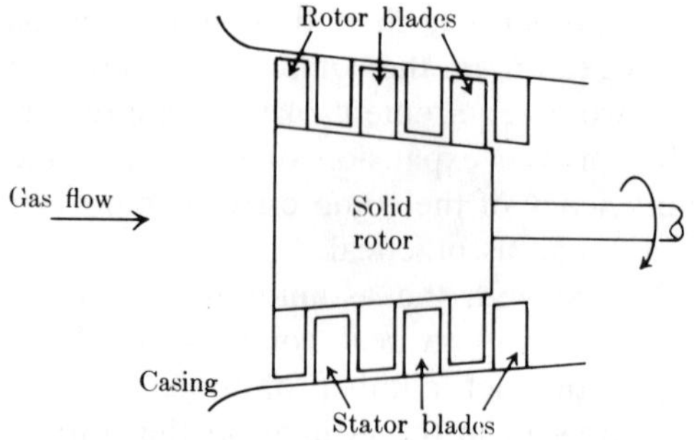

FIGURE 9.10
Axial-flow compressor.

Centrifugal and axial-flow compressors generally handle very large volumes of gases in small equipment, so the heat transfer from the gases is negligible. Thus, their performance is well described by the equations for adiabatic compressors (see Eq. 9.15). Efficiencies (Sec. 9.5) normally run from 80 to 90 percent. Although these compressors are lighter, simpler, and cheaper than PD compressors of equal capacity, they are difficult to build for high-pressure ratios (i.e., greater than about 5) or for low flow rates.

Several types of compressor have characteristics intermediate between those of PD compressors and centrifugal or axial-flow compressors. These include the lobe type, sliding vane, and liquid piston. For all these, the discussion of compressor work and compressor efficiency applies just as well as the above-mentioned types. Descriptions of the mechanical arrangements of these can be found in texts on pumps and compressors [2].

This discussion of compressors has been largely from the mechanical viewpoint; this viewpoint is helpful in understanding how these devices actually function. In Chap. 5 we showed that for any steady-flow compressor or turbine in which changes in potential and kinetic energy are negligible,

$$\frac{dW_{\text{a.o.}}}{dm} = h_{\text{in}} - h_{\text{out}} + \frac{dQ}{dm} \tag{4.40}$$

We may readily find Eq. 9.15 by substituting the relation for an isentropic process for a perfect gas in Eq. 4.40. We may find Eq. 9.13 by substituting the isothermal relation for a perfect gas in Eq. 4.40 and using the entropy balance to solve for dQ/dm (Prob. 9.24). Thus, we may find exactly the same results by a mechanical view of what happens inside the compressor or by a thermodynamic view of the compressor as the system from the outside.

9.5 COMPRESSOR EFFICIENCIES

In defining the efficiency of a pump (Sec. 9.1), we compared the useful work to the total work. For an incompressible fluid, this is done most easily by means of Bernoulli's equation, which is restricted to constant-density fluids. The definition for a pump could be restated as

$$\text{Efficiency} = \frac{\text{work required for best possible device doing job}}{\text{work actually required by device}} \tag{9.17}$$

This statement can be used for compressors as well. In the case of an adiabatic compressor, the best possible device is a reversible, adiabatic compressor for which the inlet and outlet entropies are the same. It is commonly called an *isentropic compressor*. Thus, we normally define

$$\text{Compressor efficiency } \eta = \frac{\text{work of isentropic compressor}}{\text{work of real compressor}} \tag{9.18}$$

Consider the steady-flow, adiabatic compressor shown in Fig. 9.11. The energy balance for this process (taking the compressor as the system and

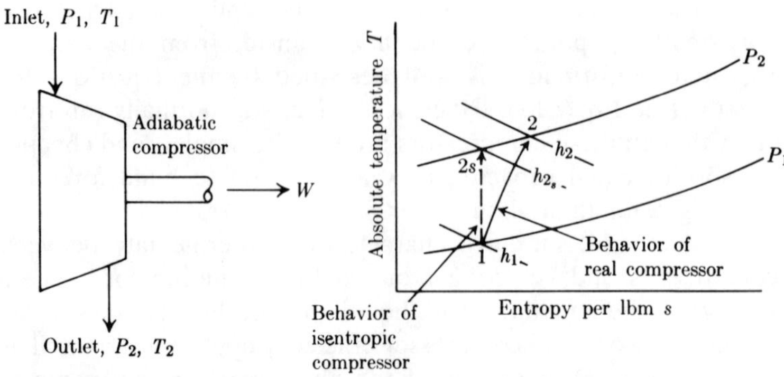

FIGURE 9.11
Compressor efficiency.

assuming that changes in kinetic and potential energies are negligible) is

$$\frac{dW}{dm} = h_{\text{in}} - h_{\text{out}} = h_1 - h_2 \tag{9.19}$$

If we wish to compare the work done by this compressor with that of a reversible compressor, we immediately see that we cannot compare it with a reversible compressor doing exactly the same thing, because the real compressor has a higher outlet entropy, temperature, and enthalpy than the outlet stream from a reversible compressor would ($2s$ in Fig. 9.11). Thus, we could compare the real compressor with a reversible one having the same outlet enthalpy, having the same outlet temperature, or having the same outlet pressure. The latter seems to be the most logical choice, since real compressors are generally regulated by controlling the outlet pressure; this is the choice which had been universally made in defining the efficiency of compressors. So Eq. 9.19 becomes

$$\eta_{\text{pump or compressor}} = \frac{W_{\text{isentropic}}}{W_{\text{real}}} = \frac{h_{2s} - h_1}{h_2 - h_1} \tag{9.20}$$

Example 9.6. An adiabatic compressor is compressing air from 20°C and 1 to 4 atm. The airflow rate is 100 kg/h, and the power required to drive the compressor is 5.3 kW. What are the efficiency of the compressor and the temperature of the outlet air? What would the outlet air temperature be if the compressor were 100 percent efficient?

Air may be assumed to be a perfect gas with $C_P = 29.3\ \text{J}/(\text{mol} \cdot \text{K})$ and $k = 1.40$. The work of an isentropic compressor doing the same job is given by Eq. 9.15:

$$-\frac{dW_{\text{a.o.}}}{dm} = \frac{293.15\ \text{K}}{29\ \text{g/mol}} \cdot \frac{8.314\ \text{J}}{\text{mol} \cdot \text{K}} \cdot \frac{1.4}{0.4} \cdot (4^{0.4/1.4} - 1) = 143\ \frac{\text{J}}{\text{g}}$$

$$\text{Po} = \dot{m}\left(-\frac{dW_{\text{a.o.}}}{dm}\right) = 10^5\ \frac{\text{g}}{\text{h}} \cdot 143\ \frac{\text{J}}{\text{g}} \cdot \frac{\text{W}\cdot\text{s}}{\text{J}} \cdot \frac{\text{h}}{3600\ \text{s}} = 3.97\ \text{kW} = 5.32\ \text{hp}$$

$$\eta = \frac{3.97\ \text{kW}}{5.3\ \text{kW}} = 0.75$$

$$\Delta T_{\text{real}} = \frac{\Delta h}{C_P} = \frac{-dW_{\text{a.o.}}/dm}{C_P}$$

$$= \frac{5.3\ \text{kW}/(100\ \text{kg/h})}{[29.3\ \text{J}/(\text{mol}\cdot\text{K})]\cdot(\text{mol}/29\ \text{g})} \cdot \frac{3600\ \text{s}}{\text{h}} \cdot \frac{\text{J}}{\text{W}\cdot\text{s}}$$

$$= 189\ \text{K} = 340°\text{F}$$

$$\therefore T_{\text{out}} = 20°\text{C} + 189\ \text{K} = 209°\text{C} = 408°\text{F}$$

For an isentropic compressor,

$$\Delta T_{\text{ideal}} = \frac{\Delta h}{C_P} = \frac{-dW_{\text{a.o.}}/dm}{C_P}$$

$$= \frac{3.97\ \text{kW}/(100\ \text{kg/h})}{[29.3\ \text{J}/(\text{mol}\cdot\text{K})]\cdot(\text{mol}/29\ \text{g})} \cdot \frac{3600\ \text{s}}{\text{h}} \cdot \frac{\text{J}}{\text{W}\cdot\text{s}}$$

$$= 142\ \text{K}$$ ■

This example illustrates the fact that the effect of the compressor inefficiency is to raise the outlet temperature of the compressor. One may look upon this compressor inefficiency as being a type of friction heating. The extra work above that which would have been required for a 100 percent efficient compressor goes to heat either the gas passing through the compressor or the surroundings.

9.6 FLUID ENGINES AND TURBINES

A pump or compressor does work on a fluid in order to increase the fluid's pressure, elevation, velocity, or internal energy. A fluid engine or turbine extracts work from a fluid by lowering its pressure, elevation, velocity, or internal energy. These definitions are the reverse of each other, so some devices could serve as pumps or as fluid engines, depending on what way they were run. Tidal power plants and pumped storage power plants use the same device as a pump for part of the day and as a turbine for another part of the day.

The PD pump shown in Fig. 9.1 can be used as a fluid engine with simple changes in valve timing. This is the form of the steam engine which supplied most of the world's mechanical power in the nineteenth and early twentieth centuries; its operation is the reverse of that of the compressor, shown diagrammatically in Fig. 9.9. This type of engine has been replaced for very large applications with turbines (cheaper and simpler) and in industry with electric motors. Most of the remaining applications are in locations where there is a sufficient fire hazard to require the distribution and application of power as steam or compressed air (spark-free) rather than as electricity, e.g., in mines and some chemical plants.

For large installations extracting power from fluids (generally water in hydroelectric plants and steam in thermal power plants), the most common device is a turbine. Water and steam turbines have the same principles of operation but very different sizes, shapes, and speeds.

A turbine consists of a wheel with blades attached to its periphery and the associated casing, etc.; for many kinds of turbines these blades are called *buckets*. The blades change the direction of the flow, which results in a force on the blades. This force turning the wheel produces power. The simplest turbine is the child's pinwheel; see Fig. 9.12. In the pinwheel a high-velocity jet strikes the blades and is slowed down. This type of device, in which the fluid undergoes its pressure reduction in the fixed nozzle (the child's lips in the figure) and flows through the turbine at practically constant pressure, is called an *impulse turbine*. Its behavior is discussed in Sec. 7.4, where it is shown that for the most efficient operation the blade speed should be one-half of the jet speed, resulting in the exit fluid's having negligible velocity.

Another kind of turbine is the *reaction turbine*, in which the fluid enters the blades with a negligible velocity and leaves at a high velocity relative to the blades. The simplest reaction turbine is the rotating garden sprinkler; see Fig. 9.13.

This is called a reaction turbine for the same reason that a rocket motor is often called a reaction motor; the force exerted is described by Newton's third law, "action equals reaction." A turbine of this type could be constructed by attaching two rockets to the ends of a shaft in place of the water jets shown in Fig. 9.13. In a reaction turbine, the pressure reduction takes place in a moving nozzle.

The simple reaction turbine is analyzed most easily by the angular momentum balance, Eq. 7.65, which, as shown in Sec. 7.7 for a steady-flow turbine, reduces to Euler's turbine equation (Eq. 7.66). To find the power produced per unit time, we multiply both sides of Eq. 7.66 by $-\omega$ (the minus sign is used here to agree with the thermodynamic sign convention that power produced by the fluid passing through the system is positive):

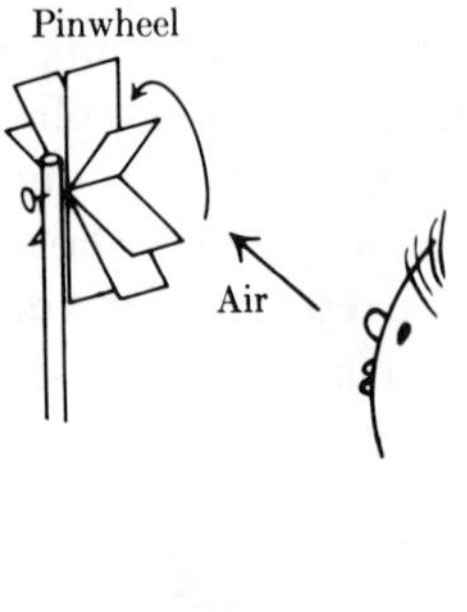

FIGURE 9.12
Child's pinwheel, the simplest impulse turbine.

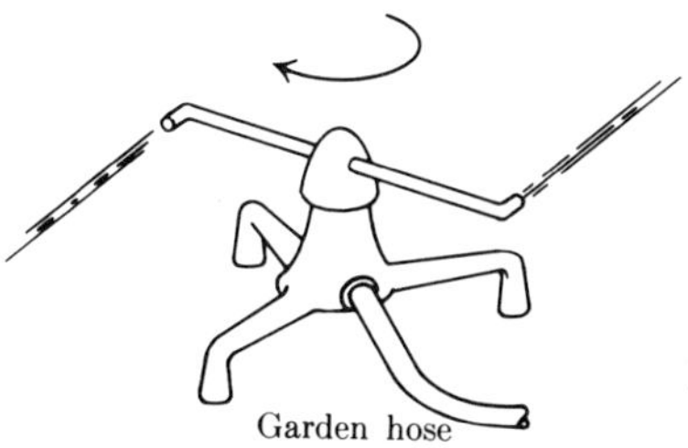

FIGURE 9.13
Rotating garden sprinkler, the simplest reaction turbine.

$$\text{Po} = \frac{dW_{\text{a.o.}}}{dt} = -\Gamma\omega = -\dot{m}\omega[(rV_t)_{\text{out}} - (rV_t)_{\text{in}}] \tag{9.21}$$

Here the fluid enters at the center ($r_{\text{in}} = 0$), so the far right term is zero. Then solving for the work per unit mass, we find

$$\frac{dW_{\text{a.o.}}}{dm} = -\omega(rV_t)_{\text{out}}$$

But ωr_{out} is the tangential velocity of the nozzle, so

$$\frac{dW_{\text{a.o.}}}{dm} = -V_{t_{\text{noz}}} V_{t_{\text{out}}}$$

The velocity $V_{t_{\text{out}}}$ is equal to $V_{t_{\text{noz}}} + V_{\text{rel}}$, where V_{rel} is the velocity of the jet as measured by an observer riding on the nozzle; so

$$\frac{dW_{\text{a.o.}}}{dm} = -V_{t_{\text{noz}}}(V_{\text{rel}} + V_{t_{\text{noz}}}) \tag{9.22}$$

In the simple garden sprinkler shown in the figure, V_{rel} is independent of $V_{t_{\text{noz}}}$; it depends on the pressure in the garden hose and the friction in the system. For constant V_{rel}, we may find the most efficient value of $V_{t_{\text{noz}}}$, that is, the one which gives the largest value of $dW_{\text{a.o.}}/dm$, by differentiating Eq. 9.22 with respect to $V_{t_{\text{noz}}}$ and setting the derivative equal to zero. The result is $V_{t_{\text{noz}}} = -\frac{1}{2}V_{\text{rel}}$; that is, the nozzle moves at one-half of the jet speed, as measured by an observer riding on the nozzle, and in the opposite direction.

Comparing the simple impulse turbine in Fig. 9.12 with the simple reaction turbine in Fig. 9.13, we see that the former is more efficient. Either may be considered to consist of a nozzle which converts internal energy and injection work ($\Delta P/\rho$) to kinetic energy and a device which uses this kinetic energy to produce work. In either case the kinetic energy of the fluid leaving the system is wasted. As shown in Sec. 7.4, we can, in principle, build an impulse turbine for which the outlet velocity (based on fixed coordinates) is zero. However, for a reaction turbine like that in Fig. 9.14, the maximum efficiency corresponds to $V_{\text{out}} = \frac{1}{2}V_{\text{rel}}$; so the outlet kinetic energy will be one-fourth of the total available kinetic energy. For this reason the pure reaction turbine of Fig. 9.13 is inefficient and is never used in industrial practice.

Most modern water and steam turbines take part of the pressure reduction in a set of fixed nozzles and part of the pressure reduction in the moving wheel. Thus, they are part impulse and part reaction. However, since pure impulse turbines are in current use, common terminology is to reserve the term "impulse turbine" for a turbine that is 100 percent impulse, with no reaction, and to call any turbine that is not 100 percent impulse a *reaction turbine*. Most reaction turbines are less than 50 percent reaction, with the remainder being impulse.

Liquid turbines are large, slow-moving devices, and gas and steam turbines are small, fast-moving devices. This is easiest to see for an impulse turbine, for which we previously showed that the optimum velocity of the blade is one-half that of the jet. So the speed of a single-stage impulse turbine rotor is set by the available jet speed. At Hoover Dam the fluid drops about 700 ft. Applying Bernoulli's equation from the water surface to the turbine nozzle, we can solve for the maximum possible jet velocity:

$$V = (2gh)^{1/2} = \left(2 \cdot 32.2\ \frac{\text{ft}}{\text{s}^2} \cdot 700\ \text{ft}\right)^{1/2} = 212\ \frac{\text{ft}}{\text{s}} = 65\ \frac{\text{m}}{\text{s}}$$

Therefore, the blade of the turbine should travel about 106 ft/s. It is convenient to build the rotor about 10 ft in diameter, so the rate of rotation (rpm) is about

$$\text{rpm} = \frac{106\ \text{ft/s}}{10\pi\ \text{ft}} \cdot \frac{60\ \text{s}}{\text{min}} = 200\ \text{rpm} \qquad \text{approximately}$$

Now consider a steam turbine; as shown by the methods in Chap. 8, the reversible, adiabatic expansion of steam through a nozzle from about 100 psia to atmospheric pressure produces a flow with a velocity of about 3000 ft/s. Thus, the blade should move at about 1500 ft/s. Here also it would be desirable to use a large-diameter wheel, but at these high rotational speeds the centrifugal force tending to pull the turbine wheel apart becomes so great that only a small-diameter wheel can survive. The largest single-stage steam-turbine wheels are about 2 or 3 ft in diameter. For a 3-ft wheel and a blade speed of 1500 ft/s, the rotational speed is

$$\text{rpm} = \frac{1500\ \text{ft/s}}{3\pi\ \text{ft}} \cdot \frac{60\ \text{s}}{\text{min}} = 15{,}000\ \text{rpm} \qquad \text{approximately}$$

The first successful steam turbine, developed by de Laval [4], was a simple one-stage impulse turbine, as described above, and it turned at about 20,000 rpm.

The turbines on most jet and gas-turbine engines are of the simple impulse variety described above and turn at about 20,000 rpm. However, this is an inconvenient speed for connecting to a generator which is producing 60-Hz current; it must run at $3600/n$ rpm, where n is any integer. The most common U.S. generator speeds are 1800, 1200, and 600 rpm. A high-speed turbine could be connected to such a generator through a gear reducer, but the more economical solution seems to be to build a turbine of lower speed. In it many wheels are attached to a common shaft; the steam flows through a nozzle, then a bladed wheel, then another nozzle, etc. Each combination of nozzle and wheel is, in effect, a separate turbine; in common terminology such a nozzle-and-wheel combination is a *stage*, and thus multiwheel turbines are called *multistage turbines*. Each stage has a small pressure reduction and, thus, a small jet velocity and a small tip speed for efficient operation. In current steam-turbine practice, the first few stages usually are pure impulse, followed by stages that are 50 percent impulse and 50 percent reaction.

For high-head, low-flow-rate, water power application, the most economical turbine is the pure impulse Pelton wheel. For high flow rates and lower heads, the friction effects in the Pelton wheel cut down its efficiency, and a radially inward-flowing, part-reaction Francis turbine seems to be the most economical. For very high flow rates and very low heads, the most economical is the Kaplan turbine, which looks quite like a ship's propeller, with adjustable blade pitch to correct for changes in head and flow rate. The Francis and Kaplan turbines can be designed to function efficiently at many more revolutions per minute than an impulse turbine with the same input head; this is an advantage, because generators of very slow speed (for example, 50 rpm) are expensive to build. Details of the various kinds of steam and gas turbines may be found elsewhere [4, 5]. Details of water turbines also may be found elsewhere [1]. There are numerous general works on pumps [6] and on pumps as used in chemical engineering [7, 8].

9.7 FLUID ENGINE AND TURBINE EFFICIENCY

Fluid engine and turbine efficiency is defined as the inverse of pump or compressor efficiency:

$$\text{Efficiency} = \frac{\text{work actually delivered}}{\text{maximum possible work}} \tag{9.23}$$

For an incompressible fluid (e.g., water), the common definition of the maximum possible work is the work that would be delivered if the fluid left the system with zero velocity and if the $\mathcal{F}$ term in Bernoulli's equation were zero. For gases (e.g., steam), the common definition of maximum work is that work which would have been obtained for zero outlet velocity and isentropic operation. Although the forms of these maximum-work definitions are different, they can both be shown to be the same, because the $\mathcal{F}$ term in Bernoulli's equation is related to the irreversible entropy increase.

9.8 SUMMARY

1. Positive-displacement pumps work by trapping a fluid in a cavity and then squeezing it out at a higher pressure; they are generally high-pressure-rise, low-flow-rate devices.
2. Centrifugal pumps work by giving the fluid kinetic energy and then converting this to injection work. They are generally high-flow-rate, low-pressure-rise devices.

3. The problem of boiling in the inlet line (or *cavitation*) limits how high any kind of pump may be placed above the reservoir on which it is drawing.
4. Compressors are pumps which pump gases over significant changes in pressure. These are normally practically adiabatic and result in a significant temperature rise for the gas. The work of compression is less for an isothermal than for an adiabatic compressor; for this reason many compressors are staged, with intercooling to decrease the work requirement.
5. Turbines work by impulse, in which a fluid is accelerated by a fixed nozzle and slowed down by moving blades, or by reaction, in which a fluid is accelerated by a moving nozzle. Most turbines are either pure impulse or part impulse and part reaction.

PROBLEMS

See the Common Units and Values for Problems and Examples inside the back cover.

9.1. How many gallons per minute should be delivered by a pump with a piston area of 10 in^2 and a piston stroke of 5 in with a speed of 1 Hz?

9.2. Calculate the hydraulic horsepower for pumping 500 gal/min from an inlet pressure of 5 psig to an outlet pressure of 30 psig.

9.3. It has been suggested that the way to solve the Los Angeles Basin smog problem is to construct fans to pump the air out of the basin and discharge it into the Mohave Desert. Let us assume that we wish to remove a layer 2000 ft thick every day from an area 70 mi by 60 mi. We also make a first guess that the pressure drop necessary from the flow is 1 psi. For this small a pressure drop, air may be considered an incompressible fluid. What hydraulic horsepower is required?

9.4. For the pump discussed in Example 9.1, calculate the temperature rise for the fluid passing through the pump. Assume that there is no heat transfer to the surroundings.

9.5. We wish to pump mercury from a sump with a PD pump. Assuming that there is no friction and that the vapor pressure of mercury is negligible, what is the maximum height above the sump at which we can locate the pump?

9.6. Repeat Example 9.3 for the pump being driven at 3600 rpm.

9.7. From the flow rate and "head" shown on Fig. 9.6, calculate the efficiency at a flow rate of 5 ft^3/s. How does this compare with the efficiency shown in Fig. 9.6?

9.8. A centrifugal pump is pumping mercury. The inlet pressure is 200 psia. The pump impeller is 2 in in diameter, and the pump is rotating at 20,000 rpm. Estimate the outlet pressure.

9.9. A centrifugal pump is tested with water and found at 1800 rpm to deliver 200 gal/min at a pressure rise of 50 psi. The mechanical efficiency is 75 percent. We wish to pump mercury in this pump at the same revolutions per minute and flow rate. Estimate the pressure rise and pump horsepower required for this operation, assuming that the pump remains 75 percent efficient.

9.10. We wish to design a centrifugal blower for air. It will take in air at 1 atm and 68°F and deliver it at a gauge pressure of 2 psig. The impeller will rotate at 3600 rpm. What is the minimum impeller diameter, assuming our simplified model?

9.11. A manufacturer has recently introduced a line of centrifugal pumps which use a gear-speed increaser to drive the impeller at 20,000 rpm with an 1800 rpm motor. One of the pumps delivers a head of 4500 ft. Estimate its impeller diameter. List the most important mechanical design problems for such a pump.

9.12. Old hands in the chemical process industries pour cold water onto the suction side of any malfunctioning centrifugal pump. Why?

9.13. Suppose that, instead of using a PD pump, in Example 9.2 we used a centrifugal pump, which for 200 gal/min had a reported NPSH requirement of 10 ft. What would be the maximum elevation above the sump at which we could place the pump?

9.14. Why is it impractical to try to build a zero-clearance compressor?

9.15. How many horsepower are required to compress 20 lbmol/h of helium from 1 atm at 70°F to 10 atm in (*a*) an isothermal compressor, (*b*) an adiabatic compressor, and (*c*) a two-stage, adiabatic compressor with optimum interstage pressure and intercooling to 68°F? For helium $k = 1.666$.

9.16. Prepare a plot of work per pound-mole versus the pressure ratio P_2/P_1 for a perfect gas, with $k = 1.4$, being compressed from an inlet condition of 68°F. Cover the pressure-ratio range of 1 to 20. Show the curves for both an adiabatic and an isothermal compressor.

9.17. Prove that the interstage pressure given by Eq. 9.16 gives the minimum work per pound-mole for a given P_{inlet} and P_{outlet}. *Hint:* Write the equation for the total work of a two-stage, intercooled compressor, and differentiate it with respect to the interstage pressure.

9.18. Show by induction how it follows from Eq. 9.16 that for a multistage compressor the optimum interstage pressures are those which result in equal pressure ratios for each stage.

9.19. Rework Example 9.5(*c*), using the optimum interstage pressure instead of the interstage pressure selected for the example.

9.20. We wish to compress air from 1 atm at 0°F to 10 atm. We will use a two-stage, intercooled, adiabatic compressor. For the intercooler we have cooling water cold enough to cool the gas to 68°F. What is the optimum interstage pressure in this case? Write out the general formula for the optimum interstage pressure in terms of the inlet temperatures to the two stages.

9.21. Show that as the number of stages of a multistage, intercooled compressor becomes very large, the work requirement approaches as a limit the work requirement of an isothermal compressor with the same inlet temperature and overall pressure ratio.

9.22. Rework Example 9.5(*c*) for a three-stage compressor, using the optimum interstage pressures.

9.23. Show that as k approaches 1, Eq. 9.15 (adiabatic compressor) approaches as a limit Eq. 9.13 (isothermal compressor). *Hint:* See Prob. 8.58.

9.24. Show the derivations of Eqs. 9.13 and 9.15 from Eq. 4.40 and the relations for the enthalpy of an ideal gas.

9.25. For a pure impulse turbine, the work per unit mass passing through the system is given by Eq. 7.54. Prepare a plot of $(dW_{\text{a.o.}}/dm)/(dW_{\text{a.o.}}/dm)_{\text{opt. blade speed}}$ versus blade speed from zero blade speed to twice the optimum blade speed. Also show the torque as a function of the blade speed on the same plot.

9.26. For a pure reaction turbine, prepare a plot of $(dW_{\text{a.o.}}/dm)/(dW_{\text{a.o.}}/dm)_{\text{opt. noz speed}}$ for nozzle speeds from zero to twice the optimum nozzle speed. Also show the torque as a function of nozzle speed on the same plot.

9.27. It has been suggested that for short-term service we could make a simple reaction turbine by attaching two solid-fuel rockets to the ends of a rotor. What would be the optimum speed for maximum power production for this type of device? Why does the answer differ from that for the garden sprinkler type shown in Fig. 9.14?

9.28. One of the highest-head water power plants in the world is that at Dixence, Switzerland, with a net head of 5330 ft. The water from this plant drives an impulse turbine (Pelton wheel) with a diameter to the middle of the blades of 10.89 ft. The wheel turns at 500 rpm [1]. What is the ratio of blade speed to jet speed for this turbine? How does this compare with the optimum discussed in Sec. 7.4?

9.29. Most PD compressors are connected to constant-speed motors, because variable-speed motors are much more expensive. This poses a problem in controlling the compressor flow rate. One way to control a compressor with a constant-speed motor is to vary the clearance volume by means of "clearance pockets," which are connected to or disconnected from the head of the compressor by remotely operated valves. For a 100 percent efficient, adiabatic compressor, what is the effect of such a pocket on the pressure ratio, flow rate, and power requirement?

9.30. See Prob. 9.29. An alternative procedure is to use an inlet or outlet valve which can be stopped in the open position by some remote controller. For a 100 percent efficient, adiabatic compressor, what is the effect of a stuck-open inlet valve on the pressure ratio, flow rate, and power requirement? Such devices are normally called *valve unloaders*.

CHAPTER 10

POTENTIAL FLOW

So far we have considered only flows which were in one direction, as in a pipe or down a straight riverbed. In the few cases in which the fluid flow was not one-dimensional, as around a sphere or in a pipe elbow or venturi meter, we have introduced experimental data to allow us to treat the problem as if it were one-dimensional. Although this one-dimensional approach adequately covers many of the practical problems in fluid mechanics, it is not satisfactory for the complicated ones, particularly for the aerodynamics problems. To solve these more complicated problems, two additional ideas are needed: potential flow and the boundary layer.

10.1 THE HISTORY OF POTENTIAL FLOW AND BOUNDARY LAYER

In the eighteenth and nineteenth centuries, there existed two schools of thought on fluid mechanics. One group, called the *hydraulicians*, looked at experimental data and attempted to generalize them into useful design equations. Their equations were generally empirical, without much theoretical content. The other group, called the *hydrodynamicists*, started with Newton's equations of motion and tried to deduce the necessary equations for fluid flow. It was quickly apparent to the hydrodynamicists that if they retained the viscous-friction terms or the change-of-density terms, then the resulting differential equations would be so cumbersome that solutions would seldom, if ever, be possible. So they ignored the viscous-friction and density-change terms by hypothesizing a *perfect fluid* with zero viscosity and constant density. For this perfect fluid they were then able to calculate the complete behavior of many kinds of flows. For flows which did not involve solid surfaces, such as deep-water waves or tides, these mathematical solutions agreed very well with observed behavior. But the hydraulicians found that the perfect-fluid solutions did not agree with observed behavior in the problems which concerned them: flow in channels, flow in pipes, forces on solid bodies caused by flow past them, etc. By 1900 the two schools had gone their separate ways, the hydrodynamicists publishing learned mathematical papers with little bearing on engineering problems and the hydraulicians solving engineering problems by trial and error, intuition, and experimental tests. A wit of the period said, "Hydrodynamicists calculate that which cannot be observed; hydraulicians observe that which cannot be calculated."

In 1904 Ludwig Prandtl [1] suggested a way to bring the two schools together by introducing a new concept, called the boundary layer. If a fluid flows past the leading edge of a flat surface, there will develop a velocity profile, as shown in Fig. 10.1. According to the laws of perfect-fluid flow, the surface should not influence the flow in any way; the velocity should be V_∞ everywhere in the flowing fluid. According to the ideas of viscous flow, there should exist a velocity gradient in the y direction extending out to infinity. Prandtl's suggestion to reconcile these views was that the flow be conceptually divided into two parts along the line shown. In the region close to the solid surface, the effects of viscosity are too large to be ignored. However, this is a fairly small region; outside it the effects of viscosity are small and can be neglected. Thus, outside this region the laws of perfect-fluid flow should be satisfactory.

Prandtl called the region where the viscous forces cannot be ignored the *boundary layer*. He arbitrarily suggested that it be considered that region in which the x component of the velocity is less than 0.99 times the free-stream velocity. Then, to obtain a complete solution to a flow problem in two or three

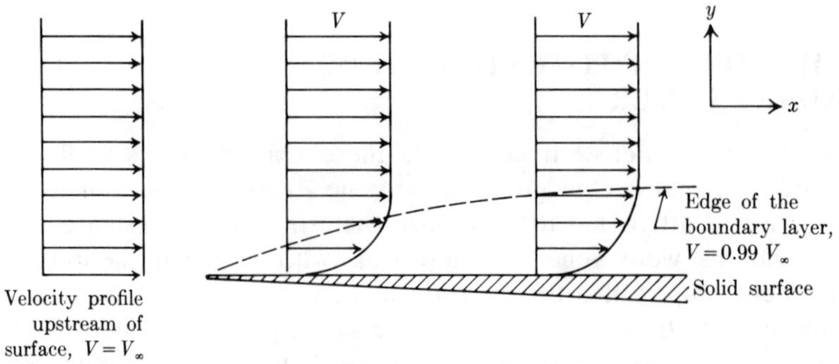

FIGURE 10.1
The idea of the boundary layer.

dimensions, one should use the viscous-flow equations inside the boundary layer and the equations of perfect-fluid flow outside the boundary layer. At the edge of the boundary layer, the pressures and velocities of the two solutions must be matched.

This division does not necessarily correspond to any physically measurable boundary. The edge of the boundary layer does not correspond to any sudden change in the flow but rather corresponds to an arbitrary mathematical definition. Even with this simplification, the calculations are very difficult, and in general only approximate mathematical solutions are possible. Nonetheless, this idea clarified numerous unexplained phenomena and provided a much better intellectual basis for discussing complicated flows than had existed previously. As a result, the boundary layer has become a standard idea in the minds of fluid mechanicians. Once it became accepted in fluid mechanics, an analogous idea was tried in heat transfer and in mass transfer, generally with useful results.

From the preceding it is clear that the ideas of perfect-fluid flow and of the boundary layer are intimately tied together. Both are generally needed for completely describing physically interesting flows, although sometimes one alone is sufficient. We consider perfect-fluid flows in this chapter and the boundary layer in the next. First we must introduce the idea of streamlines.

10.2 STREAMLINES

In one-dimensional flow, the direction of flow at every point in the flow is the same, although the velocity may not be the same at every point (e.g., laminar flow in a tube). In two- and three-dimensional flows, the velocity and direction both change from place to place. For unsteady (i.e., time-varying) flows, they also change from one instant to the next. For steady flow we can map out the velocity and direction at any point; see Fig. 10.2, in which the velocity at any point is represented by an arrow showing the relative velocity and direction of the flow.

If we follow the history of a fluid particle starting at A, we see that it moves, not in a straight line, but rather along a curve, whose direction at any point is tangent to the flow direction. Such a curve, showing the path of any fluid particle in steady flow, is called a *streamline*. Obviously, there is a streamline passing through every point in the flow; so if all the streamlines

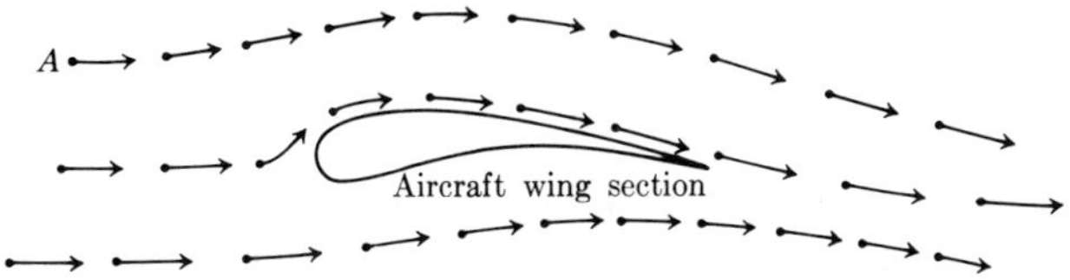

FIGURE 10.2
Point values of the flow velocity and direction for steady, two-dimensional flow.

were drawn in Fig. 10.2, the entire flow area would be printed black. Therefore, it is common practice to draw only a few streamlines, from which the intermediate ones can be readily interpolated. In steady flow there is no flow across (i.e., perpendicular to) a streamline.

For unsteady-flow problems, the direction and magnitude of the velocities in Fig. 10.2 are changing with time, so the direction and velocity of the streamline through A is changing with time. A fluid particle which was on the streamline through A at time t_1 may not be on the new streamline through A at time $t_1 + \Delta t$. To handle such problems, two other concepts are introduced: the *streak line*, which is the line made by a dye injected into a fluid at one point and which thus marks the position of all the particles of fluid that have passed that point, and the *path line*, which gives the instantaneous velocity and direction of a single particle of fluid at various times. In steady flow, streak lines, path lines, and streamlines are the same. Since we deal only with steady flow, we refer only to streamlines in the rest of this chapter.

If we use the alternative view of a streamline—a line across which there is no flow—then it is clear that the boundaries of solid objects immersed in the flow must be streamlines. For real fluid flows, the fluid adjacent to the boundary of a solid body does not move relative to the body; it clings to the wall. Thus, in real fluids the wall is a streamline of zero velocity. In the theory of perfect-fluid flow, the imaginary perfect fluid has no tendency to cling to walls, because it has no viscosity. Thus, the streamline adjacent to a solid body in perfect-fluid flow is one with finite velocity. This leads to the idea that we may divide a perfect-fluid flow along a streamline and substitute a solid body for the flow on one side of the streamline without changing the mathematical character of the flow on the other side of the streamline. Thus, to compute the flow around some solid body in perfect-fluid theory, we need only find the flow which has a streamline with the same shape as the solid body and then conceptually substitute the solid body for that part of the flow; this does not affect the rest of the flow. Several examples of this procedure will be shown.

10.3 POTENTIAL FLOW

In the region outside the boundary layer, where the fluid may be assumed to have no viscosity, the mathematical solution takes on the form known as *potential flow*. This flow is analogous to the flow of heat in a temperature field or to the flow of charge in an electrostatic field. The basic equations of heat conduction (Fourier's law) are

$$q_x = -k \frac{\partial T}{\partial x}, \qquad q_y = -k \frac{\partial T}{\partial y}, \qquad q_z = -k \frac{\partial T}{\partial z} \tag{10.1}$$

Here q_x is the flow of heat per unit time per unit area in the x direction, k is the thermal conductivity, and T is the temperature. An energy balance for some arbitrary region in space (analogous to the procedure shown in finding Eq. 3.32) yields

$$\rho C_v \frac{\partial T}{\partial t} = \frac{\partial[-k(\partial T/\partial x)]}{\partial x} + \frac{\partial[-k(\partial T/\partial y)]}{\partial y} + \frac{\partial[-k(\partial T/\partial z)]}{\partial z} \tag{10.2}$$

Here ρ is the density and C_V the heat capacity. For constant k this simplifies to

$$-\frac{\rho C_V}{k}\frac{\partial T}{\partial t} = \frac{\partial^2 T}{\partial x^2} + \frac{\partial^2 T}{\partial y^2} + \frac{\partial^2 T}{\partial z^2} \tag{10.3}$$

For steady state, the left term is zero, so that the steady-state heat conduction equation becomes

$$0 = \frac{\partial^2 T}{\partial x^2} + \frac{\partial^2 T}{\partial y^2} + \frac{\partial^2 T}{\partial z^2} \tag{10.4}$$

This is *Laplace's equation*, for which solutions are known for many geometries [2].

Similarly, in an electrostatic field the flow of charge is given by

$$J_x = -\frac{1}{\rho}\frac{\partial E}{\partial x} \qquad J_y = -\frac{1}{\rho}\frac{\partial E}{\partial y} \qquad J_x = -\frac{1}{\rho}\frac{\partial E}{\partial z} \tag{10.5}$$

Here J_x is the x component of the current density, E is the potential, and ρ is the resistivity. For steady state these also lead to Laplace's equation in the form

$$0 = \frac{\partial^2 E}{\partial x^2} + \frac{\partial^2 E}{\partial y^2} + \frac{\partial^2 E}{\partial z^2} \tag{10.6}$$

Because the flow of heat and electric charge obey Laplace's equation (under certain restrictions), the hydrodynamicists introduced a similar formulation for the flow of a liquid; they defined a *velocity potential* ϕ by the equations

$$\begin{aligned} V_x &= x \text{ component of velocity} = \frac{-\partial \phi}{\partial x} \\ V_y &= y \text{ component of velocity} = \frac{-\partial \phi}{\partial y} \\ V_z &= z \text{ component of velocity} = \frac{-\partial \phi}{\partial z} \end{aligned} \tag{10.7}$$

By applying the steady-state mass balance for a constant-density fluid (Eq. 3.33), we find that this definition also leads to Laplace's equation:

$$0 = \frac{\partial^2 \phi}{\partial x^2} + \frac{\partial^2 \phi}{\partial y^2} + \frac{\partial^2 \phi}{\partial z^2} \tag{10.8}$$

From Eq. 10.7 it is clear that ϕ must have the dimension of square feet or meters per second.

The advantage of the formulation of flow problems in terms of the velocity potential is the great simplification which this formulation allows. If we are trying to determine the general solution to some steady-flow problem, then we will have $V_x = f_1(x, y, z)$, $V_y = f_2(x, y, z)$, and $V_z = f_3(x, y, z)$, three unknown functions of three independent variables. If the problem can be formulated in terms of the velocity potential, then we can find all three of these functions from $\phi = \phi(x, y, z)$; so the problem is reduced from that of finding three functions to that of finding one.

What physical meaning should one attach to the velocity potential? For the flow of an ideal, frictionless fluid, the velocity potential has no physical meaning whatever. To illustrate this, consider the steady flow of a frictionless, constant-density fluid in a horizontal pipe; see Fig. 10.3. (Such a frictionless fluid, once started in motion by some external force, would continue moving forever, because there is no force to stop it.) For such a frictionless fluid, the velocity is uniform over the cross section perpendicular to the flow. From Bernoulli's equation we can see that there is no change with distance of pressure, velocity, or elevation, and by straightforward arguments we can show that there is no change of temperature, refractive index, dielectric constant, or any other measurable property. But from Eq. 10.7 we know that, because V_x is constant, there is a steady decrease of ϕ in the x direction. Thus the velocity potential for a perfect fluid ϕ is not a function of *any measurable physical property* of the fluid.

We need not be disturbed by this lack of physical significance of ϕ. There are other such quantities in engineering, such as $i = (-1)^{1/2}$. Clearly, there can be no physical interpretation of imaginary voltages, currents, etc.; nonetheless, the treatment of alternating currents is easier if one uses i. We should take a similar view of ϕ; it has no real physical meaning but is a useful mathematical device for solving some problems.

An alternative meaning of ϕ appears in the study of the flow of real, viscous fluids through porous media. In Chap. 12 we will see that the $V^2/2$ term in Bernoulli's equation is often negligible and that the friction-loss term is of the laminar form

$$\mathscr{F} = \frac{\text{viscosity} \cdot \text{velocity} \cdot \text{length}}{\text{permeability} \cdot \text{density}} \tag{10.9}$$

Here the permeability k is property of the porous medium (discussed at greater length in Chap. 12). If we make this substitution, Bernoulli's equation becomes

$$\Delta\left(\frac{P}{\rho} + gz\right) = -\frac{\mu}{k\rho} V_x \, \Delta x \tag{10.10}$$

If we now multiply through by $k\rho/(\mu \, \Delta x)$, take the limit of both sides as Δx

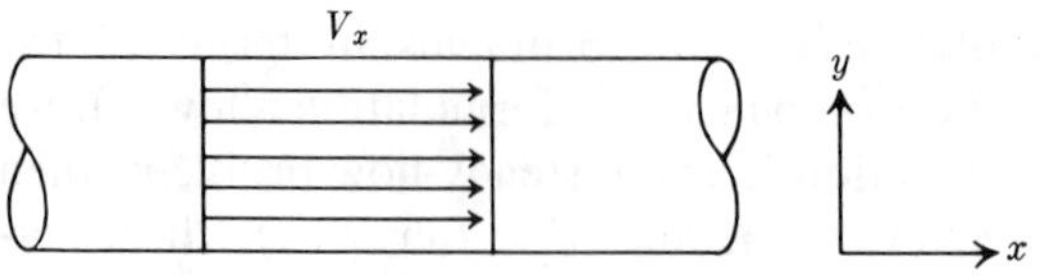

FIGURE 10.3
Flow of a frictionless fluid in a horizontal pipe.

becomes infinitesimal, and rearrange, we find

$$\frac{k}{\mu}\frac{d(P+\rho g z)}{dx} = -V_x \tag{10.11}$$

This is the same as Eq. 10.7 if

$$\phi = \frac{k}{\mu}(P+\rho g z) \tag{10.12}$$

This interpretation is intuitively quite satisfying. For the flow of a real, viscous fluid through a uniform, porous medium, the density, viscosity, and permeability are constant; so for constant elevation the velocity in the x direction $-\partial\phi/\partial x$ is proportional to the negative pressure gradient $-\partial P/\partial x$. Although this porous-medium meaning of potential flow is intuitively satisfying and has considerable practical use in petroleum reservoir engineering, groundwater hydrology, and the study of filters and packed beds, it is not the chief application of potential flow. The chief application is for imaginary perfect fluids, for which ϕ has no intuitive meaning at all.

These potential-flow systems are compared in Table 10.1. All four potential flows can be expressed in compact vector notation by

$$\text{Vector velocity or heat flux or electric flux} = -(\text{constant})\ \text{grad(potential)} \tag{10.13}$$

where "grad" represents the two- or three-dimensional operation of taking partial derivatives of the potential and multiplying these by appropriate unit vectors; see Bird et al. [3, App. A] for a clear treatment of the vector calculus approach to these problems. One consequence of Eq. 10.13 is that the flow is always perpendicular to the equipotential lines. Thus, in fluid mechanics terminology, the steady-flow streamlines are always perpendicular to the lines of constant ϕ.[†]

To gain some feeling for the idea of a potential flow, we will show what kind of flows are described by various choices of ϕ. We restrict our attention to two-dimensional flows, because they are mathematically much easier than three-dimensional flows. In general, ϕ will be $\phi = \phi(x, y)$, but not every such function satisfies Laplace's equation, so not every such function represents a potential flow. You may verify that $\phi = x^2$, $\phi = x^2 + y^2$, $\phi = e^x$, and $\phi = \sin x$

[†] This property is contrary to our experience in watching balls roll down hills. They start from rest, rolling perpendicular to the contour lines, but unless the contour lines are straight and parallel, the balls eventually cross them at some other angle. Balls do this because they have inertia and try to keep going straight when the hill curves. Flows which obey Laplace's equation generally involve no inertia. Electricity and heat have no inertia. In viscous flow in a porous medium, the inertia term $V^2/2$ is so small that it can be ignored. In perfect-fluid flow, the inertia can be significant, but the nonphysical character of ϕ and the irrotational character (described in Sec. 10.4) allow this flow with inertia to fit an inertia-free formula.

TABLE 10.1
Comparison of systems obeying the Laplace equation $\partial^2\phi/\partial x^2 + \partial^2\phi/\partial y^2 + \partial^2\phi/\partial z^2 = 0$

System	What is flowing	ϕ	Lines of constant ϕ
Steady-state temperatures	Heat (i.e., thermal energy)	Temperature	Isotherms
Steady-state electric field	Charge (i.e., electrons in opposite direction)	Potential (voltage)	Equipotentials
Steady-state perfect fluid	Perfect fluid (zero viscosity, constant density)	No physical meaning whatever	Equipotentials
Steady-state viscous flow in porous medium	Real fluid	$\frac{k}{\mu}(P + \rho g z)$	Equipotentials (or, for constant elevation, isobars; or, for constant pressure, contour lines)

do not satisfy Laplace's equation, so they cannot represent potential flows, because they violate the mass balance for a constant-density fluid (see Prob. 10.4).

To illustrate some functions which do satisfy Laplace's equation, we map out the flows described by the equations

$$\phi_1 = Ax \tag{10.14}$$

$$\phi_2 = Ax + By \tag{10.15}$$

$$\phi_3 = C \ln (x^2 + y^2)^{1/2} \tag{10.16}$$

Here A and B have dimensions of velocity (feet or meters per second) and C has dimensions of velocity times distance (square feet or meters per second). For all these functions, you may verify that Laplace's equation is satisfied.

For ϕ_1:
$$V_x = \frac{-\partial \phi}{\partial x} = -A \qquad V_y = \frac{-\partial \phi}{\partial y} = 0 \tag{10.17}$$

and ϕ_1 describes a uniform, steady flow of velocity A in the $-x$ direction. This velocity is the same over the entire region described; this might be the description of a wind blowing over the ocean at a steady, uniform velocity of A; see Fig. 10.4.

For ϕ_2:
$$V_x = \frac{-\partial \phi}{\partial x} = -A \qquad V_y = \frac{-\partial \phi}{\partial y} = -B \tag{10.18}$$

This flow is shown in Fig. 10.5. From these two examples clearly any equation of the form $\phi = Ax + By + C$ represents a uniform, constant-velocity flow with velocity $(A^2 + B^2)^{1/2}$, making the angle arctan (B/A) with the x axis. Such uniform, constant-velocity flows are not of much practical interest alone, but later we will see how they are combined with other flows to solve more interesting problems.

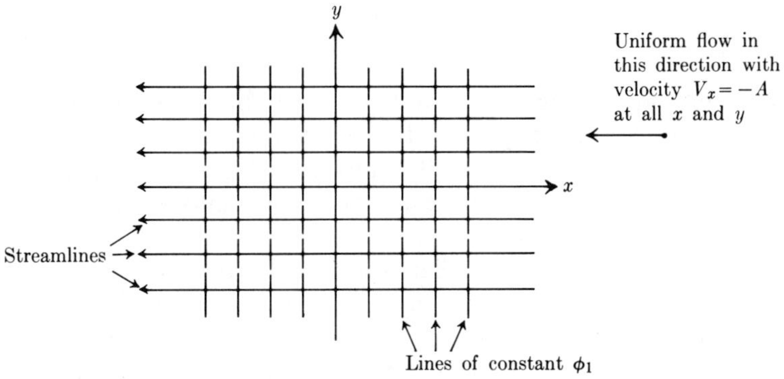

FIGURE 10.4
Flow described by the velocity potential $\phi_1 = Ax$.

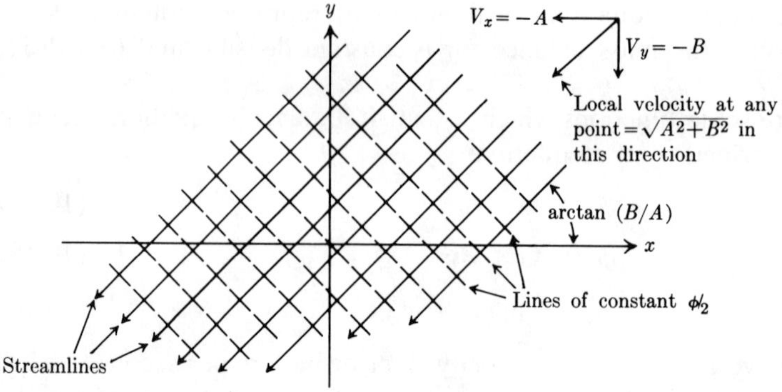

FIGURE 10.5
Flow described by the velocity potential $\phi_2 = Ax + By$.

For many potential functions it is easier to work in plane polar coordinates than in rectangular coordinates. In polar coordinates Eq. 10.7 has the form

$$\text{Radial velocity } V_r = \frac{-\partial \phi}{\partial r}$$
$$\text{Tangential velocity } V_\theta = r\frac{d\theta}{dt} = r\omega = \frac{-1}{r}\frac{\partial \phi}{\partial \theta} \tag{10.19}$$

and Laplace's equation takes the form

$$\frac{\partial \phi}{r\,\partial r} + \frac{\partial^2 \phi}{\partial r^2} + \frac{\partial^2 \phi}{r^2\,\partial \theta^2} = 0 \tag{10.20}$$

In polar coordinates ϕ_3 is expressed as

$$\phi_3 = C \ln r \tag{10.21}$$

and the velocity components are

$$V_r = \frac{-C}{r} \qquad V_\theta = 0 \tag{10.22}$$

Thus, the streamlines are radially inward lines, and the lines of constant potential are circles, as shown in Fig. 10.6. If C is positive, the flow is radially inward; if it is negative, the flow is radially outward. This flow has practical significance in the petroleum industry; it describes the flow into an oil well in a thin, horizontal stratum; see Fig. 10.7.

Equation 10.22 shows that the radial-flow velocity becomes infinite at $r = 0$; thus, this equation cannot describe any real flow at $r = 0$. In Fig. 10.7 it describes the flow from out in the oil-bearing stratum up to the well. Inside the well the flow turns and moves in the direction perpendicular to the plane of Fig. 10.7 and is clearly not described by Eq. 10.22.

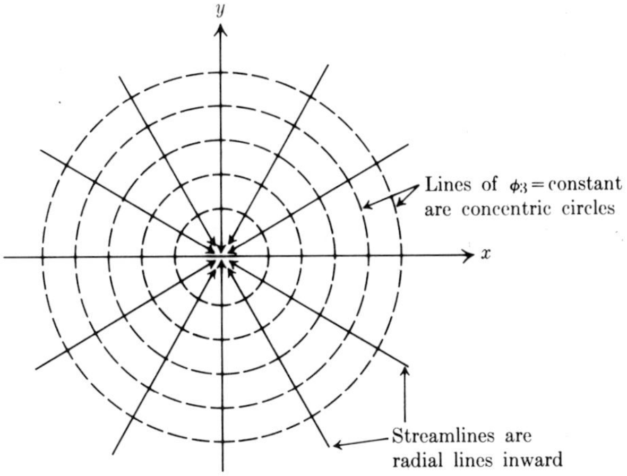

FIGURE 10.6
Flow described by the velocity potential $\phi_3 = C \ln r$.

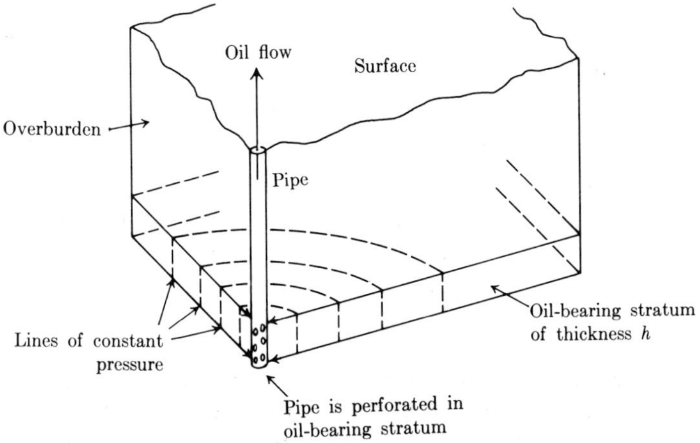

FIGURE 10.7
Flow from a thin, horizontal stratum into an oil well.

From Eq. 10.22 we can calculate the value of C for any known flow up the well. If the oil-bearing stratum is h ft thick and is producing Q ft^3/h of oil, then the steady-state flow inward across any cylindrical surface surrounding the well is Q. The radial velocity is

$$V_r = \frac{-Q}{\text{flow area}} = \frac{-Q}{h2\pi r} \tag{10.23}$$

Substituting in Eq. 10.22, we find

$$\frac{-C}{r} = \frac{-Q}{h2\pi r} \qquad \text{or} \qquad C = \frac{Q}{2\pi h} \tag{10.24}$$

In terms of ideal frictionless fluids, this flow alone has limited significance; but it has the same conceptual description. It is commonly referred to as a *sink* if the flow is radially inward or a *source* if the flow is radially outward.

An interesting and useful property of solutions of Laplace's equation is that if ϕ_1 and ϕ_2 are both individually solutions of Laplace's equation, then their sum must also be a solution, because if

$$\frac{\partial^2 \phi_1}{\partial x^2} + \frac{\partial^2 \phi_1}{\partial y^2} = 0 \qquad \text{and} \qquad \frac{\partial^2 \phi_2}{\partial x^2} + \frac{\partial^2 \phi_2}{\partial y^2} = 0$$

then

$$\frac{\partial^2 \phi_1}{\partial x^2} + \frac{\partial^2 \phi_2}{\partial x^2} + \frac{\partial^2 \phi_1}{\partial y^2} + \frac{\partial^2 \phi_2}{\partial y^2} = 0 = \frac{\partial^2(\phi_1 + \phi_2)}{\partial x^2} + \frac{\partial^2(\phi_1 + \phi_2)}{\partial y^2} \tag{10.25}$$

We can illustrate this property by adding the two potential functions ϕ_2 and ϕ_3, discussed previously, to find

$$\phi_4 = Ax + C \ln (x^2 + y^2)^{1/2} \tag{10.26}$$

which has velocity components

$$V_x = -\left(A + \frac{Cx}{x^2 + y^2}\right) \qquad V_y = \frac{-Cy}{x^2 + y^2} \tag{10.27}$$

and is sketched in Fig. 10.8.

There are several physical systems which display the flow pattern in Fig. 10.8. The flow can be in the direction of the arrows shown or in the reverse direction (to reverse the direction of the flow, we need only reverse the signs of A and C in Eq. 10.26):

1. If there is a steady flow from right to left in a thin, porous stratum (as would exist in a horizontal oil stratum with a linear pressure gradient) and some of the fluid is being withdrawn from a well in a direction perpendicular to the page, then the flow pattern is as shown, and the equipotential lines are isobars.
2. If the flow direction is reversed, then this is the pattern of a fluid being injected into a stratum in which there is steady flow from left to right.
3. The lines of flow direction marked A and B in Fig. 10.8 close on the point at which there is zero flow (that is, $y = 0$ and $x = -C/A$). If we divide the flow along curve AB and consider only the flow to the left of this curve, this represents the potential flow outside some two-dimensional body shaped like AB. If the flow is from left to right, then this is quite similar to the flow over the leading edge of an airplane wing or the upstream side of a rounded bridge abutment.

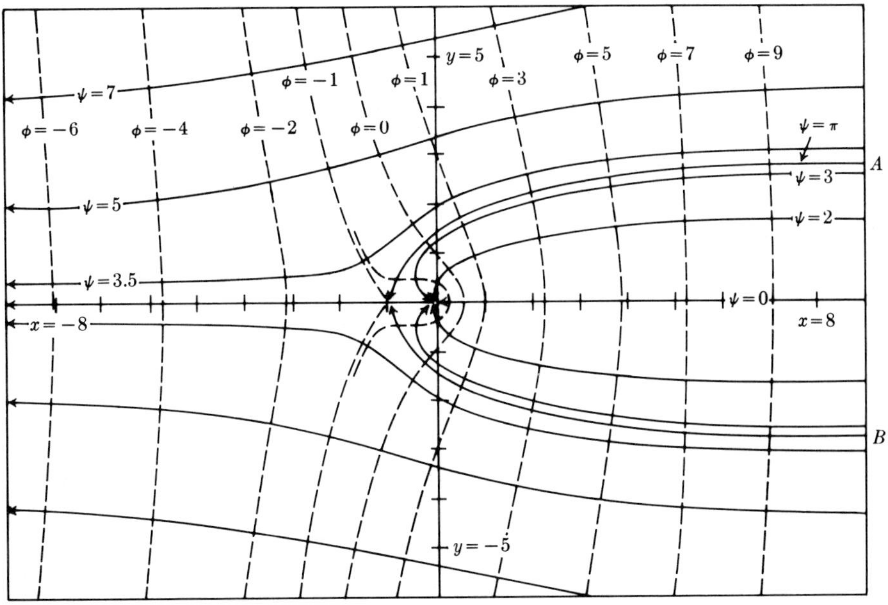

FIGURE 10.8
Flow described by the velocity potential $\phi_4 = Ax + C \ln (x^2 + y^2)^{1/2}$. Drawn to scale for $A = 1$ and $C = 1$. The markings $\psi = 7$, etc., on the streamlines are discussed in Sec. 10.5.

The latter interpretation is the one normally sought in the study of perfect-fluid flows; we wish to find the flow pattern around some arbitrary body. This is normally done by judicious combinations of steady flows, sources, sinks, etc. When a combination is found that produces a streamline with the shape of the body in question, the flow outside the streamline is a representation of the flow around the body. The flow inside that line (i.e., inside line AB in Fig. 10.8) normally has no meaning and is ignored.

Similar flow maps for a wide variety of functions and their corresponding geometries are given by Kirchhoff [4].

10.4 IRROTATIONAL FLOW

Equations 10.7, which define the velocity potential, have an interesting consequence: A flow which obeys them must be irrotational. If $\phi = \phi(x, y)$ has continuous derivatives, then the order of differentiation is immaterial, and

$$\frac{\partial^2 \phi}{\partial x \, \partial y} = \frac{\partial^2 \phi}{\partial y \, \partial x} \tag{10.28}$$

But by substituting Eqs. 10.7 in Eq. 10.28 we find

$$\frac{\partial V_x}{\partial y} = \frac{\partial V_y}{\partial x} \quad \text{or} \quad \frac{\partial V_y}{\partial x} - \frac{\partial V_x}{\partial y} = 0 \tag{10.29}$$

which is the defintion of a two-dimensional irrotational flow.

How Eq. 10.29 is related to rotation may be seen by viewing a body of fluid rotating in two-dimensional, rigid-body rotation about the origin; see Fig. 10.9. If, as shown, the fluid is rotating in rigid-body rotation with angular velocity ω, then at any point the velocity in polar coordinates is given by

$$r = \text{const} \qquad \frac{d\theta}{dt} = \omega = \text{const} \tag{10.30}$$

$$x = r\cos\theta\,, \qquad y = r\sin\theta \tag{10.31}$$

$$V_x = \frac{dx}{dt} = -r\sin\theta\,\frac{d\theta}{dt} = -r\,\frac{y}{r}\,\omega = -y\omega$$

$$V_y = \frac{dy}{dt} = r\cos\theta\,\frac{d\theta}{dt} = r\,\frac{x}{r}\,\omega = x\omega \tag{10.32}$$

so

$$\frac{\partial V_y}{\partial x} - \frac{\partial V_x}{\partial y} = \omega + \omega = 2\omega \tag{10.33}$$

These calculations were carried out for any rigid-body rotation, so we see that $\partial V_y/\partial x - \partial V_x/\partial y$ is exactly twice the angular velocity. This quantity is given the name *vorticity* in theoretical fluid mechanics:

$$\text{Vorticity } \zeta = 2\omega = \frac{\partial V_y}{\partial x} - \frac{\partial V_x}{\partial y} \qquad \text{[two-dimensional flow]} \tag{10.34}$$

For an irrotational flow, the vorticity is zero everywhere.

We see above that for simple, rigid-body rotation $\partial V_y/\partial x - \partial V_x/\partial y$ is not zero. Thus it is impossible to find any potential function ϕ which, when substituted in Eqs. 10.7, will describe such a flow. This does not mean that there can be no potential flows which have circular motion. Only those circular motions which have zero vorticity are irrotational and hence can be potential flows. For a flow to be irrotational, the two derivatives $\partial V_y/\partial x$ and $\partial V_x/\partial y$ must

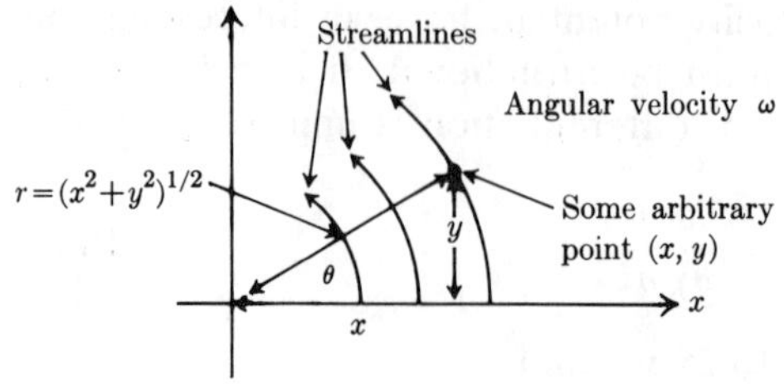

FIGURE 10.9
Solid-body rotation (sometimes called *forced vortex*).

be equal. This is illustrated by the potential flow described by

$$\phi = -xy \qquad V_x = \frac{-\partial \phi}{\partial x} = y \qquad V_y = \frac{-\partial \phi}{\partial y} = x \tag{10.35}$$

This flow satisfied Laplace's equation and does have the irrotational property (Eq. 10.29); it is sketched in Fig. 10.10.

On this flow we have marked four particles of fluid at A, B, C, and D. At time t_1 these are at the corners of a cross. Now if we follow them to time $t_1 + \Delta t$, we see that they have deformed into a flattened figure. Line AC is rotating in a clockwise direction, because the x component of the velocity increases in the y direction, and point A moves to the right faster than point C does. However, line BD is rotating counterclockwise, because the y component of the velocity increases in the x direction. We may show that these lines are rotating in the opposite directions at the same speed, so that even though the fluid is being deformed by the flow, it has no net rotation.

The flow shown in Fig. 10.10 is representative of the flow into a square corner. This may be seen by noting that the lines $y = x$ and $y = -x$ are both streamlines; thus, there is no flow across them.

To show that this irrotational property can exist in a flow in which the streamlines are circles, consider the potential flow described by

$$\phi = -A \arctan \frac{y}{x} = -A\theta \tag{10.36}$$

Here both the rectangular and polar forms of ϕ are shown. In polar coordinates the velocity components are

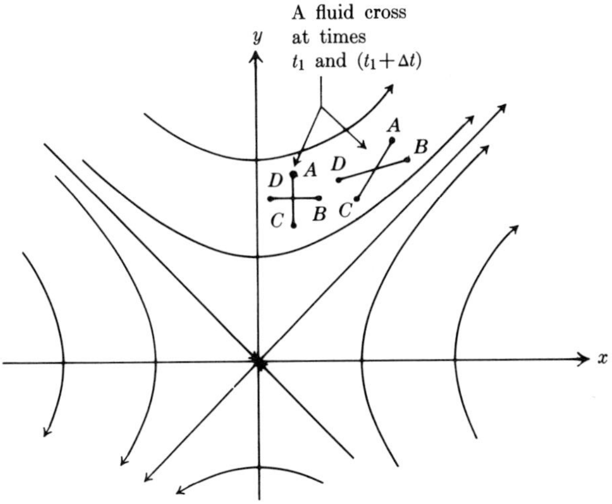

FIGURE 10.10
Flow described by the velocity potential $\phi = -xy$.

$$V_r = \frac{-\partial \phi}{\partial r} = 0 \qquad V_\theta = \frac{-1}{r}\frac{\partial \phi}{\partial \theta} = \frac{A}{r} \tag{10.37}$$

So at any point the velocity component toward the origin V_r is zero; thus, the streamlines are circles, and the equipotential lines are rays passing through the origin. This flow is sketched in Fig. 10.11.

In this figure if we mark a fluid cross $ABCD$ at time t_1 and then look at it again at time $t_1 + \Delta t$, line BD has rotated in the counterclockwise direction, but line AC has rotated in the clockwise direction, because the fluid at C is moving much faster than the fluid at A. Thus, although the streamlines in this flow are all circles, the individual particles of fluid are not rotating. We can demonstrate this by placing a float on such a flow and observing that it moves in a circle but maintains its x-y orientation, just as a compass needle would if the compass were moved in a circle.

The flow shown in Fig. 10.11 is called a *free vortex*; it is not common alone in nature. However, the combination of a free vortex with the sink shown in Fig. 10.6, obtained by adding the potential functions, produces

$$\phi = -A\theta + C \ln r \tag{10.38}$$

which describes a flow in which fluid spirals into a central sink; see Fig. 10.12.

Figure 10.12 is a fair description of most of the flow into the eye of a tornado or of the flow spiraling inward toward the drain of a bathtub. In both cases the flow does not extend into the origin but turns and moves in the $+z$ or $-z$ direction over some small region near the origin. Thus, Eq. 10.38 and Fig. 10.12 are satisfactory descriptions only of those regions in the flow far enough from the origin to have negligible velocities in the z direction.

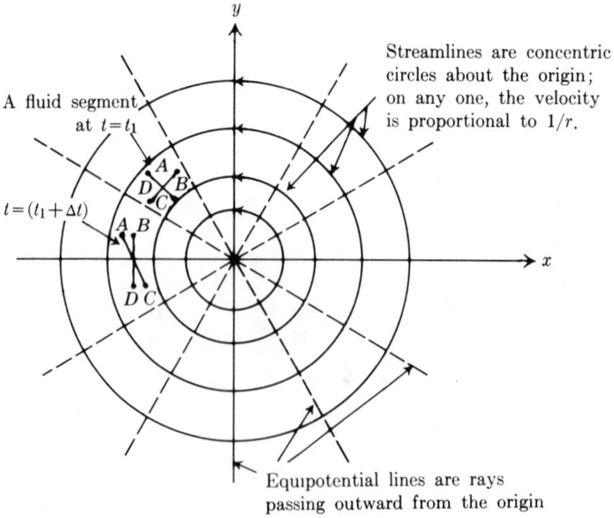

FIGURE 10.11
Flow described by the velocity potential $\phi = -A\theta$.

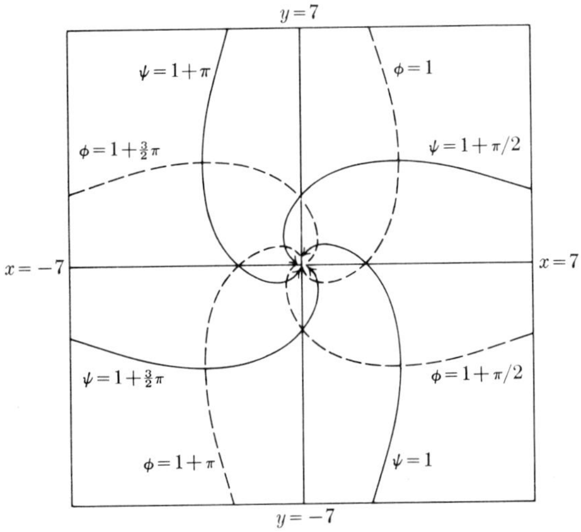

FIGURE 10.12
Flow described by the velocity potential $\phi = -A\theta + C \ln r$. Drawn to scale for $A = 1$ and $C = 1$. The markings $\psi = 1 + \pi$, etc., are discussed in Sec. 10.5. Each streamline and potential line make an infinite number of circles about the origin as r approaches zero; this drawing only shows them entering the region in which they begin their infinite number of circles.

The important idea of an irrotational flow is that at any point in the fluid the angular velocity about any axis is zero. This is shown for zero angular velocity about any axis perpendicular to the *xy* plane in Figs. 10.10 and 10.11. And it can be shown that for any three-dimensional flow which obeys Laplace's equation the angular velocity is zero about any axis.

10.5 STREAM FUNCTION

So far in mapping out potential flow directions we have merely noted that the streamline at any point must be perpendicular to the equipotential line, and we have sketched such streamlines. Now we introduce a more formal method of showing the flow directions at any point. From Eq. 3.33 we know that for a steady, incompressible, two-dimensional flow

$$\frac{\partial V_x}{\partial x} + \frac{\partial V_y}{\partial y} = 0 \tag{3.33}$$

We now arbitrarily define a function ψ, called the *stream function*, or *Lagrange stream function*, by the equations

$$V_x = \frac{-\partial \psi}{\partial y} \qquad V_y = \frac{\partial \psi}{\partial x} \tag{10.39}$$

If ψ has continuous derivatives, then we may substitute Eq. 10.39 in Eq. 3.33:

$$\frac{\partial V_x}{\partial x} + \frac{\partial V_y}{\partial y} = \frac{-\partial^2 \psi}{\partial y \, \partial x} + \frac{\partial^2 \psi}{\partial x \, \partial y} = 0 \tag{10.40}$$

So any flow that satisfies this equation automatically satisfies the two-dimensional, steady-flow, incompressible-material balance. From this equation we see that ψ must also have the dimension of square feet or meters per second. Comparing Eqs. 10.7 and 10.39, we find

$$V_x = \frac{-\partial \phi}{\partial x} = \frac{-\partial \psi}{\partial y} \tag{10.41}$$

$$V_y = \frac{-\partial \phi}{\partial y} = \frac{\partial \psi}{\partial x} \tag{10.42}$$

It is proved in App. H that Eqs. 10.41 and 10.42 can be satisfied only if for every x and y the curve of constant ϕ and the curve of constant ψ passing through that point are perpendicular. This is illustrated by Fig. 10.13 (see also Prob. 10.7).

Since streamlines also have the property of being everywhere perpendicular to equipotentials, all the streamlines which we have drawn in Figs. 10.3, 10.4, 10.5, 10.7, 10.8, and 10.10 are lines of constant ψ.

The stream function has an intuitive explanation, shown in Fig. 10.14. If, as discussed above, curves of constant ψ are streamlines, then there can be no flow across a curve of constant ψ. In Fig. 10.14 the entire flow which is passing between the curves $\psi = 1$ and $\psi = 2$ at A must also be passing between them at B. However, the space between them at B is less than that at A, so there is less area available to the flow at B; hence, by mass balance the velocity must be greater. From Eq. 10.39 we see that the velocity in the x direction can be greater at B than at A only if the curves of constant ψ are closer together at B; that is, $-\partial\psi/\partial y$ is greater at B than at A. We may thus think of the curves of constant ψ as being the boundaries of flow channels; as they squeeze together, the flow between them must go faster.

If we imagine the flow shown in Fig. 10.14 as being h ft deep in the z direction, then the flow passing between lines $\psi = 2$ and $\psi = 1$ at A is

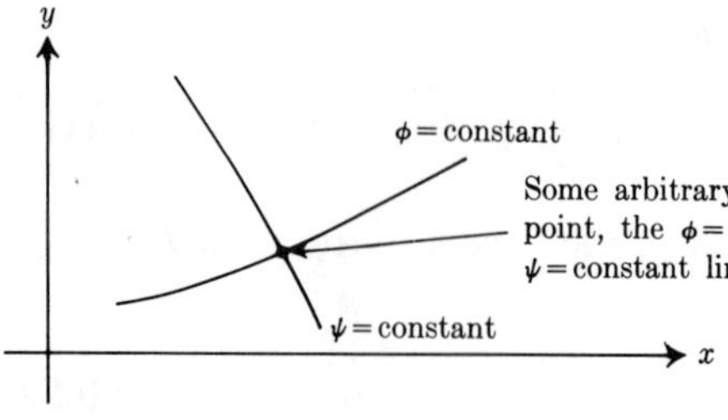

FIGURE 10.13
Relation between lines of constant stream function and constant velocity potential.

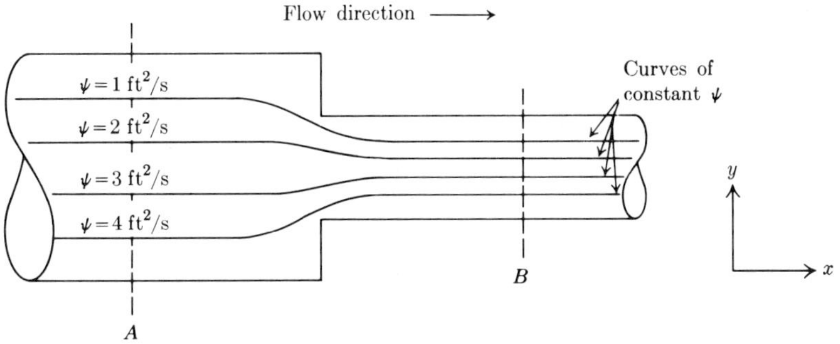

FIGURE 10.14
Stream function for steady flow in a channel transition.

$$Q = \int_{\psi = 2\ \mathrm{ft}^2/\mathrm{s}}^{\psi = 1\ \mathrm{ft}^2/\mathrm{s}} V_x\, dA = \int_{\psi = 2\ \mathrm{ft}^2/\mathrm{s}}^{\psi = 1\ \mathrm{ft}^2/\mathrm{s}} V_x h\, dy = h \int_{\psi = 2\ \mathrm{ft}^2/\mathrm{s}}^{\psi = 1\ \mathrm{ft}^2/\mathrm{s}} \left(-\frac{\partial \psi}{\partial y} \right)_x dy$$

$$= h \int_{\psi = 1\ \mathrm{ft}^2/\mathrm{s}}^{\psi = 2\ \mathrm{ft}^2/\mathrm{s}} d\psi = h(\psi)_{1\ \mathrm{ft}^2/\mathrm{s}}^{2\ \mathrm{ft}^2/\mathrm{s}} = h\ \mathrm{ft}^2/\mathrm{s} \qquad (10.43)$$

(Here h has the dimension of a length, so that the dimension of the volumetric flow rate is indeed correct.) Thus, if on a flow field curves of constant ψ are drawn at equal intervals of ψ, then the volumetric flow between each two such successive curves must be the same for each two (e.g., in Fig. 10.14 the volumetric flow between $\psi = 1$ and $\psi = 2$ is the same as the volumetric flow between $\psi = 2$ and $\psi = 3$, etc.).

In Fig. 10.14 the curves of constant ψ are numbered from the top down, so that $\partial \psi / \partial y$ is negative and hence the flow is in the positive x direction. If they were numbered from the bottom up, then $\partial \psi / \partial y$ would be positive and the flow would be in the negative x direction. This is a purely arbitrary arrangement; we must number them this way because of the signs in the definitions in Eq. 10.39. We could have reversed those signs and still had a satisfactory stream function, with the numbering direction of the stream function curves reversed. German writers usually use a stream function which has the signs so reversed; when reading German texts, we must be careful to observe this sign difference.

If we know the potential function which describes some flow, we can compute the stream function, and conversely. The calculation is based on the following general integration property of partial derivatives. If $A = A(B, C)$, where A, B, and C are any mathematical functions, then

$$A = \left[\int \left(\frac{\partial A}{\partial B} \right)_C dB \right]_{C = \mathrm{const}} + f_1(C) \qquad (10.44)$$

Here $f_1(C)$ is some function of C alone.

If we now let A be ψ and if we know ϕ, we can immediately write the necessary partial derivatives and integrate. To illustrate the procedure, we will find the ψ which corresponds to Eq. 10.26:

$$\phi_4 = Ax + C \ln (x^2 + y^2)^{1/2}$$

Using Eq. 10.41 and 10.42, we find

$$V_x = -\frac{\partial \phi}{\partial x} = -\frac{\partial \psi}{\partial y} = -\left(A + \frac{Cx}{x^2 + y^2}\right) \tag{10.45}$$

$$V_y = -\frac{\partial \phi}{\partial y} = +\frac{\partial \psi}{\partial x} = -\frac{Cy}{x^2 + y^2} \tag{10.46}$$

Letting B in Eq. 10.44 be y in Eqs. 10.45 and 10.46, we find

$$\psi = \int \left(A + \frac{Cx}{x^2 + y^2}\right) dy + f_1(x) = Ay + C \arctan \frac{y}{x} + f_1(x) \tag{10.47}$$

Here the integration has been performed by treating x as a constant, as required by Eq. 10.44.

Now we differentiate Eq. 10.47 with respect to x at constant y:

$$\frac{\partial \psi}{\partial x} = 0 - \frac{Cy}{x^2 + y^2} + \frac{df_1(x)}{dx} \tag{10.48}$$

Note that since f_1 depends on x alone, the derivative at the right is not a partial derivative. Comparing Eq. 10.48 with Eq. 10.46, we see that $df_1(x)/dx$ must be zero; so, by simple integration, $f_1(x)$ must be a constant. Thus, for this ϕ we have

$$\psi = Ay + C \arctan \frac{y}{x} + \text{const} \tag{10.49}$$

One may show (Prob. 10.8) that if we choose to make the B in Eq. 10.44 the x in Eqs. 10.45 and 10.46, we obtain the same result. By plotting various lines of constant ψ given by Eq. 10.49, we see that the lines of constant ψ given by Eq. 10.49 are the streamlines in Fig. 10.8.

For the potential function

$$\phi = C \ln (x^2 + y^2)^{1/2} = C \ln r \tag{10.16}$$

it is easier to work in polar coordinates. The stream function in polar coordinates takes the form

$$V_r = \frac{-1}{r} \frac{\partial \psi}{\partial \theta} \tag{10.50}$$

$$V_\theta = \frac{\partial \psi}{\partial r} \tag{10.51}$$

Thus, from Eq. 10.16 we have

$$V_r = \frac{-\partial \phi}{\partial r} = \frac{-1}{r} \frac{\partial \psi}{\partial \theta} = \frac{C}{r} \tag{10.52}$$

$$V_\theta = \frac{-1}{r} \frac{\partial \phi}{\partial \theta} = \frac{\partial \psi}{\partial r} = 0 \tag{10.53}$$

So letting B in Eq. 10.44 be θ and C be r, we find

$$\psi = \int (-C)\, d\theta + f_1(r) = -C\theta + f_1(r) \tag{10.54}$$

$$\left(\frac{\partial \psi}{\partial r}\right)_\theta = \frac{df_1(r)}{dr} \tag{10.55}$$

but from Eq. 10.53 we know that this is zero, so $f_1(r)$ must be a constant and

$$\psi = -C\theta + \text{const} \tag{10.56}$$

So the streamlines are rays passing through the origin at various angles, as shown in Fig. 10.6.

An interesting property of the stream function and the velocity potential is that if for a given flow $\phi = f_1(x, y)$ and $\psi = f_2(x, y)$, then there is another flow given by $\phi = f_2(x, y)$ and $\psi = f_1(x, y)$. This second flow has exactly the same map as the first, except that the labels are reversed; the streamlines on one are equipotentials on the other, and conversely. For example, Figs. 10.6 and 10.11 bear this relation one to the other. This property is also a property of the real and imaginary parts of any analytic complex function. Because the stream function and velocity potential have this property in common with the real and imaginary parts of an analytic complex function, they can be manipulated by the rules for complex functions. In particular, they obey the rules of *conformal mapping*, a complex-function procedure widely used in both heat flow and potential flow of fluids [5].

The discussion has all been for the application of the stream function to potential flows. But sometimes it can be used to advantage for flows which are not irrotational and for which a potential function cannot exist (see Probs. 10.10 and 11.4). This is possible because the stream function is used in these cases simply as a way of combining the mass balance with the other pertinent equations.

The entire discussion of the stream function has been for two-dimensional flow. The definition of a satisfactory stream function for three-dimensional flow is more difficult. However, if the flow is symmetric about some axis, e.g., uniform flow around some body of revolution, then it is possible to define a different stream function which is convenient for that problem. This three-dimensional stream function is called *Stokes stream function* [6] to distinguish it from the Lagrange stream function, which is discussed in this chapter.

10.6 BERNOULLI'S EQUATION FOR TWO-DIMENSIONAL, PERFECT-FLUID, IRROTATIONAL FLOWS

The velocity-potential stream-function methods shown in the preceding sections allow us to calculate the flow velocity and direction at any point in a two-dimensional, perfect-fluid, irrotational flow. Sometimes this is all the information sought. More often the desired information is the force on some body immersed in the flow, e.g., the lift and drag of a wing section or the drag on a particle settling through a fluid.

In Chap. 5 we derived Bernoulli's equation from the energy balance equation. Since the energy balance has no one-dimensional restriction on it, the same approach must apply to two- and three-dimensional flows. However, in our derivation of Bernoulli's equation, we restricted our attention to systems with only one flow in and out. How can we apply this idea to a two-dimensional flow field in which there is a continuously varying velocity over some region of space? In Fig. 10.15 such a region is shown with no sources, sinks, or solid bodies, but with streamlines.

At A we draw a closed curve (e.g., a circle) around a streamline, perpendicular to the flow. Then we draw streamlines from every point in that closed curve, in the direction of flow. Since streamlines cannot cross or separate except at a source, a sink, or the edge of a solid body, we can draw a closed curve farther downstream (e.g., at B), through which all the streamlines that pass through the curve at A must also pass. This curve at B probably will not have the same shape as the curve at A, but it will exist. Such a set of streamlines, which form a closed curve in any plane perpendicular to the flow, is called a *stream tube*. Let us choose such a tube from A to B as our system. Then for steady flow there is only flow into or out of the stream tube at ends A and B.

In deriving Bernoulli's equation, we assumed that the flow into and out of the system was of uniform velocity, etc. In general this will not be true of any stream tube, because the velocity may be different from one streamline to the next. However, if we make our stream tube smaller and smaller, then the nonuniform flow across its entrance (and exit) becomes negligible. In the limit the stream tube can be thought of as being so small that it shrinks down to just a streamline. For such a stream tube, Bernoulli's equation, as derived in Chap. 5, is obviously applicable. This result is true for any kind of incompressible flow; if the flow is frictionless (i.e., an ideal fluid), then the friction term may

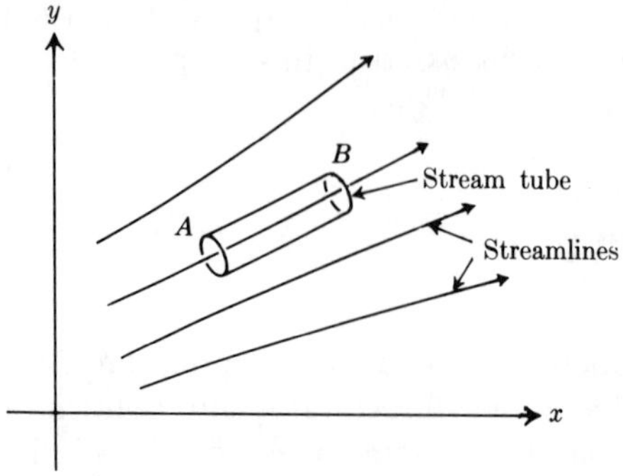

FIGURE 10.15
Stream tube.

be dropped and Bernoulli's equation integrated from some point in the streamline to any adjacent point, to find

$$d\left(\frac{P}{\rho} + gz + \frac{V^2}{2}\right) = 0 \qquad \text{or} \qquad \frac{P}{\rho} + gz + \frac{V^2}{2} = \text{const}$$

[along a streamline, frictionless-fluid flow] (10.57)

Here V is the velocity relative to our fixed coordinate system, given by

$$V = (V_x^2 + V_y^2)^{1/2} = (V_r^2 + V_\theta^2)^{1/2} \tag{10.58}$$

This result allows us to evaluate the pressure on the surface of any body immersed in the flow and hence the force, if only we know the pressure at some upstream position and the total velocity field.†

10.7 FLOW AROUND A CYLINDER

To illustrate the idea of potential flow and how to use it to calculate forces, let us calculate the pressure distribution on the surface of a cylinder which is immersed in a flow perpendicular to it. If this is a very long cylinder, then there will be negligible change in the flow in the direction of the cylinder's axis, and so the flow will be practically two-dimensional. To find the flow field, we must make a judicious combination of a steady flow, a source, and a sink. Consider first a source and a sink with equal flow rates located some distance A apart on the x axis; See Fig. 10.16. The flow between them is given by

$$\phi = C \ln (x^2 + y^2)^{1/2} - C \ln [(x - A)^2 + y^2]^{1/2} \tag{10.59}$$

Figure 10.16 is a fair representation of the flow between an injection and a production well in a porous medium and hence has some application in petroleum reservoir engineering and hydrology. If C in Eq. 10.59 is a constant, then, as A gets smaller and smaller, the flow at any point (except a point directly between the source and the sink) must become smaller and smaller, approaching zero, because more and more of the flow will take the direct path from the source to the sink. If we replace C in Eq. 10.59 with C/A, then the total flow increases just as the distance between the sink and source A decreases. In this case the flow at any point does not go to zero as A goes to zero; we find the value of ϕ as A goes to zero by *L'Hospital's rule*:

$$\phi_{A=0} = \frac{\lim_{A\to 0} \dfrac{d}{dA}\left[\dfrac{C}{2} \ln \dfrac{x^2 + y^2}{(x - A)^2 + y^2}\right]}{\lim_{A\to 0} \dfrac{d}{dA}(A)} = \frac{Cx}{x^2 + y^2} \tag{10.60}$$

† It is proved in various hydrodynamics texts that for *irrotational flow from a single reservoir* the constant in Eq. 10.57 is the same for all streamlines, whereas for rotational flows it is not the same but is only constant along one particular streamline.

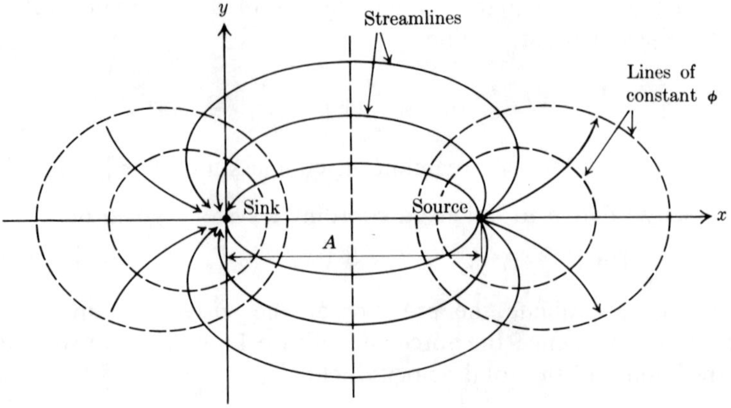

FIGURE 10.16
Flow represented by Eq. 10.59.

This limiting case of a source and a sink at zero distance apart is called a *doublet*. If we now combine this doublet flow with a uniform flow given by $\phi = Dx$, we find

$$\phi = Dx + \frac{Cx}{x^2 + y^2} \tag{10.61}$$

At this point it is convenient to switch to polar coordinates, so Eq. 10.61 becomes

$$\phi = Dr\cos\theta + C\,\frac{r\cos\theta}{r^2} = \left(\frac{C}{r} + Dr\right)\cos\theta \tag{10.62}$$

Therefore,

$$V_r = \frac{-\partial\phi}{\partial r} = \left(\frac{C}{r^2} - D\right)\cos\theta \tag{10.63}$$

$$V_\theta = \frac{-1}{r}\,\frac{\partial\phi}{\partial\theta} = \left(\frac{C}{r^2} + D\right)\sin\theta \tag{10.64}$$

This flow is sketched in Fig. 10.17, in which one of the streamlines is a circle. From Eq. 10.63, this is the circle for which $V_r = 0$; that is, $r = (C/D)^{1/2}$. Thus, this is an ideal-fluid flow which has a circular streamline. In ideal-flow theory, we can substitute a solid body for any streamline without affecting the flow outside that streamline; so the flow for $r > (C/D)^{1/2}$ is the same as the ideal-fluid flow which would exist outside a circular cylinder oriented perpendicular to the flow.

Now we can apply Bernoulli's equation to find the pressure at any point on the surface of the cylinder. Let us assume that far to the right in Fig. 10.17 the flow is undisturbed by the cylinder, so that it is moving from right to left at a uniform velocity V_0 with a uniform pressure P_0 and that the changes in

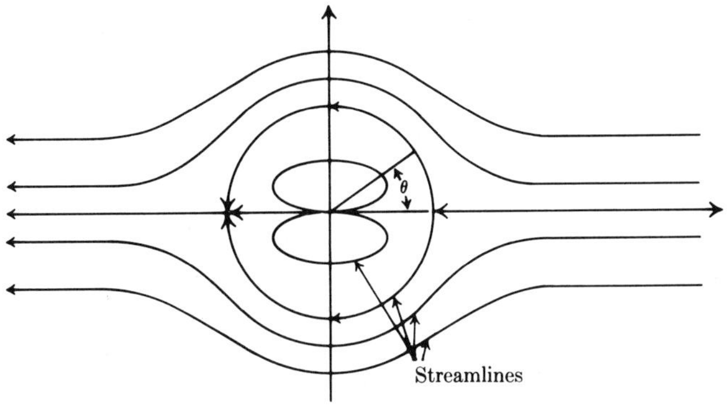

FIGURE 10.17
Streamlines for Eq. 10.62.

elevation in the entire flow are negligible. Then from Eqs. 10.57 and 10.58 we know that the pressure at any point in the flow is given by

$$P = P_0 + \rho\left(\frac{V_0^2}{2} - \frac{V^2}{2}\right) \tag{10.65}$$

Here the velocity is given by Eq. 10.58. Along the surface of the cylinder $V_r = 0$; so $V = V_\theta$. Substituting from Eq. 10.64 for $r = (C/D)^{1/2}$, we find

$$V_\theta = 2D \sin\theta \tag{10.66}$$

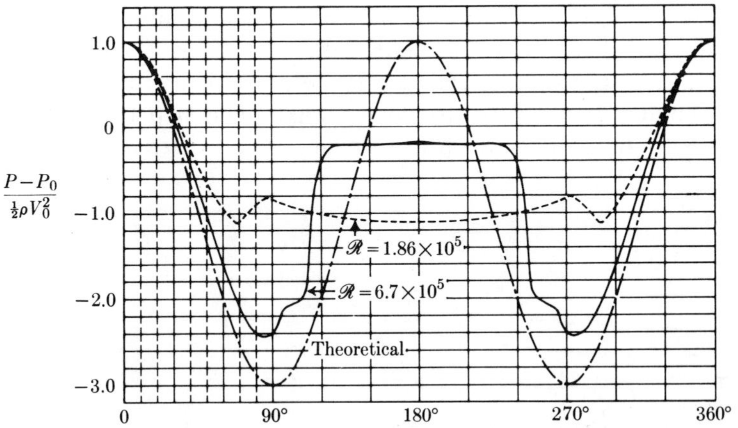

FIGURE 10.18
Pressure on the surface of a cylinder; comparison of the perfect-fluid calculation and experimental observations. The theoretical curve is Eq. 10.67; the other curves correspond to Reynolds numbers of 1.86×10^5 and 6.7×10^5. The reason for the great difference between them is given in Sec. 11.6. [From H. Muttray, "Die experimentalen Tatsachen des Widerstandes ohne Auftrieb" (The experimental data of drag without lift), in *Handbuch der Experimentalphysik Hydro- und Aero-Dynamik*, Leipzig, 1932, p. 316; based on data by Flaschbart.]

and comparing Eq. 10.62 with the steady flow shown in Fig. 10.4, we see that D is identical with V_0. So we substitute and find

$$P = P_0 + \rho \frac{V_0^2}{2} (1 - 4 \sin^2\theta) \tag{10.67}$$

Now that we have found the perfect-fluid solution for the pressure at various points on the surface of a cylinder, we must inquire whether nature really behaves this way. Figure 10.18 shows the pressure at various values of θ calculated from Eq. 10.67 as well as the measured pressures at the same angles at two different flow rates. For both flow rates there is fair agreement between the observed pressures and Eq. 10.67 along the front of the cylinder (0 to 90° and 270 to 360°), but there is very poor agreement for the back of the cylinder. The explanation is in terms of separation, described next.

10.8 SEPARATION

Figure 10.17 shows the perfect-fluid solution for flow around a cylinder. The flow splits at the upstream face, flows smoothly around the cylinder, and rejoins at the downstream face. Equation 10.64 shows that at $\theta = 0°$ and $\theta = 180°$ the flow velocity is zero; these are the points where the flow divides, hence the flow has no velocity here. Such zero-velocity points are commonly called *stagnation points*.

The actual flow pattern of a real fluid flowing around a cylinder can be observed by putting dye markers or bits of lint in a fluid. The observed pattern is sketched in Fig. 10.19. To the right (upstream of the cylinder) the flow is very similar to that described by Eq. 10.62 for a perfect fluid. However, downstream, instead of closing together behind the cylinder as a perfect fluid would, the flow pulls away from the cylinder at the points marked A, leaving the back of the cylinder covered by an eddying wake. This departure of the streamline from the body around which it is flowing is called *separation*.

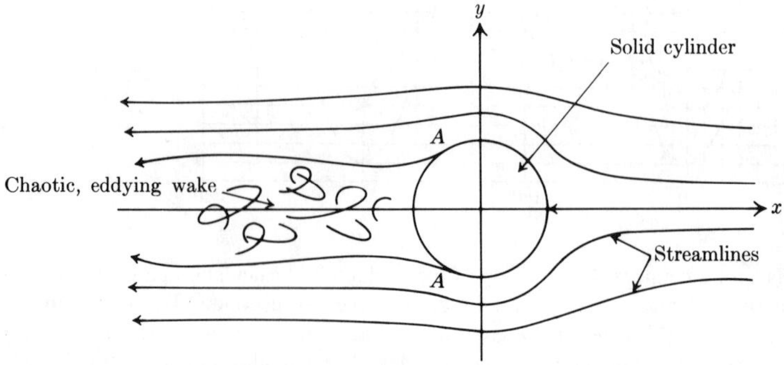

FIGURE 10.19
Real fluid flow around a cylinder.

Separation is caused by friction. In a real fluid, the friction in the "boundary layer" near the wall of the cylinder causes the fluid to go slower than the corresponding perfect fluid would. On the front of the cylinder, this has relatively little effect on the flow pattern, because there the fluid is speeding up and decreasing in pressure. However, on the rear of the cylinder, the perfect-fluid flow is one in which the fluid is slowing down and increasing in pressure. For the fluid to reach the rear stagnation point, it cannot have lost any momentum due to the effects of friction. Since the real fluid has lost some momentum due to friction, it does not have enough momentum to overcome the "adverse pressure gradient" on the rear of the cylinder, so it flows away. This is illustrated in Fig. 10.20.

For the designer of an aircraft, a racing car, or a structure, separation is very troublesome. Comparing the pressures before and behind the cylinder in Fig. 10.18, we see that the average pressure behind the cylinder, although high enough to cause the flow to separate, is not nearly as high as the pressure that would have existed if the flow had not separated. Thus, the cylinder has a large net pressure force acting on it. For an aircraft this would be a drag force, which would require the expenditure of power to overcome. This power is ultimately used up in friction heating in the eddying wake behind the cylinder. Much of the ingenuity of modern aircraft designers has gone into designing special wing structures (flaps, slots, spoilers, jets) to prevent separation [7].

For a *bluff body* like a cylinder or a flat plate perpendicular to the flow, separation is almost inevitable. But for a *streamlined body* such as an airplane wing or a fish, separation normally does not occur, and the flow pattern is very much like that shown by perfect-fluid theory. Figure 10.21 shows a comparison of the predicted and observed pressures on a "streamlined" body; the agreement between the predictions of perfect-fluid theory and experimental results are quite good except at the rear. Thus, aeronautical engineers can predict the lift of a "streamlined" wing with fair accuracy by using perfect-fluid theory. However, perfect-fluid theory predicts zero drag, whereas all real fluid flows show drag; so perfect-fluid theory alone is of little use in drag predictions.

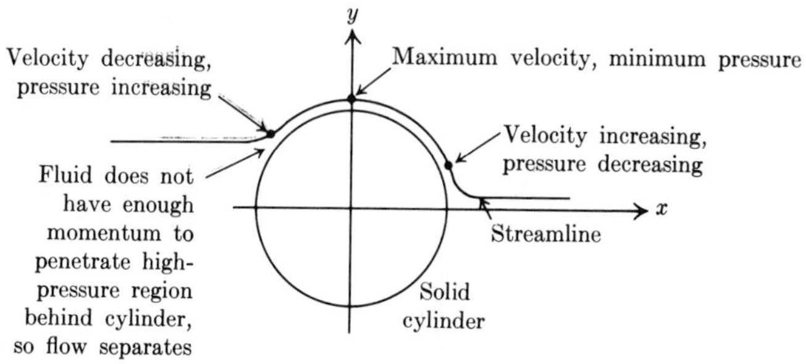

FIGURE 10.20
Separation in flow around a cylinder.

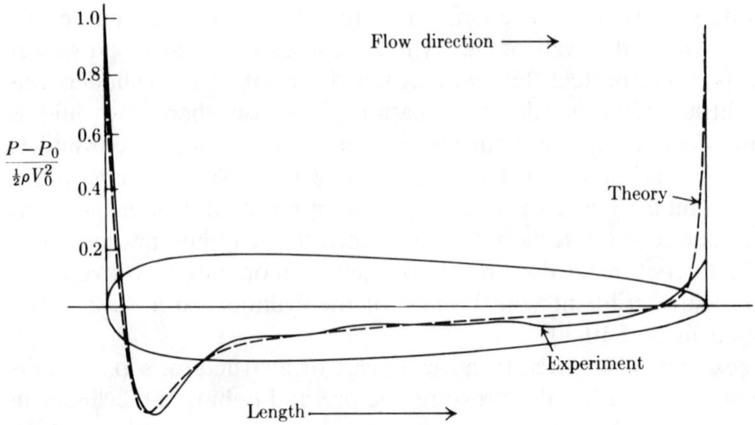

FIGURE 10.21
Comparison of experimental pressures with those calculated from perfect-fluid theory for a streamlined body. [From H. Muttray, "Die experimentalen Tatsachen des Widerstandes ohne Auftrieb" (The experimental data of drag without lift), in *Handbuch der Experimentalphysik Hydro- und Aero-Dynamik*, Leipzig, 1932, p. 316; based on data by Fuhrmann.]

One may define a streamlined body as one for which the real-fluid streamlines cling to the body all along its length without separation; normally such shapes are blunt in front and gradually taper in back.

The development of supercomputers has made it possible to apply the idea of potential flow outside the boundary layer and viscous flow inside the boundary layer to very complex structures and situations. Currently commercial aircraft designs are "tested" with supercomputers using the combination of potential flow and the boundary layer, much more quickly and cheaply than they could be tested with models in wind tunnels or in flight tests [8].

10.9 SUMMARY

1. The mathematics of two- and three-dimensional flows is much more difficult than that of one-dimensional flows. To simplify the mathematics, the concept of a perfect fluid with zero viscosity and constant density was invented.
2. This perfect fluid obeys the laws of potential flow, just as heat conduction and electrostatic fields do.
3. This potential-flow theory also describes the flow of a viscous fluid in a porous medium, which has considerable practical significance in petroleum reservoir engineering, hydrology, filters, etc.
4. Perfect-fluid solutions give fair descriptions of many flows, except near solid boundaries.
5. To handle complex flows involving solid boundaries, Prandtl introduced the idea of using perfect-fluid theory far from the solid surface and taking

viscosity into account only in a thin "boundary layer" near the surface of the solid.

6. Potential flows are irrotational.
7. For many real flows, the streamlines separate from the body around which they are flowing. This results in the formation of eddying wakes, low pressure behind the body, and large drag forces. This is not predictable by perfect-fluid theory, but it can be approached through boundary-layer theory.

PROBLEMS

See the Common Units and Values for Problems and Examples inside the back cover.

10.1. Describe the heat conduction and electrostatic fields that correspond to Figs. 10.4, 10.6, 10.8, and 10.16.

10.2. The flow described by Fig. 10.6 may be thought of as an oil well drawing fluid from a horizontal stratum. The oil stratum is 10 ft thick, and the flow is 100 ft^3/h of a fluid with a density of 50 lbm/ft^3 and a viscosity of 10 cP. The permeability of the stratum is 10^{-11} ft^2. The pressure at the inside of the well (radius 0.25 ft) is 1000 psia. Prepare a sketch of the pressure versus distance from the center of the well.

10.3. Map out the following potential flows. In each case verify that Laplace's equation is satisfied, calculate the equation of the stream function, and indicate to what physical situation these flows might correspond.

(*a*) $\phi = A(x^2 - y^2)$

(*b*) $\phi = (Ar^n \cos n\theta)/n$ for $n = \frac{1}{2}, \frac{2}{3}, 1, \frac{3}{2}, 2$

(*c*) $\phi = Bx + C \ln (x^2 + y^2)^{1/2} - C \ln [(x - A)^2 + y^2]^{1/2}$

10.4. Map out the flows predicted by $\phi = x^2$, $\phi = x^2 + y^2$, $\phi = -e^x$, and $\phi = \sin x$, and show that each does indeed result in a flow in which mass is not conserved (for a constant-density fluid), as discussed in Sec. 10.3.

10.5. Show the derivation of Eqs. 10.19 and 10.20. *Hint:* These transformations are made in terms of the chain rules of advanced calculus, as shown in any text on advanced calculus. For example, Eq. 10.20 is worked out in detail by Kaplan [9].

10.6. Show that the vorticity (Eq. 10.34) in polar coordinates is given by $\zeta = V_0/r + \partial V_\theta/\partial r - (1/r)(\partial V_r/\partial\theta)$. Then, using the polar coordinate form, show that the flow described by Eq. 10.30 is not irrotational, but that the flow described by Eq. 10.36 is irrotational.

10.7. Show that Eqs. 10.41 and 10.42 can be simultaneously satisfied only if both ϕ and ψ each satisfy Laplace's equation (Eq. 10.4). *Hint:* Differentiate both sides of Eq. 10.41 with respect to y at constant x and both sides of Eq. 10.42 with respect to x at constant y, and then subtract one equation from the other.

10.8. Show that if we choose B in Eq. 10.44 to be x in Eqs. 10.45 and 10.46, we find the same ψ shown in Eq. 10.49. *Hint:* Remember that $\arctan x = \pi/2 - \arctan (1/x)$.

10.9. Find the stream function which corresponds to Eqs. 10.15, 10.35, 10.36, and 10.38.

10.10. Although the stream function is used most often with the potential function, it can be used for viscous flows for which no potential function exists. For example, the laminar flow of a viscous fluid along a sloping flat plate is given by the equation $V_x = [(\rho g d^2 \cos\beta)/(2\mu)][1 - (y/d)^2]$, as described by Bird et al. [3, p. 39]. See Fig. 10.22. Calculate the stream function for this flow. Show that this flow cannot be represented by any potential function.

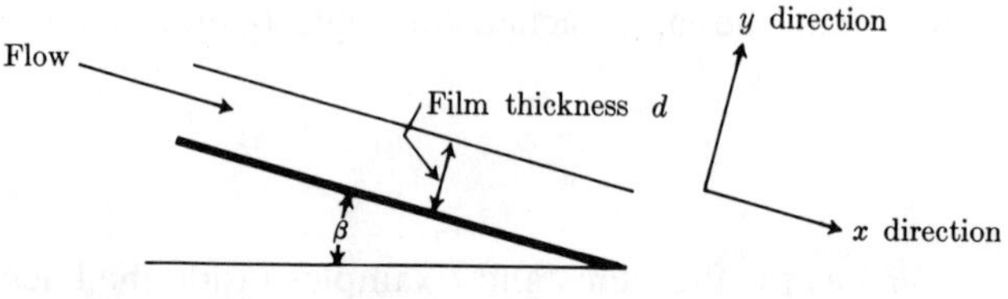

FIGURE 10.22

10.11. Show that one cannot calculate ϕ from the ψ calculated in Prob. 10.10 by the method shown in Sec. 10.5 for finding ϕ from ψ. To do this, calculate ϕ from ψ, and show that the resulting function does not give back the correct values of $\partial\phi/\partial x$ and $\partial\psi/\partial y$ on differentiation. This cannot be done because Eqs. 10.45 and 10.46 hold only (as may be shown) if both ϕ and ψ satisfy Laplace's equation, and the ψ found in Prob. 10.10 does not satisfy Laplace's equation.

10.12. The potential flow of a frictionless fluid is described by the equation $\phi = (10/\text{s})(x^2 - y^2)$. The fluid has a density of 60 lbm/ft^3. The pressure at $x = 0$, $y = 0$ is 20 psia. Calculate the pressure at $x = 1$ ft and $y = 1$ ft.

10.13. A tornado is described by Eq. 10.38. If at a distance 1 mi from the eye of the tornado (i.e., the origin of the coordinate system) the radial velocity is 1 mi/h and the tangential velocity is 1 mi/h, what are the radial and tangential velocity components 50 ft from the center of the tornado? If the pressure is 14.7 psia 1 mi from the center of the tornado, what is the pressure 50 ft from the center?

10.14. Figure 10.16 represents the flow from an injection well to a production well in a porous medium. How much of the flow between the two wells passes outside the streamlines which pass through the points $(A/2, A/2)$ and $(A/2, -A/2)$?

10.15. Figure 10.8 is drawn for $A = 1$ and $C = 1$. Sketch this figure (*a*) for $A = 1$ and $C = 10$ and (*b*) for $A = 10$ and $C = 1$.

10.16. Figure 10.12 is drawn for $A = 1$ and $C = 1$. Sketch this figure (*a*) for $A = 1$ and $C = 10$ and (*b*) for $A = 10$ and $C = 1$.

CHAPTER 11

THE BOUNDARY LAYER

The topic of the boundary layer is introduced in Sec. 10.1. The topic is a very large one. It is an active field of fluid mechanics research, so new results are constantly being published. Here we cannot hope to cover the entire topic; rather, we intend to show by a few examples what types of solution are obtainable and to impart some feeling for the results of the boundary-layer approach. More terminology is introduced than is necessary for the subject actually treated. This terminology is in common use in the boundary-layer literature; it is introduced here to show the student how these common terms relate to the other subjects treated in this book.

11.1 PRANDTL'S BOUNDARY-LAYER EQUATIONS

Ludwig Prandtl, the father of boundary-layer theory, after making the conceptual division of the flow discussed in Sec. 10.1, set out to calculate the flow in the boundary layer. He chose as his starting point the Navier-Stokes equations (Sec. 7.9) and simplified them by dropping the terms he considered unimportant. His simplifications are as follows:

1. The solid surface is taken as the x axis, the boundary layer beginning at the origin; see Fig. 11.1.
2. Gravity is unimportant compared with the other forces acting, so the gravity term can be dropped.

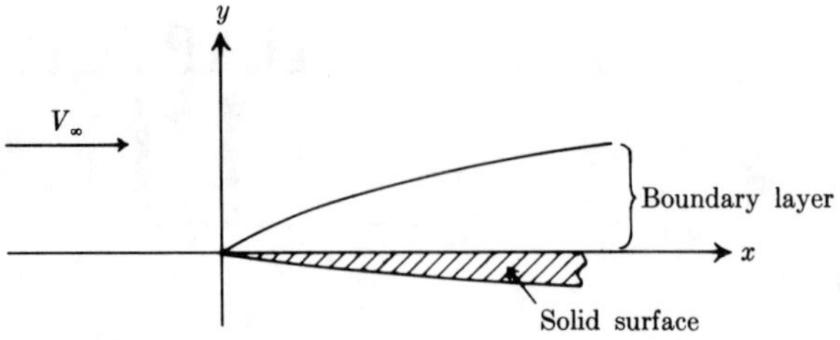

FIGURE 11.1

3. The flow is two-dimensional in the x and y directions. This means that V_z (the z component of the velocity) is zero, as are all derivatives with respect to z. These simplifications make the z momentum balance all zeros, so it can be dropped from the list of equations to be solved.
4. Although there is some flow in the y direction within the boundary layer, it is slow enough compared with the flow in the x direction for us not to need to consider the y-directed momentum balance. This does not mean that V_y is zero, but it does mean that $\partial P/\partial y$ is negligible.
5. The $\mu(\partial^2 V_x/\partial x^2)$ term in the momentum balance is small compared with the $\mu(\partial^2 V_x/\partial y^2)$ term and may be dropped.

Making these simplifications in Eq. 7.75, dividing by ρ, and replacing μ/ρ by ν (the kinematic viscosity), Prandtl found

$$\frac{\partial V_x}{\partial t} + V_x \frac{\partial V_x}{\partial x} + V_y \frac{\partial V_x}{\partial y} = -\frac{1}{\rho}\frac{\partial P}{\partial x} + \nu \frac{\partial^2 V_x}{\partial y^2} \tag{11.1}$$

This equation and the two-dimensional, constant-density material balance

$$\frac{\partial V_x}{\partial x} + \frac{\partial V_y}{\partial y} = 0 \tag{11.2}$$

are referred to as *boundary-layer equations* or *Prandtl's boundary-layer equations*.

In boundary-layer flows, as in flow in a pipe, the flow can be laminar or turbulent. Prandtl's equations, as shown above, are limited to laminar flows because of the form of the fluid friction term. The analogous form for turbulent flows can be derived [1, p. 563] but is of little use.

11.2 THE STEADY-FLOW, LAMINAR BOUNDARY LAYER ON A FLAT PLATE PARALLEL TO THE FLOW

As an example of the use of the boundary-layer equations, we consider the simplest possible boundary-layer problem, the steady flow of a constant-density, newtonian fluid past a flat plate placed parallel to the flow at velocities low enough for the flow to be laminar everywhere. This flow is shown schematically in Fig. 11.2.

To find a complete description of this flow, we must find a function $V_x = V_x(x, y)$ which satisfies Eqs. 11.1 and 11.2 together with the conditions that at $y = 0$ the velocity components V_x and V_y are zero (i.e., no flow along or across the solid surface) and that as y becomes large, V_x becomes the same as the perfect-fluid flow (see Chap. 10) outside the boundary layer.

For the system shown in Fig. 11.2, the perfect-fluid flow is described by

$$\phi = -V_\infty x \tag{11.3}$$

so that outside the boundary layer $V_x = V_\infty$ and $V_y = 0$ for all x and y. Thus, the second condition is that as y becomes large, V_x must approach V_∞.

From the perfect-fluid solution we find $\partial P/\partial x = 0$. This is true outside the boundary layer; inside the boundary layer it may not be exactly true, but according to Prandtl's third assumption, there is no change in pressure in the y direction inside the boundary layer. Therefore, at any point in the boundary layer the pressure is the same as the pressure at that x in the perfect-fluid flow outside the boundary layer. Boundary-layer experts describe this assumption by saying that the pressure in the perfect-fluid flow outside the boundary layer is *impressed* on the boundary layer. Thus, according to this assumption, the $\partial P/\partial x$ term in Eq. 11.1 can be dropped, giving

$$V_x \frac{\partial V_x}{\partial x} + V_y \frac{\partial V_x}{\partial y} = \nu \frac{\partial^2 V_x}{\partial y^2} \tag{11.4}$$

This problem of determining the flow velocity at every point near the plate is a "boundary-value problem," i.e., a set of partial differential equations with specified values on the boundaries. To see how it was solved, we consider first a simpler boundary-value problem. Suppose that an infinite fluid at rest is adjacent to an infinite plane wall at rest. At time zero the wall is suddenly set in motion in the x direction with velocity V_0. If the flow is laminar throughout, then from the momentum balance (Prob. 11.1) for any particle of fluid we find

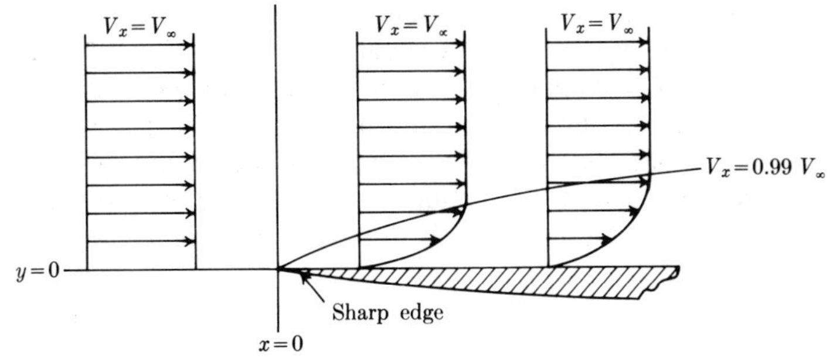

FIGURE 11.2

$$\frac{\partial V_x}{\partial t} = \nu \frac{\partial^2 V_x}{\partial y^2} \tag{11.5}$$

together with a set of boundary conditions. This problem is an approximate description of the flow near the flat side of a ship which starts suddenly.

This boundary-value problem can be completely and rigorously solved by the method of separation of variables [2] or by Laplace transforms [3]. The solution is

$$V_x = V_0\left[1 - \operatorname{erf} \frac{y}{2(\nu t)^{1/2}}\right] \tag{11.6}$$

Here erf is Gauss' error function, defined by

$$\operatorname{erf} x = \frac{2}{\pi^{1/2}} \int_0^x e^{-\lambda^2}\, d\lambda$$

where λ is a dummy variable. The values of erf are presented in various mathematical tables.

The interesting point about Eq. 11.6 is that even though V_x depends on both y and t, the solution indicates that it does not depend on them separately but rather depends only on a fixed combination of them $y/[2(\nu t)^{1/2}]$. If we introduce a new variable $A = y/[2(\nu r)^{1/2}]$, then the velocity, instead of being a function of y and t, becomes a function of A alone. This is a great simplification. Mathematically it means that the partial differential equation (Eq. 11.5) can be replaced with an ordinary differential equation. Graphically it means that instead of representing V_x by a plot of V_x versus y with lines of constant t, we can represent V_x as a single curve on a plot of V_x versus A. This simplification results from the character of the differential equation and its boundary values.

Now we reconsider the problem of the boundary layer on the flat plate. Blasius [1, p. 135] observed that although Eqs. 11.4 and 11.5 are not exactly the same, they have a similar form. He also noted the physical similarity between the situations described; if we consider the fluid to be at rest and the boundary layer to be formed by a sharp-edged plate moving through it at constant velocity (i.e., Fig. 11.2 as seen by an observer riding with the fluid), then this situation physically has much in common with the flow described by Eq. 11.5. Therefore Blasius assumed that the solution would be of the same form: that V_x, instead of depending on x and y separately, would depend only on some combination of them. Comparing the physical situation with that in the problem described by Eq. 11.5, he decided that the t in Eq. 11.6 is the time that the fluid has "known" that the plate is moving. For the boundary-layer problem this would be the distance x from the leading edge of the plate divided by the free-stream velocity V_∞. Making this substitution, he defined

$$\eta = y\left(\frac{V_\infty}{\nu x}\right)^{1/2} \tag{11.7}$$

By comparison with Eq. 11.6, he assumed he could write

$$\frac{V_x}{V_\infty} = \text{some function of} \left[y\left(\frac{V_\infty}{x\nu}\right)^{1/2} \right] \tag{11.8}$$

In the case of the other boundary-value problem mentioned, we can demonstrate mathematically that such a substitution is correct; here we cannot make such a demonstration, so the resulting solution rests on this additional assumption. This assumption converts the set of two partial differential equations to a single, ordinary differential equation, which Blasius was able to solve in numerical form. The details of this calculation are shown by Schlichting [1, p. 135]; see Prob. 11.4. The result is in the form of a curve of V_x/V_∞ versus η, shown in Fig. 11.3.

Blasius' solution for the laminar boundary layer on a flat plate, shown in Fig. 11.3, rests on a considerable string of assumptions and simplifications. However, it has been tested by numerous investigators and found to represent the experimental data very well (note that Fig. 11.3 shows the comparison between Blasius' solution and Nikuradse's experimental data). Thus, these assumptions and simplifications seem to be justified.

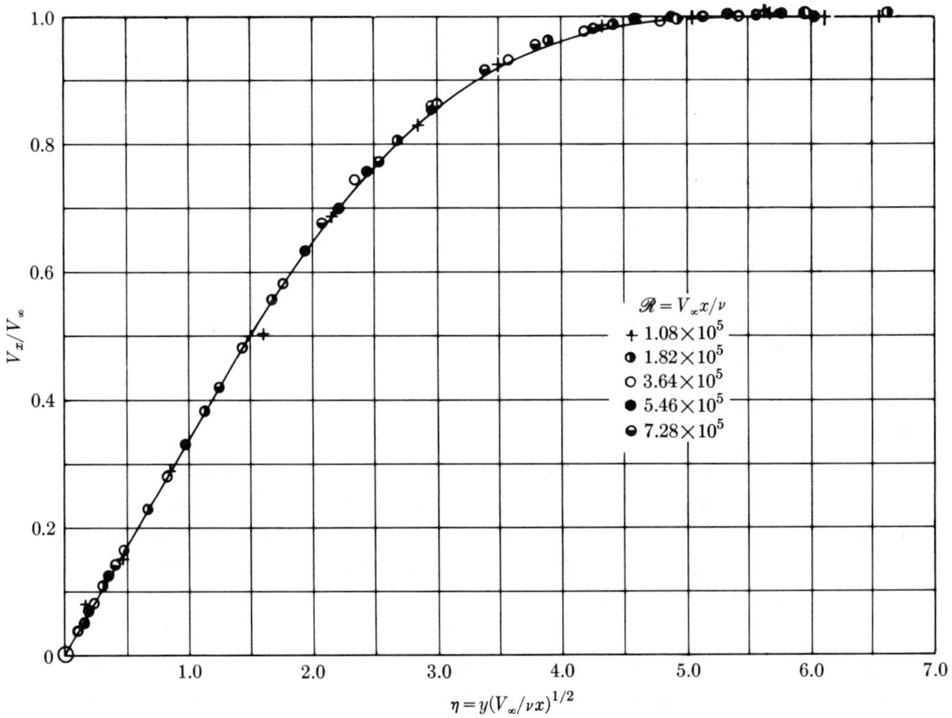

FIGURE 11.3
Blasius' solution for the laminar boundary layer on a flat plate and Nikuradse's experimental tests of same. [From J. Nikuradse, "Laminar Reibungsschichten an der laengsangestroemten Platte" (Laminar friction layers on plates with parallel flow), *Monograph Zentrale fuer Wiss. Berichtwesen*, Berlin (1942).

Blasius' solution to this problem is one of the landmarks of fluid mechanics. From a detailed study of this solution we can observe most of the principal terms and ideas in boundary-layer theory. Although this solution was found for a flat plate, it provides a guide to the solution for the flow over gently curving surfaces. Thus, although an airplane's wing surface is not flat, most of the calculated values for the laminar boundary layer on a flat plate will be qualitatively true for the laminar flow over an airplane wing, and much of the quantitative information in Blasius' solution is approximately correct for gently curving surfaces such as airplane wings and ship hulls.

From Fig. 11.3 we see that if the boundary layer is defined as that layer out to $V_x/V_\infty = 0.99$, then the laminar boundary layer extends out to a distance of $\eta \approx 5$. If we let δ be the thickness of the boundary layer, then

$$\delta \approx 5\left(\frac{\nu x}{V_\infty}\right)^{1/2} \tag{11.9}$$

This shows that the laminar boundary layer grows proportionally to the square root of the distance from the front of the plate and that the parabolic shape sketched for it in Fig. 11.2 is correct.

Example 11.1. Calculate the boundary-layer thickness for (*a*) a point on an airplane wing 2 ft from the leading edge when the plane is flying 200 mi/h through air and (*b*) a point 2 ft from the bow of a ship when the ship is moving 10 mi/h through water.

(*a*) For air

$$\delta = 5\left(\frac{1.61 \times 10^{-4}\ \text{ft}^2/\text{s} \cdot 2\ \text{ft}}{200\ \text{mi/h} \cdot 5280\ \text{ft/mi} \cdot \text{h}/3600\ \text{s}}\right)^{1/2} = 5.23 \times 10^{-3}\ \text{ft} = 1.60\ \text{mm}$$

(*b*) For water

$$\delta = 5\left(\frac{1.08 \times 10^{-5}\ \text{ft}^2/\text{s} \cdot 2\ \text{ft}}{10\ \text{mi/h} \cdot 5280\ \text{ft/mi} \cdot \text{h}/3600\ \text{s}}\right)^{1/2} = 6.07 \times 10^{-3}\ \text{ft}$$
$$= 0.07\ \text{in} = 1.8\ \text{mm}$$

■

In Sec. 10.1 we discussed Prandtl's assumption that the effect of the transition from zero velocity at a solid object to the free-stream velocity took place over a very thin layer of fluid. From Example 11.1 we see that this is certainly a very good assumption for laminar flow of air and water and typical ship and airplane velocities.

From Blasius' boundary-layer solution we can calculate the drag on any part of a flat plate. Because the solution is based on laminar flow of a newtonian fluid, we know that the shear stress at any point is given by

$$\tau = \mu \frac{dV_x}{dy} \tag{1.5}$$

Differentiating both sides of Eq. 11.7 with respect to V_x at constant x, we find

$$\left(\frac{d\eta}{dV_x}\right)_x = \left(\frac{V_\infty}{\nu x}\right)^{1/2}\left(\frac{dy}{dV_x}\right)_x \tag{11.10}$$

which can be inverted and rearranged to

$$\left(\frac{dV_x}{dy}\right)_x = V_\infty\left(\frac{V_\infty}{\nu x}\right)^{1/2}\frac{d(V_x/V_\infty)_x}{d\eta} \tag{11.11}$$

We want the value of $d(V_x/V_\infty)_x/d\eta$ at the surface of the plate, i.e., at $y = 0$ or, therefore, $\eta = 0$. From the slope of the curve in Fig. 11.3 this is 0.332. Thus, the shear stress at any point on the surface of the plate is given by

$$\tau_0 = 0.332\mu V_\infty\left(\frac{V_\infty}{\nu x}\right)^{1/2} \tag{11.12}$$

In Sec. 6.13 we showed that the drag force on numerous bodies could be represented in terms of plot of drag coefficient versus Reynolds number. If we now define a local drag coefficient for some small part of a flat plate as

$$C_f' = \frac{\tau_0}{\frac{1}{2}\rho V_\infty^2} \tag{11.13}$$

then from Eq. 11.12 we can calculate the drag coefficient corresponding to Blasius' solution:

$$C_f' = \frac{0.332}{\frac{1}{2}}\cdot\frac{\mu}{\rho}\cdot\frac{V_\infty}{V_\infty^2}\cdot\left(\frac{V_\infty}{\nu x}\right)^{1/2} = 0.664\left(\frac{\nu}{V_\infty x}\right)^{1/2} \tag{11.14}$$

Here the term in parentheses on the far right has the same form as $1/\mathcal{R}$. We have seen Reynolds numbers in which the length was a pipe diameter and those in which it was a particle diameter. In boundary-layer theory, the natural length to use seems to be the length measured from the leading edge of the solid body. Thus,

$$\mathcal{R}_x = \text{Reynolds number based on leading-edge distance} = \frac{V_\infty x}{\nu} \tag{11.15}$$

Equation 11.14 says that, according to Blasius' solution, a plot of local drag coefficients C_f' versus the Reynolds number should be given by

$$C_f' = \frac{0.662}{\mathcal{R}_x^{1/2}} \tag{11.16}$$

Figure 11.4 shows such a plot of experimental local drag coefficients. Those for which the flow is laminar agree very well with Blasius' solution. However, if the flow is turbulent, the result is quite different. We discuss turbulent boundary layers in Secs. 11.3 and 11.5.

At the leading edge of the plate ($x = 0$) the Reynolds number is zero; so, according to Eq. 11.16, the drag coefficient should be infinite. This is physically unreal and leads to the conclusion that Blasius' solution is not correct in the

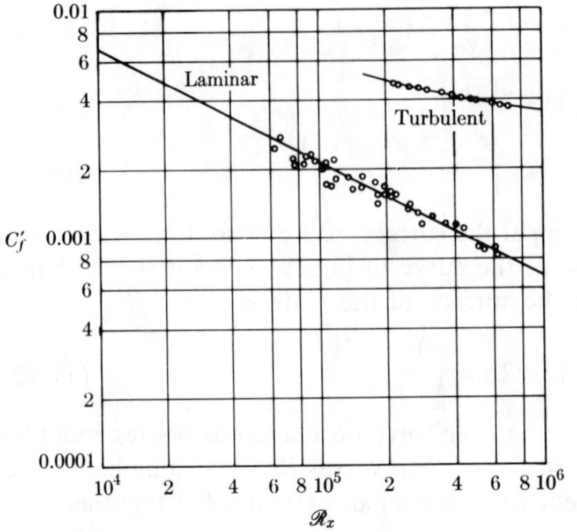

FIGURE 11.4
Local drag coefficient for a flat plate. Experimental data are compared with Blasius' solution (Eq. 11.16) and with Prandtl's equation (Eq. 11.36). [From H. W. Liepmann and S. Dahwan, "Direct measurements of local skin friction in low-speed and high-speed flow," *Proc. First U.S. Natl. Congr. Appl. Mech*, ASME, New York, 1952, p. 873. Reproduced with the permission of the publisher.]

small region nearest the leading edge [1, p. 141]. This is a minor flaw, which is of little practical concern.

The drag coefficient defined above C_f', which gives the local drag force, is less convenient for practical calculations than one that gives the drag force on an entire plate. That force for a plate of width W is

$$F = W \int_0^x \tau_0 \, dx \tag{11.17}$$

Now we define a new drag coefficient for the entire surface:

$$C_f = \frac{F}{\frac{1}{2}\rho V_\infty^2 A} = \frac{1}{A}\int C_f' \, dA \tag{11.18}$$

For Blasius' laminar boundary-layer solution, this becomes

$$C_f = \frac{0.332 W \mu V_\infty (V_\infty/\nu)^{1/2} \int_0^x dx/x^{1/2}}{0.5 \rho V_\infty^2 W x} = \frac{1.328}{\mathscr{R}_x^{1/2}} \tag{11.19}$$

Example 11.2. A 1-m^2 flat plate is towed behind a ship by a long, thin wire which does not disturb the flow. The boundary layers on both sides of the plate are laminar. The ship's velocity is 15 km/h. What force is required to tow the plate?

The Reynolds number is

$$\mathscr{R}_x = \frac{15 \times 10^3\ \text{m/h} \cdot 1\ \text{m}}{1.004 \times 10^{-6}\ \text{m}^2/\text{s}} \cdot \frac{\text{h}}{3600\ \text{s}} = 4.15 \times 10^6$$

so

$$C_f = \frac{1.328}{(4.15 \times 10^6)^{1/2}} = 6.52 \times 10^{-4}$$

and

$$F = 6.52 \times 10^{-4} \cdot \frac{1}{2} \cdot 998.2\ \frac{\text{kg}}{\text{m}^3} \cdot \left(\frac{15{,}000}{3600} \cdot \frac{\text{m}}{\text{s}}\right)^2 \cdot 2\ \text{m}^2 \cdot \frac{\text{N} \cdot \text{s}^2}{\text{kg} \cdot \text{m}}$$

$$= 11.3\ \text{N} = 2.54\ \text{lbf}$$ ■

In addition to the boundary-layer thickness δ, two other thicknesses occur frequently in the boundary-layer literature: the *displacement thickness* δ^* and the *momentum thickness* θ. To see the meaning of the displacement thickness, consider the streamlines for the laminar boundary layer on a flat plate, as sketched in Fig. 11.5.

To get around the layer of slow-moving fluid in the boundary layer, the streamlines in the entire flow are diverted away from the solid surface. To see how far, we make a material balance around the section marked in Fig. 11.5 for a width into the paper of W. The upper and lower boundaries are along streamlines, so there is no flow across them. Thus, the material balance is

$$\rho V_\infty y W = \rho W \int_0^{y+\delta^*} V_x\, dy \tag{11.20}$$

Dividing out ρW and subtracting $V_\infty \delta^*$ from both sides, we rearrange to get

$$-V_\infty \delta^* = -V_\infty(y + \delta^*) + \int_0^{y+\delta^*} V_x\, dy \tag{11.21}$$

which then can be rearranged to

$$\delta^* = \frac{\int_0^{y+\delta^*} (V_\infty - V_x)\, dy}{V_\infty} = \int_0^{y+\delta^*} \left(1 - \frac{V_x}{V_\infty}\right) dy \tag{11.22}$$

Here the displacement thickness δ^* is the distance that the streamlines are

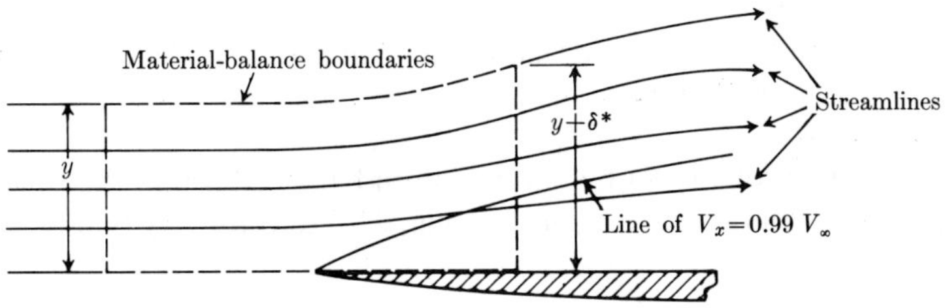

FIGURE 11.5
Displacement thickness.

moved in the direction perpendicular to the plate. Since $V_\infty - V_x$ is 0.01 at the edge of the boundary layer and rapidly goes to zero as y increases, the upper limit of integration in Eq. 11.22 is normally shown as infinity, although any large number will do. Equation 11.22 is correct for any constant-density boundary layer, whether laminar or turbulent.

For Blasius' solution we can perform the integration in Eq. 11.2 graphically on Fig. 11.3 by noting that the region above and to the left of the solid curve is

$$\int_0^\infty \left(1 - \frac{V_x}{V_\infty}\right) d\left[y\left(\frac{V_\infty}{\nu x}\right)^{1/2}\right] = \delta^*\left(\frac{V_\infty}{\nu x}\right)^{1/2} \tag{11.23}$$

By graphical integration this region has area 1.72; so

$$\delta^* = 1.72\left(\frac{\nu x}{V_\infty}\right)^{1/2} \tag{11.24}$$

Comparing Eq. 11.24 with Eq. 11.9, we see that the displacement thickness for a laminar boundary layer is 1.72/5, or about one-third of the boundary-layer thickness.

The solid body slows down a layer of the fluid without bringing any but the fluid actually touching the wall completely to rest. From all the fluid slowed down it extracts momentum in the form of a drag force on the solid plate. This same amount of momentum and hence the same force could be obtained by stopping completely some layer of the oncoming stream; the thickness of such a layer is called the momentum thickness θ. If such a layer were stopped by a plate W wide, the force required would be $W\theta\rho V_\infty^2$. The force actually exerted by the plate is given by Eq. 11.17, so

$$\theta = \frac{1}{\rho V_\infty^2} \int_0^x \tau_0 \, dx \tag{11.25}$$

$$\tau_0 = \rho V_\infty^2 \frac{d\theta}{dx} \tag{11.26}$$

which is used later in the study of turbulent boundary layers. For Blasius' laminar boundary-layer solution, we can substitute for τ_0 in Eq. 11.25, finding

$$\theta = \frac{1}{\rho V_\infty^2} 0.322\mu V_\infty \left(\frac{V_\infty}{\nu}\right)^{1/2} \int_0^x \frac{dx}{x^{1/2}} = \frac{0.664}{\mathcal{R}_x^{1/2}} \tag{11.27}$$

Comparing Eqs. 11.27 and 11.9, we see that for the laminar boundary layer on a flat plate the momentum thickness is 0.664/5, or about one-eighth of the boundary-layer thickness.

We can also show (Prob. 11.9) by a momentum balance around the same region we made the material balance around in Fig. 11.5 that for any boundary layer

$$\theta = \int_0^\infty \frac{V_x}{V_\infty}\left(1 - \frac{V_x}{V_\infty}\right) dy \tag{11.28}$$

This equation also is used in treating turbulent boundary layers.

Blasius' steady-flow, laminar, flat-plate, boundary-layer solution is a numerical solution of his simplification of Prandtl's boundary-layer equations, which are a simplified, one-dimensional momentum balance and a mass balance. This type of solution is known in the boundary-layer literature as an *exact solution*. Exact solutions can be found for only a very limited number of cases. Therefore, approximate methods are available for making reasonable estimates of the behavior of laminar boundary layers (Prob. 11.8).

11.3 TURBULENT BOUNDARY LAYERS

Like the flow in a pipe (Sec. 6.2), the flow in a boundary layer can be laminar or turbulent. In a pipe the transition takes place at a Reynolds number of about 2000, although it may be delayed to higher Reynolds numbers by taking extreme care to avoid pipe roughness or vibration. In a constant-diameter pipe, the flow has the same character over the entire length of the flow, except for a small region near the entrance. The same is not true of a boundary-layer flow. In a boundary-layer flow, the characteristic dimension is the distance from the leading edge. As we have seen, the appropriate Reynolds number for boundary-layer calculations is based on this length. As in pipe flow, the Reynolds number furnishes the criterion for transition from laminar to turbulent flow in a boundary layer. For a flat plate, the transition takes place for Reynolds numbers from 3.5×10^5 to 2.8×10^6. The transition is strongly influenced by turbulence in the stream outside the boundary layer and by roughness of the surface. A typical boundary layer on a smooth surface of sufficient length might look like Fig. 11.6. The figure shows not only laminar, transition, and turbulent boundary layers but also a "laminar sublayer" beneath the turbulent boundary layer (to be discussed later).

For laminar boundary layers, as for laminar flow in a pipe, it was possible to calculate the flow behavior from a set of plausible assumptions and then to show experimentally that the flow behaved as calculated. For turbulent boundary layers, as for turbulent flow in pipes, no one is yet able to calculate the flow behavior without starting with experimental data. However, from experimental measurements it has been possible to make some generalizations, which can then be used to extrapolate to other conditions.

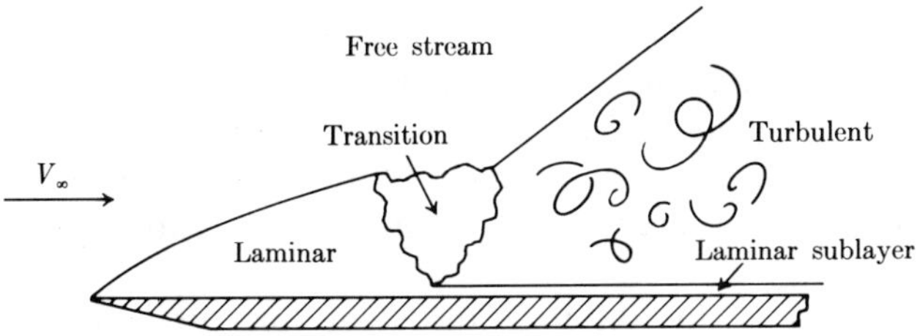

FIGURE 11.6
Laminar-turbulent transition in a boundary layer.

In a turbulent flow, the velocity at any point fluctuates randomly with time. One may speak of any such velocity as consisting of a time-average component and a fluctuating component; so at any point

$$V = V_{av} + \nu \tag{11.29}$$

These are defined such that V_{av} is the average reading over some time interval of a velocity meter at the point

$$V_{av} = \frac{1}{t}\int_0^t V\,dt \tag{11.30}$$

and such that the average of ν over some time interval is zero, because it is positive and negative for equal parts of the total time. In all the following discussions, the velocities are the time-average velocities.

Most of the results available for turbulent boundary layers have been found by measuring time-average velocities at various points in flow in pipes or over flat plates and by attempting to generalize the velocity profiles. For various experimental reasons it is easier to make such measurements in pipes, so most of the results are pipe results. Now we consider the turbulent flow in pipes for one section, and then we return to the turbulent boundary layer.

11.4 TURBULENT FLOW IN PIPES

As discussed in Sec. 6.4, turbulent flow differs from laminar flow in that the principal cause of the shear stresses between adjacent layers of the fluid is the interchange of masses of fluid between adjacent layers of fluid moving at different velocities. This gives rise to additional stresses, called *Reynolds stresses*. The most dramatic effect of these stresses is the large increase of friction heating in turbulent flow over that found in laminar flow. The other dramatic effect is the change in shape of the velocity profile from laminar to turbulent flow. This is shown in Fig. 11.7, where the experimental turbulent velocity profiles measured by Nikuradse are compared with the profile for laminar flow.

As discussed in Sec. 6.3, the velocity profile for laminar flow in a tube is parabolic. For turbulent flow it is much closer to *plug flow*, i.e., to a uniform velocity over the entire pipe cross section. Furthermore, as seen from Fig. 11.7, as the Reynolds number is increased, the velocity profile approaches closer and closer to plug flow. At the wall the turbulent eddies disappear; so the shear stress at the wall for both laminar and turbulent flow of newtonian fluids is given by $\tau_0 = \mu\, dV_x/dy$. Although it is very difficult experimentally to measure velocity gradients very close to the wall, it is clear from Fig. 11.7 that at the wall the velocity gradient is steeper for turbulent flow. Hence, the shear stress and friction heating must be larger for turbulent flow than for laminar flow.

Prandtl showed that each of the different turbulent-flow curves in Fig. 11.7 could be represented fairly well by an equation of the form

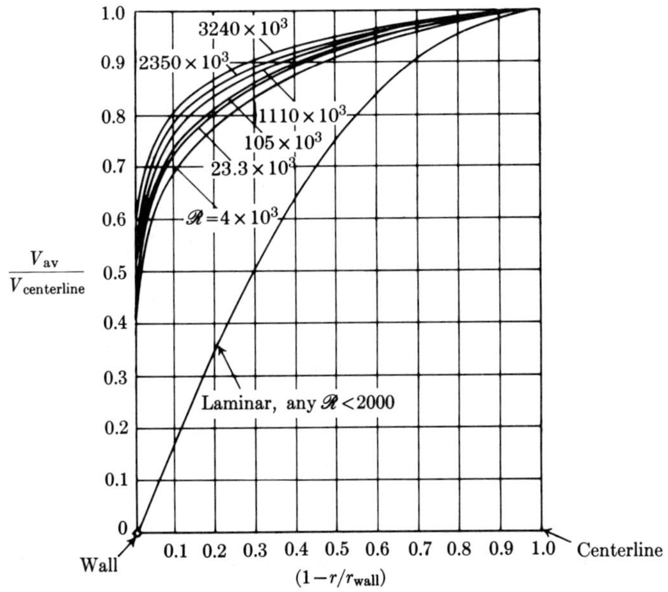

FIGURE 11.7
Velocity distribution in laminar and turbulent flow in smooth, circular pipes. [From J. Nikuradse, "Gesetzmaessigkeiten der turbulenten Stroemung in glatten Rohren" (Regularities of turbulent flow in smooth tubes), *Forschungsheft 356* (1932). Reproduced by permission of the publisher.]

$$\frac{V_x}{V_{x\ \text{centerline}}} = \left(1 - \frac{r}{r_{\text{wall}}}\right)^n \tag{11.31}$$

For the curves in Fig. 11.7, the value of n that gives the best representation of the experimental curves varies from $\frac{1}{6}$ for the lowest Reynolds number to $\frac{1}{10}$ for the highest Reynolds number. Prandtl selected $\frac{1}{7}$ as the best average, deducing "Prandtl's $\frac{1}{7}$ power velocity distribution rule." This is not an exact rule, because if it were a general rule, then all the curves in Fig. 11.7 would be identical. Furthermore, it cannot be correct very near the wall of the tube, because there it predicts that dV/dy is infinite and hence that the shear stress is infinite. Nonetheless, it is widely used because it is simple and, as we will see in Sec. 11.5, because it gives useful results.

It is possible to find more complex correlations for the velocity distribution in a pipe which do not have the limitations of Prandtl's $\frac{1}{7}$ power rule. In Fig. 11.7 the Reynolds number appears as a parameter in the velocity distribution plot. In trying to produce a universal velocity distribution rule, it seems logical to change the coordinates in Fig. 11.7 so that the Reynolds number enters either explicitly or implicitly in one of the coordinates, in the hope of getting all the data onto one curve.

The most successful method of doing this has been to define a new quantity called the *friction velocity*

$$u^* = \left(\frac{\tau_{\text{wall}}}{\rho}\right)^{1/2} = V_{x,\text{ av}}\left(\frac{f}{2}\right)^{1/2} \tag{11.32}$$

where f is the friction factor used in Chap. 6. This is not a physical velocity, which one could measure at any point in the flow, but is a combination of terms that has the dimensions of a velocity and so is called a velocity. Using this "velocity" as a parameter, we can prepare a universal plot of pipe velocities, as shown in Fig. 11.8.

Figure 11.8 shows that in making up the universal velocity distribution it was necessary to introduce two combinations of variables which are in common use in the fluid mechanics literature. The ratio of the local velocity to the friction velocity is called u^+ (spoken of as "u plus"). This is also the ratio of (the local time-average velocity/the average velocity in the entire flow) times the square root of $(2/f)$. The combination of the distance from the pipe wall and the friction velocity divided by the kinematic viscosity is called y^+. This is the product of a kind of Reynolds number based on distance from the wall rather than on pipe diameter and $\sqrt{f/2}$.

Figure 11.8 shows that the flow can be divided conceptually into three zones: a *laminar sublayer* nearest the pipe wall, in which the shear stress is principally due to viscous shear; a turbulent core in the middle of the pipe, in which the shear stress is principally due to turbulent Reynolds stresses; and a layer between them, called the *buffer layer*, in which both viscous and Reynolds stresses are of the same order of magnitude. Good experimental measurements are difficult to make in the laminar sublayer and buffer layer, so there is some controversy over the best location for the boundaries shown in Fig. 11.8. Deissler [4] and coworkers place the buffer layer at a y^+ of 5 to 26, and Schlichting and coworkers place it at a y^+ of 5 to about 70. Furthermore, current work seems to indicate that the location of the edge of these layers is not fixed in place but fluctuates up and down; so these values indicate only the mean locations of these edges [5]. Thus, Fig. 11.8 may be too simple a picture of the actual behavior. Nonetheless, it provides a reasonable conceptual model and is able to correlate most of the available data with reasonable accuracy.

Figure 11.8 is for smooth pipes. As shown in Fig. 6.10, increasing the roughness of the pipe wall in turbulent flow generally leads to an increase in the friction factor. In Fig. 11.8 we see that increasing the friction factor will increase y^+ and decrease u^+; so increasing the roughness while holding everything else constant will move a point on the curve downward and to the right. Schlichting [1, p. 621] presents a plot like Fig. 11.8 with a smooth-pipe line identical with that in Fig. 11.8 and other lines below and to the right of it for various relative roughnesses.

Example 11.3. Water is flowing in a 3-in ID smooth pipe, with an average velocity of 10 ft/s. How far from the wall are the edge of the laminar sublayer and the edge of the buffer layer? What is the average velocity at each of those points?

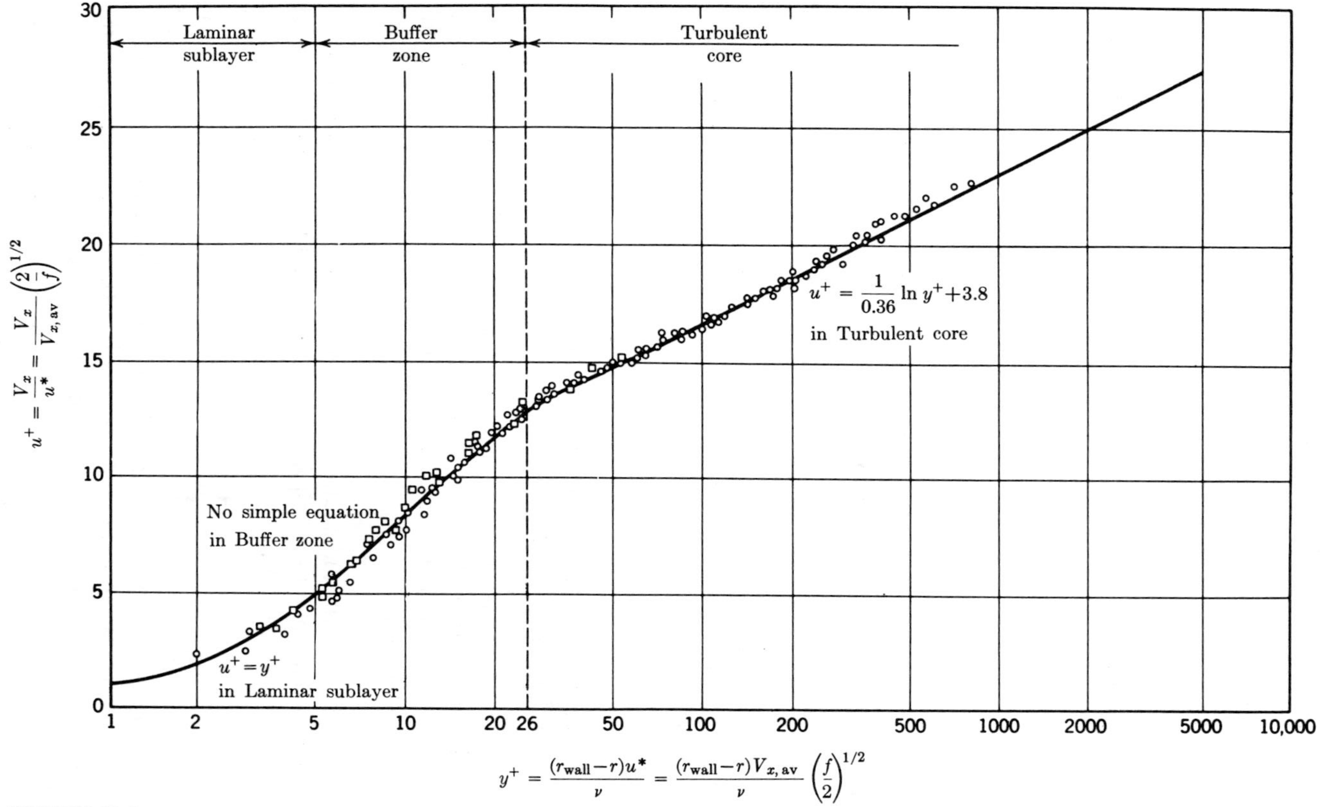

FIGURE 11.8
Universal velocity distribution for turbulent flow in smooth tubes. [After Deissler; reproduced from R. B. Bird, W. E. Stewart, and E. N. Lightfoot, *Transport Phenomena*, Wiley, New York, 1960. Reproduced by permission of the publisher.]

Here we have

$$\mathcal{R} = \frac{10\ \text{ft/s} \cdot 0.25\ \text{ft}}{1.08 \times 10^{-5}\ \text{ft}^2/\text{s}} = 2.3 \times 10^5$$

From Fig. 6.10 for smooth pipes we have $f = 0.0037$, so

$$u^* = 10\ \frac{\text{ft}}{\text{s}} \left(\frac{0.0037}{2}\right)^{1/2} = 0.44\ \frac{\text{ft}}{\text{s}} = 0.13\ \frac{\text{m}}{\text{s}}$$

From Fig. 11.8 at the edge of the laminar sublayer we have $u^+ \approx 5$ and $y^+ \approx 5$, so

$$V_x = u^+ u^* = 5 \cdot 0.44\ \text{ft/s} = 2.6\ \text{ft/s} = 0.79\ \text{m/s}$$

$$r_{\text{wall}} - r = \frac{y^+ \nu}{u^*} = \frac{5 \cdot 1.08 \times 10^{-5}\ \text{ft}^2/\text{s}}{0.44\ \text{ft/s}}$$

$$= 1.2 \times 10^{-4}\ \text{ft} = 1.4 \times 10^{-3}\ \text{in} = 0.037\ \text{mm}$$

At the edge of the buffer layer we have $u^+ \approx 12$ and $y^+ \approx 26$, so

$$V_x = 5.2\ \text{ft/s} = 1.59\ \text{m/s} \qquad r_{\text{wall}} - r = 7 \times 10^{-3}\ \text{in} = 0.18\ \text{mm}$$ ■

This example illustrates why there are so few experimental data in the laminar sublayer and buffer layer; these layers are extremely thin and have very steep velocity gradients.

11.5 THE STEADY, TURBULENT BOUNDARY LAYER ON A FLAT PLATE

There are no known analytical solutions for turbulent boundary layers that are analogous to Blasius' solution for the laminar boundary layer on a flat plate. Prandtl, to describe the steady, turbulent boundary layer on a flat plate, made the following assumptions:

1. The average velocity in the x direction at any point has the same kind of distribution as that found in a pipe and is represented by Prandtl's $\frac{1}{7}$ power rule (Eq. 11.31) in the form

$$\frac{V_x}{V_\infty} = \left(\frac{y}{\delta}\right)^{1/7} \qquad (11.33)$$

   This presupposes that the velocity profiles at any x are similar to each other; this is the kind of assumption made by Blasius when he assumed that the velocity was a function of η, not of x and y separately.
2. Over the Reynolds number range of 3×10^3 to 3×10^5, the friction factor plot for smooth pipes can be approximated (see Prob. 11.13) by

$$f = \frac{0.0791}{\mathcal{R}^{1/2}} \qquad (11.34)$$

Blasius has shown that this equation fits the smooth-pipes curve in Fig. 6.10 quite well. Prandtl assumed that it could be taken over directly for determining the shear stress at the surface of the plate, understanding the length term in the Reynolds number given above to be *twice* the thickness of the boundary layer.

Combining Eqs. 11.33 and 11.34, as shown in Prob. 11.14, Prandtl found

$$\delta = 0.37x\left(\frac{\nu}{V_\infty x}\right)^{1/5} \tag{11.35}$$

From this equation and several other relations (Prob. 11.15) we can compute the drag coefficient for a turbulent boundary layer

$$C_f' = \frac{0.0576}{\mathcal{R}_x^{1/5}} \tag{11.36}$$

$$C_f = \frac{0.072}{\mathcal{R}_x^{1/5}} \tag{11.37}$$

These two equations are based on some very severe assumptions, and their use can be justified only by experimental verification. Most tests have indicated that they give a very good representation of experimental data. For example, Fig. 11.4 shows a comparison of experimental data on the local drag coefficient with Eq. 11.36; the agreement is excellent. Other experimental data can be adduced to show that these equations are at least satisfactory for engineering purposes. Equation 11.37 assumes that the boundary layer is turbulent from the beginning of the plate ($x = 0$) to the end of the plate. Such a situation can exist if the beginning of the plate is artificially roughened. However, more commonly (Fig. 11.6) the first part of the plate has a laminar boundary layer, which then makes a transition to turbulent flow farther down the plate. To calculate the drag on such a plate, we calculate the separate contributions from the laminar and turbulent parts of the boundary layer.

Equation 11.35 also indicates that turbulent boundary layers grow with distance as x to the $\frac{4}{5}$ power, compared with the $\frac{1}{2}$ power for laminary boundary layers. Thus, for the same distance, the boundary layer will be larger and growing faster if it is turbulent rather than laminar.

Example 11.4. A speedboat is towing a smooth plate 1 ft wide and 20 ft long through still water at a speed of 50 ft/s. Determine the boundary-layer thickness at the end of the plate and the drag on the plate.

At the end of the plate,

$$\mathcal{R}_x = \frac{50\ \text{ft/s} \cdot 20\ \text{ft}}{1.08 \times 10^{-5}\ \text{ft}^2/\text{s}} = 0.93 \times 10^8$$

so from Eq. 11.35

$$\delta = \frac{0.37 \cdot 20\ \text{ft}}{(0.93 \times 10^8)^{1/5}} = 0.189\ \text{ft} = 2.3\ \text{in} = 58\ \text{mm}$$

As a first approximation, we assume that the entire boundary layer is turbulent so that Eq. 11.37 applies. Then

$$F = C_f \tfrac{1}{2} \rho V_\infty^2 A$$

$$= \frac{0.072}{(0.93 \times 10^8)^{1/5}} \cdot \frac{1}{2} \cdot 62.3 \frac{\text{lbm}}{\text{ft}^3} \cdot \left(50 \frac{\text{ft}}{\text{s}}\right)^2 \cdot (2 \cdot 20 \text{ ft} \cdot 1 \text{ ft}) \cdot \frac{\text{lbf} \cdot \text{s}^2}{32.2 \text{ lbm} \cdot \text{ft}}$$

$$= 178 \text{ lbf} = 790 \text{ N}$$

Now, to ascertain how large an error we made by assuming that the entire boundary layer was turbulent, we assume that transition from laminar to turbulent flow takes place at an $\mathscr{R}_x$ of 10^6. From the above, this corresponds to a distance of $\frac{1}{100}$ of the length of the plate; so the boundary layer over the first 0.2 ft presumably is laminar. For this area the drag due to a laminar boundary layer is given by Eq. 11.19 as

$$F = \frac{1.328}{(10^6)^{1/2}} \cdot \frac{1}{2} \cdot 62.3 \frac{\text{lbm}}{\text{ft}^3} \cdot \left(50 \frac{\text{ft}}{\text{s}}\right)^2 \cdot (2 \cdot 0.2 \text{ ft} \cdot 1 \text{ ft}) \cdot \frac{\text{lbf} \cdot \text{s}^2}{32.2 \text{ lbm} \cdot \text{ft}}$$

$$= 1.3 \text{ lbf} = 5.8 \text{ N}$$

In the calculation which assumed that the entire boundary layer was turbulent, this leading area contributed a force of

$$F = \frac{0.072}{(10^6)^{1/5}} \cdot \frac{1}{2} \cdot 62.3 \frac{\text{lbm}}{\text{ft}^3} \cdot \left(50 \frac{\text{ft}}{\text{s}}\right)^2 \cdot (2 \cdot 0.2 \text{ ft} \cdot 1 \text{ ft}) \cdot \frac{\text{lbf} \cdot \text{s}^2}{32.2 \text{ lbm} \cdot \text{ft}}$$

$$= 4.4 \text{ lbf} = 19.5 \text{ N}$$

So the calculation done by assuming a completely turbulent boundary layer gives a drag force which is too high by $4.4 - 1.3 = 3.1$ lbf. This error is small compared with the uncertainties introduced by the approximate nature of Eq. 11.37. ■

11.6 THE SUCCESSES OF BOUNDARY-LAYER THEORY

The foregoing shows that even in the simplest possible boundary-layer problems the mathematics is formidable. For calculating the boundary layer around an airplane or a ship, the mathematics is beyond our current abilities. We must resort to approximations and simplifications, mostly based on the results of the simple cases like those shown. Nonetheless, the success of boundary-layer theory in explaining the behavior of ships, airplanes, projectiles, etc., has been very good.

In Sec. 10.8 we discussed separation. In almost every case, separation results in a great increase in drag, which is normally a very undesirable result. From boundary-layer theory it is possible to make some good estimates of when separation will or will not occur. In particular, it can be shown that turbulent boundary layers are less likely to separate than laminar ones. One may visualize why by considering the flow around a cylinder, as shown in Fig. 10.20. There separation occurs because the fluid flowing near the surface has lost momentum due to drag and cannot penetrate the high-pressure layer behind the cylinder. The effect of turbulence is to transfer momentum from the free stream into the boundary layer by means of turbulent eddies. Thus, the layer near the wall is not slowed as much and does not separate as soon.

The most dramatic example of this occurs in flow around a sphere. A laminar boundary layer separates well ahead of the maximum diameter perpendicular to flow, while a turbulent one clings on much further. This is illustrated in Fig. 11.9, which shows two bowling balls moving at the same velocity in water. One ball has a patch of sand glued to its leading surface, making the boundary layer turbulent. This holds the separation back much further. The turbulent boundary layer has a higher skin-friction drag than the laminar one but has a much smaller low-pressure wake behind it. Since this low-pressure wake is the principal source of the total drag on the ball, the net drag is greatly reduced by the patch of sand. This principle was discovered long before the idea of the boundary layer by golfers, who found that old, rough balls went farther than new, smooth ones. This discovery led to the invention of the dimpled golf ball which is rough enough to have a turbulent boundary layer but still smooth enough to putt well. Its performance was enjoyed by many but understood by no one until the invention of boundary-layer theory [6].

In Fig. 6.22 we showed that the drag coefficient for a sphere moving in a still liquid drops dramatically at a particle Reynolds number of about 3×10^5.

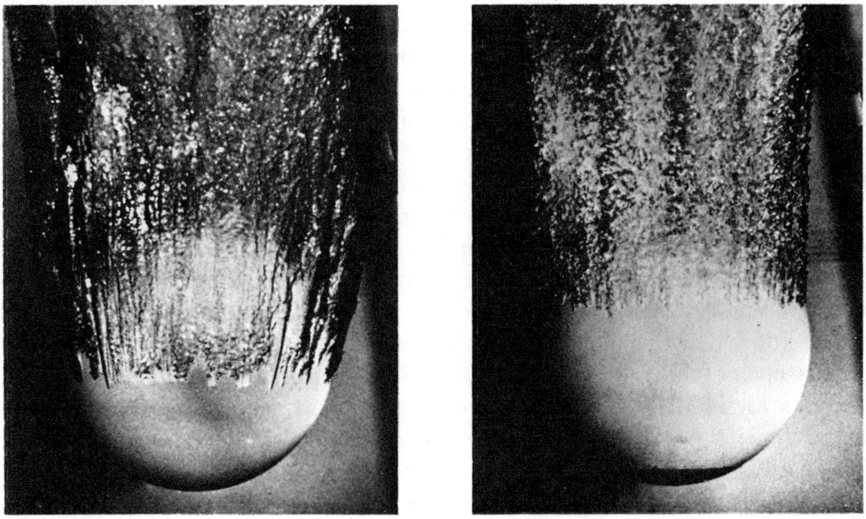

FIGURE 11.9
Two 8.5-in bowling balls entering still water at 25 ft/s. The one on the left is smooth; the one on the right has a patch of sand on its nose. [Official photograph, U.S. Navy.]

Perceptive readers will recognize this as being the lower limit of the transition from laminar to turbulent boundary layers on a flat plate and will conclude that this must be the result of the transition of the boundary layer on the sphere from laminar to turbulent flow. This has been verified experimentally. The function of the dimples on the golf ball is to make this transition occur at a lower Reynolds number.

The lift obtainable from an airplane wing is a strong function of the plane's angle to the horizontal, called the *angle of attack*. The higher the angle of attack, the greater the lift, up to the point where the flow separates on the top of the wing, causing the lift to decrease dramatically. This phenomenon in airplanes is called *stall*. A stall close to the ground is almost certain to cause a crash. Considerable efforts have been directed at preventing this separation so that airplanes can lift more or have smaller wings.

Normally stall is a problem only on takeoff and landing. One can increase the lift by going faster, but high takeoff and landing speed require longer runways than low ones. Thus, there is considerable incentive to find ways to increase the low-speed lift of a wing so that the airplane can take off and land from short runways. In most airplanes the wing shape is modified during takeoff and landing via slats and flaps, which allow some high-pressure air to flow from the bottom of the wing into the boundary layer on the top of the wing, increasing its velocity and making separation less likely. Other airplanes use small obstructions on the surface of the wing to increase the turbulence there and thus to prevent stall. The function of all these devices is referred to as *boundary-layer control*; their success indicates that the boundary-layer viewpoint has been very fruitful in aeronautical engineering.

For many years mariners were amazed at the swimming speed of dolphins, who can swim for hours at high speeds. Careful studies showed, e.g., that the drag coefficient of a dolphin must be less than one-half that of the best torpedo designs produced by our navies. Apparently, dolphins can do this because they have a specially designed resilient skin, which damps out turbulence and keeps the boundary layer laminar in circumstances in which a smooth but rigid surface would have a turbulent boundary layer [7].

In discussing friction in pipes in Chap. 6, we restricted ourselves to the case of flow well downstream of the pipe entrance. Much higher pressure losses per foot are observed in the "entrance region" of a pipe. When the flow enters a pipe, a boundary layer begins to grow at the wall. From what we have seen here, we know that the shear stress is highest at the front of a plate (where the boundary layer is thin). Thus, we would expect the pressure loss per unit length to be greatest at the inlet and to decrease with distance down the pipe. At some distance down the pipe, the boundary layers from the opposite walls grow together, filling the pipe. Thereafter the flow is no longer of the boundary-layer type and can be treated as a pipe flow. The distance for the boundary layers to grow together, forming "fully developed" flow, is a function of the pipe diameter and the Reynolds number [8].

Other cases of the success of boundary-layer theory can be cited. The same kind of idea has been taken over into heat-transfer and mass-transfer theory with generally useful results.

Notice also that this entire chapter seems to be a mixture of mathematics and sweeping assumptions. This is typical of boundary-layer studies. The entire problem of describing the flow around some complicated structure like an airplane is far beyond our current abilities, even with supercomputers. However, by using judicious guesswork we can reduce the important parts of the problem to mathematically manageable form. The real genius of Ludwig Prandtl was his ability to guess correctly which terms he could drop out of his equations and still have the calculated result agree with experimental data. This inspired guesswork usually can be done only by engineers who have a very good understanding of what every term in an equation means physically. It is strongly recommended that students try to develop this understanding.

11.7 SUMMARY

1. Prandtl started with the Navier-Stokes equations and discarded enough terms to make his boundary-layer equations, which are the working form of the momentum and continuity equations for boundary-layer problems.
2. By assuming that the velocity was a function of $y[V_\infty/(x\nu)]^{1/2}$, Blasius was able to solve Prandtl's equations for the steady-flow laminar boundary layer on a flat plate. He found that the laminar boundary-layer thickness is proportional to the square root of the length down the plate.
3. The solution for the turbulent boundary layer is based on results obtained in pipes. Using pipe results, Prandtl calculated that the thickness of a turbulent boundary layer is proportional to the length of the plate to the $\frac{4}{5}$ power.
4. Only a very small number of boundary-layer problems can be solved in any closed or simple mathematical form. However, the boundary-layer viewpoint has been very fruitful in the field of flow around solid bodies and in several other fields.

PROBLEMS

See the Common Units and Values for Problems and Examples inside the back cover.

11.1. Derive Eq. 11.5 in two ways: (*a*) by writing a momentum balance for a small element of fluid and (*b*) by canceling the zero terms in the x-directed Navier-Stokes equation (Eq. 7.79). For laminar flow in this geometry, you may assume that the y and z components of the velocity are everywhere zero.

11.2. List the boundary values for Eq. 11.5.

11.3. Show that the substitution of $A = y/[2(\nu t)^{1/2}]$ converts Eq. 11.5 from a partial differential equation with y and t as variables to an ordinary differential equation with A as a variable.

11.4. Blasius' laminar boundary-layer solution (Fig. 11.3) is an excellent example of the use of the stream function to simplify and solve the partial differential equations of fluid mechanics. To solve Eqs. 11.4 and 11.2 together, he substituted for V_x and V_y from Eqs. 10.41 and 10.42.

(*a*) Show that this leads to

$$\frac{\partial \psi}{\partial y}\frac{\partial^2 \psi}{\partial y\,\partial x} - \frac{\partial \psi}{\partial x}\frac{\partial^2 \psi}{\partial y^2} = \nu \frac{\partial^3 \psi}{\partial y^3}$$

in which, instead of having two dependent variables V_x and V_y, we have only one, ψ.

(*b*) We wish ψ to depend on the dimensionless group η, but it cannot depend on η alone, because ψ has the dimensions of length squared divided by time and η is dimensionless. Therefore, it must depend not only on η, but also on some combination of variables x, y, ν, and V_∞. Several such combinations have the required dimensions; Blasius found that the mathematics was simplest if he made the choice $\psi = (\nu x V_\infty)^{1/2} f(\eta)$, where $f(\eta)$ is an unknown function, to be determined. Show that on the basis of this choice

$$V_x = -\frac{\partial \psi}{\partial y} = \frac{-\partial \psi}{\partial \eta}\frac{\partial \eta}{\partial y} = V_\infty f'(\eta)$$

where f' stands for $df/d\eta$ and that

$$V_y = \frac{\partial \psi}{\partial x} = -0.5\left(\frac{\nu V_\infty}{x}\right)^{1/2}(\eta f' - f)$$

Hint: For V_y differentiate directly for $\partial f/\partial x$, and note that $\partial f/\partial x = (df/d\eta)(\partial \eta/\partial x)$.

(*c*) Show that if we proceed to compute the necessary second and third derivatives in the same way as in (*b*) and then substitute them in the first equation in this problem, we find

$$\left(\frac{-V_\infty^2}{2x}\right)\eta f' f'' + \left(\frac{V_\infty^2}{2x}\right)(\eta f' - f)f'' = \nu\left(\frac{V_\infty^2}{2\nu}\right)f'''$$

which can be simplified to $ff'' + 2f''' = 0$. This is now an ordinary differential equation involving only $f(\eta)$ and its derivatives with respect to η. Although it appears simple, it is nonlinear, and no analytical solution of it has been found. However, it has been solved numerically by using an infinite-series method, and tabulated values are available [1, p. 135]. The solid curve in Fig. 11.3 is based on this infinite-series solution.

11.5. Assuming that the transition from a laminar to a turbulent boundary layer takes place at a Reynolds number of 10^6, what is the maximum thickness for the laminar boundary layer on a flat plate for (*a*) air flowing at 10 ft/s, (*b*) water flowing at 10 ft/s, and (*c*) glycerin flowing at 10 ft/s ($\nu_{\text{glyc}} = 8.07 \times 10^{-3}\ \text{ft}^2/\text{s}$)?

11.6. Calculate the values of the Reynolds number based on length for the two parts of Example 11.1. Boundary layers on flat plates are laminar only up to Reynolds numbers of 3.5×10^5 to 2.8×10^6. Are these values exceeded for the boundary layers in Example 11.1?

11.7. From Blasius' solution (Fig. 11.3) we wish to find V_y (the y component of the velocity) at any point. Starting with the mass balance equation, Eq. 11.2, show that

$$V_y = -\left[\int_0^y \left(\frac{\partial V_x}{\partial x}\right)_y dy\right]_{\text{all at constant } x} = \frac{V_\infty}{2(V_\infty x/\nu)^{1/2}} \int_0^\eta \frac{d(V_x/V_\infty)}{d\eta}\, d\eta$$

Then rearrange to

$$\frac{V_y}{V_\infty}\left(\frac{V_\infty x}{\nu}\right)^{1/2} = \frac{1}{2}\int_0^\eta \frac{d(V_x/V_\infty)}{d\eta}\,\eta\,d\eta$$

and integrate graphically, noting that $d(V_x/V_\infty)/d\eta$ is the slope of Fig. 11.3. Schlichting gives a table in which the far right column gives these slopes to five significant figures; he also gives a plot of the result of this graphical integration [1, pp. 137, 139].

11.8. For boundary layers on curved surfaces, the pressure will change with distance. This greatly complicates the solution of the boundary-layer equations compared with that on a flat plate (in which $\partial P/\partial x$ was zero), and so very few "exact" solutions are known for such boundary layers. Some estimate of the behavior of such boundary layers is given by several methods. To illustrate, we apply them to the laminar boundary layer on a flat plate, where we can compare the results with Blasius' "exact" solution. These methods begin by assuming a velocity profile of the form $V_x/V_\infty = f(y/\delta)$, where δ is the boundary-layer thickness.

(*a*) Show that any satisfactory assumed function $f(y/\delta)$ must have the following properties: When $y/\delta = 0$, then $f = 0$; when $y/\delta = 1$, then $f = 1$. In addition, for best results it is desirable that secondary conditions be met: When $y/\delta = 0$, then $d^2f/dy^2 = 0$; when $y/\delta = 1$, then $df/dy = 0$ and $d^2f/dy^2 = 0$. Show graphically the meaning of these conditions.

(*b*) Several useful choices of $f(y/\delta)$ are $f = y/\delta$, $f = \frac{3}{2}y/\delta - \frac{1}{2}(y/\delta)^3$, and $f = \sin(\pi/2)(y/\delta)$. Show which of these conditions are satisfied by these functions.

(*c*) Below is the approximate boundary-layer calculation for the assumed function $f = y/\delta$. Repeat the calculation for the other assumed functions shown in part (*b*). Substituting $V_x/V_\infty = y/\delta$ in Eq. 11.28 and integrating only from 0 to δ, we find

$$\theta = \int_0^\delta \frac{y}{\delta}\left(1 - \frac{y}{\delta}\right) dy = \frac{1}{\delta^2}\left[\frac{y^2\delta}{2} - \frac{y^3}{3}\right]_0^\delta = \frac{\delta}{6}$$

and using Newton's law of viscosity in Eq. 11.26, we find

$$\tau_0 = \mu\,\frac{dV_x}{dy} = \mu\,\frac{V_\infty}{\delta} = \rho V_\infty^2\,\frac{d\theta}{dx}$$

But, as shown above, $\theta = \delta/6$, and therefore

$$\frac{d\theta}{dx} = \frac{1}{6}\frac{d\delta}{dx} \qquad \text{or} \qquad \delta\,\frac{d\delta}{dx} = \frac{6\nu x}{V_\infty}$$

Separating variables and integrating from $x = 0$ and $\delta = 0$ to $x = x$ and $\delta = \delta$, we find

$$\frac{\delta^2}{2} = \frac{6\nu x}{V_\infty} \qquad \text{or} \qquad \delta = \left(\frac{12\nu x}{V_\infty}\right)^{1/2} = 3.46\left(\frac{\nu x}{V_\infty}\right)^{1/2}$$

Comparing this result with Blasius' "exact" solution for the same boundary layer (Eq. 11.9), we see that this approximate solution gives a boundary-layer thickness which is about 3.46/5, or approximately 70 percent of the correct one.

Computing the drag coefficient from Eq. 11.13, we get

$$C_f' = \frac{\mu V_\infty/\delta}{\frac{1}{2}\rho V_\infty^2} = \frac{2\nu}{V_\infty}\left(\frac{V_\infty}{12\nu x}\right)^{1/2} = 0.577\,\frac{\nu}{V_\infty x}$$

Comparing this value with the drag coefficient based on Blasius' solution (Eq. 11.14), we find that this approximate solution gives a drag coefficient of 0.577/0.664, or about 87 percent of the correct solution. In the same way, all the other properties of the boundary layer can be computed for this assumed velocity profile or any other. The striking thing here is that this very simple assumed profile gives reasonably accurate results with an expenditure of much less effort than Blasius' solution requires. For more complicated flows, this saving in effort can be vast. More on these approximate methods is given by Schlichting [1, chap. 10].

11.9. Derive Eq. 11.28 by making an x momentum balance around the boundaries shown in Fig. 11.5 and solving for the force on the plate. Then eliminate the displacement thickness by means of Eq. 11.22, and equate the force on the plate to $W\theta\rho V_\infty^2$.

11.10. For Prandtl's $\frac{1}{7}$ power velocity rule, calculate the ratio of the maximum velocity at the center of the pipe to the average velocity in the entire pipe. Also calculate the ratio of the kinetic energy of the fluid to the kinetic energy it would have if it were all flowing at the average velocity.

11.11. Any velocity distribution equation which is to represent the flow in the neighborhood of the wall must satisfy the following requirements: $V_x = 0$ at $y = 0$; dV_x/dy is finite and nonzero at $y = 0$. As we saw in the text, Prandtl's $\frac{1}{7}$ power rule satisfies the first condition but not the second. Which of the following kinds of functions satisfy both conditions? (*a*) $V_x = A + By$, (*b*) $V_x = By^n$, where n is some power other than 1, (*c*) $V_x = A \sin y$, and (*d*) $V_x = A \exp y^n$, where n is any power.

11.12. Show from the definitions of u^+, y^+, and the friction factor that the limiting value of the relation of u^+ to y^+ as y^+ approaches zero must be $u^+ = y^+$.

11.13. Show that Eq. 11.34 is approximately correct over the Reynolds number range of 3×10^3 to 3×10^5 by calculating the friction factor from it and comparing it with the friction factor for smooth pipes in Fig. 6.10 for several $\mathcal{R}$ values in this range.

11.14. Show that if we make the assumptions shown in Sec. 11.5, we find Prandtl's equation for the drag on a flat plate with a turbulent boundary layer. The procedure is as follows:

1. From the $\frac{1}{7}$ power distribution rule (Eq. 11.33), deduce the ratio of the momentum thickness to the boundary-layer thickness, using Eq. 11.28. Here the integration is from 0 to δ rather than from 0 to infinity.
   Answer: $\theta = \frac{7}{72}\delta$.
2. In Eq. 11.34 the velocity in the Reynolds number and the velocity in the expression for the friction factor are average pipe velocities. From Prandtl's $\frac{1}{7}$ power rule it can be shown that for circular pipes (Prob. 11.10) this average velocity is 0.817 times the maximum velocity. Prandtl rounded this to 0.8. Making this substitution and recalling from Chap. 6 that

$$f = \frac{\tau_0}{\rho} \frac{1}{2} V_{x,\,\mathrm{av}}^2$$

convert Eq. 11.4 to

$$\frac{\tau_0}{\rho} \frac{1}{2} V_\infty^2 = 0.225\left(\frac{\nu}{V_\infty \delta}\right)^{1/4}$$

3. Combine this result with Eq. 11.27 and the answer from part 1 to obtain $\frac{7}{72}\, d\delta/dx = 0.0225[\nu/(V_\infty \delta)]^{1/4}$, which may be integrated from $\delta = 0$ at $x = 0$ to $\delta = \delta(x)$ at any x, obtaining Eq. 11.35.

11.15. Starting with Eq. 11.35, derive Eqs. 11.36 and 11.37. Use Eq. 11.26 for the shear stress as a function of the momentum thickness and the result obtained in Prob. 11.14 that $\theta = \frac{7}{72}\delta$. The two different drag coefficients are defined in Eqs. 11.13 and 11.18.

11.16. In Example 7.14 we considered the laminar flow of a newtonian fluid between two parallel plates and showed that well downstream from the entrance the velocity distribution was parabolic. At the entrance to such a pair of plates, the flow will be initially plane, and boundary layers will grow from the walls, eventually meeting in the center, as sketched in Fig. 11.10. Show that if we make the simple possible assumptions—that the growing boundary layers do not interact with each other and that the fluid between the boundary layers has a constant velocity—then the distance downstream required for the boundary layers to grow together (the "entrance length") will be given by $L_e/h = 0.01\mathcal{R}$. These assumptions are gross simplifications; the worse one says that the fluid in the center does not speed up. By material balance we may show that it must reach a velocity of twice the entrance velocity when the layers meet. More complicated analyses which take this into account [1, p. 186] lead to an approximate formula for parallel plates of $L_e/h = 0.04\mathcal{R}$. To see the magnitude of this entrance length, calculate it for air flowing at 10 ft/s between plates 12.0 in apart. (Here $\mathcal{R}$ is the Reynolds number based on distance between plates, not on the distance from the leading edge.)

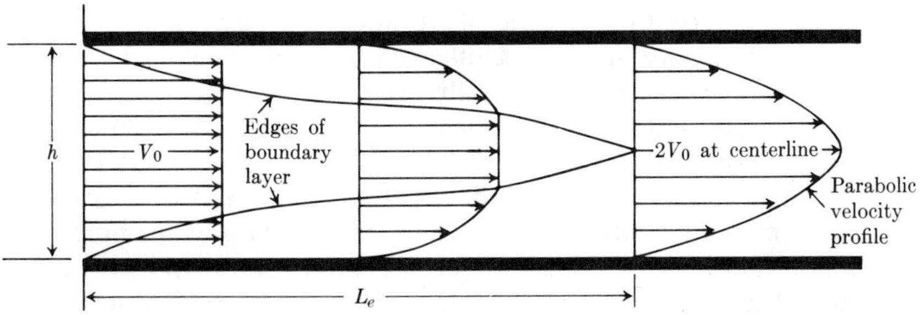

FIGURE 11.10
Entrance length.

CHAPTER 12

FLOW THROUGH POROUS MEDIA

A *porous medium* is a continuous solid phase which has many void spaces, or pores, in it. Examples are sponges, cloths, wicks, paper, sand and gravel, filters, bricks, plaster walls, many naturally occurring rocks (e.g., sandstones and some limestones), and the packed beds used for distillation, absorption, etc. In many such porous solids, the void spaces are not connected, so there is no possibility that fluid will flow through them. For example, expanded polystyrene hot-drink cups, life preservers, and iceboxes have many pores, but because of the "closed cell" structure of the plastic these pores are not interconnected. Thus these porous media form excellent barriers to fluid flow. A pile of sand, however, has fewer pores than an expanded polystyrene drinking cup, but its pores are all connected so that fluids can flow easily through it. The porous media with no interconnected pores are described as impermeable to fluid flow and those with interconnected pores as permeable (we give a mathematical definition of permeability in Sec. 12.1). The flow of fluids in *permeable* porous media is of great practical significance in groundwater hydrology, oil and gas production, filters, packed absorption, distillation columns, and fluidized beds.

To view the similarities and differences between this kind of flow and the flows we discussed previously, let us consider the gravity flow of water through some vessel (see Fig. 12.1). In a vessel of this type, as we saw in Chaps. 5 and 6, the flow could be described by Bernoulli's equation. There are no pumps or

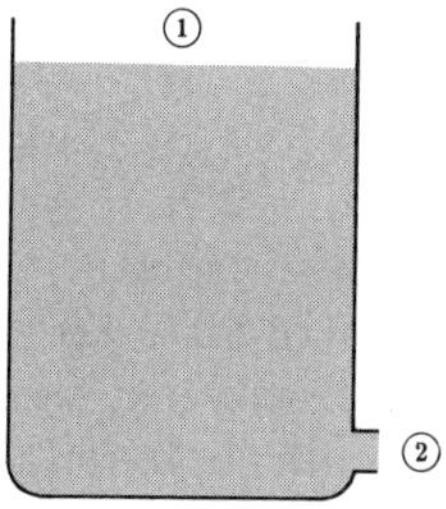

FIGURE 12.1

turbines, and the pressure change from point 1 to point 2 is negligible. So Bernoulli's equation simplifies to

$$g\,\Delta z + \frac{\Delta V^2}{2} = -\mathscr{F} \tag{12.1}$$

In Chap. 5 we considered numerous cases in which the friction term in Eq. 12.1 was negligible. In Chap. 6 we considered how we could calculate the friction term (normally based on generalizations of experimental data) in those cases in which the friction term was not negligible.

Now suppose the entire vessel in Fig. 12.1 is filled with some porous solid, such as sand (this vessel now resembles the sand filters frequently used to clarify muddy water). Equation 12.1 still describes the situation exactly as it did before, because it is based on the steady-flow energy balance for a constant-density fluid and it has no built-in assumption that the flow is occurring in an open vessel rather than a porous medium. The significant differences that we will see, if we compare the two situations, are as follows:

1. In most porous-medium flows, the friction term is much, much larger than it would be in the analogous flow in an empty vessel, and it is not directly calculable from the results in Chap. 6.
2. For most porous-medium flows, even though V_2 does not equal V_1, both velocities are so small that ΔV^2 is negligible compared with $\mathscr{F}$.
3. If the tank in Fig. 12.1 does not contain sand and is originally full of one fluid, e.g., air, when we admit a second fluid, e.g., water, it will quickly flush out all the first fluid. However, if the tank contains sand and its voids are originally full of air, then admitting water will not flush out all the air. Some significant part of the air (probably 10 to 30 percent) will be trapped permanently in the pores. This kind of behavior is of great significance in groundwater hydrology and oil recovery.

In this chapter we examine the friction term in Bernoulli's equation for flow in a porous medium as well as the phenomenon of incomplete displacement of one fluid from a porous medium by another. We also examine competitive countercurrent flow in porous media and look briefly at filtration and fluidization.

12.1 FLUID FRICTION IN POROUS MEDIA

Consider a porous medium consisting of sand or some porous rock or glass beads or macaroni or cotton cloth contained in a pipe; see Fig. 12.2. If we attach this pipe to the apparatus for the pressure-drop experiment shown in Fig. 6.1 and run exactly the same tests on it that we described there for a pipe, we find results of the same form as those shown in Fig. 6.2, except that the abrupt transitions region on Fig. 6.2 will be replaced with a smooth curve for a porous-medium flow. From these results we guess that the two end parts of the curve correspond to laminar and turbulent flows; this is experimentally verifiable.[†]

For flow in a pipe, we were able to calculate the laminar-flow portion of the curve from a simple force balance and to make a simple correlation for the turbulent-flow portion of the curve. For porous media this has been successfully done only for media consisting of uniformly sized, spherical particles. Here we examine that solution, because it provides useful insights into flow in more complex media and allows us to discuss many of the terms in common use in the porous-media literature. We begin with some definitions.

At any one cross section perpendicular to the flow, the average velocity may be based on the entire cross-sectional area of the pipe, in which case it is called the *superficial velocity* V_s

$$V_s = \frac{Q}{A_{\text{pipe}}} = \frac{\dot{m}}{\rho A_{\text{pipe}}} \tag{12.2}$$

Or it may be based on the area actually open to the flowing fluid, in which case it is called the *interstitial velocity* V_I

$$V_I = \frac{Q}{\varepsilon A_{\text{pipe}}} = \frac{\dot{m}}{\varepsilon \rho A_{\text{pipe}}} \tag{12.3}$$

where ε is the *porosity*, or *void fraction*:

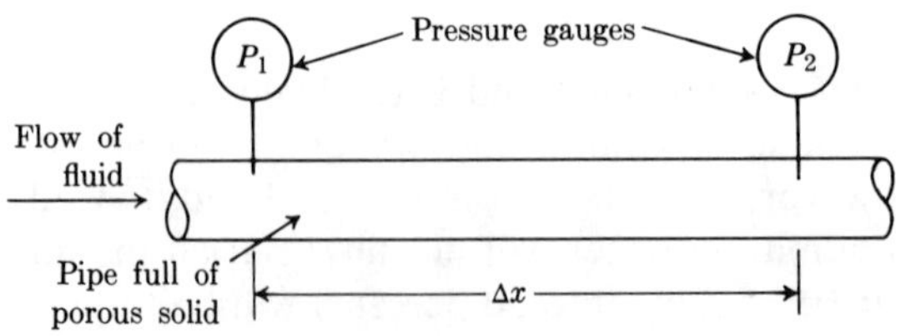

FIGURE 12.2

[†] The term "laminar" literally means in "shells" or "laminae." This is not an accurate description of flow in a porous medium, in which the width of the individual flow channels changes from point to point. A better term is "streamline flow," indicating that the individual fluid particles follow streamlines, which do not cross or mix, as they would in turbulent flow. However, the name "laminar" is more widely used and is used here.

$$\varepsilon = \frac{\text{total volume of system} - \text{volume of solids in system}}{\text{total volume of system}}$$

$$= \frac{A\,\Delta x\,(1 - \text{average fraction of cross section by solids})}{A\,\Delta x}$$

$$= \text{average fraction of cross section not occupied by solids} \tag{12.4}$$

From a theoretical standpoint, the interstitial velocity is the more important; it determines the kinetic energy and the fluid forces and whether the flow is turbulent or laminar. From a practical standpoint, the superficial velocity is generally more useful; it shows the flow rate in terms of readily measured variables. Both see common use.

Previously it was indicated that for noncircular conduits Fig. 6.10 (the friction factor plot) could be used if we replaced the diameter in both the friction factor and the Reynolds number with 4 times the hydraulic radius (HR). The hydraulic radius is the cross-sectional area perpendicular to flow, divided by the wetted perimeter. For a uniform duct this is a constant. For a packed bed it varies from point to point. But if we multiply both the cross-sectional area and the perimeter by the length of the bed, it becomes

$$\text{HR for porous medium} = \frac{\text{volume open to flow}}{\text{total wetted surface}} \tag{12.5}$$

For a porous medium made of equally sized spherical particles,

$$\text{HR} = \frac{\text{volume of bed} \cdot \varepsilon}{\text{no. spherical particles} \cdot \text{surface area of one particle}}$$

but

$$\text{No. particles} = \frac{\text{volume of bed} \cdot (1 - \varepsilon)}{\text{volume of one particle}}$$

so

$$\text{HR} = \frac{\text{volume of bed} \cdot \varepsilon}{\text{volume of bed} \cdot (1 - \varepsilon) \cdot \text{surface/volume}}$$

$$= \frac{\varepsilon}{(1 - \varepsilon)[\pi D_p^2 / (\frac{1}{6}\pi D_p^3)]}$$

$$= \frac{D_p}{6} \cdot \frac{\varepsilon}{1 - \varepsilon} \tag{12.6}$$

Here D_p is the particle diameter. If we now insert 4 times this definition of the hydraulic radius into the definitions of the pipe flow friction factor and the pipe flow Reynolds number, we find

$$f = \mathscr{F}\,\frac{4D_p\varepsilon / [6(1 - \varepsilon)]}{4\,\Delta x} \cdot \frac{2}{V_1^2} = \frac{\mathscr{F}}{3} \cdot \frac{D_p}{\Delta x} \cdot \frac{\varepsilon}{1 - \varepsilon} \cdot \frac{1}{V_1^2} \tag{12.7}$$

$$\mathcal{R} = \frac{V_I 4\{D_p \varepsilon / [6(1-\varepsilon)]\} \rho}{\mu} = \frac{2 D_p \varepsilon V_I \rho}{3\mu(1-\varepsilon)} \tag{12.8}$$

It is customary to replace V_I in these equations with V_s/ε, so

$$f = \frac{\mathcal{F}}{3} \cdot \frac{D_p}{\Delta x} \cdot \frac{\varepsilon^3}{1-\varepsilon} \cdot \frac{1}{V_s^2} \tag{12.9}$$

and

$$\mathcal{R} = \frac{2 D_p V_s \rho}{3\mu(1-\varepsilon)} \tag{12.10}$$

As in the case of flow in pipes, there are several different friction factors in common usage for flowing porous media, all differing by a constant. The choice between these is completely arbitrary; in this text we drop the $\frac{1}{3}$ in Eq. 12.9 and the $\frac{2}{3}$ in Eq. 12.10 to find our working forms of the friction factor and Reynolds number for porous media:

$$f_{\text{porous medium}} = \frac{D_p}{\Delta x} \frac{\varepsilon^3}{(1-\varepsilon)} \frac{1}{V_s^2} = f_{\text{PM}} \tag{12.11}$$

$$\mathcal{R}_{\text{porous medium}} = \frac{D_p V_s \rho}{\mu(1-\varepsilon)} = \mathcal{R}_{\text{PM}} \tag{12.12}$$

Having made definitions, we now inquire whether the pipe flow friction factor plot will predict the pressure drop for flow in porous media. For the laminar-flow region in pipes, Poiseuille's equation may be rewritten as $f = 16/\mathcal{R}$. Here the f and $\mathcal{R}$ are consistent with the definitions given in Eqs. 12.9 and 12.10. When we convert to the definitions given in Eqs. 12.11 and 12.12, this becomes $f_{\text{PM}} = 72/\mathcal{R}_{\text{PM}}$ (see Prob. 12.1). There is one obvious error in this derivation, namely, the tacit assumption that the flow is in the x direction. Actually, the flow is a zigzag; it must detour around one particle and then around another. If we assume that this zigzag proceeds with an average angle of 45° to the x axis, then the actual flow path is $\sqrt{2}$ times as long as the flow path shown in Eq. 12.11 and the actual interstitial velocity is $\sqrt{2}$ times the interstitial velocity used in Eq. 12.11. If we make these changes, we conclude that to agree with the pipe friction factor plot, laminar flow in a porous medium made of uniform-sized spheres should be described by $f_{\text{PM}} = 144/\mathcal{R}_{\text{PM}}$ (see Prob. 12.2). Experimental data indicate that the constant is about 150; i.e., the 45° assumption made above is slightly incorrect, so that for laminar flow we find experimentally

$$f_{\text{PM}} = \frac{150}{\mathcal{R}_{\text{PM}}}$$

or, rearranged,

$$\mathcal{F} = 150 \frac{V_s \mu (1-\varepsilon)^2}{D_p^2 \varepsilon^3} \frac{\Delta x}{\rho} \qquad \text{[laminar flow]} \tag{12.13}$$

Equation 12.13 is known as the *Blake-Kozeny equation* or the *Kozeny-Carman equation*; it describes the experimental data for steady flow of newtonian fluids through beds of uniform-size spheres satisfactorily for $\mathscr{R}_{\text{PM}}$ less than about 10.

Example 12.1. Figure 12.3 shows a water softener in which water trickles by gravity over a bed of spherical ion-exchange resin particles, each 0.05 in in diameter. The bed has a porosity of 0.33. Calculate the volumetric flow rate of water.

Applying Bernoulli's equation from the top surface of the fluid to the outlet of the packed bed and ignoring the kinetic-energy term and the pressure drop through the support screen, which are both small, we find

$$g(\Delta z) = -\mathscr{F}$$

Substituting from Eq. 12.13 and solving for V_s, we find

$$\begin{aligned} V_s &= \frac{g(-\Delta z)D_p^2\varepsilon^3\rho}{150\mu(1-\varepsilon)^2\Delta x} \\ &= \frac{32.2\text{ ft/s}\cdot 1.25\text{ ft}\cdot(0.05\text{ ft}/12)^2\cdot 0.33^3\cdot 62.3\text{ lbm/ft}^3}{150\cdot 1\text{ cP}\cdot 0.67^2\cdot 1\text{ ft}\cdot 6.72\times 10^{-4}\text{ lbm/(ft}\cdot\text{s}\cdot\text{cP)}} \\ &= 0.035\text{ ft/s} = 0.011\text{ m/s} \end{aligned}$$

Therefore,

$$Q = AV_s = \left(\frac{2}{12}\text{ ft}\right)^2\cdot\frac{\pi}{4}\cdot 0.035\ \frac{\text{ft}}{\text{s}} = 0.00075\ \frac{\text{ft}^3}{\text{s}} = 21\ \frac{\text{cm}^3}{\text{s}}$$

Before accepting this as the correct solution, we check the Reynolds number, finding

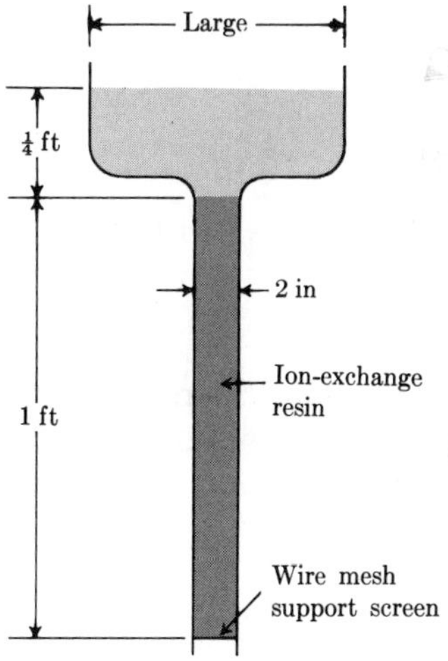

FIGURE 12.3

$$\mathcal{R}_{\text{PM}} = \frac{(0.5\ \text{ft}/12) \cdot 0.035\ \text{ft/s} \cdot 62.3\ \text{lbm/ft}^3}{1\ \text{cP} \cdot 0.67 \cdot 6.72 \times 10^{-4}\ \text{lbm/(ft} \cdot \text{s} \cdot \text{cP)}} = 20.2$$

This is slightly above the value of 10, for which we can safely use Eq. 12.13. Looking ahead to Fig. 12.4, we see that this estimated velocity is probably high by 10 to 15 percent. ■

If there had been no porous medium in the lower part of the apparatus in Fig. 12.3, then the exit velocity would have been given by Torricelli's equation, equal to about 9 ft/s. Here the calculated velocity is $\frac{1}{250}$ as large. Fluid friction effects in porous media are large!

We saw that for fully turbulent flow in a pipe the friction factor was constant for a given relative roughness but varied considerably for different relative roughnesses. For pipes the relative roughness can vary over a wide range, as can the friction factor in fully turbulent flow. However, for porous media made of uniform spherical particles there can be little variation in relative roughness. Here the roughness does not consist of rough spots on the surface of the individual spheres, but consists of the constantly changing shape of the individual flow channels, as they wend their way between the individual particles. The height of a typical obstruction is about the diameter of a single particle; the width of a typical flow channel is about one-half the diameter of a single particle. Therefore, the relative roughness is generally about 2. Referring to the friction factor plot for pipes (Fig. 6.10), we see that this relative roughness is very much larger than that ever encountered in pipes; so we would expect that the friction factor for completely turbulent flow in porous media should be very much larger than any friction factor ever encountered in a pipe. This is experimentally verifiable. The pressure drop for highly turbulent flow in a porous medium made up of uniform, spherical particles is reasonably well described by

$$f_{\text{PM}} = 1.75$$

which can be rearranged to

$$\mathcal{F} = 1.75\,\frac{V_s^2\,\Delta x}{D_p} \cdot \frac{1-\varepsilon}{\varepsilon^3} \tag{12.14}$$

This equation is known as the *Burke-Plumber equation*; it is satisfactory for $\mathcal{R}_{\text{PM}}$ greater than about 1000.

Example 12.2. We now wish to apply a sufficient pressure difference to the water flowing through the packed bed in Fig. 12.3 for the water superficial velocity to be 1 ft/s. What pressure gradient is required?

Applying Bernoulli's equation as before, we find

$$\frac{\Delta P}{\rho} + g\,\Delta z = -\mathcal{F}$$

Here, however, the gravity term is negligible compared with the others, so substituting from Eq. 12.14, we find

$$\frac{\Delta P}{\Delta x} = \frac{1.75 \rho V_s^2}{D_P} \cdot \frac{1-\varepsilon}{\varepsilon^3}$$

$$= \frac{1.75 \cdot 62.3 \text{ lbm/ft}^3 \cdot (1 \text{ ft/s})^2 \cdot 0.67}{(0.05 \text{ ft}/12) \cdot 0.33^3 \cdot 32.2 \text{ lbm} \cdot \text{ft}/(\text{lbf} \cdot \text{s}^2) \cdot 144 \text{ in}^2/\text{ft}^2}$$

$$= 105 \text{ psi/ft} = 2.37 \text{ MPa/m}$$

Here we may check the Reynolds number to see whether this is indeed turbulent flow. ■

This startingly high pressure drop illustrates the fact that turbulent flows very seldomly occur in porous media composed of particles this small. Since this particle size is typical of those encountered in most soils or underground aquifers or petroleum reservoirs and in most industrial filters, we see why almost all flows of fluids in the earth or in industrial filters are laminar.

For the transition region from laminar to turbulent flow in pipes, there is no simple friction-loss correlation, because the flow may be laminar or turbulent or oscillate between the two. This is not the case in a porous medium, because the flow does not switch all at once from laminar to turbulent. The reason is that the flow is not through one channel, but through a large number of parallel channels, of varying sizes. As the flow rate is increased from a low value, it is originally entirely laminar, and then the largest channels switch to turbulent flow. As the flow rate is further increased, more and more channels become turbulent, until at very high flow rates there is turbulent flow in all the channels. This leads to a smooth transition from all-laminar to all-turbulent flow.

Thus, in a plot of f_{PM} versus $\mathcal{R}_{\text{PM}}$, there is one smooth curve for transition from laminar to turbulent flow; Fig. 12.4 is such a plot. Ergun [1] showed that if we add the right-hand sides of the Kozeny-Carman and Burke-Plumber equations, the result fits the data in the transition region reasonably well, i.e.,

$$f_{\text{PM}} = 1.75 + \frac{150}{\mathcal{R}_{\text{PM}}}$$

which can be rearranged to

$$\mathcal{F} = 1.75 \frac{V_s^2}{D_p} \frac{(1-\varepsilon)\,\Delta x}{\varepsilon^3} + 150 \frac{V_s \mu (1-\varepsilon)^2\,\Delta x}{D_p^2 \varepsilon^3 \rho} \qquad (12.15)$$

which is known as the *Ergun equation*. It fits the data satisfactorily for all Reynolds numbers because the second term becomes negligible at high Reynolds numbers, giving Eq. 12.14. At small Reynolds numbers, the second term becomes so large that the first term is negligible in comparison, giving Eq. 12.13 (see Prob. 12.3).

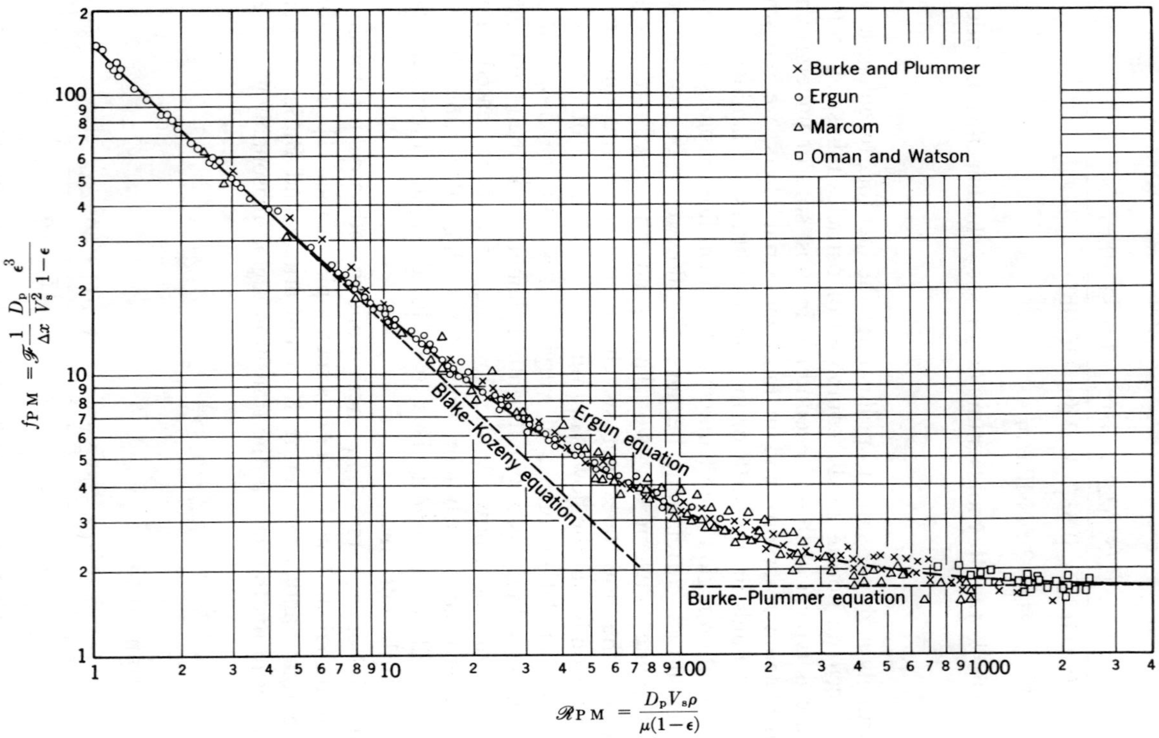

FIGURE 12.4
Pressure-drop data for flow through porous media. [From R. B. Bird, W. E. Stewart, and E. N. Lightfoot, *Transport Phenomena*, Wiley, New York, 1960, who redrew it from S. Ergun, "Fluid flow through packed columns," *CEP 48:* 89–94 (1952). Reproduced by permission of the publisher.]

The foregoing is all for beds of uniformly sized spherical particles. For other shapes of uniform particles, some efforts have been made to relate the results to those shown here through defining an empirical "sphericity" factor [2, p. 45 *et seq*]. However, results have not been successful enough to allow one to calculate the behavior of such porous media accurately without experimental test, because even for a completely uniform set of nonspherical particles the porosity is a strong function of how the bed is assembled, and if the particles are loose, the porosity can be significantly altered by simply shaking the bed, etc.

There has been very little progress in calculating the flow of water, oil, or gas in naturally occurring rocks without experimental test because naturally occurring rocks are much, much less uniform than the beds of uniform sized spheres described by Fig. 12.4. Thus, in the study of groundwater movement and in petroleum reservoir engineering, it is customary to simplify Eq. 12.13 to

$$\mathcal{F} = \frac{V_s}{k} \cdot \frac{\mu}{\rho} \Delta x \tag{12.16}$$

where k is called the permeability. Equation 12.16 is *Darcy's equation*, and the unit of permeability is called the *darcy*:

$$1 \text{ darcy} = \frac{(1 \text{ cm/s}) \cdot \text{cP}}{\text{atm/cm}} = 0.99 \times 10^{-8} \text{ cm}^2 = 1.06 \times 10^{-11} \text{ ft}^2$$

Example 12.3. In a test of a horizontal-flow filter with compressed air at 1 atm, the following data were obtained: Filter area, 1 ft^2; pressure difference across filter, 2 psi; length of filter medium in flow direction, $\frac{1}{2}$ in; flow rate, 1 ft^3/min.

Calculate the permeability of the filter medium, and estimate the pressure drop necessary to force 1 ft^3/min of water through it.

The permeability is given by Eq. 12.16, rearranged:

$$k = \frac{V_s \mu \,\Delta x}{\rho \mathcal{F}} = \frac{V_s \mu \,\Delta x}{\Delta P} = \frac{Q \mu \,\Delta x}{A \,\Delta P}$$

$$= \frac{1 \text{ ft}^3/\text{min} \cdot 0.018 \text{ cP} \cdot \text{ft}/24}{1 \text{ ft}^2 \cdot 2 \text{ lbf/in}^2} \cdot \frac{\text{ft}^2}{144 \text{ in}^2} \cdot 2.09 \times 10^{-5} \frac{\text{lbf} \cdot \text{s}}{\text{ft}^2 \cdot \text{cP}} \cdot \frac{\text{min}}{60 \text{ s}} \cdot \frac{\text{darcy}}{1.06 \times 10^{-11} \text{ ft}^2}$$

$$= 0.080 \text{ darcy}$$

For water we could use this value directly in Eq. 12.16. However, it is easier to solve Eq. 12.16 for the pressure difference and then apply it twice, once to the airflow and once to the water flow, and take the ratio. All the terms cancel except

$$\frac{\Delta P_{\text{water}}}{\Delta P_{\text{air}}} = \frac{\mu_{\text{water}}}{\mu_{\text{air}}} = \frac{1.0 \text{ cP}}{0.018 \text{ cP}} = 55.5$$

so that the required pressure drop for water is 111 psi = 765 kPa. ■

The velocities encountered in groundwater and petroleum reservoir flow are generally small enough for the kinetic energy terms in Bernoulli's equation to be neglected. Furthermore, the flow is almost always laminar, so that $\mathcal{F}$ is described by Darcy's equation (Eq. 12.16). Thus, Bernoulli's equation for this situation becomes

$$\Delta\left(\frac{P}{\rho} + gz\right) = -\frac{\mu}{k} \cdot \frac{V\,\Delta x}{\rho} \tag{12.17}$$

which for constant ρ, k, μ, and g may be rewritten

$$\frac{d(P + \rho g z)}{dx} = -\frac{\mu}{k} V_x \tag{12.18}$$

If we write a similar equation for the y direction, differentiate each with respect to x and y, respectively, and then add them, we find

$$\frac{\partial^2(P + \rho g z)}{\partial x^2} + \frac{\partial^2(P + \rho g z)}{\partial y^2} = -\frac{\mu}{k}\left(\frac{\partial V_x}{\partial x} + \frac{\partial V_y}{\partial y}\right) \tag{12.19}$$

but from Eq. 3.33, the constant-density mass balance equation, the right-hand side of Eq. 12.19 is zero; so

$$\frac{\partial^2 \phi}{\partial x^2} + \frac{\partial^2 \phi}{\partial y^2} = 0 \qquad \phi = P + \rho g z \tag{12.20}$$

This is Laplace's equation, which describes "potential flow." It is widely used in heat flow and electrostatic field problems; an enormous number of solutions to Laplace's equation are known for various geometries. These can be used to predict the two-dimensional flow in oil fields, underground water flow, etc. The same method can be used in three dimensions, but solutions are more difficult. The solutions to the two-dimensional Laplace equation for common problems in petroleum reservoir engineering are summarized by Muskat [3]. The analogous solutions for groundwater flow are shown in the numerous texts on hydrology, e.g., Todd [4]. See Chap. 10 for more on potential flow.

12.2 TWO-FLUID COCURRENT FLOW IN POROUS MEDIA

So far we have assumed that all the pore space in the porous medium was occupied by the same fluid, such as air or water or oil. However, there are very important problems in which two immiscible fluids are present in the same pore space, e.g., the simultaneous flow of oil and gas or of oil and water, which occurs in petroleum reservoirs, or the air-blowing of a filter cake to drive out a valuable filtrate.

If in the experimental apparatus shown in Fig. 12.2 we fill all the pores with water and then force air through the system, the fraction of water in the outlet stream behaves as shown in Fig. 12.5. Initially only water will flow out of

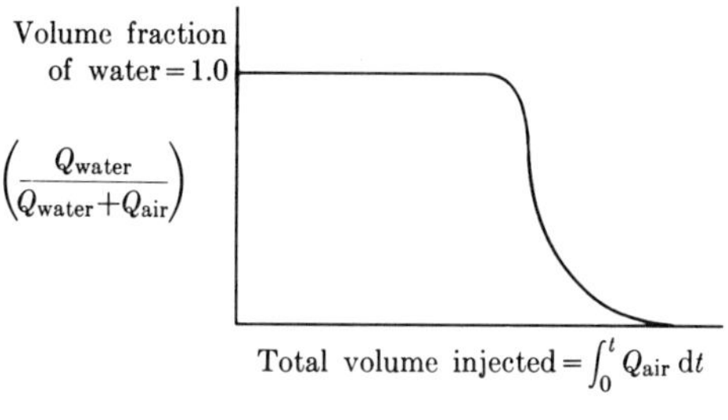

FIGURE 12.5

the downstream end of the apparatus. Its volumetric flow rate will equal the volumetric flow rate of air entering the apparatus. Then air will "break through" and appear at the outlet end. For a brief period, both air and water will emerge from the outlet, the volume fraction of water steadily decreasing. Finally, no more water will emerge (except that which is being removed as water vapor in the air, due to evaporation inside the apparatus). If, after we have reached the point of no more water flow in Fig. 12.5, we open the apparatus and determine the amount of water present inside, we find that 10 to 30 percent of the pores in the sample are full of water.

Why does this water not flow out? It is held in place by surface forces. This can be made plausible by comparing the surface of 1 gal of fluid in a cylindrical container with 1 gal of fluid in a typical sandstone. An ordinary 1-gal paint can has a surface area of 1.5 ft^2. A gallon of fluid contained in the pores of a sandstone made of spherical grains of diameter 0.01 in and porosity 0.3 has 216 ft^2 of surface area. One may think of the fluid in the pores of such a sandstone as being spread out as a film of average thickness 0.007 in = 0.18 mm.

If only one fluid is present, then this is all solid-fluid surface and the interaction is simply one of adhesion at the surface. If two fluids are present, then in addition to two kinds of solid-fluid interface there will be a fluid-fluid interface. The pressure difference which can exist across such a fluid-fluid interface is of the form†

$$\text{Pressure difference} = \frac{\sigma}{r} \tag{12.21}$$

Here σ is the surface tension, and r is the radius of the pore space. For a water-air interface σ is about 4×10^{-4} lbf/in. For 1 gal of water in an ordinary 1-gal can, r is about 3 in, so the pressure difference between the air and the water is about 2×10^{-4} psi, which is negligible in most problems. If, however, the surface is inside a pore of radius 0.001 in, then the pressure difference is

† This assumes a "contact angle" of 0°. See Scheidegger [5] for a discussion of this equation without this simplifying assumption.

about 0.8 psi, which frequently is not negligible (see Chap. 17 for more on surface tension effects).

These surface forces prevent the complete displacement of one phase from a porous medium by another. The displacing fluid (air in the above example) tends to move first into the largest of the available flow channels in the porous medium and thus bypass some of the displaced fluid. When this bypassed displaced fluid, which is still flowing, is reduced in quantity so much that it breaks up into droplets surrounded by the displacing fluid rather than moving as a continuous filament, then it stops moving (in petroleum terminology it becomes *immobile*). If we examine microscopically a sand which has had water displaced from it by air, as described above, we see that the retained water exists, not as continuous filaments, but as small layers or drops, normally held in the junctions between various grains of the sand.

From this physical description we concluded that a particle of fluid stops moving when the displacing force (which equals the pressure gradient times the length of the droplet times its cross-sectional area perpendicular to the flow) is balanced by the surface force (which equals the surface tension divided by the radius of the drop times its cross-sectional area). Equating these, we find that the fluid particle should stop moving when

$$\frac{\Delta P}{\Delta x} LA = \frac{\sigma}{r} A \tag{12.22}$$

or

$$\frac{\Delta P}{\Delta x} \frac{Lr}{\sigma} = 1 \tag{12.23}$$

For a naturally occurring stone, it is impossible to measure L (the length of the drop) or r in Eq. 12.23. Furthermore, this analysis only applies to spherical droplets, whereas microscopic observation shows that the drops are seldom spherical. However, the basic idea is sound. To show this, we note that Lr has the same dimension as the permeability; low permeabilities go with low values of L and r and high permeabilities with high values of L and r. Thus, we could reasonably expect that the fraction of the void space full of displaced fluid, when the displaced fluid stops moving, should be some function of the dimensionless group

$$\frac{k}{\sigma} \frac{\Delta P}{\Delta x} = \text{capillary number} \tag{12.24}$$

Figure 12.6 shows a correlation of measured *residual saturation* (residual saturation is the fraction of pore space occupied by displaced fluid when the displaced fluid stops flowing) as a function of this capillary number. For high permeabilities (e.g., large pores, as might exist in a pile of bricks or coarse gravel) the residual saturation is very low, perhaps 2 or 3 percent, whereas for low permeabilities (e.g., a very fine-grained sandstone or shale) the residual saturation is very high, perhaps 30 to 60 percent. The scatter of the data shows

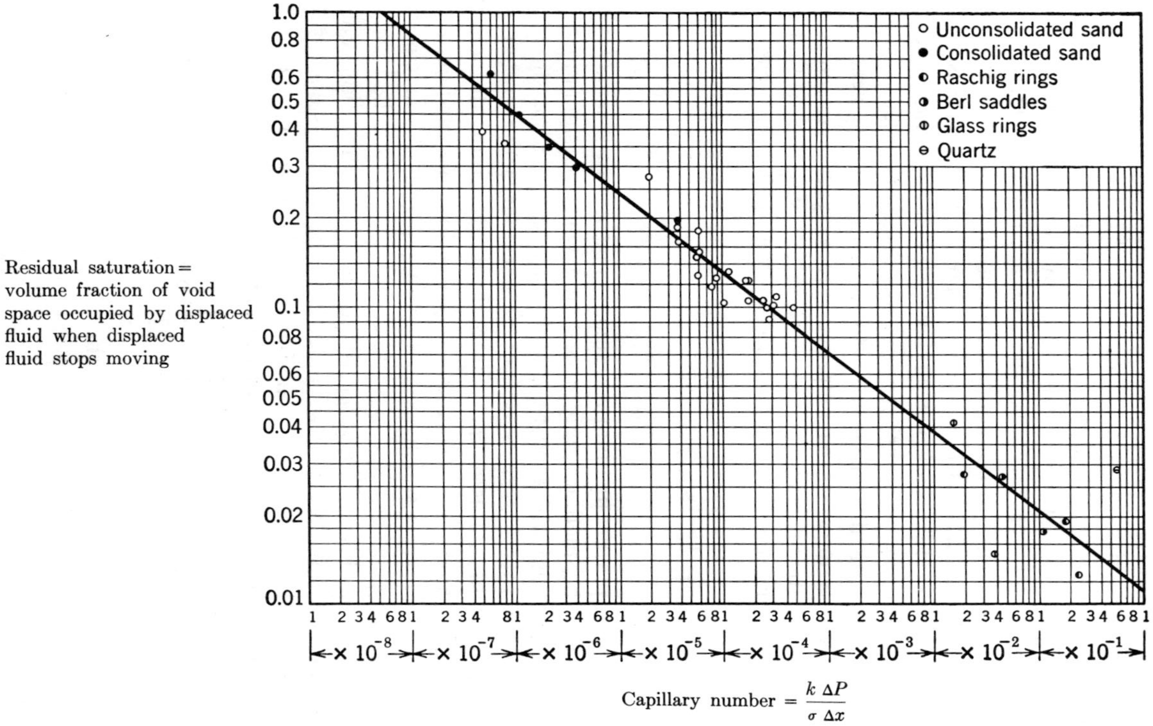

FIGURE 12.6
Residual saturation as a function of capillary number. [From L. E. Brownell and D. L. Katz, "Flow of fluids through porous media," *CEP 43:* 601–612 (1950). Reproduced by permission of the publisher.]

that this correlation should be used only for order-of-magnitude estimates of the fraction of immobile fluid in a porous medium.

This kind of two-phase *cocurrent* flow is very important in oil fields, where the two phases may be oil and water or oil and natural gas; in hydrology, where the two phases flowing (near the earth's surface) are air and water; or in filters, where air is blowing residual liquid out of the filter cake. The treatment here is much simplified; for more detailed treatments see Todd [4] or Scheidegger [5]. These flows are complicated by the effect of viscosity. If the displacing fluid is more viscous than the displaced fluid, then viscosity damps out any unevenness in the boundary between displacing and displaced fluids. But if the displacing fluid is less viscous, then viscosity increases such unevenness, and the less viscous displacing fluid "fingers" through the displaced fluid, leading to early breakthrough and much less complete displacement than if the displacing fluid is more viscous than the displaced fluid. Unfortunately in cases of industrial significance (recovery of petroleum, air-blowing of filter cakes) the displacing fluid is often less viscous, so that this fingering is a very serious limitation on the effectiveness of the processes [6].

12.3 COUNTERCURRENT FLOW IN POROUS MEDIA

When the particles in a porous medium are small, then if a fluid is moving in one direction through the medium, any other fluid which is simultaneously moving through the medium will almost certainly move in the same direction. However, when the particles are large (say, $\frac{1}{4}$ in or larger), it is entirely possible for two immiscible fluids to move in opposite directions through the same medium at the same time. This is the basis for the gas-liquid and liquid-liquid contacting devices which use a bed of particles to increase the surface of contact between the fluids. Such devices, usually called *packed towers*, are commonly used for absorption, distillation, humidification, etc.

The important feature of the two-fluid flow in such systems is the competition of the fluids for the area available to flow. Normally the denser fluid (e.g., water) runs down the surface of the particles by gravity, while the less dense fluid (e.g., air) flows upward because it is introduced at the bottom of the column at a higher pressure than that at which it is withdrawn from the top. Typical pressure-difference results for such a system are shown in Fig. 12.7.

In Fig. 12.7 the pressure decrease from the bottom to the top of the tower is plotted versus the gas superficial mass flow rate up the column for various liquid superficial flow rates down the column. First consider curve *A*. This is for flow of gas only; its slope (on a log-log plot) is 1.8, indicating that the flow is in the transition region on Fig. 12.4, in which f_{PM} is proportional to $\mathcal{R}_{PM}$ to the -0.2 power. Curve *B* is for no liquid flow, but for the packing having been wetted and drained. The slope of the curve is the same as for curve *A*, but at each flow rate the pressure drop is slightly higher, because some of the flow

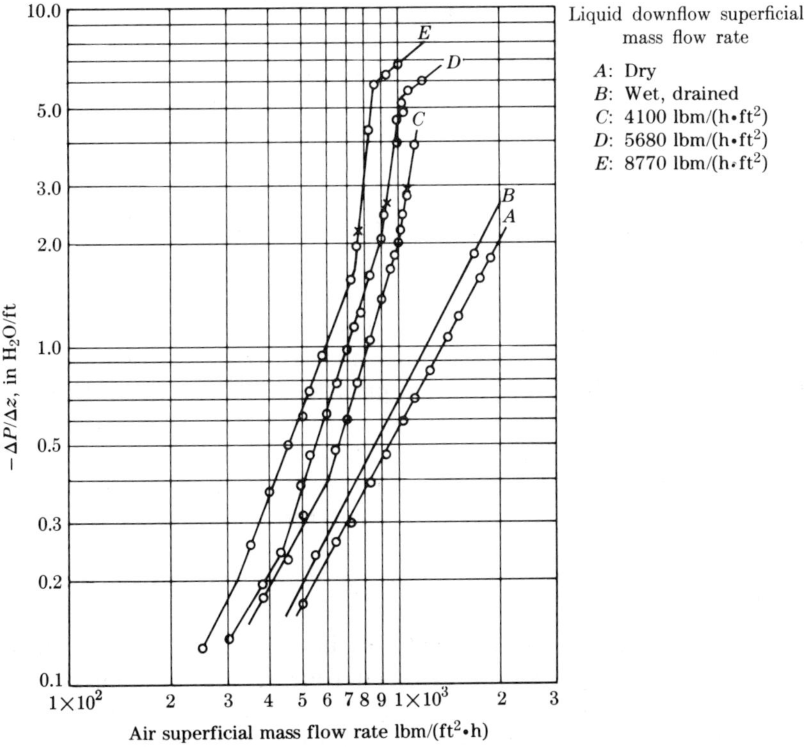

FIGURE 12.7
Pressure drop in competitive countercurrent air-water flow in a porous medium, which consisted of raschig rings, which are thin-walled cylinders 1 in long and 1 in in diameter, with both ends open. [From B. J. Lerner and E. S. Grove, Jr., "Critical conditions of two-phase flow in packed columns," *Ind. Eng. Chem. 53:* 216 (1951). Copyright 1951 by the American Chemical Society. Reproduced by permission of the publisher.]

passages are now blocked by retained liquid. With some of the passages blocked, the gas's interstitial velocity must increase because less area is available for flow. That raises the pressure drop due to friction. Curve *C* shows a typical curve for the competitive flow of gas and liquid. At low gas flow rates, the form is similar to curves *A* and *B*, but the pressure drop is higher, because more of the passages are blocked by liquid. However, at an airflow rate of about $600\,\text{lbm}/(\text{h}\cdot\text{ft}^2)$ the curve turns sharply upward. Visual observation indicates that at this point fluid begins to be held up in passages in which it previously flowed downward. This blocks these passages to the flow of the gas. Continued flow-rate increase causes more and more of these passages to be thus blocked, and the pressure rises steeply; this behavior is sometimes called *loading*.

Curves *D* and *E* show similar behavior, but they also show that for higher liquid flow rates a new region can be entered, in which the steep rise in the pressure-drop curve moderates. In this region, called *flooding*, the liquid fills the column, and it becomes the continuous phase instead of the dispersed one. The gas rises through it as bubbles rather than as a continuous gas stream.

Because devices with this kind of flow are widely used in distillation and absorption, the empirical methods for estimating their performance are summarized in books on those topics [7].

12.4 SIMPLE FILTER THEORY

Filters are widely used to purify gases and liquids or to separate valuable products from gases or liquids. We can learn something about their behavior by applying the results previously found in this chapter. There are two kinds of filters: surface filters, in which the collected particles form a coherent cake on the filter surface (e.g., coffee filters, colanders, most industrial baghouses, most industrial thickeners, plate-and-frame or drum filters), and depth filters, in which the particles are collected throughout the entire depth of the filter (e.g., cigarette filters, automobile oil filters, most home furnace filters).

A. Surface Filters

The flow through a filter is shown schematically in Fig. 12.8. A slurry (a fluid containing suspended solids) flows through a filter medium (most often a cloth, but sometimes paper, porous metal or a bed of sand). The solid particles in the slurry deposit on the face of the filter medium, forming the "filter cake." The liquid, free from solids, flows through both cake and filter medium. Applying Bernoulli's equation from point 1 to point 3, we see that there is no change in elevation. There is a slight change in velocity (due to the solid particles left behind in the filter cake in all cases and due to the pressure drop if the fluid is a gas), but this is generally negligible, and there is no pump or compressor work; so $\Delta P/\rho = -\mathcal{F}$. The flow is laminar in almost all filters, so that the pressure drop due to friction is given by Eq. 12.16. Solving Eq. 12.16 for the superficial velocity, we find

$$V_s = \frac{Q}{A} = \frac{-\Delta P}{\mu}\,\frac{k}{\Delta x} \tag{12.25}$$

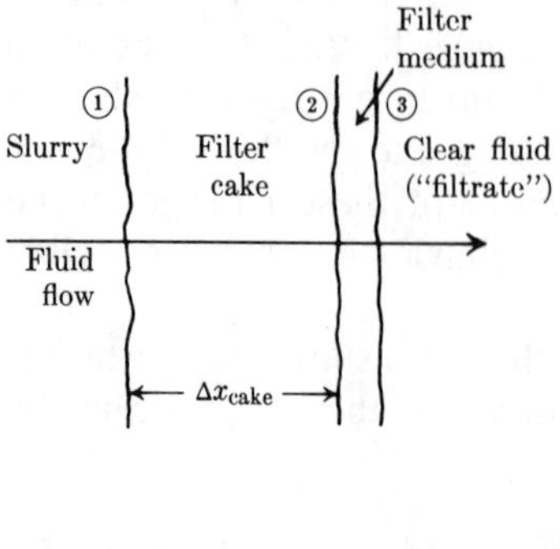

FIGURE 12.8
Flow in a filter.

Here there are two resistances in series with the same flow rate through them. Letting the subscript "FM" indicate "filter medium," we write Eq. 12.25 twice and equate the identical flow rates (see Fig. 12.8):

$$V_s = \frac{P_1 - P_2}{\mu}\left(\frac{k}{\Delta x}\right)_{\text{cake}} = \frac{P_2 - P_3}{\mu}\left(\frac{k}{\Delta x}\right)_{\text{FM}} \tag{12.26}$$

Solving for P_2, we get

$$P_2 = P_1 - \mu V_s\left(\frac{\Delta x}{k}\right)_{\text{cake}} = P_3 + \mu V_s\left(\frac{\Delta x}{k}\right)_{\text{FM}} \tag{12.27}$$

and then solving this equation for V_s, we get

$$V_s = \frac{P_1 - P_3}{\mu[(\Delta x/k)_{\text{cake}} + (\Delta x/k)_{\text{FM}}]} = \frac{Q}{A_{\text{filter}}} \tag{12.28}$$

This equation describes the instantaneous flow rate through a filter; it is analogous to Ohm's law for two resistors in series, so the $\mu\,\Delta x/k$ terms are called the *cake resistance* and the *cloth resistance*.

The resistance of the filter medium is normally assumed to be a constant independent of time, so $(\Delta x/k)_{\text{FM}}$ is replaced with a constant a. If the filter cake is uniform, then its instantaneous resistance is proportional to its instantaneous thickness. However, this thickness is related to the volume of filtrate which has passed through the cake by the material balance:

$$\begin{aligned}\Delta x_{\text{cake}} &= \frac{\text{mass of cake}}{\text{area}} \cdot \frac{1}{\rho_{\text{cake}}} \\ &= \frac{1}{\rho_{\text{cake}}} \cdot \frac{\text{volume of filtrate}}{\text{area}} \cdot \frac{\text{mass of solids}}{\text{volume of filtrate}}\end{aligned} \tag{12.29}$$

Customarily we define

$$W = \frac{\text{mass of solids}}{\text{volume of filtrate}} \cdot \frac{1}{\rho_{\text{cake}}} \tag{12.30}$$

so that

$$\Delta x_{\text{cake}} = \frac{V}{A} W \tag{12.31}$$

Here V is the volume of filtrate. Substituting Eq. 12.31 for the cake thickness in Eq. 12.28, we find

$$\frac{Q}{A} = \frac{1}{A}\frac{dV}{dt} = \frac{P_1 - P_3}{\mu[VW/(kA) + a]} \tag{12.32}$$

For many industrial filtrations, the filter is supplied by a centrifugal pump or blower at practically constant pressure, so $P_1 - P_3$ is a constant, and Eq. 12.32 may be rearranged and integrated to

$$\left(\frac{V}{A}\right)^2 \frac{\mu W}{2k} + \frac{V}{A}\mu a = (P_1 - P_3)t \tag{12.33}$$

For many filtrations the resistance a of the filter medium is negligible compared with the cake resistance (except perhaps for the very first part of the filtration, in which the cake is forming), so that the second term of Eq. 12.33 may be dropped. In such cases the volume of filtrate is proportional to the square root of the time of filtration.

Because of this buildup of cake, the filter must be cleaned at regular intervals; the optimum time between cleanings is discussed in Chen [8].

For a few industrial filtrations, the filter is supplied by a positive-displacement pump, which is practically a constant-flow-rate device. Such a pump feeds the filter at a pressure which is steadily increasing during the filtration. Equation 12.32 shows that for constant k and negligible a the pressure increases linearly with time, because the cake thickness increases linearly with time.

In many real filtration problems, the k in Eq. 12.32 is not constant but is a function of the pressure. This occurs because many filtrates, such as the iron hydroxides and aluminum hydroxides used in water clarification, are weak-structured gels or flocs. In the loose state they have a relatively low flow resistance, but under pressure they collapse and form denser structures which have a higher flow resistance. The common practice in describing such cakes is to write

$$\text{Cake specific resistance} = \frac{1}{k} = \alpha P^s \tag{12.34}$$

Substituting Eq. 12.34 in Eq. 12.32 and letting P_3 be zero (i.e., atmospheric pressure), we find

$$\frac{Q}{A} = \frac{1}{A}\frac{dV}{dt} = \frac{P}{\mu(\alpha P^s VW/A + a)} \tag{12.35}$$

If a is negligible (the usual case), then Eq. 12.35 indicates that at a given cake thickness increasing the pressure (1) will linearly increase the flow rate if s is 0 (as for sand), (2) will have no effect at all on the flow rate if s is 1 (as for some gelatinous hydroxides), and (3) will have some intermediate effect if s is between 0 and 1. There are cases in which s is less than 1 at low pressures and greater than 1 at higher pressures, so that there is a pressure which gives maximum flow rate with lower flow rates for lower or higher pressures. Much of the art of filtration consists of selecting additives, "filter aids," which lower α or s [9].

B. Depth Filters

Surface filters are used when the concentration of solids to be removed from the fluid is high; the solids form a cake which actually does the filtering. Once the cake forms, the filter medium serves only to support the cake, but does practically no filtering. Depth filters are used when the concentration of solids is low and the goal is to produce a very clean fluid stream at the outlet. Often the depth filter is the final cleanup step, following a surface filter which takes out most of the particles to be removed. This is the case with the depth filters which make the final cleanup of air going to operating rooms or to microchip fabrication facilities. While surface filters can be cleaned by removing the cake from the surface, most depth filters are thrown away when they become loaded with solids, e.g., cigarette filters and automobile oil filters.

The fluid mechanics of depth filters do not lend themselves to as simple a mathematical treatment as that shown above for surface filters [10].

12.5 FLUIDIZATION

The previous treatment of flow in porous media will help us understand fluidization, which plays a major role in many chemical processes. Figure 12.9 shows an apparatus for creating a fluidized bed. A granular material such as sand is resting on some sort of support screen in a tube. What would happen if a fluid, such as air or water, were to flow upward through this material?

The behavior will be described by Bernoulli's equation, with the laminar-flow friction heating term given by Eq. 12.13:

$$\frac{\Delta P}{\rho} + g\,\Delta z = -\mathscr{F} \tag{12.36}$$

Here we have dropped the $\Delta V^2/2$ term, which is negligible. Now we can multiply through by ρ and assume that we are dealing with air, in which $\rho g\,\Delta z$ is also small and can be dropped. Thus, we see that the pressure difference across the bed of granular material $P_2 - P_1$ is linearly proportional to the volumetric flow rate.

Now if we steadily increase the fluid flow rate, what will happen when the pressure force on the entire bed ($P_2 - P_1$ times the cross-sectional area) is slightly greater than the weight of the bed? If the bed is made up of some solid,

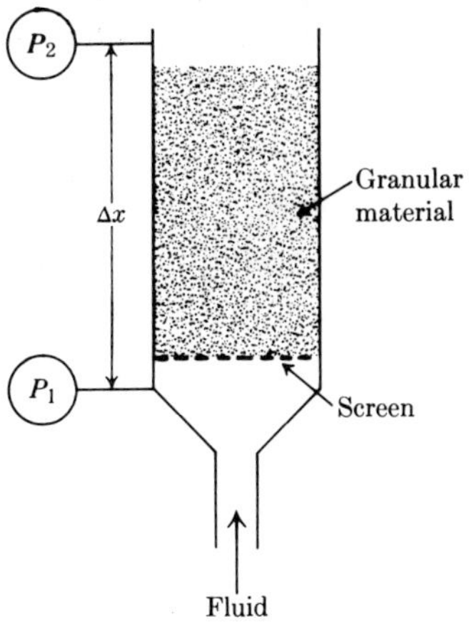

FIGURE 12.9
A fluidized-bed demonstration.

porous material such as a block of sandstone, then the entire block will be expelled from the tube, exactly as a bullet is expelled from a gun by the high-pressure gases behind it. However, if the solid is a mass of individual particles, such as sand grains, their resistance to flow can be decreased by increasing their porosity. Substituting Eq. 12.13 in Eq. 12.36, we see that for gas-solid systems

$$-\Delta P = g\,\Delta z\,(1-\varepsilon)\rho_{\text{particles}}$$

$$= \frac{150 V_s \mu (1-\varepsilon)^2\,\Delta x}{(D_p)^2 \varepsilon^3} \qquad \text{[gas-solid systems]} \qquad (12.37)$$

As V_s increases, ε may increase and hold ΔP constant (Δx will also increase, but its effect is much less than the effect of a change in ε). Thus, the experimental result for such a test is shown in Fig. 12.10. For velocities less than the *minimum fluidizing velocity* V_{mf}, the bed behaves as a packed bed. However, as the velocity is increased past V_{mf}, not only does the bed expand (Δx increases), but also the particles move apart and are able to slide past each other, and the entire particle-fluid mass becomes a fluid which can be poured from one vessel to another and pumped, etc. As the velocity is further increased, the bed becomes more and more expanded, and the solid content becomes more and more dilute. Finally, the velocity becomes as large as the terminal settling velocity of the individual particles, so the particles are blown out of the system. Thus, the velocity range for which a fluidized bed can exist is from V_{mf} (whose value can be calculated from Eq. 12.37 for spherical particles and analogous equations for nonspherical ones) to V_t (whose value can be calculated as shown in Sec. 6.14).

The above description of the particles in the bed increasing their value of ε to accommodate an increasing flow rate, as sketched in the upper part of Fig. 12.10, is actually observed if the fluidizing fluid is a liquid. However, if the fluidizing fluid is a gas (by far the more common case industrially), the behavior is more complex. In that case as the fluid flow rate is increased past V_{mf}, the increased gas flow forms bubbles, which contain virtually no particles and which rise through the bed of fluidized particles, very much as bubbles rise

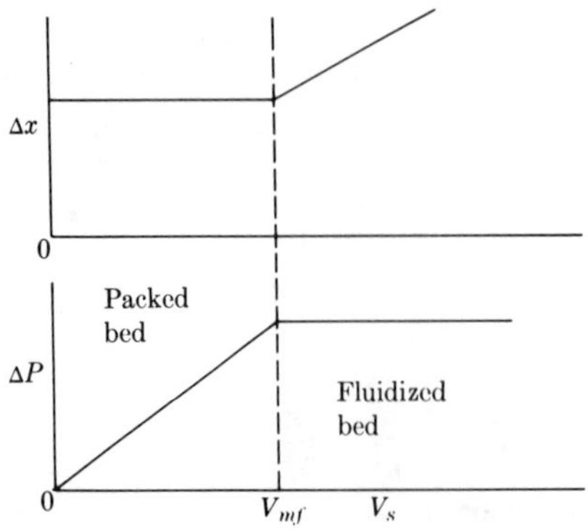

FIGURE 12.10
Transition from a packed to a fluidized bed.

through liquids [11]. These bubbles greatly complicate all the aspects of the behavior of fluidized beds and of the chemical reactions carried out in them.

These fluidized beds have proved very useful in chemical technology when one wishes to move a granular solid through a series of processing steps in a continuous fashion. Their most dramatic application is in *fluidized-bed catalytic cracking*, which is a standard petroleum refining operation, but Zenz and Othmer [12] list dozens of other applications. This brief treatment only shows how they are formed; more on their properties and uses can be found elsewhere [2, Chap. 1; 12; 13].

If one continues to increase the velocity of the fluidizing stream, eventually it will exceed the settling velocity of the largest particle in the bed, and then the entire bed will be conveyed upward. This ceases to be a fluidized bed and becomes a pneumatic transport pipe, which is widely used to move granular solids such as grain, portland cement, or plastic pellets [14]. As a rule of thumb, fluidized beds operate with gas superficial velocities of 1 to 3 ft/s (0.3 to 1 m/s); pneumatic transport operates with superficial velocities of 30 to 60 ft/s (10 to 20 m/s).

12.6 SUMMARY

1. Flow through a porous medium can be laminar, turbulent, or intermediate. The transition from laminar to turbulent flow is smoother and more reproducible than for flow in a pipe. Generally, the flow in filters, the flow of groundwater, and the flow in petroleum and natural-gas fields are laminar.
2. For a porous media made up of equal-size spherical particles, we can calculate the pressure drop with fair accuracy by using a modified friction factor plot.
3. For most filter cakes and naturally occurring rocks, it is impossible to calculate the pressure drop based on nonflow measurements, but the pressure drop can be correlated very satisfactorily in the laminar-flow region by Darcy's equation.
4. Two-phase flow in porous media is strongly influenced by surface forces; these generally lead to incomplete displacement of one phase by another in a porous medium.
5. Countercurrent flow of two immiscible fluids in porous media is strongly influenced by the competition of the two fluids for the available pore space.
6. Fluidized beds are formed when the fluid friction and pressure plus buoyant forces are equal and opposite to the gravity force on the particles.

PROBLEMS

See the Common Units and Values for Problems and Examples inside the back cover.

12.1. Show that $f = 16/\mathscr{R}$ is equivalent to $f_{\text{PM}} = 72/\mathscr{R}_{\text{PM}}$.

12.2. Show that if one assumes that $V_I = \sqrt{2}(V_I)_{x\text{ dir}}$ and that the length of the flow path is $\sqrt{2}\,\Delta x$, then the friction factor should be twice as high as shown in $f_{\text{PM}} = 72/\mathscr{R}_{\text{PM}}$.

12.3. Show the relative sizes of the two terms in the Ergun equation (Eq. 12.15) at $\mathscr{R}_{\text{PM}}$ values of 0.1, 1, 10, 100, 1000, and 10,000.

12.4. Calculate $\mathscr{R}_{\text{PM}}$ in Example 12.2.

12.5. Calculate the permeability of the bed of ion-exchange particles in Example 12.1.

12.6. For the flows in Examples 12.1 and 12.2, calculate the magnitudes of the $\Delta V^2/2$ terms omitted in Bernoulli's equation, and compare these with the magnitude of the $\mathscr{F}$ terms.

12.7. For the apparatus in Example 12.1, estimate the residual saturation of air if the apparatus is originally full of air and then filled with water so that water is allowed to percolate through as shown in Fig. 12.3.

12.8. In Fig. 12.7, what value would the abscissa have if the packing were filled completely with water which was not moving? From this value what can you say about the fraction of the pore space occupied by liquid in the "flooded" regime at the top of curves D and E?

12.9. Calculate the $\mathscr{R}_{\text{PM}}$ range for curve A on Fig. 12.7. Here use $D_p = 1$ in and $\varepsilon = 0.5$. Does this Reynolds number correspond to the laminar, transition, or turbulent range of flow rates?

12.10. The results of a constant-pressure filtration test are shown in a table of volume of filtrate per unit area of filter versus time. The data are presumed to agree with Eq. 12.33. Show how to plot these data so that they will form a straight-line plot and how to find the values of k and a from this line.

12.11. Equation 12.37 correctly predicts the minimum fluidizing velocity V_{mf} for beds of spherical particles, but it does not correctly predict the relationship of V_s to ε after the bed has started to expand. Why not?

12.12. Write the equation analogous to Eq. 12.37 for water flowing upward through a bed of sand. Note that in this case the $g\,\Delta z$ term in Eq. 12.36 cannot be neglected.

12.13. For spherical sand particles with $D_p = 0.03$ in and $\rho_{\text{particles}} = 150\ \text{lbm/ft}^3$, estimate the minimum fluidizing velocity for air and for water. Assume $\varepsilon = 0.3$, and note that in the case of the water you must rederive Eq. 12.37, taking into account the buoyant force on the particles.

CHAPTER 13

MODELS, DIMENSIONAL ANALYSIS, AND SIMILITUDE

13.1 MODELS

A model is an intellectual construct which represents reality and which can be manipulated to predict the consequences of future actions. Most of engineering is the application of mathematical models to practical problems. For example, $F = ma$ is a mathematical model of the relation between force, mass, and acceleration. Using it, engineers have been spectacularly successful in predicting the behavior of real physical systems. Much more complex mathematical models are regularly used; as the size and power of our computers have grown, the size and complexity of the mathematical models we can use also have grown.

But still for many problems we do not yet have mathematical models in which we have so much confidence that we will risk large sums of money or human lives, without first testing their predictions with some kind of test of a physical model. For example, although we have made great strides in computational power, we still will not build a major new aircraft based on calculations alone, without wind tunnel tests of smaller-scale models which verify the computations. We know a great deal about chemical reactors, but not enough that we will build a full-scale plant to produce a chemical by a novel reaction scheme without some bench-scale or pilot plant tests to verify that our mathematical models are reliable. We still cannot predict the behavior of complex structures under earthquake loads with total confidence, so we resort to testing of physical models there are as well. For problems like these, in which we still do not have completely reliable mathematical models, engineers have developed some valuable techniques for making useful predictions without a complete answer. These methods played a greater role in the precomputer era than they do now. For that reason, to some extent this chapter is more part of an engineer's historical, cultural background than of an engineer's current technical tool kit. However, the methods described here often offer a useful physical insight into problems, which will complement the insight that mathematical models and computer solutions provide.

These methods are mostly based on studies using physical model tests. If we cannot calculate the behavior of a new shape of ship hull or airplane or a new type of chemical reactor, then we must build it and test it. If we can build and test a small model of the finished product instead of a full-sized airplane or ship or reactor, we will save time, money, and possibly the lives of test pilots. (Have your failures on a small scale, in private; have your successes on a large scale, in public!) The enormous progress of the aircraft industry from the first powered flight in 1903 to the present is largely due to the fact that engineers have learned to test new designs by using small-scale models and to employ the test results for designing full-scale airplanes. Similarly, the progress of the chemical industry is largely due to the chemical engineer's ability to "scale up" bench-size or pilot-size plants to commercial plants with confidence that the resulting plants will perform as predicted. With the advent of big computers, we can do much more by calculation than we could 20 years ago. For that reason we do more computing (mathematical model testing) and less physical model testing than in the past. Nonetheless, really important engineering decisions are often made on the basis of a combination of computations and physical model tests, with the physical model tests serving to confirm the computations before the full-scale airplane, boat, or chemical plant is built.

However, just making and testing a scale model of the airplane, boat, or chemical reactor is not the whole story, as indicated by the following quotation from J. B. S. Haldane [1]:

> The most obvious differences between different animals are differences of size, but for some reason the zoologists have paid singularly little attention to them. In a large textbook of zoology before me I find no indication that the eagle is larger than the sparrow, or the hippopotamus bigger than the hare, though some grudging admissions are made in the case of the mouse and the whale. But yet it is easy to show that a hare could not be as large as a hippopotamus, or a whale as small as a herring. For every type of animal there is a most convenient size, and a large change in size inevitably carries with it a change of form.
>
> Let us take the most obvious of possible cases, and consider a giant man sixty feet high—about the height of Giant Pope and Giant Pagan in the illustrated *Pilgrim's Progress* of my childhood. These monsters were not only ten times as

high as Christian, but ten times as wide and ten times as thick, so that their total weight was a thousand times his, or about eighty to ninety tons. Unfortunately the cross sections of their bones were only a hundred times those of Christian, so that every square inch of giant bone had to support ten times the weight borne by a square inch of human bone. As the human thigh-bone breaks under about ten times the human weight, Pope and Pagan would have broken their thighs every time they took a step. This was doubtless why they were sitting in the picture I remember. But it lessens one's respect for Christian and Jack the Giant Killer.

To turn to zoology, suppose that a gazelle, a graceful little creature with long thin legs, is to become large; it will break its bones unless it does one of two things. It may make its legs short and thick, like the rhinoceros, so that every pound of weight has still about the same area of bone to support it. Or it can compress its body and stretch out its legs obliquely to gain stability, like the giraffe. I mention these two beasts because they happen to belong to the same order as the gazelle, and both are quite successful mechanically, being remarkably fast runners.†

13.2 DIMENSIONLESS NUMBERS

If, as shown in the preceding quotation, holding the shape constant while increasing the size does not guarantee equal performance by differently sized equipment or animals, what does? In the case of the bones of the giant, presumably one should keep the ratio of stress to crushing strength constant. To do this, as the height of the giant is increased, the crushing strength of the giant's bones could be increased. Alternatively, the average density of the giant's body could be lowered, to keep the same bone stress while the height increased; or the giant could go to a planet with a lower acceleration of gravity. Combining all these possibilities, we see that if the giant keeps constant the ratio

$$\frac{\text{Height} \cdot \text{average body density} \cdot \text{acceleration of gravity}}{\text{Crushing strength of bones}}$$

then the giant can become any height with the same relative resistance to bone failure as a human has. This ratio is dimensionless, a pure number.

Thus, we might suspect that dimensionless ratios like the one above would be important in predicting the behavior of a large piece of equipment from tests of a small model. Experience indicates that this is certainly the case. These dimensionless numbers have also proved invaluable in correlating, interpreting, and comparing experimental data. For example, if you were asked to compare a business venture in which you invest \$4000 for a return of \$400 per year with one in which you invest \$6000 for a return of \$650 per year,

† From J. B. S. Haldane, *Possible Worlds*, Harper & Row, New York, 1928; reprinted in J. R. Newman, *The World of Mathematics*, Simon and Schuster, New York, 1956, p. 952 *et seq*. Quoted by permission of the publisher.

you would certainly make the comparison through the most common of all dimensionless numbers, the percentage.

Furthermore, there is often a real benefit in understanding if we can show our experimental or computational results in dimensionless form. The behavior of nature does not depend on the system of dimensions that we humans use to describe that behavior. Thus, the results of our observations, if they are correct, can be expressed in a form which is totally independent of the system of units we use. If they cannot, then we should wonder whether they are correct. Finding the best way to reduce observations made in any system of units to dimensionless form is always a good test of the experimental data, and often it is a good test of our understanding of that data. The application of dimensionless analysis shown in Fig. 6.10 reduces a problem with six dimensioned variables to one with three dimensionless variables.

13.3 FINDING THE DIMENSIONLESS NUMBERS

How do we go about finding which dimensionless numbers are important for a given type of model study or for changing our experimental findings from the dimensioned form in which we made the observations to a dimensionless form? Three general methods are in common use: the governing-equation method, the method of force ratios, and Buckingham's π method.

A. The Method of Governing Equations

Suppose that our problem concerns a complicated fluid flow system in which we suspect that Bernoulli's equation, along with other equations, would apply. Then we can write Bernoulli's equation in differential form (without pump or compressor work) and integrate to find

$$\frac{P}{\rho} + gz + \frac{V^2}{2} + \mathcal{F} = \text{const} \tag{13.1}$$

Let us assume that the $\mathcal{F}$ term is of the form given by the Poiseuille equation

$$\mathcal{F} = \frac{(V_{\text{av}} D_0^2 \pi / 4)[\Delta x \, \mu (128/\pi)]}{D_0^4 \rho} \tag{6.12}$$

Each term in Eq. 13.1 has the same dimensions; therefore, if we divide through by any one of them, the result will be a dimensionless equation.

Let us divide by $V^2/2$. The result is

$$\frac{2P}{\rho V^2} + \frac{2gz}{V^2} + \frac{\mu}{\rho V D} 64 \frac{\Delta x}{D} = \frac{\text{const}}{V^2} \tag{13.2}$$

The first term on the left is important enough to be given a name in fluid mechanics, the *pressure coefficient*; it is also sometimes called 1/(Euler number)^2. It appears in problems in which there are significant changes in velocity and pressure between different parts of the system. For example, Eq. 6.53 may be rewritten as

$$C_d = \frac{F}{A} \cdot \frac{2}{\rho V^2} = \frac{2P_{\text{av}}}{\rho V^2} \tag{13.3}$$

so this pressure coefficient is equivalent to the drag coefficient previously discussed.

The second term on the left in Eq. 13.2 is also important enough to have a name,[†] 2/Froude number, or

$$\text{Froude number} = \frac{V^2}{gz} \tag{13.4}$$

In problems involving changes in velocity and free surfaces, the Froude number plays a very important role, e.g., in ship-model studies and open-channel flow. We saw in Chap. 7 that it is the key parameter in describing hydraulic jumps.

The third term on the left-hand side of Eq. 13.2 is 1/(Reynolds number) times 64(length/diameter). The Reynolds number is important whenever viscous forces are important and there are significant changes in velocity. We have seen its use in Chaps. 6 and 11.

This example shows that simply by dividing Bernoulli's equation by one of the terms, we can find three of the most frequently used dimensionless groups in fluid mechanics. Similarly, we saw in Chap. 8 that dividing both sides of Eq. 8.14 by part of one side leads naturally to the Mach number, which is also very frequently used.

These simple examples do not show all the possibilities of this method; for more on it, see Kline [2, chap. 4]. Interesting variants of this method are shown by Bird et al. [3] and Hellums and Churchill [4].

B. The Method of Force Ratios

The method of force ratios, discussed here, is often referred to as the *method of similitude* or the *method of similarity*.

Most dimensionless groups in fluid mechanics may be thought of as ratios of lengths or ratios of forces. For example, in pipe flow we saw two length ratios, $\Delta x/D$ and ε/D, and two dimensionless groups (which, as we will see below, are expressible as force ratios), the Reynolds number and the friction factor. The length ratios are obviously important in model studies; a scale model has the same length/width or length/height as the original. The force ratios are important in model studies because for two different-size models to encounter the same kind of fluid behavior, the influences of gravity, viscosity, compressibility, etc., must be in the same proportion for both. The dimension-

[†] Some textbooks call the square root of the value shown here the Froude number.

less groups generally may be thought of in some other way (e.g., the Mach number as a ratio of velocities), but they also can be seen as force ratios. This is easiest to see for the pressure coefficient. The pressure force exerted by a fluid on some planar body is

$$\text{Pressure force} = \int_{\text{all surface}} P\,dA = P_{\text{av}}A = F_P \tag{13.5}$$

Similarly, the force required to stop a unit volume of flow (which a body inserted into the flow does) is given by $F = ma$. Multiplying both sides by dx, we get

$$F\,dx = ma\,dx = m\,\frac{dV}{dt}\,dx = mV\,dV \tag{13.6}$$

But m is the mass of a unit volume equal to ρL^3, where L is the length of one side of the unit volume. Substituting this and integrating, we find

$$\int F\,dx = F_{\text{av}}\,\Delta x = \int \rho L^3 V\,dV = L^3\rho\,\Delta\left(\frac{V^2}{2}\right) \tag{13.7}$$

Here the final velocity is zero, so we may replace $\Delta(V^2/2)$ in this equation with $-V^2/2$. Solving Eq. 13.7 for the average force, we have

$$-F_{\text{av}} = \frac{L^3}{\Delta x}\,\rho\,\frac{V^2}{2} \tag{13.8}$$

Here the minus sign indicates that the required force is in the direction opposite to that of the initial velocity of the fluid. Now $L^3/\Delta x$ has the same dimensions as an area, so we may replace it with some area A. Thus, the force to stop the fluid, which is commonly called the *inertia force*, is

$$\text{Inertia force} = A\rho\,\frac{V^2}{2} = F_I \tag{13.9}$$

Dividing Eq. 13.5 by Eq. 13.9, we find

$$\frac{F_P}{F_I} = \frac{A\,\Delta P}{A\rho V^2/2} = \frac{\Delta P}{\rho V^2/2} = \text{pressure coefficient} \tag{13.10}$$

We can find the appropriate force ratios systematically by making up a list of forces and their dimensions.

To find the viscous force, we consider the shear force exerted by a fluid flowing past a surface in laminar flow. From Chap. 1 we know that the shear stress τ is given by

$$\tau = \frac{F}{A} = \mu\,\frac{dV}{dy} \tag{1.5}$$

Solving for the force gives

$$F = \mu A\,\frac{dV}{dy} \tag{13.11}$$

However, at the wall $V=0$ and $y=0$, so this becomes

$$\text{Viscous force} = \frac{\mu AV}{L} = F_V \tag{13.12}$$

Similarly, the force of gravity on a unit volume of fluid is

$$\text{Gravity force} = g\rho L^3 = F_G \tag{13.13}$$

and the surface tension force on a unit length of fluid surface is

$$\text{Surface tension force} = \sigma L = F_S \tag{13.14}$$

The elastic force for a spring is given by Hooke's law as $F = kx$, where k is the spring constant and x is the displacement. If we apply the same equation to the one-dimensional compression of a fluid, we find that the *spring constant* is $A\,dP/dx$; so

$$\text{Elastic force} = A\,\frac{dP}{dx}\,\Delta x = \frac{A^2\,dP\,\Delta x}{A\,dx} = A^2\,\frac{dP}{dV}\,\Delta x \tag{13.15}$$

But $V = m/\rho$, so for a constant mass $dV = -(m/\rho^2)\,d\rho$. Substituting gives

$$-\text{Elastic force} = \frac{A^2\rho^2}{m}\,\Delta x\,\frac{dP}{d\rho} \tag{13.16}$$

The minus sign here indicates that the force exerted by the fluid is in the direction opposite to the direction of compression; we may drop the minus sign and substitute $m = \rho V$ to get

$$\text{Elastic force} = \frac{A^2\rho^2}{\rho V}\,\Delta x\,\frac{dP}{d\rho} = A^2\,\frac{\Delta x}{V}\,\rho\,\frac{dP}{d\rho} \tag{13.17}$$

but $V/\Delta x$ has the dimensions of an area, so we may call it an area;

$$\text{Elastic force} = A\rho\,\frac{dP}{d\rho} = F_E \tag{13.18}$$

From this list of six kinds of forces we may make up a table of all the possible ratios, Table 13.1. From this table we see that we can combine the six kinds of forces into 15 force ratios, of which eight are important enough in fluid mechanics to have the common names shown. An entirely analogous table is also shown by Kline [2, p. 50] for the dimensionless ratios important in heat transfer, in which, instead of using force ratios, we use the ratios of energy quantities. A list of dimensionless groups used in chemical engineering fluid mechanics, made up in a way similar to Table 13.1, is given by Sakiadis [5].

The merit of Table 13.1 is that in using it we can quickly estimate which ratios are likely to be important for a given problem.

Example 13.1. Using Table 13.1, estimate which dimensionless ratios are probably important for (*a*) steady laminar flow in a horizontal pipe, (*b*) completely turbulent steady flow in a horizontal pipe, (*c*) resistance to an

TABLE 13.1
Dimensionless numbers obtained as force ratios[†]

	F_I	F_V	F_P	F_E	F_S	F_G
	$A\rho \dfrac{V^2}{2}$	$\dfrac{\mu A V}{L}$	$A\,\Delta P$	$A\rho \dfrac{dP}{d\rho}$	σL	$g\rho L^3$
F_I	$\dfrac{F_I}{F_I}=1$	$\dfrac{F_V}{F_I}=\dfrac{2\mu}{LV\rho}$ **1**	$\dfrac{F_P}{F_I}=\dfrac{2\,\Delta P}{\rho V^2}$ **2**	$\dfrac{F_E}{F_I}=\dfrac{2(dP/d\rho)}{V^2}$ **3**	$\dfrac{F_S}{F_I}=\dfrac{2\sigma}{L\rho V^2}$ **4**	$\dfrac{F_G}{F_I}=\dfrac{2gL}{V^2}$ **5**
F_V		$\dfrac{F_V}{F_V}=1$	$\dfrac{F_P}{F_V}=\dfrac{\Delta PL}{\mu V}$ **6**	$\dfrac{F_E}{F_V}=\dfrac{\rho(dP/d\rho)L}{\mu V}$	$\dfrac{F_S}{F_V}=\dfrac{\sigma}{\mu V}$	$\dfrac{F_G}{F_V}=\dfrac{g\rho L^2}{\mu V}$
F_P	These numbers are merely reciprocals of numbers shown above and to the right.		$\dfrac{F_P}{F_P}=1$	$\dfrac{F_E}{F_P}=\rho\dfrac{dP/d\rho}{\Delta P}$	$\dfrac{F_S}{F_P}=\dfrac{\sigma}{L\,\Delta P}$ **7**	$\dfrac{F_G}{F_P}=\dfrac{g\rho L}{\Delta P}$
F_E				$\dfrac{F_E}{F_E}=1$	$\dfrac{F_S}{F_E}=\dfrac{\sigma}{L\rho(dP/d\rho)}$	$\dfrac{F_G}{F_E}=\dfrac{gL}{dP/d\rho}$
F_S					$\dfrac{F_S}{F_S}=1$	$\dfrac{F_G}{F_S}=\dfrac{g\rho L^2}{\sigma}$ **8**
F_G						$\dfrac{F_G}{F_G}=1$

[†] *After Kline* [2, *p.* 39].

1: 2/(Reynolds number)
2: Pressure coefficient = 1/(Euler number)^2
3: 2/(Mach number)^2 = Cauchy number
4: 2/Weber number
5: 2/Froude number
6: Stokes number
7: Capillary number
8: Eötvös number

airplane in steady flight, (*d*) resistance to a ship in steady motion, (*e*) resistance to a submerged submarine in steady motion, and (*f*) the rise of a fluid in a capillary tube.

(*a*) For steady laminar flow in a pipe, we assume that the only important forces are the pressure force and the viscous force (in fact, we know that they are equal and opposite). Then from Table 13.1 we conclude that the only important force ratio is probably the Stokes number. We also assume that the length/diameter ratio is important and hence that the Stokes number is some function of that ratio. If we assume that the Stokes number is a constant times the ratio of length to diameter, we can solve for ΔP per length:

$$\frac{\Delta P}{\text{length}} = (\text{const})\left(\frac{\mu V}{LD}\right) \tag{13.19}$$

Here the L in the Stokes number is length perpendicular to the flow direction (in this case, the diameter), so that the right-hand side of the equation is $\rho V/D^2$. This result is the same as Poiseuille's equation (written for the average velocity).

Dimensional analysis will not tell us the value of the constant in Eq. 13.19, but it will suggest to us that for steady laminar flow in horizontal pipes a plot of pressure drop per unit length versus viscosity times average velocity divided by diameter squared might give the same straight line for all fluids, pipes, and velocities (which is experimentally verifiable).

(*b*) For completely turbulent flow, we assume that viscous forces are negligible compared with the pressure and inertia forces, so that the only important force ratio should be the pressure coefficient. The length/diameter ratio should be important as well as the pipe roughness. If we assume that for a constant pipe relative roughness the pressure coefficient is a constant times the length/diameter ratio, we can solve for the pressure drop per unit length:

$$\frac{\Delta P}{\Delta x} = \frac{\text{const}}{D} \rho \frac{V^2}{2} \qquad \text{[for a given pipe relative roughness]} \tag{13.20}$$

Comparing this with the definition of the friction factor, we see that the constant here is $4f$. So far we have not included the effect of variable pipe roughness. It is plausible to assume that the constant in this equation is a function of pipe roughness, so that our final form is

$$\frac{\Delta P}{\Delta x} = \text{some function of } \frac{\varepsilon}{D} \text{ and } \frac{\rho V^2}{2D} \tag{13.21}$$

It is also plausible to assume that this function is linear,

$$\frac{\Delta P}{\Delta x} = \text{const} \cdot \frac{\varepsilon}{D} \cdot \frac{\rho V^2}{2D}$$

but experimental data (Fig. 6.10) indicate that nature disagrees with this plausible assumption. However, Eq. 13.21 is a good description of the right-hand side of the friction factor plot; the friction factor for very high Reynolds numbers depends on the roughness alone, not on the Reynolds number.

(*c*) For an airplane the forces which may be significant are the pressure force, inertia force, and elastic force. At low velocities the elastic force is probably negligible compared with the others, so we conclude that the pressure coefficient (normally called a drag coefficient for airplanes) should be a function of the Reynolds number and geometry. At high velocities the viscous forces are probably negligible compared with the elastic forces, so the pressure coefficient should depend on the Mach number and geometry. Both assumptions are experimentally verifiable.

We might make either of two plausible assumptions about intermediate velocities: The two ranges overlap, i.e., there is a range of velocities in which both the Reynolds number and the Mach number affect the pressure coefficient; or these two ranges do not meet, i.e., there is a range of velocities in which neither the Reynolds number nor the Mach number affects the pressure coefficient, which depends on the geometry alone. From theoretical speculation alone we cannot decide between these two possibilities. Experiments indicate that the latter is correct; over a large range of velocities, the pressure coefficient is independent of the Reynolds and Mach numbers; it depends only on the shape of the airplane and its angle of attack relative to the oncoming airstream.

(*d*) For a ship the forces which may be significant are the pressure force, gravity force, viscous force, and inertia force. The gravity force enters this list because of the bow wave thrown up by ships; because of this wave the ship seems to be steadily traveling "uphill." Then, from Table 13.1, we assume that the pressure coefficient should be a function of the Froude number and the Reynolds number as well as the shape of the ship. This is experimentally verifiable.

(*e*) For a submerged submarine the gravity force is no longer important, because it causes no bow wave. Therefore, we should drop the Froude number from the list obtained in (*d*).

(*f*) For capillary rise the significant forces are gravity and surface tension of the fluid. Therefore, from Table 13.1 we see that the important force ratio is $g\rho L^2/\sigma$, the Eötvös number. Here L^3 in the gravity force indicates three perpendicular directions. The L in the surface tension force cancels one of them, but the remaining L^2 refers not to one dimension squared, but rather to two perpendicular dimensions. Setting this force ratio equal to a constant and solving for one of these dimensions, we find

$$L_1 = \text{const} \cdot \frac{\sigma}{gL_2} \tag{13.22}$$

If the geometry we are discussing is a right cylinder with vertical axis and if we choose L_1 as the height of capillary rise for perfect wetting of the surface by the fluid, then L_2 is a characteristic length at right angles, e.g., the radius of the tube. In that case it can be shown (see Chap. 17) that the constant in this equation is a dimensionless 2. ■

C. Buckingham's π Method

Another systematic approach to finding the dimensionless numbers is the method of Buckingham [6], often referred to as the *π theorem* or *Buckingham's π theorem*. It states that if there is some relationship where A (the dependent variable) is a function of $B_1, B_2, \ldots, B_n$ (the independent variables), then the relationships can be written

$$A = f(B_1, B_2, \ldots, B_n) \tag{13.23}$$

or

$$f(A, B_1, B_2, B_3, \ldots, B_n) = 0 \tag{13.24}$$

Furthermore, if the quantities contain among them k independent dimensions, then it is possible to rewrite Eq. 13.24 as

$$f(\pi_1, \pi_2, \ldots, \pi_m) = 0 \tag{13.25}$$

in which the π's are dimensionless quantities and the number of π's is $n + 1 - k$. Here the π's may be dimensionless numbers like the Reynolds number or may be ratios of dimensions, like the L/D which appears in equations for the friction effect in pipe flow. This part of Buckingham's theorem shows how to decide how many such dimensionless groupings we should seek for a given problem.

The proof is given by Buckingham [5]; in brief, if a relation such as Eq. 13.24 exists and if the final relation contains more than one term (i.e., is not of the form $A = 0$), then each term must have the same dimensions. Thus, we know that some relation of the form of Eq. 13.24 exists, and we know that k independent equations exist among the dimensions of the $n + 1$ quantities in Eq. 13.24. This means that we can, in principle, eliminate k unknowns among these equations; hence, there are $n + 1 - k$ independent, dimensionless π's.

There are several restrictions in this logic:

1. The list of independent dimensions should not contain redundant dimensions. For example, if it contains length and force, then it should not contain energy (energy is dimensionally equivalent to the product of length and force). If we wish to include dimensions which are redundant in this sense, then the conversion factor for converting the redundant equations to each other should be included in the B's. For example, if we wish to include length, force, and thermal energy in our list of dimensions, then we should add to the list of B's the conversion factor between mechanical and thermal energy, which is $778\ \text{ft} \cdot \text{lbf/Btu} = 4.184\ \text{J/cal} = 1$. Thus, we can add a redundant dimension and a conversion factor, and the number of π's, equal to $n + 1 - k$, will not change, because both n and k increase by 1. The other common example of this approach is the inclusion of force, mass, length, and time as independent dimensions. This is permissible *if* one adds $g_c = 32.2\ \text{lbm} \cdot \text{ft}/(\text{lbf} \cdot \text{s}^2) = \text{kg} \cdot \text{m}/(\text{N} \cdot \text{s}^2) = 1$ to the list of B's.
2. If two dimensions occur only in a specific ratio, then they are not independent and must be treated as one dimension. Suppose that our list of A's and B's consists of two velocities V_1 and V_2 and two forces F_1 and F_2. By simple application of Buckingham's theorem we conclude that $n + 1$ equals 4 and that k equals 3 (length, time, force); so there should be one π. But to conclude that there is only one π here is incorrect. Since length and time appear in our list of variables only in the combination length/time, there are really only two independent dimensions, force and length/time; so k is 2,

and there are two π's, which are presumably V_1/V_2 and F_1/F_2. In this case it is obvious that the dimensions of length and of time are not independent. In less obvious cases, the recommended procedure is to find from the list the largest number of variables that cannot form a dimensionless group. In this example that number is 2; we can select one velocity and one force, and they cannot be converted to a dimensionless group by using the other members of the variable list. Any three from the list can form a dimensionless group, either V_1/V_2 or F_1/F_2. This largest number of uncombinable variables is equal to the number of independent dimensions. It can never be more than the total number of dimensions; as shown above, it may be less.

Example 13.2. We believe that some force is a function of a velocity, a density, a viscosity, and two lengths. How many dimensionless π's should be required to correlate the data for this problem?

Here we have three choices of the set of dimensions, all of which give the same result:

1. We may choose as our dimensions length, force, and time, in which case we must express the density in dimensions of force times $(\text{time})^2$ divided by $(\text{length})^4$ and the viscosity in force times time divided by $(\text{length})^2$. In this case $n + 1$ equals 6, and k equals 3. Can we find three parameters which cannot be combined into any dimensionless group? Yes, the force, one of the lengths, and the density cannot be so combined; the three dimensions are independent, and the number of π's is $6 - 3 = 3$.
2. We may choose as our dimensions length, mass, and time, in which case we must express the force in dimensions of mass times length divided by $(\text{time})^2$ and the viscosity in mass divided by $(\text{length} \cdot \text{time})$. Here again $n + 1$ equals 6, and k equals 3. The same group of a length, the force, and the density again cannot be internally combined into any dimensionless group; so the three dimensions are independent, and the number of π's is 3.
3. We may choose force, mass, length, and time as our dimensions, in which case we must add g_c to our list of parameters. Now we can select a length, the force, the density, and g_c as the four parameters which cannot be combined into any dimensionless group; so all four dimensions are independent, and the number of π's is now $7 - 4 = 3$. ■

Having decided how many π's to look for, how do we look for them? Buckingham's theorem also provides an algorithm for selecting π's:

1. Select k variables which cannot be combined internally into a dimensionless group. These are the *repeating variables*.
2. Then the combination of the repeating variables with any one of the remaining, nonrepeating variables can form a dimensionless group; make up $n + 1 - k$ such groups from the variables. This can often be done by

inspection; it can be done systematically by the algorithm shown in Example 13.3.

Example 13.3. For the variables shown in Example 13.2, find a set of three π's.

This problem can be worked in three ways, just as Example 13.2 could. Here we work with only the first of the three parts of Example 13.2. The other two are Probs. 13.4 and 13.5.

In this choice of dimensions the independent dimensions are length, force, and time. We now construct a table of the variables, indicating their dimensions.

Variable	Description	Dimensions
A	A force F_1	F
B_1	Velocity V	$\frac{L}{t}$
B_2	Density ρ	$\frac{Ft^2}{L^4}$
B_3	Viscosity μ	$\frac{Ft}{L^2}$
B_4	Length L_1	L
B_5	Length L_2	L

As our repeating variables, we may select any three which cannot be combined internally to form a dimensionless group, for example, A, B_2, and B_4 (a force, the density, and a length). Then the three dimensionless π's may be formed from the following groups: A, B_2, B_4, B_1; A, B_2, B_4, B_3; and A, B_2, B_4, B_5.

The first π must be some product of powers (plus or minus) of A, B_2, B_4, B_1. We form this product, using a, b, c, and d to indicate the unknown powers:

$$\pi_1 = A^a B_2^b B_4^c B_1^d \tag{13.26}$$

However, because the π's are dimensionless, the dimensions of this equation are

$$L^0 t^0 F^0 = F^a \left(\frac{Ft^2}{L^4}\right)^b L^c \left(\frac{L}{t}\right)^d \tag{13.27}$$

This can be true only if the following three simultaneous equations are satisfied:

Equation for length:	$0 = -4b + c + d$
Equation for time:	$0 = 2b - d$
Equation for force:	$0 = a + b$

By straightforward algebra we may show that this set of equations has the solution

$$a = -\frac{d}{2} \qquad b = \frac{d}{2} \qquad c = d$$

Now we may arbitrarily select a value for d; we select 2 as the lowest value of d which avoids fractional exponents. Then our first π is

$$\pi_1 = A^{-1}B_2^1B_4^2B_1^2 = \frac{\rho L^2V^2}{F}$$

If the force here is a pressure force, then L^2/F is equivalent to 1/(a pressure) and π_1 is 2/(pressure coefficient). If we had chosen $d = 1$ above, our resulting π_1 would be the square root of the π_1 we found here.

By equally straightforward mathematics (Prob. 13.3) we can set up the systems of equations for π_2 and π_3 and solve for them:

$$\pi_2 = \frac{\mu^2}{F\rho} \qquad \text{and} \qquad \pi_3 = \frac{L_1}{L_2}$$

Here π_3 is the ratio of two lengths, and π_2 can be shown to be equal to 2/(Reynolds number · pressure coefficient)^2. Examining this set of π's, most experienced engineers would select as their dimensionless groups π_3, $1/\pi_1$, and $(\pi_2/\pi_1)^{1/2}$, that is, the length ratio, the pressure coefficient, and the Reynolds number. The justification for this procedure is that a different selection of the repeating variables will give a different set of π's. However, the different sets are all products or ratios of the first set to some power. For example, if the repeating variables in Example 13.3 are V, ρ, and L_1, then the same procedure leads to $\pi_1 = 2/(\text{pressure coefficient})$, $\pi_2 = 1/(\text{Reynolds number})$, and $\pi_3 = L_1/L_2$ (see Prob. 13.6). ∎

13.4 JUDGMENT, GUESSWORK, AND CAUTION

Dimensional analysis and physical-model studies are needed only in those problems which currently cannot be completely or rigorously solved with mathematical models. Therefore, we cannot hope to obtain a complete and certain solution by these methods. Furthermore, as the examples show, applying these methods requires judgment and good guesswork. Because the results are only tentative, they should be applied with caution.

However, the gain in simplifying a problem may be very great. As stated by Kline,†

---

† From S. J. Kline, *Similitude and Approximation Theory*, McGraw-Hill, New York, 1965, p. 17. Quoted by permission of the publisher.

> The use of dimensionless parameters reduces the number of independent coordinates required. A convenient way to realize the importance of such a reduction is to recall that a function of one independent coordinate can be recorded on a single line; two independent coordinates a page; three require a book; and four, a library.

The gain in increased understanding of the problem by seeing it in its dimensionless form may be even more valuable than the gain in simplifying the problem.

13.5 SUMMARY

1. Although many flow situations can be represented by a simple, closed, analytical equation (e.g., laminar flow in a pipe), many others cannot (e.g., turbulent flow in a pipe).
2. In the latter case the experimental data can often be correlated and simplified by the use of dimensionless ratios (e.g., the friction factor, Reynolds number, relative roughness plot).
3. In designing full-size equipment from tests on small-scale physical models, it is necessary to "scale up," keeping the values of the pertinent dimensionless groups the same for model and full-size equipment.
4. The common methods of finding the pertinent dimensionless groups are the method of governing equations, the method of force ratios, and Buckingham's π method.
5. All these methods require judgment and good guesswork. The results must be applied with caution.

PROBLEMS

See the Common Units and Values for Problems and Examples inside the back cover.

13.1. Estimate which of the dimensionless force ratios in Table 13.1 should be important for the following kinds of flow: (*a*) breakup of a jet of liquid into droplets, (*b*) formation of a vortex in the free surface of the drain of a bathtub, (*c*) the same as in (*b*) but for a bathtub full of molasses, (*d*) the shape of a bubble rising in a viscous liquid, (*e*) the breakaway of a bubble from a submerged horizontal orifice, (*f*) the simultaneous horizontal flow of two fluids through a porous medium at low flow rates (e.g., the flow of oil and gas into an oil well, and (*g*) the same as in (*f*) but flow in the vertical direction.

13.2. In the example in the text in which the list of variables consists of two velocities and two forces, show what the possible forms of the final equation are if you assume that there is only one π and that there are two π's.

13.3. Find π_2 and π_3 in Example 13.3.

13.4. Rework Example 13.3, using L, m, and t as the independent dimensions.

13.5. Rework Example 13.3, using L, m, t, and F as independent dimensions (and adding g_c to the list of variables).

13.6. Rework Example 13.3, using V, ρ, and L_1 as the repeating variables.

13.7. Rework Example 13.3 by the method of force ratios.

13.8. Make a list of the dimensionless groups which appear in the common introductory course in thermodynamics. Indicate whether these are ratios, and if so, of what.

13.9. An airplane scale model is to be tested. The problem of interest is one in which the landing behavior is to be tested. Since this is a low-speed problem, the Reynolds number is to be held constant between the model and full-scale airplane. The model is to be one-tenth of the length of the airplane with the same shape. The landing speed of the full-size airplane is 60 mi/h. How fast should the air in the wind tunnel move past the model to have the same Reynolds number? Suppose we could test the model in water. What should the water velocity be for the same Reynolds numbers?

13.10. We are designing a new type of racing boat hull. In the model tests in a towing basin we will use a model one-tenth the size of the actual boat. We know that the drag of the boat (its resistance to moving through the water) is a function of the Reynolds and Froude numbers. Can we simultaneously hold both of these constant between the model and the real boat in our tests? If not, how can we test the model?

CHAPTER 14

GAS-LIQUID FLOW

In preceding chapters, as in most fluid mechanics problems, the flowing fluid was composed of one homogeneous phase, such as water, oil, air, or steam. However, many interesting and important flows involve the simultaneous flow of two quite different materials through the same conduit, such as the flow in a coffee percolator (Fig. 2.19), the flow in most vaporizers or boilers or condensers, and the flow in the carburetor of an automobile (Fig. 5.31). These flows can be of gas-liquid mixtures, liquid-liquid mixtures, gas-solid mixtures, or liquid-solid mixtures. Although there are industrially important examples of each of these combinations, the most important and interesting seems to be the gas-liquid case, which we discuss in this chapter. Brief summaries of what is known about gas-solid and liquid-solid flows are given by Sakiadis [1].

In all these flows the influence of gravity is much greater than in the one-phase flows previously considered. For laminar or turbulent flow of water in a tube, the velocity distribution and friction effect ($\mathscr{F}$) would be the same on earth as in the zero-gravity environment of an earth satellite. Neither the velocity distribution nor $\mathscr{F}$ is influenced at all by changing the pipe's position from horizontal to vertical. Gravity does not significantly influence the flow pattern or $\mathscr{F}$, because it works equally on each particle of the fluid. This is not the case for two-phase flows, because normally the phases have different densities and thus are affected to different extents by gravity. Not all the flows described in the following sections would be the same in a zero-gravity environment as on earth, and as shown in Secs. 14.1 and 14.2, horizontal and vertical two-phase flows have very different velocity distributions and $\mathscr{F}$'s.

14.1 VERTICAL, UPWARD GAS-LIQUID FLOW

Many of the salient features of multiphase flow can be illustrated by considering the simultaneous gas-liquid flow up a vertical pipe with the apparatus sketched in Fig. 14.1. Assume that at first only water flows; then from Bernoulli's equation

$$P_1 - P_2 = \rho g(z_2 - z_1) + \rho \mathcal{F} \tag{14.1}$$

For zero flow rate we can calculate $P_1 - P_2 = \rho g \, \Delta z = 8.7$ psi. $= 60$ kPa. As we start the flow rate of water only, $P_1 - P_2$ will increase, because of the increase in $\mathcal{F}$ with increasing velocity.

Now let us hold the water flow rate constant at some modest average velocity, such as 2 ft/s, and slowly increase the air velocity from zero to some large value. This will cause $\mathcal{F}$ to increase, since the overall linear velocity is increased. However, now there will be bubbles of gas in the pipe; the density in Eq. 14.1 is no longer the density of water but is the average density of the gas-liquid mixture in the vertical pipe. At low flow rates the density goes down much faster than $\mathcal{F}$ goes up, so $P_1 - P_2$ decreases steadily as we increase the gas flow rate. Finally, a point is reached where further increase in the gas velocity causes $\mathcal{F}$ to increase faster than ρ decreases; $P_1 - P_2$ will increase with an increasing gas flow rate. A typical plot of experimental data for such a system is shown in Fig. 14.2.

Figure 14.2 has the shape described above, with a distinct kink in it near a flow rate ratio of 10. This is typical of such systems and indicates that the change described above does not take place smoothly. If the system is made of glass, so that the flow pattern can be observed, it will be seen that several distinctly different flow patterns are formed as the airflow rate is increased.

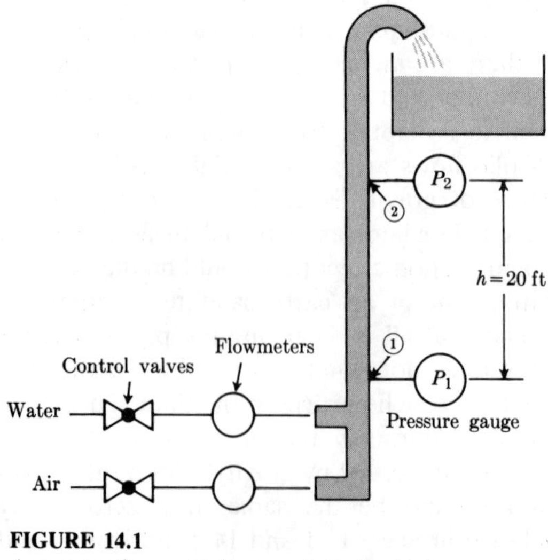

FIGURE 14.1
Apparatus for vertical, upward gas-liquid flow.

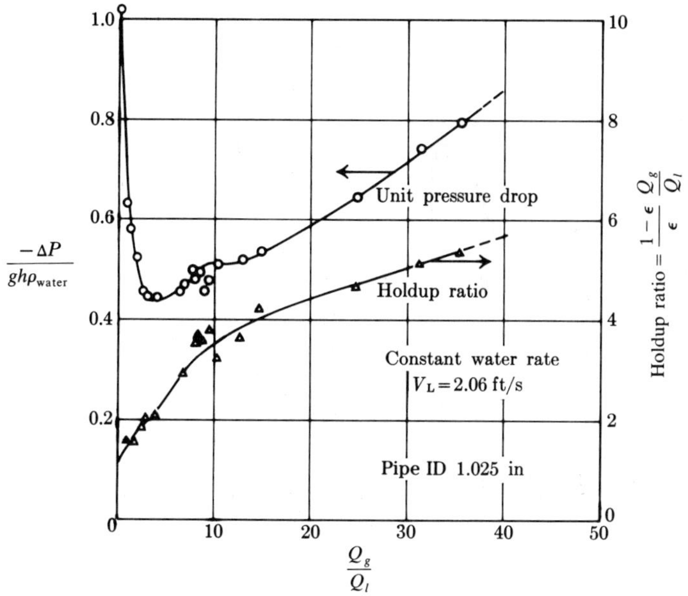

FIGURE 14.2
Typical experimental results for vertical, upward air-water flow in the apparatus shown in Fig. 14.1, with constant liquid flow rate. Here V_L is Q_l/A. [From G. W. Govier, B. A. Radford, and J. S. C. Dunn, "The upward vertical flow of air-water mixtures," *Can. J. Chem. Eng. 35:* 58–70 (1957). Reproduced by permission of the publisher.]

This is illustrated in Fig. 14.3. At the lowest airflow rates, small bubbles rise through the liquid. As the airflow rate is increased, large single bubbles are formed, which practically fill the tube, driving slugs of liquid between them. At higher rates these slugs become frothy, and finally at high gas flow rates the liquid is present either as an annular film on the walls or as a mist in the gas.

Two additional ideas are widely used to correlate the results of experiments like that shown in Fig. 14.1: *holdup* and *slip*. One can measure what fraction of the tube is occupied by gas, ε, and what fraction by liquid, $1 - \varepsilon$, in such a flow (experimentally this is done by placing two quick-closing valves in the tube, establishing the flow, and then closing the valves simultaneously and measuring the amounts of gas and liquid so trapped). One might assume that the ratio of gas to liquid, $\varepsilon/(1 - \varepsilon)$, would be the same as the flow rate ratio of the two streams Q_g/Q_l, but experimental evidence indicates the contrary. In Fig. 14.2 are shown the experimental values for the *holdup ratio*, which is the ratio of the liquid-gas volume ratio actually present in the pipe to the liquid-gas volume ratio in the stream passing through [2]. This holdup ratio is always greater than 1 for vertical, upward flow, because some of the liquid is always falling back by gravity and thus having to make several upward trips in order to get out.

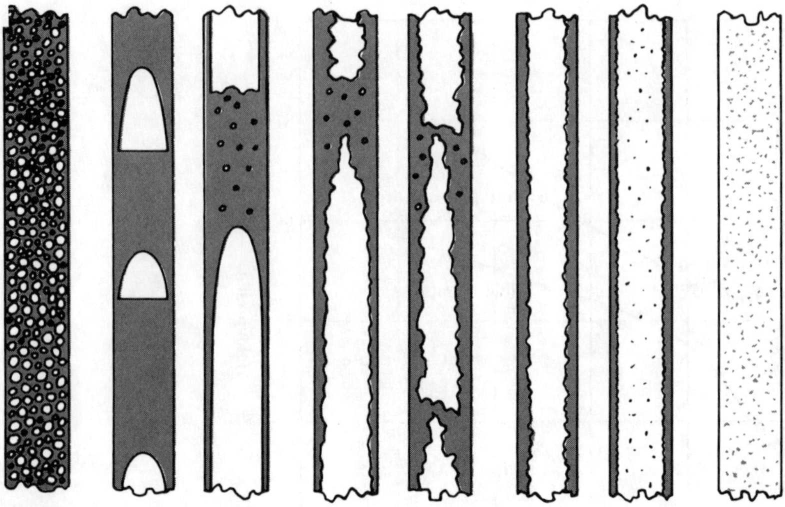

FIGURE 14.3
Two-phase flow patterns in vertical tubes. The liquid flow rate is upward at a small, constant velocity. The gas flow rate upward increases steadily from left to right. The "annular" pattern shown is often referred to as *climbing film flow*. [From D. J. Nicklin and J. F. Davidson, "The onset of instability in two-phase slug flow," in *Symposium in Two-Phase Flow*, Institution of Mechanical Engineers, London, 1962. Reproduced by permission of the publisher.]

We may define an *average velocity of the gas in the tube* as

$$V_{\text{av, gas}} = \frac{Q_g}{A_{\text{tube}}\varepsilon} \tag{14.2}$$

and analogously

$$V_{\text{av, liq}} = \frac{Q_l}{A_{\text{tube}}(1-\varepsilon)} \tag{14.3}$$

The difference between these two is defined as the *slip velocity*:

$$V_{\text{slip}} = V_{\text{av, gas}} - V_{\text{av, liq}} = \frac{1}{A_{\text{tube}}}\left(\frac{Q_g}{\varepsilon} - \frac{Q_l}{1-\varepsilon}\right) \tag{14.4}$$

If the holdup ratio were exactly the same as the volumetric flow ratio, then the term in parentheses on the right in Eq. 14.4 would be exactly zero. For actual upward flows the slip velocity is always positive, because the gas mostly moves upward, while the liquid partly moves upward and partly flows backward downhill because of gravity, lowering its net average upward velocity.

Example 14.1. From the data shown in Fig. 14.2 for $Q_g/Q_l = 10$, calculate the values of $\mathcal{F}$, ε, and V_{slip}.

We start by reading that at this value of Q_g/Q_l the holdup ratio is approximately 3.5, so that

$$\frac{1-\varepsilon}{\varepsilon} = \frac{3.5}{Q_g/Q_l} = \frac{3.5}{10} = 0.35 \qquad \varepsilon = 0.741$$

In Eq. 14.1 we have

$$\rho = \rho_{air}\,\varepsilon + \rho_{water}(1-\varepsilon)$$

but since we have $\rho_{air} \ll \rho_{water}$, we may neglect the first term and use $\rho = \rho_{water}(1-\varepsilon)$. Solving Eq. 14.1 for $\mathcal{F}$, we get

$$\mathcal{F} = \frac{-\Delta P}{\rho} - gh$$

and from Fig. 14.2 we read $-\Delta P/(gh\rho_{water}) = 0.5$, so

$$\mathcal{F} = \frac{0.5gh\rho_{water}}{(1-\varepsilon)\rho_{water}} - gh = gh\left(\frac{0.5}{0.259} - 1\right)$$
$$= 0.93gh$$

Thus, at this flow rate the conversion of injection work to internal energy by friction heating is 93 percent as large as the conversion of injection work to potential energy by increasing the elevation of the fluid. Finally, from Eqs. 14.2 and 14.3

$$V_{av,\,liq} = \frac{Q_l}{A_{tube}} \cdot \frac{1}{1-\varepsilon} = \frac{2.06\text{ ft/s}}{1-0.741} = 7.95\ \frac{\text{ft}}{\text{s}} = 2.42\ \frac{\text{m}}{\text{s}}$$

$$V_{av,\,gas} = \frac{Q_g}{A_{tube}\,\varepsilon} = \frac{Q_g}{Q_l} \cdot \frac{Q_l}{A_{tube}} \cdot \frac{1}{\varepsilon} = \frac{10 \cdot 2.06\text{ ft/s}}{0.741} = 27.8\ \frac{\text{ft}}{\text{s}} = 8.48\ \frac{\text{m}}{\text{s}}$$

$$V_{slip} = 27.8 - 7.9 = 19.9\text{ ft/s} = 6.07\text{ m/s}$$ ■

Vertical cocurrent flow of this type is extremely common in engineering. It occurs in almost all flowing and gas-lifted oil wells and in gas-lift pumps. Much of the work done on this type of flow has been connected with boiling of liquids in vertical tubes. In a vertical tube in which a liquid (e.g., water) is being boiled, it is entirely possible to have all the flow patterns shown in Fig. 14.3 present in the same tube at the same time. In that case the fluid enters the bottom of the tube as all liquid. Then as it passes up the tube, more and more of it is converted to a vapor by boiling, so that at the top of the tube it may be all vapor (this is generally avoided in boilers, because it results in a dangerously high metal temperature at the top of the tube). Thus, the various patterns shown in Fig. 14.3 exist at various levels in the same tube, with the bottom of the tube corresponding to the left of Fig. 14.3 and the top corresponding to the right.

Because the various flow patterns shown in Fig. 14.3 are so different from each other, it is unlikely that there will ever be a completely successful single description of the friction effect and holdup, of vertical, gas-liquid flows comparable, e.g., to the friction factor plot for pipe flow. The various flow forms are so different physically that they cannot be expected to obey the same mathematical relationships. Several such overall correlations are available, e.g., that of Hughmark and Pressburg [3], in which for a variety of fluids they correlated liquid holdup $1 - \varepsilon$ with the group

$$\left(\frac{\dot{m}_l}{\dot{m}_g}\right)^{0.9} \cdot \frac{\mu_l^{0.19} \sigma^{0.205} \rho_g^{0.70}}{(Q_l/A + Q_g/A)^{0.435} \rho_l^{0.72}}$$

They found that they could represent all the available data for the holdup in such flow on a plot of $1 - \varepsilon$ versus this group with an average error of about 12 percent.

The other approach is to try to find separate equations for the different flow patterns shown in Fig. 14.3. This approach requires some kind of "flow map" that indicates which flow pattern will be observed for the conditions of interest and then suitable relations for all the individual flow patterns. There has been considerable progress in this approach [4, 5]. One major complication in trying to correlate the behavior of such flows is the fact that for a given choice of fluids, pipe size, and flow rates the quantities ε and $\mathscr{F}$ are strongly influenced by the design of the gas-liquid mixer. If in Fig. 14.1 we switch from the simple pipe tee mixer to one in which the liquid is introduced through porous walls, this makes no change in our observations in the slug flow region, but it makes a significant change in the observations in the annular and mist flow regions [6, p. 248]. This effect becomes unimportant for very long pipes, but is important for short ones.

14.2 HORIZONTAL GAS-LIQUID FLOW

If the apparatus shown in Fig. 14.1 is turned so that the pipe is horizontal, then the behavior will be quite different from that for the vertical case. The observed flow patterns are shown on Fig. 14.4. For a constant liquid flow rate, as we increase the gas flow rate from zero, at first the gas flows in bubbles along the top of the pipe. Then the bubbles grow in size and length. Finally they become so large that they coalesce into a continuous stream of gas flowing over a continuous layer of liquid. Further increase in gas velocity causes the gas to raise waves on the liquid surface. These waves grow until they eventually reach across the tube, in which case they are propelled as slugs of liquid with interspersed slugs of gas. Then this pattern switches to the annular pattern observed in vertical gas-liquid flow and finally to a mist flow.

Knowledge of the flow pattern can be quite important, If, for example, we are vaporizing a liquid in a horizontal tube under high heat fluxes, then in annular flow the tube wall will always be covered with liquid and presumably

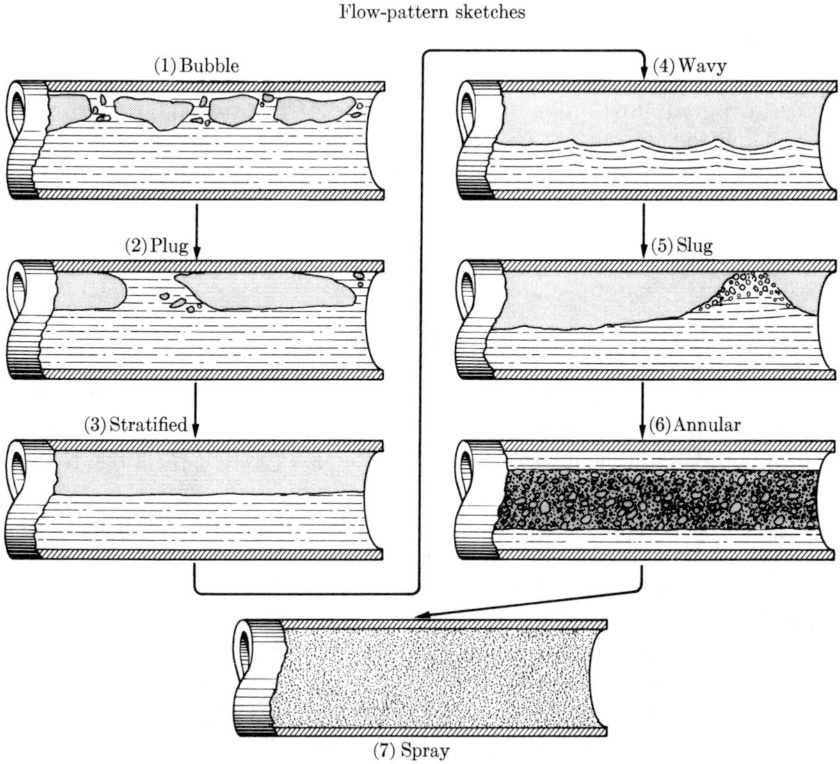

FIGURE 14.4
Flow patterns in horizontal gas-liquid flows. These are for constant liquid flow rates, with the gas flow rate increasing as shown by the arrows. [From R. S. Brodkey, *The Phenomena of Fluid Motions*, Addison-Wesley, Reading, Mass., 1967, as redrawn from G. Alves, "Concurrent liquid-gas flow in a pipeline contactor," *CEP 50:* 449–456 (1954). Reproduced by permission of the publisher.]

be safe from excessive metal temperatures. But if the flow is stratified, then the top part of the tube will be covered by gas, which is much less effective in conducting heat away from the surface; so the inside metal temperatures may become very high, even to the point of melting the tube. Although several correlations have been proposed for determining what kind of flow patterns will exist [6, pp. 199–277], none is now known that is universally applicable. Furthermore, the transitions from one kind of flow to another do not occur at sharply defined conditions but may take place over a range of conditions, and as in vertical gas-liquid flow for a given set of fluids and flow rates, the flow pattern can be completely changed and the value of $\mathscr{F}$ doubled by simply changing the type of gas-liquid mixer.

All available experimental data indicate that $\mathscr{F}$ is always higher for two-phase horizontal flow than for single-phase flow under similar conditions.

This is principally due to the movement of the two phases relative to each other in the tube, which does not contribute to flow along the tube but does contribute to the conversion of other forms of energy to internal energy. For horizontal gas-liquid flow, numerous empirical correlations have been proposed, of which the most widely used is that of Lockhart and Martinelli [1]. The comparison of those correlations with experiment has been extensively reviewed by Scott [6, pp. 199–277], indicating that it is not reliable to more than about ±50 percent in most cases but that it is as good as any other single correlation.

14.3 TWO-PHASE FLOW WITH BOILING

When a liquid at its boiling point flows in a pipe, the pressure decrease due to friction will cause the pressure of the liquid to fall below its saturation pressure, and the liquid will boil. This type of flow, called *flashing flow*, is important in the design of boilers, steam condensate lines, etc.

The principal difference between this type of flow and those discussed above is that as the pressure falls, more and more vapor is formed, so the volumetric flow rates, average velocities, and pressure drops per foot are not constant for an entire pipe but vary with length. The velocity and pressure gradient increase rapidly as the amount of vapor increases. Furthermore, in this type of flow it is very common to have the kind of choked condition found in high-velocity gas flows (Sec. 8.3). However, the observed sonic velocities for gas-liquid mixtures are much lower than that for gas alone, so that this choking occurs at velocities much lower than the sonic velocity of the vapor alone. This kind of flow is of great interest in the design of steam boilers and vaporizing furnaces of all kinds [1].

14.4 SUMMARY

1. In gas-liquid flows, numerous different flow patterns are possible, depending on the gas and liquid flow rates and properties and on the direction of the flow relative to the direction of gravity.
2. For such flows the pressure drop due to friction heating of the fluids $\mathcal{F}$ is always greater than that for single-phase flows under comparable circumstances.
3. Numerous empirical correlations are available for predicting the pressure drop, holdup, and slip velocity for such systems. These are not nearly as reliable as the pressure-drop correlations for single-phase flow.

PROBLEMS

See the Common Units and Values for Problems and Examples inside the back cover.

14.1. From the data given in Fig. 14.2, prepare a plot showing the average density of the fluid in the tube ρ_{av}, $\rho_{av} g(z_2 - z_1)$, and $\rho_{av}\mathcal{F}$ versus Q_g/Q_l.

14.2. From the data in Fig. 14.2, prepare a plot of slip velocity versus Q_g/Q_l.

14.3. For the flow discussed in Example 14.1, calculate the values of $\mathcal{F}$ which would exist if only the liquid were flowing and if only the gas were flowing. Compare these with the observed value of $\mathcal{F}$ for the simultaneous two-phase flow.

14.4. If we wished to perform the same liquid-lifting task as shown in Example 14.1, we could use the apparatus sketched in Fig. 14.5. For such an apparatus, with the flow rates given in Example 14.1, calculate the necessary pressures at the inlets to the turbine and to the pump (assume that these pressures are the same). Compare these with the necessary pressure at the base of the gas-liquid column in Example 14.1. Assume that the overall vertical elevation change is 20 ft for both gas and liquid in both cases.

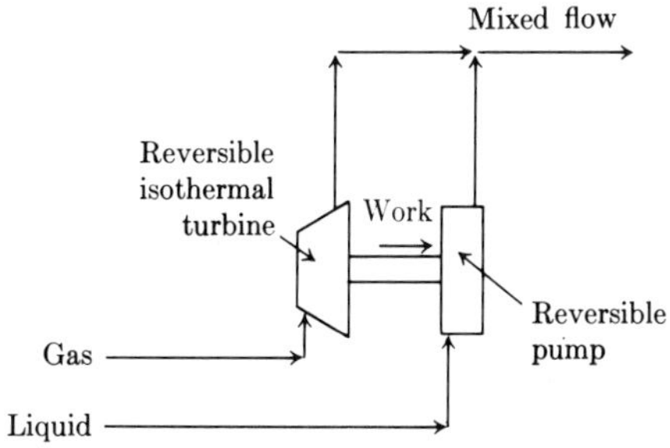

FIGURE 14.5

14.5. For the coffee percolator sketched in Fig. 2.19, on the basis of Bernoulli's equation estimate the maximum possible mass flow rate per unit area of the steam-water mixture in the riser tube.

14.6. Figure 14.6 shows the flow diagram of an air-lift or gas-lift pump. In it a fluid is pumped from the lower to the upper reservoir by way of a vertical pipe, into the bottom of which air or gas is introduced. Explain why such pumps are widely used in the processing of the highly radioactive solutions in nuclear fuel processing.

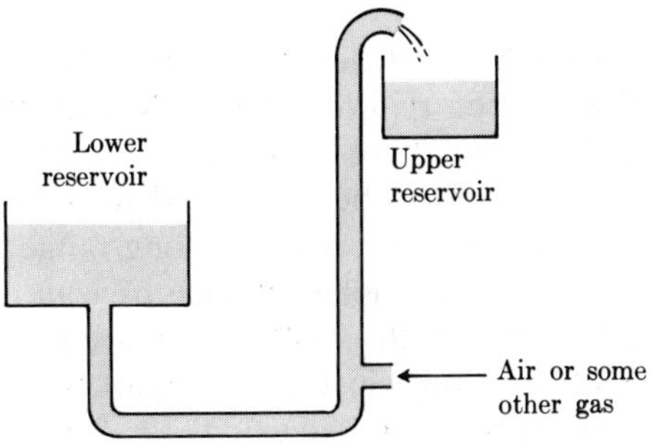

FIGURE 14.6
Air lift or gas lift.

14.7. For the pump in Fig. 14.6, sketch a plot of liquid flow rate versus airflow rate for constant geometry and fluid properties. *Hint:* See Fig. 14.2.

CHAPTER 15

NONNEWTONIAN FLUIDS

This topic was introduced in Sec. 1.5C. In this brief chapter some comparisons are made between the behavior of newtonian and nonnewtonian fluids in pipe flow, and references are given for the student who wishes to pursue the subject further.

15.1 THE ROLE OF STRUCTURE IN NONNEWTONIAN BEHAVIOR

Almost all nonnewtonian fluids contain suspended particles or dissolved molecules which are large compared with the size of typical fluid molecules (a typical polymer molecule may be many thousand times as large as a water molecule). Most nonnewtonian behavior is believed to be associated with the "long-range structure" due to such larger constituents, where "long-range" implies long compared with the diameter of a small molecule such as water. For example, a Bingham fluid is assumed to have a three-dimensional elastic structure, which will resist small shearing stresses but which comes apart when subjected to a stress higher than its yield strength. Pseudoplastic fluids (by far the most common type of nonnewtonian fluid) mostly have dissolved or dispersed particles (e.g., dissolved long-chain molecules), which have a random orientation in the fluid at rest but which line up when the fluid is sheared. They offer more resistance to deformation in the random position, so the viscosity drops as they become aligned. Dilatant fluids are almost all slurries of solid particles in which there is barely enough liquid to keep the solid particles from touching each other. Their behavior is explained by assuming that at low shear rates the fluid between the particles is able to lubricate the sliding of one particle past another but that at high shear rates this lubrication breaks down.

Thixotropic fluids are assumed to have alignable particles, as pseudoplastic fluids do (most thixotropic fluids are pseudoplastic), but with a finite time required for the particles to become aligned with the flow. An additional factor in thixotropic behavior is probably the existence of weak bonds between molecules (e.g., hydrogen bonds or entanglements of polymer chains). The bonds are gradually destroyed by shearing (some authors suggest that ordinary pseudoplastic fluids are really thixotropic fluids whose particles align or whose bonds break much faster than can be observed on currently available viscometers). Rheopectic fluids are rare and generally only show rheopectic behavior under very mild shearing. It has been suggested that mild shearing may help particles in the fluid to fit together better, thus forming a tighter structure and increasing the viscosity. Viscoelastic fluids normally contain long-chain molecules, which can exist in coiled or extended forms and which can connect one to another. When stretched, these molecules straighten out, but when the flow stops, they tend to revert to their coiled position, causing the elastic behavior.

These descriptions are in accord with most observed behavior of these fluids and thus offer a mental picture of what may be going on within the fluid. However, they are by no means rigorous descriptions of the microscopic internal behavior of such fluids, and they may be modified by further studies of nonnewtonian fluids.

15.2 MEASUREMENT AND DESCRIPTION OF NONNEWTONIAN FLUIDS

Much of the past and present research in nonnewtonian fluids has consisted of measuring their stress-rate-of-strain curves (such as Fig. 1.5) and trying to find mathematical descriptions of these curves. The study of the flow behavior of materials is called *rheology* (from Greek words meaning "the study of flow"), and diagrams like Fig. 1.5 are often called *rheograms*.

As shown in Sec. 1.5, the basic definition of viscosity is in terms of the sliding-plate experiment shown in Fig. 1.4. For newtonian fluids it was shown in Sec. 6.3 (Example 6.2) that the viscosity could be determined easily by a capillary-tube viscometer. It can be shown both theoretically and experimentally that the viscosity determined by such a viscometer for a newtonian liquid is exactly the same as the viscosity one would determine on a sliding-plate viscometer. Since capillary-tube viscometers are cheap and simple to operate, they are widely used in industry for newtonian fluids.

For nonnewtonial fluids which are not time-dependent or viscoelastic, it is possible to convert capillary-tube viscometer measurements to the equivalent sliding-plate measurements, but this involves some mathematical manipulations. For time-dependent (e.g., thixotropic) fluids, this does not seem to be possible. Thus, most studies of the behavior of nonnewtonian fluids use some variant of the sliding-plate viscometer. The most common is the concentric-cylinder viscometer; see Fig. 15.1. (Cone-and-plate viscometers are also widely used, but they are not discussed here [1, p. 517].)

In such a device a motor-driven cylindrical cup is rotated at a constant speed. The fluid being tested is in the thin, annular region between the cup and the bob. The shear stress generated by the fluid on the wall of the bob tends to turn the bob, but this turning motion is resisted by the torsion wire which supports the bob. The bob takes up a position where the torque exerted by the torsion wire is equal and opposite to the torque supplied by the fluid shear on its surface; from its position, as indicated by a pointer and scale and the calibration of the torsion wire, one can readily compute the shear stress at the wall.

This device is really the sliding-plate device wrapped around a cylinder. Mathematical corrections are needed to make the readings of this viscometer correspond exactly to those of the sliding-plate viscometer [2], but these are generally small; see Prob. 15.11. This type of viscometer is suited to newtonian or nonnewtonian fluids with or without time dependence. Several other comparable viscometer types are known [2].

The experimental data from a viscometer like that shown in Fig. 15.1 are normally represented on a plot such as Fig. 1.5. For newtonian fluids the stress-rate-of-strain behavior is described by Newton's law of viscosity, Eq. 1.5. In reading the nonnewtonian literature, observe that most authors use μ as the symbol for viscosity only of newtonian fluids and use η as the symbol for viscosity of nonnewtonian fluids.

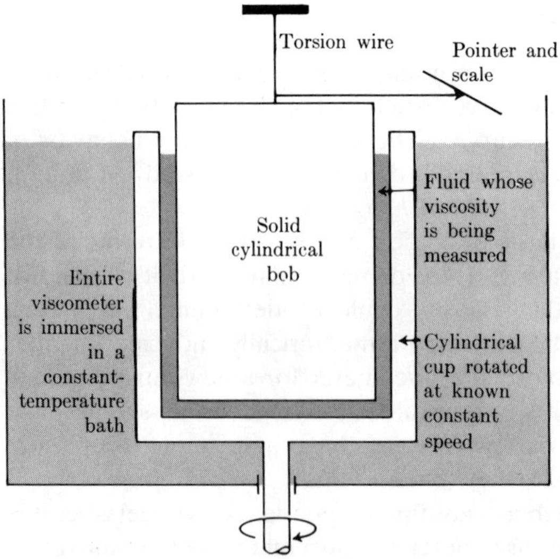

FIGURE 15.1
Concentric cylinder or "cup and bob" viscometer.

The data on a plot such as Fig. 1.5 can be used more easily if they can be represented by an equation. The Bingham fluid can be easily represented by

$$T \le \tau_{\text{yield}} \qquad \frac{dV}{dy} = 0 \qquad \tau \ge \tau_{\text{yield}} \qquad \tau = \tau_{\text{yield}} + \mu_0 \frac{dV}{dy} \tag{15.1}$$

where μ_0 is the slope of the curve on Fig. 1.5.

In many cases the experimental curves for both dilatant and pseudoplastic fluids can be reasonably well represented by the *power law*, also called the *Ostwald-de Waele equation*:

$$\tau = K\left(\frac{dV}{dy}\right)^n \tag{15.2}$$

Here K and n are constants whose values are determined by fitting experimental data. For newtonian fluids $n = 1$ and $K = \mu$. For pseudoplastic fluids n is less than 1, and for dilatant fluids it is greater than 1. The power law has little theoretical basis; its virtues are that it represents a considerable amount of experimental data with reasonable accuracy and that it leads to relatively simple mathematics. Many other equations have been used to represent these stress-strain rate curves. Some of the simpler ones are those of Ellis [3]

$$\tau = \frac{dV/dy}{A + B\tau^C} \tag{15.3}$$

Reiner-Phillipoff [3]

$$\tau = \left[A + \frac{B - A}{1 + (\tau/C)^2}\right]\frac{dV}{dy} \tag{15.4}$$

and Powell-Eyring [4, p. 372]

$$\tau = A\frac{dV}{dy} + \frac{1}{B}\sinh^{-1}\left(\frac{1}{C}\frac{dV}{dy}\right) \tag{15.5}$$

In each of these three equations A, B, and C are constants determined from experimental data. The Ellis and Reiner-Phillipoff equations are based simply on looking at experimental-data plots and deciding what form of equation would give the best fit (Probs. 15.1 and 15.2); the Powell-Eyring equation results from Eyring's theories of the structure of the liquid state. Since these three equations each contain three adjustable constants, compared with the two adjustable constants of the power-law equation, they can fit experimental data somewhat better than the power law but at the expense of greater mathematical complexity in their use. These three equations are about equal in ability to represent wide ranges of experimental data accurately and in extrapolating experimental data. Such equations are frequently called *constitutive equations* or *rheological equations of state*.

For time-dependent fluids (thixotropic or rheopectic) there are no simple relations now available for showing the stress-strain-rate-time dependence. Figure 15.2 is a typical stress-time curve for a thixotropic fluid, showing lines of

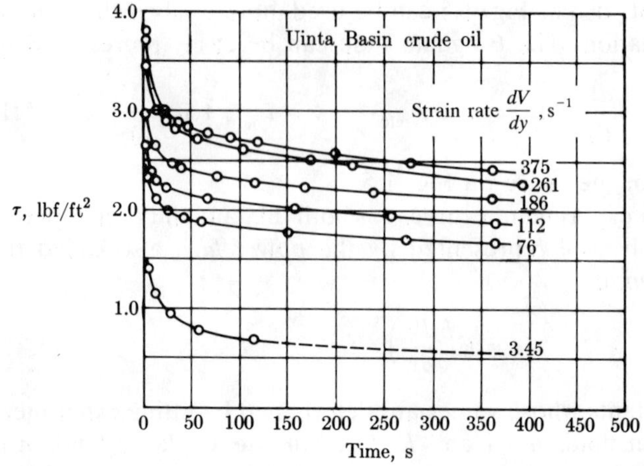

FIGURE 15.2
Stress-time curve for various strain rates for a typical thixotropic fluid, obtained in an apparatus like that shown in Fig. 15.1. [Courtesy of the late E. B. Christiansen.]

constant strain rate. The change with time occurs mostly in the first 60 s, after which the change with time is minor. For most engineering applications it would be safe to treat this fluid as a simple pseudoplastic fluid with properties corresponding to those of the right-hand side of Fig. 15.2.

For viscoelastic fluids no simple relations are known at all, and current thought is that it may never be possible to describe these fluids adequately by simple scalar equations, only by tensorial equations [5].

15.3 LAMINAR FLOW OF NONNEWTONIAN FLUIDS IN CIRCULAR TUBES

Most fluids with pronounced nonnewtonian behavior have such high viscosities that their flow is laminar in most industrially interesting situations. We saw in Sec. 6.3 that for any fluid the shear stress at any point in a horizontal circular pipe is given by

$$\tau = \frac{-r(P_1 - P_2)}{2\,\Delta x} = \frac{r}{2}\frac{dP}{dx} \tag{6.3}$$

For laminar flow of newtonian fluids we substituted Newton's law of viscosity for the shear stress and integrated twice to find Poiseuille's equation.

For nonnewtonian fluids we can experimentally determine a plot like Fig. 1.5 of the shear stress as a function of dV/dy (which equals dV/dr for circular pipe flow). From this plot we can find the equivalent of Poiseuille's equation by two graphical integrations. This is tedious and has prompted much of the work of trying to find equations that will represent the data in curves such as Fig. 1.5. If the data can be fit by the power law (Eq. 15.2), then the two integrations can be easily performed (Prob. 15.4), yielding

$$V = \left(\frac{1}{2K} \frac{-dP}{dx} \right)^{1/n} \cdot \frac{n}{n+1} \cdot \left(r_W^{(n+1)/n} - r^{(n+1)/n} \right) \tag{15.6}$$

$$Q = \frac{n\pi}{3n+1} \left(\frac{1}{2K} \frac{-dP}{dx} \right)^{1/n} r_W^{(3n+1)/n} = \frac{n\pi D^3}{8(3n+1)} \left(\frac{D}{4K} \frac{-dP}{dx} \right)^{1/n} \tag{15.7}$$

where r_W is the radius of the tube or pipe. It is also possible to integrate several other of the shear-stress-strain-rate equations to find analytical solutions for laminar flow in a circular tube [4, p. 377]. Closed-form solutions for the flow of power-law fluids in a variety of other geometries are shown by Bird et al. [1, p. 176].

The laminar flow of various kinds of fluids in circular pipes can be easily compared by plotting $(D/4)(-dP/dx)$ versus $32Q/(\pi D^3) = 8V_{av}/D$, as shown in Fig. 15.3. This plot (or its equivalent on logarithmic paper) is very widely used in nonnewtonian flow calculations and publications. Its merit can be seen by rewriting Poiseuille's equation (Eq. 6.8) in the form

$$\frac{D}{4} \frac{-dP}{dx} = \mu \frac{32Q}{\pi D^3} = \mu \frac{8V_{av}}{D} \tag{15.8}$$

From Eq. 15.8 we see that for newtonian fluids this plot must be a straight line through the origin with a slope equal to the viscosity, as in Fig. 1.5. The left-hand side of Eq. 15.8 (and the ordinate of Fig. 15.3) is exactly equal to the shear stress at the wall of the pipe, as may be seen by comparison with Eq. 6.3. The abscissa is related to the shear rate at the wall by

$$\left(\frac{dV}{dr} \right)_w = \frac{8V_{av}}{D} \left[\frac{3}{4} + \frac{1}{4} \frac{d \ln (8V_{av}/D)}{d \ln (D/4)(-dP/dx)} \right] \tag{15.9}$$

This equation, due to Rabinowitsch and Mooney [4, p. 377; 6] is derivable for

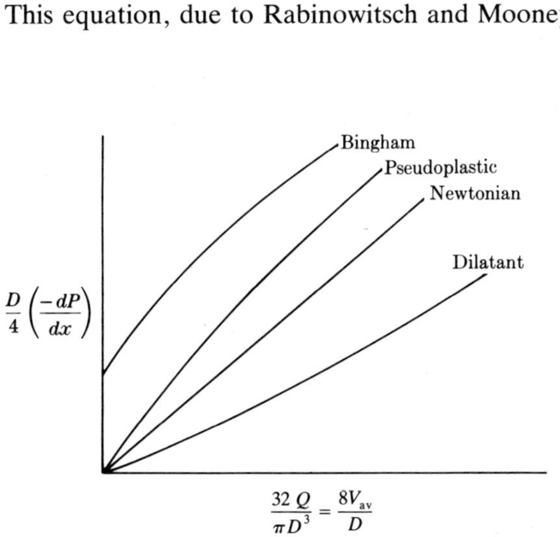

FIGURE 15.3

the laminar flow of any homogeneous, non-time-dependent fluid in a circular pipe. It has been shown experimentally to work quite well for slurries, which are not homogeneous, so its applicability is quite broad. For newtonian fluids the term in brackets on the right in Eq. 15.9 is equal to 1, so the abscissa is exactly equal to the shear rate at the wall (Prob. 15.5). For more complicated fluids the term in brackets is either a constant (for power-law fluids) or some relatively simple function of the shear rate. Thus, Fig. 15.3 is the same kind of figure as Fig. 1.5, except for a scale factor or some scale-changing function.

Just as the curve for a specific fluid at a given temperature must be the same in Fig. 1.5, independent of the kind or size of viscometer used, so also the curve for a given fluid at a given temperature in laminar flow must be the same in Fig. 15.3, independent of the size of tube in which the fluid is flowing. Figure 15.4 shows a set of experimental data for a lime-water slurry flowing in four tubes of different diameters. As indicated above, the laminar-flow data all lie on one curve. The steeply rising parts at the upper right of the curve are for turbulent flow, discussed in Sec. 15.4.

Example 15.1. It is desired to pump 15 gal/min of 23 percent lime slurry in a 1-in pipe. What is the required pressure gradient?

From App. A.3 we have

$$V_{\text{av}} = \frac{15 \text{ gal/min}}{2.69 \text{ (gal/min)/(ft/s)}} = 5.6 \frac{\text{ft}}{\text{s}} = 1.7 \frac{\text{m}}{\text{s}}$$

so that

$$\frac{1}{4}\left(\frac{8V_{\text{av}}}{\pi D}\right) = \frac{1}{4}\left(\frac{8}{\pi}\right)\left(\frac{5.6 \text{ ft/s}}{1.049 \text{ ft}/12}\right) = \frac{42.7}{\text{s}}$$

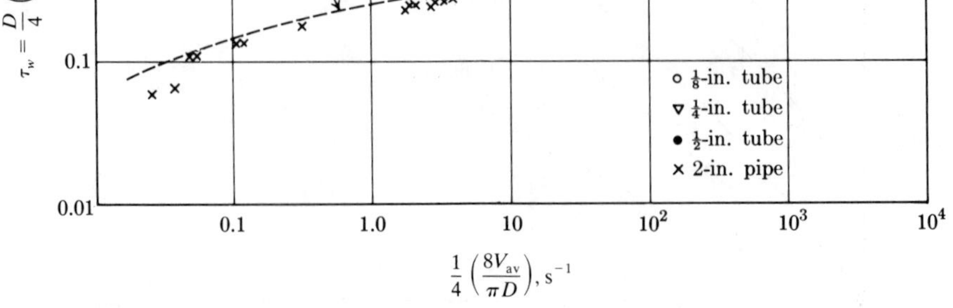

FIGURE 15.4
Data of Alves et al. [6], replotted by E. B. Christiansen.

From Fig. 15.4 we have

$$\frac{D}{4}\left(\frac{-dP}{dx}\right) = 0.45\ \frac{\text{lbf}}{\text{ft}^2} = 21.6\ \text{Pa}$$

$$\frac{-dP}{dx} = 0.45\ \frac{\text{lbf}}{\text{ft}^2} \cdot \frac{4}{1.049\ \text{ft}/12} \cdot \frac{\text{ft}^2}{144\ \text{in}^2} = 0.143\ \frac{\text{psi}}{\text{ft}} = 3.23\ \frac{\text{kPa}}{\text{m}} \quad \blacksquare$$

Example 15.2. We wish to double the flow rate in Example 15.1. How much must we increase the pressure gradient? How much would we have to increase it if the fluid were newtonian?

From Poiseuille's equation we know that if the fluid were newtonian (in laminar flow), we would have to double the pressure gradient to double the flow rate. From Fig. 15.4 we can measure the slope of the curve at $\frac{1}{4}[8V_{\text{av}}/(\pi D)] = 42.7$ and find that it is approximately 0.13; so we must multiply the pressure gradient by only $2^{0.13} = 1.094$ to double the flow rate. ■

15.4 TURBULENT FLOW OF NONNEWTONIAN FLUIDS IN PIPES

For turbulent flow of newtonian fluids in pipes, the experimental pressure-gradient data are represented by a friction factor–Reynolds number plot (Fig. 6.10). It seems logical to do the same for nonnewtonian fluids, but in so doing we must redefine the Reynolds number.

For newtonian fluids the viscosity is independent of the shear stress, so there is no ambiguity as to which value of the viscosity to use in the Reynolds number. However, for a nonnewtonian fluid the viscosity is a strong function of the shear stress; and from Eq. 6.3 we see that the shear stress decreases linearly with the distance from the wall, becoming zero at the tube center. Thus, there is no obvious choice of the correct viscosity to use in calculating the Reynolds number. Numerous theories and methods have been proposed for determining the critierion for laminar-turbulent transition and the proper Reynolds number to use in Fig. 6.10. The following method is the simplest and the most widely used.

Poiseuille's equation, Eq. 6.8, can be rewritten as Eq. 6.19:

$$\mathcal{R} = \frac{16}{f} \tag{6.19}$$

If we accept this as the definition of the laminar-flow Reynolds number, then for any constitutive equation which can be integrated twice to give the nonnewtonian equivalent of the Poiseuille equation, Eq. 15.9 can be used to define a working Reynolds number. For example, for power-law fluids (Eq. 15.7) this leads (Prob. 15.7) to

$$\mathcal{R} = \frac{8\rho V_{\text{av}}^{2-n} D^n}{K[2(3n+1)/n]^n} \tag{15.10}$$

Figure 15.5 shows a plot of the friction factor versus the Reynolds number as defined in Eq. 15.10. Because the Reynolds number has been defined by Eq. 15.10, the laminar-flow data must fall on the line shown. For flow at Reynolds numbers greater than 2000, two possible kinds of behavior are known. All slurries and many polymer solutions are represented by the solid curves in Fig 15.5. These do not seem to significantly suppress the turbulent behavior of the fluid. However, some polymer solutions and polymer melts, particularly those which show distinct viscoelastic behavior (such as rubber cement) obey the curves shown dotted at the right in Fig. 15.5. Visual observation [7] indicates that for these fluids the turbulence in the fluid is much less than it would be for a newtonian fluid at the same Reynolds number.

The decrease in the friction factor for polymer solutions compared with newtonian fluids can be quite startling. Dissolving as little as 5 ppm of some polymers in water produces a solution with only 60 percent of the friction factor of water at high Reynolds numbers [1, p. 88]. Such pressure-loss-reduction additives are in current large-scale industrial use [8, 9].

In this brief chapter we have not discussed elastic effects in fluid flow, which occur in many polymer melts and solutions of polymers. Some of these

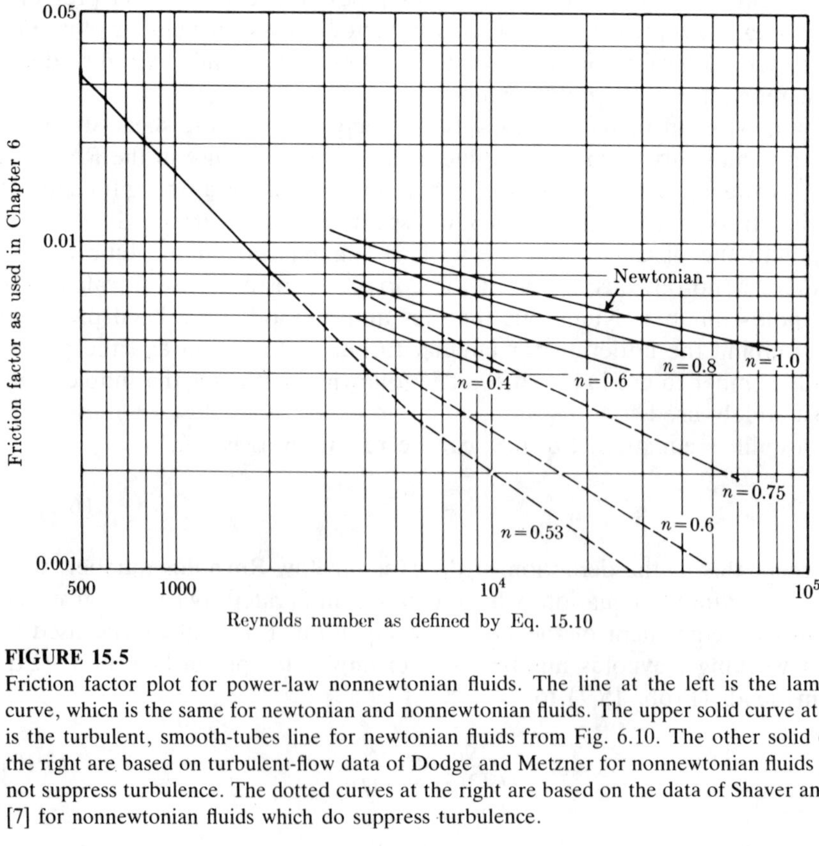

FIGURE 15.5
Friction factor plot for power-law nonnewtonian fluids. The line at the left is the laminar-flow curve, which is the same for newtonian and nonnewtonian fluids. The upper solid curve at the right is the turbulent, smooth-tubes line for newtonian fluids from Fig. 6.10. The other solid curves at the right are based on turbulent-flow data of Dodge and Metzner for nonnewtonian fluids which do not suppress turbulence. The dotted curves at the right are based on the data of Shaver and Merrill [7] for nonnewtonian fluids which do suppress turbulence.

effects are quite bizarre and startling; their theoretical explanation is one of the current major challenges in nonnewtonian fluid mechanics [1, chap. 2].

15.5 SUMMARY

1. Although intuitive explanations and numerous equations are available to describe the behavior of nonnewtonian fluids, no general, universally applicable theory or equation has been developed yet. For time-dependent and viscoelastic fluids, our knowledge consists mostly of descriptions of observed behavior.
2. For laminar flow of nonnewtonian fluids in circular pipes, we can readily calculate the behavior from pipe flow data in pipes of other sizes or from data from any kind of viscometer.
3. For turbulent flow the friction factors for nonnewtonian fluids are generally less than those for newtonian fluids. Some polymer solutions have surprisingly low friction factors.
4. The behavior of nonnewtonian fluids is currently a very active research topic. More detailed summaries of results to date can be found [1, 10].

PROBLEMS

See the Common Units and Values for Problems and Examples inside the back cover.

15.1. For pseudoplastic fluids (the most common types of nonnewtonian fluid) the fluid frequently appears to be a newtonian fluid with very high viscosity μ_0 at low shear rates and then again to be a newtonian fluid of lower viscosity μ_∞ at higher shear rates, with a transition between, as sketched in Fig. 15.6. Show that the Reiner-Phillipoff equation corresponds to this behavior, and show what constants in that equation corresponds to μ_0 and μ_∞.

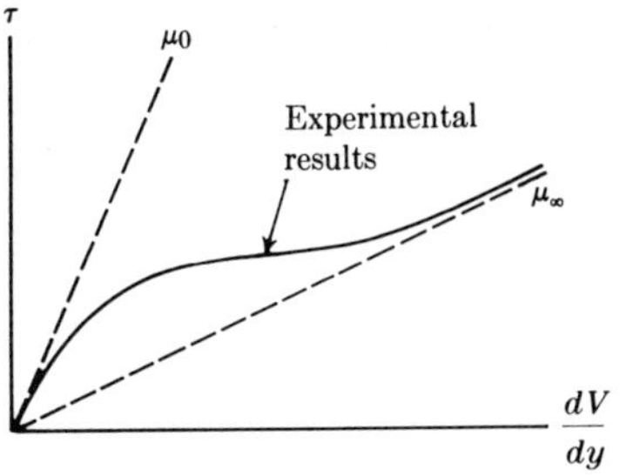

FIGURE 15.6

15.2. Show that the Ellis equation corresponds, practically, to a newtonian fluid at low shear rates and a power-law fluid at high shear rates. Show what constant or combination of constants in the Ellis equation corresponds to μ_0 in Fig. 15.6.

15.3. The data for 200 s in Fig. 15.2 can be reasonably represented by a power-law expression. Find the constants in that expression. *Hint:* The power law can be represented as a straight line with slope n on log paper.

15.4. Derive Eqs. 15.6 and 15.7. Also show that for a newtonian fluid ($n = 1$ and $K = \mu$) these become the same as Poiseuille's equation.

15.5. Show that for a newtonian fluid the term in brackets at the right in Eq. 15.9 is exactly equal to 1 and that for a power-law fluid it is exactly equal to $\frac{3}{4} + \frac{1}{4n}$.

15.6. We wish to represent the data in Fig. 15.4 by the power-law equation, Eq. 15.2. Select a set of constants which do this satisfactorily, taking into account the relation between wall shear rate and the dimensions of Fig. 15.4 by means of Eq. 15.9.

15.7. Derive Eq. 15.10. Note that many authors further substitute $K = K'[4n/(3n+1)]$, so that the final expression becomes $\mathscr{R} = D^n V_{\text{av}}^{2-n}/(8^{n-1}K')$.

15.8. In turbulent flow the pressure drop is proportional to the flow rate to the 1.8 to 2.0 power. This means that a plot of pressure drop versus flow rate on log paper should have a slope of 1.8 to 2.0. Is this observed in the turbulent parts of Fig. 15.4?

15.9. The upper right-hand end of the laminar-flow curve in Fig. 15.4 can be reasonably well approximated by $\tau = 0.04(\text{lbf} \cdot \text{s}^{0.4}/\text{ft}^2)(dV/dy)^{0.4}$. Using this approximation, calculate the value of the Reynolds number defined in Eq. 15.10 for the laminar-turbulent transition in each of the three pipe diameters. The density of the slurry may be estimated at 95 lbm/ft^3. Check the dimensions of the Reynolds number.

15.10. Sketch the equivalent of Fig. 15.4 for a newtonian fluid, including the laminar and turbulent regions. Indicate the slopes of the various lines.

15.11. In the viscometer shown in Fig. 15.1, in addition to the shear on the cylindrical walls, there is a shear stress experienced by the bottom of the cylindrical bob.

(*a*) Assuming that the fluid is newtonian, show the ratio of this torque to the torque on the walls, under the assumption that the distance from the bottom of the bob to the bottom of the cup is the same as the distance from the cylindrical wall of the bob to the cylindrical wall of the cup.

(*b*) On the basis of this result, comment on the best choice of length/diameter for the bob.

(*c*) How is the ratio of end torque to the cylindrical torque influenced by increasing the distance from the bottom of the bob to the bottom of the cup while all other dimensions are held constant?

(*d*) One way to eliminate this problem is shown in Fig. 15.7. Explain how this eliminates the problem.

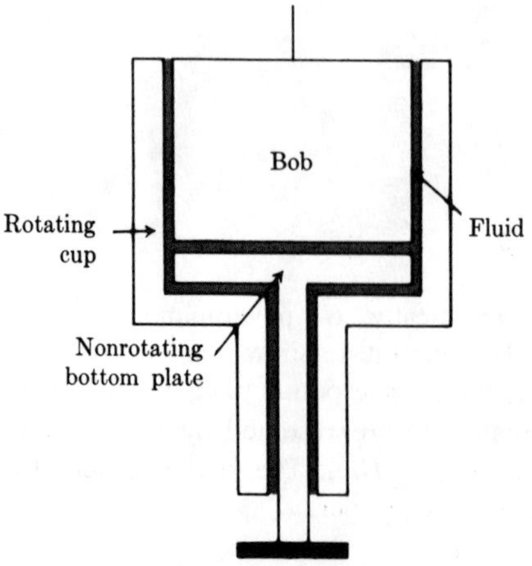

FIGURE 15.7

CHAPTER 16

TURBULENCE

The detailed behavior of turbulent flows is so complex that in spite of considerable research effort over the past 50 years we do not now possess a comprehensive theory of turbulence or a simple conceptual model of how it works in detail. Most of what we know consists of qualitative observations, measurements of various properties of turbulent flows, and some definitions and correlations of these measurements. This chapter is largely devoted to explaining what can be directly measured in turbulent flows and to exhibiting and discussing some of the experimental results.

Even though we cannot provide a detailed mathematical description of turbulent flow, we can provide several physical descriptions, which help the student form an intuitive picture of turbulent flow.

The first observation we can make is that turbulence dies out. If we violently stir a bowl of soup, the soup will have obvious turbulent eddies in it, but after a few minutes they will be gone and the soup will be motionless. This happens because of viscosity. Large tubulent eddies transfer energy to smaller ones, and then to even smaller ones, until the size of the eddy is small enough that viscosity slows and stops it. Thus the total store of turbulent kinetic energy which we put into the soup with a spoon is said to "decay" into internal energy by viscous friction heating. If we want the mass of fluid to have turbulence which does not die out, then we must continue to put in turbulent kinetic energy as fast as viscosity is causing that energy to decay to heat.

The most common way for turbulent kinetic energy to enter flow is by a shear layer. When we stir the soup, often we use a circular motion and induce a circular flow and/or circular eddies. This is also common in vessels with rotating mixers. But for flows in pipes, ducts, around airplanes or ships, or in the atmosphere, there is no such rotating element. Still we observe that such rotating eddies occur in flows like the flow between two plates, in which no solid surface is rotating. Figure 16.1(*a*) shows one plate sliding relative to another, with fluid in between. If, instead of fluid, the space between the plates had been filled with cylindrical rods aligned at right angles to the flow, the plates would set the rods in rotation, in the clockwise direction shown. So, too, this "shear flow" sets the fluid between the plates into rotational motion as shown. However, the whole fluid cannot form one rotating cell, so the fluid tends to form parallel threads of rotation, called *vortex threads*. If the viscosity is high enough or the velocity low enough, these rotational threads will be quickly stopped by viscosity. But for high velocity or low viscosity (actually for high Reynolds number), they will persist.

If the vortex threads were permanently attached to rigid masses of fluid, then they would behave as the solid cylindrical rods described above. But they are not; fluid flows through them, just as fluid flows through the isolated, stationary vortex that forms on draining a sink. The fluid in a vortex thread, being fluid, deforms in response to the forces acting on it. Each vortex thread influences the behavior of its neighbors, because at their boundaries vortex threads are generally moving in opposite directions. As a result, they twist and kink and split and divide. If they were visible, then a turbulent fluid flow would look like a wriggling mass of interlaced spaghetti, with many different diameters and lengths of spaghetti pieces present and with large ones constantly being formed, turning into smaller and smaller ones, and the smallest ones eventually disappearing.

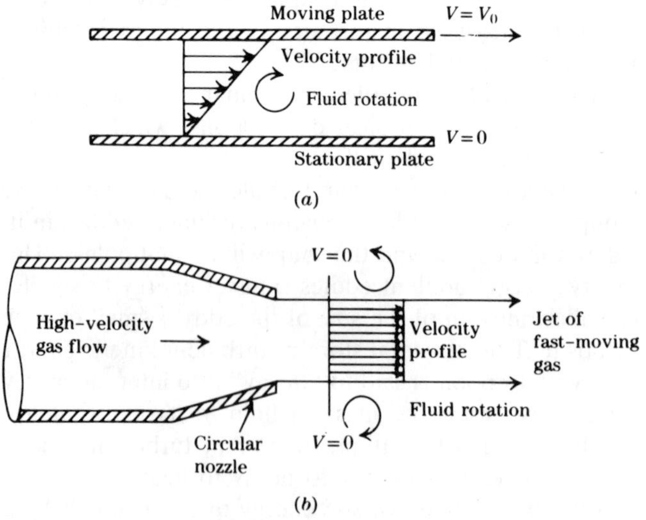

FIGURE 16.1
Fluid rotation produced by (*a*) flow between moving plates and (*b*) a free jet.

If the above intuitive picture is correct, then no turbulence can be formed except by large-scale mechanical stirring or by some kind of shearing, analogous to that shown in Fig. 16.1(*a*). This is indeed observed. The turbulence in pipes, ducts, and channels is produced by the shear layers at the walls of the duct. Turbulence is steadily fed into the main flow in the duct and ultimately is consumed in viscous heating in the main flow. The turbulence in the wake of ships and airplanes is caused by the shear layer adjacent to the surface of the airplane or ship, and the turbulence decays with time, due to viscosity, after the airplane or ship passes.

We can also have shear layers not involving solid surfaces, as shown in Fig. 16.1(*b*). The jet of hot gas leaving a jet engine is moving very rapidly, relative to the surrounding air; the boundary between the jet of hot gas and the surrounding air forms such a shear layer, which sets the surrounding air into rotation. If the jet is circular, under some conditions it can form circular vortex threads, which form "smoke rings." Similarly the hot smoke from a cigarette, rising in a still room due to buoyancy, sets up a shear layer between itself and the surrounding air.

The large eddies in the atmosphere which form major storms are formed not by shear layers, but by the rotation of the earth, interacting with north-south flows caused by solar heating of the tropics. They do not decay rapidly as long as they are over warm oceans where solar heat drives them; they decay rapidly over land.

Except for rotating systems like tropical storms or stirred soup bowls, if there is no such shear layer, then there can be no turbulence. You have probably observed, on calm days, that jet airliners leave a condensation trail high in the sky. The trail itself is wrinkled and convoluted, because of the turbulence put into it by the airplane's wake and the turbulence due to the interaction of the rapid jet exhaust and the stationary surrounding air. But after the plane passes, the condensation trail does not grow or disperse; it remains practically constant in size and in place. The reason is that such trails are only formed and visible in areas of the atmosphere high above the ground, which have practically no turbulence; if there were turbulence, the condensation trails would disappear. There is only turbulence in the atmosphere where it is caused by wind blowing over the surface or when vertical turbulence is caused by solar heating of the ground and the rising of heated air or the sinking of cooled air. For a flow of the atmosphere, well away from the ground, uninfluenced by rising heated air, there is negligible turbulence.

Closer to the ground, turbulence can be fed into the atmosphere by the shear layer at the ground or by hot air rising from the solar-heated ground, as in the example of the hot plume from the cigarette in a still room. You have probably observed that the plumes from smokestacks in the lower atmosphere continually expand as they flow in the downwind direction. This expansion is caused by the turbulence in the lower atmosphere. At night, when the ground is cold, there is very little turbulence in the lower atmosphere.

Turbulence is inherently three-dimensional. If there is a pressure-driven laminar flow between two parallel plates, all the velocity will be in one direction and all the velocity gradients will be in the direction perpendicular to the parallel plates. There is no velocity or velocity gradient in the direction perpendicular to these two directions. If we increase the Reynolds number enough for the flow to become turbulent, then we will observe fluctuating turbulent velocity components in all three directions.

In liquids or gases at atmospheric pressure, the smallest eddies, which are busily converting turbulent kinetic energy to internal energy by viscous heating, are thousands or millions of times as large as the space between molecules, or the mean free path. Thus treating turbulence as a phenomenon in a continuous phase and ignoring the existence of molecules and atoms do not introduce any serious error [1, p. 8].

The largest eddies in a flow generally will be as large as the boundaries of the system or the size of the disturbance causing the turbulence. So in turbulent flow in a pipe or duct there will be some eddies practically as large as the pipe diameter. In the wake of ships and airplanes, the largest eddies will be of the order of magnitude of the size of the ship or airplane. In the atmosphere, because of thermal stratification, there will generally be a layer adjacent to the ground, called the *mixing layer*, up to which turbulence initiated from the ground will reach; above this there will be substantially no turbulence. The largest eddies will be the size of the height of this mixing layer. (We have all seen days in which there are scattered clouds, all well above the ground and which all have the same elevation at their base. On these days the mixing layer reaches up to the cloud bases; below the cloud bases the air is mildly turbulent, and above it in the air layer containing the clouds there is very little turbulence.) On the scale of major storms in the atmosphere the flow is not turbulent, while on the scale of individual parts of the storms it generally is. Thus the huge eddy of a hurricane, seen from an earth satellite, has little flow and eddying across the main directions of flow, but the cloud edges are ragged, indicating local turbulence. In the oceans turbulence mostly enters through wind-driven waves at the surface and penetrates only a few wavelengths down. There is large-scale buoyant motion in the ocean, caused by solar heating and temperature and salinity differences. These and the large-scale ocean currents like the Gulf Stream can generate large-scale turbulence at their edges. Away from the surface and from the edges of these large-scale flows, there is very little turbulence in the oceans [2].

16.1 WHY STUDY TURBULENCE?

The principal goal of turbulence research is to place turbulent flows on as sound a footing as we now have for laminar flows. For laminar flows we can start with Newton's laws of motion (generally in the form of the Navier-Stokes equations in Sec. 7.9), and from a description of the flow boundaries and the fluid properties we deduce a complete description of the flow. This can be done analytically only for very simple flows, but with the advent of supercomputers it is possible to do it numerically for very complex flows. Furthermore, for laminar flows, if we can find the velocity distribution as described above, then we can generally describe the heat transfer and mass transfer in the flow by mathematical analysis without recourse to experimental measurement.

For turbulent flows we are not so fortunate. In general, we cannot calculate such velocity distributions or heat transfer and mass transfer from basic laws but must depend on experimental measurements. For widely used systems such as the flow inside a straight, cylindrical pipe or the perpendicular flow across the outside of a cylinder, enough experimental data are available, and these data have been correlated sufficiently well by the methods of dimensional analysis (Chap. 13), to allow us to predict the results of any given experiment with considerable accuracy. However, although this may be satisfactory for the practical engineer, it is unsettling to the theorist. Furthermore, when an extrapolation outside the range of experimental data is required or application is desired to a shape for which no experimental data are available, the need for a turbulence theory becomes clear.

Turbulence theories are not so far advanced that they allow us to extrapolate experimental data or to calculate flows around new shapes. Rather, turbulence specialists have concentrated on trying to reproduce the existing experimental data from some kind of comprehensive theory. This has not yet been accomplished. However, the partial results and partial understandings of turbulent flow have been useful in predicting the results of some experiments, e.g., turbulent boundary layers, as discussed in Sec. 11.5.

One of the first historical examples in which turbulence research proved very useful was in the comparison of wind tunnel tests with the corresponding results in free flight. The early experimental work in this field indicated that results from one wind tunnel did not necessarily agree with those from another wind tunnel or with the results for free flight. These differences ultimately were explained by the study of the differences in turbulence between various wind tunnels. Some of the first careful measurements of turbulence were made to explain these contradictory wind tunnel results [3].

Heat-transfer and mass-transfer studies look to fluid mechanics for an understanding of turbulence, because it would be very useful in those fields. Even our limited understanding of turbulence has already been of some value in those fields. The mixing of fluids play a significant part in many processes, e.g., the fuel-oxidizer mixing in all combustion processes, the mixing of reactants in most chemical reactors, and the blending of ingredients for foods, plastics, etc. If the fluids being mixed have low viscosities, then turbulence greatly speeds the mixing. Thus, a knowledge of the detailed structure of turbulence is a prerequisite of any scientific understanding of the mixing of low-viscosity fluids. Turbulence plays a major role in low-altitude meteorological processes [4]. Modern theories of astronomy on a galactic scale indicate that turbulence is very important in the evolution of galaxies [5].

Thus, although the study of turbulence is difficult and the large efforts expended have not resulted in general or comprehensive results, the potential benefits of such a thorough knowledge of turbulence are great enough to justify the effort.

16.2 TURBULENCE MEASUREMENTS

As discussed in Sec. 11.3, it is common in discussing turbulence to regard any velocity as consisting of an average component[†] and a fluctuating component:

$$V_x = \overline{V_x} + v_x \tag{16.1}$$

Here V_x is the instantaneous value of the x component of the velocity, $\overline{V_x}$ is the time average of this velocity over some reasonable period, and v_x is the instantaneous fluctuation of this velocity from its time-average value.

Here we must distinguish between time averages and position averages. A time average of some arbitrary function of time ϕ is given by

$$\bar{\phi} = \underset{t \to \infty}{\text{limit}} \frac{1}{t} \int_0^t \phi(t)\, dt \tag{16.2}$$

which is the average value of the reading of some kind of ϕ meter located at some fixed point. We can also define position averages as, e.g., the average velocity across the cross section of a pipe, as defined in Sec. 6.3. All average quantities in this chapter are time averages, as defined by Eq. 16.2.

Although the time average given above is shown as the limit as times goes to infinity, in practice it is only necessary to make measurements over a period which is long compared with the frequency of the fluctuations. Thus, for turbulence measurements in pipes, the average value found by Eq. 16.2 for t equal to a few seconds is numerically equal to that found for any longer period.

According to these definitions, the average value of v_x must be zero, because it is negative for as much of the time as it is positive.

The fluctuations in turbulent flow in pipes and channels are mostly so fast that ordinary fluid flow-measuring devices do not detect them at all; those devices record only the values associated with $\overline{V_x}$. For example, the pressures in turbulent pipe flow fluctuate with a very high frequency. However, ordinary pressure gauges do not respond to such high frequencies, so they show a steady average pressure for such a flow. Similarly, the velocities indicated by venturi meters, orifice meters, pitot tubes, etc., for turbulent pipe flows generally do

[†] Throughout the turbulence literature (and this chapter) a bar over a symbol indicates the average value of that quantity (most often a time average, but not always). This is not to be confused with the bar over a symbol indicating a vector quantity. In turbulence literature, vectors are normally indicated by boldface symbols.

The division into average and fluctuating components can be made not only for velocities and velocity components, but also for pressure, temperature, etc.

not show the fluctuating component at all; their response is simply too slow. Therefore, a pitot tube in a turbulent pipe flow simply reads $\overline{V_x}$ (subject to slight corrections due to turbulent fluctuations).

To measure v_x, we need a flowmeter, which is much more responsive to rapid changes in the flow rate. The most successful flowmeter and the one which has produced most of the world's measurements of turbulence is the hot-wire anemometer, shown in Fig. 16.2. In such a device, a fluid flows over a very thin electrically heated wire, whose temperature is much higher than that of the fluid. The wire is generally made of platinum or tungsten, both metals which show a significant increase in electric resistance with increasing temperature. By suitable electric circuitry we can measure the fluctuating resistance (and hence the fluctuating temperature) at constant heat input or the fluctuating heat input required to hold the wire's temperature constant.

Over a wide range of flow rates, the heat removed from a hot wire by a fluid flowing at right angles to it is described by

$$\text{Heat removal rate} = A + B(\text{fluid velocity})^{1/2} \tag{16.3}$$

where A and B are experimental constants. Thus, after suitable calibration we can use a hot-wire anemometer to measure fluid velocities. Furthermore, it can be shown experimentally that because the wire is very small (the best are about 0.005 mm in diameter, or one-tenth of the average diameter of a human hair), it can respond very rapidly to changing flow rates. Some hot-wire anemometers can follow velocity fluctuations that are as fast as 50,000 cycles per second, or hertz (Hz).

The hot-wire anemometer shown in Fig. 16.2 is very sensitive to flow perpendicular to the wire but much less sensitive to flow in the direction parallel to the wire (this flow is much less effective for heat transfer). Thus, by using arrays of hot-wire anemometers placed at different angles to the flow, we can determine both the fluid velocity and its direction.

Once the basic scheme for hot-wire anemometry was worked out, numerous variants on it were developed [6]. Generally, hot-wire anemometers are delicate, temperamental instruments which require expensive electronic circuitry and considerable care and training to use.

One alternative procedure for studying turbulence is to inject some tracer into a turbulent flow and to measure its concentration distribution at some point downstream in the flow. (This was the method used by Reynolds in his

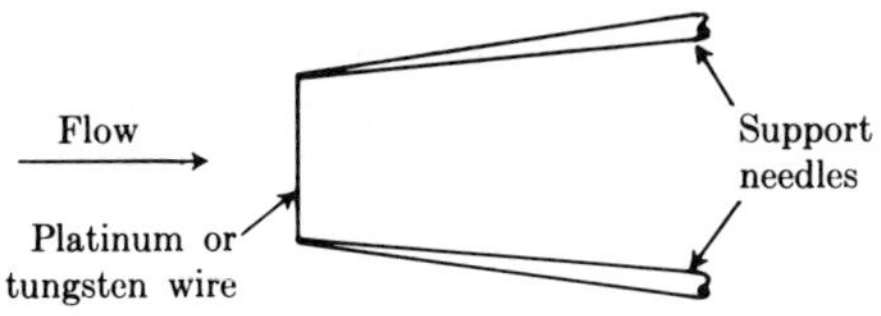

FIGURE 16.2
Hot-wire anemometer. The wire is usually about 0.005 mm in diameter and about 1 mm long.

pioneering exploration of turbulence, described in Sec. 6.2.) As a tracer we can use a dye, which is injected at a point and then photographed downstream, or an electrically conducting solution (e.g., salt solution in water), whose presence is detected downstream by electric conductivity probes. Both methods have the drawback that the "tracer" fluid they introduce may disturb the flow. Another approach is to suspend in the fluid small particles of the same density as that of the fluid and then to record their trajectories photographically or by laser-Doppler measurements. The latter method is currently a strong competitor to hot-wire anemometry [7].

16.3 FREE AND CONFINED TURBULENT FLOWS

The turbulent flow in a pipe is quite different in character from the turbulent flow in a wind tunnel or in the lower atmosphere. In the atmosphere or in the central section of a wind tunnel, the nearest wall is so far away that it has little influence on the flow. This kind of flow, substantially uninfluenced by nearby walls, is called *free turbulence*. In a typical long pipe, the flow is strongly influenced by the nearby presence of a wall. This kind of flow is often called *shear turbulence* or *wall turbulence*.

In the lower atmosphere or in a wind tunnel, the turbulence at any point generally has the same properties in all directions; this is called *isotropic turbulence*.[†] Turbulence in pipes is generally not isotropic.

In a pipe the turbulence character does not change as one moves downstream at a constant radial position. This type of turbulence is called *homogeneous turbulence*. Normally this is not the case in the atmosphere or in a wind tunnel, where the turbulence, although isotropic at a point, tends to become less intense with distance downstream from the source of the disturbance.

16.4 TURBULENT KINETIC ENERGY

A fluid flowing in turbulent flow possesses more kinetic energy per unit mass than the same fluid moving at the same average velocity in nonturbulent flow. Furthermore, the more intense the turbulence, the greater the amount of "turbulent kinetic energy."

The turbulent kinetic energy is fed into the flow by some kind of external work, generally a pressure gradient (i.e., injection work) in a pipe or a fan or a blower in a wind tunnel or the sun-driven work of thermal convection currents in the lower atmosphere. This work may first cause an increase in kinetic energy in the flow direction or merely a gradient in the velocity which will lead

† The same term is used with an analogous meaning in the study of matter. Water is isotropic; it has the same properties in all directions. Wood is nonisotropic; it has different strengths with the grain and across the grain.

to turbulence. The turbulent kinetic energy leaves the flow by viscous conversion to internal energy.

In a wind tunnel the turbulent kinetic energy is put in by the blower or an inlet screen, downstream from a blower, and no further turbulent kinetic energy is put into the flow as it moves through the tunnel (because the tunnel is so short and wide that the walls have little effect in the center where the tests are made). Thus, the amount of turbulence decreases with distance from the blower in a wind tunnel; the turbulence is said to "decay." Similarly, if we measure the turbulence in the wake of a ship or airplane, we find that the turbulence is strong close to the ship or airplane but decays farther from the ship or plane, eventually vanishing far enough downstream. In the latter case, the turbulent energy is supplied by the engines, which drive the ship or airplane, overcoming drag.

At the other extreme in "fully developed" turbulent flow well downstream from the entrance of a pipe, turbulent kinetic energy is steadily being fed into the flow by the work being expended, overcoming the shear (frictional) resistance at the pipe walls. This kinetic energy is being steadily converted to internal energy by viscous friction in the turbulent flow, so that the rate of addition of turbulent kinetic energy to the flow exactly balances the rate of destruction of kinetic energy by viscous friction, and the amount present stays constant as the fluid moves down the pipe. Thus, the turbulence in a pipe does not decay with distance, as does the turbulence in a wind tunnel or the wake behind a ship or plane.

16.5 EXPERIMENTAL AND MATHEMATICAL DESCRIPTIONS OF TURBULENT FLOWS

The ultimate description we would like to have of turbulent flow would be an explicit expression for $\overline{V_x}$, $\overline{V_y}$, $\overline{V_z}$ and ν_x, ν_y, and ν_z as functions of time and position. Then we could predict the average and the fluctuating velocities at any point and any time. Currently it seems impossible to make such a description; the problem is much too complex. The next best thing is a statistical description of the flow, i.e., what fraction of the time V, ν_x, ν_y, etc., have certain values. So far most of the experimental and theoretical work done on turbulence has been directed at these statistical properties of the flow. Below we give a set of definitions which are widely used in the turbulence literature to describe such statistical properties of the flow and some experimental values of the quantities so defined.

A. Turbulent Intensity

Turbulent intensity, or simply *intensity*, is a measure of how strong, violent, or intense the turbulence is. (In the older literature it is often called *level of turbulence* or *degree of turbulence*.) Turbulent intensity is defined by

$$x - \text{turbulent intensity} = \left(\overline{\nu_x^2}\right)^{1/2} \tag{16.4}$$

Here $\overline{(v_x^2)}^{1/2}$ is the root mean square (rms) of the x component of the fluctuating component of the velocity. The relative intensity is defined by[†]

$$x - \text{relative intensity} = T_x = \frac{\overline{(v_x^2)}^{1/2}}{\bar{V}} \tag{16.5}$$

Here $\bar{V}$ is the average of the absolute magnitude of the vector velocity, equal to $(\overline{V_x^2 + V_y^2 + V_z^2})^{1/2}$. We may define the y and z components of the intensity by replacing the subscript x in Eq. 16.4 with a subscript y or z. If Eq. 16.5 defines the x component of the relative intensity, then the entire relative intensity must be given by

$$T = \left[\frac{1}{3}\left(\frac{\overline{(v_x^2)}}{\overline{V^2}} + \frac{\overline{(v_y^2)}}{\overline{V^2}} + \frac{\overline{(v_z^2)}}{\overline{V^2}}\right)\right]^{1/2} = \left[\frac{1}{3}(T_x^2 + T_y^2 + T_z^2)\right]^{1/2} \tag{16.6}$$

As discussed in Sec. 16.2, the average value of v_x is zero, because it is positive as often as it is negative. However, $\overline{(v_x^2)}$, the rms value of v_x, is not zero, because squaring before averaging removes the minus signs.

Example 16.1. A turbulent flow is described by the equations[‡]

$$V_x = 10\ \text{ft/s} + (1\ \text{ft/s})(\sin t) \qquad V_y = V_z = 0$$

Calculate $\overline{V_x}$, v_x, $\overline{v_x}$, $\overline{(v_x^2)}^{1/2}$, and T_x. By inspection,

$$\overline{V_x} = 10\ \text{ft/s} \qquad v_x = (1\ \text{ft/s})(\sin t)$$

$$\overline{v_x} = \lim_{t\to\infty} \frac{1}{t}\int_0^t 1\,\frac{\text{ft}}{\text{s}} \cdot \sin t\, dt = 1\,\frac{\text{ft}}{\text{s}} \lim_{t\to\infty}\left(-\frac{\cos t}{t}\right) = 0$$

$$\overline{(v_x^2)} = \lim_{t\to\infty} \frac{1}{t}\int_0^t \left(\frac{1\ \text{ft}}{\text{s}}\right)^2 (\sin t)^2\, dt$$

$$= \left(\frac{1\ \text{ft}}{\text{s}}\right)^2 \lim_{t\to\infty} \frac{1}{t}\left(-\frac{1}{2}\cos t \sin t + \frac{1}{2}t\right) = \frac{1}{2}\,\frac{\text{ft}^2}{\text{s}^2}$$

$$\overline{(v_x^2)}^{1/2} = (\tfrac{1}{2}\ \text{ft}^2/\text{s}^2)^{1/2} = 0.707\ \text{ft/s} = 0.22\ \text{m/s}$$

$$\bar{V} = (\overline{V_x^2} + 0 + 0)^{1/2} = \overline{V_x} = 10\ \text{ft/s} = 3.05\ \text{m/s}$$

$$T_x = \frac{0.707\ \text{ft/s}}{10\ \text{ft/s}} = 0.0707$$

■

[†] Some writers call T_x, which we call the relative intensity, simply the intensity; others call it the absolute intensity.

[‡] No real turbulent flow can be described by equations this simple. These equations serve only to show the relations between the various defined quantities for turbulent flow.

By suitable electronic circuitry, a hot-wire anemometer can be made to read directly the rms fluctuating velocity. Thus, with suitable equipment we can directly read the turbulent intensity in the direction perpendicular to the anemometer wire. A typical set of experimental results for turbulent-intensity measurements in a confined channel are shown in Fig. 16.3. From this figure we see that (1) the intensity varies with location and is generally greatest in a layer near the wall, (2) the intensity falls off to zero right at the wall (this must be, because the solid boundary stops any motion at right angles to the flow direction, and there is zero velocity in the flow direction right at the wall), (3) the maximum observed relative intensity T_x is about 0.23 (measured relative intensities seldom exceed this value), (4) the intensity in the flow direction is greater than the intensity in the direction at right angles to the flow, and (5) near the center of the channel the intensities in the x and y directions are approaching each other (for a very large channel they become practically equal at the center, so that the turbulence is practically isotropic).

The measurements shown in Fig. 16.3 are typical of those made in a channel where no effort has been made to keep the turbulence level low. For wind tunnels we desire as low a turbulent intensity as possible; good ones have $T \approx 0.0005$.

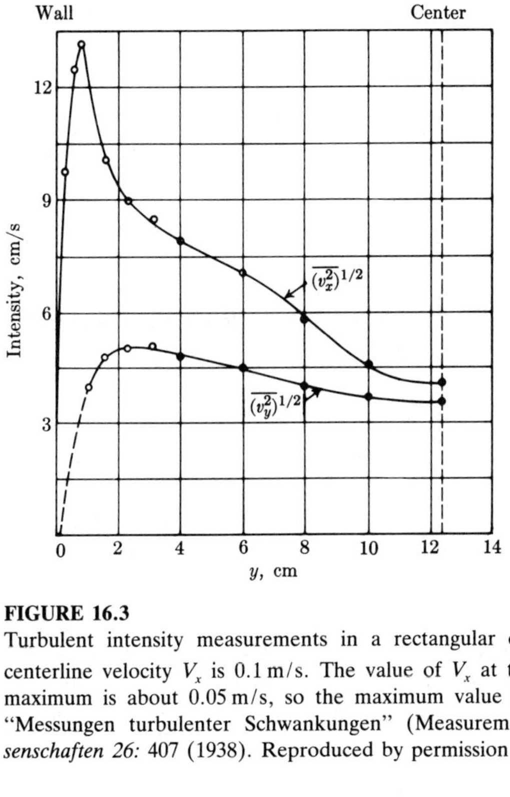

FIGURE 16.3
Turbulent intensity measurements in a rectangular channel 1 m wide and 0.24 m high. The centerline velocity V_x is 0.1 m/s. The value of V_x at the point near the wall where $(\overline{v_x^2})^{1/2}$ is a maximum is about 0.05 m/s, so the maximum value of T is about 0.23. [From H. Reichardt, "Messungen turbulenter Schwankungen" (Measurements of turbulent oscillations), *Naturwissenschaften 26:* 407 (1938). Reproduced by permission of the publisher.]

B. Correlation Coefficient and Scale

The intensity is a measure of "how much" turbulence is present relative to the average flow. However, this amount of turbulence could be made up of many small eddies going back and forth very frequently or of a few large eddies going back and forth less often. Therefore, to characterize the turbulence, we need some measure of the average size of an eddy. In advance, we may say that for pipe flows the average eddy is smaller than the pipe diameter and is usually measured in fractions of an inch. Eddies in the lower atmosphere are usually measured in tens of feet. The giant eddy of a hurricane is up to 100 mi across. For galactic eddies the average size is measured in light-years.

Unfortunately, there is no very simple measurement we can make of the size of an individual eddy, nor is there even a very direct measurement of the average size of the eddies passing a given point. Therefore, two terms (the correlation coefficient and the scale of turbulence) have been defined as expressions of the average size of eddies. The *correlation coefficient* (borrowed from statistics, where it is widely used) is a measure of how much of the time two variables coincide with each other. The correlation coefficient of two arbitrary functions of time $\phi_1(t)$ and $\phi_2(t)$ is

$$\text{Correlation coefficient } R = \frac{\overline{\phi_1 \phi_2}}{\left(\overline{\phi_1^2}\right)^{1/2} \left(\overline{\phi_2^2}\right)^{1/2}} \tag{16.7}$$

This is the average of the product of the two functions, divided by the product of their rms averages.

Example 16.2. Determine the correlation coefficients for the following sets of functions: (*a*) $\phi_1(t) = t$ and $\phi_2(t) = t$ and (*b*) $\phi_1(t) = \sin t$ and $\phi_2(t) = \cos t$.
(*a*) We have

$$R = \frac{\overline{t^2}}{\left(\overline{t^2}\right)^{1/2} \left(\overline{t^2}\right)^{1/2}}$$

Taking the averages as shown by Eq. 16.2, we see that R equal 1. Thus, for two functions which are the same, R equals 1.
(*b*) We have

$$R = \frac{\overline{\sin t \cos t}}{\left(\overline{\sin^2 t}\right)^{1/2} \left(\overline{\cos^2 t}\right)^{1/2}}$$

Here the numerator is zero (Prob. 16.2), so R equals 0. Thus, for two functions which are 90° out of phase with each other, the correlation coefficient is zero. ■

The correlation coefficient can take values from +1 to −1. Here it is illustrated for simple analytic functions, where its value is obvious. In the study of turbulence it is generally applied to randomly fluctuating variables. It can be shown that if $\phi_1(t)$ and $\phi_2(t)$ are any two randomly fluctuating variables whose average values are zero and whose instantaneous values are not related in any way, then their correlation coefficient is zero.

Returning now to the problem of defining the size of an eddy, let us consider the correlation coefficient between the values of ν_x at two points:

$$R_{\nu_2} = \frac{\overline{\nu_{x_1}\nu_{x_2}}}{(\nu_{x_1}^2)^{1/2}(\nu_{x_2}^2)^{1/2}} \tag{16.8}$$

Here the ν_x's are the fluctuating x components of the velocity at two points. Each is some function of time. If the two points are the same point, then $\nu_{x_1} = \nu_{x_2}$ and Eq. 16.7 will lead to $R_{\nu_x} = 1$. But if the two points are very far away from each other, then we should expect no systematic relation between the two ν_x's; we should expect R_{ν_x} to be zero. In the intermediate case we expect that if the two points are separated by a distance much smaller than the size of an average eddy, then both points generally will be within the same eddy and hence R_{ν_x} will be nearly 1. As the distance between them increases, we should expect R_{ν_x} to get smaller and smaller, finally reaching zero when the distance between the points is larger than the size of the largest eddy. To measure the correlation coefficient, we place one hot-wire anemometer at a fixed place in a turbulent flow and then move another hot-wire anemometer from place to place in the flow. By suitable electronic circuitry it is possible to measure both the individual rms fluctuating components and the average value of the product of the two fluctuating components. From these measures we can find the correlation coefficient for any distance between the two anemometers.

Figure 16.4 shows a typical set of experimental measurements of the correlation coefficient in a flow in a pipe. The data in the figure were taken at

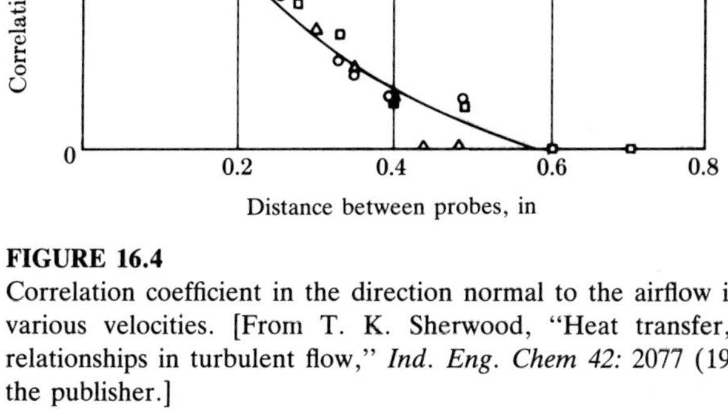

FIGURE 16.4

Correlation coefficient in the direction normal to the airflow in a pipe of 4-in inside diameter at various velocities. [From T. K. Sherwood, "Heat transfer, mass transfer and fluid friction relationships in turbulent flow," *Ind. Eng. Chem 42:* 2077 (1950). Reproduced by permission of the publisher.]

three different velocities, which varied by a factor of almost 3; the resulting correlation coefficients are substantially the same. This is typical of such measurements; for the same geometry the correlation coefficient is practically independent of fluid velocity.

Figure 16.4 is typical of correlation coefficient measurements in a pipe. In free turbulence one often measures a curve as sketched in Fig. 16.5. The negative correlation coefficient shown in this figure is not an experimental error. Rather it reflects the fact that in free turbulence generally two adjacent eddies will, on average, be moving in opposite directions. Thus, when the probes are placed far enough apart to be normally present in two adjacent eddies, the correlation coefficient will be negative.

From the experimental results in Fig. 16.4 we may infer the size of the eddies. Observationally it is clear that not all eddies are the same size. Nonetheless, it is convenient to have a numerical length for an "average" eddy rather than a cumbersome plot like Fig. 16.4. There are numerous ways we could decide on an appropriate length from a figure like Fig. 16.4; the one which has been most widely accepted is the *scale of turbulence*, defined by

$$\text{Scale } L = \int_0^\infty R(y)\, dy \tag{16.9}$$

This is simply the area between the curve and the x axis on Fig. 16.4. By graphical integration the scale of the turbulence represented on Fig. 16.4 is 0.22 in = 5.6 mm. This is a reasonable estimate of the size of an "average eddy" in this particular set of flows.

Although the correlation coefficient shown here is the most commonly used in the turbulence literature, Eq. 16.7, which defines the correlation coefficient, can be applied to various combinations of functions. Thus, we sometimes see references to the correlation coefficient between v_x and v_y at some point or to the correlation between v_x at time t and v_x at time $t + \Delta t$. The latter is often referred to as an *autocorrelation*.

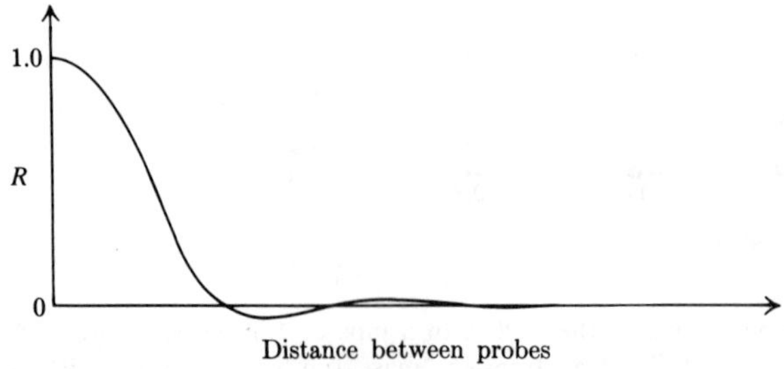

FIGURE 16.5
Typical correlation coefficient measurement for free turbulence.

C. Spectrum

Another measurable physical property of a turbulent flow is the distribution of frequencies of turbulent oscillations. Via suitable electronic filters it is possible to separate the output signal from a hot-wire anemometer into various frequency ranges.

If n is the frequency of oscillation in hertz (cycles per second) and we record first the value of $\overline{\nu_x^2}$ for the entire range of frequencies and then the value of $\Delta\overline{\nu_x^2}$ for some frequency range Δn, we can form the ratio

$$f(n) = \frac{1}{\overline{\nu_x^2}} \frac{\Delta\overline{\nu_x^2}}{\Delta n} \tag{16.10}$$

In experimental practice we must always use a finite value of Δn, but in principle we can take the limit as Δn approaches zero, finding

$$f(n) = \frac{1}{\overline{\nu_x^2}} \frac{d\overline{\nu_x^2}}{dn} \tag{16.11}$$

This $f(n)$ is thus the fraction of the total value of $\overline{\nu_x^2}$ per hertz. Here $\overline{\nu_x^2}$ is twice the x component of the turbulent kinetic energy per unit mass, so that this fraction (Eq. 16.11) is really the fraction of the x component of the turbulent kinetic energy per hertz (cycle per second). A typical experimental measurement of this fraction (of the total turbulent kinetic energy, rather than of the x component) is shown in Fig. 16.6, which also shows the function

$$F(n) = \int_{n=0}^{n=n} f(n)\, dn = \frac{1}{\overline{\nu_x^2}} \int_{n=0}^{n=n} \frac{d\overline{\nu_x^2}}{dn}\, dn \tag{16.12}$$

This function, which is simply the area under the $f(n)$ curve up to a given value of n, shows what fraction of the turbulent kinetic energy is contained in oscillations of lower frequency. By definition, the value of $F(n)$ must approach 1 as n becomes very large.

A curve like Fig. 16.6 is called a *turbulent kinetic-energy spectrum*, analogous to the energy spectra of light which appear in texts on optics. From Fig. 16.6 it is apparent that for this flow one-half of the turbulent kinetic energy is contained in velocity fluctuations which have frequencies between 0 and 2 Hz, and 90 percent of the turbulent kinetic energy is contained in fluctuations having frequencies between 0 and 8 Hz. As a general rule, there are more small (high-frequency) eddies than large (low-frequency) eddies, but most of the turbulent kinetic energy is in the large eddies.

Experimentally it is observed [8] that at higher overall velocities the spectrum curve is shifted to higher frequencies. Thus, for high-velocity flow in a pipe we should expect more of the turbulent kinetic energy to be contained in oscillations of higher frequency, whereas in low-velocity flow of the lower atmosphere we should expect more of the turbulent kinetic energy to be contained in oscillations of lower frequency.

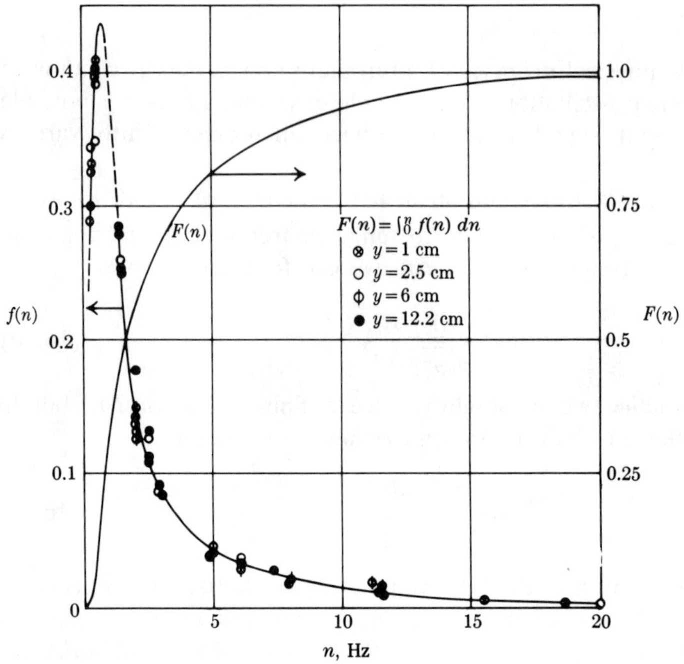

FIGURE 16.6
Spectrum of kinetic energy for the system shown in Fig. 16.3. [From H. Motzfeld, "Frequenzanalyse turbulenter Schwankungen" (Frequency analysis of turbulent oscillations), *Zeitschrift fuer Angewandte Mathematik und Mechanik 18*: 362–365 (1938). Reproduced by permission of the publisher.]

It can also be shown [8] that there exists a relatively simple mathematical relationship between the spectrum of turbulence and the correlation coefficient, so that a detailed, accurate measurement of one enables the calculation of the other.

16.6 REYNOLDS STRESSES

In Sec. 6.4 we discussed how turbulent fluctuations lead to shear stresses in addition to those due to simple viscous shear. These additional stresses are called *Reynolds stresses*. It can be shown mathematically [9, p. 559] that the Reynolds stress components for a general three-dimensional flow (see Sec. 7.7) are given by

$$\sigma_{xx} = -\rho\overline{v_x^2} \tag{16.13}$$

$$\tau_{xy} = -\rho\overline{v_x v_y} \tag{16.14}$$

with analogous equations for τ_{xy} and τ_{yz}. Using these Reynolds stress components, we can form the equivalent of Eq. 7.74 for turbulent flow.

The most interesting of the Reynolds stresses are the shear stresses. From Eq. 16.14 we see that these require the fluctuations to be in two directions. If there is no velocity gradient and the turbulence is isotropic, then in general there will be no such correlation, so that the Reynolds shearing stress will be zero. But, if there is a velocity gradient (as exists in any flow near a wall or at the boundary of a free jet), then, as shown below, there is always such a correlation and there will be a Reynolds stress.

Figure 16.7 shows a representation of a flow near a solid wall. The value of $\overline{V_x}$ is shown; $\overline{V_y}$ is 0. Now consider a small mass of fluid which is carried by an eddy upward from A to B. This mass of fluid moves to a larger y over some finite time, so it must have a positive y velocity. However, $\overline{V_y}$ is zero, so for this eddy ν_y is positive. Before this mass of fluid began to move in the y direction, its x velocity was V_A. When it arrives at B, it probably has not had time to change its x velocity, so it still has an x velocity V_A. This is less than the average x velocity at that point (V_B), so this represents a negative x fluctuation of the velocity at y or a negative ν_x.

Now some time later a mass of fluid from C is brought to B by an eddy. By similar arguments, ν_y is negative and ν_x is positive. Thus, for both kinds of fluctuations $\overline{\nu_x \nu_y}$ is negative, and there is, indeed, a correlation between the two velocity fluctuations at right angles to each other and, hence, a significant Reynolds stress.

The magnitude and character of these Reynolds stresses can be visualized through the concept of the eddy viscosity, first introduced by Boussinesq [1, p. 23]. He suggested that we retain the form of Newton's law of viscosity

$$\tau = \mu \frac{dV_x}{dy} \tag{1.5}$$

which holds only for laminar flow and make it fit the experimental turbulent-flow data by introducing a new quantity, called the *eddy viscosity* ε,[†]

FIGURE 16.7
Diagram showing how Reynolds stresses arise in a turbulent flow with a velocity gradient.

[†] The eddy viscosity ε is really an eddy *kinematic* viscosity, but its common name is simply *eddy viscosity*.

$$\tau = (\mu + \rho\varepsilon)\frac{dV_x}{dy} \tag{16.15}$$

To avoid confusion in discussions involving the eddy viscosity, the "ordinary" viscosity is often called the *molecular viscosity*.

This definition of ε has the disadvantage that ε is not a simple property of the fluid, such as the molecular viscosity, but is also a function of the flow rate and position in the flow. It has the advantage that it lets us easily formulate the ratio of the Reynolds stresses to viscous stresses. In addition, in calculations of heat and mass transfer, we may introduce a similar eddy thermal conductivity and eddy diffusivity. Under some circumstances these three eddy properties are identical, and under all circumstances they are at least of the same order of nagnitude. So this approach helps to apply fluid flow data to the solution of problems in heat and mass transfer.†

From a measured average velocity profile ($\overline{V}_x$ versus y or r) in a flow and information on the viscosity and density of the fluid, we can calculate the eddy viscosity for any point in the flow. A typical plot of velocity versus position is shown in Fig. 11.7. Figure 16.8 shows the eddy viscosity divided by the kinematic viscosity for flow in smooth pipes, calculated from a figure like Fig. 11.7.

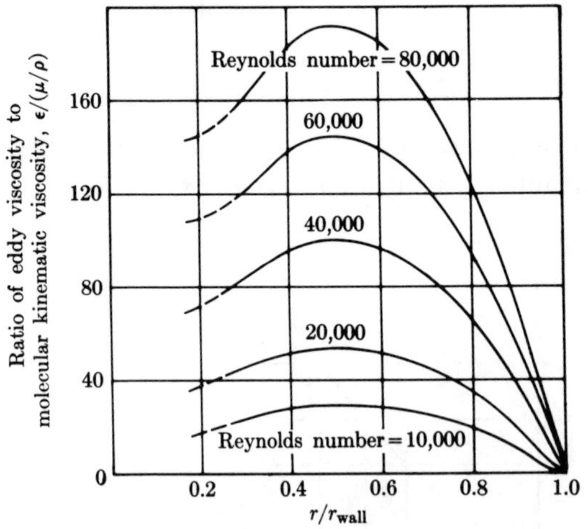

FIGURE 16.8
Calculated ratio of eddy to kinematic viscosity for fully developed flow in smooth pipes. [From W. H. Corcoran and B. H. Sage, "Role of eddy conductivity in thermal transport," *AIChEJ.*, *2*, 251–258 (1956). Reproduced by permission of the publisher.]

† This approach to solving problems in heat and mass transfer through fluid mechanics is discussed in various texts under the name "Reynolds analogy."

From Fig. 16.8 we see that in ordinary pipe flow for regions away from the wall the eddy viscosity is typically about 100 times the molecular viscosity (i.e., the Reynolds stresses are about 100 times the stresses due to molecular viscosity), that the eddy viscosity is a strong function of position and Reynolds number, and that it is difficult to calculate values of the eddy viscosity near the center of the pipe. From Eq. 16.15 we see that the sum of the eddy and molecular viscosities is equal to $\tau/(d\overline{V}_x/dy)$; at the center of the pipe both quantities are zero. To obtain the correct limit in this ratio as both numerator and denominator approach zero requires more precise experimental measurements of $\overline{V}_x$ and y than are currently available. We may infer from Fig. 16.8 that in this type of pipe flow the heat transfer and mixing will be of the order of 100 times the heat transfer and mixing due to molecular thermal conductivity and molecular diffusion.

16.7 TURBULENCE THEORIES

Most of the best known theoretical fluid mechanic experts of the twentieth century have attempted to deduce a comprehensive theory of turbulence. The resulting theories each provide some insight into the relations that must exist between various quantities in turbulent flow, but all contain undefined constants which must be measured experimentally to make the theory fit the observations. The theories of Prandtl and von Karman are well summarized by Schlichting [9, chap. 19]. That of G. I. Taylor is summarized by Dryden [10]. The theories of Kolmogoroff are discussed by Hinze [1] and Corrsin [5]. The problem of calculating the velocity distribution in turbulent flow in a pipe from the various theories is discussed by Bird et al. [11].

Many of these theories are quite complex mathematically. The more involved mathematics led one fluid mechanics expert to comment that these theories "confirm one's suspicion that the aim of the statistical theory of turbulence is full employment for mathematicians" [12].

16.8 SUMMARY

1. In analyzing and measuring turbulence, it is customary to divide the flow conceptually into steady and fluctuating components.
2. Historically, most turbulence measurements have been made with hot-wire anemometers. In recent years laser-Doppler anemometers have been widely used as well.
3. Free turbulence has a different experimental character from that of turbulence near a solid wall or a free jet.
4. The readily measured experimental properties of a turbulent flow are the time-average velocity and the intensity, scale, and spectrum of the turbulence.
5. Reynolds shear stresses arise out of the correlation of turbulent fluctuations

in two directions. These are rare in free turbulence but are almost always present in turbulent flow near a wall or the edge of a free jet.

PROBLEMS

See the Common Values for Problems and Examples inside the back cover.

16.1. Select two functions $\phi_1(t)$ and $\phi_2(t)$ whose correlation coefficient is -1.

16.2. Show that $\overline{\sin t \cos t} = 0$.

16.3. Show the relationship between the kinetic energy of a stream with velocity $V = \bar{V} + v$ and the kinetic energy of a stream with velocity $V = \bar{V}$.

16.4. From Fig. 11.7 estimate the value of $\varepsilon/(\mu/\rho)$ at a Reynolds number of 80,000 and r/r_{wall} of 0.6. Compare your calculation with the result shown in Fig. 16.7.

CHAPTER 17

SURFACE FORCES

The introduction to this topic begins with Sec. 1.5E. There we discussed how surface forces arise. Such forces are present in any system in which there is a two-phase interface, i.e., solid-liquid, solid-gas, liquid-gas, or liquid-liquid. Thus, they are present in all the examples treated so far in this text, but in those examples they are generally small and can be neglected without measurable error. But they must be taken into account in very small systems or in systems in which other forces are small or zero. To see why surface forces are important in small systems, consider the pressure difference due to surface tension in a spherical drop or bubble.

Figure 17.1 shows such a drop, cut in half along its equator. If the system is at rest, then the surface force, which tends to shrink the bubble, is equal and opposite to the pressure-difference force, which tends to expand it. Here $\Delta P = P_{\text{inside}} - P_{\text{outside}}$. Equating these two forces and solving for ΔP, we find

$$\Delta P = \frac{4\sigma}{D} \tag{17.1}$$

As long as D is large, this pressure difference is negligible. For example, if this is a drop of water at 68°F with a 1-in diameter, then

$$\Delta P = \frac{4 \cdot 0.00043 \text{ lbf/in}}{1 \text{ in}} = 0.00172 \text{ psi} = 11.9 \text{ Pa}$$

which is negligible for most engineering applications. But if the drop has a diameter of 10^{-5} in, then the pressure difference is 172 psi = 1.19 MPa.

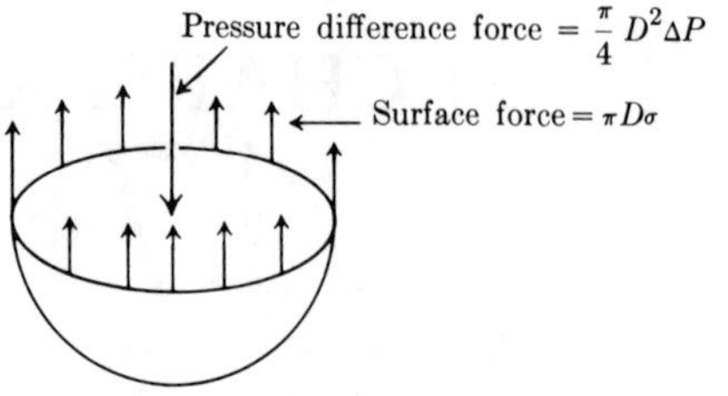

FIGURE 17.1
Pressure difference due to surface forces

We may appreciate why surface forces become important in small systems by observing that the surface forces are proportional to the diameter of the body, whereas the pressure forces are proportional to the projected area (i.e., proportional to the diameter squared) and the gravity and inertia forces are proportional to the volume (i.e., the diameter cubed). For other shapes some other dimension replaces the diameter, but the idea is the same. Hence if we hold the shape of a system constant and increase all its dimensions, the inertia and gravity forces grow most quickly, the pressure force at an intermediate rate, and the surface forces most slowly.

Surface tension forces are also important in problems in which the other forces are negligible. For example, in an ordinary liquid storage tank on earth, the shape and position of the liquid are determined by gravity and pressure forces, as discussed in Chap. 2. However, in rockets and earth satellites, which have zero gravity, the position and shape of a liquid in a tank are largely determined by surface forces [1, 2].

17.1 SURFACE TENSION AND SURFACE ENERGY

Surface tension, as discussed in Sec. 1.5E, has the dimensions of force/length, in pound-force per inch or newtons per meter. (Historically surface tensions have been reported in handbooks in dynes per centimeter; 1 dyn/cm = 0.001 N/m = 5.7×10^{-6} lbf/in.) However, if we multiply the top and bottom of this ratio by length, we obtain force · length/length^2 (ft · lbf/ft^2 or J/m^2), which is equivalent to energy/area. Thus, we may conclude that the surface tension is connected with the surface energy per unit area. However, it is an experimental fact that if we prepare a film, as shown in Fig. 1.10, and stretch it by increasing the weight of the frame, the film cools. Some of the internal energy of the fluid is used to create a new surface. We may visualize this as the fluid below the surface making a new surface by pushing molecules from the bulk into the surface. Thus, the surface energy is not merely the energy put into the system by the external force moving through the distance but also the energy derived from the bulk fluid, which suffers a decrease in internal energy. If the process is performed slowly, so that heat may flow in to hold the temperature constant, then the increase in surface energy is equal to the work done ($F\,dx = \sigma\,dA$) plus the heat added. It may be shown [3, p. 1] from free-energy arguments that the correct expression for the surface energy is

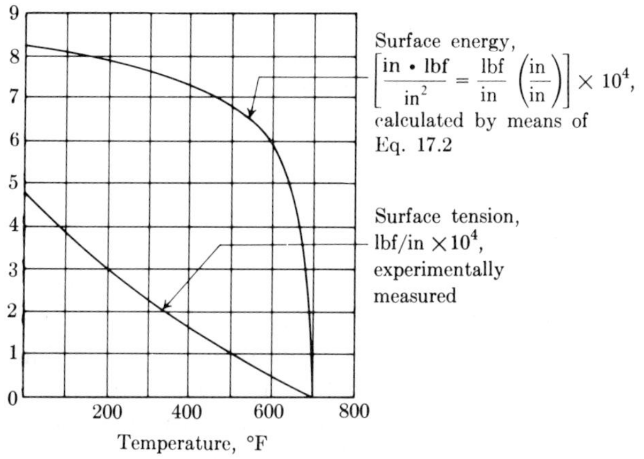

FIGURE 17.2
Surface tension and surface energy of water as a function of temperature.

$$\frac{E_s}{A} = \sigma - T\,\frac{d\sigma}{dT} \tag{17.2}$$

This equation is equivalent to the statement that the surface tension is exactly equal to the Helmholz free energy per unit area. Thus, we may make use of the thermodynamic equilibrium statement that isolated systems tend toward the condition of lowest free energy to show that the stable state of a system is the one with minimum free energy, including the contribution of the surface free energy.

In most fluid mechanics problems, it is simplest to think of surface tension as a force per unit length, but in problems involving several liquids and a solid it is sometimes more convenient to work in terms of surface energy. It is hard to visualize a solid's having surface tension but easier to visualize its having surface energy.

We may compare the relative magnitudes of the surface tension and the surface energy in Fig. 17.2, which shows the measured surface tension of water as a function of temperature, and the surface energy calculated from these surface tension measurements by Eq. 17.2 (see Prob. 17.1). The measured surface tension decreases steadily with increasing temperature, becoming zero at the critical temperature, where the surface of separation between gas and liquid vanishes.

17.2 WETTING AND CONTACT ANGLE

Before discussing how we measure surface tensions, we must discuss wetting. Fluids wet some solids and do not wet others. Figure 17.3 shows some of the possible wetting behaviors of a drop of liquid placed on a horizontal, solid surface (the remainder of the surface is covered with air, so two fluids are present).

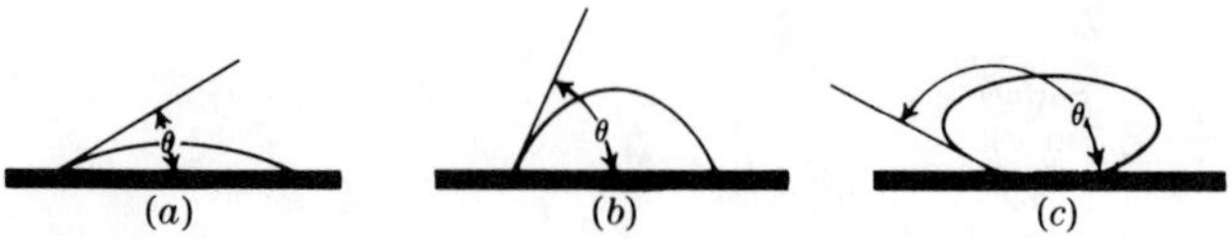

FIGURE 17.3
Wetting and contact angle.

Figure 17.3(*a*) represents the case of a liquid which wets a solid surface well, e.g., water on a very clean glass or mercury on very clean copper. The angle θ shown is the angle between the edge of the liquid surface and the solid surface, measured inside the liquid. This angle is called the *contact angle* and is a measure of the quality of the wetting. For perfect wetting, in which the liquid spreads as a thin film over the surface of the solid, θ is zero.

Figure 17.3(*b*) shows the case of moderate wetting, in which θ is less than 90° but not zero. This might be observed for water on dirty glass or mercury on a slightly oxidized copper.

Figure 17.3(*c*) represents the case of no wetting. If there were exactly zero wetting, θ would be 180°. However, the gravity force on the drop flattens the drop, so that a 180° angle is never observed. This might represent water on Teflon or mercury on clean glass.

We normally say that a liquid wets a surface if θ is less than 90° and does not wet it if θ is more than 90°. Values of θ less than 20° are considered strong wetting, and values of θ greater than 140° are examples of strong nonwetting.

In most of the methods of measuring surface and interfacial tensions described in the next section, we assume that the liquid wets some part of the apparatus perfectly, that is, $\theta = 0$.

17.3 MEASUREMENT OF SURFACE TENSION

One of the simplest measurements of surface tension is by means of capillary rise. A small-diameter glass tube is inserted in a bath of liquid; see Fig. 17.4. The fluid is assumed to wet the surface of the tube perfectly, so that the contact angle θ is 0. For small-diameter tubes, the free surface of the liquid in the tube is practically a hemisphere, so the film pulls up uniformly around the perimeter, and the net surface force upward is

$$F_s = \pi D \sigma \tag{17.3}$$

This is opposed by the gravity force on the column of fluid, which is equal to the weight of the fluid which is above the free surface and which equals

$$F_g = \frac{\pi}{4} D^2 h g \rho_l \tag{17.4}$$

Here ρ_l is the density of the liquid. Equating these forces and solving for the surface tension, we find

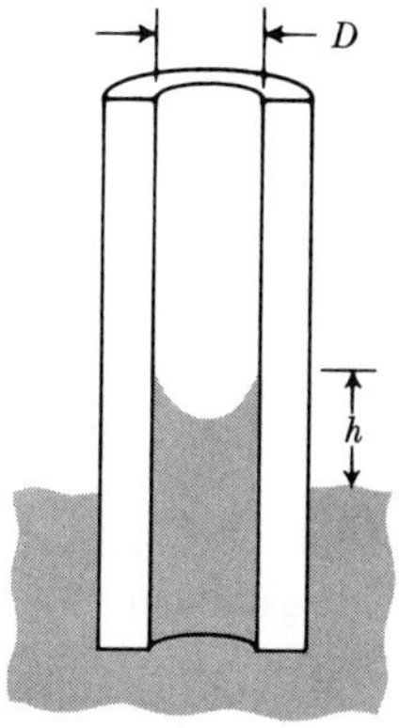

FIGURE 17.4
Capillary rise.

$$\sigma = \frac{hg\rho_l D}{4} \tag{17.5}$$

This equation omits the buoyant force due to the air, which is generally small compared with the fluid's weight. However, these omissions are not as serious as the difficulty of knowing the small diameter of the tube accurately and the difficulty of getting the inside of the tube very clean so that there will be perfect wetting and θ will be zero.

To solve these problems, we sometimes use the drop-weight method; see Fig. 17.5. In this method we allow drops to fall slowly (one every 2 to 5 min) from the tip of a burette or hypodermic needle. The drops are caught and weighed. If the liquid wets the burette perfectly, then at the instant that the drop breaks away its weight must be exactly equal to the surface force holding it up, or

$$V_{\text{drop}} \rho_l g = \pi D \sigma \tag{17.6}$$

or

$$\sigma = \frac{V_{\text{drop}} \rho_l g}{\pi D} \tag{17.7}$$

Again, we neglect the buoyant force due to the air, but this is negligible in most cases. The drop-weight method solves the problem of measuring D

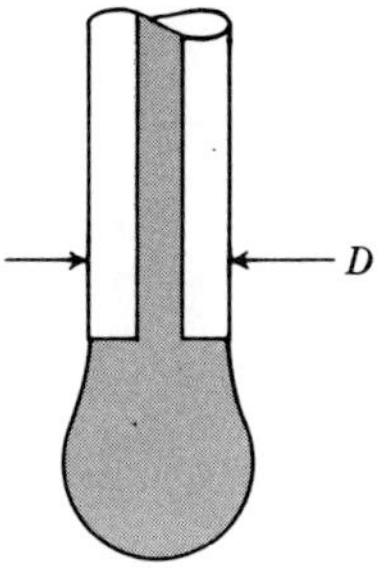

FIGURE 17.5
Drop-weight method of measuring surface tension.

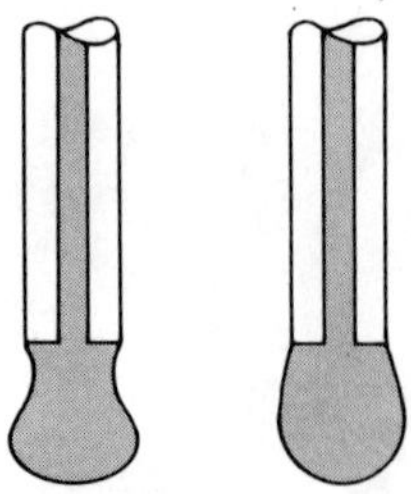

FIGURE 17.6
Difficulties with drop-weight method.

accurately, because measuring the outside diameter of a small cylinder is much easier than measuring the inside diameter. Furthermore, the glass surface, which must be very clean, is an exterior surface, which is easier to clean than an interior surface. The difficulty with this method is that the liquid surface is not always vertical but may take one of the shapes shown in Fig. 17.6. In either of these cases not all the surface tension force acts in the vertical direction, so the volume of the drop at breakaway will be less than that shown by Eq. 17.6. Experimentally, we must multiply the calculated surface tension by a correction factor whose value is generally about 1.5 to take into account this shape change [3, p. 45].

The most common routine laboratory method of measuring surface tension involves the Du Nouy tensiometer. As sketched in Fig. 17.7, a small ring is fabricated from thin platinum wire and then immersed in the liquid. The ring is drawn out of the liquid by applying a force F on a hanger or stirrup. This force is applied through a balance (normally a torsion balance), and the force required to remove the ring is measured. If the fluid wets the ring perfectly ($\theta = 0$) and if the films at the inside and outside of the ring point vertically downward at the instant of breakaway, then

$$F = 2(\pi D \sigma) \qquad \text{or} \qquad \sigma = \frac{F}{2\pi D} \tag{17.8}$$

Experimentally it is found that the assumptions of perfect wetting, etc., are not exactly correct and that the surface tension calculated by Eq. 17.8 is too large by a factor of about 1.1. However, with this kind of apparatus, the cleaning problem is much easier than the case of the other two, and the dimensions are easily measured. For these and other reasons of convenience, this type of apparatus is the most common laboratory device for measuring surface tension.

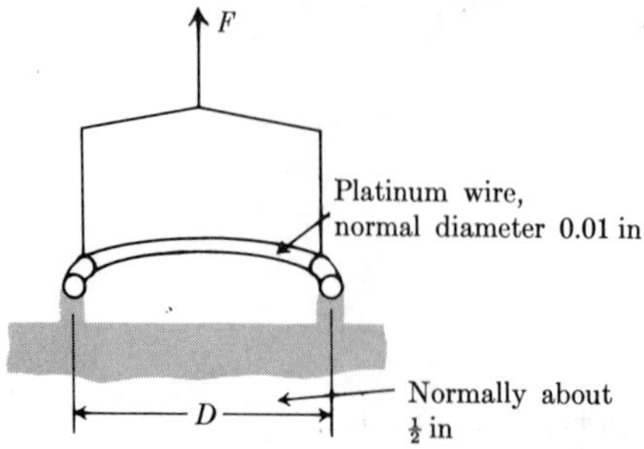

FIGURE 17.7
Du Nouy tensiometer.

17.4 INTERFACIAL TENSION

Surface forces exist at both gas-liquid and liquid-liquid interfaces. The latter are called *interfacial tensions*. They can be measured most easily with a Du Nouy tensiometer (Fig. 17.7), if the denser fluid wets the ring. In that case the force required to pull the ring from the lower fluid up into the upper fluid depends on the interfacial tension.

In general, interfacial tensions are greater for liquid pairs with low mutual solubilities than for those with high ones. Thus, hexane-water (very low mutual solubility) has an interfacial tension two-thirds that of air-water, whereas butanol-water (reasonably large mutual solubility) has an interfacial tension only a few percent of that of air-water. For miscible liquid pairs such as ethanol-water, there can be no interfacial tension because there can be no interface.

In all surface and interfacial tension measurements, we must take extreme care to keep the surfaces clean. Many impurities tend to collect at interfaces, and very small quantities of them can make large changes in interfacial properties. Soaps, detergents, and wetting agents are prime examples of *surface-active agents*. These normally consist of tadpole-shaped molecules with a polar, water-soluble head and a nonpolar, oil-soluble tail. Here "oil" means any organic liquid not miscible with water. Small quantities of detergent are soluble in either water or oil, but their preferred position is at the water-oil interface, where they can put their water-soluble heads in the water and their oil-soluble tails in the oil. Thus the concentration of such agents at water-oil interfaces is much higher than in the bulk fluids surrounding the interfaces. Their high concentration there changes the chemical and physical properties of the interfaces. The principal function of such soaps and detergents is to disperse oils and fats into microscopic droplets, called *micelles*, and to keep them from coalescing. Thus, soaps disperse oils in water and allow them to be washed off surfaces.

17.5 FORCES DUE TO CURVED SURFACES

A plane surface exerts forces in the plane of the surface, as shown in Fig. 1.10. A curved surface can also exert a net force at right angles to itself, as shown in Fig. 17.8.

Here we see a small piece of liquid surface perpendicular to the z axis, whose projection on the xy plane is the small rectangle Δx by Δy. The forces exerted by the surface along its four edges are $F = \sigma \, \Delta y$ for the two edges perpendicular to the x axis and $F = \sigma \, \Delta x$ for the two edges perpendicular to the y axis. Assuming that this piece of surface is symmetric about the x and y axes, we can see that the x and y components of the force on the surface are zero, because the plus and minus parts cancel.

To find the z component of this force, we sum the z components of the four surface forces on the edges to find

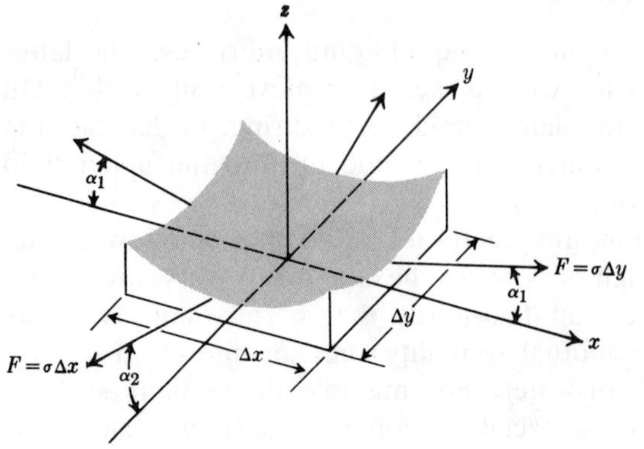

FIGURE 17.8
Forces due to curved surfaces.

$$F_{z_{\text{total}}} = \sum_{4 \text{ sides}} F_z = 2\sigma\, \Delta y \sin \alpha_1 + 2\sigma\, \Delta x \sin \alpha_2 \tag{17.9}$$

For small angles α_1 and α_2, we have

$$\sin \alpha_1 = \frac{1}{2}\frac{\Delta x}{R_1} \qquad \sin \alpha_2 = \frac{1}{2}\frac{\Delta y}{R_2} \tag{17.10}$$

where R_1 and R_2 are the radii of curvature of the surface in the xz and yz planes, respectively. We can visualize these radii by imagining a circle drawn tangent to the surface at the origin; its radius is the radius of curvature in that plane (see Fig. 17.9).

Substituting Eq. 17.10 in Eq. 17.9, we find

$$F_z = 2\sigma\left(\Delta y \,\frac{1}{2}\frac{\Delta x}{R_1} + \Delta x \,\frac{1}{2}\frac{\Delta y}{R_2}\right) = \sigma A\left(\frac{1}{R_1} + \frac{1}{R_2}\right) \tag{17.11}$$

Here A is the area of the surface with sides Δx and Δy.

Example 17.1. Evaluate the force per unit area due to surfaces in the shape of (*a*) a sphere, (*b*) a cylinder, and (*c*) a plane.

(*a*) Let the sphere have radius r and have its center at the origin. Then in the xy plane it forms a circle whose equation is

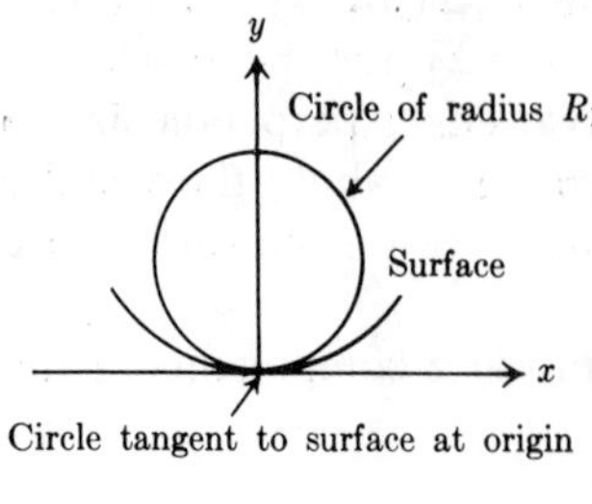

FIGURE 17.9

$$x^2 + y^2 = r^2 \qquad \frac{dy}{dx} = -\frac{x}{y} \qquad \frac{d^2y}{dx^2} = \frac{x^2 + y^2}{y^3}$$

In the calculus it is proved that the radius of curvature of any curve on the *xy* plane is given by

$$R = \frac{[1 + (dy/dx)^2]^{3/2}}{d^2y/dx^2}$$

So here

$$R_1 = \frac{(1 + x^2/y^2)^{3/2}}{(x^2 + y^2)/y^3} = (x^2 + y^2)^{1/2} = r$$

Similarly, for the *yz* plane we have $R_2 = r$, so

$$\frac{F}{A} = \sigma\left(\frac{1}{r} + \frac{1}{r}\right) = \frac{2\sigma}{r} = \frac{4\sigma}{D}$$

which is the result found from Eq. 17.1.

(*b*) Let the cylinder have radius r and its axis along the x axis. Then its intersection with the *xz* plane is a straight line perpendicular to the z axis, for which dz/dx and d^2z/dx^2 are both zero. So

$$R_1 = \frac{(1 + 0)^{3/2}}{0} = \infty$$

In the *yz* plane its intersection is a circle with radius r, so by a similar argument to that used in part (*a*) we have $R_2 = r$ and

$$\frac{F}{A} = \sigma\left(\frac{1}{r} + \frac{1}{\infty}\right) = \frac{\sigma}{r}$$

(*c*) Let the plane be parallel to the *xy* plane, so that its intersections with both the *xz* and *yz* planes are straight lines perpendicular to the z axis, for which, by the argument in part (*b*), we have $R_1 = R_2 = \infty$. Thus,

$$\frac{F}{A} = \sigma\left(\frac{1}{\infty} + \frac{1}{\infty}\right) = 0$$

■

For static surfaces the force described by Eq. 17.11 is normally balanced by a pressure-difference force. If the fluid is moving, this force can accelerate the fluid, causing oscillations, breakup of jets, etc.

For a fluid at rest with both sides exposed to the same pressure, Eq. 17.11 indicates that $[1/R_1 + 1/R_2]$ must be zero. This is obviously true of a plane surface, but it is also true of many complicated surfaces, such as that shown in Fig. 17.10. This is the shape taken by an open-ended soap film attached to two circular wire loops. Here the center of curvature in the *xy* plane is on the z axis, and the center of curvature in the *xz* plane is outside the film. Since these are on opposite sides of the film, R_1 and R_2 have opposite signs. To make $[1/R_1 + 1/R_2]$ equal zero, we must make them have the same absolute value; we can show (Prob. 17.10) that this is the description of a catenoid curve.

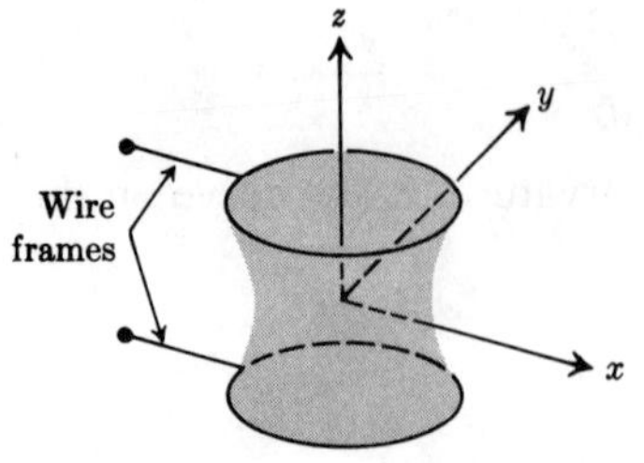

FIGURE 17.10
Soap bubble suspended between two wire frames.

17.6 SOME EXAMPLES OF SURFACE FORCE EFFECTS

In many industrial devices, bubbles are formed by forcing a gas through an orifice into a stagnant pool of liquid. For gas being forced through a circular, horizontal orifice, this situation is shown in Fig. 17.11.

If the bubble is spherical, then the buoyant force on the bubble is

$$F_{\text{buoy}} = \pi D^3 \frac{(\rho_l - \rho_g)g}{6} \tag{17.12}$$

If the liquid wets the orifice, including the vertical part of the hole, as shown in Fig. 17.11, then the surface force at the bottom of the hole acts vertically downward with magnitude

$$R_{\text{surf}} = \pi D_0 \sigma \tag{17.13}$$

The bubble breaks away from the orifice when the buoyant force exceeds the surface force. Assuming that breakaway begins when these forces are just equal, we can equate these two forces and solve for the bubble diameter:

$$D = \left[\frac{6 D_0 \sigma}{(\rho_l - \rho_g)g}\right]^{1/3} \tag{17.14}$$

which is called *Tate's law*. This treatment ignores the possibility that the bubbles may interact with the preceding or following bubbles and neglects the momentum of the gas flowing into the bubble and the momentum given to the fluid around the bubble. Nevertheless, it can be shown experimentally that Eq. 17.14 gives an excellent prediction of the observed bubble diameter for low gas flow rates through an orifice. The criterion for the departure of the experimen-

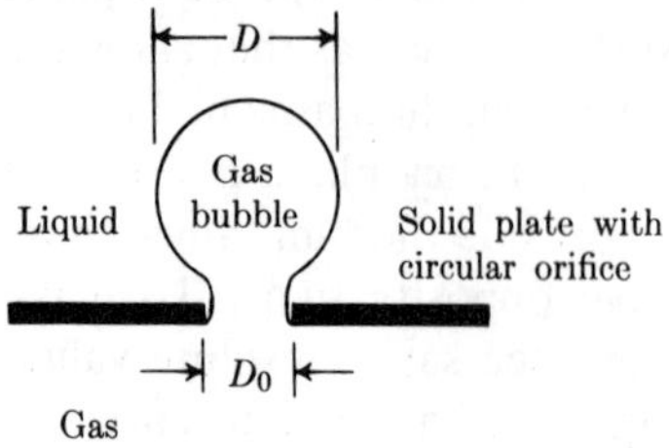

FIGURE 17.11
Gas bubble growing at a circular orifice.

tal data from this simple equation is discussed by Soo [4]. A more complex treatment, which takes into account the momentum of the gas and the liquid, and which is accurate to higher velocities, is given by Hayes et al. [5]. This behavior is greatly complicated by the interaction between the growing bubble and the gas chamber from which it grows [6].

As shown in Fig. 17.11, this treatment assumes that the gas does not wet the orifice and that instead a thin film of liquid wets the inside of the orifice. This is commonly observed for gas-liquid systems and for some liquid-liquid systems. However, if the fluid flowing through the orifice wets the surface of the orifice, then much larger bubbles result [7].

An unconfined jet of liquid will break up into drops; this is observable in the jet leaving a faucet or a garden hose (see Fig. 17.12). Here a cylindrical jet of liquid is leaving a nozzle. As the liquid falls, it speeds up, because it is being accelerated by gravity. This causes the jet to decrease in cross-sectional area to satisfy the material balance. Finally the jet breaks up into liquid droplets. This breakup is caused by surface forces; the cylindrical column of fluid can rearrange into a system with less surface area by changing over into spherical droplets.

This breakup is possible for a cylinder whose length is four-ninths that of the diameter (Prob. 17.13), but such a cylinder is metastable: To pass from the cylindrical shape to the spherical one, it would have to pass through intermediate states with more surface. Rayleigh [8] analyzed the problem as follows. He assumed that there existed such a cylinder of length L (Fig. 17.13) which had superimposed on its cylindrical surface a cosine wave disturbance of wavelength L. Thus the radius at any point is given by

$$r = r_0 + a \cos \frac{2\pi}{L} z \tag{17.15}$$

In his analysis a is always assumed very small, so that the curvature shown in Fig. 17.13 is quite exaggerated. By assuming that a was very small he could show (Prob. 17.14) that if L was greater than $2\pi r_0$, a small disturbance of this form resulted in a decrease in the surface area compared with the undisturbed

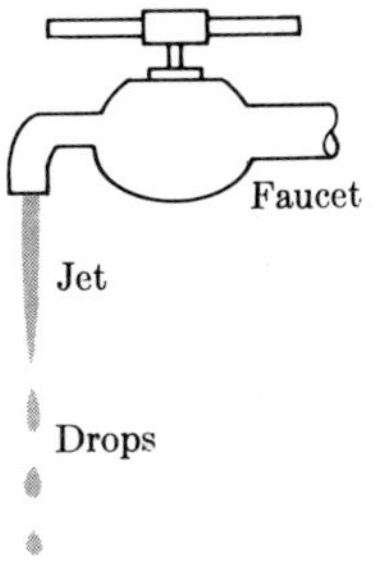

FIGURE 17.12
Breakup of liquid jet.

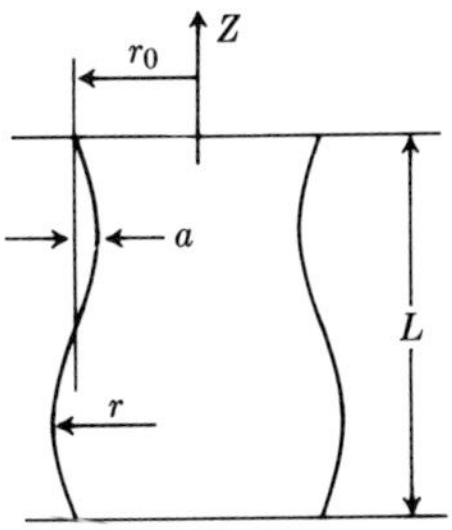

FIGURE 17.13
Small displacements on a cylinder.

cylinder. It can be shown by thermodynamic reasoning that systems proceed to states of lower free energy whenever possible and that a state of lower surface area is a state of lower free energy; hence, a fluid cylinder longer than $2\pi r_0$ is unstable and will break down under a small disturbance. This can be experimentally verified [9] with cylindrical soap bubbles. When their length is less than $2\pi r_0$, they are stable; for any length longer than this, they break down as a result of any minor vibration.

Rayleigh also showed in the same paper that although any disturbance with wavelength greater than $2\pi r_0$ would grow and break up a cylindrical jet of liquid, the fastest-growing disturbances should be those with wavelength $9r_0$. One may demonstrate this breakup in a kitchen sink by adjusting the flow from the faucet so that it has the smallest flow which will have a continuous column of fluid reach the bottom of the sink. Then tapping the faucet will cause the faucet to vibrate, introducing such disturbances and causing the jet to break up into droplets.

This simple analysis indicates only what size jet is unstable and how the surface forces cause the breakup of the jets. It does not indicate how fast such a breakup proceeds. Rayleigh's analysis of the most favored wavelength for breakup was based on the assumption that the viscosity of the fluid was negligible. For thick jets of liquids such as water, the experimental data [10] indicate that the breakup occurs most frequently at lengths corresponding $10r_0$ to $12r_0$ rather than the $9r_0$ predicted by Rayleigh.

For a very viscous fluid like maple syrup, the breakup is very slow (which is why we can pour a long, thin stream of maple syrup onto pancakes without its breaking into droplets). Jet breakup is very important in such processes as spray drying, vaporization of liquid fuels in combustion, spray painting, and insecticide spraying. More on this subject can be found elsewhere [10, 11, 12].

In all the preceding we assumed that the surface tension was uniform on the entire surface. If there is a gradient in the surface tension, then the surface will tend to flow in the direction of the higher surface tension and to drag adjacent liquid with it, causing bulk fluid motion. Such gradients in the surface tension can arise from differences in composition, differences in temperature, or differences in electric charge. An easily observed example of surface flow caused by concentration gradients is the formation of "wine tears" (Fig.

FIGURE 17.14
Wine tears.

17.14). To see these, place any liquor or wine with more than 20 percent alcohol in a clean glass. Because the liquid wets the glass, a film of liquid is dragged up the side of the glass. Alcohol is constantly being evaporated from all the exposed surface of the liquid. In the bulk liquid in the glass, the concentration is kept practically constant, because alcohol diffuses in from below to replace that being evaporated. In the films along the walls, there is no comparable source of alcohol resupply, so the concentration of alcohol there falls. In alcohol-water solutions, the surface tension increases with increasing water concentration, so the surface tension is higher on the film upon the walls of the glass. This causes the film to flow up the walls and drag fluid with it from below. At the top of this film the fluid accumulates, forming the "tears," which run back down the glass.

The same kind of motion can occur at the interface between two immiscible phases when a chemical reaction is taking place between components present in both phases or there is diffusion of one substance from one phase to another. This type of behavior occurs in several chemical engineering systems and is called *interfacial turbulence* or the *Marangoni effect* [3].

Consider a gas or air bubble which is at rest in a converging tube, as shown in Fig. 17.15. The tube is assumed small enough for the ends of the bubble to be hemispherical with radii r_1 and r_2. The bubble is assumed at rest, so we may assume a uniform pressure P_i inside the bubble. From Eq. 17.1 we can then calculate the pressures at points 1 and 2, finding

$$P_1 = P_i - \frac{2\sigma}{r_1} \qquad P_2 = P_i - \frac{2\sigma}{r_2}$$

$$P_1 - P_2 = 2\sigma\left(\frac{1}{r_2} - \frac{1}{r_1}\right) \tag{17.16}$$

Thus, in the absence of viscous forces, this bubble can stay in the converging tube only when there is a pressure difference, as indicated by Eq. 17.16, from its large end to its small end. It can only be driven into the converging direction by a pressure difference greater than that computed by Eq. 17.16. If it is driven into the converging direction, the values of the radii will decrease, until it finds a location where the surface tension forces will exactly balance the pressure-difference force, and then the bubble will remain in place.

This ability of a bubble to resist a pressure gradient, which would tend to move it, can be very annoying in laboratory glassware and in the small-diameter lines used for measuring instruments. Such bubbles regularly form,

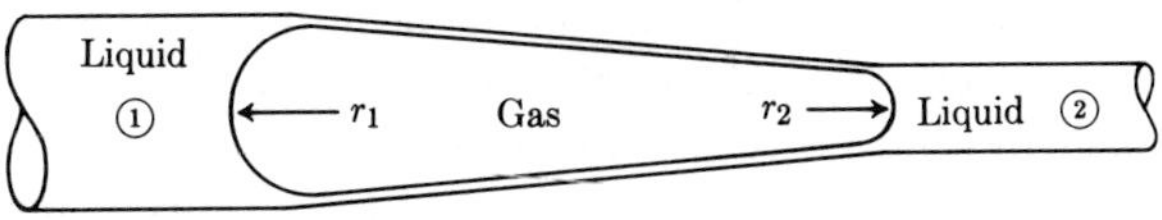

FIGURE 17.15
Jamin effect.

often from the evolution of dissolved air from water, and block the tubing. The same effect has great technical and economic significance in two-phase flow in porous media, such as groundwater flow in the presence of air and oil flow in the presence of water or gas. Such flows may be conceived of as occurring in a series of interconnecting, irregularly shaped tubes of very small diameter. The flow normally occurs as a result of a pressure difference. When one of the fluids breaks up into globules or bubbles, it can become trapped, just as the bubble in Fig. 17.15 is. It is experimentally observed in such systems that once one of the phases becomes discontinuous (i.e., breaks up into bubbles or drops), it stops moving, and no amount of flushing with the other fluid will make it move. Although other factors are involved, this surface tension factor is one of the major causes of this result. This effect is known in the petroleum engineering literature as the *Jamin effect* [14].

17.7 SUMMARY

1. Surface forces are likely to be important in small systems and in systems in which other forces are small or negligible.
2. We may think of surface tension as a force per unit length or as an energy per unit area. The surface tension is exactly equal to the surface Helmholz free energy per unit area.
3. When solid surfaces are involved in surface phenomena, it is necessary to take into account the wetting or nonwetting properties of the solid-fluid boundaries. These are normally expressed by the contact angle.
4. Surface forces are also present at the interfaces between immiscible liquids; these are called interfacial tensions. Such tensions are strongly influenced by impurities in the fluids, called surface-active agents.
5. The study of bubbles, drops, and interfaces between fluids generally requires the study of the surface forces involved.

PROBLEMS

See the Common Units and Values for Problems and Examples inside the back cover.

17.1. Calculate the surface energy of water at 200°F from the surface tension curve in Fig. 17.2. Compare your result with the surface energy curve in Fig. 17.2.

17.2. From the data shown in Fig. 17.2 calculate the surface Helmholz free energy per unit of surface at 200°F.

17.3. Some automatic dishwashers add a "wetting agent" to the final rinse to prevent the formation of droplet marks on glassware. Explain how these agents work.

17.4. The surface tension of a liquid was measured with a capillary-rise tube and found to be A. Later tests show that this liquid does not wet the glass perfectly but makes a contact angle of $\theta = 30°$. Estimate the true value of the surface tension of this liquid.

17.5. As discussed in Sec. 17.4, we may use a Du Nouy tensiometer to measure the interfacial tension of immiscible liquid pairs in the normal way, if the denser fluid wets the ring. Sketch how to set up this instrument to measure such tensions if the less dense fluid wets the ring.

17.6. Calculate the correction factor to be applied to Eq. 17.7 for the case in which the film at the surface is slanted 10° from the vertical. Does it make any difference whether the film slants inward or outward?

17.7. A common statement in chemistry laboratory manuals is that "20 drops from a burette equals approximately 1 cm^3." Assuming this applies only to dilute aqueous solutions whose surface tension is approximately equal to that of water, calculate the diameter of the tip of a standard burette.

17.8. Experiments [3, p. 45] indicate that the surface tension calculated from drop weights on burettes by means of Eq. 17.7 must be multiplied by a factor which ranges up to 1.5 to agree with those obtained by the most reliable methods. Assuming that this is entirely due to the shape of the droplet just before breakaway, calculate the angle that the drop's surface makes with the vertical.

17.9. Two perfectly flat glass plates are assembled as shown in Fig. 17.16. The space between the plates has the form of a wedge with zero thickness on one side and thickness B on the other side; $A \gg B$. The lower edges of the plates are now immersed in a pan of water. Calculate the shape of the water layer drawn up between the plates by surface tension in terms of A, B, and σ. This result is quite easy to verify experimentally; the glass plates used for microscope slides are easy to use.

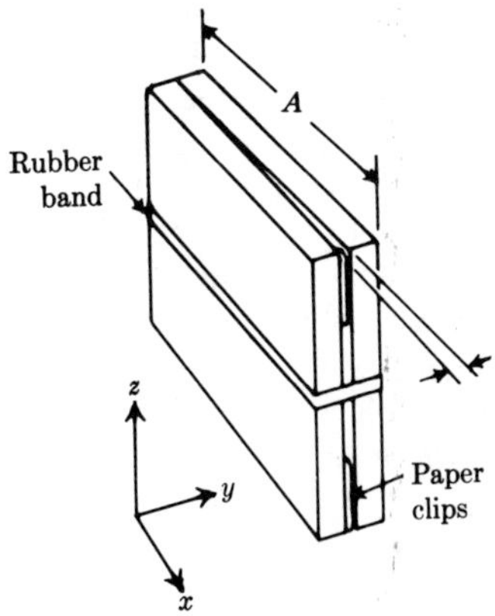

FIGURE 17.16

17.10. Show that the surface shown in Fig. 17.10 is a catenoid surface (one described by rotating a catenary curve about the z axis). *Hint:* See Fig. 17.17. From the condition that the film exerts no force in the direction normal to it, we know that the two radii of curvature at right angles to each other and normal to the film are equal and of opposite sign. These are R_1 and R_2. Equate these, and then show that $R_2 = y[1 + (dy/dz)^2]^{1/2}$. Then show that the resulting differential equation is satisfied by $y = (1/a)\cosh(z/a)$, which is the equation of a catenary curve. Here a is an arbitrary constant.

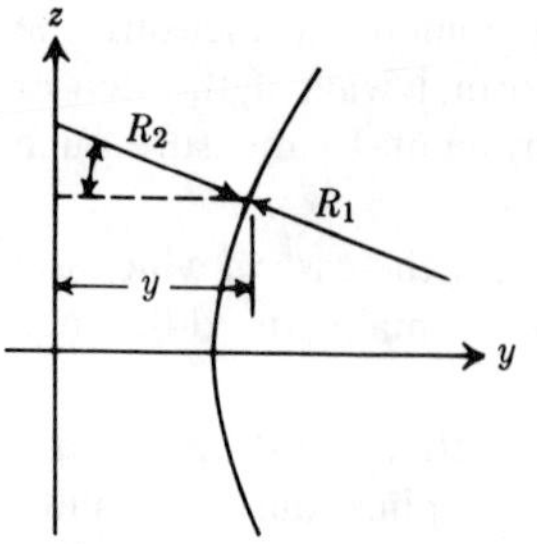

FIGURE 17.17

17.11. One can prepare a closed-end cylindrical soap bubble by blowing between two solid rings, as shown in Fig. 17.18. Because of the pressure inside the bubble, which is necessary to hold the long surface in the cylindrical shape, the top and bottom surfaces will bow out into spherical segments. If the rings have diameter D, what is the radius of the spherical segments?

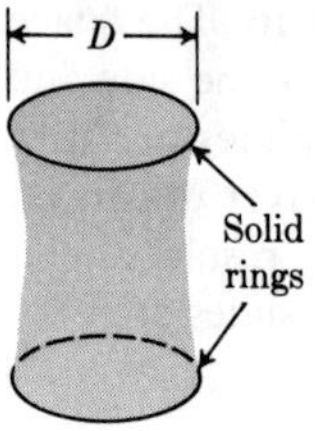

FIGURE 17.18

17.12. Two flat, rectangular pieces of glass of the type used to make lantern slides (3.25×4.25 in) are held parallel and vertical, about $\frac{1}{32}$ in apart. Their lower edges are immersed in water. The water rises in the space between them because of surface tension; the plates pull together. After they have pulled together, the plates can be easily separated by being slid parallel to their surface, but they are very difficult to separate by being pulled perpendicular to the surface. Why? What is the magnitude of the force pulling them together? Does this force change if they are removed from the water and placed on a horizontal table?

17.13. (*a*) Consider a cylindrical column of length L and diameter D_c. Show that if this column were rearranged into a sphere of equal volume, the ratio of the surface of the new sphere to the cylindrical surface of the cylinder (total surface less the surface of the ends) would equal $S_s/S_{cyl} = (\frac{3}{2})^{2/3}(D_c/L)^{1/3}$ and hence that such a rearrangement causes a decrease in surface if L is greater than $4D_c/9$, no change if L equals $4D_c/9$, and an increase if L is less than $4D_c/9$.

(*b*) If we take the area of the ends into account in calculating the surface of the cylinder, what minimum length is required so that the cylinder can rearrange into a sphere with a smaller surface area?

17.14. Below is an excerpt from Rayleigh's classic paper [8] on the instability of jets, in which he describes the maximum stable length† (see Fig. 17.13).

† Here we have changed the symbols in the original paper to match those used in this book.

"Let us, then, taking the axis of z along the axis of the cylinder, suppose that at time t, the surface of the cylinder is of the form

$$r = r_0 + a \cos \frac{2\pi}{L} z \tag{1}$$

where a is a small quantity variable with the time . . .

"If we denote the surface corresponding (on the average) to the unit length along the axis by A, we readily find

$$A = 2\pi r_0 + \frac{1}{2} \pi r_0 \left(\frac{2\pi}{L}\right)^2 a^2 \tag{2}$$

"In this, however, we have to substitute for r_0 (which is not strictly constant) its value obtained from the condition that V, the volume enclosed per unit of length, is given. We have

$$V = \pi r_0^2 + \tfrac{1}{2}\pi a^2 \tag{3}$$

whence

$$r_0 = \sqrt{\frac{V}{\pi}}\left(1 - \frac{1}{4}\frac{\pi a^2}{V}\right) \tag{4}$$

"Using this in (2), we get with sufficient approximation

$$A = 2\sqrt{\pi V} + \frac{a\pi^2}{2r_0}\left[\left(\frac{2\pi r_0}{L}\right)^2 - 1\right] \tag{5}$$

or, if A_0 be the value of A for the undisturbed condition,

$$A - A_0 = \frac{\pi a^2}{2}\left[\left(\frac{2\pi r_0}{L}\right)^2 - 1\right] \tag{6}$$

"From this we infer that, if $2\pi r_0/L > 1$, the surface is greater after displacement than before."[†]

And conversely, if $2\pi r_0/L < 1$, the surface is less after displacement.

Fill in the missing steps in the derivation. *Hint:* In finding Eq. (1), he has started with the equation for the length of an element on the surface of the cylinder, parallel to the axis, and then used the binomial theorem to perform the necessary integration. Similarly, to obtain Eq. (4) from Eq. (3), he has used the binomial theorem. In both cases this is the form of $(1 + x)^{1/2} = 1 + \frac{1}{2}x +$ other terms. When x is small, the other terms may be neglected, which he has done here.

17.15. It is frequently stated that the breakup of a jet of fluid is a way of preparing a high-surface area of the fluid for vaporization, chemical reaction, etc. However, the analysis in Sec. 17.6 indicates that the breakup occurs because the fluid is decreasing in surface area. How are these ideas to be reconciled?

[†] From Lord Rayleigh (John William Strutt), "On the instability in jets," *Proc. London Math. Soc. 10:* 4–13 (1879); reprinted in *Scientific Papers of Lord Rayleigh*, Dover, New York, 1964, p. 362. Quoted by permission of the publisher.

17.16. Calculate the pressure difference required to make an air bubble move through a water-wet orifice whose diameter is 0.001 in.

17.17. If we dip a wire ring into a soap solution and hold it as shown in Fig. 17.19, we have a vertical soap film. From a force balance on a small section of the film, we see that it is acted on by gravity force downward, and if it is to stay in place, it must be acted on by an upward surface force. How can this upward surface force be generated? From this consideration, can we conclude that it is impossible to form such a film from an absolutely pure liquid? This topic is discussed in detail by Ross [15].

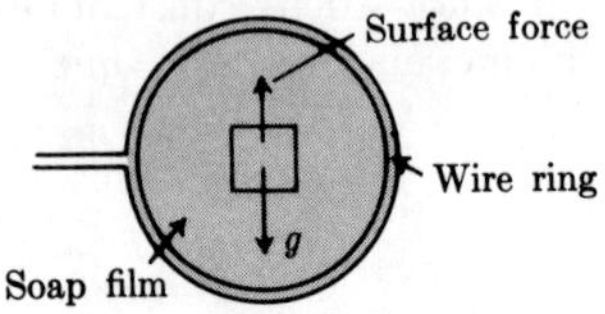

FIGURE 17.19

APPENDIX A

TABLES AND CHARTS

A.1 VISCOSITIES OF VARIOUS FLUIDS AT 1-ATM PRESSURE

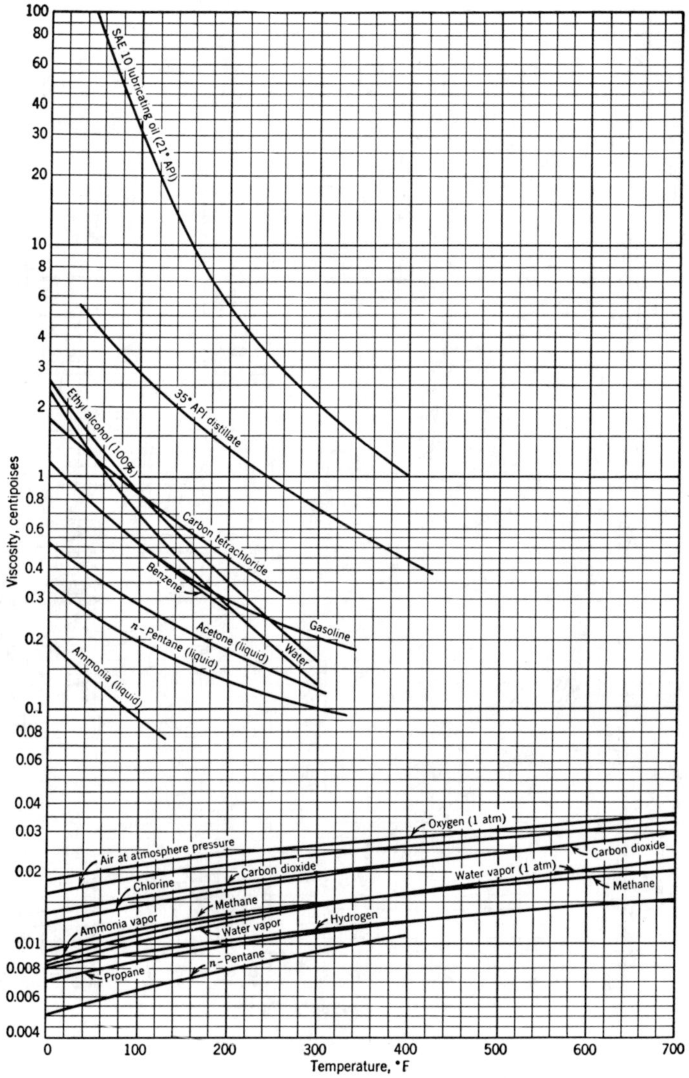

From G. G. Brown et al., *Unit Operations*, Wiley, New York, 1951, p. 586. Reproduced by permission of the publisher.

A.2 PRESSURE-ENTHALPY DIAGRAM FOR FREON-12 REFRIGERANT

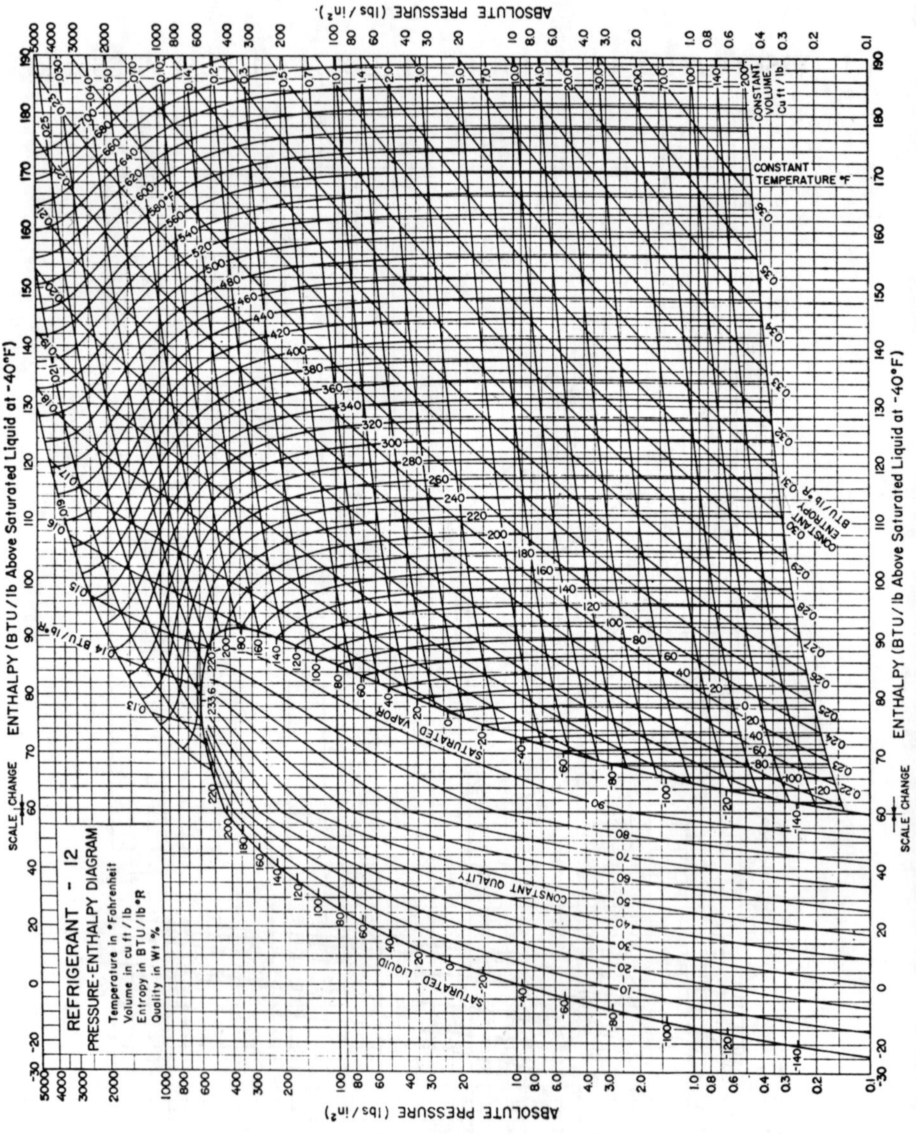

In the examples in Chapter 4 this plot is read to 4 significant figures. The student is not expected to do this without the tables which accompany this figure, which can be obtained from du Pont.

Courtesy of E. I. du Pont de Nemours and Company, Inc.

A.3 STEEL PIPE DIMENSIONS: CAPACITIES AND WEIGHTS

Nominal pipe size, in	Outside diam., in	Schedule no.	Wall thickness, in	Inside diam., in	Cross-sectional area metal, in^2	Inside sectional area, ft^2	Circumference, ft, or surface, ft^2/ft, of length		Capacity at 1 ft/s velocity		Weight of pipe, lb/ft
							Outside	Inside	U.S. gal/min	lb/h water	
$\frac{1}{8}$	0.405	40	0.068	0.269	0.072	0.00040	0.106	0.0705	0.179	89.5	0.25
		80	0.095	0.215	0.093	0.00025	0.106	0.0563	0.112	56.0	0.32
$\frac{1}{4}$	0.540	40	0.088	0.364	0.125	0.00072	0.141	0.0954	0.323	161.5	0.43
		80	0.119	0.302	0.157	0.00050	0.141	0.0792	0.224	112.0	0.54
$\frac{3}{8}$	0.675	40	0.091	0.493	0.167	0.00133	0.177	0.1293	0.596	298.0	0.57
		80	0.126	0.423	0.217	0.00098	0.177	0.1110	0.440	220.0	0.74
$\frac{1}{2}$	0.840	40	0.109	0.622	0.250	0.00211	0.220	0.1630	0.945	472.5	0.85
		80	0.147	0.546	0.320	0.00163	0.220	0.1430	0.730	365.0	1.09
		160	0.187	0.466	0.384	0.00118	0.220	0.1220	0.529	264.5	1.31
$\frac{3}{4}$	1.050	40	0.113	0.824	0.333	0.00371	0.275	0.2158	1.665	832.5	1.13
		80	0.154	0.742	0.433	0.00300	0.275	0.1942	1.345	672.5	1.48
		160	0.218	0.614	0.570	0.00206	0.275	0.1610	0.924	462.0	1.94
1	1.315	40	0.133	1.049	0.494	0.00600	0.344	0.2745	2.690	1,345.0	1.68
		80	0.179	0.957	0.639	0.00499	0.344	0.2505	2.240	1,120.0	2.17
		160	0.250	0.815	0.837	0.00362	0.344	0.2135	1.625	812.5	2.85
$1\frac{1}{4}$	1.660	40	0.140	1.380	0.669	0.01040	0.435	0.362	4.57	2,285.0	2.28
		80	0.191	1.278	0.881	0.00891	0.435	0.335	3.99	1,995.0	3.00
		160	0.250	1.160	1.107	0.00734	0.435	0.304	3.29	1,645.0	3.77
$1\frac{1}{2}$	1.990	40	0.145	1.610	0.799	0.01414	0.498	0.422	6.34	3,170.0	2.72
		80	0.200	1.500	1.068	0.01225	0.498	0.393	5.49	2,745.0	3.64
		160	0.281	1.338	1.429	0.00976	0.498	0.350	4.38	2,190.0	4.86

(*continued*)

A.3 (*continued*)

Nominal pipe size, in	Outside diam., in	Schedule no.	Wall thickness, in	Inside diam., in	Cross-sectional area metal, in^2	Inside sectional area, ft^2	Circumference, ft, or surface, ft^2/ft, of length		Capacity at 1 ft/s velocity		Weight of pipe, lb/ft
							Outside	Inside	U.S. gal/min	lb/h water	
2	2.375	40	0.154	2.067	1.075	0.02330	0.622	0.542	10.45	5,225.0	3.66
		80	0.218	1.939	1.477	0.02050	0.622	0.508	9.20	4,600.0	5.03
		160	0.343	1.689	2.190	0.01556	0.622	0.442	6.97	3,485.0	7.45
2½	2.875	40	0.203	2.469	1.704	0.3322	0.753	0.647	14.92	7,460.0	5.80
		80	0.276	2.323	2.254	0.02942	0.753	0.609	13.20	6,600.0	7.67
		160	0.375	2.125	2.945	0.02463	0.753	0.557	11.07	5,535.0	10.0
3	3.500	40	0.216	3.068	2.228	0.05130	0.917	0.804	23.00	11,500.0	7.58
		80	0.300	2.900	3.016	0.04587	0.917	0.760	20.55	10,275.0	10.3
		160	0.437	2.626	4.205	0.03761	0.917	0.688	16.90	8,450.0	14.3
3½	4.000	40	0.226	3.548	2.680	0.06870	1.047	0.930	30.80	15,400.0	9.11
		80	0.318	3.364	3.678	0.06170	1.047	0.882	27.70	13,850.0	12.5
4	4.500	40	0.237	4.026	3.173	0.08840	1.178	1.055	39.6	19,800.0	10.8
		80	0.337	3.826	4.407	0.07986	1.178	1.002	35.8	17,900.0	15.0
		120	0.437	3.626	5.578	0.07170	1.178	0.950	32.2	16,100.0	19.0
		160	0.531	3.438	6.621	0.06447	1.178	0.901	28.9	14,450.0	22.6
5	5.563	40	0.258	5.047	4.304	0.1390	1.456	1.322	62.3	31,150.0	14.7
		80	0.375	4.813	6.112	0.1263	1.456	1.263	57.7	28,850.0	20.8
		120	0.500	4.563	7.953	0.1136	1.456	1.197	51.0	25,500.0	27.1
		160	0.625	4.313	9.696	0.1015	1.456	1.132	45.5	22,750.0	33.0
6	6.625	40	0.280	6.065	5.584	0.2006	1.734	1.590	90.0	45,000.0	19.0
		80	0.432	5.761	8.405	0.1810	1.734	1.510	81.1	40,500.0	28.6
		120	0.562	5.501	10.71	0.1650	1.734	1.445	73.9	36,950.0	36.4
		160	0.718	5.189	13.32	0.1469	1.734	1.360	65.8	32,900.0	45.3
8	8.625	20	0.250	8.125	6.570	0.3601	2.258	2.130	161.5	80,750.0	22.4
		30	0.277	8.071	7.260	0.3553	2.258	2.115	159.4	79,700.0	24.7
		40	0.322	7.981	8.396	0.3474	2.258	2.090	155.7	77,850.0	28.6
		60	0.406	7.813	10.48	0.3329	2.258	2.050	149.4	74,700.0	35.7
		80	0.500	7.625	12.76	0.3171	2.258	2.000	142.3	71,150.0	43.4

		100	0.593	7.439	14.96	0.3018	2.258	1.947	135.3	67,650.0	50.9
		120	0.718	7.189	17.84	0.2819	2.258	1.883	126.5	63,250.0	60.7
		140	0.812	7.001	19.93	0.2673	2.258	1.835	120.0	60,000.0	67.8
		160	0.906	6.813	21.97	0.2532	2.258	1.787	113.5	56,750.0	74.7
10	10.75	20	0.250	10.250	8.24	0.5731	2.814	2.685	257.0	128,500.0	28.1
		30	0.307	10.136	10.07	0.5603	2.814	2.655	252.0	126,000.0	34.3
		40	0.365	10.020	11.90	0.5475	2.814	2.620	246.0	123,000.0	40.5
		60	0.500	9.750	16.10	0.5185	2.814	2.550	233.0	116,500.0	54.8
		80	0.593	9.564	18.92	0.4989	2.814	2.503	224.0	112,000.0	64.4
		100	0.718	9.314	22.63	0.4732	2.814	2.440	212.0	106,000.0	77.0
		120	0.843	9.064	26.24	0.4481	2.814	2.373	201.0	100,500.0	89.2
		140	1.000	8.750	30.63	0.4176	2.814	2.290	188.0	93,750.0	105.0
		160	1.125	8.500	34.02	0.3941	2.814	2.230	177.0	88,500.0	116.0
12	12.75	20	0.250	12.250	9.82	0.8185	3.338	3.31	367.0	183,500.0	33.4
		30	0.330	12.090	12.87	0.7972	3.338	3.17	358.0	179,000.0	43.8
		40	0.406	11.938	15.77	0.7773	3.338	3.13	349.0	174,500.0	53.6
		60	0.562	11.626	21.52	0.7372	3.338	3.05	331.0	165,500.0	73.2
		80	0.687	11.376	26.03	0.7058	3.338	2.98	317.0	158,500.0	88.6
		100	0.843	11.064	31.53	0.6677	3.338	2.90	299.0	149,500.0	108.0
		120	1.000	10.750	36.91	0.6303	3.338	2.82	283.0	141,500.0	126.0
		140	1.125	10.500	41.08	0.6013	3.338	2.75	270.0	135,000.0	140.0
		160	1.312	10.126	47.14	0.5592	3.338	2.66	251.0	125,500.0	161.0
14	14.0	10	0.250	13.500	10.80	0.9940	3.665	3.54	446.0	223,000.0	36.8
		20	0.312	13.376	13.42	0.9750	3.665	3.51	438.0	219,000.0	45.7
		30	0.375	13.250	16.05	0.9575	3.665	3.47	430.0	215,000.0	54.6
		40	0.437	13.126	18.61	0.9397	3.665	3.44	422.0	211,000.0	63.3
		60	0.593	12.814	24.98	0.8956	3.665	3.36	402.0	201,000.0	85.0
		80	0.750	12.500	31.22	0.8522	3.665	3.28	382.0	191,000.0	107.0
		100	0.937	12.126	38.45	0.8020	3.665	3.18	360.0	180,000.0	131.0
		120	1.062	11.876	43.17	0.7693	3.665	3.11	345.0	172,500.0	147.0
		140	1.250	11.500	50.07	0.7213	3.665	3.01	324.0	162,000.0	171.0
		160	1.406	11.188	55.63	0.6827	3.665	2.93	306.0	153,000.0	190.0
16	16.0	10	0.250	15.500	12.37	1.3104	4.189	4.06	587.0	293,500.0	42.1
		20	0.312	15.376	15.38	1.2895	4.189	4.03	578.0	289,000.0	52.3
		30	0.375	15.250	18.41	1.2680	4.189	4.00	568.0	284,000.0	62.6
		40	0.500	15.000	24.35	1.2272	4.189	3.93	550.0	275,000.0	82.8

(continued)

A.3 (*continued*)

Nominal pipe size, in	Outside diam., in	Schedule no.	Wall thickness, in	Inside diam., in	Cross-sectional area metal, in^2	Inside sectional area, ft^2	Circumference, ft, or surface, ft^2/ft, of length		Capacity at 1 ft/s velocity		Weight of pipe, lb/ft
							Outside	Inside	U.S. gal/min	lb/h water	
		60	0.656	14.688	31.62	1.1766	4.189	3.85	528.0	264,000.0	108.0
		80	0.843	14.314	40.14	1.1175	4.189	3.76	500.0	250,000.0	137.0
		100	1.031	13.938	48.48	1.0596	4.189	3.65	474.0	237,000.0	165.0
		120	1.218	13.564	56.56	1.0035	4.189	3.56	450.0	225,000.0	193.0
		140	1.437	13.126	65.74	0.9397	4.189	3.44	422.0	211,000.0	224.0
		160	1.562	12.876	70.85	0.9043	4.189	3.37	405.0	202,500.0	241.0
18	18.0	10	0.250	17.50	13.94	1.6703	4.712	4.59	748.0	374,000.0	47.4
		20	0.312	17.376	17.34	1.6468	4.712	4.55	738.0	369,000.0	59.0
		30	0.437	17.126	24.11	1.5993	4.712	4.49	717.0	358,500.0	82.0
		40	0.562	16.876	30.79	1.5533	4.712	4.42	697.0	348,500.0	105.0
		60	0.718	15.564	38.98	1.4964	4.712	4.34	670.0	335,000.0	133.0
		80	0.937	16.126	50.23	1.4183	4.712	4.23	635.0	317,500.0	171.0
		100	1.156	15.688	61.17	1.3423	4.712	4.11	602.0	301,000.0	208.0
		120	1.343	15.314	70.28	1.2791	4.712	4.02	573.0	286,500 0	239.0
		140	1.562	14.876	80.66	1.2070	4.712	3.90	540.0	270,000.0	275.0
		160	1.750	14.500	89.34	1.1467	4.712	3.80	514.0	257,000.0	304.0
20	20.0	10	0.250	19.500	15.51	2.0740	5.236	5.11	930.0	465,000.0	52.8
		20	0.375	19.250	23.12	2.0211	5.236	5.05	902.0	451,000.0	78.6
		30	0.500	19.000	30.63	1.9689	5.236	4.98	883.0	441,500.0	105.0
		40	0.593	18.814	36.15	1.9305	5.236	4.94	866.0	433,000.0	123.0
		60	0.812	18.376	48.95	1.8317	5.236	4.81	826.0	413,000.0	167.0
		80	1.031	17.938	61.44	1.7550	5.236	4.70	787.0	393,500.0	209.0
		100	1.250	17.500	73.63	1.6703	5.236	4.59	750.0	375,000.0	251.0
		120	1.500	17.000	87.18	1.5762	5.236	4.46	707.0	353,500.0	297.0
		140	1.750	16.500	100.3	1.4849	5.236	4.32	665.0	332,500.0	342.0
		160	1.937	16.126	109.9	1.4183	5.236	4.22	635.0	317,500.0	374.0

The schedule number corresponds roughly to 10^3 allowable pressure per allowable stress. Thus, for a material with an allowable stress of 10,000 lb/in^2 a schedule 40 pipe would have an allowable pressure of 400 lb/in^2. [From *Chemical Engineers' Handbook*, by Perry, Chilton, and Kirkpatrick. Copyright © 1963, McGraw-Hill, Inc. Used by permission of the publisher.]

A.4 FLOW OF WATER THROUGH SCHEDULE 40 STEEL PIPE

Discharge		Pressure drop per 100 ft and velocity in schedule 40 pipe for water at 60°F															
gal/min	ft^3/s	Velocity, ft/s	Press. drop, lb/in^2	Velocity, ft/s	Press. drop, lb/in^2	Velocity, ft/s	Press. drop, lb/in^2	Velocity, ft/s	Press. drop, lb/in^2	Velocity, ft/s	Press. drop, lb/in^2	Velocity, ft/s	Press. drop, lb/in^2	Velocity, ft/s	Press. drop, lb/in^2	Velocity, ft/s	Press. drop, lb/in^2
		$\frac{1}{8}$ in		$\frac{1}{4}$ in		$\frac{3}{8}$ in		$\frac{1}{2}$ in		$\frac{3}{4}$ in							
0.2	0.000446	1.13	1.86	0.616	0.359												
0.3	0.000668	1.69	4.22	0.924	0.903	0.504	0.159	0.317	0.061								
0.4	0.000891	2.26	6.98	1.23	1.61	0.672	0.345	0.422	0.086								
0.5	0.00111	2.82	10.5	1.54	2.39	0.840	0.539	0.528	0.167	0.301	0.033						
0.6	0.00134	3.39	14.7	1.85	3.29	1.01	0.751	0.633	0.240	0.361	0.041						
0.8	0.00178	4.52	25.0	2.46	5.44	1.34	1.25	0.844	0.408	0.481	0.102						
												1 in		$1\frac{1}{4}$ in		$1\frac{1}{2}$ in	
1	0.00223	5.65	37.2	3.08	8.28	1.68	1.85	1.06	0.600	0.602	0.155	0.371	0.048				
2	0.00446	11.29	134.4	6.16	30.1	3.36	6.58	2.11	2.10	1.20	0.526	0.743	0.164	0.429	0.044		
3	0.00668			9.25	64.1	5.04	13.9	3.17	4.33	1.81	1.09	1.114	0.336	0.644	0.090	0.473	0.043
4	0.00891			12.33	111.2	6.72	23.9	4.22	7.42	2.41	1.83	1.49	0.565	0.858	0.150	0.630	0.071
5	0.01114					8.40	36.7	5.28	11.2	3.01	2.75	1.86	0.835	1.073	0.223	0.788	0.104
		2 in		$2\frac{1}{2}$ in													
6	0.01337	0.574	0.044			10.08	51.9	6.33	15.8	3.61	3.84	2.23	1.17	1.29	0.309	0.946	0.145
8	0.01782	0.765	0.073			13.44	91.1	8.45	27.7	4.81	6.60	2.97	1.99	1.72	0.518	1.26	0.241
10	0.02228	0.956	0.108	0.670	0.046			10.56	42.4	6.02	9.99	3.71	2.99	2.15	0.774	1.58	0.361
15	0.03342	1.43	0.224	1.01	0.094	3 in				9.03	21.6	5.57	6.36	3.22	1.63	2.37	0.755
20	0.04456	1.91	0.375	1.34	0.158	0.868	0.056			12.03	37.8	7.43	10.9	4.29	2.78	3.16	1.28
								$3\frac{1}{2}$ in		4 in							
25	0.05570	2.39	0.561	1.68	0.234	1.09	0.083	0.812	0.041			9.28	16.7	5.37	4.22	3.94	1.93
30	0.06684	2.87	0.786	2.01	0.327	1.30	0.114	0.974	0.056			11.14	23.8	6.44	5.92	4.73	2.72
35	0.07798	3.35	1.05	2.35	0.436	1.52	0.151	1.14	0.704	0.882	0.041	12.99	32.2	7.51	7.90	5.52	3.64
40	0.08912	3.83	1.35	2.68	0.556	1.74	0.192	1.30	0.095	1.01	0.052	14.85	41.5	8.59	10.24	6.30	4.65
45	0.1003	4.30	1.67	3.02	0.668	1.95	0.239	1.46	0.117	1.13	0.064			9.67	12.80	7.09	5.85
												5 in					
50	0.1114	4.78	2.03	3.35	0.839	2.17	0.288	1.62	0.142	1.26	0.076			10.74	15.66	7.88	7.15
60	0.1337	5.74	2.87	4.02	1.18	2.60	0.406	1.95	0.204	1.51	0.107			12.89	22.2	9.47	10.21
70	0.1560	6.70	3.84	4.69	1.59	3.04	0.540	2.27	0.261	1.76	0.143	1.12	0.047			11.05	13.71
80	0.1782	7.65	4.97	5.36	2.03	3.47	0.687	2.60	0.334	2.02	0.180	1.28	0.060			12.62	17.59
90	0.2005	8.60	6.20	6.03	2.53	3.91	0.861	2.92	0.416	2.27	0.224	1.44	0.074			14.20	22.0
														6 in			
100	0.2228	9.56	7.59	6.70	3.09	4.34	1.05	3.25	0.509	2.52	0.272	1.60	0.099	1.11	0.036	15.78	26.9
125	0.2785	11.97	11.76	8.38	4.71	5.43	1.61	4.06	0.769	3.15	0.415	2.01	0.135	1.39	0.055	19.72	41.4
150	0.3342	14.36	16.70	10.05	6.69	6.51	2.24	4.87	1.08	3.78	0.580	2.41	0.190	1.67	0.077		
175	0.3899	16.75	22.3	11.73	8.97	7.60	3.00	5.68	1.44	4.41	0.774	2.81	0.253	1.94	0.102		
200	0.4456	19.14	28.8	13.42	11.68	8.68	3.87	6.49	1.85	5.04	0.985	3.21	0.323	2.22	0.130		
																8 in	
225	0.5013	...	...	15.09	14.63	9.77	4.83	7.30	2.32	5.67	1.23	3.61	0.401	2.50	0.162	1.44	0.043
250	0.557	...	...	...	...	10.85	5.93	8.12	2.84	6.30	1.46	4.01	0.495	2.78	0.195	1.60	0.051
275	0.6127	...	...	...	...	11.94	7.14	8.93	3.40	6.93	1.79	4.41	0.583	3.05	0.234	1.76	0.061
300	0.6684	...	...	...	...	13.00	8.36	9.74	4.02	7.56	2.11	4.81	0.683	3.33	0.275	1.92	0.072
325	0.7241	...	...	...	...	14.12	9.89	10.53	4.09	8.19	2.47	5.21	0.797	3.61	0.320	2.08	0.083

(continued)

A.4 *(continued)*

Discharge		Pressure drop per 100 ft and velocity in schedule 40 pipe for water at 60°F															
gal/min	ft^3/s	Velocity, ft/s	Press. drop, lb/in^2	Velocity, ft/s	Press. drop, lb/in^2	Velocity, ft/s	Press. drop, lb/in^2	Velocity, ft/s	Press. drop, lb/in^2	Velocity, ft/s	Press. drop, lb/in^2	Velocity, ft/s	Press. drop, lb/in^2	Velocity, ft/s	Press. drop, lb/in^2	Velocity, ft/s	Press. drop, lb/in^2
350	0.7798			...	...	...	...	11.36	5.41	8.82	2.84	5.62	0.919	3.89	0.367	2.24	0.095
375	0.8355			...	...	...	...	12.17	6.18	9.45	3.25	6.02	1.05	4.16	0.416	2.40	0.108
400	0.8912			...	...	...	...	12.98	7.03	10.08	3.68	6.42	1.19	4.44	0.471	2.56	0.121
425	0.9469			...	...	...	...	13.80	7.89	10.71	4.12	6.82	1.33	4.72	0.529	2.73	0.136
450	1.003	10 in		...	...	...	...	14.61	8.80	11.34	4.60	7.22	1.48	5.00	0.590	2.89	0.151
475	1.059	1.93	0.054			...	...	...	...	11.97	5.12	7.62	1.64	5.27	0.653	3.04	0.166
500	1.114	2.03	0.059			...	...	...	...	12.60	5.65	8.02	1.81	5.55	0.720	3.21	0.182
550	1.225	2.24	0.071			...	...	...	...	13.85	6.79	8.82	2.17	6.11	0.861	3.53	0.219
600	1.337	2.44	0.083			...	...	...	...	15.12	8.04	9.63	2.55	6.66	1.02	3.85	0.258
650	1.448	2.64	0.097	12 in		...	...	...	...	...	...	10.43	2.98	7.22	1.18	4.17	0.301
700	1.560	2.85	0.112	2.01	0.047			...	...	...	...	11.23	3.43	7.78	1.35	4.49	0.343
750	1.671	3.05	0.127	2.15	0.054	14 in		...	...	...	...	12.03	3.92	8.33	1.55	4.81	0.392
800	1.782	3.25	0.143	2.29	0.061			...	...	...	...	12.83	4.43	8.88	1.75	5.13	0.443
850	1.894	3.46	0.160	2.44	0.068	2.02	0.042	...	...	...	...	13.64	5.00	9.44	1.96	5.45	0.497
900	2.005	3.66	0.179	2.58	0.075	2.13	0.047	...	...	...	...	14.44	5.58	9.99	2.18	5.77	0.554
950	2.117	3.86	0.198	2.72	0.083	2.25	0.052			...	...	15.24	6.21	10.55	2.42	6.09	0.613
1,000	2.228	4.07	0.218	2.87	0.091	2.37	0.057	16 in		...	...	16.04	6.84	11.10	2.68	6.41	0.675
1,100	2.451	4.48	0.260	3.15	0.110	2.61	0.068			...	...	17.65	8.23	12.22	3.22	7.05	0.807
1,200	2.674	4.88	0.306	3.44	0.128	2.85	0.080	2.18	0.042	...	...	...	...	13.33	3.81	7.70	0.948
1,300	2.896	5.29	0.355	3.73	0.150	3.08	0.093	2.36	0.048	...	...	...	...	14.43	4.45	8.33	1.11
1,400	3.119	5.70	0.409	4.01	0.171	3.32	0.107	2.54	0.055					15.55	5.13	8.98	1.28
1,500	3.342	6.10	0.466	4.30	0.195	3.56	0.122	2.72	0.063	18 in				16.66	5.85	9.62	1.46
1,600	3.565	6.51	0.527	4.59	0.219	3.79	0.138	2.90	0.071					17.77	6.61	10.26	1.65
1,800	4.010	7.32	0.663	5.16	0.276	4.27	0.172	3.27	0.088	2.58	0.050			19.99	8.37	11.54	2.08
2,000	4.456	8.14	0.808	5.73	0.339	4.74	0.209	3.63	0.107	2.87	0.060	20 in		22.21	10.3	12.82	2.55
2,500	5.570	10.17	1.24	7.17	0.515	5.93	0.321	4.54	0.163	3.59	0.091					16.03	3.94
3,000	6.684	12.20	1.76	8.60	0.731	7.11	0.451	5.45	0.232	4.30	0.129	3.46	0.075	24 in		19.24	5.59
3,500	7.798	14.24	2.38	10.03	0.982	8.30	0.607	6.35	0.312	5.02	0.173	4.04	0.101			22.44	7.56
4,000	8.912	16.27	3.08	11.47	1.27	9.48	0.787	7.26	0.401	5.74	0.222	4.62	0.129	3.19	0.052	25.65	9.80
4,500	10.03	18.31	3.87	12.90	1.60	10.67	0.990	8.17	0.503	6.46	0.280	5.20	0.162	3.59	0.065	28.87	12.2
5,000	11.14	20.35	4.71	14.33	1.95	11.85	1.21	9.08	0.617	7.17	0.340	5.77	0.199	3.99	0.079	...	...
6,000	13.37	24.41	6.74	17.20	2.77	14.23	1.71	10.89	0.877	8.61	0.483	6.93	0.280	4.79	0.111	...	...
7,000	15.60	28.49	9.11	20.07	3.74	16.60	2.31	12.71	1.18	10.04	0.652	8.08	0.376	5.59	0.150	...	...
8,000	17.82	...	...	22.93	4.84	18.96	2.99	14.52	1.51	11.47	0.839	9.23	0.488	6.38	0.192	...	...
9,000	20.05	...	...	25.79	6.09	21.34	3.76	16.34	1.90	12.91	1.05	10.39	0.608	7.18	0.242	...	...
10,000	22.28	...	...	28.66	7.46	23.71	4.61	18.15	2.34	14.34	1.28	11.54	0.739	7.98	0.294	...	...
12,000	26.74	...	...	34.40	10.7	28.45	6.59	21.79	3.33	17.21	1.83	13.85	1.06	9.58	0.416	...	...
14,000	31.19	...	...	...	...	33.19	8.89	25.42	4.49	20.08	2.45	16.16	1.43	11.17	0.562	...	...
16,000	35.65	...	...	...	...	...	...	29.05	5.83	22.95	3.18	18.47	1.85	12.77	0.723	...	...
18,000	40.10	...	...	...	...	...	...	32.68	7.31	25.82	4.03	20.77	2.32	14.36	0.907	...	...
20,000	44.56	...	...	...	...	...	...	36.31	9.03	28.69	4.93	23.08	2.86	15.96	1.12	...	...

From Crane Technical Paper No. 410, reproduced by permission of the Crane Company.

A.5 COMPRESSIBLE-FLOW TABLES FOR $k = 1.4$

$\mathcal{M}$	$\frac{P}{P_R}$	$\frac{\rho}{\rho_R}$	$\frac{T}{T_R}$	$\frac{A}{A^*}$	$\frac{V}{c^*}$	$\mathcal{M}$	$\frac{P}{P_R}$	$\frac{\rho}{\rho_R}$	$\frac{T}{T_R}$	$\frac{A}{A^*}$	$\frac{V}{c^*}$
0	1.0000	1.0000	1.0000	∞	0	0.25	0.9575	0.9694	0.9877	2.4027	0.27217
0.01	0.9999	1.0000	1.0000	57.8738	0.01095	0.26	0.9541	0.9670	0.9867	2.3173	0.28294
0.02	0.9997	0.9998	0.9999	28.9421	0.02191	0.27	0.9506	0.9645	0.9856	2.2385	0.29361
0.03	0.9994	0.9996	0.9998	19.3005	0.03286	0.28	0.9470	0.9619	0.9846	2.1656	0.30435
0.04	0.9989	0.9992	0.9997	14.4815	0.04381	0.29	0.9433	0.9592	0.9835	2.0979	0.31504
0.05	0.9983	0.9988	0.9995	11.5014	0.05476	0.30	0.9395	0,9564	0.9823	2.0351	0.32572
0.06	0.9975	0.9982	0.9993	9.6659	0.06570	0.31	0.9355	0.9535	0.9811	1.9765	0.33637
0.07	0.9966	0.9976	0.9990	8.2915	0.97664	0.32	0.9315	0.9506	0.9799	1.9219	0.34701
0.08	0.9955	0.9968	0.9987	7.2616	0.08758	0.33	0.9274	0.9476	0.9787	1.8707	0.35762
0.09	0.9944	0.9960	0.9984	6.4613	0.09851	0.34	0.9231	0.9445	0.9774	1.8229	0.36822
0.10	0.9930	0.9950	0.9980	5.8218	0.10944	0.35	0.9188	0.9413	0.9761	1.7780	0.37879
0.11	0.9916	0.9940	0.9976	5.2992	0.12035	0.36	0.9143	0.9380	0.9747	1.7358	0.38935
0.12	0.9900	0.9928	0.9971	4.8643	0.13126	0.37	0.9098	0.9347	0.9733	1.6961	0.39988
0.13	0.9883	0.9916	0.9966	4.4969	0.14217	0.38	0.9052	0.9313	0.9719	1.6587	0.41039
0.14	0.9864	0.9903	0.9961	4.1824	0.15306	0.39	0.9004	0.9278	0.9705	1.6234	0.42087
0.15	0.9844	0.9888	0.9955	3.9103	0.16395	0.40	0.8956	0.9243	0.9690	1.5901	0.43133
0.16	0.9823	0.9873	0.9949	3.6727	0.17482	0.41	0.8907	0.9207	0.9675	1.5587	0.44177
0.17	0.9800	0.9857	0.9943	3.4635	0.18569	0.42	0.8857	0.9170	0.9659	1.5289	0.45218
0.18	0.9776	0.9840	0.9936	3.2779	0.19654	0.43	0.8807	0.9132	0.9643	1.5007	0.46257
0.19	0.9751	0.9822	0.9928	3.1123	0.20739	0.44	0.8755	0.9094	0.9627	1.4740	0.47293
0.20	0.9725	0.9803	0.9921	2.9635	0.21822	0.45	0.8703	0.9055	0.9611	1.4487	0.48326
0.21	0.9697	0.9783	0.9913	2.8293	0.22904	0.46	0.8650	0.9016	0.9594	1.4246	0.49357
0.22	0.9668	0.9762	0.9904	2.7076	0.23984	0.47	0.8596	0.8976	0.9577	1.4048	0.50385
0.23	0.9638	0.9740	0.9895	2.5968	0.25063	0.48	0.8541	0.8935	0.9560	1.3801	0.51410
0.24	0.9607	0.9718	0.9886	2.4956	0.26141	0.49	0.8486	0.8894	0.9542	1.3595	0.52433

(*continued*)

A.5 *(continued)*

$\mathcal{M}$	$\frac{P}{P_R}$	$\frac{\rho}{\rho_R}$	$\frac{T}{T_R}$	$\frac{A}{A^*}$	$\frac{V}{c^*}$
0.50	0.8430	0.8852	0.9524	1.3398	0.53452
0.51	0.8374	0.8809	0.9506	1.3212	0.54469
0.52	0.8317	0.8766	0.9487	1.3034	0.55483
0.53	0.8259	0.8723	0.9468	1.2865	0.56493
0.54	0.8201	0.8679	0.9449	1.2703	0.57501
0.55	0.8142	0.8634	0.9430	1.2550	0.58506
0.56	0.8082	0.8589	0.9410	1.2403	0.59507
0.57	0.8022	0.8544	0.9390	1.2263	0.60505
0.58	0.7962	0.8498	0.9370	1.2130	0.61501
0.59	0.7901	0.8451	0.9349	1.2003	0.62492
0.60	0.7840	0.8405	0.9328	1.1882	0.63481
0.61	0.7778	0.8357	0.9307	1.1767	0.64465
0.62	0.7716	0.8310	0.9286	1.1657	0.65443
0.63	0.7654	0.8262	0.9265	1.1552	0.66427
0.64	0.7591	0.8213	0.9243	1.1452	0.67402
0.65	0.7528	0.8164	0.9221	1.1356	0.68374
0.66	0.7465	0.8115	0.9199	1.1265	0.69342
0.67	0.7401	0.8066	0.9176	1.1179	0.70307
0.68	0.7338	0.8016	0.9153	1.1097	0.71268
0.69	0.7274	0.7966	0.9131	1.1018	0.72225
0.70	0.7209	0.7916	0.9107	1.0944	0.73179
0.71	0.7145	0.7865	0.9084	1.0873	0.74129
0.72	0.7080	0.7814	0.9061	1.0806	0.75076
0.73	0.7016	0.7763	0.9037	1.0742	0.76019
0.74	0.6951	0.7712	0.9013	1.0681	0.76958
0.75	0.6886	0.7660	0.8989	1.0624	0.77894
0.76	0.6821	0.7609	0.8964	1.0570	0.78825
0.77	0.6756	0.7557	0.8940	1.0519	0.79753
0.78	0.6691	0.7505	0.8915	1.0471	0.80677
0.79	0.6625	0.7452	0.8890	1.0425	0.81597
0.80	0.6560	0.7400	0.8865	1.0382	0.82514
0.81	0.6495	0.7347	0.8840	1.0342	0.83426
0.82	0.6430	0.7295	0.8815	1.0305	0.84335
0.83	0.6365	0.7242	0.8789	1.0270	0.85239
0.84	0.6300	0.7189	0.8763	1.0237	0.86140
0.85	0.6235	0.7136	0.8737	1.0207	0.87037
0.86	0.6170	0.7083	0.8711	1.0179	0.87929
0.87	0.6106	0.7030	0.8685	1.0153	0.88818
0.88	0.6041	0.6977	0.8659	1.0129	0.89703
0.89	0.5977	0.6924	0.8632	1.0108	0.90583
0.90	0.5913	0.6870	0.8606	1.0089	0.91460
0.91	0.5849	0.6817	0.8579	1.0071	0.92332
0.92	0.5785	0.6764	0.8552	1.0056	0.93201
0.93	0.5721	0.6711	0.8525	1.0043	0.94065
0.94	0.5658	0.6658	0.8498	1.0031	0.94925
0.95	0.5595	0.6604	0.8471	1.0022	0.95781
0.96	0.5532	0.6551	0.8444	1.0014	0.96633
0.97	0.5469	0.6498	0.8416	1.0008	0.97481
0.98	0.5407	0.6445	0.8389	1.0003	0.98325
0.99	0.5345	0.6392	0.8361	1.0001	0.99165
1.00	0.5283	0.6339	0.8333	1.0000	1.00000

$\mathcal{M}$ or $\mathcal{M}_x$	$\frac{P}{P_R}$	$\frac{\rho}{\rho_R}$	$\frac{T}{T_R}$	$\frac{A}{A^*}$	$\frac{V}{c^*}$	$\mathcal{M}_y$	$\frac{P_y}{P_x}$	$\frac{\rho_y}{\rho_x}$	$\frac{T_y}{T_x}$	$\frac{P_{R_y}}{P_{R_x}}$	$\frac{P_x}{P_{R_y}}$
1.00	0.5283	0.6339	0.8333	1.000	1.00000	1.000	1.000	1.000	1.000	1.000	0.5283
1.01	0.5221	0.6287	0.8306	1.000	1.00831	0.9901	1.023	1.017	1.007	1.000	0.5221
1.02	0.5160	0.6234	0.8278	1.000	1.01658	0.9805	1.047	1.033	1.013	1.000	0.5160
1.03	0.5099	0.6181	0.8250	1.001	1.02481	0.9712	1.071	1.050	1.020	1.000	0.5100
1.04	0.5039	0.6129	0.8222	1.001	1.03300	0.9620	1.095	1.067	1.026	0.9999	0.5039
1.05	0.4979	0.6077	0.8193	1.002	1.04114	0.9531	1.120	1.084	1.033	0.9999	0.4980
1.06	0.4919	0.6024	0.8165	1.003	1.04925	0.9444	1.144	1.101	1.089	0.9997	0.4920
1.07	0.4860	0.5972	0.8137	1.004	1.05731	0.9360	1.169	1.118	1.046	0.9996	0.4861
1.08	0.4800	0.5920	0.8108	1.005	1.06533	0.9277	1.194	1.135	1.052	0.9994	0.4803
1.09	0.4742	0.5869	0.8080	1.006	1.07331	0.9196	1.219	1.152	1.059	0.9992	0.4746
1.10	0.4684	0.5817	0.8052	1.008	1.08124	0.9118	1.245	1.169	1.065	0.9989	0.4689
1.11	0.4626	0.5766	0.8023	1.010	1.08913	0.9041	1.271	1.186	1.071	0.9986	0.4632
1.12	0.4568	0.5714	0.7994	1.011	1.09699	0.8966	1.297	1.203	1.078	0.9982	0.4576
1.13	0.4511	0.5663	0.7966	1.013	1.10479	0.8892	1.323	1.221	1.084	0.9978	0.4521
1.14	0.4455	0.5612	0.7937	1.015	1.11256	0.8820	1.350	1.238	1.090	0.9973	0.4467
1.15	0.4398	0.5562	0.7908	1.017	1.12029	0.8750	1.376	1.255	1.097	0.9967	0.4413
1.16	0.4343	0.5511	0.7879	1.020	1.12797	0.8682	1.403	1.272	1.103	0.9961	0.4360
1.17	0.4287	0.5461	0.7851	1.022	1.13561	0.8615	1.430	1.290	1.109	0.9953	0.4307
1.18	0.4232	0.5411	0.7822	1.025	1.14321	0.8549	1.458	1.307	1.115	0.9946	0.4255
1.19	0.4178	0.5361	0.7793	1.026	1.15077	0.8485	1.485	1.324	1.122	0.9937	0.4204
1.20	0.4124	0.5311	0.7764	1.030	1.15828	0.8422	1.513	1.342	1.128	0.9928	0.4154
1.21	0.4070	0.5262	0.7735	1.033	1.16565	0.8360	1.541	1.359	1.134	0.9918	0.4104
1.22	0.4017	0.5213	0.7706	1.037	1.17319	0.8300	1.570	1.376	1.141	0.9907	0.4055
1.23	0.3964	0.5164	0.7677	1.040	1.18057	0.8241	1.598	1.394	1.147	0.9896	0.4006
1.24	0.3912	0.5115	0.7648	1.043	1.18792	0.8183	1.627	1.411	1.153	0.9884	0.3958
1.25	0.3861	0.5067	0.7619	1.047	1.19523	0.8126	1.656	1.429	1.159	0.9871	0.3911
1.26	0.3809	0.5019	0.7590	1.050	1.20249	0.8071	1.686	1.446	1.166	0.9857	0.3865
1.27	0.3759	0.4971	0.7561	1.054	1.20972	0.8017	1.715	1.463	1.172	0.9842	0.3819
1.28	0.3708	0.4923	0.7532	1.058	1.21690	0.7963	1.745	1.481	1.178	0.9827	0.3774
1.29	0.3658	0.4876	0.7503	1.062	1.22404	0.7911	1.775	1.498	1.185	0.9811	0.3729

(*continued*)

A.5 *(continued)*

$\mathcal{M}$ or $\mathcal{M}_x$	$\frac{P}{P_R}$	$\frac{\rho}{\rho_R}$	$\frac{T}{T_R}$	$\frac{A}{A^*}$	$\frac{V}{c^*}$	$\mathcal{M}_y$	$\frac{P_y}{P_x}$	$\frac{\rho_y}{\rho_x}$	$\frac{T_y}{T_x}$	$\frac{P_{R_y}}{P_{R_x}}$	$\frac{P_x}{P_{R_y}}$
1.30	0.3609	0.4829	0.7474	1.066	1.23114	0.7860	1.805	1.516	1.191	0.9794	0.3685
1.31	0.3560	0.4782	0.7445	1.071	1.23819	0.7809	1.835	1.533	1.197	0.9776	0.3642
1.32	0.3512	0.4736	0.7416	1.075	1.24521	0.7760	1.866	1.551	1.204	0.9758	0.3599
1.33	0.3464	0.4690	0.7387	1.080	1.25218	0.7712	1.897	1.568	1.210	0.9738	0.3557
1.34	0.3417	0.4644	0.7358	1.084	1.25912	0.7664	1.928	1.585	1.216	0.9718	0.3516
1.35	0.3370	0.4598	0.7329	1.089	1.26601	0.7618	1.960	1.603	1.223	0.9697	0.3475
1.36	0.3323	0.4553	0.7300	1.094	1.27286	0.7572	1.991	1.620	1.229	0.9676	0.3435
1.37	0.3277	0.4508	0.7271	1.099	1.27968	0.7527	2.023	1.638	1.235	0.9653	0.3395
1.38	0.3232	0.4463	0.7242	1.104	1.28645	0.7483	2.055	1.655	1.242	0.9630	0.3356
1.39	0.3187	0.4418	0.7213	1.109	1.29318	0.7440	2.087	1.672	1.248	0.9607	0.3317
1.40	0.3142	0.4374	0.7184	1.115	1.29987	0.7397	2.120	1.690	1.255	0.9582	0.3280
1.41	0.3098	0.4330	0.7155	1.120	1.30652	0.7355	2.153	1.707	1.261	0.9557	0.3242
1.42	0.3055	0.4287	0.7126	1.126	1.31313	0.7314	2.186	1.724	1.268	0.9531	0.3205
1.43	0.3012	0.4244	0.7097	1.132	1.31970	0.7274	2.219	1.742	1.274	0.9504	0.3169
1.44	0.2969	0.4201	0.7069	1.138	1.32623	0.7235	2.253	1.759	1.281	0.9476	0.3133
1.45	0.2927	0.4158	0.7040	1.144	1.33272	0.7196	2.286	1.776	1.287	0.9448	0.3098
1.46	0.2886	0.4116	0.7011	1.150	1.33917	0.7157	2.320	1.793	1.294	0.9420	0.3063
1.47	0.2845	0.4074	0.6982	1.156	1.34558	0.7120	2.354	1.811	1.300	0.9390	0.3029
1.48	0.2804	0.4032	0.6954	1.163	1.35195	0.7083	2.389	1.828	1.307	0.9360	0.2996
1.49	0.2764	0.3991	0.6925	1.169	1.35828	0.7047	2.423	1.845	1.314	0.9329	0.2962
1.50	0.2724	0.3950	0.6897	1.176	1.36458	0.7011	2.458	1.862	1.320	0.9298	0.2930
1.51	0.2685	0.3909	0.6868	1.183	1.37083	0.6976	2.493	1.879	1.327	0.9266	0.2898
1.52	0.2646	0.3869	0.6840	1.190	1.37705	0.6941	2.529	1.896	1.334	0.9233	0.2866
1.53	0.2608	0.3829	0.6811	1.197	1.38322	0.6907	2.564	1.913	1.340	0.9200	0.2835
1.54	0.2570	0.3789	0.6783	1.204	1.38936	0.6874	2.600	1.930	1.347	0.9166	0.2804
1.55	0.2533	0.3750	0.6754	1.212	1.39546	0.6841	2.636	1.947	1.354	0.9132	0.2773
1.56	0.2496	0.3710	0.6726	1.219	1.40152	0.6809	2.673	1.964	1.361	0.9097	0.2744
1.57	0.2459	0.3672	0.6698	1.227	1.40755	0.6777	2.709	1.981	1.367	0.9061	0.2714

1.58	0.2423	0.3633	0.6670	1.234	1.41353	0.6746	2.746	1.998	1.374	0.9026	0.2685
1.59	0.2388	0.3595	0.6642	1.242	1.41948	0.6715	2.783	2.015	1.381	0.8989	0.2656
1.60	0.2353	0.3557	0.6614	1.250	1.42539	0.6684	2.820	2.032	1.388	0.8952	0.2628
1.61	0.2318	0.3520	0.6586	1.258	1.43127	0.6655	2.857	2.049	1.395	0.8915	0.2600
1.62	0.2284	0.3483	0.6558	1.267	1.43710	0.6625	2.895	2.065	1.402	0.8877	0.2573
1.63	0.2250	0.3446	0.6530	1.275	1.44290	0.6596	2.933	2.082	1.409	0.8838	0.2546
1.64	0.2217	0.3409	0.6502	1.284	1.44866	0.6568	2.971	2.099	1.416	0.8799	0.2519
1.65	0.2184	0.3373	0.6475	1.292	1.45439	0.6540	3.010	2.115	1.423	0.8760	0.2493
1.66	0.2151	0.3337	0.6447	1.301	1.46008	0.6512	3.048	2.132	1.430	0.8720	0.2467
1.67	0.2119	0.3302	0.6419	1.310	1.46573	0.6485	3.087	2.148	1.437	0.8680	0.2442
1.68	0.2088	0.3266	0.6392	1.319	1.47135	0.6458	3.126	2.165	1.444	0.8640	0.2417
1.69	0.2057	0.3232	0.6364	1.328	1.47693	0.6431	3.165	2.181	1.451	0.8598	0.2392
1.70	0.2026	0.3197	0.6337	1.338	1.48247	0.6405	3.205	2.198	1.458	0.8557	0.2368
1.71	0.1996	0.3163	0.6310	1.347	1.48798	0.6380	3.245	2.214	1.466	0.8516	0.2344
1.72	0.1966	0.3129	0.6283	1.357	1.49345	0.6355	3.285	2.230	1.473	0.8474	0.2320
1.73	0.1936	0.3095	0.6256	1.367	1.49889	0.6330	3.325	2.247	1.480	0.8431	0.2296
1.74	0.1907	0.3062	0.6229	1.376	1.50429	0.6305	3.366	2.263	1.487	0.8389	0.2273
1.75	0.1878	0.3029	0.6202	1.386	1.50966	0.6281	3.406	2.279	1.495	0.8346	0.2251
1.76	0.1850	0.2996	0.6175	1.397	1.51499	0.6257	3.447	2.295	1.502	0.8302	0.2228
1.77	0.1822	0.2964	0.6148	1.407	1.52029	0.6234	3.488	2.311	1.509	0.8259	0.2206
1.78	0.1794	0.2931	0.6121	1.418	1.52555	0.6210	3.530	2.327	1.517	0.8215	0.2184
1.79	0.1767	0.2900	0.6095	1.428	1.53078	0.6188	3.571	2.343	1.524	0.8171	0.2163
1.80	0.1740	0.2868	0.6068	1.439	1.53598	0.6165	3.613	2.359	1.532	0.8127	0.2142
1.81	0.1714	0.2837	0.6041	1.450	1.54114	0.6143	3.655	2.375	1.539	0.8082	0.2121
1.82	0.1688	0.2806	0.6015	1.461	1.54626	0.6121	3.698	2.391	1.547	0.8038	0.2100
1.83	0.1662	0.2776	0.5989	1.472	1.55136	0.6099	3.740	2.407	1.554	0.7993	0.2080
1.84	0.1637	0.2745	0.5963	1.484	1.55642	0.6078	3.783	2.422	1.562	0.7948	0.2060
1.85	0.1612	0.2715	0.5936	1.495	1.56145	0.6057	3.826	2.438	1.569	0.7902	0.2040
1.86	0.1587	0.2686	0.5910	1.507	1.56644	0.6036	3.870	2.454	1.577	0.7857	0.2020
1.87	0.1563	0.2656	0.5884	1.519	1.57140	0.6016	3.913	2.469	1.585	0.7811	0.2001
1.88	0.1539	0.2627	0.5859	1.531	1.57633	0.5996	3.957	2.485	1.592	0.7765	0.1982
1.89	0.1516	0.2598	0.5833	1.543	1.58123	0.5976	4.001	2.500	1.600	0.7720	0.1963

(*continued*)

A.5 (*continued*)

$\mathcal{M}$ or $\mathcal{M}_x$	$\frac{P}{P_R}$	$\frac{\rho}{\rho_R}$	$\frac{T}{T_R}$	$\frac{A}{A^*}$	$\frac{V}{c^*}$	$\mathcal{M}_y$	$\frac{P_y}{P_x}$	$\frac{\rho_y}{\rho_x}$	$\frac{T_y}{T_x}$	$\frac{P_{R_y}}{P_{R_x}}$	$\frac{P_x}{P_{R_y}}$
1.90	0.1492	0.2570	0.5807	1.555	1.58609	0.5956	4.045	2.516	1.608	0.7674	0.1945
1.91	0.1470	0.2542	0.5782	1.568	1.59092	0.5937	4.089	2.531	1.616	0.7627	0.1927
1.92	0.1447	0.2514	0.5756	1.580	1.59572	0.5918	4.134	2.546	1.624	0.7581	0.1909
1.93	0.1425	0.2486	0.5731	1.593	1.60049	0.5899	4.179	2.562	1.631	0.7535	0.1891
1.94	0.1403	0.2459	0.5705	1.606	1.60523	0.5880	4.244	2.577	1.639	0.7488	0.1873
1.95	0.1381	0.2432	0.5680	1.619	1.60993	0.5862	4.270	2.592	1.647	0.7442	0.1856
1.96	0.1360	0.2405	0.5655	1.633	1.61460	0.5844	4.315	2.607	1.655	0.7395	0.1839
1.97	0.1339	0.2378	0.5630	1.646	1.61925	0.5826	4.361	2.622	1.663	0.7349	0.1822
1.98	0.1318	0.2352	0.5605	1.660	1.62386	0.5808	4.407	2.637	1.671	0.7302	0.1806
1.99	0.1298	0.2326	0.5580	1.674	1.62844	0.5791	4.453	2.652	1.679	0.7255	0.1789
2.00	0.1278	0.2300	0.5556	1.688	1.63299	0.5774	4.500	2.667	1.688	0.7209	0.1773
2.01	0.1258	0.2275	0.5531	1.702	1.63751	0.5757	4.547	2.681	1.696	0.7162	0.1757
2.02	0.1239	0.2250	0.5506	1.716	1.64201	0.5740	4.594	2.696	1.704	0.7115	0.1741
2.03	0.1220	0.2225	0.5482	1.730	1.64647	0.5723	4.641	2.711	1.712	0.7069	0.1726
2.04	0.1201	0.2200	0.5458	1.745	1.65090	0.5707	4.689	2.725	1.720	0.7022	0.1710
2.05	0.1182	0.2176	0.5433	1.760	1.65530	0.5691	4.736	2.740	1.729	0.6975	0.1695
2.06	0.1164	0.2152	0.5409	1.775	1.65967	0.5675	4.784	2.755	1.737	0.6928	0.1680
2.07	0.1146	0.2128	0.5385	1.790	1.66402	0.5659	4.832	2.769	1.745	0.6882	0.1665
2.08	0.1128	0.2104	0.5361	1.806	1.66833	0.5643	4.881	2.783	1.754	0.6835	0.1651
2.09	0.1111	0.2081	0.5337	1.821	1.67262	0.5628	4.929	2.798	1.762	0.6789	0.1636
2.10	0.1094	0.2058	0.5313	1.837	1.67687	0.5613	4.978	2.812	1.770	0.6742	0.1622
2.11	0.1077	0.2035	0.5290	1.853	1.68110	0.5598	5.027	2.826	1.779	0.6696	0.1608
2.12	0.1060	0.2013	0.5266	1.869	1.68530	0.5583	5.077	2.840	1.787	0.6619	0.1594
2.13	0.1043	0.1990	0.5243	1.885	1.68947	0.5568	5.126	2.854	1.796	0.6603	0.1580
2.14	0.1027	0.1968	0.5219	1.902	1.69362	0.5554	5.176	2.868	1.805	0.6557	0.1567

From NASA Report 1135, 1953; reproduced by permission of the publisher.

A.6 FLUID DENSITIES

For gases at low pressures, the density is given to satisfactory accuracy by the perfect-gas law

$$\rho = \frac{MP}{RT} \tag{A.1}$$

(Most authors use the terms "ideal gas" and "perfect gas" to mean exactly the same thing. Some authors distinguish between them. Those who make this distinction indicate that both ideal and perfect gases obey Eq. A.1, but that a perfect gas has a heat capacity which is constant, independent of temperature, while an ideal gas has a heat capacity which changes with changing temperature. This distinction plays some role in some calculation methods for high-temperature gas processes, but is generally unimportant and is made less often in current books than in older ones. Here we treat the terms "ideal gas" and "perfect gas" as identical and mean only a gas whose behavior is represented by Eq. A.1.) Values of M for common gases are shown in App. A.7, and R is a universal constant whose values in various units are shown in App. A.9. For higher pressures one must account for departures from the perfect-gas law. For many gases there are published data on the behavior of that specific gas, e.g., the steam tables. For quick estimates the best procedure is to use the compressibility factor z, defined so that

$$\rho = \frac{MP}{zRT} \tag{A.2}$$

Clearly, z is identically 1 for perfect gases. For real gases it can be shown that z for all gases is given *approximately* by

$$z = z(P_R, T_R) \tag{A.3}$$

where $P_R = P/P_{\text{crit}}$ and $T_R = T/T_{\text{crit}}$. This relation is shown graphically in App. A.8. This simple relationship is accurate to plus or minus a few percent for values of z near 1, but may have errors of ± 10 percent for values of z less than about 0.6. If greater accuracy is needed, either find specific data for the gas in question, or see Reid et al. [1] for more details on estimating the value of z. Values of T_{crit} and P_{crit} for some common gases are given in App. A.7.

For any fluid (or solid) the density can be written as a Taylor series:

$$\rho = \rho_0 + \frac{\partial \rho}{\partial T}(T - T_0) + \frac{\partial \rho}{\partial P}(P - P_0) + \frac{\partial^2 \rho}{\partial T^2}\left[\frac{(T-T_0)^2}{2}\right] + \frac{\partial^2 \rho}{\partial P^2}\left[\frac{(P-P_0)^2}{2}\right]$$
$$+ \frac{\partial^2 \rho}{\partial P\,\partial T}(P - P_0)(T - T_0) + \cdots \tag{A.4}$$

This equation is correct for all conditions, if one uses an infinite series of terms. For liquids at temperatures well below the critical temperature (say, 200°F below the critical) and for solids, one may neglect all but the first three terms on the right and write the equation as

$$\rho = \rho_0\left[1 + \frac{1}{\rho_0}\left(\frac{\partial\rho}{\partial T}\right)(T - T_0) + \frac{1}{\rho_0}\left(\frac{\partial\rho}{\partial P}\right)(P - P_0)\right]$$

$$= \rho_0[1 - \alpha(T - T_0) + \beta(P - P_0)] \tag{A.5}$$

Here

$$\alpha = -\frac{1}{\rho_0}\frac{\partial\rho}{\partial T} = \text{coefficient of thermal expansion} \tag{A.6}$$

$$\beta = -\frac{1}{\rho_0}\frac{\partial\rho}{\partial P} = \text{isothermal compressibility} = \frac{1}{\text{bulk modulus}} \tag{A.7}$$

Values of the density, isothermal compressibility, and coefficient of thermal expansion for some common fluids are listed in App. A.10. Notice that for most fluids the effect of a change in temperature is more significant than the effect of a change in pressure. Normally a temperature decrease of 1°F will have the same effect as a pressure increase of 100 psi.

Generally the bulk modulus is practically constant over a wide range of pressures, but the coefficient of thermal expansion increases with increasing temperature. Therefore, the single value given in App. A.10 should not be used for temperature changes above 100°F. For a more complete set of values which includes the effect of increasing temperature on α, see Forsythe [2] and Lange [3]. The behavior of liquids near their critical states can be best estimated from App. A.8.

REFERENCES

1. R. C. Reid, J. M. Prausnitz, and T. K. Sherwood, *The Properties of Liquids and Gases*, 3d ed., McGraw-Hill, New York, 1977, chap. 3.
2. W. E. Forsythe, *Smithsonian Physical Tables*, 9th ed., Smithsonian Institute, Washington, 1954, p. 153. This table is widely reproduced in other texts.
3. N. A. Lange, *Handbook of Chemistry*, McGraw-Hill, New York, 1967, p. 1682.

A.7 SOME PROPERTIES OF GASES

Gas	Molecular weight, lbm/(lbmol)	T_{crit}, °R	P_{crit}, atm
Hydrogen	2	59.9	12.8
Helium	4	9.47	2.26
Nitrogen	28	227	33.5
Oxygen	32	278	47.9
Air (79% N_2, 21% O_2)	29	240	37
Water	18	1165	218.5
Carbon monoxide	28	239	34.5
Carbon dioxide	44	548	73.0
Freon-12	121	694	40.7
Typical gasoline	≈100	≈1000	≈25
Methane	16	343	45.8

More extensive tables are found in *International Critical Tables*, McGraw-Hill, New York, 1926, vol. 3, pp. 248–249.

A.8 COMPRESSIBILITY FACTOR

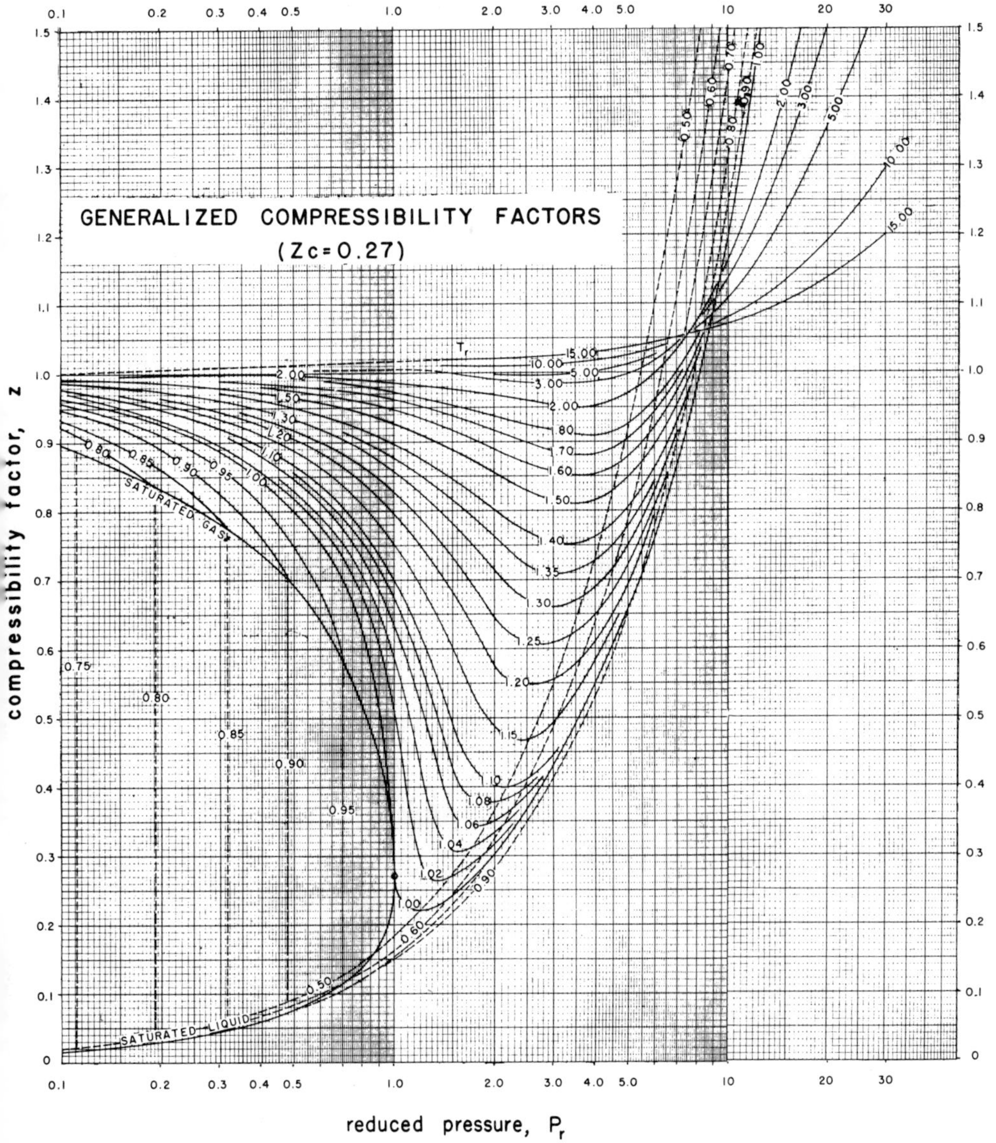

From O. A. Hougen et al., *Chemical Process Principles*, Wiley, New York, 1959. Reproduced by permission of the publishers.

A.9 VALUES OF THE UNIVERSAL GAS CONSTANT

$$R = \frac{10.73(\mathrm{lbf/in^2})\mathrm{ft^3}}{\mathrm{lbmol \cdot °R}} = \frac{0.7302\ \mathrm{atm \cdot ft^3}}{\mathrm{lbmol \cdot °R}}$$

$$= \frac{8.314\ \mathrm{m^3 \cdot Pa}}{\mathrm{mol \cdot K}} = \frac{0.08206\ \mathrm{L \cdot atm}}{\mathrm{mol \cdot K}} = \frac{0.08314\ \mathrm{L \cdot bar}}{\mathrm{mol \cdot K}}$$

$$= \frac{1.987\ \mathrm{Btu}}{\mathrm{lbmol \cdot °R}} = \frac{1.987\ \mathrm{cal}}{\mathrm{mol \cdot K}} = \frac{1.987\ \mathrm{kcal}}{\mathrm{kgmol \cdot K}} = \frac{8.314\ \mathrm{J}}{\mathrm{mol \cdot K}}$$

A.10 SOME PROPERTIES OF LIQUIDS

Liquid	Density,† $\mathrm{lbm/ft^3}$	$10^3\alpha(1/\mathrm{F°})$	$10^5\beta(1/\mathrm{psi})$
Hydrogen	4.4	34	11
Helium	9.1	15	48
Typical gasoline	43	0.7	0.7
Benzene	54.6	0.67	0.7
Water	62.3	0.11	0.3
Carbon tetrachloride	99.2	0.67	0.7
Mercury	845	0.10	0.3

† Measured at 1 atm and 60°F, except for hydrogen and helium, which are at their 1-atm boiling points, 20 and 2.1 K, respectively.

APPENDIX B

PROOF THAT FOR A FLUID AT REST THE PRESSURE IS THE SAME IN ALL DIRECTIONS†

The laws of fluid mechanics work perfectly well in any gravity situation, including the zero gravity of an earth satellite. The following proof is shown for zero gravity, because that makes it simple. The result is then extended for a finite gravity field. The proof rests on the definition of a fluid: "A fluid, when subject to any shear stress, moves."

Consider a prism of fluid, as shown in Fig. B.1. For the fluid to be at rest, there can be no shear forces on any surfaces of the prism. Furthermore, because it is at rest, there are no unbalanced forces; i.e., the sum of the forces in any direction is zero. For the sum of the forces in the z direction to be zero, the pressure force on $ABCE$ must equal the z component of the pressure force on $BCFD$, or

$$P_{\text{bottom}}\,\Delta x\,\Delta y = P_{\text{sloping face}}\,\Delta y\,(\text{length } BD)\cos\theta \tag{B.1}$$

But length BD is exactly $\Delta x/\cos\theta$, so Eq. B.1 reduces to

$$P_{\text{bottom}} = P_{\text{sloping face}} \tag{B.2}$$

† See Sec. 1.6.

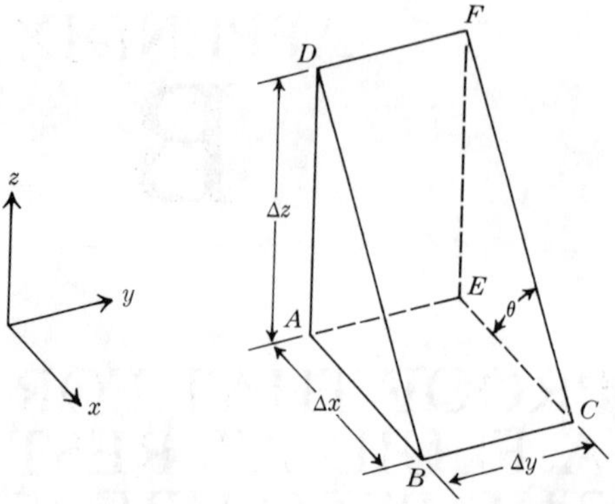

FIGURE B.1

Equation B.2 is true for any angle θ. It shows that the pressure in any direction is the same as the pressure vertically upward and hence that the pressure is the same in all directions.

In the foregoing it is shown that for the zero-gravity situation the pressure in an entire body of fluid at rest is the same in all directions. To apply the same reasoning to the prism in the case in which there is significant gravity, one includes in Eq. B.1 a term for the force of gravity on the fluid element. One then lets the size of the prism decrease, i.e., $\Delta x \to 0$, and observes that the gravity terms is proportional to Δx^3 whereas the pressure terms are proportional to Δx^2. So in the limit (i.e., at a point) the same argument holds as that given above for any value of the acceleration of gravity.

APPENDIX C

THE HYDRAULIC JUMP EQUATIONS†

We start with the continuity equation

$$V_1 z_1 = V_2 z_2 \tag{7.32}$$

and the one-dimensional momentum balance

$$0 = l\rho z_1 V_1 (V_1 - V_2) + \frac{l\rho g}{2}(z_1^2 - z_2^2) \tag{7.35}$$

We then divide Eq. 7.35 by $l\rho g/2$ and substitute for V_2 from Eq. 7.32 to find

$$0 = \frac{2V_1^2 z_1}{g}\left(1 - \frac{z_1}{z_2}\right) + z_1^2 - z_2^2 \tag{C.1}$$

we next multiply the far right-hand term in Eq. C.1 by z_2^2/z_2^2 and regroup, getting

$$0 = \frac{2V_1^2 z_1}{g}\left(1 - \frac{z_1}{z_2}\right) - z_2^2\left(1 - \frac{z_1^2}{z_2^2}\right) \tag{C.2}$$

† See Sec. 7.3.

and we factor $(1 - z_1^2/z_2^2)$ into $(1 + z_1/z_2)(1 - z_1/z_2)$ and divide by $(1 - z_1/z_2)$ to find

$$0 = \frac{2V_1^2 z_1}{g} - z_2^2\left(1 + \frac{z_1}{z_2}\right) \tag{C.3}$$

This can be multiplied out to give

$$0 = z_2^2 + z_2 z_1 - \frac{2V_1^2 z_1}{g} \tag{C.4}$$

which is a standard quadratic equation with solution

$$z_2 = \frac{-z_1}{2} \pm \sqrt{\left(\frac{z_1}{2}\right)^2 + \frac{2V_1^2 z_1}{g}} \tag{C.5}$$

APPENDIX D

PROPERTIES OF A PERFECT GAS

In several derivations in the text, we use the properties of a perfect gas. All these properties are derived here.

D.1 DEFINITIONS

A perfect gas is one whose pressure, density, and absolute temperature are related by

$$\frac{P}{\rho} = \frac{RT}{M} \tag{D.1}$$

Here R is the universal gas constant, whose values in various units are shown in App. A.9. The density in this equation is the mass density, or mass/volume. If we wish the molar density, moles/volume, we simply drop the M from Eq. D.1 and from all the equations in this appendix. There is rarely any question whether the mass density or molar density is being used in any of the equations in this book.

It is shown in any standard text on thermodynamics that for a perfect gas the enthalpy per unit mass h and the internal energy per unit mass u are functions of temperature alone; they do not depend on pressure. (The same is *not* true for real gases or solids or liquids.) Thus, we can define the heat capacity at constant pressure C_P and the heat capacity at constant volume C_V as follows:

$$C_P = \frac{dh}{dT} \qquad dh = C_P\, dT \tag{D.2}$$

$$C_V = \frac{du}{dT} \qquad du = C_V\, dT \tag{D.3}$$

For any material the enthalpy per unit mass is defined by $h = u + P/\rho$. Combining these definitions with Eqs. D.1, D.2, and D.3, we find

$$dh = du + d\left(\frac{P}{\rho}\right) \tag{D.4}$$

$$C_P\, dT = C_V\, dT + \frac{R}{M}\, dT \tag{D.5}$$

$$C_P = C_V + \frac{R}{M} \tag{D.6}$$

We now introduce the definition of k, using Eq. D.6:

$$k = \frac{C_P}{C_V} = \frac{C_V + R/M}{C_V} = 1 + \frac{R}{MC_V} \tag{D.7}$$

Rearranging produces

$$\frac{R}{MC_V} = k - 1 \tag{D.8}$$

or

$$C_V = \frac{R}{M(k-1)} \tag{D.9}$$

and

$$C_P = C_V + \frac{R}{M} = \frac{Rk}{M(k-1)} \tag{D.10}$$

D.2 ISENTROPIC RELATIONS

In any standard text on thermodynamics one may find this relation among entropy per unit mass, pressure, enthalpy per unit mass, temperature, and density:

$$dh = T\, ds + \frac{1}{\rho}\, dP \tag{D.11}$$

(This is true for any substance—solid, liquid, or gas.) For a constant-entropy process (i.e., an isentropic process), the first term on the right of the equals sign is zero. We then substitute for dh from Eq. D.2 and substitute for C_P from Eq. D.10 to find

$$\frac{Rk}{M(k-1)}\, dT_s = \frac{1}{\rho}\, dP_s \tag{D.12}$$

The subscript reminds us that this equation applies to isentropic processes only. Now we replace ρ by its equivalent from Eq. D.1:

$$\frac{Rk}{M(k-1)}\, dT_s = \frac{RT}{PM}\, dP_s \tag{D.13}$$

Canceling the R's and M's and rearranging, we find

$$\frac{k}{k-1}\,\frac{dT_s}{T} = \frac{dP_s}{P} \tag{D.14}$$

which is readily integrated to

$$\frac{k}{k-1}\ln T_s\big]_{T_1}^{T_2} = \ln P_s\big]_{P_1}^{P_2} \tag{D.15}$$

or

$$\left(\frac{T_2}{T_1}\right)_s^{k/(k-1)} = \left(\frac{P_2}{P_1}\right)_s \tag{D.16}$$

Substituting $\rho RT/M$ for P_1 and for P_2 and canceling R's and M's, we find

$$\left(\frac{T_2}{T_1}\right)_s^{k/(k-1)} = \left(\frac{T_2}{T_1}\right)_s\left(\frac{\rho_2}{\rho_1}\right)_s \tag{D.17}$$

or

$$\left(\frac{\rho_2}{\rho_1}\right)_s = \left(\frac{T_2}{T_1}\right)_s^{k/(k-1)}\left(\frac{T_1}{T_2}\right)_s = \left(\frac{T_2}{T_1}\right)_s^{k/(k-1)-1} = \left(\frac{T_2}{T_1}\right)_s^{1/(k-1)} \tag{D.18}$$

Returning now to Eq. D.17, we substitute $PM/(R\rho)$ for T_1 and for T_2 and cancel, finding

$$\left(\frac{P_2}{P_1}\right)_s = \left(\frac{P_2}{P_1}\,\frac{\rho_1}{\rho_2}\right)_s^{k/(k-1)} \tag{D.19}$$

Taking both sides to the $(k-1)/k$ power produces

$$\left(\frac{P_2}{P_1}\right)_s^{(k-1)/k} = \left(\frac{P_2}{P_1}\right)_s\left(\frac{\rho_1}{\rho_2}\right)_s \tag{D.20}$$

Therefore

$$\left(\frac{\rho_1}{\rho_2}\right)_s = \left(\frac{P_2}{P_1}\right)_s^{(k-1)/k}\left(\frac{P_1}{P_2}\right)_s = \left(\frac{P_2}{P_1}\right)_s^{-1/k} \tag{D.21}$$

This we rearrange to

$$\left(\frac{\rho_1}{\rho_2}\right)_s^k = \left(\frac{P_2}{P_1}\right)_s^{-1} = \left(\frac{P_1}{P_2}\right)_s \tag{D.22}$$

or

$$\left(\frac{P}{\rho^k}\right)_s = \text{const} \qquad \text{(D.23)}$$

Now we differentiate:

$$\frac{dP_s}{\rho^k} + P_s\left(\frac{-k\,d\rho_s}{\rho^{k+1}}\right) = 0 \qquad \text{(D.24)}$$

or

$$\frac{dP_s}{d\rho_s} = \left(\frac{\partial P}{\partial \rho}\right)_s = \frac{kP_s\rho_s^k}{\rho_s^{k+1}} = \frac{kP_s}{\rho_s} \qquad \text{(D.25)}$$

D.3 ENTROPY CHANGE

Solving Eq. D.11 for ds, we find

$$ds = \frac{dh}{T} - \frac{dP}{\rho T} \qquad \text{(D.26)}$$

Substituting from Eqs. D.3 and D.1 produces

$$ds = C_P\,\frac{dT}{T} - \left(\frac{R}{M}\,\frac{dP}{P}\right) \qquad \text{(D.27)}$$

Then, inserting Eq. D.11 and factoring, we get

$$ds = \frac{R}{M}\left(\frac{k}{k-1}\,\frac{dT}{T} - \frac{dP}{P}\right) \qquad \text{(D.28)}$$

which integrates to

$$\frac{M(s_2 - s_1)}{R} = \ln\left(\frac{T_2}{T_1}\right)^{k/(k-1)} - \ln\frac{P_2}{P_1} = \ln\left(\frac{T_2}{T_1}\right)^{k/(k-1)}\left(\frac{P_1}{P_2}\right) \qquad \text{(D.29)}$$

APPENDIX E

THE AREA RATIO†

Starting with Eq. 8.20

$$\frac{A_1}{A^*} = \frac{\rho^* V^*}{\rho_1 V_1} \tag{8.20}$$

we substitute from Eq. D.18:

$$\frac{\rho^*}{\rho_1} = \left(\frac{T^*}{T_1}\right)^{1/(k-1)} \tag{E.1}$$

and

$$\frac{V^*}{V_1} = \frac{\mathscr{M}^* c^*}{\mathscr{M}_1 c_1} = \frac{1}{\mathscr{M}_1} \frac{c^*}{c_1} \tag{E.2}$$

because $\mathscr{M}^* \equiv 1$. But, from Eq. 8.11,

$$\frac{c^*}{c_1} = \left(\frac{T^*}{T_1}\right)^{1/2} \tag{E.3}$$

† See Sec. 8.2, Eq. 8.21.

Substituting Eqs. E.1, E.2, and E.3 in Eq. 8.20, we find

$$\frac{A_1}{A^*} = \frac{1}{\mathcal{M}_1}\left(\frac{T^*}{T_1}\right)^{1/2}\left(\frac{T^*}{T_1}\right)^{1/(k-1)} = \frac{1}{\mathcal{M}_1}\left(\frac{T^*}{T_1}\right)^{(k+1)/[2(k-1)]} \quad \text{(E.4)}$$

However,

$$\frac{T^*}{T_1} = \frac{T_R/T_1}{T_R/T^*} = \frac{\mathcal{M}_1^2[(k-1)/2]+1}{1^2[(k-1)/2]+1} \quad \text{(E.5)}$$

so

$$\frac{A}{A^*} = \frac{1}{\mathcal{M}_1}\left\{\frac{\mathcal{M}_1^2[(k-1)/2]+1}{(k-1)/2+1}\right\}^{(k+1)/[2(k-1)]} \quad \text{(8.21)}$$

APPENDIX F

NORMAL SHOCK WAVES†

In this appendix we derive the equations for a normal shock wave in a perfect gas, starting with the continuity equation

$$\rho_x V_x = \rho_y V_y \tag{8.25}$$

the momentum balance

$$V_x - V_y = \frac{P_y}{\rho_y V_y} - \frac{P_x}{\rho_x V_x} \tag{8.26}$$

and the energy balance

$$\frac{T_R}{T_1} = \mathcal{M}_1^2\left(\frac{k-1}{2}\right) + 1 \tag{8.16}$$

From the perfect-gas law and the equation for the speed of sound in a perfect gas,

$$\frac{P}{\rho} = \frac{RT}{M} = \frac{c^2}{k} \tag{F.1}$$

† See Sec. 8.5.

Substituting this twice in Eq. 8.26, we find

$$V_x - V_y = \frac{c_y^2}{kV_y} - \frac{c_x^2}{kV_x} \tag{F.2}$$

Now we write Eq. 8.16 twice: once for any arbitrary state 1 and once for the critical state, at which $\mathcal{M} = 1$. For these states T_R is the same, because the flow is assumed adiabatic, so dividing one energy balance by the other, we find

$$\frac{T_R/T_1}{T_R/T^*} = \frac{T^*}{T_1} = \frac{\mathcal{M}_1^2[(k-1)/2]+1}{(k-1)/2+1} \tag{F.3}$$

but $T^*/T_1 = c^{*2}/c_1^2$ and $\mathcal{M}_1^2 = V_1^2 c_1^2$, so

$$\frac{c^{*2}}{c_1^2} = \frac{(V_1^2/c_1^2)[(k-1)/2]+1}{(k-1)/2+1} \tag{F.4}$$

Multiplying through by c_1^2, simplifying the denominator, and solving for c_1^2 give

$$c_1^2 = c^{*2}\,\frac{k+1}{2} - V_1^2\,\frac{k-1}{2} \tag{F.5}$$

Here we have let 1 be an arbitrary state, so we can let it be either x or y. Thus, we use Eq. F.5 twice to eliminate c_x^2 and c_y^2 from Eq. F.2:

$$V_x - V_y = \frac{c^{*2}\left(\dfrac{k+1}{2}\right) - V_y^2\left(\dfrac{k-1}{2}\right)}{kV_y} - \frac{c^{*2}\left(\dfrac{k+1}{2}\right) - V_x^2\left(\dfrac{k-1}{2}\right)}{kV_x} \tag{F.6}$$

Now we multiply both sides by $2kV_x/(k+1)$, finding

$$\frac{2kV_x}{k+1}(V_x - V_y) = \frac{V_x}{V_y}\left[c^{*2} - V_y^2\left(\frac{k-1}{k+1}\right)\right] - c^{*2} + V_x^2\left(\frac{k-1}{k+1}\right) \tag{F.7}$$

Regrouping gives

$$\frac{2kV_x}{k+1}(V_x - V_y) = c^{*2}\,\frac{V_x - V_y}{V_y} + \frac{k-1}{k+1}\,V_x(V_x - V_y) \tag{F.8}$$

Canceling $V_x - V_y$, multiplying by V_y, rearranging, and simplifying, we find

$$V_x V_y = c^{*2} \tag{F.9}$$

which is known as the *Prandtl relation* or the *Prandtl-Meyer relation*. Now, returning to Eq. F.4, we multiply by $(c_1/c^*)^2[(k-1)/2+1]$, finding

$$\frac{k-1}{2} + 1 = \left(\frac{V_1}{c^*}\right)^2\left(\frac{k-1}{2}\right) + \left(\frac{c_1}{c^*}\right)^2 \tag{F.10}$$

Substituting for $(c_1/c^*)^2$ from Eq. F.4 and simplifying, we find

$$\frac{k+1}{2} = \left(\frac{V_1}{c^*}\right)^2\left(\frac{k-1}{2}\right) + \frac{(k+1)/2}{\mathcal{M}_1^2(k-1)/2+1} \tag{F.11}$$

Multiplying by $(k-1)/2$ and rearranging give

$$\left(\frac{V_1}{c^*}\right)^2 = \frac{k+1}{k-1}\left[1 - \frac{1}{\mathcal{M}_1^2(k-1)/2+1}\right] = \frac{\mathcal{M}_1^2(k+1)}{\mathcal{M}_1^2(k-1)+2} \tag{F.12}$$

Equation F.12 is not specific for shock waves but applies to any steady, adiabatic flow of a perfect gas. The reader may check a few V/c^* values in App. A.5 to see that they have indeed been made up from Eq. F.12.

We may now use Eq. F.12 twice in Eq. F.9, finding

$$\frac{\mathcal{M}_x^2(k+1)}{\mathcal{M}_x^2(k-1)+2}\,\frac{\mathcal{M}_y^2(k+1)}{\mathcal{M}_y^2(k-1)+2} = 1 \tag{F.13}$$

from which

$$\mathcal{M}_y^2 = \frac{\mathcal{M}_x^2(k-1)+2}{\mathcal{M}_x^2(k+1)^2}\,[\mathcal{M}_y^2(k-1)+2] \tag{F.14}$$

Collecting $\mathcal{M}_y^2$'s and putting them over a common denominator gives

$$\mathcal{M}_y^2\,\frac{\mathcal{M}_x^2(k+1)^2 - (k-1)[\mathcal{M}_x^2(k-1)+2]}{\mathcal{M}_x^2(k+1)^2} = 2\,\frac{\mathcal{M}_x^2(k-1)+2}{\mathcal{M}_x^2(k+1)^2} \tag{F.15}$$

Dividing out the term on the left and simplifying, we find

$$\mathcal{M}_y^2 = \frac{1+\mathcal{M}_x^2(k-1)/2}{k\mathcal{M}_x^2-(k-1)/2} \tag{F.16}$$

Now that we have a relationship between the upstream and downstream Mach numbers, we can compute the other relations easily. Writing Eq. 8.16 twice gives

$$\frac{T_R/T_x}{T_R/T_y} = \frac{T_y}{T_x} = \frac{\mathcal{M}_x^2(k-1)/2+1}{\mathcal{M}_y^2-(k-1)/2+1} \tag{F.17}$$

Substituting for $\mathcal{M}_y$ from Eq. F.16, we get

$$\frac{T_y}{T_x} = \frac{\mathcal{M}_x^2\left(\dfrac{k-1}{2}\right)+1}{\dfrac{k-1}{2}\,\dfrac{1+\mathcal{M}_x^2(k-1)/2}{k\mathcal{M}_x^2-(k-1)/2}+1} \tag{F.18}$$

To find the velocity ratio, we write

$$\frac{V_x}{V_y} = \frac{V_x^2}{V_yV_x} = \frac{V_x^2}{c^{*2}} = \frac{\mathcal{M}_x^2(k+1)}{\mathcal{M}_x^2(k-1)+2} \tag{F.19}$$

Here we have substituted for V_xV_y from Eq. F.9 and then for V_x/c^* from Eq. (F.12).

From Eq. 8.25 we know that

$$\frac{\rho_y}{\rho_x} = \frac{V_x}{V_y} = \frac{\mathcal{M}_x^2(k+1)}{\mathcal{M}_x^2(k-1)+2} \tag{F.20}$$

Finally, we have P_y/P_x from

$$\frac{P_y}{P_x} = \frac{\rho_y}{\rho_x}\frac{T_x}{T_y} = \frac{\left[\dfrac{\mathcal{M}_x^2(k+1)}{\mathcal{M}_x^2(k-1)+2}\right]\left\{\dfrac{\left(\dfrac{k-1}{2}\right)\left[1+\mathcal{M}_x^2\left(\dfrac{k-1}{2}\right)\right]}{k\mathcal{M}_x^2-(k-1)/2}+1\right\}}{\mathcal{M}_x^2\left(\dfrac{k-1}{2}\right)+1} \tag{F.21}$$

which after some algebra simplifies to

$$\frac{P_y}{P_x} = \frac{2k}{k+1}\mathcal{M}_x^2 - \frac{k-1}{k+1} \tag{F.22}$$

Thus, we now have all the pertinent property ratios in terms of $\mathcal{M}_x$, which allows us to tabulate these ratios in App. A.5 and thus to solve normal shock wave problems very conveniently.

APPENDIX G

EQUATIONS FOR ADIABATIC, ZERO-CLEARANCE, ISENTROPIC PERFECT-GAS COMPRESSORS†

Starting with

$$PV^k = \text{constant} \tag{D.24}$$

we write

$$V = V_1\left(\frac{P_1}{P}\right)^{1/k} = \frac{nRT}{P}\left(\frac{P_1}{P}\right)^{1/k} \tag{G.1}$$

Then

$$\int_{P_1}^{P_2} V\,dP = \frac{nRT_1}{P_1}\,P_1^{1/k}\int_{P_1}^{P_2}\left(\frac{1}{P}\right)^{1/k} dP = \frac{nRT_1}{P_1^{(k-1)/k}}\int_{P_1}^{P_2}\left(\frac{1}{P}\right)^{1/k} dP \tag{G.2}$$

† See Sec. 9.3.

But

$$\int_{P_1}^{P_2} \left(\frac{1}{P}\right)^{1/k} dP = \int_{P_1}^{P_2} P^{-1/k}\, dP = \frac{1}{1-1/k}\, P^{1-1/k}\Big]_{P_1}^{P_2}$$

$$= \frac{k}{k-1}\left(P_2^{(k-1)/k} - P_1^{(k-1)/k}\right) \tag{G.3}$$

So

$$\int_{P_1}^{P_2} V\, dP = \frac{k}{k-1}\, nRT_1\, \frac{P_2^{(k-1)/k} - P_1^{(k-1)/k}}{P_1^{(k-1)/k}}$$

$$= \frac{k}{k-1}\, nRT_1\left[\left(\frac{P_2}{P_1}\right)^{(k-1)/k} - 1\right]$$

(9.15)

APPENDIX H

PROOF THAT THE CURVES OF CONSTANT ϕ AND OF CONSTANT ψ ARE PERPENDICULAR[†]

Starting with

$$V_x = -\left(\frac{\partial \phi}{\partial x}\right)_y = -\left(\frac{\partial \psi}{\partial y}\right)_x \qquad (10.41)$$

and

$$V_y = -\left(\frac{\partial \phi}{\partial y}\right)_x = \left(\frac{\partial \psi}{\partial x}\right)_y \qquad (10.42)$$

we introduce the identity proved in all calculus books

$$\left(\frac{\partial A}{\partial B}\right)_C = -\left(\frac{\partial A}{\partial C}\right)_B \left(\frac{\partial C}{\partial B}\right)_A \qquad (H.1)$$

which is true for any A, B, and C. Using this identity, we find

$$\left(\frac{\partial \phi}{\partial x}\right)_y = -\left(\frac{\partial \phi}{\partial y}\right)_x \left(\frac{\partial y}{\partial x}\right)_\phi \qquad (H.2)$$

† See Sec. 10.5.

$$\left(\frac{\partial \psi}{\partial y}\right)_x = -\left(\frac{\partial \psi}{\partial x}\right)_y \left(\frac{\partial x}{\partial y}\right)_\psi \tag{H.3}$$

Substituting these in Eq. 10.41 produces

$$\left(\frac{\partial \phi}{\partial y}\right)_x \left(\frac{\partial y}{\partial x}\right)_\phi = \left(\frac{\partial \psi}{\partial y}\right)_x \left(\frac{\partial x}{\partial y}\right)_\psi \tag{H.4}$$

Substituting for $(\partial \phi / \partial y)_x$ from Eq. 10.42 and canceling, we get

$$-\left(\frac{\partial y}{\partial x}\right)_\phi = \left(\frac{\partial x}{\partial y}\right)_\psi \tag{H.5}$$

or

$$\left(\frac{\partial y}{\partial x}\right)_\phi \left(\frac{\partial x}{\partial y}\right)_\psi = -1 \tag{H.6}$$

Equation H.6 says that the product of the slopes of a curve of constant ϕ and a curve of constant ψ is -1. This is the condition that the two curves be perpendicular, proved in all calculus books. This was shown for any x and y, not for any specific values, which indicates that at any point x, y the lines of constant ϕ and constant ψ passing through that point are perpendicular.

REFERENCES

CHAPTER 1

1. M. Reiner, "The flow of matter," *Sci. Am. 201:* 122–138 (December 1959).
2. R. B. Bird, E. N. Stewart, and W. E. Lightfoot, *Transport Phenomena*, Wiley, New York, 1960, chap. 1.
3. T. C. Schneirla and G. Piel, "The army ant," *Sci. Am. 178:* 16–23 (June 1948).
4. J. T. Davies and E. K. Rideal, *Interfacial Phenomena*, Academic Press, New York, 1961.
5. C. V. Boys, *Soap Bubbles and the Forces Which Mould Them*, Doubleday, Garden City, N.Y., 1959 (1st ed. 1902). This interesting, informative book is highly recommended. Reading time is about 3 hours.
6. A. S. Lodge, *Elastic Fluids*, Academic Press, New York, 1964.
7. M. Jakob, *Heat Transfer*, vol. 1, Wiley, New York, 1958, p. 618.
8. M. H. Zimmerman, "How sap moves in trees," *Sci. Am. 208*(3): 133–142 (1963).
9. Noel de Nevers, "The poundal per square foot, the pascal and SI units," *Eng. Educ. 78*(2): 137 (1987).
10. ASTM, *Metric Practice Guide*, ASTM Publ. E 380-72^{e}, Philadelphia, 1973.
11. Richard Leavitt, "Kyoto," in *Fodor's Japan and Korea*, 1982, Alan Tucker, ed., Fodor's Modern Guides, New York, 1982.

CHAPTER 2

1. L. Prandtl and O. G. Tietjens, *Fundamentals of Hydro- and Aeromechanics*, L. Rosenhead, transl., Dover, New York, 1957.
2. *ASME Boiler and Pressure Vessel Code*, American Society of Mechanical Engineers, New York, 1980.
3. V. L. Streeter, *Fluid Mechanics*, 8th ed., McGraw-Hill, New York, 1985.

CHAPTER 3

1. G. Gamow, "Modern cosmology," in *The New Astronomy*, Simon and Schuster, New York, 1959, p. 14.
2. N. de Nevers and J. R. Morris, "Rollback modelling: Basic and modified," *J. Air Pollut. Control Assoc. 25:* 943–947 (1974).
3. ACGIH, *Threshold Limit Values and Biological Exposure Indices for 1988–1989*, American Conference of Governmental and Industrial Hygienists, Cincinnati, Ohio, 1988.

CHAPTER 4

1. Sandborn C. Brown, *Count Rumford*, Doubleday, Garden City, New York, 1962.
2. J. P. Joule, "On the mechanical equivalent of heat," originally presented to the Royal Society, June 21, 1849. See *The Scientific Papers of James Prescott Joule*, Physical Society of London, 1884. This short paper is highly recommended as an example of the care and the thought which must go into even a relatively simple experiment if the results are to be accurate to ±1 percent. Note also that in Joule's time the distinction between "temperature" and "heat" had not become clear, so he occasionally interchanged them.
3. W. H. Giedt, *Thermophysics*, preliminary edition, Van Nostrand, Princeton, N.J., 1964, p. 13.
4. G. Van Wylen and R. E. Sonntag, *Fundamentals of Classical Thermodynamics*, Wiley, New York, 1965.
5. M. W. Zemansky and H. C. Van Ness, *Basic Engineering Thermodynamics*, McGraw-Hill, New York, 1966, p. 273.

CHAPTER 5

1. Horace Lamb, *Hydrodynamics*, 6th ed., Dover, New York, 1945, p. 99.
2. B. C. Sakiadis, "Fluid and particle mechanics," in *Perry's Chemical Engineers' Handbook*, 6th ed., R. H. Perry and D. W. Green, eds., McGraw-Hill, New York, 1984, pp. 5–12.
3. "Flows of fluids through valves, fittings, and pipes," *Tech. Paper No. 410*, Crane Company, 475 N. Gary Ave., Carol Stream, Ill., 1957, p. A-19.
4. Kermit Fischer, "How to predict calibration of variable-area flow meters," *Chem. Eng. 59*(6): 180–184 (1952). You must read this paper carefully to see that it is discussing corrections of only a few percent to the simple equations presented in this text.
5. R. Davies, ed., *Cavitation in Real Liquids*, Elsevier, New York, 1964.
6. J. W. Holl and G. M. Wood, eds., *Cavitation Research Facilities and Techniques*, ASME, New York, 1964.
7. C. E. Kindsvater and R. W. Carter, "Discharge characteristics of rectangular thin-plate weirs," *Trans. Am. Soc. Civil Eng. 124:* 772–822 (1959). This long paper, with comments by others, shows that using a coefficient of 0.67 in Eq. 5.40 will reproduce most of the published test data, over a wide range of conditions to ±5 percent.

CHAPTER 6

1. O. Reynolds, "An experimental investigation of the circumstances which determine whether the motion of water shall be direct or sinuous and of the law of resistance in parallel channels," *Phil. Trans. Roy. Soc. 174* (1883). The history of this work and its relation to others are discussed in Rouse and Ince [2, p. 206].
2. H. Rouse and S. Ince, *History of Hydraulics*, Iowa Institute of Hydraulic Research, Iowa City, 1957.
3. J. Nikuradse, "Stroemungsgesetze in rauhen Rohren" ("Flow laws in rough tubes"), *VDI-Forsch.-Arb. Ing.-Wes. Heft 361* (1933).

4. L. W. Moody, "Friction factors for pipe flow," *Trans. ASME 66* (1944).
5. C. F. Colebrook, "Turbulent flow in pipes, with particular reference to the transition region between the smooth and rough pipe laws," *J. Inst. Civ. Eng. 11:* 133–155 (1938–9).
6. H. Schlichting, *Boundary Layer Theory*, 7th ed., J. Kestin, transl., McGraw-Hill, New York, 1979, p. 615.
7. *ASHRAE Handbook of Fundamentals*, ASHRAE, Atlanta, 1972.
8. Don J. Wood, "An explicit friction factor relationship," *Civil. Eng. 36*(12): 60–61 (1966).
9. "Flows of fluids through valves, fittings, and pipes," *Tech. Paper No. 410*, Crane Company, 475 N. Gary Ave., Carol Stream, Ill., 1957 (regularly reprinted).
10. M. P. O'Brian and G. H. Hickox, *Applied Fluid Mechanics*, McGraw-Hill, New York, 1937, p. 211.
11. R. B. Bird, W. E. Stewart, and E. N. Lightfoot, *Transport Phenomena*, Wiley, New York, 1960.
12. B. C. Sakiadis, "Fluid and particle mechanics," in *Perry's Chemical Engineers' Handbook*, 6th ed., R. H. Perry, D. W. Green, and J. O. Maloney, eds., McGraw-Hill, New York, 1984, pp. 5–29.
13. H. Cross, "Analysis of flow in networks of conduits or conductors," *Univ. Illinois Bull. 86* (1946). This method is described in detail in most texts on hydraulics.
14. Brice Carnahan, H. A. Luther, and J. O. Wilkes, *Applied Numerical Methods*, Wiley, New York, 1969, p. 310.
15. M. S. Peters and K. D. Timmerhaus, *Plant Design and Economics for Chemical Engineers*, 3d ed., McGraw-Hill, New York, 1980.
16. D. F. Boucher and G. E. Alves, "Fluid and particle mechanics," in *Chemical Engineers' Handbook*, 4th ed., R. H. Perry, C. H. Chilton, and S. D. Kirkpatrick, eds., McGraw-Hill, New York, 1963, pp. 5–30.
17. I. Newton, *Philos. Nat. Primo. Math. 2* (1710), sec. 2, cited in L. B. Torobin and W. H. Gauvin, *Can. J. Chem. Eng. 37:* 129 (1959).
18. A. C. Charters and R. N. Thomas, "The aerodynamic performance of small spheres from subsonic to high supersonic velocities," *J. Aeron. Sci. 12:* 468 (1945).
19. Horace Lamb, *Hydrodynamics*, 6th ed., Dover, New York, 1945, p. 598.
20. D. B. Steinman, "Suspension bridges: The aerodynamic stability problem and its solution," *Am. Sci. 42:* 397 (1954).
21. R. Von Mises, *Theory of Flight*, Dover, New York, 1945, pp. 139 *et seq*. This book from before the jet era gives a thorough, readable treatment of basic aerodynamics.
22. G. G. Brown, A. S. Foust, and D. L. Katz, *Unit Operations*, Wiley, New York, 1950, p. 142.
23. W. Thompson (Lord Kelvin), "On the economy of metal in conductors of electricity," *Report to the British Association for the Advancement of Science 1881*, John Murray, London, 1882, p. 526.
24. R. R. Hughes and E. R. Gilliland, "The mechanics of drops," *Chem. Eng. Prog. 48:* 497 (1952).

CHAPTER 7

1. John Parmakian, *Waterhammer Analysis*, Prentice-Hall, Englewood Cliffs, N.J., 1955.
2. W. Ley, *Rockets, Missiles and Space Travel*, Viking, New York, 1957. This is an excellent history of the rocket business, which can be read and understood by high school graduates. The author was involved in much of the early German rocket work and has many amusing tales to tell about it.
3. G. P. Sutton, *Rocket Propulsion Elements*, Wiley, New York, 1959. This is, as the subtitle suggests, "An introduction to the engineering of rockets."
4. A. C. Adler, "Vertical takeoff," *Int. Sci. Tech. 48*(12): 50–58 (1965).
5. D. B. Shepherd, *Principles of Turbomachinery*, Macmillan, New York, 1956.

6. H. Schlichting, *Boundary Layer Theory*, 7th ed., J. Kestin, transl., McGraw-Hill, New York, 1979.
7. L. Prandtl and O. G. Tietjens, *Fundamentals of Hydro- and Aeromechanics*, L. Rosenhead, transl., Dover, New York, 1957, p. 251.
8. R. B. Bird, W. E. Stewart, and E. N. Lightfoot, *Transport Phenomena*, Wiley, New York, 1960.
9. B. Lewis and G. Von Elbe, *Combustion, Flames and Explosions of Gases*, 2d ed., Academic Press, New York, 1961, p. 206.

CHAPTER 8

1. H. Semat, *Fundamentals of Physics*, Reinhart, New York, 1951, p. 404.
2. R. S. Westfall, "Newton and the fudge factor," *Science 179:* 751–758 (1973).
3. Horace Lamb, *Hydrodynamics*, 6th ed., Dover, New York, 1945, p. 478.
4. *Handbook of Chemistry and Physics*, 47th ed., R. C. Weast and S. M. Selby, eds., Chemical Rubber, Cleveland, 1964, p. F-9. There is an extensive table of isothermal compressibilities here.
5. A. H. Shapiro, *The Dynamics and Thermodynamics of Compressible Fluid Flow*, vol. 1, Ronald Press, New York, 1958.
6. V. L. Streeter and E. B. Wylie, *Fluid Mechanics*, 6th ed., McGraw-Hill, New York, 1975, p. 362.
7. O. Levenspied, "The discharge of gases from a reservoir through a pipe," *AIChE J. 23:* 402–403 (1977).
8. C. E. Lapple, "Isothermal and adiabatic flow of compressible fluids," *Trans. AIChE 39:* 385–432 (1943).
9. D. L. Katz, Cornell, and Kobayashi, *Handbook of Natural Gas Engineering*, McGraw-Hill, New York, 1959, p. 305.
10. G. Taylor, "The formation of a blast wave by a very intense explosion," *Proc. Roy. Soc. Ser. A 201:* 159–187 (1950).
11. A. Stodola, *Steam and Gas Turbines*, Louis C. Loewenstein, transl., McGraw-Hill, New York, 1927, p. 558.
12. H. W. Liepmann and A. Roshko, *Elements of Gas Dynamics*, Wiley, New York, 1957.
13. J. H. Keenan, F. G. Keyes, P. G. Hill, and J. G. Moore, *Steam Tables; Thermodynamic Properties of Water Including Vapor, Liquid and Solid Phases*, Wiley, New York, 1969.
14. R. D. Zucker, *Fundamentals of Gas Dynamics*, Matrix Publishers, Champagne, Ill., 1977, p. 175.
15. National Advisory Committee for Aeronautics, *Report 1135, Equations, Tables and Charts for Compressible Flow*, Government Printing Office, Washington, 1953.

CHAPTER 9

1. R. L. Daughterty and J. B. Franzini, *Fluid Mechanics with Engineering Applications*, 6th ed., McGraw-Hill, New York, 1965, p. 527.
2. D. G. Shepherd, *Principles of Turbomachinery*, Macmillan, New York, 1956.
3. R. P. Genereaux, et al., "Transport and storage of fluids," in *Perry's Chemical Engineers' Handbook*, 6th ed., R. H. Perry and D. W. Green, eds., McGraw-Hill, New York, 1984, pp. 6–18.
4. A. Stodola, *Steam and Gas Turbines*, Louis C. Loewenstein, transl., McGraw-Hill, New York, 1927, p. 558.
5. B. G. A. Skrotski, "Steam turbines," *Power 106*(6): s-1, s-42 (1962).
6. I. J. Karassik, W. C. Krutzsch, and W. H. Fraser, eds., *Pump Handbook*, McGraw-Hill, New York, 1976.

7. Jay Matley (ed.), *Fluid Movers; Pumps, Compressors, Fans and Blowers* (a series of papers from *Chemical Engineering* on this topic), McGraw-Hill, New York, 1979.
8. T. G. Hicks and T. W. Edwards, *Pump Application Engineering*, McGraw-Hill, New York, 1971.

CHAPTER 10

1. L. Prandtl, "Ueber Fluessigkeitsbewegung mit kleiner Reibung" ("Concerning fluid movements with small friction"), *Verhlandl III Intern. Math.-Kongr.*, Heidelberg (1904); also reprinted in *Vier Abhandlungen zur Hydrodynamik*, Goetingen, (1927).
2. H. S. Carslaw and J. C. Jaeger, *Conduction of Heat in Solids*, 2d ed., Oxford University Press, London, 1959.
3. R. B. Bird, W. E. Lightfoot, and E. N. Stewart, *Transport Phenomena*, Wiley, New York, 1960.
4. R. H. Kirchhoff, *Potential Flows; Computer Graphic Solutions*, Marcel Dekker, New York, 1985.
5. L. Milne-Thompson, *Theoretical Hydrodynamics*, Macmillan, New York, 1955.
6. Horace Lamb, *Hydrodynamics*, 6th ed., Dover, New York, 1945.
7. D. C. Hazen and C. J. Lynch, "Better airplane wings," *Int. Sci. Technol. 50*(2): 38–46 (1966).
8. E. N. Tinoco and A. W. Chen, "CFD applications to engine/airframe integration," *Prog. Astronaut. Aeronaut. 102:* 219–255 (1986).
9. W. Kaplan, *Advanced Calculus*, Addison-Wesley, Reading, Mass., 1953, p. 90.

CHAPTER 11

1. H. Schlichting, *Boundary Layer Theory*, 7th ed., J. Kestin, transl., McGraw-Hill, New York, 1979.
2. R. V. Churchill, *Fourier Series and Boundary Value Problems*, McGraw-Hill, New York, 1941, p. 120.
3. R. V. Churchill, *Modern Operational Mathematics in Engineering*, McGraw-Hill, New York, 1944, p. 11. This same boundary-value problem arises as a heat-transfer problem when a semi-infinite slab, originally all at one temperature, suddenly has its surface temperature raised. This heat-transfer problem is solved in almost every book on heat transfer.
4. R. B. Bird, W. E. Stewart, and E. N. Lightfoot, *Transport Phenomena*, Wiley, New York, 1960, p. 164.
5. P.A. Longwell, *Mechanics of Fluid Flow*, McGraw-Hill, New York, 1966, p. 316.
6. R. D. Mehta, "Aerodynamics of sports balls," *Annu. Rev. Fluid Mech. 17:* 151–189 (1985).
7. M. O. Kramer, "Hydrodynamics of the dolphin," *Adv. Hydrosci. 2:* 11–130 (1965).
8. B. C. Sakiadis, "Fluid and particle mechanics," in *Perry's Chemical Engineers' Handbook*, 6th ed., R. H. Perry and D. W. Green, eds., McGraw-Hill, New York, 1984, pp. 5–34.

CHAPTER 12

1. S. Ergun, "Fluid flow through packed columns," *Chem. Eng. Prog. 48:* 89–94 (1952).
2. Max Leva, *Fluidization*, McGraw-Hill, New York, 1959.
3. M. Muskat, *Physical Principles of Oil Production*, McGraw-Hill, New York, 1949.
4. D. K. Todd, *Ground Water Hydrology*, Wiley, New York, 1960, p. 53.
5. A. E. Scheidegger, *The Physics of Flow through Porous Media*, Macmillan, New York, 1960.
6. R. A. Wooding and H. J. Morel-Seytoux, "Multiphase fluid flow through porous media," *Annu. Rev. Fluid Mech. 8:* 233–274 (1976).

7. R. E. Treybal, *Mass Transfer Operations*, 3d ed., McGraw-Hill, New York, 1980, p. 200.
8. N. H. Chen, "Liquid-Solid filtration: Generalized design and optimization calculations," *Chem. Eng. 85*(17): 97–101 (1978).
9. R. J. Akers, "Filtration pretreatment," in *Filtration, Principles and Practices*, 2d ed., M. J. Matteson and Clyde Orr, eds., Marcel Dekker, New York, 1987.
10. M. I. Tillery, "High efficiency air filtration," in *Filtration, Principles and Practices*, 2d ed., M. J. Matteson and Clyde Orr, eds., Marcel Dekker, New York, 1987.
11. J. F. Davidson, D. Harrison, and J. R. F. Guedes de Carvalho, "On the liquidlike behavior of fluidized beds," *Annu. Rev. Fluid Mech. 9:* 55–86 (1977).
12. F. A. Zenz and D. F. Othmer, *Fluidization and Fluid-Particle Systems*, Rheinhold, New York, 1960.
13. J. F. Davidson and D. Harrison, *Fluidized Particles*, Cambridge University Press, London, 1963.
14. A. J. Stepanoff, *Gravity Flow of Bulk Solids and Transport of Solids in Suspension*, Wiley, New York, 1969.

CHAPTER 13

1. J. B. S. Haldane, "On being the right size," in *Possible Worlds and Other Essays*, Harper & Row, New York, 1927; reprinted in J. R. Newman, *The World of Mathematics*, Simon and Schuster, New York, 1956. This short essay is an absolute masterpiece, which everyone should read for pleasure and for information.
2. S. J. Kline, *Similitude and Approximation Theory*, McGraw-Hill, New York, 1965.
3. R. B. Bird, W. E. Stewart, and E. N. Lightfoot, *Transport Phenomena*, Wiley, New York, 1960, p. 107.
4. J. D. Hellumns and S. W. Churchill, "Simplification of the mathematical description of boundary and initial value problems," *AIChE J. 10:* 110–114 (1964).
5. B. C. Sakiadis, "Fluid and particle mechanics," in *Perry's Chemical Engineers' Handbook*, 6th ed. R. H. Perry, D. W. Green, and J. O. Maloney, eds., McGraw-Hill, New York, 1984, pp. 5–62.
6. E. Buckingham, "Model experiments and the form of empirical equations," *Trans. ASME 37:* 263–296 (1915). See also P. W. Buckingham, *Dimensional Analysis*, Harvard University Press, Cambridge, Mass., 1921.

CHAPTER 14

1. B. C. Sakiadis, "Fluid and particle mechanics," in *Perry's Chemical Engineers' Handbook* 6th ed., R. H. Perry, D. W. Green, and J. O. Maloney, eds., McGraw-Hill, New York, 1984, pp. 5–46.
2. G. W. Govier, B. A. Radford, and J. S. C. Dunn, "The upward vertical flow of air-water mixtures," *Can. J. Chem. Eng. 35:* 58–70 (1957). See also later articles by Govier and his coworkers in the same journal: *36:* 195–202 and *38:* 62–66.
3. G. A. Hughmark and B. S. Pressburg, "Holdup and pressure drop with gas-liquid flow in a vertical pipe," *AIChE J. 7:* 677–682 (1961).
4. J. M. Delhaye, "Two-phase flow patterns," in *Two-Phase Flow and Heat Transfer in the Power and Process Industries*, Hemisphere-McGraw-Hill, New York, 1981.
5. D. Barnea and Y. Taitel, "Flow pattern transition in two-phase gas-liquid flows," in *Encyclopedia of Fluid Mechanics*, vol. 3, *Gas-Liquid Flows*, N. P. Cheremisinoff, ed., Gulf Publishing, Houston, 1986.
6. D. S. Scott, "Properties of cocurrent gas-liquid flow," in *Advances in Chemical Engineering*, 4th ed., T. B. Drew, J. W. Hoopes, Jr., and T. Vermeulen, eds., Academic Press, New York, 1963.

CHAPTER 15

1. R. B. Bird, R. C. Armstrong, and Ole Hassager, *Dynamics of Polymeric Liquids*, 2d ed., vol. 1: *Fluid Mechanics*, Wiley, New York, 1986.
2. W. L. Wilkinson, *Non-Newtonian fluids*, Pergamon, New York, 1960, chap. 2.
3. R. B. Bird, W. E. Stewart, and E. N. Lightfoot, *Transport Phenomena*, Wiley, New York, 1960, p. 14.
4. P. Longwell, *Mechanics of Fluid Flow*, McGraw-Hill, New York, 1966.
5. W. R. Schowalter, *Mechanics of Non-Newtonian Fluids*, Pergamon, New York, 1978.
6. G. E. Alves, D. F. Boucher, and R. L. Pigford, "Pipeline design for non-Newtonian solutions and suspensions," *Chem. Eng. Prog. 48:* 385–393 (1952).
7. R. G. Shaver and E. W. Merrill, "Turbulent flow of pseudoplastic polymer solutions in straight cylindrical tubes," *AIChE J. 5:* 181–188 (1959).
8. J. G. Savins, "Drag reduction characteristics of solutions of macromolecules in turbulent pipe flow," *Soc. Petrol Eng. J. 4:* 203–214 (1964).
9. N. S. Berman, "Drag reduction by polymers," *Annu. Rev. Fluid Mech. 10:* 47–64 (1978).
10. A. H. P. Skelland, *Non-Newtonian Flow and Heat Transfer*, McGraw-Hill, New York, 1967.

CHAPTER 16

1. J. O. Hinze, *Turbulence*, 2d ed., McGraw-Hill, New York, 1975.
2. A. E. Gargett, "Ocean turbulence," *Annu. Rev. Fluid Mech. 21:* 419–451 (1989).
3. H. L. Dryden and A. M. Kuethe, "Effect of turbulence on wind tunnel measurements," *NASA Rep. No. 342* (1929).
4. C. B. H. Priestly, *Turbulent Transfer in the Lower Atmosphere*, Chicago University Press, Chicago, 1959.
5. S. Corrsin, "Turbulent flow," *Am. Sci. 49:* 300–325 (1961).
6. A. E. Perry, *Hot-Wire Anemometry*, Clarendon Press, Oxford, 1982.
7. F. Durst, A. Melling, and J. H. Whitelaw, *Principles and Practices of Laser-Doppler Anemometry*, 2d ed. Academic Press, New York, 1981.
8. G. I. Taylor, "The spectrum of turbulence," *Proc. Roy. Soc. Ser. A 164:* 476–490 (1938).
9. H. Schlichting, *Boundary Layer Theory*, J. Kestin, transl., 7th ed., McGraw-Hill, New York, 1979.
10. H. L. Dryden, "A review of the statistical theory of turbulence," *Q. Appl. Math. 1:* 7–42 (1943). This paper and most of G. I. Taylor's basic papers are contained in S. K. Friedlander and L. Topper, eds., *Turbulence—Classical Papers on Statistical Theory*, Interscience, New York, 1961.
11. R. B. Bird, W. E. Lightfoot, and E. N. Stewart, *Transport Phenomena*, Wiley, New York, 1960, p. 160.
12. A. A. Townsend, *J. Fluid Mech. 10:* 635–636 (1961).

CHAPTER 17

1. W. J. Masica, J. D. Derdul, and D. A. Petrash, "Hydrostatic stability of the liquid-vapor interface in a low-acceleration field," *NASA Tech. Note*, NASA TN-2444 (1964).
2. V. N. Serebryakov, "Control of the dynamics of a two-phase liquid-gas weightless medium with the aid of surface effects," available in *NASA Tech. Transl.*, NASA TT F-469 (1967).
3. J. T. Davies and E. K. Rideal, *Interfacial Phenomena*, Academic Press, New York, 1961.
4. S. L. Soo, *Fluid Dynamics of Multiphase Systems*, Blaisdell, Waltham, Mass., 1967, p. 100.
5. W. B. Hayes, Ill, B. W. Hardy, and C. D. Holland, "Formation of gas bubbles at submerged orifices," *AIChE J. 5:* 319–324 (1959).

6. Yongjue Park, A. L. Tyler, and Noel de Nevers, "The chamber-orifice interaction in the formation of bubbles," *Chem. Eng. Sci. 32:* 907–916 (1977).
7. J. E. Halligan and L. E. Burkhart, "Determination of the profile of a growing droplet," *AIChE J. 14:* 411–414 (1968).
8. Lord Rayleigh (John William Strutt), "On the instability of jets," *Proc. London Math. Soc. 10:* 4–13 (1879); reprinted in *Scientific Papers of Lord Rayleigh*, Dover, New York, 1964.
9. C. V. Boys, *Soap Bubbles and the Forces Which Mould Them*, Doubleday, New York, 1959, p. 75; first published in 1902.
10. W. E. Meyer and W. E. Ranz, "Sprays," *Encyclopaedia of Chemical Technology*, 1st ed., vol. 12, R. E. Kirk and D. F. Othmer, eds., Interscience, New York, 1954, pp. 703–721.
11. R. A. Castleman, Jr., "The mechanism of the atomization of liquids," *Natl. Bur. Std. J. Res. 6:* 369–376 (1931).
12. E. Giffen and A. Muraszew, *The Atomization of Liquid Fuels*, Wiley, New York, 1953.
13. C. V. Sternling and L. E. Scriven, "Interfacial turbulence: Hydrodynamic instability and the Marangoni effect," *AIChE J. 5:* 514–523 (1959).
14. M. Muskat, *Physical Principles of Oil Production*, McGraw-Hill, New York, 1949, pp. 290–298.
15. S. Ross, "Mechanisms of foam stabilization and antifoam action," *Chem. Eng. Progr. 63:* 41–47 (September 1967).

ANSWERS TO SELECTED PROBLEMS

1.3 102 lbm/ft^3
1.13 1.1×10^{12} gal; 1.14 mi^3
1.15 1 toof = 32.2 ft; 1 dnoces = 0.176 s
1.18 9660 ft/s
1.20 4.6×10^5
1.22 5.7×10^{-6} lbf/in; 0.001 N/m

2.1 7×10^{-4} in
2.3 998.2 kgf/m^3
2.7 16,000 lbf/in^2 gauge
2.10 2.68 in
2.16 97,000 ft
2.18 −0.00356°F/ft; −0.00536°F/ft
2.20 1.2×10^{19} lbm
2.22 88.3 kPa gauge
2.24 4.2×10^7 lbf; 2.1×10^7 lbf; 2.1×10^7 lbf
2.26 26.0×10^6 lbf
2.28 31,200 lbf; 49,000 lbf
2.30 87 wt %
2.31 62.5 lbf
2.40 1.08×10^6 ft^3
2.42 $\Delta h = \Delta z\, \rho_w/(\rho_{Hg} - \rho_w)$
2.44 10.7 in, 2.8 in

2.46 5.3 ft, 0.008 ft
2.48 0.47 in H_2O = 0.017 psi
2.50 1215 lbf/in^2 gauge; −1.7%
2.55 14.3 ft/s^2; up
2.57 1.78°
2.59 0.15 lbf; toward the axis of rotation
2.62 $t = Pr/(2\sigma)$

3.7 224,400 gal/min; 0.476 ft/s
3.11 2.55 ft^3/s, 159 lbm/s, 13 ft/s
3.13 47.4 min; 1.33×10^{-3} atm
3.15 Rising; 72 mm/h
3.17 0.0051 lbm/min

4.3 3.73 ft · lbf
4.6 289°C
4.9 228 Btu/lbm added
4.11 2.8×10^{-4} kWh/lbm
4.13 −412 kJ/kg
4.15 1.602 MW
4.19 160°F
4.21 14.8 Btu
4.24 2.18×10^4 kJ
4.27 0.29
4.29 3500 ft/s
4.32 1.1×10^{14} Btu/h; 2.9 lbm/h
4.36 2.2 Btu/lbm

5.1 1.3 Btu/lbm; 11 F°, 1.3 F°
5.3 31.94 kPa, 0 J/kg; 28.7 kPa, 3.20 J/kg; 0 kPa, 32.0 J/kg
5.7 55.6 ft^3/s
5.9 60.6 ft^3/s
5.12 127 ft/s
5.14 11.3 ft/s
5.16 57.4 ft/s
5.18 5.58 m/s
5.20 112 ft/s
5.22 17.3 mi/h
5.24 0.46 ft/s
5.28 4.72 ft^3/s
5.30 123 ft^3/s
5.36 63 ft/s, 41 ft/s
5.38 62 rpm
5.40 36.5 s
5.42 $10h^{1/2}$ $s/ft^{1/2}$

6.3 2.84 ft/s; $2.08 \times 10^{-4}(lbf/in^2)/ft$
6.7 0.0013 ft^3/s
6.9 0.114 lbf

6.13 20 lbf/in^2 per 1000 ft; 35 to 40 lbf/in^2 per 1000 ft
6.17 21 lbf/in^2; 19 lbf/in^2
6.19 1600 gal/min
6.21 40 gal/min; from the second tank to the first tank
6.27 31 gal/min; 1600; 87 gal/min; 4500
6.37 0.125 lbf/in^2
6.39 1040 gal/min
6.41 0.0166 ft/ft
6.45 2.2×10^{-6} ft^3/s
6.49 7.4-in sides
6.51 300 ft/s
6.53 47 ft/s
6.55 62 gal/min; right to left
6.57 423 gal/min; 192 gal/min; vessel 2 to vessel 1
6.62 +16%
6.72 219 ft/yr
6.74 1.5 ft/s
6.76 384 ft/s, 31.6 s, 10,500 ft

7.1 2×10^{-24} ft/s; 0
7.4 4000 N
7.7 1.333
7.11 828 lbf
7.13 446 lbf
7.17 3.25 kN/(kg/s)
7.19 2470 lbf; yes!
7.24 16.7 ft; 6.0 ft/s
7.39 19,500 m/s

8.1 810 psi/ft; 1.0 psi/ft
8.3 10,800 ft/s
8.5 2.8×10^5 psi
8.7 334 ft/s
8.9 226°R
8.11 226°R, 3.6 psia, 0.006 lbm/ft^3
8.13 501°R; 45.2 psia; 1931 ft/s
8.15 420°R; 7.16 psia; 2950 ft/s; 0.0063 lbm/ft^3
8.17 6.3 psia; 440°R; 1.17
8.19 639°R; 6.48 psia; 3060 ft/s
8.21 0.69 lbm/s
8.23 3.7 lbm/s
8.29 3.03 psia; 0.026 lbm/ft^3
8.31 37,700 lbf
8.48 165 lbm/s
8.50 1467 ft/s; 631°R; 18 psia
8.54 0.48; 1.0
8.56 $\mathscr{M} = 1.71$; lowest $P = 0.612 P_{R_1}$

9.1 13 gal/min
9.3 2×10^8 hp

9.5 2.5 ft
9.7 0.84
9.9 680 psi; 106 hp
9.11 6 in
9.13 11.4 ft
9.15 19 hp; 31 hp; 24 hp
9.19 2859 Btu/(lbmol)

10.12 14.8 psia
10.13 −155 ft/s; −155 ft/s; 14.3 psia
10.14 $\frac{1}{2}$

11.5 0.08 ft; 0.005 ft; 4.0 ft
11.10 1.22; 1.06
11.16 17.3 ft

12.5 887 darcies
12.7 0.12
12.9 1900 to 7600; all turbulent
12.13 0.64 ft/s; 0.0067 ft/s

13.3 $\mu^2/(\rho F)$; L_1/L_2
13.9 600 mi/h; 40 mi/h

15.3 $K = 0.54\ \mathrm{lbf/ft^2}$; $n = 0.27$
15.6 $K = 0.2\ \mathrm{lbf/ft^2}$; $n = 0.2$
15.9 2100; 2000; 3100

17.7 0.087 in
17.9 $z = 2\sigma A/(g\rho Bx)$
17.16 0.8 psi

INDEX

A
absolute pressure, 17
adiabatic flame temperature, 116
adiabatic throttle, 112
adjusted length, 206
aeronautical engineering, 266
angle of attack, 404
angular-momentum balance, 270
API gravity, 8
archimedes principle, 47

B
balance equation, 76
barometric equation, 33
basic equation of fluid statics, 33
Bernoulli's equation, 139
Bingham fluid, Bingham plastic, 10
Blasius solution, 389
bluff body, 381
bottle filling, 117
boundary layer, 355, 385
boundary-layer equations, 386
boundary-layer thickness, 390
bourdon tube, 154
box model, 88
British thermal unit, 104
Buckingham's π theorem, 442
buffer layer, 398
bulk density, 7
bulk modulus, 292, 522
buoyancy, 46
buoyant force, 47

C
calorie, 104
capillary number, 422
capillary viscometer, 186
carburetor, 173
cavitation, 161
centipoise, 12
centistoke, 12
centrifugal pumps, 334
centroid, 42
choking, 304
closed system, 78
coefficient of discharge, 154
coefficient of thermal expansion, 522
coefficient of viscosity, 9
compressibility factor, 521
compressive stress, 1
compressor, 113, 339
compressor efficiencies, 345
conservation of energy, 94
constitutive equations, 461
contact angle, 421, 491
continuity equation, 76
control volume, 78
converging-diverging nozzle, 316
correlation coefficient, 480
corrosion allowance, 75
Couette flow, 287
creeping flow, 224
critical state, 297

D
darcy, 30, 419
datum, 95
de Laval nozzle, 316
density, 6
depth filters, 428
diffuser, 335
dilatant fluid, 11

dimensionless numbers, 435
discharge, 84
displacement thickness, 393
doublet, 378
draft tube, 71
drag coefficient, 222
drag forces, 222, 268

E
economic pipe diameter, 218
eddy viscosity, 485
electrostatic energy, 97
energy, 94
energy balance, 99
enlargements and contractions, 208
enthalpy, 109
enthalpy balance, 122
entrance region, 404, 409
equation of motion, 241
equivalent lengths, 206
escape velocity, 125
Euler equation, 277
Euler's turbine equation, 271
eulerian viewpoint, 275
exact solution, 395
extensive property, 78

F
filters, 426
first law of thermodynamics, 94
fitting losses, 206
flashing flow, 456
flow energy, 108
flow work, 108
fluid, 1
fluidization, 429
force, 18
friction factor, 189, 414
friction heat loss, 142
friction heating, 94, 140
friction loss, 141
friction velocity, 397
Froude number, 259, 437

G
gases, 5
gauge pressure, 17
general continuity equation, 90

H
head form of Bernoulli's equation, 143
heat, 98
heat balance, 122
holdup, 451
hoop stress, 64
hot-wire anemometer, 475
hydraulic horsepower, 332
hydraulic jump, 257
hydraulic radius, 211
hydraulicians, 355
hydraulics, 3
hydrodynamicists, 355
hydrology, 3

I
impact tube, 151
impulse turbine, 348
incompressible flow, 142
injection work, 107
intensity of turbulence, 477
intensive property, 79
intercooling, 342
interfacial tensions, 495
internal energy, 95, 103
interstitial velocity, 412
intrusion work, 108
irrotational flow, 367
isentropic compressor, 345
isothermal compressibility, 292, 522

J
jets, 248

K
kilocalorie, 104
kilogram-force, 21
kilogram-mass, 21
kinematic viscosity, 12
kinetic energy, 96

L
lagrangian viewpoint, 275
laminar flow, 180
laminar sublayer, 398
Laplace equation, 359, 420
lift coefficient, 228
liquids, 5
lost work, 142

M
Mach number, 295
magnetic energy, 97
manometer, 50
mass, 18
mass balance, 76
mass flow rate, 83
material balance, 80
mechanical energy balance, 143
minimum fluidizing velocity, 430
mixing layer, 472
models, 433
moment of momentum equation, 271
momentum, 242
momentum balance, 241
momentum flux, 29
momentum thickness, 393

N
natural gas, 174
Navier-Stokes equations, 277
net positive suction head (NPSH), 338
Newtonian fluid, 9
Newton's equation of motion, 241
Newton's law of viscosity, 9
nonnewtonian fluid, 9, 458
non-uniform flows, 166
normal shock waves, 311
no-slip condition, 183
nozzle-choking, 304
nozzles and diffusers, 316
nuclear energy, 98

O
open system, 78, 107
orifice meter, 157

P
packed towers, 424
particle density, 7
particle dynamics, 3
particle Reynolds number, 224
pathline, 358
peizometer, 54
perfect fluid, 356
perfect gas law, 36, 521
Pi theorem, 442
pitot-static tube, 152
pitot tube, 151, 320
poise, 12
Poiseuille equation, 184
porosity, 412
porous medium, 410
positive-displacement pumps, 329
potential energy, 96
potential flow, 355, 420
poundal, 22
pound-force, 21
pound-mass, 21
power, 114
power law, 461
Prandtl's boundary-layer equations, 386
pressure, 15
pressure coefficient, 436
pressure drop, 179
pressure measurement, 49
pressure recovery, 144
principle of conservation of mass, 76
propane, 174
pseudoplastic fluid, 10
pump, 113
pycnometer, 6

R
radiation, 99
rarefaction, 304
rate of shear deformation, 9
rate of strain, 9
ratio of specific heats, 38, 293
reaction turbine, 348
relative roughness, 190
relative velocities, 262, 314
reservoir condition, 295
reservoir mechanics, 3
residual saturation, 422
resistance coefficient, 208
Reynolds analogy, 199
Reynolds experiment, 180
Reynolds number, 181, 391, 414, 437
Reynolds stress, 189, 396, 484
rheograms, 459
rheology, 459
rheopectic fluid, 11
rigid body motion, 59
rigid body rotation, 61
rockets, 253
rotameter, 159

S
scale (of turbulence), 482
second law of thermodynamics, 131
separation, 380
sharp-edged weir, 166
shear rate, 9
shear stress, 2
shock waves, 311
similarity, 437
similitude, 437
simple heater, 115
sink, 366
siphon, 161
slip, 451
slip velocity, 452
slug, 22
slurry, 426
solid, 1
source, 366
specific gravity, 7
specific impulse, 30, 255
specific volume, 106
specific weight, 34
spectrum of turbulence, 483
speed of sound, 290
square-edged orifice, 159
staging, 342
stagnation condition, 295
stagnation point, 380
stagnation tube, 151
stall, 404
standard atmosphere, 39
starting and stopping flows, 259
static tube, 152
steady state, 80
Stokes derivative, 275
Stokes law, 224
Stokes stream function, 375
streakline, 358
stream function, 371
stream tube, 376
streamline flow, 412
streamlined body, 381
streamlines, 357
substantive derivative, 275
suction lift, 333
sudden expansion, 144, 255
superficial velocity, 412
surface active agents, 495
surface energy, 97, 490
surface filters, 426
surface tension, 13, 490
surroundings, 77
system, 77

T
Tate's law, 498
tensile stress, 1
terminal velocity, 226
thin-walled pressure vessels, 64
thixotropic fluid, 11
throat, 153
thrust, 254
Torricelli's equation, 148
total condition, 295
transition region, 181
turbine, 113, 348
turbulent flow, 180

U
units and conversion factors, 18
unsteady state, 84

V
vectors, 242
velocity potential, 359
vena contracta, 209
venturi meter, 153
viscoelastic fluid, 11
viscosity, 8
void fraction, 412
volume (specific), 106
volumetric flow rate, 84
vortex threads, 470
vorticity, 368

W
water hammer, 165, 262
weight, 18
weir, 166
wetting, 491
Weymouth equation, 310
work, 98